The R Book

The R Book

Michael J. Crawley
Imperial College London at Silwood Park, UK

John Wiley & Sons, Ltd

Other Wiley Editorial Offices

John Wiley & Sons Inc., 111 River Street, Hoboken, NJ 07030, USA

Jossey-Bass, 989 Market Street, San Francisco, CA 94103-1741, USA

Wiley-VCH Verlag GmbH, Boschstr. 12, D-69469 Weinheim, Germany

John Wiley & Sons Australia Ltd, 42 McDougall Street, Milton, Queensland 4064, Australia

John Wiley & Sons (Asia) Pte Ltd, 2 Clementi Loop #02-01, Jin Xing Distripark, Singapore 129809

John Wiley & Sons Canada Ltd, 6045 Freemont Blvd, Mississauga, ONT, L5R 4J3

Wiley also publishes its books in a variety of electronic formats. Some content that appears in print may not be
available in electronic books.

Anniversary Logo Design: Richard J. Pacifico

British Library Cataloguing in Publication Data

A catalogue record for this book is available from the British Library

ISBN-13: 978-0-470-51024-7

Typeset in 10/12pt Times by Integra Software Services Pvt. Ltd, Pondicherry, India
Printed and bound in Great Britain by Antony Rowe Ltd, Chippenham, Wiltshire
This book is printed on acid-free paper responsibly manufactured from sustainable forestry in
which at least two trees are planted for each one used for paper production.

Contents

Preface

R is a high-level language and an environment for data analysis and graphics. The design of R was heavily influenced by two existing languages: Becker, Chambers and Wilks' S and Sussman's Scheme. The resulting language is very similar in appearance to S, but the underlying implementation and semantics are derived from Scheme. This book is intended as an introduction to the riches of the R environment, aimed at beginners and intermediate users in disciplines ranging from science to economics and from medicine to engineering. I hope that the book can be read as a text as well as dipped into as a reference manual. The early chapters assume absolutely no background in statistics or computing, but the later chapters assume that the material in the earlier chapters has been studied. The book covers data handling, graphics, mathematical functions, and a wide range of statistical techniques all the way from elementary classical tests, through regression and analysis of variance and generalized linear modelling, up to more specialized topics such as spatial statistics, multivariate methods, tree models, mixed-effects models and time series analysis. The idea is to introduce users to the assumptions that lie behind the tests, fostering a critical approach to statistical modelling, but involving little or no statistical theory and assuming no background in mathematics or statistics.

Why should you switch to using R when you have mastered a perfectly adequate statistical package already? At one level, there is no point in switching. If you only carry out a very limited range of statistical tests, and you don't intend to do more (or different) in the future, then fine. The main reason for switching to R is to take advantage of its unrivalled coverage and the availability of new, cutting edge applications in fields such as generalized mixed-effects modelling and generalized additive models. The next reason for learning R is that you want to be able to understand the literature. More and more people are reporting their results in the context of R, and it is important to know what they are talking about. Third, look around your discipline to see who else is using R: many of the top people will have switched to R already. A large proportion of the world's leading statisticians use R, and this should tell you something (many, indeed, contribute to R, as you can see below). Another reason for changing to R is the quality of back-up and support available. There is a superb network of dedicated R wizards out there on the web, eager to answer your questions. If you intend to invest sufficient effort to become good at statistical computing, then the structure of R and the ease with which you can write your own functions are major attractions. Last, and certainly not least, the product is free. This is some of the finest integrated software in the world, and yet it is yours for absolutely nothing.

Although much of the text will equally apply to S-PLUS, there are some substantial differences, so in order not to confuse things I concentrate on describing R. I have made no attempt to show where S-PLUS is different from R, but if you have to work in S-PLUS, then try it and see if it works.

Acknowledgements

S is an elegant, widely accepted, and enduring software system with outstanding conceptual integrity, thanks to the insight, taste, and effort of John Chambers. In 1998, the Association for Computing Machinery (ACM) presented him with its Software System Award, for 'the S system, which has forever altered the way people analyze, visualize, and manipulate data'. R was inspired by the S environment that was developed by John Chambers, and which had substantial input from Douglas Bates, Rick Becker, Bill Cleveland, Trevor Hastie, Daryl Pregibon and Allan Wilks.

R was initially written by Ross Ihaka and Robert Gentleman at the Department of Statistics of the University of Auckland in New Zealand. Subsequently, a large group of individuals contributed to R by sending code and bug reports. John Chambers graciously contributed advice and encouragement in the early days of R, and later became a member of the core team. The current R is the result of a collaborative effort with contributions from all over the world.

Since mid-1997 there has been a core group known as the 'R Core Team' who can modify the R source code archive. The group currently consists of Doug Bates, John Chambers, Peter Dalgaard, Robert Gentleman, Kurt Hornik, Stefano Iacus, Ross Ihaka, Friedrich Leisch, Thomas Lumley, Martin Maechler, Duncan Murdoch, Paul Murrell, Martyn Plummer, Brian Ripley, Duncan Temple Lang, Luke Tierney, and Simon Urbanek.

R would not be what it is today without the invaluable help of the following people, who contributed by donating code, bug fixes and documentation: Valerio Aimale, Thomas Baier, Roger Bivand, Ben Bolker, David Brahm, Göran Broström, Patrick Burns, Vince Carey, Saikat DebRoy, Brian D'Urso, Lyndon Drake, Dirk Eddelbuettel, Claus Ekström, John Fox, Paul Gilbert, Frank E. Harrell Jr, Torsten Hothorn, Robert King, Kjetil Kjernsmo, Roger Koenker, Philippe Lambert, Jan de Leeuw, Uwe Ligges, Jim Lindsey, Patrick Lindsey, Catherine Loader, Gordon Maclean, John Maindonald, David Meyer, Eiji Nakama, Jens Oehlschaegel, Steve Oncley, Richard O'Keefe, Hubert Palme, Roger D. Peng, Jose C. Pinheiro, Tony Plate, Anthony Rossini, Jonathan Rougier, Deepayan Sarkar, Guenther Sawitzki, Marc Schwartz, Detlef Steuer, Bill Simpson, Gordon Smyth, Adrian Trapletti, Terry Therneau, Rolf Turner, Bill Venables, Gregory R. Warnes, Andreas Weingessel, Morten Welinder, James Wettenhall, Simon Wood and Achim Zeileis. I have drawn heavily on the R help pages in writing this book, and I am extremely grateful to all the R contributors who wrote the help files.

Special thanks are due to the generations of graduate students on the annual GLIM course at Silwood. It was their feedback that enabled me to understand those aspects of R that are most difficult for beginners, and highlighted the concepts that require the most detailed explanation. Please tell me about the errors and omissions you find, and send suggestions for changes and additions to m.crawley@imperial.ac.uk.

M.J. Crawley
Ascot
September 2006

1

Getting Started

Installing R

I assume that you have a PC or an Apple Mac, and that you want to install R on the hard disc. If you have access to the internet then this could hardly be simpler. First go to the site called CRAN (this stands for Comprehensive R Archive Network). You can type its full address,

```
http://cran.r-project.org/
```

or simply type CRAN into Google and be transported effortlessly to the site. Once there, you need to 'Download and Install R' by running the appropriate precompiled binary distributions. Click to choose between Linux, Mac OS and Windows, then follow the (rather different) instructions. You want the 'base' package and you want to run the setup program which will have a name like R*.exe (on a PC) or R*.dmg (on a Mac). When asked, say you want to 'Run' the file (rather than 'Save' it). Then just sit back and watch. If you do not have access to the internet, then get a friend to download R and copy it onto a CD or a memory stick for you.

Running R

To run R, just click on the R icon. If there is no icon, go to Programs, then to R, then click on the R icon. The first thing you see is the version number of R and the date of your version. It is a good idea to visit the CRAN site regularly to make sure that you have got the most up-to-date version of R. If you have an old version, it is best to uninstall your current version before downloading the new one.

The header explains that there is no warranty for this free software, and allows you to see the list of current contributors. Perhaps the most important information in the header is found under

citation()

which shows how to cite the R software in your written work. The R Development Core Team has done a huge amount of work and we, the R user community, should pay them due credit whenever we publish work that has used R.

The R Book   Michael J. Crawley
© 2007 John Wiley & Sons, Ltd

Below the header you will see a blank line with a > symbol in the left hand margin. This is called the **prompt** and is R's way of saying 'What now?'. This is where you type in your commands, as introduced on p. 9. When working, you will sometimes see + at the left-hand side of the screen instead of >. This means that the last command you typed is incomplete. The commonest cause of this is that you have forgotten one or more brackets. If you can see what is missing (e.g. a final right-hand bracket) then just type the missing character and press enter, at which point the command will execute. If you have made a mistake, then press the Esc key and the command line prompt > will reappear. Then use the Up arrow key to retrieve your last command, at which point you can correct the mistake, using the Left and Right arrow keys.

Getting Help in R

The simplest way to get help in R is to click on the Help button on the toolbar of the RGui window. Alternatively, if you are connected to the internet, you can type CRAN in Google and search for the help you need at CRAN. However, if you know the name of the function you want help with, you just type a question mark ? at the command line prompt followed by the name of the function. So to get help on read.table, just type

?read.table

Sometimes you cannot remember the precise name of the function, but you know the subject on which you want help (e.g. data input in this case). Use the help.search function (without a question mark) with your query in double quotes like this:

help.search("data input")

and (with any luck) you will see the names of the R functions associated with this query. Then you can use ?read.table to get detailed help.

Other useful functions are find and apropos. The find function tells you what package something is in:

find(lowess)

```
[1] "package:stats"
```

while apropos returns a character vector giving the names of all objects in the search list that match your (potentially partial) enquiry:

apropos(lm)

```
[1]  ". __C__anova.glm"    ". __C__anova.glm.null"  ". __C__glm"
[4]  ". __C__glm.null"     ". __C__lm"              ". __C__mlm"
[7]  "anova.glm"           "anova.glmlist"          "anova.lm"
[10] "anova.lmlist"        "anova.mlm"              "anovalist.lm"
[13] "contr.helmert"       "glm"                    "glm.control"
[16] "glm.fit"             "glm.fit.null"           "hatvalues.lm"
[19] "KalmanForecast"      "KalmanLike"             "KalmanRun"
[22] "KalmanSmooth"        "lm"                     "lm.fit"
[25] "lm.fit.null"         "lm.influence"           "lm.wfit"
[28] "lm.wfit.null"        "model.frame.glm"        "model.frame.lm"
[31] "model.matrix.lm"     "nlm"                    "nlminb"
[34] "plot.lm"             "plot.mlm"               "predict.glm"
```

```
[37]  "predict.lm"      "predict.mlm"      "print.glm"
[40]  "print.lm"        "residuals.glm"    "residuals.lm"
[43]  "rstandard.glm"   "rstandard.lm"     "rstudent.glm"
[46]  "rstudent.lm"     "summary.glm"      "summary.lm"
[49]  "summary.mlm"     "kappa.lm"
```

Online Help

The is a tremendous amount of information about R on the web, but your first port of call is likely to be CRAN at

> `http://cran.r-project.org/`

Here you will find a variety of R manuals:

- *An Introduction to R* gives an introduction to the language and how to use R for doing statistical analysis and graphics.

- A draft of the *R Language Definition* documents the language *per se* – that is, the objects that it works on, and the details of the expression evaluation process, which are useful to know when programming R functions.

- *Writing R Extensions* covers how to create your own packages, write R help files, and use the foreign language (C, C + +, Fortran, . . .) interfaces.

- *R Data Import/Export* describes the import and export facilities available either in R itself or via packages which are available from CRAN.

- *R Installation and Administration*, which is self-explanatory.

- *R: A Language and Environment for Statistical Computing* (referred to on the website as 'The R Reference Index') contains all the help files of the R standard and recommended packages in printable form.

(These manuals are also available in R itself by choosing Help/Manuals (in PDF) from the menu bar.) There are also answers to *Frequently Asked Questions* (FAQs) and *R News*, a newsletter which contains interesting articles, book reviews and news of forthcoming releases. The most useful part of the site, however, is the Search facility which allows you to investigate the contents of most of the R documents, functions, and searchable mail archives.

Worked Examples of Functions

To see a worked **example** just type the function name (linear models, lm, in this case)

example(lm)

and you will see the printed and graphical output produced by the lm function.

Demonstrations of R Functions

These can be useful for seeing the range of things that R can do. Here are some for you to try:

```
demo(persp)
demo(graphics)
demo(Hershey)
demo(plotmath)
```

Libraries in R

To use one of the libraries (listed in Table 1.1), simply type the library function with the name of the library in brackets. Thus, to load the spatial library type

```
library(spatial)
```

Table 1.1. Libraries used in this book that come supplied as part of the base package of R.

lattice	lattice graphics for panel plots or trellis graphs
MASS	package associated with Venables and Ripley's book entitled *Modern Applied Statistics using S-PLUS*
mgcv	generalized additive models
nlme	mixed-effects models (both linear and non-linear)
nnet	feed-forward neural networks and multinomial log-linear models
spatial	functions for kriging and point pattern analysis
survival	survival analysis, including penalised likelihood

Contents of Libraries

It is easy to use the help function to discover the contents of library packages. Here is how you find out about the contents of the spatial library:

```
library(help=spatial)
```

```
           Information on package "spatial"
Package:       spatial
Description:  Functions for kriging and point pattern analysis.
```

followed by a list of all the functions and data sets. You can view the full list of the contents of a library using objects with search() like this. Here are the contents of the spatial library:

```
objects(grep("spatial",search()))
```

```
 [1]  "anova.trls"   "anovalist.trls"   "correlogram"    "expcov"
 [5]  "gaucov"       "Kaver"            "Kenvl"          "Kfn"
 [9]  "plot.trls"    "ppgetregion"      "ppinit"         "pplik"
[13]  "ppregion"     "predict.trls"     "prmat"          "Psim"
```

```
[17]  "semat"      "sphercov"      "SSI"             "Strauss"
[21]  "surf.gls"   "surf.ls"       "trls.influence" "trmat"
[25]  "variogram"
```

Then, to find out how to use, say, Ripley's K (Kfn), just type

?Kfn

Installing Packages and Libraries

The base package does not contain some of the libraries referred to in this book, but downloading these is very simple. Run the R program, then from the command line use the **install.packages** function to download the libraries you want. You will be asked to highlight the mirror nearest to you for fast downloading (e.g. London), then everything else is automatic. The packages used in this book are

```
install.packages("akima")
install.packages("chron")
install.packages("lme4")
install.packages("mcmc")
install.packages("odesolve")
install.packages("spdep")
install.packages("spatstat")
install.packages("tree")
```

If you want other libraries, then go to CRAN and browse the list called 'Packages' to select the ones you want to investigate.

Command Line versus Scripts

When writing functions and other multi-line sections of input you will find it useful to use a text editor rather than execute everything directly at the command line. I always use Word for this, because it is so easy to keep a copy of all the output and graphics produced by R using Copy and Paste. Other people prefer to use R's own built-in editor. It is accessible from the RGui menu bar. Click on File then click on New script. At this point R will open a window entitled Untitled - R Editor. You can type and edit in this, then when you want to execute a line or group of lines, just highlight them and press Ctrl + R (the Control key and R together). The lines are automatically transferred to the command window and executed.

By pressing Ctrl + S you can save the contents of the R Editor window in a file that you will have to name. It will be given a .R file extension automatically. In a subsequent session you can click on File/Open script . . . when you will see all your saved .R files and can select the one you want to open.

Data Editor

There is a data editor within R that can be accessed from the menu bar by selecting Edit/Data editor. . . . You provide the name of the matrix or dataframe containing the material you want to edit (this has to be a dataframe that is active in the current R session, rather than one which is stored on file), and a **Data Editor** window appears. Alternatively, you can do this from the command line using the fix function (e.g. fix(data.frame.name)). Suppose you want to edit the bacteria dataframe which is part of the **MASS** library:

```
library(MASS)
attach(bacteria)
fix(bacteria)
```

The window has the look of an Excel spreadsheet, and you can change the contents of the cells, navigating with the cursor or with the arrow keys. My preference is to do all of my data preparation and data editing in Excel itself (because that is what it is good at). Once checked and edited, I save the data from Excel to a tab-delimited text file (*.txt) that can be imported to R very simply using the function called read.table (p. 98). One of the most persistent frustrations for beginners is that they cannot get their data imported into R. Things that typically go wrong at the data input stage and the necessary remedial actions are described on p. 98.

Changing the Look of the R Screen

The default settings of the command window are inoffensive to most people, but you can change them if you don't like them. The **Rgui Configuration Editor** under Edit/GUI preferences . . . is used to change the look of the screen. You can change the colour of the input line (default is red), the output line (default navy) or the background (default white). The default numbers of rows (25) and columns (80) can be changed, and you have control over the font (default Courier New) and font size (default 10).

Significance Stars

If these worry you, then turn them off. Significance stars are shown by default next to the *p* values in the output of statistical models.

```
gg<-read.table("c:\\temp\\Gain.txt",header=T)
attach(gg)
names(gg)
```

```
[1] "Weight" "Sex" "Age" "Genotype" "Score"
```

This is what the default output looks like for an analysis of covariance:

```
model<-lm(Weight~Age+Sex)
summary(model)
```

```
Coefficients:
```

	Estimate	Std. Error	t value	Pr(| t |)	
(Intercept)	8.17156	0.33118	24.674	< 2e-16	***
Age	0.29958	0.09185	3.262	0.00187	**
Sexmale	-0.83161	0.25980	-3.201	0.00224	**

```
- - -
Signif. codes: 0 '***' 0.001 '**' 0.01 '*' 0.05 '.' 0.1 ' ' 1
```

```
Residual standard error: 1.006 on 57 degrees of freedom
Multiple R-Squared: 0.2681, Adjusted R-squared: 0.2425
F-statistic: 10.44 on 2 and 57 DF,  p-value: 0.0001368
```

Here is the output with the significance stars turned off:

```
options(show.signif.stars=FALSE)
summary(model)
```

```
Coefficients:
              Estimate  Std. Error  t value   Pr(>|t|)
(Intercept)   8.17156     0.33118    24.674    < 2e—16
Age           0.29958     0.09185     3.262    0.00187
Sexmale      -0.83161     0.25980    -3.201    0.00224
```

```
Residual standard error: 1.006 on 57 degrees of freedom
Multiple R-Squared: 0.2681,  Adjusted R-squared: 0.2425
F-statistic: 10.44 on 2 and 57 DF,  p-value: 0.0001368
```

You decide.

Disappearing Graphics

To stop multiple graphs whizzing by, use

```
par(ask=TRUE)
```

then each graph will stay on the screen until you press the Enter key. You can pause execution to mimic a slide show effect. You need to specify the number of seconds delay that you want between each action, using the Sys.sleep function (see p. 102).

Good Housekeeping

To see what variables you have created in the current session, type

```
objects()
```

```
[1] "colour.factor"  "colours"      "dates"     "index"
[5] "last.warning"   "nbnumbers"    "nbtable"   "nums"
[9] "wanted"         "x"            "xmat"      "xv"
```

To see which libraries and dataframes are attached:

```
search()
```

```
 [1] ".GlobalEnv"         "nums"                "nums"
 [4] "package:methods"    "package:stats"       "package:graphics"
 [7] "package:grDevices"  "package:utils"       "package:data sets"
[10] "Autoloads"          "package:base"
```

Linking to Other Computer Languages

Advanced users can employ the functions .C and .Fortran to provide a standard interface to compiled code that has been linked into R, either at build time or via dyn.load. They are primarily intended for compiled C and Fortran code respectively, but the .C function can be used with other languages which can generate C interfaces, for example C + +. The .Internal and .Primitive interfaces are used to call C code compiled into R at build time. Functions .Call and .External provide interfaces which allow compiled code (primarily compiled C code) to manipulate R objects.

Tidying Up

At the end of a session in R, it is good practice to remove (rm) any variables names you have created (using, say, x <-5.6) and to detach any dataframes you have attached earlier in the session (see p. 18). That way, variables with the same names but different properties will not get in each other's way in subsequent work:

```
rm(x,y,z)
detach(worms)
```

This command does not make the dataframe called worms disappear; it just means that the variables within worms, such as Slope and Area, are no longer accessible directly by name. To get rid of everything, including all the dataframes, type

```
rm(list=ls())
```

but be absolutely sure that you really want to be as draconian as this before you execute the command.

2

Essentials of the R Language

There is an enormous range of things that R can do, and one of the hardest parts of learning R is finding your way around. I suggest that you start by looking down all the chapter names at the front of this book (p. v) to get an overall idea of where the material you want might be found. The Index for this book has the names of R functions in **bold**. The dataframes used in the book are listed by name, and also under 'dataframes'.

Alternatively, if you know the name of the function you are interested in, then you can go directly to R's help by typing the exact name of the function after a question mark on the command line (p. 2).

Screen prompt

The screen prompt > is an invitation to put R to work. You can either use one of the many facilities which are the subject of this book, including the built-in functions discussed on p. 11, or do ordinary calculations:

```
> log(42/7.3)
```

```
[1] 1.749795
```

Each line can have at most 128 characters, so if you want to give a lengthy instruction or evaluate a complicated expression, you can continue it on one or more further lines simply by ending the line at a place where the line is obviously incomplete (e.g. with a trailing comma, operator, or with more left parentheses than right parentheses, implying that more right parentheses will follow). When continuation is expected, the prompt changes from > to +

```
> 5+6+3+6+4+2+4+8+
+       3+2+7
```

```
[1] 50
```

Note that the + continuation prompt does not carry out arithmetic plus. If you have made a mistake, and you want to get rid of the + prompt and return to the > prompt, then either press the Esc key or use the Up arrow to edit the last (incomplete) line.

The R Book   Michael J. Crawley
© 2007 John Wiley & Sons, Ltd

Two or more expressions can be placed on a single line so long as they are separated by semi-colons:

2+3; 5*7; 3-7

```
[1]  5
[1]  35
[1]  -4
```

From here onwards and throughout the book, the prompt character > will be omitted. The material that you should type on the command line is shown in Arial font. Just press the Return key to see the answer. The output from R is shown in Courier New font, which uses absolute rather than proportional spacing, so that columns of numbers remain neatly aligned on the page or on the screen.

Built-in Functions

All the mathematical functions you could ever want are here (see Table 2.1). The log function gives logs to the base e ($e = 2.718282$), for which the antilog function is exp

log(10)

```
[1]  2.302585
```

exp(1)

```
[1]  2.718282
```

If you are old fashioned, and want logs to the base 10, then there is a separate function

log10(6)

```
[1]  0.7781513
```

Logs to other bases are possible by providing the log function with a second argument which is the base of the logs you want to take. Suppose you want log to base 3 of 9:

log(9,3)

```
[1]  2
```

The trigonometric functions in R measure angles in radians. A circle is 2π radians, and this is 360°, so a right angle (90°) is $\pi/2$ radians. R knows the value of π as pi:

pi

```
[1]  3.141593
```

sin(pi/2)

```
[1]  1
```

cos(pi/2)

```
[1]  6.123032e-017
```

Notice that the cosine of a right angle does not come out as exactly zero, even though the sine came out as exactly 1. The e-017 means 'times 10^{-17}'. While this is a very small

Table 2.1. Mathematical functions used in R.

Function	Meaning
log(x)	log to base e of x
exp(x)	antilog of x (e^x)
log(x,n)	log to base n of x
log10(x)	log to base 10 of x
sqrt(x)	square root of x
factorial(x)	$x!$
choose(n,x)	binomial coefficients $n!/(x!\ (n-x)!)$
gamma(x)	$\Gamma(x)$, for real x $(x-1)!$, for integer x
lgamma(x)	natural log of $\Gamma(x)$
floor(x)	greatest integer $< x$
ceiling(x)	smallest integer $> x$
trunc(x)	closest integer to x between x and 0 trunc(1.5) $= 1$, trunc(-1.5) $= -1$ trunc is like floor for positive values and like ceiling for negative values
round(x, digits=0)	round the value of x to an integer
signif(x, digits=6)	give x to 6 digits in scientific notation
runif(n)	generates n random numbers between 0 and 1 from a uniform distribution
cos(x)	cosine of x in radians
sin(x)	sine of x in radians
tan(x)	tangent of x in radians
acos(x), asin(x), atan(x)	inverse trigonometric transformations of real or complex numbers
acosh(x), asinh(x), atanh(x)	inverse hyperbolic trigonometric transformations of real or complex numbers
abs(x)	the absolute value of x, ignoring the minus sign if there is one

number it is clearly not exactly zero (so you need to be careful when testing for exact equality of real numbers; see p. 77).

Numbers with Exponents

For very big numbers or very small numbers R uses the following scheme:

1.2e3	means 1200 because the e3 means 'move the decimal point 3 places to the right'
1.2e-2	means 0.012 because the e-2 means 'move the decimal point 2 places to the left'
3.9+4.5i	is a complex number with real (3.9) and imaginary (4.5) parts, and i is the square root of -1.

Modulo and Integer Quotients

Integer quotients and remainders are obtained using the notation %/% (percent, divide, percent) and %% (percent, percent) respectively. Suppose we want to know the integer part of a division: say, how many 13s are there in 119:

```
119 %/% 13
```

```
[1]   9
```

Now suppose we wanted to know the remainder (what is left over when 119 is divided by 13): in maths this is known as **modulo**:

```
119 %% 13
```

```
[1]   2
```

Modulo is very useful for testing whether numbers are odd or even: odd numbers have modulo 2 value 1 and even numbers have modulo 2 value 0:

```
9 %% 2
```

```
[1]   1
```

```
8 %% 2
```

```
[1]   0
```

Likewise, you use modulo to test if one number is an exact multiple of some other number. For instance to find out whether 15 421 is a multiple of 7, ask:

```
15421 %% 7 == 0
```

```
[1]   TRUE
```

Rounding

Various sorts of rounding (rounding up, rounding down, rounding to the nearest integer) can be done easily. Take 5.7 as an example. The 'greatest integer less than' function is floor

```
floor(5.7)
```

```
[1]   5
```

and the 'next integer' function is ceiling

```
ceiling(5.7)
```

```
[1]   6
```

You can round to the nearest integer by adding 0.5 to the number then using floor. There is a built-in function for this, but we can easily write one of our own to introduce the notion of function writing. Call it rounded, then define it as a function like this:

```
rounded<-function(x) floor(x+0.5)
```

Now we can use the new function:

```
rounded(5.7)
```

```
[1]   6
```

```
rounded(5.4)
```

```
[1]   5
```

Infinity and Things that Are Not a Number (NaN)

Calculations can lead to answers that are plus infinity, represented in R by `Inf`, or minus infinity, which is represented as `-Inf`:

```
3/0
```

```
[1]  Inf
```

```
-12/0
```

```
[1]  -Inf
```

Calculations involving infinity can be evaluated: for instance,

```
exp(-Inf)
```

```
[1]  0
```

```
0/Inf
```

```
[1]  0
```

```
(0:3)^Inf
```

```
[1]  0  1  Inf  Inf
```

Other calculations, however, lead to quantities that are not numbers. These are represented in R by `NaN` ('not a number'). Here are some of the classic cases:

```
0/0
```

```
[1]  NaN
```

```
Inf-Inf
```

```
[1]  NaN
```

```
Inf/Inf
```

```
[1]  NaN
```

You need to understand clearly the distinction between `NaN` and `NA` (this stands for 'not available' and is the missing-value symbol in R; see below). The function `is.nan` is provided to check specifically for `NaN`, and `is.na` also returns `TRUE` for `NaN`. Coercing `NaN` to logical or integer type gives an `NA` of the appropriate type. There are built-in tests to check whether a number is finite or infinite:

```
is.finite(10)
```

```
[1]  TRUE
```

```
is.infinite(10)
```

```
[1]  FALSE
```

```
is.infinite(Inf)
```

```
[1]  TRUE
```

Missing values NA

Missing values in dataframes are a real source of irritation because they affect the way that model-fitting functions operate and they can greatly reduce the power of the modelling that we would like to do.

Some functions do not work with their default settings when there are missing values in the data, and **mean** is a classic example of this:

```
x<-c(1:8,NA)
```

```
mean(x)
```

```
[1]  NA
```

In order to calculate the mean of the non-missing values, you need to specify that the NA are to be removed, using the na.rm=TRUE argument:

```
mean(x,na.rm=T)
```

```
[1]  4.5
```

To check for the location of missing values within a vector, use the function is.na(x) rather than x !="NA". Here is an example where we want to find the locations (7 and 8) of missing values within a vector called vmv:

```
vmv<-c(1:6,NA,NA,9:12)
```

```
vmv
```

```
[1]  1  2  3  4  5  6  NA  NA  9  10  11  12
```

Making an index of the missing values in an array could use the **seq** function,

```
seq(along=vmv)[is.na(vmv)]
```

```
[1]  7  8
```

but the result is achieved more simply using which like this:

```
which(is.na(vmv))
```

```
[1]  7  8
```

If the missing values are genuine counts of zero, you might want to edit the NA to 0. Use the is.na function to generate subscripts for this

```
vmv[is.na(vmv)]<- 0
```

```
vmv
```

```
[1]  1  2  3  4  5  6  0  0  9  10  11  12
```

or use the ifelse function like this

```
vmv<-c(1:6,NA,NA,9:12)
```

```
ifelse(is.na(vmv),0,vmv)
```

```
[1]  1  2  3  4  5  6  0  0  9  10  11  12
```

Assignment

Objects obtain values in R by assignment ('*x gets* a value'). This is achieved by the **gets arrow** <- which is a composite symbol made up from 'less than' and 'minus' with no space between them. Thus, to create a scalar constant x with value 5 we type

```
x<-5
```

and not x = 5. Notice that there is a potential ambiguity if you get the spacing wrong. Compare our x<-5, '*x* gets 5', with x < -5 which is a logical question, asking 'is x less than minus 5?' and producing the answer TRUE or FALSE.

Operators

R uses the following operator tokens:

+ - */%% ^	arithmetic
> >= < <= == !=	relational
! & |	logical
~	model formulae
<- ->	assignment
$	list indexing (the 'element name' operator)
:	create a sequence

Several of the operators have different meaning inside model formulae. Thus * indicates the main effects plus interaction, : indicates the interaction between two variables and ^ means all interactions up to the indicated power (see p. 332).

Creating a Vector

Vectors are variables with one or more values of the same type: logical, integer, real, complex, string (or character) or raw. A variable with a single value (say 4.3) is often known as a scalar, but in R a scalar is a vector of length 1. Vectors could have length 0 and this can be useful in writing functions:

```
y<-4.3
z<-y[-1]
length(z)
[1]  0
```

Values can be assigned to vectors in many different ways. They can be generated by R: here the vector called *y* gets the sequence of integer values 10 to 16 using : (colon), the sequence-generating operator,

```
y <- 10:16
```

You can type the values into the command line, using the concatenation function c,

```
y <- c(10, 11, 12, 13, 14, 15, 16)
```

or you can enter the numbers from the keyboard one at a time using scan:

y <- scan()

1: 10

2: 11

3: 12

4: 13

5: 14

6: 15

7: 16

8:

Read 7 items

pressing the Enter key instead of entering a number to indicate that data input is complete. However, the commonest way to allocate values to a vector is to read the data from an external file, using read.table (p. 98). Note that read.table will convert character variables into factors, so if you do *not* want this to happen, you will need to say so (p. 100). In order to refer to a vector by name with an R session, you need to attach the dataframe containing the vector (p. 18). Alternatively, you can refer to the dataframe name and the vector name within it, using the element name operator $ like this: dataframe$y

One of the most important attributes of a vector is its length: the number of numbers it contains. When vectors are created by calculation from other vectors, the new vector will be as long as the longest vector used in the calculation (with the shorter variable(s) recycled as necessary): here A is of length 10 and B is of length 3:

A<-1:10
B<-c(2,4,8)
A*B

[1] 2 8 24 8 20 48 14 32 72 20

Warning message: longer object length is not a multiple of shorter object length in: A * B

The vector B is recycled three times in full and a warning message in printed to indicate that the length of A is not a multiple of the length of B (11×4 and 12×8 have not been evaluated).

Named Elements within Vectors

It is often useful to have the values in a vector labelled in some way. For instance, if our data are counts of 0, 1, 2, ... occurrences in a vector called counts

(counts<-c(25,12,7,4,6,2,1,0,2))

[1] 25 12 7 4 6 2 1 0 2

so that there were 25 zeros, 12 ones and so on, it would be useful to name each of the counts with the relevant number 0 to 8:

names(counts)<-0:8

Now when we inspect the vector called counts we see both the names and the frequencies:

counts

```
 0   1  2  3  4  5  6  7  8
25  12  7  4  6  2  1  0  2
```

If you have computed a table of counts, and you want to *remove* the names, then use the as.vector function like this:

(st<-table(rpois(2000,2.3)))

```
  0    1    2    3    4    5   6   7  8  9
205  455  510  431  233  102  43  13  7  1
```

as.vector(st)

```
[1]  205  455  510  431  233  102  43  13  7  1
```

Vector Functions

One of R's great strengths is its ability to evaluate functions over entire vectors, thereby avoiding the need for loops and subscripts. Important vector functions are listed in Table 2.2.

Table 2.2. Vector functions used in R.

Operation	Meaning
max(x)	maximum value in x
min(x)	minimum value in x
sum(x)	total of all the values in x
mean(x)	arithmetic average of the values in x
median(x)	median value in x
range(x)	vector of $\min(x)$ and $\max(x)$
var(x)	sample variance of x
cor(x,y)	correlation between vectors x and y
sort(x)	a sorted version of x
rank(x)	vector of the ranks of the values in x
order(x)	an integer vector containing the permutation to sort x into ascending order
quantile(x)	vector containing the minimum, lower quartile, median, upper quartile, and maximum of x
cumsum(x)	vector containing the sum of all of the elements up to that point
cumprod(x)	vector containing the product of all of the elements up to that point
cummax(x)	vector of non-decreasing numbers which are the cumulative maxima of the values in x up to that point
cummin(x)	vector of non-increasing numbers which are the cumulative minima of the values in x up to that point
pmax(x,y,z)	vector, of length equal to the longest of x, y or z, containing the maximum of x, y or z for the ith position in each

Table 2.2. (Continued)

Operation	Meaning
pmin(x,y,z)	vector, of length equal to the longest of x, y or z, containing the minimum of x, y or z for the ith position in each
colMeans(x)	column means of dataframe or matrix x
colSums(x)	column totals of dataframe or matrix x
rowMeans(x)	row means of dataframe or matrix x
rowSums(x)	row totals of dataframe or matrix x

Summary Information from Vectors by Groups

One of the most important and useful vector functions to master is tapply. The 't' stands for 'table' and the idea is to apply a function to produce a table from the values in the vector, based on one or more grouping variables (often the grouping is by factor levels). This sounds much more complicated than it really is:

```
data<-read.table("c:\\temp\\daphnia.txt",header=T)
attach(data)
names(data)
```

```
[1]  "Growth.rate"  "Water"  "Detergent"  "Daphnia"
```

The response variable is Growth.rate and the other three variables are factors (the analysis is on p. 479). Suppose we want the mean growth rate for each detergent:

```
tapply(Growth.rate,Detergent,mean)
```

```
BrandA  BrandB  BrandC  BrandD
  3.88    4.01    3.95    3.56
```

This produces a table with four entries, one for each level of the factor called Detergent. To produce a two-dimensional table we put the two grouping variables in a list. Here we calculate the median growth rate for water type and daphnia clone:

```
tapply(Growth.rate,list(Water,Daphnia),median)
```

```
        Clone1  Clone2  Clone3
Tyne     2.87    3.91    4.62
Wear     2.59    5.53    4.30
```

The first variable in the list creates the rows of the table and the second the columns. More detail on the tapply function is given in Chapter 6 (p. 183).

Using with rather than attach

Advanced R users do not routinely employ attach in their work, because it can lead to unexpected problems in resolving names (e.g. you can end up with multiple copies of the same variable name, each of a different length and each meaning something completely different). Most modelling functions like lm or glm have a data= argument so attach is unnecessary in those cases. Even when there is no data= argument it is preferable to wrap the call using with like this

```
with(data, function(. . .))
```

The with function evaluates an R expression in an environment constructed from data. You will often use the with function other functions like tapply or plot which have no built-in data argument. If your dataframe is part of the built-in package called datasets (like OrchardSprays) you can refer to the dataframe directly by name:

```
with(OrchardSprays,boxplot(decrease~treatment))
```

Here we calculate the number of 'no' (not infected) cases in the bacteria dataframe which is part of the MASS library:

```
library(MASS)
with(bacteria,tapply((y=="n"),trt,sum))
```

```
placebo  drug  drug+
     12    18     13
```

and here we plot brain weight against body weight for mammals on log-log axes:

```
with(mammals,plot(body,brain,log="xy"))
```

without attaching either dataframe. Here is an unattached dataframe called reg.data:

```
reg.data<-read.table("c:\\temp\\regression.txt",header=T)
```

with which we carry out a linear regression and print a summary

```
with (reg.data, {
             model<-lm(growth~tannin)
             summary(model) } )
```

The linear model fitting function lm knows to look in reg.data to find the variables called growth and tannin because the with function has used reg.data for constructing the environment from which lm is called. Groups of statements (different lines of code) to which the with function applies are contained within curly brackets. An alternative is to define the data environment as an argument in the call to lm like this:

```
summary(lm(growth~tannin,data=reg.data))
```

You should compare these outputs with the same example using attach on p. 388. Note that whatever form you choose, you still need to get the dataframe into your current environment by using read.table (if, as here, it is to be read from an external file), or from a library (like MASS to get bacteria and mammals, as above). To see the names of the dataframes in the built-in package called datasets, type

```
data()
```

but to see all available data sets (including those in the installed packages), type

```
data(package = .packages(all.available = TRUE))
```

Using **attach** in This Book

I use **attach** throughout this book because experience has shown that it makes the code easier to understand for beginners. In particular, using **attach** provides simplicity and brevity, so that we can

- refer to variables by name, so x rather than dataframe$x

- write shorter models, so lm(y~x) rather than lm(y~x,data=dataframe)

- go straight to the intended action, so plot(y~x) not with(dataframe,plot(y~x))

Nevertheless, readers are encouraged to use **with** or **data=** for their own work, and to avoid using **attach** wherever possible.

Parallel Minima and Maxima: **pmin** and **pmax**

Here are three vectors of the same length, x, y and z. The parallel minimum function, pmin, finds the minimum from any one of the three variables for each subscript, and produces a *vector* as its result (of length equal to the longest of x, y, or z):

x

```
[1]  0.99822644  0.98204599  0.20206455  0.65995552  0.93456667  0.18836278
```

y

```
[1]  0.51827913  0.30125005  0.41676059  0.53641449  0.07878714  0.49959328
```

z

```
[1]  0.26591817  0.13271847  0.44062782  0.65120395  0.03183403  0.36938092
```

pmin(x,y,z)

```
[1]  0.26591817  0.13271847  0.20206455  0.53641449  0.03183403  0.18836278
```

so the first and second minima came from z, the third from x, the fourth from y, the fifth from z, and the sixth from x. The functions min and max *produce scalar* results.

Subscripts and Indices

While we typically aim to apply functions to vectors as a whole, there are circumstances where we want to select only some of the elements of a vector. This selection is done using **subscripts** (also known as **indices**). Subscripts have square brackets [2] while functions have round brackets (2). Subscripts on vectors, matrices, arrays and dataframes have one set of square brackets [6], [3,4] or [2,3,2,1] while subscripts on lists have double square brackets [[2]] or [[i,j]] (see p. 65). When there are two subscripts to an object like a matrix or a dataframe, the first subscript refers to the row number (the rows are defined as **margin** no. 1) and the second subscript refers to the column number (the columns are margin no. 2). There is an important and powerful convention in R such that *when a subscript appears as a blank it is understood to mean 'all of'*. Thus

- [,4] means all rows in column 4 of an object
- [2,] means all columns in row 2 of an object.

There is another indexing convention in R which is used to extract named components from objects using the $ operator like this: **model$coef** or **model$resid** (p. 363). This is known as 'indexing tagged lists' using the element names operator $.

Working with Vectors and Logical Subscripts

Take the example of a vector containing the 11 numbers 0 to 10:

x<-0:10

There are two quite different kinds of things we might want to do with this. We might want to *add up* the values of the elements:

sum(x)

```
[1]  55
```

Alternatively, we might want to *count* the elements that passed some logical criterion. Suppose we wanted to know how many of the values were less than 5:

sum(x<5)

```
[1]  5
```

You see the distinction. We use the vector function **sum** in both cases. But **sum(x)** adds up the values of the *x*s and **sum(x<5)** counts up the number of cases that pass the logical condition '*x* is less than 5'. This works because of *coercion* (p. 25). Logical TRUE has been coerced to numeric 1 and logical FALSE has been coerced to numeric 0.

That is all well and good, but how do you add up the values of just some of the elements of *x*? We specify a logical condition, but we don't want to count the number of cases that pass the condition, we want to add up all the values of the cases that pass. This is the final piece of the jigsaw, and involves the use of *logical subscripts*. Note that when we counted the number of cases, the counting was applied to the entire vector, using **sum(x<5)**. To find the sum of the values of *x* that are less than 5, we write:

sum(x[x<5])

```
[1]  10
```

Let's look at this in more detail. The logical condition **x<5** is either true or false:

x<5

```
[1]    TRUE    TRUE  TRUE  TRUE  TRUE  FALSE  FALSE  FALSE  FALSE
[10]  FALSE  FALSE
```

You can imagine false as being numeric 0 and true as being numeric 1. Then the vector of subscripts [x<5] is five 1s followed by six 0s:

1*(x<5)

```
[1]  1 1 1 1 1 0 0 0 0 0 0
```

Now imagine multiplying the values of *x* by the values of the logical vector

```
x*(x<5)
```

```
[1]  0  1  2  3  4  0  0  0  0  0  0
```

When the function **sum** is applied, it gives us the answer we want: the sum of the values of the numbers $0 + 1 + 2 + 3 + 4 = 10$.

```
sum(x*(x<5))
```

```
[1]  10
```

This produces the same answer as **sum(x[x<5])**, but is rather less elegant.

Suppose we want to work out the sum of the three largest values in a vector. There are two steps: first sort the vector into descending order. Then add up the values of the first three elements of the sorted array. Let's do this in stages. First, the values of y:

```
y<-c(8,3,5,7,6,6,8,9,2,3,9,4,10,4,11)
```

Now if you apply **sort** to this, the numbers will be in ascending sequence, and this makes life slightly harder for the present problem:

```
sort(y)
```

```
[1]  2  3  3  4  4  5  6  6  7  8  8  9  9  10  11
```

We can use the reverse function, **rev** like this (use the Up arrow key to save typing):

```
rev(sort(y))
```

```
[1]  11  10  9  9  8  8  7  6  6  5  4  4  3  3  2
```

So the answer to our problem is $11 + 10 + 9 = 30$. But how to compute this? We can use specific subscripts to discover the contents of any element of a vector. We can see that 10 is the second element of the sorted array. To compute this we just specify the subscript [2]:

```
rev(sort(y))[2]
```

```
[1]  10
```

A range of subscripts is simply a series generated using the colon operator. We want the subscripts 1 to 3, so this is:

```
rev(sort(y))[1:3]
```

```
[1]  11  10  9
```

So the answer to the exercise is just

```
sum(rev(sort(y))[1:3])
```

```
[1]  30
```

Note that we have not changed the vector *y* in any way, nor have we created any new space-consuming vectors during intermediate computational steps.

Addresses within Vectors

There are two important functions for finding addresses within arrays. The function which is very easy to understand. The vector y (see above) looks like this:

y

```
[1]  8  3  5  7  6  6  8  9  2  3  9  4  10  4  11
```

Suppose we wanted to know which elements of y contained values bigger than 5. We type

which(y>5)

```
[1]  1  4  5  6  7  8  11  13  15
```

Notice that the answer to this enquiry is *a set of subscripts*. We don't use subscripts inside the which function itself. The function is applied to the whole array. To see the values of y that are larger than 5, we just type

y[y>5]

```
[1]  8  7  6  6  8  9  9  10  11
```

Note that this is a shorter vector than y itself, because values of 5 or less have been left out:

length(y)

```
[1]  15
```

length(y[y>5])

```
[1]  9
```

To extract every nth element from a long vector we can use seq as an index. In this case I want every 25th value in a 1000-long vector of normal random numbers with mean value 100 and standard deviation 10:

```
xv<-rnorm(1000,100,10)
xv[seq(25,length(xv),25)]
```

```
 [1]  100.98176  91.69614  116.69185   97.89538  108.48568  100.32891   94.46233
 [8]  118.05943  92.41213  100.01887  112.41775  106.14260   93.79951  105.74173
[15]  102.84938  88.56408  114.52787   87.64789  112.71475  106.89868  109.80862
[22]   93.20438  96.31240   85.96460  105.77331   97.54514   92.01761   97.78516
[29]   87.90883  96.72253   94.86647   90.87149   80.01337   97.98327   92.77398
[36]  121.47810  92.40182   87.65205  115.80945   87.60231
```

Finding Closest Values

Finding the value in a vector that is closest to a specified value is straightforward using which. Here, we want to find the value of xv that is closest to 108.0:

which(abs(xv-108)==min(abs(xv-108)))

```
[1]  332
```

The closest value to 108.0 is in location 332. But just how close to 108.0 is this 332nd value? We use 332 as a subscript on xv to find this out xv[332]

```
[1]  108.0076
```

Thus, we can write a function to return the closest value to a specified value (*sv*)

```
closest<-function(xv,sv){
xv[which(abs(xv-sv)==min(abs(xv-sv)))] }
```

and run it like this:

closest(xv,108)

```
[1]  108.0076
```

Trimming Vectors Using Negative Subscripts

Individual subscripts are referred to in square brackets. So if *x* is like this:

x<- c(5,8,6,7,1,5,3)

we can find the 4th element of the vector just by typing

x[4]

```
[1]  7
```

An extremely useful facility is to use negative subscripts to drop terms from a vector. Suppose we wanted a new vector, *z*, to contain everything but the first element of *x*

z <- x[-1]

z

```
[1]  8  6  7  1  5  3
```

Suppose our task is to calculate a trimmed mean of *x* which ignores both the smallest and largest values (i.e. we want to leave out the 1 and the 8 in this example). There are two steps to this. First, we sort the vector *x*. Then we remove the first element using x[-1] and the last using x[-length(x)]. We can do both drops at the same time by concatenating both instructions like this: -c(1,length(x)). Then we use the built-in function mean:

trim.mean <- function (x) mean(sort(x)[-c(1,length(x))])

Now try it out. The answer should be mean(c(5,6,7,5,3)) $= 26/5 = 5.2$:

trim.mean(x)

```
[1]  5.2
```

Suppose now that we need to produce a vector containing the numbers 1 to 50 but omitting all the multiples of seven (7, 14, 21, etc.). First make a vector of all the numbers 1 to 50 including the multiples of 7:

vec<-1:50

Now work out how many numbers there are between 1 and 50 that are multiples of 7

```
(multiples<-floor(50/7))
```

```
[1]   7
```

Now create a vector of the first seven multiples of 7 called *subscripts*:

```
(subscripts<-7*(1:multiples))
```

```
[1]   7  14  21  28  35  42  49
```

Finally, use negative subscripts to drop these multiples of 7 from the original vector

```
vec[-subscripts]
```

```
[1]    1   2   3   4   5   6   8   9  10  11  12  13  15  16  17  18  19  20  22  23  24  25
[23]  26  27  29  30  31  32  33  34  36  37  38  39  40  41  43  44  45  46  47  48  50
```

Alternatively, you could use modulo seven %%7 to get the result in a single line:

```
vec[-(1:50*(1:50%%7==0))]
```

```
[1]    1   2   3   4   5   6   8   9  10  11  12  13  15  16  17  18  19  20  22  23  24  25
[23]  26  27  29  30  31  32  33  34  36  37  38  39  40  41  43  44  45  46  47  48  50
```

Logical Arithmetic

Arithmetic involving logical expressions is very useful in programming and in selection of variables. If logical arithmetic is unfamiliar to you, then persevere with it, because it will become clear how useful it is, once the penny has dropped. The key thing to understand is that logical expressions evaluate to either true or false (represented in R by TRUE or FALSE), and that R can coerce TRUE or FALSE into numerical values: 1 for TRUE and 0 for FALSE. Suppose that x is a sequence from 0 to 6 like this:

```
x<-0:6
```

Now we can ask questions about the contents of the vector called x. Is x less than 4?

```
x<4
```

```
[1]   TRUE   TRUE   TRUE   TRUE   FALSE   FALSE   FALSE
```

The answer is yes for the first four values (0, 1, 2 and 3) and no for the last three (4, 5 and 6). Two important logical functions are all and any. They check an entire vector but return a single logical value: TRUE or FALSE. Are all the x values bigger than 0?

```
all(x>0)
```

```
[1]   FALSE
```

No. The first x value is a zero. Are any of the x values negative?

```
any(x<0)
```

```
[1]   FALSE
```

No. The smallest x value is a zero. We can use the answers of logical functions in arithmetic. We can count the true values of (x<4), using sum

```
sum(x<4)
```

```
[1]  4
```

or we can multiply (x<4) by other vectors

```
(x<4)*runif(7)
```

```
[1]  0.9433433  0.9382651  0.6248691  0.9786844  0.0000000  0.0000000
     0.0000000
```

Logical arithmetic is particularly useful in generating simplified factor levels during statistical modelling. Suppose we want to reduce a five-level factor called **treatment** to a three-level factor called t2 by lumping together the levels a and e (new factor level 1) and c and d (new factor level 3) while leaving b distinct (with new factor level 2):

```
(treatment<-letters[1:5])
```

```
[1]  "a"  "b"  "c"  "d"  "e"
```

```
(t2<-factor(1+(treatment=="b")+2*(treatment=="c")+2*(treatment=="d")))
```

```
[1]  1  2  3  3  1
```

```
Levels:  1  2  3
```

The new factor t2 gets a value 1 as default for all the factors levels, and we want to leave this as it is for levels a and e. Thus, we do not add anything to the 1 if the old factor level is a or e. For old factor level b, however, we want the result that $t2 = 2$ so we add 1 (treatment=="b") to the original 1 to get the answer we require. This works because the logical expression evaluates to 1 (TRUE) for every case in which the old factor level is b and to 0 (FALSE) in all other cases. For old factor levels c and d we want the result that $t2 = 3$ so we add 2 to the baseline value of 1 if the original factor level is either c (2*(treatment=="c")) or d (2*(treatment=="d")). You may need to read this several times before the penny drops. Note that 'logical equals' is a double = sign without a space between the two equals signs. You need to understand the distinction between:

x <- y   x is assigned the value of y (x *gets* the values of y);

x = y   in a function or a list x is set to y unless you specify otherwise;

x == y   produces TRUE if x is exactly equal to y and FALSE otherwise.

Evaluation of combinations of TRUE and FALSE

It is important to understand how combinations of logical variables evaluate, and to appreciate how logical operations (such as those in Table 2.3) work when there are missing values, NA. Here are all the possible outcomes expressed as a logical vector called x:

```
x <- c(NA, FALSE, TRUE)
```

```
names(x) <- as.character(x)
```

Table 2.3. Logical operations.

Symbol	Meaning
!	logical NOT
&	logical AND
|	logical OR
<	less than
<=	less than or equal to
>	greater than
>=	greater than or equal to
==	logical equals (double =)
!=	not equal
&&	AND with IF
||	OR with IF
xor(x,y)	exclusive OR
isTRUE(x)	an abbreviation of identical(TRUE,x)

To see the logical combinations of & (logical AND) we can use the outer function with x to evaluate all nine combinations of NA, FALSE and TRUE like this:

```
outer(x, x, "&")
```

```
         <NA>   FALSE    TRUE
<NA>       NA   FALSE      NA
FALSE   FALSE   FALSE   FALSE
 TRUE      NA   FALSE    TRUE
```

Only TRUE & TRUE evaluates to TRUE. Note the behaviour of NA & NA and NA & TRUE. Where one of the two components is NA, the result will be NA if the outcome is ambiguous. Thus, NA & TRUE evaluates to NA, but NA & FALSE evaluates to FALSE. To see the logical combinations of | (logical OR) write

```
outer(x, x, "|")
```

```
         <NA>   FALSE    TRUE
<NA>       NA      NA    TRUE
FALSE      NA   FALSE    TRUE
 TRUE    TRUE    TRUE    TRUE
```

Only FALSE | FALSE evaluates to FALSE. Note the behaviour of NA | NA and NA | FALSE.

Repeats

You will often want to generate repeats of numbers or characters, for which the function is rep. The object that is named in the first argument is repeated a number of times as specified in the second argument. At its simplest, we would generate five 9s like this:

```
rep(9,5)
```

```
[1]  9  9  9  9  9
```

You can see the issues involved by a comparison of these three increasingly complicated uses of the rep function:

rep(1:4, 2)

```
[1]  1  2  3  4  1  2  3  4
```

rep(1:4, each = 2)

```
[1]  1  1  2  2  3  3  4  4
```

rep(1:4, each = 2, times = 3)

```
[1]  1  1  2  2  3  3  4  4  1  1  2  2  3  3  4  4  1  1  2  2  3  3  4  4
```

In the simplest case, the *entire* first argument is repeated (i.e. the sequence 1 to 4 is repeated twice). You often want each *element* of the sequence to be repeated, and this is accomplished with the each argument. Finally, you might want each number repeated and the whole series repeated a certain number of times (here 3 times).

When each element of the series is to be repeated a different number of times, then the second argument must be a vector of the same length as the vector comprising the first argument (length 4 in this example). So if we want one 1, two 2s, three 3s and four 4s we would write:

rep(1:4,1:4)

```
[1]  1  2  2  3  3  3  4  4  4  4
```

In the most complex case, there is a different but irregular repeat of each of the elements of the first argument. Suppose that we need four 1s, one 2, four 3s and two 4s. Then we use the concatenation function c to create a vector of length 4 c(4,1,4,2)) which will act as the second argument to the rep function:

rep(1:4,c(4,1,4,2))

```
[1]  1  1  1  1  2  3  3  3  3  4  4
```

Generate Factor Levels

The function gl ('generate levels') is useful when you want to encode long vectors of factor levels: the syntax for the three arguments is this:

gl('up to', 'with repeats of', 'to total length')

Here is the simplest case where we want factor levels up to 4 with repeats of 3 repeated only once (i.e. to total length = 12):

gl(4,3)

```
[1]  1  1  1  2  2  2  3  3  3  4  4  4
Levels:  1  2  3  4
```

Here is the function when we want that whole pattern repeated twice:

gl(4,3,24)

```
[1]  1  1  1  2  2  2  3  3  3  4  4  4  1  1  1  2  2  2  3  3  3  4  4  4
Levels:  1  2  3  4
```

If the total length is not a multiple of the length of the pattern, the vector is truncated:

```
gl(4,3,20)
```

```
[1]  1  1  1  2  2  2  3  3  3  4  4  4  1  1  1  2  2  2  3  3
Levels:  1  2  3  4
```

If you want text for the factor levels, rather than numbers, use labels like this:

```
gl(3,2,24,labels=c("A","B","C"))
```

```
[1]  A  A  B  B  C  C  A  A  B  B  C  C  A  A  B  B  C  C  A  A  B  B  C  C
Levels:  A  B  C
```

Generating Regular Sequences of Numbers

For regularly spaced sequences, often involving integers, it is simplest to use the colon operator. This can produce ascending or descending sequences:

```
10:18
```

```
[1]  10  11  12  13  14  15  16  17  18
```

```
18:10
```

```
[1]  18  17  16  15  14  13  12  11  10
```

```
-0.5:8.5
```

```
[1]  -0.5  0.5  1.5  2.5  3.5  4.5  5.5  6.5  7.5  8.5
```

When the interval is not 1.0 you need to use the seq function. In general, the three arguments to seq are: initial value, final value, and increment (or decrement for a declining sequence). Here we want to go from 0 up to 1.5 in steps of 0.2:

```
seq(0,1.5,0.2)
```

```
[1]  0.0  0.2  0.4  0.6  0.8  1.0  1.2  1.4
```

Note that seq stops *before* it gets to the second argument (1.5) if the increment does not match exactly (our sequence stops at 1.4). If you want to seq downwards, the third argument needs to be negative

```
seq(1.5,0,-0.2)
```

```
[1]  1.5  1.3  1.1  0.9  0.7  0.5  0.3  0.1
```

Again, zero did not match the decrement, so was excluded and the sequence stopped at 0.1. Non-integer increments are particularly useful for generating x values for plotting smooth curves. A curve will look reasonably smooth if it is drawn with 100 straight line segments, so to generate 100 values of x between min(x) and max(x) you could write

```
x.values<-seq(min(x),max(x),(max(x)-min(x))/100)
```

If you want to create a sequence of the same length as an existing vector, then use along like this. Suppose that our existing vector, x, contains 18 random numbers from a normal distribution with a mean of 10.0 and a standard deviation of 2.0:

```
x<-rnorm(18,10,2)
```

and we want to generate a sequence of the same length as this (18) starting at 88 and stepping down to exactly 50 for x[18]

```
seq(88,50,along=x)
```

```
 [1] 88.00000  85.76471  83.52941  81.29412  79.05882  76.82353
     74.58824  72.35294
 [9] 70.11765  67.88235  65.64706  63.41176  61.17647  58.94118
     56.70588  54.47059
[17] 52.23529  50.00000
```

This is useful when you do not want to go to the trouble of working out the size of the increment but you do know the starting value (88 in this case) and the final value (50). If the vector is of length 18 then the sequence will involve 17 decrements of size:

```
(50-88)/17
```

```
[1] -2.235294
```

The function **sequence** (spelled out in full) is slightly different, because it can produce a *vector* consisting of sequences

```
sequence(5)
```

```
[1] 1 2 3 4 5
```

```
sequence(5:1)
```

```
[1] 1 2 3 4 5 1 2 3 4 1 2 3 1 2 1
```

```
sequence(c(5,2,4))
```

```
[1] 1 2 3 4 5 1 2 1 2 3 4
```

If the argument to **sequence** is itself a sequence (like 5:1) then several sequences are concatenated (in this case a sequence of 1 to 5 is followed by a sequence of 1 to 4 followed by a sequence of 1 to 3, another of 1 to 2 and a final sequence of 1 to 1 (= 1). The successive sequences need not be regular; the last example shows sequences to 5, then to 2, then to 4.

Variable Names

- Variable names in R are case-sensitive so x is not the same as X.
- Variable names should not begin with numbers (e.g. $1x$) or symbols (e.g. $\%x$).
- Variable names should not contain blank spaces: use *back.pay* (not *back pay*).

Sorting, Ranking and Ordering

These three related concepts are important, and one of them (**order**) is difficult to understand on first acquaintance. Let's take a simple example:

```
houses<-read.table("c:\\temp \\houses.txt",header=T)
attach(houses)
names(houses)
```

```
[1]  "Location"  "Price"
```

Now we apply the three different functions to the vector called Price,

```
ranks<-rank(Price)
sorted<-sort(Price)
ordered<-order(Price)
```

and make a dataframe out of the four vectors like this:

```
view<-data.frame(Price,ranks,sorted,ordered)
view
```

	Price	ranks	sorted	ordered
1	325	12.0	95	9
2	201	10.0	101	6
3	157	5.0	117	10
4	162	6.0	121	12
5	164	7.0	157	3
6	101	2.0	162	4
7	211	11.0	164	5
8	188	8.5	188	8
9	95	1.0	188	11
10	117	3.0	201	2
11	188	8.5	211	7
12	121	4.0	325	1

Rank

The prices themselves are in no particular sequence. The ranks column contains the value that is the rank of the particular data point (value of Price), where 1 is assigned to the lowest data point and length(Price) – here 12 – is assigned to the highest data point. So the first element, Price = 325, is the highest value in Price. You should check that there are 11 values smaller than 325 in the vector called Price. Fractional ranks indicate ties. There are two 188s in Price and their ranks are 8 and 9. Because they are tied, each gets the average of their two ranks $(8 + 9)/2 = 8.5$.

Sort

The sorted vector is very straightforward. It contains the values of Price sorted into ascending order. If you want to sort into descending order, use the reverse order function rev like this: y<-rev(sort(x)). Note that sort *is potentially very dangerous*, because it uncouples values that might need to be in the same row of the dataframe (e.g. because they are the explanatory variables associated with a particular value of the response variable). It is bad practice, therefore, to write x<-sort(x), not least because there is no 'unsort' function.

Order

This is the most important of the three functions, and much the hardest to understand on first acquaintance. The order function returns an integer vector containing the permutation that will sort the input into ascending order. You will need to think about this one. The lowest value of Price is 95. Look at the dataframe and ask yourself what is the subscript in the original vector called Price where 95 occurred. Scanning down the column, you find it in row number 9. This is the first value in ordered, ordered[1]. Where is the next smallest value (101) to be found within Price? It is in position 6, so this is ordered[2]. The third smallest Price (117) is in position 10, so this is ordered[3]. And so on.

This function is particularly useful in sorting dataframes, as explained on p. 113. Using order with subscripts is a much safer option than using sort, because with sort the values of the response variable and the explanatory variables could be uncoupled with potentially disastrous results if this is not realized at the time that modelling was carried out. The beauty of order is that we can use order(Price) as a subscript for Location to obtain the price-ranked list of locations:

Location[order(Price)]

```
[1] Reading     Staines      Winkfield Newbury
[5] Bracknell   Camberley    Bagshot     Maidenhead
[9] Warfield    Sunninghill  Windsor     Ascot
```

When you see it used like this, you can see exactly why the function is called order. If you want to reverse the order, just use the rev function like this:

Location[rev(order(Price))]

```
[1] Ascot       Windsor      Sunninghill Warfield
[5] Maidenhead  Bagshot      Camberley   Bracknell
[9] Newbury     Winkfield    Staines     Reading
```

The sample Function

This function shuffles the contents of a vector into a random sequence while maintaining all the numerical values intact. It is extremely useful for randomization in experimental design, in simulation and in computationally intensive hypothesis testing. Here is the original y vector again:

y

```
[1]  8  3  5  7  6  6  8  9  2  3  9  4  10  4  11
```

and here are two samples of y:

sample(y)

```
[1]  8  8  9  9  2  10  6  7  3  11  5  4  6  3  4
```

sample(y)

```
[1]  9  3  9  8  8  6  5  11  4  6  4  7  3  2  10
```

The order of the values is different each time that sample is invoked, but the same numbers are shuffled in every case. This is called *sampling without replacement*. You can specify the size of the sample you want as an optional second argument:

```
sample(y,5)
```

```
[1]  9  4  10  8  11
```

```
sample(y,5)
```

```
[1] 9  3  4  2  8
```

The option **replace=T** allows for *sampling with replacement*, which is the basis of boot-strapping (see p. 320). The vector produced by the **sample** function with **replace=T** is the same length as the vector sampled, but some values are left out at random and other values, again at random, appear two or more times. In this sample, 10 has been left out, and there are now three 9s:

```
sample(y,replace=T)
```

```
[1]  9  6  11  2  9  4  6  8  8  4  4  4  3  9  3
```

In this next case, the are two 10s and only one 9:

```
sample(y,replace=T)
```

```
[1]  3  7  10  6  8  2  5  11  4  6  3  9  10  7  4
```

More advanced options in **sample** include specifying different probabilities with which each element is to be sampled (**prob=**). For example, if we want to take four numbers at random from the sequence 1:10 without replacement where the probability of selection (p) is 5 times greater for the middle numbers (5 and 6) than for the first or last numbers, and we want to do this five times, we could write

```
p <- c(1, 2, 3, 4, 5, 5, 4, 3, 2, 1)
```

```
x<-1:10
```

```
sapply(1:5,function(i) sample(x,4,prob=p))
```

```
      [,1]  [,2]  [,3]  [,4]  [,5]
[1,]    8     7     4    10     8
[2,]    7     5     7     8     7
[3,]    4     4     3     4     5
[4,]    9    10     8     7     6
```

so the four random numbers in the first trial were 8, 7, 4 and 9 (i.e. column 1).

Matrices

There are several ways of making a matrix. You can create one directly like this:

```
X<-matrix(c(1,0,0,0,1,0,0,0,1),nrow=3)
```

```
X
```

```
      [,1]  [,2]  [,3]
[1,]    1     0     0
[2,]    0     1     0
[3,]    0     0     1
```

where, by default, the numbers are entered columnwise. The class and attributes of X indicate that it is a matrix of three rows and three columns (these are its dim attributes)

class(X)

```
[1]  "matrix"
```

attributes(X)

```
$dim
```

```
[1]  3  3
```

In the next example, the data in the vector appear row-wise, so we indicate this with byrow=T:

vector<-c(1,2,3,4,4,3,2,1)
V<-matrix(vector,byrow=T,nrow=2)
V

```
     [,1] [,2] [,3] [,4]
[1,]    1    2    3    4
[2,]    4    3    2    1
```

Another way to convert a vector into a matrix is by providing the vector object with two dimensions (rows and columns) using the dim function like this:

dim(vector)<-c(4,2)

We can check that vector has now become a matrix:

is.matrix(vector)

```
[1]  TRUE
```

We need to be careful, however, because we have made no allowance at this stage for the fact that the data were entered row-wise into vector:

vector

```
     [,1] [,2]
[1,]    1    4
[2,]    2    3
[3,]    3    2
[4,]    4    1
```

The matrix we want is the transpose, t, of this matrix:

(vector<-t(vector))

```
     [,1] [,2] [,3] [,4]
[1,]    1    2    3    4
[2,]    4    3    2    1
```

Naming the rows and columns of matrices

At first, matrices have numbers naming their rows and columns (see above). Here is a 4 × 5 matrix of random integers from a Poisson distribution with mean = 1.5:

```
X<-matrix(rpois(20,1.5),nrow=4)
X
```

```
     [,1] [,2] [,3] [,4] [,5]
[1,]    1    0    2    5    3
[2,]    1    1    3    1    3
[3,]    3    1    0    2    2
[4,]    1    0    2    1    0
```

Suppose that the rows refer to four different trials and we want to label the rows 'Trial.1' etc. We employ the function rownames to do this. We could use the paste function (see p. 44) but here we take advantage of the prefix option:

```
rownames(X)<-rownames(X,do.NULL=FALSE,prefix="Trial.")
X
```

```
          [,1] [,2] [,3] [,4] [,5]
Trial.1      1    0    2    5    3
Trial.2      1    1    3    1    3
Trial.3      3    1    0    2    2
Trial.4      1    0    2    1    0
```

For the columns we want to supply a vector of different names for the five drugs involved in the trial, and use this to specify the colnames(X):

```
drug.names<-c("aspirin", "paracetamol", "nurofen", "hedex", "placebo")
colnames(X)<-drug.names
X
```

```
          aspirin  paracetamol  nurofen  hedex  placebo
Trial.1         1            0        2      5        3
Trial.2         1            1        3      1        3
Trial.3         3            1        0      2        2
Trial.4         1            0        2      1        0
```

Alternatively, you can use the dimnames function to give names to the rows and/or columns of a matrix. In this example we want the rows to be unlabelled (NULL) and the column names to be of the form 'drug.1', 'drug.2', etc. The argument to dimnames has to be a list (rows first, columns second, as usual) with the elements of the list of exactly the correct lengths (4 and 5 in this particular case):

```
dimnames(X)<-list(NULL,paste("drug.",1:5,sep=""))
X
```

```
     drug.1  drug.2  drug.3  drug.4  drug.5
[1,]      1       0       2       5       3
[2,]      1       1       3       1       3
[3,]      3       1       0       2       2
[4,]      1       0       2       1       0
```

Calculations on rows or columns of the matrix

We could use subscripts to select parts of the matrix, with a blank meaning 'all of the rows' or 'all of the columns'. Here is the mean of the rightmost column (number 5), mean(X[,5])

```
[1]  2
```

calculated over all the rows (blank then comma), and the variance of the bottom row,

var(X[4,])

```
[1]  0.7
```

calculated over all of the columns (a blank in the second position). There are some special functions for calculating summary statistics on matrices:

rowSums(X)

```
[1]  11  9  8  4
```

colSums(X)

```
[1]  6  2  7  9  8
```

rowMeans(X)

```
[1]  2.2  1.8  1.6  0.8
```

colMeans(X)

```
[1]  1.50  0.50  1.75  2.25  2.00
```

These functions are built for speed, and blur some of the subtleties of dealing with NA or NaN. If such subtlety is an issue, then use apply instead (p. 68). Remember that columns are margin no. 2 (rows are margin no. 1):

apply(X,2,mean)

```
[1]  1.50  0.50  1.75  2.25  2.00
```

You might want to sum groups of rows within columns, and rowsum (singular and all lower case, in contrast to rowSums, above) is a very efficient function for this. In this case we want to group together row 1 and row 4 (as group A) and row 2 and row 3 (group B). Note that the grouping vector has to have length equal to the number of rows:

group=c("A","B","B","A")

rowsum(X, group)

```
   [,1]  [,2]  [,3]  [,4]  [,5]
A    2     0     4     6     3
B    4     2     3     3     5
```

You could achieve the same ends (but more slowly) with tapply or aggregate:

tapply(X, list(group[row(X)], col(X)), sum)

```
   1  2  3  4  5
A  2  0  4  6  3
B  4  2  3  3  5
```

Note the use of row(X) and col(X), with row(X) used as a subscript on group.

aggregate(X,list(group),sum)

```
   Group.1  V1  V2  V3  V4  V5
1        A   2   0   4   6   3
2        B   4   2   3   3   5
```

Suppose that we want to shuffle the elements of each column of a matrix independently. We apply the function sample to each column (margin no. 2) like this:

apply(X,2,sample)

```
      [,1]  [,2]  [,3]  [,4]  [,5]
[1,]    1     1     2     1     3
[2,]    3     1     0     1     3
[3,]    1     0     3     2     0
[4,]    1     0     2     5     2
```

apply(X,2,sample)

```
      [,1]  [,2]  [,3]  [,4]  [,5]
[1,]    1     1     0     5     2
[2,]    1     1     2     1     3
[3,]    3     0     2     2     3
[4,]    1     0     3     1     0
```

and so on, for as many shuffled samples as you need.

Adding rows and columns to the matrix

In this particular case we have been asked to add a row at the bottom showing the column means, and a column at the right showing the row variances:

X<-rbind(X,apply(X,2,mean))
X<-cbind(X,apply(X,1,var))
X

```
      [,1]  [,2]  [,3]  [,4]  [,5]      [,6]
[1,]   1.0   0.0  2.00  5.00     3   3.70000
[2,]   1.0   1.0  3.00  1.00     3   1.20000
[3,]   3.0   1.0  0.00  2.00     2   1.30000
[4,]   1.0   0.0  2.00  1.00     0   0.70000
[5,]   1.5   0.5  1.75  2.25     2   0.45625
```

Note that the number of decimal places varies across columns, with one in columns 1 and 2, two in columns 3 and 4, none in column 5 (integers) and five in column 6. The default in R is to print the minimum number of decimal places consistent with the contents of the column as a whole.

Next, we need to label the sixth column as 'variance' and the fifth row as 'mean':

colnames(X)<-c(1:5,"variance")
rownames(X)<-c(1:4,"mean")
X

```
     1     2     3     4   5   variance
1   1.0   0.0  2.00  5.00   3   3.70000
2   1.0   1.0  3.00  1.00   3   1.20000
3   3.0   1.0  0.00  2.00   2   1.30000
4   1.0   0.0  2.00  1.00   0   0.70000
```

```
mean  1.5  0.5  1.75  2.25  2  0.45625
```

When a matrix with a single row or column is created by a subscripting operation, for example row <- mat[2,], it is by default turned into a vector. In a similar way, if an array with dimension, say, $2 \times 3 \times 1 \times 4$ is created by subscripting it will be coerced into a $2 \times 3 \times 4$ array, losing the unnecessary dimension. After much discussion this has been determined to be a *feature* of R. To prevent this happening, add the option drop = FALSE to the subscripting. For example,

```
rowmatrix <- mat[2, , drop = FALSE]
colmatrix <- mat[, 2, drop = FALSE]
a <- b[1, 1, 1, drop = FALSE]
```

The drop = FALSE option should be used defensively when programming. For example, the statement

```
somerows <- mat[index,]
```

will return a vector rather than a matrix if index happens to have length 1, and this might cause errors later in the code. It should be written as

```
somerows <- mat[index , , drop = FALSE]
```

The sweep function

The sweep function is used to 'sweep out' array summaries from vectors, matrices, arrays or dataframes. In this example we want to express a matrix in terms of the departures of each value from its column mean.

```
matdata<-read.table("c: \\temp \\sweepdata.txt")
```

First, you need to create a vector containing the parameters that you intend to sweep out of the matrix. In this case we want to compute the four column means:

```
(cols<-apply(matdata,2,mean))
   V1      V2      V3      V4
 4.60   13.30    0.44  151.60
```

Now it is straightforward to express all of the data in matdata as departures from the relevant column means:

```
sweep(matdata,2,cols)
        V1      V2      V3      V4
1     -1.6    -1.3   -0.04   -26.6
2      0.4    -1.3    0.26    14.4
3      2.4     1.7    0.36    22.4
4      2.4     0.7    0.26   -23.6
5      0.4     4.7   -0.14   -15.6
6      4.4    -0.3   -0.24     3.4
7      2.4     1.7    0.06   -36.6
8     -2.6    -0.3    0.06    17.4
9     -3.6    -3.3   -0.34    30.4
10    -4.6    -2.3   -0.24    14.4
```

Note the use of margin = 2 as the second argument to indicate that we want the sweep to be carried out on the columns (rather than on the rows). A related function, scale, is used for centring and scaling data in terms of standard deviations (p. 191).

You can see what sweep has done by doing the calculation long-hand. The operation of this particular sweep is simply one of subtraction. The only issue is that the subtracted object has to have the same dimensions as the matrix to be swept (in this example, 10 rows of 4 columns). Thus, to sweep out the column means, the object to be subtracted from matdata must have the each column mean repeated in each of the 10 rows of 4 columns:

```
(col.means<-matrix(rep(cols,rep(10,4)),nrow=10))
```

```
        [,1]  [,2]  [,3]   [,4]
 [1,]   4.6  13.3  0.44  151.6
 [2,]   4.6  13.3  0.44  151.6
 [3,]   4.6  13.3  0.44  151.6
 [4,]   4.6  13.3  0.44  151.6
 [5,]   4.6  13.3  0.44  151.6
 [6,]   4.6  13.3  0.44  151.6
 [7,]   4.6  13.3  0.44  151.6
 [8,]   4.6  13.3  0.44  151.6
 [9,]   4.6  13.3  0.44  151.6
[10,]   4.6  13.3  0.44  151.6
```

Then the same result as we got from sweep is obtained simply by

```
matdata-col.means
```

Suppose that you want to obtain the subscripts for a columnwise or a row-wise sweep of the data. Here are the row subscripts repeated in each column:

```
apply(matdata,2,function (x) 1:10)
```

```
        V1  V2  V3  V4
 [1,]    1   1   1   1
 [2,]    2   2   2   2
 [3,]    3   3   3   3
 [4,]    4   4   4   4
 [5,]    5   5   5   5
 [6,]    6   6   6   6
 [7,]    7   7   7   7
 [8,]    8   8   8   8
 [9,]    9   9   9   9
[10,]   10  10  10  10
```

Here are the column subscripts repeated in each row:

```
t(apply(matdata,1,function (x) 1:4))
```

```
     [,1]  [,2]  [,3]  [,4]
1      1     2     3     4
2      1     2     3     4
3      1     2     3     4
4      1     2     3     4
5      1     2     3     4
6      1     2     3     4
7      1     2     3     4
8      1     2     3     4
9      1     2     3     4
10     1     2     3     4
```

Here is the same procedure using **sweep**:

sweep(matdata,1,1:10,function(a,b) b)

```
        [,1]  [,2]  [,3]  [,4]
[1,]      1     1     1     1
[2,]      2     2     2     2
[3,]      3     3     3     3
[4,]      4     4     4     4
[5,]      5     5     5     5
[6,]      6     6     6     6
[7,]      7     7     7     7
[8,]      8     8     8     8
[9,]      9     9     9     9
[10,]    10    10    10    10
```

sweep(matdata,2,1:4,function(a,b) b)

```
        [,1]  [,2]  [,3]  [,4]
[1,]      1     2     3     4
[2,]      1     2     3     4
[3,]      1     2     3     4
[4,]      1     2     3     4
[5,]      1     2     3     4
[6,]      1     2     3     4
[7,]      1     2     3     4
[8,]      1     2     3     4
[9,]      1     2     3     4
[10,]     1     2     3     4
```

Arrays

Arrays are numeric objects with dimension attributes. We start with the numbers 1 to 25 in a vector called array:

```
array<-1:25
is.matrix(array)
```

```
[1]  FALSE
```

```
dim(array)
```

```
NULL
```

The vector is not a matrix and it has no (NULL) dimensional attributes. We give the object dimensions like this (say, with five rows and five columns):

```
dim(array)<-c(5,5)
```

Now it does have dimensions and it is a matrix:

```
dim(array)
```

```
[1]  5  5
```

```
is.matrix(array)
```

```
[1]  TRUE
```

When we look at array it is presented as a two-dimensional table (but note that it is *not* a table object; see p. 187):

```
array
```

```
      [,1] [,2] [,3] [,4] [,5]
[1,]    1    6   11   16   21
[2,]    2    7   12   17   22
[3,]    3    8   13   18   23
[4,]    4    9   14   19   24
[5,]    5   10   15   20   25
```

```
is.table(array)
```

```
[1]  FALSE
```

Note that the values have been entered into array in columnwise sequence: this is the default in R. Thus a vector is a one-dimensional array that lacks any dim attributes. A matrix is a two-dimensional array. Arrays of three or more dimensions do not have any special names in R; they are simply referred to as three-dimensional or five-dimensional arrays. You should practise with subscript operations on arrays until you are thoroughly familiar with them. Mastering the use of subscripts will open up many of R's most powerful features for working with dataframes, vectors, matrices, arrays and lists. Here is a three-dimensional array of the first 24 lower-case letters with three matrices each of four rows and two columns:

```
A<-letters[1:24]
dim(A)<-c(4,2,3)
A
```

```
, , 1
      [,1] [,2]
[1,]   "a"  "e"
[2,]   "b"  "f"
[3,]   "c"  "g"
[4,]   "d"  "h"

, , 2
      [,1] [,2]
[1,]   "i"  "m"
[2,]   "j"  "n"
```

```
[3,]    "k"    "o"
[4,]    "l"    "p"

, , 3
        [,1]  [,2]
[1,]    "q"    "u"
[2,]    "r"    "v"
[3,]    "s"    "w"
[4,]    "t"    "x"
```

We want to select all the letters a to p. These are all the rows and all the columns of tables 1 and 2, so the appropriate subscripts are [,,1:2]

A[,,1:2]

```
, , 1

        [,1]  [,2]
[1,]    "a"    "e"
[2,]    "b"    "f"
[3,]    "c"    "g"
[4,]    "d"    "h"

, , 2
        [,1]  [,2]
[1,]    "i"    "m"
[2,]    "j"    "n"
[3,]    "k"    "o"
[4,]    "l"    "p"
```

Next, we want only the letters q to x. These are all the rows and all the columns from the third table, so the appropriate subscripts are [,,3]:

A[,,3]

```
        [,1]  [,2]
[1,]    "q"    "u"
[2,]    "r"    "v"
[3,]    "s"    "w"
[4,]    "t"    "x"
```

Here, we want only c, g, k, o, s and w. These are the third rows of all three tables, so the appropriate subscripts are [3,,]:

A[3,,]

```
        [,1]  [,2]  [,3]
[1,]    "c"    "k"    "s"
[2,]    "g"    "o"    "w"
```

Note that when we drop the whole first dimension (there is just one row in A[3,,]) the shape of the resulting matrix is altered (two rows and three columns in this example). This is a *feature* of R, but you can override it by saying drop = F to retain all three dimensions:

A[3,,,drop=F]

```
, , 1

        [,1]  [,2]
[1,]    "c"   "g"

, , 2

        [,1]  [,2]
[1,]    "k"   "o"

, , 3

        [,1]  [,2]
[1,]    "s"   "w"
```

Finally, suppose we want all the rows of the second column from table 1, A[,2,1], the first column from table 2, A[,1,2], and the second column from table 3, A[,2,3]. Because we want all the rows in each case, so the first subscript is blank, but we want different column numbers (cs) in different tables (ts) as follows:

```
cs<-c(2,1,2)
ts<-c(1,2,3)
```

To get the answer, use sapply to concatenate the columns from each table like this:

```
sapply (1:3, function(i) A[,cs[i],ts[i]])

        [,1]  [,2]  [,3]
[1,]    "e"   "i"   "u"
[2,]    "f"   "j"   "v"
[3,]    "g"   "k"   "w"
[4,]    "h"   "l"   "x"
```

Character Strings

In R, character strings are defined by double quotation marks:

```
a<-"abc"
b<-"123"
```

Numbers can be characters (as in *b*, above), but characters cannot be numbers.

```
as.numeric(a)
```

```
[1]  NA
Warning message:
NAs introduced by coercion
as.numeric(b)
```

```
[1]  123
```

One of the initially confusing things about character strings is the distinction between the length of a character object (a vector) and the numbers of characters in the strings comprising that object. An example should make the distinction clear:

```
pets<-c("cat","dog","gerbil","terrapin")
```

Here, pets is a vector comprising four character strings:

length(pets)

```
[1]   4
```

and the individual character strings have 3, 3, 6 and 7 characters, respectively:

nchar(pets)

```
[1]  3  3  6  7
```

When first defined, character strings are not factors:

class(pets)

```
[1]  "character"
```

is.factor(pets)

```
[1]  FALSE
```

However, if the vector of characters called pets was part of a dataframe, then R would coerce all the character variables to act as factors:

df<-data.frame(pets)
is.factor(df$pets)

```
[1]   TRUE
```

There are built-in vectors in R that contain the 26 letters of the alphabet in lower case (letters) and in upper case (LETTERS):

letters

```
[1]  "a" "b" "c" "d" "e" "f" "g" "h" "i" "j" "k" "l" "m" "n" "o" "p"
[17] "q" "r" "s" "t" "u" "v" "w" "x" "y" "z"
```

LETTERS

```
[1]  "A" "B" "C" "D" "E" "F" "G" "H" "I" "J" "K" "L" "M" "N" "O" "P"
[17] "Q" "R" "S" "T" "U" "V" "W" "X" "Y" "Z"
```

To discover which number in the alphabet the letter n is, you can use the which function like this:

which(letters=="n")

```
[1]   14
```

For the purposes of printing you might want to suppress the quotes that appear around character strings by default. The function to do this is called noquote:

noquote(letters)

```
[1] a  b  c  d  e  f  g  h  i  j  k  l  m  n  o  p  q  r  s  t  u  v  w  x  y  z
```

You can amalgamate strings into vectors of character information:

c(a,b)

```
[1]  "abc"  "123"
```

This shows that the concatenation produces a vector of two strings. It does *not* convert two 3-character strings into one 6-charater string. The R function to do that is **paste**:

```
paste(a,b,sep="")
```

```
[1]  "abc123"
```

The third argument, sep="", means that the two character strings are to be pasted together without any separator between them: the default for paste is to insert a single blank space, like this:

```
paste(a,b)
```

```
[1]  "abc 123"
```

Notice that you do *not* lose blanks that are within character strings when you use the sep="" option in paste.

```
paste(a,b,"a longer phrase containing blanks",sep="")
```

```
[1]  "abc123a longer phrase containing blanks"
```

If one of the arguments to paste is a vector, each of the elements of the vector is pasted to the specified character string to produce an object of the same length as the vector:

```
d<-c(a,b,"new")
e<-paste(d,"a longer phrase containing blanks")
e
```

```
[1]  "abc a longer phrase containing blanks"
[2]  "123 a longer phrase containing blanks"
[3]  "new a longer phrase containing blanks"
```

Extracting parts of strings

We being by defining a phrase:

```
phrase<-"the quick brown fox jumps over the lazy dog"
```

The function called substr is used to extract substrings of a specified number of characters from a character string. Here is the code to extract the first, the first and second, the first, second and third, ... (up to 20) characters from our phrase

```
q<-character(20)
for (i in 1:20) q[i]<- substr(phrase,1,i)
q
```

```
[1]   "t"                  "th"                 "the"
[4]   "the "               "the q"              "the qu"
[7]   "the qui"            "the quic"           "the quick"
[10]  "the quick "         "the quick b"        "the quick br"
[13]  "the quick bro"      "the quick brow"     "the quick brown"
[16]  "the quick brown "   "the quick brown f"  "the quick brown fo"
[19]  "the quick brown fox "
```

The second argument in substr is the number of the character at which extraction is to begin (in this case always the first), and the third argument is the number of the character at which extraction is to end (in this case, the *i*th). To split up a character string into individual characters, we use strsplit like this

```
strsplit(phrase,split=character(0))
```

```
[[1]]
[1]    "t" "h" "e" " " "q" "u" "i" "c" "k" " " "b" "r" "o" "w" "n" " "
[17]   "f" "o" "x" " " "j" "u" "m" "p" "s" " " "o" "v" "e" "r"
[31]   " " "t" "h" "e" " " "l" "a" "z" "y" " " "d" "o" "g"
```

The table function is useful for counting the number of occurrences of characters of different kinds:

```
table(strsplit(phrase,split=character(0)))
```

a	b	c	d	e	f	g	h	i	j	k	l	m	n	o	p	q	r	s	t	u	v	w	x	y	z
8	1	1	1	3	1	1	2	1	1	1	1	1	1	4	1	1	2	1	2	2	1	1	1	1	1

This demonstrates that all of the letters of the alphabet were used at least once within our phrase, and that there were 8 blanks within phrase. This suggests a way of counting the number of words in a phrase, given that this will always be one more than the number of blanks:

```
words<-1+table(strsplit(phrase,split=character(0)))[1]
words
```

```
9
```

When we specify a particular string to form the basis of the split, we end up with a list made up from the components of the string that *do not contain the specified string*. This is hard to understand without an example. Suppose we split our phrase using 'the':

```
strsplit(phrase,"the")
```

```
[[1]]

[1]  ""   " quick brown fox jumps over "   " lazy dog"
```

There are three elements in this list: the first one is the empty string "" because the first three characters within phrase were exactly 'the' ; the second element contains the part of the phrase between the two occurrences of the string 'the'; and the third element is the end of the phrase, following the second 'the'. Suppose that we want to extract the characters between the first and second occurrences of 'the'. This is achieved very simply, using subscripts to extract the second element of the list:

```
strsplit(phrase,"the")[[1]] [2]
```

```
[1]   " quick brown fox jumps over "
```

Note that the first subscript in double square brackets refers to the number within the list (there *is* only one list in this case) and the second subscript refers to the second element within this list. So if we want to know how many characters there are between the first and second occurrences of the word "the" within our phrase, we put:

```
nchar(strsplit(phrase,"the")[[1]] [2])
```

```
[1]  28
```

It is easy to switch between upper and lower cases using the toupper and tolower functions:

```
toupper(phrase)
```

```
[1]    "THE QUICK BROWN FOX JUMPS OVER THE LAZY DOG"
```

tolower(toupper(phrase))

```
[1]    "the quick brown fox jumps over the lazy dog"
```

The match Function

The match function answers the question 'Where do the values in the second vector appear in the first vector?'. This is impossible to understand without an example:

first<-c(5,8,3,5,3,6,4,4,2,8,8,8,4,4,6)
second<-c(8,6,4,2)
match(first,second)

```
[1]  NA  1  NA  NA  NA  2  3  3  4  1  1  1  3  3  2
```

The first thing to note is that match produces a vector of subscripts (index values) and that these are subscripts within the *second* vector. The length of the vector produced by match is the length of the *first* vector (15 in this example). If elements of the first vector do not occur anywhere in the second vector, then match produces NA.

Why would you ever want to use this? Suppose you wanted to give drug A to all the patients in the first vector that were identified in the second vector, and drug B to all the others (i.e. those identified by NA in the output of match, above, because they did *not* appear in the second vector). You create a vector called drug with two elements (A and B), then select the appropriate drug on the basis of whether or not match(first,second) is NA:

drug<-c("A","B")
drug[1+is.na(match(first,second))]

```
[1] "B"  "A"  "B"  "B"  "B"  "A"  "A"  "A"  "A"  "A"  "A"  "A"  "A"  "A"  "A"
```

The match function can also be very powerful in manipulating dataframes to mimic the functionality of a relational database (p. 127).

Writing functions in R

Functions in R are objects that carry out operations on *arguments* that are supplied to them and return one or more values. The syntax for writing a function is

function (argument list) body

The first component of the function declaration is the keyword function, which indicates to R that you want to create a function. An argument list is a comma-separated list of formal arguments. A formal argument can be a symbol (i.e. a variable name such as x or y), a statement of the form symbol = expression (e.g. pch=16) or the special formal argument . . . (triple dot). The body can be any valid R expression or set of R expressions. Generally, the body is a group of expressions contained in curly brackets { }, with each expression on a separate line. Functions are typically assigned to symbols, but they need not be. This will only begin to mean anything after you have seen several examples in operation.

Arithmetic mean of a single sample

The mean is the sum of the numbers $\sum y$ divided by the number of numbers $n = \sum 1$ (summing over the number of numbers in the vector called y). The R function for n is length(y) and for $\sum y$ is sum(y), so a function to compute arithmetic means is

```
arithmetic.mean<-function(x)   sum(x)/length(x)
```

We should test the function with some data where we know the right answer:

```
y<-c(3,3,4,5,5)
```

```
arithmetic.mean(y)
```

```
[1]   4
```

Needless to say, there is a built-in function for arithmetic means called mean:

```
mean(y)
```

```
[1]   4
```

Median of a single sample

The median (or 50th percentile) is the middle value of the sorted values of a vector of numbers:

```
sort(y)[ceiling(length(y)/2)]
```

There is slight hitch here, of course, because if the vector contains an even number of numbers, then there *is* no middle value. The logic is that we need to work out the arithmetic average of the two values of y on either side of the middle. The question now arises as to how we know, in general, whether the vector y contains an odd or an even number of numbers, so that we can decide which of the two methods to use. The trick here is to use modulo 2 (p. 12). Now we have all the tools we need to write a general function to calculate medians. Let's call the function med and define it like this:

```
med<-function(x) {
odd.even<-length(x)%%2
if (odd.even == 0) (sort(x)[length(x)/2]+sort(x)[1+ length(x)/2])/2
else sort(x)[ceiling(length(x)/2)]
}
```

Notice that when the if statement is true (i.e. we have an even number of numbers) then the expression immediately following the if function is evaluated (this is the code for calculating the median with an even number of numbers). When the if statement is false (i.e. we have an odd number of numbers, and odd.even == 1) then the expression following the else function is evaluated (this is the code for calculating the median with an odd number of numbers). Let's try it out, first with the odd-numbered vector y, then with the even-numbered vector y[-1], after the first element of y (y[1] = 3) has been dropped (using the negative subscript):

```
med(y)
```

```
[1]   4
```

```
med(y[-1])
```

```
[1]   4.5
```

Again, you won't be surprised that there is a built-in function for calculating medians, and helpfully it is called median.

Geometric mean

For processes that change multiplicatively rather than additively, neither the arithmetic mean nor the median is an ideal measure of central tendency. Under these conditions, the appropriate measure is the geometric mean. The formal definition of this is somewhat abstract: the geometric mean is the nth root of the product of the data. If we use capital Greek pi ($\prod$) to represent multiplication, and $\hat{y}$ (pronounced y-hat) to represent the geometric mean, then

$$\hat{y} = \sqrt[n]{\prod y}.$$

Let's take a simple example we can work out by hand: the numbers of insects on 5 plants were as follows: 10, 1, 1000, 1, 10. Multiplying the numbers together gives 100 000. There are five numbers, so we want the fifth root of this. Roots are hard to do in your head, so we'll use R as a calculator. Remember that roots are fractional powers, so the fifth root is a number raised to the power $1/5 = 0.2$. In R, powers are denoted by the $\char`\^$ symbol:

```
100000^0.2
```

```
[1] 10
```

So the geometric mean of these insect numbers is 10 insects per stem. Note that two of the data were exactly like this, so it seems a reasonable estimate of central tendency. The arithmetic mean, on the other hand, is a hopeless measure of central tendency, because the large value (1000) is so influential: it is given by $(10 + 1 + 1000 + 1 + 10)/5 = 204.4$, and none of the data is close to it.

```
insects<-c(1,10,1000,10,1)
mean(insects)
```

```
[1] 204.4
```

Another way to calculate geometric mean involves the use of logarithms. Recall that to multiply numbers together we add up their logarithms. And to take roots, we divide the logarithm by the root. So we should be able to calculate a geometric mean by finding the antilog (exp) of the average of the logarithms (log) of the data:

```
exp(mean(log(insects)))
```

```
[1]  10
```

So a function to calculate geometric mean of a vector of numbers x:

```
geometric<-function (x) exp(mean(log(x)))
```

and testing it with the insect data

```
geometric(insects)
```

```
[1]  10
```

The use of geometric means draws attention to a general scientific issue. Look at the figure below, which shows numbers varying through time in two populations. Now ask yourself which population is the more variable. Chances are, you will pick the upper line:

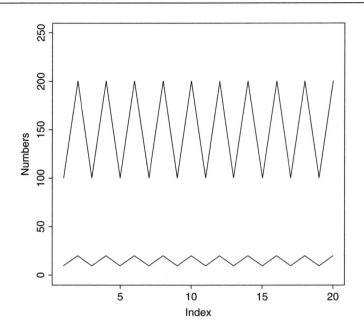

But now look at the scale on the *y* axis. The upper population is fluctuating 100, 200, 100, 200 and so on. In other words, it is doubling and halving, doubling and halving. The lower curve is fluctuating 10, 20, 10, 20, 10, 20 and so on. It, too, is doubling and halving, doubling and halving. So the answer to the question is that they are equally variable. It is just that one population has a higher mean value than the other (150 vs. 15 in this case). In order not to fall into the trap of saying that the upper curve is more variable than the lower curve, it is good practice to graph the logarithms rather than the raw values of things like population sizes that change multiplicatively, as below.

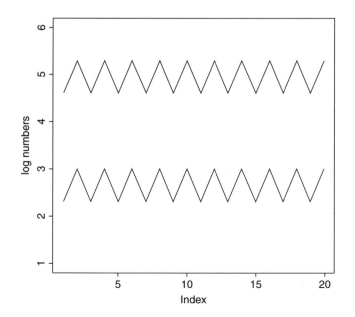

Now it is clear that both populations are equally variable. Note the change of scale, as specified using the ylim=c(1,6) option within the plot function (p. 181).

Harmonic mean

Consider the following problem. An elephant has a territory which is a square of side $= 2$ km. Each morning, the elephant walks the boundary of this territory. He begins the day at a sedate pace, walking the first side of the territory at a speed of 1 km/hr. On the second side, he has sped up to 2 km/hr. By the third side he has accelerated to an impressive 4 km/hr, but this so wears him out, that he has to return on the final side at a sluggish 1 km/hr. So what is his average speed over the ground? You might say he travelled at 1, 2, 4 and 1 km/hr so the average speed is $(1+2+4+1)/4 = 8/4 = 2$ km/hr. But that is wrong. Can you see how to work out the right answer? Recall that velocity is defined as distance travelled divided by time taken. The distance travelled is easy: it's just $4 \times 2 = 8$ km. The time taken is a bit harder. The first edge was 2 km long, and travelling at 1 km/hr this must have taken 2 hr. The second edge was 2 km long, and travelling at 2 km/hr this must have taken 1 hr. The third edge was 2 km long and travelling at 4 km/hr this must have taken 0.5 hr. The final edge was 2 km long and travelling at 1 km/hr this must have taken 2 hr. So the total time taken was $2+1+0.5+2 = 5.5$ hr. So the average speed is not 2 km/hr but $8/5.5 = 1.4545$ km/hr. The way to solve this problem is to use the **harmonic mean**.

The harmonic mean is the reciprocal of the average of the reciprocals. The average of our reciprocals is

$$\frac{1}{1} + \frac{1}{2} + \frac{1}{4} + \frac{1}{1} = \frac{2.75}{4} = 0.6875.$$

The reciprocal of this average is the harmonic mean

$$\frac{4}{2.75} = \frac{1}{0.6875} = 1.4545.$$

In symbols, therefore, the harmonic mean, $\tilde{y}$ (y-curl), is given by

$$\tilde{y} = \frac{1}{(\Sigma(1/y))/n} = \frac{n}{\Sigma(1/y)}.$$

An R function for calculating harmonic means, therefore, could be

harmonic<-function (x) 1/mean(1/x)

and testing it on our elephant data gives

harmonic(c(1,2,4,1))

```
[1] 1.454545
```

Variance

A measure of variability is perhaps the most important quantity in statistical analysis. The greater the variability in the data, the greater will be our uncertainty in the values of parameters estimated from the data, and the less will be our ability to distinguish between competing hypotheses about the data.

The variance of a sample is measured as a function of 'the sum of the squares of the difference between the data and the arithmetic mean'. This important quantity is called the 'sum of squares':

$$SS = \Sigma(y - \bar{y})^2.$$

Naturally, this quantity gets bigger with every new data point you add to the sample. An obvious way to compensate for this is to measure variability as the average of the squared departures from the mean (the 'mean square deviation'.). There is a slight problem, however. Look at the formula for the sum of squares, SS, above and ask yourself what you need to know before you can calculate it. You have the data, y, but the only way you can know the sample mean, $\bar{y}$, is to calculate it from the data (you will never know $\bar{y}$ in advance).

Degrees of freedom

To complete our calculation of the variance we need the **degrees of freedom** (d.f.) This important concept in statistics is defined as follows:

$$\text{d.f.} = n - k,$$

which is the sample size, n, minus the number of parameters, k, estimated from the data. For the variance, we have estimated one parameter from the data, $\bar{y}$, and so there are $n - 1$ degrees of freedom. In a linear regression, we estimate two parameters from the data, the slope and the intercept, and so there are $n - 2$ degrees of freedom in a regression analysis.

Variance is denoted by the lower-case Latin letter s squared: s^2. The square root of variance, s, is called the standard deviation. We always calculate variance as

$$\text{variance} = s^2 = \frac{\text{sum of squares}}{\text{degrees of freedom}}.$$

Consider the following data, y:

```
y<-c(13,7,5,12,9,15,6,11,9,7,12)
```

We need to write a function to calculate the sample variance: we call it **variance** and define it like this:

```
variance<-function(x) sum((x − mean(x))^2)/(length(x)-1)
```

and use it like this:

```
variance(y)
```

```
[1] 10.25455
```

Our measure of variability in these data, the variance, is thus 10.25455. It is said to be an unbiased estimator because we divide the sum of squares by the degrees of freedom $(n - 1)$ rather than by the sample size, n, to compensate for the fact that we have estimated one parameter from the data. So the variance is *close* to the average squared difference between the data and the mean, especially for large samples, but it is not exactly equal to the mean squared deviation. Needless to say, R has a built-in function to calculate variance called **var**:

```
var(y)
```

```
[1] 10.25455
```

Variance Ratio Test

How do we know if two variances are significantly different from one another? We need to carry out Fisher's F test, the ratio of the two variances (see p. 224). Here is a function to print the p value (p. 290) associated with a comparison of the larger and smaller variances:

```
variance.ratio<-function(x,y) {
   v1<-var(x)
   v2<-var(y)
   if (var(x) > var(y)) {
   vr<-var(x)/var(y)
   df1<-length(x)-1
   df2<-length(y)-1}
else { vr<-var(y)/var(x)
   df1<-length(y)-1
   df2<-length(x)-1}
2*(1-pf(vr,df1,df2)) }
```

The last line of our function works out the probability of getting an F ratio as big as vr or bigger by chance alone if the two variances were really the same, using the cumulative probability of the F distribution, which is an R function called pf. We need to supply pf with three *arguments*: the size of the variance ratio (vr), the number of degrees of freedom in the numerator (9) and the number of degrees of freedom in the denominator (also 9).

Here are some data to test our function. They are normally distributed random numbers but the first set has a variance of 4 and the second a variance of 16 (i.e. standard deviations of 2 and 4, respectively):

```
a<-rnorm(10,15,2)
b<-rnorm(10,15,4)
```

Here is our function in action:

```
variance.ratio(a,b)
```

```
[1] 0.01593334
```

We can compare our p with the p-value given by the built-in function called var.test

```
var.test(a,b)
```

```
F test to compare two variances
```

```
data: a and b
F = 0.1748, num df = 9, denom df = 9, p-value = 0.01593
alternative hypothesis: true ratio of variances is not equal to 1
95 percent confidence interval:
0.04340939 0.70360673
```

```
sample estimates:
ratio of variances
         0.1747660
```

Using Variance

Variance is used in two main ways: for establishing measures of unreliability (e.g. confidence intervals) and for testing hypotheses (e.g. Student's *t* test). Here we will concentrate on the former; the latter is discussed in Chapter 8.

Consider the properties that you would like a measure of unreliability to possess. As the variance of the data increases, what would happen to the unreliability of estimated parameters? Would it go up or down? Unreliability would go up as variance increased, so we would want to have the variance on the top (the numerator) of any divisions in our formula for unreliability:

$$\text{unreliability} \propto s^2.$$

What about sample size? Would you want your estimate of unreliability to go up or down as sample size, *n*, increased? You would want unreliability to go down as sample size went up, so you would put sample size on the bottom of the formula for unreliability (i.e. in the denominator):

$$\text{unreliability} \propto \frac{s^2}{n}.$$

Finally, consider the units in which unreliability is measured. What are the units in which our current measure is expressed? Sample size is dimensionless, but variance is based on the sum of squared differences, so it has dimensions of mean squared. So if the mean was a length in cm, the variance would be an area in cm^2. This is an unfortunate state of affairs. It would make good sense to have the dimensions of the unreliability measure and of the parameter whose unreliability it is measuring the same. That is why all unreliability measures are enclosed inside a big square root term. Unreliability measures are called *standard errors*. What we have just worked out is the *standard error of the mean*,

$$se_{\bar{y}} = \sqrt{\frac{s^2}{n}},$$

where s^2 is the variance and *n* is the sample size. There is no built-in R function to calculate the standard error of a mean, but it is easy to write one:

```
se<-function(x) sqrt(var(x)/length(x))
```

You can refer to functions from within other functions. Recall that a confidence interval (*CI*) is '*t* from tables times the standard error':

$$CI = t_{\alpha/2, df} \times se.$$

The R function qt gives the value of Student's t with $1 - \alpha/2 = 0.975$ and degrees of freedom $df = \text{length(x)-1}$. Here is a function called ci95 which uses our function se to compute 95% confidence intervals for a mean:

```
ci95<-function(x) {
    t.value<- qt(0.975,length(x)-1)
    standard.error<-se(x)
    ci<-t.value*standard.error
    cat("95% Confidence Interval = ", mean(x) -ci, "to ", mean(x) +ci,"\n") }
```

We can test the function with 150 normally distributed random numbers with mean 25 and standard deviation 3:

```
x<-rnorm(150,25,3)
ci95(x)
```

```
95% Confidence Interval = 24.76245 to 25.74469
```

If we were to repeat the experiment, we can be 95% certain that the mean of the new sample would lie between 24.76 and 25.74.

We can use the se function to investigate how the standard error of the mean changes with the sample size. First we generate one set of data from which we shall take progressively larger samples:

```
xv<-rnorm(30)
```

Now in a loop take samples of size 2, 3, 4, ..., 30:

```
sem<-numeric(30)
sem[1]<-NA
for(i in 2:30) sem[i]<-se(xv[1:i])
plot(1:30,sem,ylim=c(0,0.8),
    ylab="standard error of mean",xlab="sample size n",pch=16)
```

You can see clearly that as the sample size falls below about $n = 15$, so the standard error of the mean increases rapidly. The blips in the line are caused by outlying values being included in the calculations of the standard error with increases in sample size. The smooth curve is easy to compute: since the values in xv came from a standard normal distribution with mean 0 and standard deviation 1, so the average curve would be $1/\sqrt{n}$ which we can add to our graph using lines:

```
lines(2:30,1/sqrt(2:30))
```

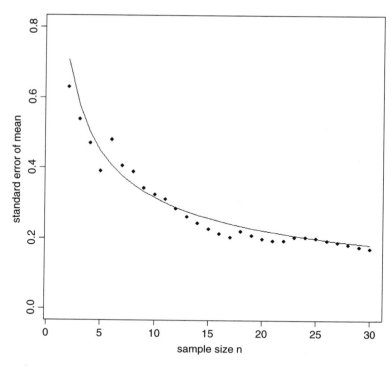

You can see that our single simulation captured the essence of the shape but was wrong in detail, especially for the samples with the lowest replication. However, our single sample was reasonably good for $n > 24$.

Error Bars

There is no function in the base package of R for drawing error bars on bar charts, although several contributed packages use the **arrows** function for this purpose (p. 147). Here is a simple, stripped down function that is supplied with three arguments: the heights of the bars (yv), the lengths (up and down) of the error bars (z) and the labels for the bars on the x axis (nn):

```
error.bars<-function(yv,z,nn){
xv<-
barplot(yv,ylim=c(0,(max(yv)+max(z))),names=nn,ylab=deparse(substitute(yv)
))
g=(max(xv)-min(xv))/50
for (i in 1:length(xv)) {
lines(c(xv[i],xv[i]),c(yv[i]+z[i],yv[i]-z[i]))
lines(c(xv[i]-g,xv[i]+g),c(yv[i]+z[i], yv[i]+z[i]))
lines(c(xv[i]-g,xv[i]+g),c(yv[i]-z[i], yv[i]-z[i]))
}}
```

Here is the error.bars function in action with the plant competition data (p. 370):

```
comp<-read.table("c:\\temp\\competition.txt",header=T)
attach(comp)
```

```
names(comp)
```

```
[1] "biomass" "clipping"
```

```
se<-rep(28.75,5)
labels<-as.character(levels(clipping))
ybar<-as.vector(tapply(biomass,clipping,mean))
```

Now the invoke the function with the means, standard errors and bar labels:

```
error.bars(ybar,se,labels)
```

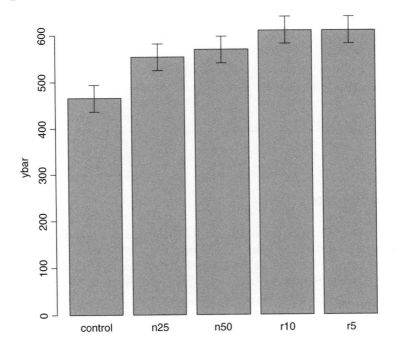

Here is a function to plot error bars on a scatterplot in both the x and y directions:

```
xy.error.bars<-function (x,y,xbar,ybar){
   plot(x, y, pch=16, ylim=c(min(y-ybar),max(y+ybar)),
      xlim=c(min(x-xbar),max(x+xbar)))
   arrows(x, y-ybar, x, y+ybar, code=3, angle=90, length=0.1)
   arrows(x-xbar, y, x+xbar, y, code=3, angle=90, length=0.1) }
```

We test it with these data:

```
x <- rnorm(10,25,5)
y <- rnorm(10,100,20)
xb <- runif(10)*5
yb <- runif(10)*20
xy.error.bars(x,y,xb,yb)
```

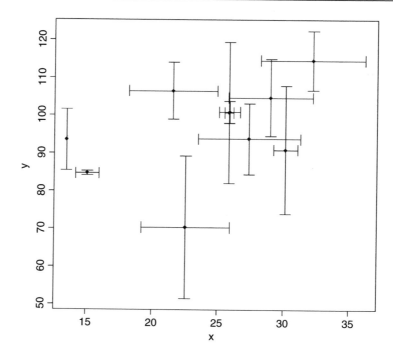

Loops and Repeats

The classic, Fortran-like loop is available in R. The syntax is a little different, but the idea is identical; you request that an index, i, takes on a sequence of values, and that one or more lines of commands are executed as many times as there are different values of i. Here is a loop executed five times with the values of i from 1 to 5: we print the square of each value:

```
for (i in 1:5) print(i^2)
```

```
[1]   1
[1]   4
[1]   9
[1]  16
[1]  25
```

For multiple lines of code, you use curly brackets {} to enclose material over which the loop is to work. Note that the 'hard return' (the Enter key) at the end of each command line is an essential part of the structure (you can replace the hard returns by semicolons if you like, but clarity is improved if you put each command on a separate line):

```
j<-k<-0
for (i in 1:5) {
j<-j+1
k<-k+i*j
print(i+j+k) }
```

```
[1]   3
[1]   9
```

```
[1]   20
[1]   38
[1]   65
```

Here we use a **for** loop to write a function to calculate factorial x (written $x!$) which is

$$x! = x \times (x-1) \times (x-2) \times (x-3) \ldots \times 2 \times 1$$

So $4! = 4 \times 3 \times 2 = 24$. Here is the function:

```
fac1<-function(x) {
    f <- 1
    if (x<2) return (1)
    for (i in 2:x) {
    f <- f*i
    f }}
```

That seems rather complicated for such a simple task, but we can try it out for the numbers 0 to 5:

```
sapply(0:5,fac1)
```

```
[1] 1  1  2  6  24  120
```

There are two other looping functions in R: **repeat** and **while**. We demonstrate their use for the purpose of illustration, but we can do much better in terms of writing a compact function for finding factorials (see below). First, the **while** function:

```
fac2<-function(x) {
f <- 1
t <- x
while(t>1) {
f <- f*t
t <- t-1 }
return(f) }
```

The key point is that if you want to use **while**, you need to set up an indicator variable (t in this case) and change its value *within* each iteration (t<-t-1). We test the function on the numbers 0 to 5:

```
sapply(0:5,fac2)
```

```
[1] 1 1 2 6 24 120
```

Finally, we demonstrate the use of the **repeat** function:

```
fac3<-function(x) {
f <- 1
t <- x
repeat {
if (t<2) break
f <- f*t
t <- t-1  }
return(f)  }
```

Because the repeat function contains no explicit limit, you need to be careful not to program an infinite loop. You must have a logical escape clause that leads to a **break** command:

```
sapply(0:5,fac3)
```

```
[1]  1  1  2  6  24  120
```

It is almost always better to use a built-in function that operates on the entire vector and hence removes the need for loops or repeats of any sort. In this case, we can make use of the cumulative product function, cumprod. Here it is in action:

```
cumprod(1:5)
```

```
[1]  1  2  6  24  120
```

This is already pretty close to what we need for our factorial function. It does not work for 0! of course, because the whole vector would end up full of zeros if the first element in the vector was zero (try 0:5 and see). The factorial of $x > 0$ is the maximum value from the vector produced by cumprod:

```
fac4<-function(x) max(cumprod(1:x))
```

This definition has the desirable side effect that it also gets 0! correct, because when x is 0 the function finds the maximum of 1 and 0 which is 1 which is 0!.

```
max(cumprod(1:0))
```

```
[1]  1
```

```
sapply(0:5,fac4)
```

```
[1]  1  1  2  6  24  120
```

Alternatively, you could adapt an existing built-in function to do the job. $x!$ is the same as $\Gamma(x+1)$, so

```
fac5<-function(x) gamma(x+1)
sapply(0:5,fac5)
```

```
[1]  1  1  2  6  24  120
```

Until recently there was no built-in factorial function in R, but now there is:

```
sapply(0:5,factorial)
```

```
[1]  1  1  2  6  24  120
```

Here is a function that uses the while function in converting a specified number to its binary representation. The trick is that the smallest digit (0 for even or 1 for odd numbers) is always at the right-hand side of the answer (in location 32 in this case):

```
binary<-function(x) {
  i<-0
  string<-numeric(32)
  while(x>0) {
    string[32-i]<-x %% 2
    x<-x%/% 2
    i<-i+1 }
```

```
first<-match(1,string)
string[first:32] }
```

The leading zeros (1 to first − 1) within the string are not printed. We run the function to find the binary representation of the numbers 15 to 17:

```
sapply(15:17,binary)
```

```
[[1]]
[1]  1  1  1  1

[[2]]
[1]  1  0  0  0  0

[[3]]
[1]  1  0  0  0  1
```

The next function uses while to generate the Fibonacci series 1, 1, 2, 3, 5, 8, . . . in which each term is the sum of its two predecessors. The key point about while loops is that the logical variable controlling their operation is altered inside the loop. In this example, we alter n, the number whose Fibonacci number we want, starting at n, reducing the value of n by 1 each time around the loop, and ending when n gets down to 0. Here is the code:

```
fibonacci<-function(n) {
   a<-1
   b<-0
   while(n>0)
     {swap<-a
     a<-a+b
     b<-swap
     n<-n-1 }
   b }
```

An important general point about computing involves the use of the swap variable above. When we replace a by $a + b$ on line 6 we lose the original value of a. If we had not stored this value in swap, we could not set the new value of b to the old value of a on line 7. Now test the function by generating the Fibonacci numbers 1 to 10:

```
sapply(1:10,fibonacci)
```

```
[1]  1  1  2  3  5  8  13  21  34  55
```

Loop avoidance

It is good R programming practice to avoid using loops wherever possible. The use of vector functions (p. 17) makes this particularly straightforward in many cases. Suppose that you wanted to replace all of the negative values in an array by zeros. In the old days, you might have written something like this:

```
for (i in 1:length(y)) { if(y[i] < 0) y[i] <- 0 }
```

Now, however, you would use logical subscripts (p. 21) like this:

```
y [y < 0] <- 0
```

The ifelse function

Sometimes you want to do one thing if a condition is true and a different thing if the condition is false (rather than do nothing, as in the last example). The **ifelse** function allows you to do this for entire vectors without using for loops. We might want to replace any negative values of y by -1 and any positive values and zero by $+1$:

```
z <- ifelse (y < 0, -1, 1)
```

Here we use **ifelse** to convert the continuous variable called Area into a new, two-level factor with values 'big' and 'small' defined by the median Area of the fields:

```
data<-read.table("c:\\temp\\worms.txt",header=T)
attach(data)
ifelse(Area>median(Area),"big","small")
```

```
 [1]    "big"    "big"  "small"  "small"    "big"  "big"  "big"  "small"  "small"
[10]  "small"  "small"    "big"    "big"  "small"  "big"  "big"  "small"    "big"
[19]  "small"  "small"
```

You should use the much more powerful function called cut when you want to convert a continuous variable like Area into many levels (p. 241).

Another use of **ifelse** is to override R's natural inclinations. The log of zero in R is $-\text{Inf}$, as you see in these 20 random numbers from a Poisson process with a mean count of 1.5:

```
y<-log(rpois(20,1.5))
y
```

```
 [1]  0.0000000  1.0986123  1.0986123  0.6931472  0.0000000  0.6931472  0.6931472
 [8]  0.0000000  0.0000000  0.0000000  0.0000000       -Inf       -Inf       -Inf
[15]  1.3862944  0.6931472  1.6094379       -Inf       -Inf  0.0000000
```

However, we want the log of zero to be represented by NA in our particular application:

```
ifelse(y<0,NA,y)
```

```
 [1]  0.0000000  1.0986123  1.0986123  0.6931472  0.0000000  0.6931472  0.6931472
 [8]  0.0000000  0.0000000  0.0000000  0.0000000         NA         NA         NA
[15]  1.3862944  0.6931472  1.6094379         NA         NA  0.0000000
```

The slowness of loops

To see how slow loops can be, we compare two ways of finding the maximum number in a vector of 10 million random numbers from a uniform distribution:

```
x<-runif(10000000)
```

First, using the vector function max:

```
system.time(max(x))
```

```
[1]  0.13  0.00  0.12  NA  NA
```

As you see, this operation took just over one-tenth of a second (0.12) to solve using the vector function max to look at the 10 million numbers in x. Using a loop, however, took more than 15 seconds:

```
pc<-proc.time()
cmax<-x[1]
```

```
for (i in 2:10000000) {
if(x[i]>cmax) cmax<-x[i] }
proc.time()-pc
```

```
[1]  15.52  0.00  15.89  NA  NA
```

The functions **system.time** and **proc.time** produce a vector of five numbers, showing the user, system and total elapsed times for the currently running R process, and the cumulative sum of user (subproc1) and system times (subproc2) of any child processes spawned by it (none in this case, so NA). It is the third number (elapsed time in seconds) that is typically the most useful.

Do not 'grow' data sets in loops or recursive function calls

Here is an extreme example of what *not* to do. We want to generate a vector containing the integers 1 to 1 000 000:

```
z<-NULL
for (i in 1:1000000){
z<-c(z,i) }
```

This took a ridiculous 4 hours 14 minutes to execute. The moral is clear: do not use concatenation $c(z,i)$ to generate iterative arrays. The simple way to do it,

```
z<-1:1000000
```

took 0.05 seconds to accomplish.

The **switch** Function

When you want a function to do different things in different circumstances, then the switch function can be useful. Here we write a function that can calculate any one of four different measures of central tendency: arithmetic mean, geometric mean, harmonic mean or median (p. 51). The character variable called measure should take one value of Mean, Geometric, Harmonic or Median; any other text will lead to the error message **Measure not included**. Alternatively, you can specify the number of the switch (e.g. 1 for Mean, 4 for Median).

```
central<-function(y, measure) {
   switch(measure,
      Mean = mean(y),
      Geometric = exp(mean(log(y))),
      Harmonic = 1/mean(1/y),
      Median = median(y),
   stop("Measure not included")) }
```

Note that you have to include the character strings in quotes as arguments to the function, but they must not be in quotes within the switch function itself.

```
central(rnorm(100,10,2),"Harmonic")
```

```
[1]  9.554712
```

```
central(rnorm(100,10,2),4)
```

```
[1]  10.46240
```

The Evaluation Environment of a Function

When a function is called or invoked a new *evaluation frame* is created. In this frame the formal arguments are matched with the supplied arguments according to the rules of **argument matching** (below). The statements in the body of the function are evaluated sequentially in this environment frame.

The first thing that occurs in a function evaluation is the matching of the formal to the actual or supplied arguments. This is done by a three-pass process:

- **Exact matching on tags**. For each named supplied argument the list of formal arguments is searched for an item whose name matches exactly.

- **Partial matching on tags**. Each named supplied argument is compared to the remaining formal arguments using partial matching. If the name of the supplied argument matches exactly with the first part of a formal argument then the two arguments are considered to be matched.

- **Positional matching**. Any unmatched formal arguments are bound to unnamed supplied arguments, in order. If there is a . . . argument, it will take up the remaining arguments, tagged or not.

- If any arguments remain unmatched an error is declared.

Supplied arguments and default arguments are treated differently. The supplied arguments to a function are evaluated in the evaluation frame of the calling function. The default arguments to a function are evaluated in the evaluation frame of the function. In general, supplied arguments behave as if they are local variables initialized with the value supplied and the name of the corresponding formal argument. Changing the value of a supplied argument within a function will not affect the value of the variable in the calling frame.

Scope

The **scoping rules** are the set of rules used by the evaluator to find a value for a symbol. A symbol can be either **bound** or **unbound**. All of the formal arguments to a function provide bound symbols in the body of the function. Any other symbols in the body of the function are either local variables or unbound variables. A local variable is one that is defined within the function, typically by having it on the left-hand side of an assignment. During the evaluation process if an unbound symbol is detected then R attempts to find a value for it: the environment of the function is searched first, then its enclosure and so on until the global environment is reached. The value of the first match is then used.

Optional Arguments

Here is a function called charplot that produces a scatterplot of x and y using solid red circles as the plotting symbols: there are two essential arguments (x and y) and two optional (*pc* and *co*) to control selection of the plotting symbol and its colour:

```
charplot<-function(x,y,pc=16,co="red"){
plot(y~x,pch=pc,col=co)}
```

The optional arguments are given their default values using = in the argument list. To execute the function you need only provide the vectors of x and y,

```
charplot(1:10,1:10)
```

to get solid red circles. You can get a different plotting symbol simply by adding a third argument

```
charplot(1:10,1:10,17)
```

which produces red solid triangles (pch=17). If you want to change only the colour (the fourth argument) then you have to specify the variable name because the optional arguments would not then be presented in sequence. So, for navy-coloured solid circles, you put

```
charplot(1:10,1:10,co="navy")
```

To change both the plotting symbol and the colour you do not need to specify the variable names, so long as the plotting symbol is the third argument and the colour is the fourth

```
charplot(1:10,1:10,15,"green")
```

which produces solid green squares. Reversing the optional arguments does not work

```
charplot(1:10,1:10,"green",15)
```

(this uses the letter g as the plotting symbol and colour no. 15). If you specify both variable names, then the order does not matter:

```
charplot(1:10,1:10,co="green",pc=15)
```

This produces solid green squares despite the arguments being out of sequence.

Variable Numbers of Arguments (. . .)

Some applications are much more straightforward if the number of arguments does not need to be specified in advance. There is a special formal name . . . (triple dot) which is used in the argument list to specify that an arbitrary number of arguments are to be passed to the function. Here is a function that takes any number of vectors and calculates their means and variances:

```
many.means <- function ( . . . ) {
    data <- list( . . . )
    n<- length(data)
    means <- numeric(n)
    vars <- numeric(n)
    for (i in 1:n) {
        means[i]<-mean(data[[i]])
        vars[i]<-var(data[[i]])
    }
    print(means)
    print(vars)
    invisible(NULL)
}
```

The main features to note are these. The function definition has . . . as its only argument. The 'triple dot' argument . . . allows the function to accept additional arguments of unspecified name and number, and this introduces tremendous flexibility into the structure and behaviour of functions. The first thing done inside the function is to create an object called data out of the list of vectors that are actually supplied in any particular case. The length of this list is the number of vectors, not the lengths of the vectors themselves (these could differ from one vector to another, as in the example below). Then the two output variables (means and vars) are defined to have as many elements as there are vectors in the parameter list. The loop goes from 1 to the number of vectors, and for each vector uses the built-in functions mean and var to compute the answers we require. It is important to note that because data is a list, we use double [[ ]] subscripts in addressing its elements.

Now try it out. To make things difficult we shall give it three vectors of different lengths. All come from the standard normal distribution (with mean 0 and variance 1) but x is 100 in length, y is 200 and z is 300 numbers long.

```
x<-rnorm(100)
y<-rnorm(200)
z<-rnorm(300)
```

Now we invoke the function:

```
many.means(x,y,z)
```

```
[1]  -0.039181830   0.003613744   0.050997841
[1]       1.146587      0.989700      0.999505
```

As expected, all three means (top row) are close to 0 and all 3 variances are close to 1 (bottom row). You can use . . . to absorb some arguments into an intermediate function which can then be extracted by functions called subsequently. R has a form of *lazy evaluation* of function arguments in which arguments are not evaluated until they are needed (in some cases the argument will never be evaluated).

Returning Values from a Function

Often you want a function to return a single value (like a mean or a maximum), in which case you simply leave the last line of the function unassigned (i.e. there is no 'gets arrow' on the last line). Here is a function to return the median value of the parallel maxima (built-in function pmax) of two vectors supplied as arguments:

```
parmax<-function (a,b) {
c<-pmax(a,b)
median(c) }
```

Here is the function in action: the unassigned last line median(c) returns the answer

```
x<-c(1,9,2,8,3,7)
y<-c(9,2,8,3,7,2)
parmax(x,y)
```

```
[1]  8
```

If you want to return two or more variables from a function you should use return with a list containing the variables to be returned. Suppose we wanted the median value of both the parallel maxima and the parallel minima to be returned:

```
parboth<-function (a,b) {
c<-pmax(a,b)
d<-pmin(a,b)
answer<-list(median(c),median(d))
names(answer)[[1]]<-"median of the parallel maxima"
names(answer)[[2]]<-"median of the parallel minima"
return(answer) }
```

Here it is in action with the same x and y data as above:

```
parboth(x,y)
```

```
$"median of the parallel maxima"
[1]  8

$"median of the parallel minima"
[1]  2
```

The point is that you make the multiple returns into a list, then return the list. The provision of multi-argument returns (e.g. return(median(c),median(d)) in the example above) has been deprecated in R and a warning is given, as multi-argument returns were never documented in S, and whether or not the list was named differs from one version of S to another.

Anonymous Functions

Here is an example of an anonymous function. It generates a vector of values but the function is not allocated a name (although the answer could be).

```
(function(x,y){ z <- 2*x^2 + y^2; x+y+z })(0:7, 1)
```

```
[1]   2   5  12  23  38  57  80  107
```

The function first uses the supplied values of x and y to calculate z, then returns the value of $x + y + z$ evaluated for eight values of x (from 0 to 7) and one value of y (1). Anonymous functions are used most frequently with apply, sapply and lapply (p. 68).

Flexible Handling of Arguments to Functions

Because of the **lazy evaluation** practised by R, it is very simple to deal with missing arguments in function calls, giving the user the opportunity to specify the absolute minimum number of arguments, but to override the default arguments if they want to. As a simple example, take a function plotx2 that we want to work when provided with either one or two arguments. In the one-argument case (only an integer $x > 1$ provided), we want it to plot z^2 against z for $z = 1$ to x in steps of 1. In the second case, when y is supplied, we want it to plot y against z for $z = 1$ to x.

```
plotx2 <- function (x, y=z^2) {
   z<-1:x
   plot(z,y,type="l") }
```

In many other languages, the first line would fail because z is not defined at this point. But R does not evaluate an expression until the body of the function actually calls for it to be evaluated (i.e. never, in the case where y is supplied as a second argument). Thus for the one-argument case we get a graph of z^2 against z and in the two-argument case we get a graph of y against z (in this example, the straight line 1:12 vs. 1:12)

```
par(mfrow=c(1,2))
plotx2(12)
plotx2(12,1:12)
```

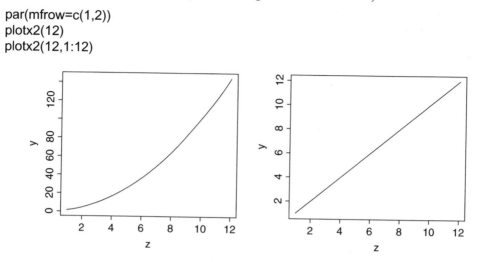

You need to specify that the type of plot you want is a line (type="l" using lower-case L, not upper-case I and not number 1) because the default is to produce a scatterplot with open circles as the plotting symbol (type="p"). If you want your plot to consist of points with lines joining the dots, then use type="b" (for 'both' lines and points). Other types of plot that you might want to specify include vertical lines from the x axis up to the value of the response variable (type="h"), creating an effect like very slim barplots or histograms, and type="s" to produce a line drawn as steps between successive ranked values of x. To plot the scaled axes, but no lines or points, use type="n" (see p. 137).

It is possible to access the actual (not default) expressions used as arguments inside the function. The mechanism is implemented via promises. You can find an explanation of promises by typing ?promise at the command prompt.

Evaluating Functions with apply, sapply and lapply

apply and sapply

The apply function is used for applying functions to the rows or columns of matrices or dataframes. For example:

```
(X<-matrix(1:24,nrow=4))
```

```
     [,1] [,2] [,3] [,4] [,5] [,6]
[1,]    1    5    9   13   17   21
[2,]    2    6   10   14   18   22
[3,]    3    7   11   15   19   23
[4,]    4    8   12   16   20   24
```

Note that placing the expression to be evaluated in parentheses (as above) causes the value of the result to be printed on the screen. This saves an extra line of code, because to achieve the same result without parentheses requires us to type

```
X<-matrix(1:24,nrow=4)
X
```

Often you want to apply a function across one of the margins of a matrix – margin 1 being the rows and margin 2 the columns. Here are the row totals (four of them):

```
apply(X,1,sum)
```

```
[1]  66  72  78  84
```

and here are the column totals (six of them):

```
apply(X,2,sum)
```

```
[1]  10  26  42  58  74  90
```

Note that in both cases, the answer produced by apply is a vector rather than a matrix. You can apply functions to the individual elements of the matrix rather than to the margins. The margin you specify influences only the shape of the resulting matrix.

```
apply(X,1,sqrt)
```

```
          [,1]      [,2]      [,3]      [,4]
[1,]  1.000000  1.414214  1.732051  2.000000
[2,]  2.236068  2.449490  2.645751  2.828427
[3,]  3.000000  3.162278  3.316625  3.464102
[4,]  3.605551  3.741657  3.872983  4.000000
[5,]  4.123106  4.242641  4.358899  4.472136
[6,]  4.582576  4.690416  4.795832  4.898979
```

```
apply(X,2,sqrt)
```

```
          [,1]      [,2]      [,3]      [,4]      [,5]      [,6]
[1,]  1.000000  2.236068  3.000000  3.605551  4.123106  4.582576
[2,]  1.414214  2.449490  3.162278  3.741657  4.242641  4.690416
[3,]  1.732051  2.645751  3.316625  3.872983  4.358899  4.795832
[4,]  2.000000  2.828427  3.464102  4.000000  4.472136  4.898979
```

Here are the shuffled numbers from each of the rows, using sample without replacement:

```
apply(X,1,sample)
```

```
      [,1]  [,2]  [,3]  [,4]
[1,]     5    14    19     8
[2,]    21    10     7    16
[3,]    17    18    15    24
[4,]     1    22    23     4
[5,]     9     2     3    12
[6,]    13     6    11    20
```

Note that the resulting matrix has 6 rows and 4 columns (i.e. it has been transposed). You can supply your own function definition within apply like this:

```
apply(X,1,function(x) x^2+x)
```

```
        [,1]   [,2]   [,3]   [,4]
[1,]       2      6     12     20
[2,]      30     42     56     72
[3,]      90    110    132    156
[4,]     182    210    240    272
[5,]     306    342    380    420
[6,]     462    506    552    600
```

This is an anonymous function because the function is not named.

If you want to apply a function to a vector then use **sapply** (rather than **apply** for matrices or margins of matrices). Here is the code to generate a **list** of sequences from 1:3 up to 1:7 (see p. 30):

```
sapply(3:7, seq)
```

```
[[1]]
[1]  1  2  3

[[2]]
[1]  1  2  3  4

[[3]]
[1]  1  2  3  4  5

[[4]]
[1]  1  2  3  4  5  6

[[5]]
[1]  1  2  3  4  5  6  7
```

The function **sapply** is most useful with complex iterative calculations. The following data show decay of radioactive emissions over a 50-day period, and we intend to use non-linear least squares (see p. 663) to estimate the decay rate a in $y = \exp(-ax)$:

```
sapdecay<-read.table("c:\\temp\\sapdecay.txt",header=T)
attach(sapdecay)
names(sapdecay)
```

```
[1] "x"  "y"
```

We need to write a function to calculate the sum of the squares of the differences between the observed (y) and predicted (yf) values of y, when provided with a specific value of the parameter a:

```
sumsq <- function(a,xv=x,yv=y)
  { yf <- exp(-a*xv)
  sum((yv-yf)^2) }
```

We can get a rough idea of the decay constant, a, for these data by linear regression of $\log(y)$ against x, like this:

```
lm(log(y)~x)
```

```
Coefficients:
(Intercept)              x
 0.04688         -0.05849
```

So our parameter *a* is somewhere close to 0.058. We generate a range of values for *a* spanning an interval on either side of 0.058:

```
a<-seq(0.01,0.2,.005)
```

Now we can use **sapply** to apply the sum of squares function for each of these values of *a* (without writing a loop), and plot the deviance against the parameter value for *a*:

```
plot(a,sapply(a,sumsq),type="l")
```

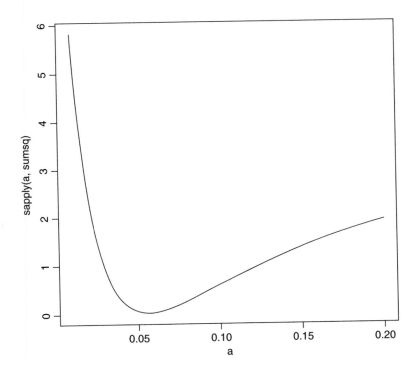

This shows that the least-squares estimate of *a* is indeed close to 0.06 (this is the value of *a* associated with the minimum deviance). To extract the minimum value of *a* we use min with subscripts (square brackets) to extract the relevant value of *a*:

```
a[min(sapply(a,sumsq))==sapply(a,sumsq)]
```

```
[1]  0.055
```

Finally, we could use this value of *a* to generate a smooth exponential function to fit through our scatter of data points:

```
plot(x,y)
xv<-seq(0,50,0.1)
lines(xv,exp(-0.055*xv))
```

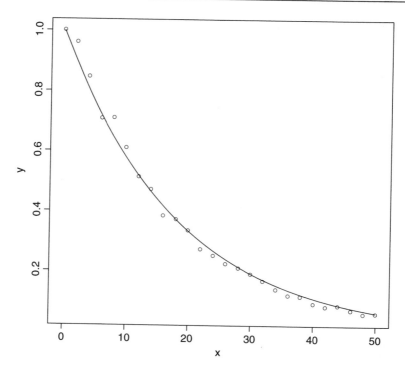

Here is the same procedure streamlined by using the **optimize** function. Write a function showing how the sum of squares depends on the value of the parameter a:

```
fa<-function(a) sum((y-exp(-a*x))^2)
```

Now use **optimize** with a specified range of values for a, here c(0.01,0.1), to find the value of a that minimizes the sum of squares:

```
optimize(fa,c(0.01,0.1))
```

```
$minimum
```

```
[1]  0.05538411
```

```
$objective
[1]  0.01473559
```

The value of a is that minimizes the sum of squares is 0.055 38 and the minimum value of the sum of squares is 0.0147. What if we had chosen a different way of assessing the fit of the model to the data? Instead of minimizing the sum of the squares of the residuals, we might want to minimize the sum of the absolute values of the residuals. We need to write a new function to calculate this quantity,

```
fb<-function(a) sum(abs(y-exp(-a*x)))
```

then use **optimize** as before:

```
optimize(fb,c(0.01,0.1))
```

```
$minimum
```

```
[1]   0.05596058
```

```
$objective
[1]   0.3939221
```

The results differ only in the fourth digit after the decimal point, and you could not choose between the two methods from a plot of the model.

Lists and lapply

We start by creating a list object that consists of three parts: character information in a vector called *a*, numeric information in a vector called *b*, and logical information in a vector called *c*:

```
a<-c("a","b","c","d")
b<-c(1,2,3,4,4,3,2,1)
c<-c(T,T,F)
```

We create our list object by using the list function to bundle these variables together:

```
list.object<-list(a,b,c)
class(list.object)
```

```
[1]  "list"
```

To see the contents of the list we just type its name:

```
list.object
```

```
[[1]]
[1]  "a"  "b"  "c"  "d"

[[2]]
[1]  1  2  3  4  4  3  2  1

[[3]]
[1]   TRUE   TRUE   FALSE
```

The function **lapply** applies a specified function to each of the elements of a list in turn (without the need for specifying a loop, and not requiring us to know how many elements there are in the list). A useful function to apply to lists is the **length** function; this asks the question how many elements comprise each component of the list. Technically we want to know the length of each of the vectors making up the list:

```
lapply(list.object,length)
```

```
[[1]]
[1]  4

[[2]]
[1]  8

[[3]]
[1]  3
```

This shows that list.object consists of three vectors ([[1]], [[2]] and [[3]]), and shows that there were four elements in the first vector, eight in the second and three in the third. But four of what, and eight of what? To find out, we apply the function class to the list:

lapply(list.object,class)

```
[[1]]
[1] "character"

[[2]]
[1] "numeric"

[[3]]
[1] "logical"
```

So the answer is there were 4 characters in the first vector, 8 numbers in the second and 3 logical values in the third vector.

Applying numeric functions to lists will only work for objects of class numeric or objects (like logical values) that can be coerced into numbers. Here is what happens when we use lapply to apply the function mean to list.object:

lapply(list.object,mean)

```
[[1]]
[1]  NA

[[2]]
[1]  2.5

[[3]]
[1]  0.6666667
```

```
Warning message:
argument is not numeric or logical: returning NA in:
mean.default(X[[1]], ...)
```

We get a warning message pointing out that the first vector cannot be coerced to a number (it is not numeric or logical), so NA appears in the output. The second vector is averaged as expected. The third vector produces the answer 2/3 because logical false (F) is coerced to numeric 0 and logical true (T) is coerced to numeric 1.

Looking for runs of numbers within vectors

The function is called rle, which stands for 'run length encoding' and is most easily understood with an example. Here is a vector of 150 random numbers from a Poisson distribution with mean 0.7:

(poisson<-rpois(150,0.7))

```
  [1]   1   1   0   0   2   1   0   1   0   1   0   0   0   0   2   1   0   0   3   1
        0   0   1   0   2   0   1   1   0   0   0   1   0   0   0   2   1
 [38]   0   0   0   1   0   0   0   2   0   0   0   1   1   0   2   1   0   0   0   2
        0   0   2   3   2   1   0   2   0   0   0   0   0   1   1   0   0
 [75]   0   0   0   1   1   1   0   0   1   0   1   2   2   0   0   2   0   0   0   0
        0   0   0   2   1   0   0   1   0   1   0   1   1   1   2   0   3
[112]   0   0   2   0   0   1   0   1   0   4   0   0   1   0   2   1   0   1   1   0
        0   1   3   3   0   0   1   1   0   1   0   0   0   0   0   1   0
[149]   2   0
```

We can do our own run length encoding on the vector by eye: there is a run of two 1s, then a run of two 0s, then a single 2, then a single 1, then a single 0, and so on. So the run lengths are 2, 2, 1, 1, 1, 1,.... The values associated with these runs were 1, 0, 2, 1, 0, 1,.... Here is the output from rle:

rle(poisson)

```
Run Length Encoding
lengths: int [1:93] 2 2 1 1 1 1 1 1 4 1...
values : num [1:93] 1 0 2 1 0 1 0 1 0 2...
```

The object produced by rle is a list of two vectors: the lengths and the values. To find the longest run, and the value associated with that longest run, we use the indexed lists like this:

max(rle(poisson)[[1]])

```
[1] 7
```

So the longest run in this vector of numbers was 7. But 7 of what? We use which to find the location of the 7 in lengths, then apply this index to values to find the answer:

which(rle(poisson)[[1]]==7)

```
[1]  55
```

rle(poisson)[[2]][55]

```
[1]  0
```

So, not surprisingly given that the mean was just 0.7, the longest run was of zeros.
    Here is a function to return the length of the run and its value for any vector:

```
run.and.value<-function (x) {
a<- max(rle(poisson)[[1]])
b<-rle(poisson)[[2]][which(rle(poisson)[[1]] == a)]
cat("length = ",a," value = ",b, "\n")}
```

Testing the function on the vector of 150 Poisson data gives

run.and.value(poisson)

```
length = 7  value = 0
```

It is sometimes of interest to know the number of runs in a given vector (for instance, the lower the number of runs, the more aggregated the numbers; and the greater the number of runs, the more regularly spaced out). We use the length function for this:

length(rle(poisson)[[2]])

```
[1]  93
```

indicating that the 150 values were arranged in 93 runs (an intermediate value, characteristic of a random pattern). The value 93 appears in square brackets [1:93] in the output of the run length encoding function.
    In a different example, suppose we had n_1 values of 1 representing 'present' and n_2 values of 0 representing 'absent', then the minimum number of runs would be 2 (a solid block of 1s and a sold block of 0s). The maximum number of runs would be $2n+1$ if they alternated (until the smaller number $n = \min(n1,n2)$ ran out). Here is a simple **runs test** based on 1000 randomizations of 25 ones and 30 zeros:

```
n1<-25
n2<-30
y<-c(rep(1,n1),rep(0,n2))
len<-numeric(10000)
for (i in 1:10000) len[i]<-length(rle(sample(y))[[2]])
quantile(len,c(0.025,0.975))
```

```
2.5%  97.5%
  21     35
```

Thus, for these data ($n_1 = 25$ and $n_2 = 30$) an aggregated pattern would score 21 or fewer runs, and a regular pattern would score 35 or more runs. Any scores between 21 and 35 fall within the realm of random patterns.

Saving Data Produced within R to Disc

It is often convenient to generate numbers within R and then to use them somewhere else (in a spreadsheet, say). Here are 1000 random integers from a negative binomial distribution with mean mu= 1.2 and clumping parameter or aggregation parameter (k) size = 1.0, that I want to save as a single column of 1000 rows in a file called nbnumbers.txt in directory 'temp' on the c: drive:

```
nbnumbers<-rnbinom(1000, size=1, mu=1.2)
```

There is general point to note here about the number and order of arguments provided to built-in functions like rnbinom. This function can have two of three optional arguments: size, mean (mu) and probability (prob) (see ?rnbinom). R knows that the unlabelled number 1000 refers to the number of numbers required because of its position, first in the list of arguments. If you are prepared to specify the names of the arguments, then the order in which they appear is irrelevant: rnbinom(1000, size=1, mu=1.2) and rnbinom(1000, mu=1.2, size=1) would give the same output. If optional arguments are not labelled, then their order is crucial: so rnbinom(1000, 0.9, 0.6) is different from rnbinom(1000, 0.6, 0.9) because if there are no labels, then the second argument *must* be size and the third argument *must* be prob.

To export the numbers I use write like this, specifying that the numbers are to be output in a single column (i.e. with third argument 1 because the default is 5 columns):

```
write(nbnumbers,"c:\\temp\\nbnumbers.txt",1)
```

Sometimes you will want to save a table or a matrix of numbers to file. There is an issue here, in that the write function transposes rows and columns. It is much simpler to use the write.table function which does *not* transpose the rows and columns

```
xmat<-matrix(rpois(100000,0.75),nrow=1000)
write.table(xmat,"c:\\temp\\table.txt",col.names=F,row.names=F)
```

but it does add made-up row names and column names unless (as here) you specify otherwise. You have saved 1000 rows each of 100 Poisson random numbers with $\lambda = 0.75$.

Suppose that you have counted the number of different entries in the vector of negative binomial numbers (above):

```
nbtable<-table(nbnumbers)
nbtable
```

```
nbnumbers
  0    1    2    3    4    5    6    7    8    9   11   15
445  248  146   62   41   33   13    4    1    5    1    1
```

and you want write this output to a file. If you want to save both the counts and their frequencies in adjacent columns, use

write.table(nbtable,"c:\\temp\\table.txt",col.names=F,row.names=F)

but if you only want to export a single column of frequencies (445, 248, .. etc) use

write.table(unclass(nbtable),"c:\\temp\\table.txt",col.names=F,row.names=F)

Pasting into an Excel Spreadsheet

Writing a vector from R to the Windows clipboard uses the function writeClipboard(x) where x is a character vector, so you need to build up a spreadsheet in Excel one column at a time. Remember that character strings in dataframes are converted to factors on input unless you protect them by as.is(name) on input. For example

writeClipboard(as.character(factor.name))

Go into Excel and press Ctrl+V, and then back into R and type

writeClipboard(as.character(numeric.variable))

Then go into Excel and Ctrl+V in the second column, and so on.

Writing an Excel Readable File from R

Suppose you want to transfer the dataframe called data to Excel:

write.table(data,"clipboard",sep="\t",col.names=NA)

Then, in Excel, just type Ctrl+V or click on the Paste icon (the clipboard).

Testing for Equality

You need to be careful in programming when you want to test whether or not two computed numbers are equal. R will assume that you mean 'exactly equal', and what *that* means depends upon machine precision. Most numbers are rounded to 53 binary digits accuracy. Typically therefore, two floating point numbers will *not* reliably be equal unless they were computed by the same algorithm, and not always even then. You can see this by squaring the square root of 2: surely these values are the same?

```
x <- sqrt(2)
x * x == 2
```

```
[1] FALSE
```

We can see by how much the two values differ by subtraction:

```
x * x - 2
```

```
[1]  4.440892e-16
```

Sets: union, intersect and setdiff

There are three essential functions for manipulating sets. The principles are easy to see if we work with an example of two sets:

```
setA<-c("a", "b", "c", "d", "e")
setB<-c("d", "e", "f", "g")
```

Make a mental note of what the two sets have in common, and what is unique to each.

The **union** of two sets is everything in the two sets taken together, but counting elements only once that are common to both sets:

```
union(setA,setB)
```

```
[1]  "a"  "b"  "c"  "d"  "e"  "f"  "g"
```

The **intersection** of two sets is the material that they have in common:

```
intersect(setA,setB)
```

```
[1]  "d"  "e"
```

Note, however, that the **difference** between two sets is order-dependent. It is the material that *is* in the first named set, that *is not* in the second named set. Thus setdiff(A,B) gives a different answer than setdiff(B,A). For our example,

```
setdiff(setA,setB)
```

```
[1]  "a"  "b"  "c"
```

```
setdiff(setB,setA)
```

```
[1]  "f"  "g"
```

Thus, it should be the case that setdiff(setA,setB) plus intersect(setA,setB) plus setdiff(setB,setA) is the same as the union of the two sets. Let's check:

```
all(c(setdiff(setA,setB),intersect(setA,setB),setdiff(setB,setA))==
   union(setA,setB))
```

```
[1]  TRUE
```

There is also a built-in function setequal for testing if two sets are equal

```
setequal(c(setdiff(setA,setB),intersect(setA,setB),setdiff(setB,setA)),
   union(setA,setB))
```

```
[1]  TRUE
```

You can use %in% for comparing sets. The result is a logical vector whose length matches the vector on the left

setA %in% setB

```
[1] FALSE FALSE FALSE TRUE TRUE
```

setB %in% setA

```
[1] TRUE TRUE FALSE FALSE
```

Using these vectors of logical values as subscripts, we can demonstrate, for instance, that setA[setA %in% setB] is the same as intersect(setA,setB):

setA[setA %in% setB]

```
[1] "d" "e"
```

intersect(setA,setB)

```
[1] "d" "e"
```

Pattern Matching

We need a dataframe with a serious amount of text in it to make these exercises relevant:

```
wf<-read.table("c:\\temp\\worldfloras.txt",header=T)
attach(wf)
names(wf)
```

```
[1] "Country" "Latitude" "Area" "Population" "Flora"
[6] "Endemism" "Continent"
```

Country

As you can see, there are 161 countries in this dataframe (strictly, 161 places, since some of the entries, such as Sicily and Balearic Islands, are not countries). The idea is that we want to be able to select subsets of countries on the basis of specified patterns within the character strings that make up the country names (factor levels). The function to do this is grep. This searches for matches to a pattern (specified in its first argument) within the character vector which forms the second argument. It returns a vector of indices (subscripts) within the vector appearing as the second argument, where the pattern was found in whole or in part. The topic of pattern matching is very easy to master once the penny drops, but it hard to grasp without simple, concrete examples. Perhaps the simplest task is to select all the countries containing a particular letter – for instance, upper case R:

as.vector(Country[grep("R",as.character(Country))])

```
[1]   "Central African Republic"   "Costa Rica"
[3]   "Dominican Republic"         "Puerto Rico"
[5]   "Reunion"                    "Romania"
[7]   "Rwanda"                     "USSR"
```

To restrict the search to countries whose *first* name begins with R use the ^ character like this:

as.vector(Country[grep("^R",as.character(Country))])

```
[1] "Reunion" "Romania" "Rwanda"
```

To select those countries with multiple names with upper case R as the first letter of their second or subsequent names, we specify the character string as 'blank R' like this:

as.vector(Country[grep(" R",as.character(Country)])

```
[1]    "Central African Republic"    "Costa Rica"
[3]    "Dominican Republic"          "Puerto Rico"
```

To find all the countries with two or more names, just search for a blank " "

as.vector(Country[grep(" ",as.character(Country))])

```
 [1]    "Balearic Islands"            "Burkina Faso"
 [3]    "Central African Republic"    "Costa Rica"
 [5]    "Dominican Republic"          "El Salvador"
 [7]    "French Guiana"               "Germany East"
 [9]    "Germany West"                "Hong Kong"
[11]    "Ivory Coast"                 "New Caledonia"
[13]    "New Zealand"                 "Papua New Guinea"
[15]    "Puerto Rico"                 "Saudi Arabia"
[17]    "Sierra Leone"                "Solomon Islands"
[19]    "South Africa"                "Sri Lanka"
[21]    "Trinidad & Tobago"           "Tristan da Cunha"
[23]    "United Kingdom"              "Viet Nam"
[25]    "Yemen North"                 "Yemen South"
```

To find countries with names ending in 'y' use the $ (dollar) symbol like this:

as.vector(Country[grep("y$",as.character(Country))])

```
[1] "Hungary" "Italy" "Norway" "Paraguay" "Sicily" "Turkey"
[7] "Uruguay"
```

To recap: the start of the character string is denoted by ^ and the end of the character string is denoted by $. For conditions that can be expressed as groups (say, series of numbers or alphabetically grouped lists of letters), use square brackets inside the quotes to indicate the range of values that is to be selected. For instance, to select countries with names containing upper-case letters from C to E inclusive, write:

as.vector(Country[grep("[C-E]",as.character(Country))])

```
 [1]    "Cameroon"                    "Canada"
 [3]    "Central African Republic"    "Chad"
 [5]    "Chile"                       "China"
 [7]    "Colombia"                    "Congo"
 [9]    "Corsica"                     "Costa Rica"
[11]    "Crete"                       "Cuba"
[13]    "Cyprus"                      "Czechoslovakia"
[15]    "Denmark"                     "Dominican Republic"
[17]    "Ecuador"                     "Egypt"
[19]    "El Salvador"                 "Ethiopia"
[21]    "Germany East"                "Ivory Coast"
[23]    "New Caledonia"               "Tristan da Cunha"
```

Notice that this formulation picks out countries like Ivory Coast and Tristan da Cunha that contain upper-case Cs in places other than as their first letters. To restrict the choice to first letters use the ^ operator before the list of capital letters:

as.vector(Country[grep("^[C-E]",as.character(Country))])

```
 [1]   "Cameroon"                  "Canada"
 [3]   "Central African Republic"  "Chad"
 [5]   "Chile"                     "China"
 [7]   "Colombia"                  "Congo"
 [9]   "Corsica"                   "Costa Rica"
[11]   "Crete"                     "Cuba"
[13]   "Cyprus"                    "Czechoslovakia"
[15]   "Denmark"                   "Dominican Republic"
[17]   "Ecuador"                   "Egypt"
[19]   "El Salvador"               "Ethiopia"
```

How about selecting the counties *not* ending with a specified patterns? The answer is simply to *use negative subscripts* to drop the selected items from the vector. Here are the countries that do not end with a letter between 'a' and 't':

as.vector(Country[-grep("[a-t]$",as.character(Country))])

```
[1] "Hungary"  "Italy"    "Norway"   "Paraguay" "Peru"     "Sicily"
[7] "Turkey"   "Uruguay"  "USA"      "USSR"     "Vanuatu"
```

You see that USA and USSR are included in the list because we specified lower-case letters as the endings to omit. To omit these other countries, put ranges for both upper- and lower-case letters inside the square brackets, separated by a space:

as.vector(Country[-grep("[A-T a-t]$",as.character(Country))])

```
[1] "Hungary"  "Italy"    "Norway"    "Paraguay" "Peru"  "Sicily"
[7] "Turkey"   "Uruguay"  "Vanuatu"
```

Dot . as the 'anything' character

Countries with 'y' as their second letter are specified by ^.y The ^ shows 'starting', then a single dot means one character of any kind, so y is the specified second character:

as.vector(Country[grep("^.y",as.character(Country))])

```
[1] "Cyprus" "Syria"
```

To search for countries with 'y' as third letter:

as.vector(Country[grep("^..y",as.character(Country))])

```
[1] "Egypt" "Guyana" "Seychelles"
```

If we want countries with 'y' as their sixth letter

as.vector(Country[grep("^. {5}y",as.character(Country))])

```
[1] "Norway" "Sicily"  "Turkey"
```

(5 'anythings' is shown by '.' then curly brackets {5} then *y*). Which are the countries with 4 or fewer letters in their names?

as.vector(Country[grep("^. {,4}$",as.character(Country))])

```
[1]   "Chad"   "Cuba"   "Iran"   "Iraq"   "Laos"   "Mali"   "Oman"
[8]   "Peru"   "Togo"   "USA"    "USSR"
```

The '.' means 'anything' while the {,4} means 'repeat up to four' anythings (dots) before $ (the end of the string). So to find all the countries with 15 or more characters in their name is just

```
as.vector(Country[grep("^. {15, }$",as.character(Country))])
```

```
[1]   "Balearic Islands"      "Central African Republic"
[3]   "Dominican Republic"    "Papua New Guinea"
[5]   "Solomon Islands"       "Trinidad & Tobago"
[7]   "Tristan da Cunha"
```

Substituting text within character strings

Search-and-replace operations are carried out in R using the functions **sub** and **gsub**. The two substitution functions differ only in that **sub** replaces only the first occurrence of a pattern within a character string, whereas **gsub** replaces all occurrences. An example should make this clear. Here is a vector comprising seven character strings, called text:

```
text <- c("arm","leg","head", "foot","hand", "hindleg", "elbow")
```

We want to replace all lower-case 'h' with upper-case 'H':

```
gsub("h","H",text)
```

```
[1] "arm" "leg" "Head" "foot" "Hand" "Hindleg" "elbow"
```

Now suppose we want to convert the first occurrence of a lower-case 'o' into an upper-case 'O'. We use **sub** for this (not **gsub**):

```
sub("o","O",text)
```

```
[1] "arm" "leg" "head" "fOot" "hand" "hindleg" "elbOw"
```

You can see the difference between **sub** and **gsub** in the following, where both instances of 'o' in foot are converted to upper case by **gsub** but not by **sub**:

```
gsub("o","O",text)
```

```
[1] "arm" "leg" "head" "fOOt" "hand" "hindleg" "elbOw"
```

More general patterns can be specified in the same way as we learned for **grep** (above). For instance, to replace the first character of every string with upper-case 'O' we use the dot notation (. stands for 'anything') coupled with ^ (the 'start of string' marker):

```
gsub("^.","O",text)
```

```
[1] "Orm" "Oeg" "Oead" "Ooot" "Oand" "Oindleg" "Olbow"
```

It is useful to be able to manipulate the cases of character strings. Here, we capitalize the first character in each string:

```
gsub("(\\w)(\\w*)", "\\U\\1\\L\\2",text, perl=TRUE)
```

```
[1] "Arm" "Leg" "Head" "Foot" "Hand" "Hindleg" "Elbow"
```

while here we convert all the characters to upper case:

```
gsub("(\\w*)", "\\U\\1",text, perl=TRUE)
```

```
[1] "ARM" "LEG" "HEAD" "FOOT" "HAND" "HINDLEG" "ELBOW"
```

Locations of the pattern within a vector of character strings using regexpr

Instead of substituting the pattern, we might want to know *if* it occurs in a string and, if so, *where* it occurs within each string. The result of regexpr, therefore, is a numeric vector (as with grep, above), but now indicating the position of the (first instance of the) pattern within the string (rather than just *whether* the pattern was there). If the pattern does not appear within the string, the default value returned by regexpr is -1. An example is essential to get the point of this:

```
text
```

```
[1] "arm" "leg" "head" "foot" "hand" "hindleg" "elbow"
```

```
regexpr("o",text)
```

```
[1] -1  -1  -1  2  -1  -1  4
```

```
attr(,"match.length")
```

```
[1] -1  -1  -1  1  -1  -1  1
```

This indicates that there were lower-case 'o's in two of the elements of text, and that they occurred in positions 2 and 4, respectively. Remember that if we wanted just the subscripts showing which elements of text contained an 'o' we would use grep like this:

```
grep("o",text)
```

```
[1] 4 7
```

and we would extract the character strings like this:

```
text[grep("o",text)]
```

```
[1] "foot" "elbow"
```

Counting how many 'o's there are in each string is a different problem again, and this involves the use of gregexpr:

```
freq<-as.vector(unlist (lapply(gregexpr("o",text),length)))
present<-ifelse(regexpr("o",text)<0,0,1)
freq*present
```

```
[1] 0 0 0 2 0 0 1
```

indicating that there are no 'o's in the first three character strings, two in the fourth and one in the last string. You will need lots of practice with these functions to appreciate all of the issues involved.

The function charmatch is for matching characters. If there are multiple matches (two or more) then the function returns the value 0 (e.g. when all the elements contain 'm'):

```
charmatch("m", c("mean", "median", "mode"))
```

```
[1] 0
```

If there is a unique match the function returns the index of the match within the vector of character strings (here in location number 2):

```
charmatch("med", c("mean", "median", "mode"))
```

```
[1] 2
```

Using %in% and which

You want to know all of the matches between one character vector and another:

stock<-c('car','van')
requests<-c('truck','suv','van','sports','car','waggon','car')

Use which to find the locations in the first-named vector of any and all of the entries in the second-named vector:

which(requests %in% stock)

```
[1] 3 5 7
```

If you want to know *what* the matches are as well as *where* they are,

requests [which(requests %in% stock)]

```
[1] "van" "car" "car"
```

You could use the match function to obtain the same result (p. 47):

stock[match(requests,stock)][!is.na(match(requests,stock))]

```
[1] "van" "car" "car"
```

but it's more clumsy. A slightly more complicated way of doing it involves sapply

which(sapply(requests, "%in%", stock))

```
van   car   car
 3     5     7
```

Note the use of quotes around the %in% function. Note that the match must be perfect for this to work ('car' with 'car' is not the same as 'car' with 'cars').

More on pattern matching

For the purposes of specifying these patterns, certain characters are called **metacharacters**, specifically \| () [{^ $ * + ? Any metacharacter with special meaning in your string may be quoted by preceding it with a backslash: \\{$ or * for instance. You might be used to specifying one or more 'wildcards' by * in DOS-like applications. In R, however, the regular expressions used are those specified by POSIX 1003.2, either extended or basic, depending on the value of the extended argument, unless perl = TRUE when they are those of PCRE (see ?grep for details).

Note that the square brackets in these class names [] are part of the symbolic names, and must be included in addition to the brackets delimiting the bracket list. For example, [[:alnum:]] means [0-9A-Za-z], except the latter depends upon the locale and the character encoding, whereas the former is independent of locale and character set. The interpretation below is that of the POSIX locale.

[:alnum:]   Alphanumeric characters: [:alpha:] and [:digit:].
[:alpha:]   Alphabetic characters: [:lower:] and [:upper:].
[:blank:]   Blank characters: space and tab.
[:cntrl:]   Control characters in ASCII, octal codes 000 through 037, and 177 (DEL).
[:digit:]   Digits: 0 1 2 3 4 5 6 7 8 9.
[:graph:]   Graphical characters: [:alnum:] and [:punct:].
[:lower:]   Lower-case letters in the current locale.
[:print:]   Printable characters: [:alnum:], [:punct:] and space.

[:punct:]   Punctuation characters:
             ! " # $ % & ( ) * +, - ./: ; <=> ? @ [\] ^ _ ' { | } ~.
[:space:]   Space characters: tab, newline, vertical tab, form feed, carriage return, space.
[:upper:]   Upper-case letters in the current locale.
[:xdigit:]   Hexadecimal digits: 0 1 2 3 4 5 6 7 8 9 A B C D E F a b c d e f.

Most metacharacters lose their special meaning inside lists. Thus, to include a literal], place it first in the list. Similarly, to include a literal ^, place it anywhere but first. Finally, to include a literal -, place it first or last. Only these and \ remain special inside character classes. To recap:

- Dot . matches any single character.

- Caret ^ matches the empty string at the beginning of a line.

- Dollar sign $ matches the empty string at the end of a line.

- Symbols \< and \> respectively match the empty string at the beginning and end of a word.

- The symbol \b matches the empty string at the edge of a word, and \B matches the empty string provided it is not at the edge of a word.

A regular expression may be *followed* by one of several repetition quantifiers:

?   The preceding item is optional and will be matched at most once.

*   The preceding item will be matched zero or more times.

+   The preceding item will be matched one or more times.

{n}   The preceding item is matched exactly n times.

{n, }   The preceding item is matched n or more times.

{,m}   The preceding item is matched up to m times.

{n,m}   The preceding item is matched at least n times, but not more than m times.

You can use the OR operator | so that "abba|cde" matches either the string "abba" or the string "cde".

Here are some simple examples to illustrate the issues involved.

```
text <- c("arm","leg","head", "foot","hand", "hindleg", "elbow")
```

The following lines demonstrate the "consecutive characters" {n} in operation:

```
grep("o{1}",text,value=T)
```

```
[1] "foot" "elbow"
```

```
grep("o{2}",text,value=T)
```

```
[1] "foot"
```

```
grep("o{3}",text,value=T)
```

```
character(0)
```

The following lines demonstrate the use of {n, } "n or more" character counting in words:

```
grep("[[:alnum:]]{4, }",text,value=T)
```

```
[1] "head" "foot" "hand" "hindleg" "elbow"
```

```
grep("[[:alnum:]]{5, }",text,value=T)
```

```
[1] "hindleg" "elbow"
```

```
grep("[[:alnum:]]{6, }",text,value=T)
```

```
[1] "hindleg"
```

```
grep("[[:alnum:]]{7, }",text,value=T)
```

```
[1] "hindleg"
```

Perl regular expressions

The perl = TRUE argument switches to the PCRE library that implements regular expression pattern matching using the same syntax and semantics as Perl 5.6 or later (with just a few differences). For details (and there are many) see ?regexp.

Perl is good for altering the cases of letters. Here, we capitalize the first character in each string:

```
gsub("(\\w)(\\w*)", "\\U\\1\\L\\2",text, perl=TRUE)
```

```
[1] "Arm" "Leg" "Head" "Foot" "Hand" "Hindleg" "Elbow"
```

while here we convert all the character to upper case:

```
gsub("(\\w*)", "\\U\\1",text, perl=TRUE)
```

```
[1] "ARM" "LEG" "HEAD" "FOOT" "HAND" "HINDLEG" "ELBOW"
```

Stripping patterned text out of complex strings

Suppose that we want to tease apart the information in these complicated strings:

```
(entries <-c ("Trial 1 58 cervicornis (52 match)", "Trial 2 60 terrestris (51 matched)",
"Trial 8 109 flavicollis (101 matches)"))
```

```
[1] "Trial 1 58 cervicornis (52 match)"
[2] "Trial 2 60 terrestris (51 matched)"
[3] "Trial 8 109 flavicollis (101 matches)"
```

The first task is to remove the material on numbers of matches including the brackets:

```
gsub(" *$", "", gsub("\\(.*\\)$", "", entries))
```

```
[1] "Trial 1 58 cervicornis"  "Trial 2 60 terrestris"
[3] "Trial 8 109 flavicollis"
```

The first argument " *$", "", removes the "trailing blanks" while the second deletes every-
thing .* between the left \\(and right \\) hand brackets "\\(.*\\)$" substituting this with nothing
"". The next job is to strip out the material in brackets and to extract that material, ignoring
the brackets themselves:

```
pos<- regexpr("\\(.*\\)$", entries)
substring(entries, first=pos+1, last=pos+attr(pos,"match.length")-2)
```

```
[1]  "52 match"  "51 matched"  "101 matches"
```

To see how this has worked it is useful to inspect the values of pos that have emerged from
the regexpr function:

```
pos
```

```
[1] 25 23 25
attr(,"match.length")
[1] 10 12 13
```

The left-hand bracket appears in position 25 in the first and third elements (note that there
are two blanks before 'cervicornis') but in position 23 in the second element. Now the
lengths of the strings matching the pattern \\(.*\\)$ can be checked; it is the number of
'anything' characters between the two brackets, plus one for each bracket: 10, 12 and 13.

Thus, to extract the material in brackets, but to ignore the brackets themselves, we need
to locate the first character to be extracted (pos+1) and the last character to be extracted
pos+attr(pos,"match.length")-2, then use the substring function to do the extracting. Note
that first and last are vectors of length 3 (= length(entries)).

Testing and Coercing in R

Objects have a type, and you can test the type of an object using an is.type function
(Table 2.4). For instance, mathematical functions expect numeric input and text-processing

Table 2.4. Functions for testing (is) the attributes of different categories of
object (arrays, lists, etc.) and for coercing (as) the attributes of an object into
a specified form. Neither operation changes the attributes of the object.

Type	Testing	Coercing
Array	is.array	as.array
Character	is.character	as.character
Complex	is.complex	as.complex
Dataframe	is.data.frame	as.data.frame
Double	is.double	as.double
Factor	is.factor	as.factor
List	is.list	as.list
Logical	is.logical	as.logical
Matrix	is.matrix	as.matrix
Numeric	is.numeric	as.numeric
Raw	is.raw	as.raw
Time series (ts)	is.ts	as.ts
Vector	is.vector	as.vector

functions expect character input. Some types of objects can be coerced into other types. A familiar type of coercion occurs when we interpret the TRUE and FALSE of logical variables as numeric 1 and 0, respectively. Factor levels can be coerced to numbers. Numbers can be coerced into characters, but non-numeric characters cannot be coerced into numbers.

```
as.numeric(factor(c("a","b","c")))
```

```
[1] 1 2 3
```

```
as.numeric(c("a","b","c"))
```

```
[1] NA NA NA
Warning message:
NAs introduced by coercion
```

```
as.numeric(c("a","4","c"))
```

```
[1] NA  4 NA
Warning message:
NAs introduced by coercion
```

If you try to coerce complex numbers to numeric the imaginary part will be discarded. Note that is.complex and is.numeric are never both TRUE.

We often want to coerce tables into the form of vectors as a simple way of stripping off their dimnames (using as.vector), and to turn matrixes into dataframes (as.data.frame). A lot of testing involves the NOT operator ! in functions to return an error message if the wrong type is supplied. For instance, if you were writing a function to calculate geometric means you might want to test to ensure that the input was numeric using the !is.numeric function

```
geometric<-function(x){
if(!is.numeric(x)) stop ("Input must be numeric")
exp(mean(log(x))) }
```

Here is what happens when you try to work out the geometric mean of character data

```
geometric(c("a","b","c"))
```

```
Error in geometric(c("a", "b", "c")) : Input must be numeric
```

You might also want to check that there are no zeros or negative numbers in the input, because it would make no sense to try to calculate a geometric mean of such data:

```
geometric<-function(x){
if(!is.numeric(x)) stop ("Input must be numeric")
if(min(x)<=0) stop ("Input must be greater than zero")
exp(mean(log(x))) }
```

Testing this:

```
geometric(c(2,3,0,4))
```

```
Error in geometric(c(2, 3, 0, 4)) : Input must be greater than zero
```

But when the data are OK there will be no messages, just the numeric answer:

```
geometric(c(10,1000,10,1,1))
```

```
[1]  10
```

Dates and Times in R

The measurement of time is highly idiosyncratic. Successive years start on different days of the week. There are months with different numbers of days. Leap years have an extra day in February. Americans and Britons put the day and the month in different places: 3/4/2006 is March 4 for the former and April 3 for the latter. Occasional years have an additional 'leap second' added to them because friction from the tides is slowing down the rotation of the earth from when the standard time was set on the basis of the tropical year in 1900. The cumulative effect of having set the atomic clock too slow accounts for the continual need to insert leap seconds (32 of them since 1958). There is currently a debate about abandoning leap seconds and introducing a 'leap minute' every century or so instead. Calculations involving times are complicated by the operation of time zones and daylight saving schemes in different countries. All these things mean that working with dates and times is excruciatingly complicated. Fortunately, R has a robust system for dealing with this complexity. To see how R handles dates and times, have a look at Sys.time():

```
Sys.time()
```

```
[1] "2005-10-23 10:17:42 GMT Daylight Time"
```

The answer is strictly hierarchical from left to right: the longest time scale (years) comes first, then month then day separated by hyphens (minus signs), then there is a blank space and the time, hours first (in the 24-hour clock) then minutes, then seconds separated by colons. Finally there is a character string explaining the time zone. You can extract the date from Sys.time() using substr like this:

```
substr(as.character(Sys.time()),1,10)
```

```
[1] "2005-10-23"
```

or the time

```
substr(as.character(Sys.time()),12,19)
```

```
[1] "10:17:42"
```

If you type

```
unclass(Sys.time())
```

```
[1] 1130679208
```

you get the number of seconds since 1 January 1970. There are two basic classes of date/times. Class POSIXct represents the (signed) number of seconds since the beginning of 1970 as a numeric vector: this is more convenient for including in dataframes. Class POSIXlt is a named list of vectors closer to human-readable forms, representing seconds, minutes, hours, days, months and years. R will tell you the date and time with the date function:

```
date()
```

```
[1] "Fri Oct 21 06:37:04 2005"
```

The default order is day name, month name (both abbreviated), day of the month, hour (24-hour clock), minute, second (separated by colons) then the year. You can convert Sys.time to an object that inherits from class POSIXlt like this:

```
date<- as.POSIXlt(Sys.time())
```

You can use the element name operator $ to extract parts of the date and time from this object using the following names: sec, min, hour, mday, mon, year, wday, yday and isdst (with obvious meanings except for mday (=day number within the month), wday (day of the week starting at 0 = Sunday), yday (day of the year after 1 January = 0) and isdst which means 'is daylight savings time in operation?' with logical 1 for TRUE or 0 for FALSE). Here we extract the day of the week (date$wday = 0 meaning Sunday) and the Julian date (day of the year after 1 January as date$yday)

```
date$wday
```

```
[1] 0
```

```
date$yday
```

```
[1] 295
```

for 23 October. Use unclass with unlist to view all of the components of date:

```
unlist(unclass(date))
```

```
sec   min   hour   mday   mon   year   wday   yday   isdst
42    17    10     23     9     105    0      295    1
```

Note that the month of October is 9 (not 10) because January is scored as month 0, and years are scored as post-1900.

Calculations with dates and times

You can do the following calculations with dates and times:

- time + number
- time − number
- time1 − time2
- time1 'logical operation' time2

where the logical operations are one of ==, !=, <, <=, '>' or >=. You can add or subtract a number of seconds or a difftime object (see below) from a date-time object, but you cannot add two date-time objects. Subtraction of two date-time objects is equivalent to using difftime (see below). Unless a time zone has been specified, POSIXlt objects are interpreted as being in the current time zone in calculations.

The thing you need to grasp is that you should convert your dates and times into POSIXlt objects *before* starting to do any calculations. Once they are POSIXlt objects, it is straightforward to calculate means, differences and so on. Here we want to calculate the number of days between two dates, 22 October 2003 and 22 October 2005:

```
y2<-as.POSIXlt("2003-10-22")
y1<-as.POSIXlt("2005-10-22")
```

Now you can do calculations with the two dates:

```
y1-y2
```

```
Time difference of 731 days
```

Note that you cannot *add* two dates. It is easy to calculate differences between times using this system. Note that the dates are separated by hyphens whereas the times are separated by colons:

```
y3<-as.POSIXlt("2005-10-22 09:30:59")
y4<-as.POSIXlt("2005-10-22 12:45:06")
y4-y3
```

```
Time difference of 3.235278 hours
```

The **difftime** function

Working out the time difference between to dates and times involves the **difftime** function, which takes two date-time objects as its arguments. The function returns an object of class **difftime** with an attribute indicating the units. How many days elapsed between 15 August 2003 and 21 October 2005?

```
difftime("2005-10-21","2003-8-15")
```

```
Time difference of 798 days
```

If you want only the number of days, for instance to use in calculation, then write

```
as.numeric(difftime("2005-10-21","2003-8-15"))
```

```
[1] 798
```

For differences in hours include the times (colon-separated) and write

```
difftime("2005-10-21 5:12:32","2005-10-21 6:14:21")
```

```
Time difference of -1.030278 hours
```

The result is negative because the first time (on the left) is before the second time (on the right). Alternatively, you can subtract one date-time object from another directly:

```
ISOdate(2005,10,21)-ISOdate(2003,8,15)
```

```
Time difference of 798 days
```

You can convert character stings into **difftime** objects using the **as.difftime** function:

```
as.difftime(c("0:3:20", "11:23:15"))
```

```
Time differences of 3.333333, 683.250000 mins
```

You can specify the format of your times. For instance, you may have no information on seconds, and your times are specified just as hours (format %H) and minutes (%M). This is what you do:

```
as.difftime(c("3:20", "23:15", "2:"), format= "%H:%M")
```

```
Time differences of 3.333333, 23.250000, NA hours
```

Because the last time in the sequence '2:' had no minutes it is marked as NA.

The strptime function

You can 'strip a date' out of a character string using the strptime function. There are functions to convert between character representations and objects of classes POSIXlt and POSIXct representing calendar dates and times. The details of the formats are system-specific, but the following are defined by the POSIX standard for strptime and are likely to be widely available. Any character in the format string other than the % symbol is interpreted literally.

%a   Abbreviated weekday name

%A   Full weekday name

%b   Abbreviated month name

%B   Full month name

%c   Date and time, locale-specific

%d   Day of the month as decimal number (01–31)

%H   Hours as decimal number (00–23) on the 24-hour clock

%I   Hours as decimal number (01–12) on the 12-hour clock

%j   Day of year as decimal number (001–366)

%m   Month as decimal number (01–12)

%M   Minute as decimal number (00–59)

%p   AM/PM indicator in the locale

%S   Second as decimal number (00–61, allowing for two 'leap seconds')

%U   Week of the year (00–53) using the first Sunday as day 1 of week 1

%w   Weekday as decimal number (0–6, Sunday is 0)

%W   Week of the year (00–53) using the first Monday as day 1 of week 1

%x   Date, locale-specific

%X   Time, locale-specific

%Y   Year with century

%Z   Time zone as a character string (output only)

Where leading zeros are shown they will be used on output but are optional on input.

Dates in Excel spreadsheets

The trick is to learn how to specify the format of your dates properly in strptime. If you had dates (and no times) in a dataframe in Excel format (day/month/year)

```
excel.dates <- c("27/02/2004", "27/02/2005", "14/01/2003",
          "28/06/2005", "01/01/1999")
```

then the appropriate format would be "%d/%m/%Y" showing the format names (from the list above) and the 'slash' separators / (note the upper case for year %Y; this is the unambiguous year including the century, 2005 rather than the potentially ambiguous 05 for which the format is %y). To turn these into R dates, write

```
strptime(excel.dates,format="%d/%m/%Y")
```

```
[1] "2004-02-27" "2005-02-27" "2003-01-14" "2005-06-28" "1999-01-01"
```

Here is another example, but with years in two-digit form (%y), and the months as abbreviated names (%b) and no separators:

```
other.dates<- c("1jan99", "2jan05", "31mar04", "30jul05")
strptime(other.dates, "%d%b%y")
```

```
[1] "1999-01-01" "2005-01-02" "2004-03-31" "2005-07-30"
```

You will often want to create POSIXlt objects from components stored in vectors within dataframes. For instance, here is a dataframe with the hours, minutes and seconds from an experiment with two factor levels in separate columns:

```
times<-read.table("c:\\temp\\times.txt",header=T)
times
```

	hrs	min	sec	experiment
1	2	23	6	A
2	3	16	17	A
3	3	2	56	A
4	2	45	0	A
5	3	4	42	A
6	2	56	25	A
7	3	12	28	A
8	1	57	12	A
9	2	22	22	B
10	1	42	7	B
11	2	31	17	B
12	3	15	16	B
13	2	28	4	B
14	1	55	34	B
15	2	17	7	B
16	1	48	48	B

```
attach(times)
```

Because the times are not in POSIXlt format, you need to paste together the hours, minutes and seconds into a character string with colons as the separator:

```
paste(hrs,min,sec,sep=":")
```

```
 [1]   "2:23:6"  "3:16:17"  "3:2:56"   "2:45:0"   "3:4:42"  "2:56:25" "3:12:28"
 [8]  "1:57:12" "2:22:22"  "1:42:7"   "2:31:17" "3:15:16"  "2:28:4"  "1:55:34"
[15]   "2:17:7"  "1:48:48"
```

Now save this object as a difftime vector called duration:

```
duration<-as.difftime (paste(hrs,min,sec,sep=":"))
```

Then you can carry out calculations like mean and variance using the tapply function:

tapply(duration,experiment,mean)

```
       A          B
2.829375   2.292882
```

Calculating time differences between the rows of a dataframe

A common action with time data is to compute the time difference between successive rows of a dataframe. The vector called duration created above is of class **difftime** and contains 16 times measured in decimal hours:

class(duration)

```
[1] "difftime"
```

duration

```
Time differences of 2.385000, 3.271389, 3.048889, 2.750000, 3.078333,
2.940278, 3.207778, 1.953333, 2.372778, 1.701944, 2.521389, 3.254444,
2.467778, 1.926111, 2.285278, 1.813333 hours
```

We can compute the differences between successive rows using subscripts, like this

duration[1:15]-duration[2:16]

```
Time differences of -0.8863889, 0.2225000, 0.2988889, -0.3283333,
0.1380556, -0.2675000, 1.2544444, -0.4194444, 0.6708333, -0.8194444,
-0.7330556, 0.7866667, 0.5416667, -0.3591667, 0.4719444 hours
```

You might want to make the differences between successive rows into part of the dataframe (for instance, to relate change in time to one of the explanatory variables in the dataframe). Before doing this, you need to decide on the row in which to put the first of the differences. Is the change in time between rows 1 and 2 related to the explanatory variables in row 1 or row 2? Suppose it is row 1 that we want to contain the first time difference (-0.886). Because we are working with differences (see p. 719) the vector of differences is shorter by one than the vector from which it was calculated:

length(duration[1:15]-duration[2:16])

```
[1] 15
```

length(duration)

```
[1] 16
```

so we need to add one 'NA' to the bottom of the vector (in row 16).

diffs<-c(duration[1:15]-duration[2:16],NA)
diffs

```
 [1]  -0.8863889   0.2225000  0.2988889  -0.3283333   0.1380556  -0.2675000
 [7]   1.2544444  -0.4194444  0.6708333  -0.8194444  -0.7330556   0.7866667
[13]   0.5416667  -0.3591667  0.4719444          NA
```

Now we can make this new vector part of the dataframe called times:

times$diffs<-diffs
times

	hrs	min	sec	experiment	diffs
1	2	23	6	A	-0.8863889
2	3	16	17	A	0.2225000
3	3	2	56	A	0.2988889
4	2	45	0	A	-0.3283333
5	3	4	42	A	0.1380556
6	2	56	25	A	-0.2675000
7	3	12	28	A	1.2544444
8	1	57	12	A	-0.4194444
9	2	22	22	B	0.6708333
10	1	42	7	B	-0.8194444
11	2	31	17	B	-0.7330556
12	3	15	16	B	0.7866667
13	2	28	4	B	0.5416667
14	1	55	34	B	-0.3591667
15	2	17	7	B	0.4719444
16	1	48	48	B	NA

There is more about dates and times in dataframes on p. 126.

Data Input

You can get numbers into R through the keyboard, from the clipboard or from an external file. For a single variable of up to 10 numbers or so, it is probably quickest to type the numbers at the command line, using the *concatenate* function c like this:

y <- c (6,7,3,4,8,5,6,2)

For intermediate sized variables, you might want to enter data from the keyboard using the scan function. For larger data sets, and certainly for sets with several variables, you should make a dataframe in Excel and read it into R using read.table (p. 98).

The scan Function

This is the function to use if you want to type (or paste) a few numbers into a vector from the keyboard.

x<-scan()

1:

At the 1: prompt type your first number, then press the Enter key. When the 2: prompt appears, type in your second number and press Enter, and so on. When you have put in all the numbers you need (suppose there are eight of them) then simply press the Enter key at the 9: prompt.

```
x<-scan()
1: 6
2: 7
3: 3
4: 4
5: 8
6: 5
7: 6
8: 2
9:
Read 8 items
```

The R Book   Michael J. Crawley
© 2007 John Wiley & Sons, Ltd

You can also use scan to paste in groups of numbers from the clipboard. In Excel, highlight the column of numbers you want, then type Ctrl+C (the accelerator keys for Copy). Now go back into R. At the 1: prompt just type Ctrl+V (the accelerator keys for Paste) and the numbers will be scanned into the named variable (*x* in this example). You can then paste in another set of numbers, or press Return to complete data entry. If you try to read in a group of numbers from a *row* of cells in Excel, the characters will be pasted into a single multi-digit number (definitely *not* what is likely to have been intended). So, if you are going to paste numbers from Excel, make sure the numbers are in columns, not in rows, in the spreadsheet. Use Edit/Paste Special/Transpose in Excel to turn a row into a column if necessary.

Data Input from Files

You can read data from a file using scan (see p. 102) but read.table is much more user-friendly. The read.table function reads a file in table format and automatically creates a dataframe from it, with cases corresponding to rows (lines) and variables to columns (fields) in the file (see p. 107). Much the simplest way to proceed is always to make your dataframe as a spreadsheet in Excel, and always to save it as a tab-delimited text file. That way you will always use read.table for data input, and you will avoid many of the most irritating problems that people encounter when using other input formats.

Saving the File from Excel

Once you have made your dataframe in Excel and corrected all the inevitable data-entry and spelling errors, then you need to save the dataframe in a file format that can be read by R. Much the simplest way is to save all your dataframes from Excel as tab-delimited text files: File/Save As . . . / then from the 'Save as type' options choose 'Text (Tab delimited)'. There is no need to add a suffix, because Excel will automatically add '.txt' to your file name. This file can then be read into R directly as a dataframe, using the read.table function like this:

```
data<-read.table("c:\\temp\\regression.txt",header=T)
```

Common Errors when Using read.table

It is important to note that read.table would fail if there were any spaces in any of the variable names in row 1 of the dataframe (the header row, see p. 107), such as Field Name, Soil pH or Worm Density, or between any of the words within the same factor level (as in many of the field names). You should replace all these spaces by dots '.' before saving the dataframe in Excel (use Edit/Replace with " " replaced by "."). Now the dataframe can be read into R. There are three things to remember:

- The whole path and file name needs to be enclosed in double quotes: "c:\\abc.txt".
- header=T says that the first row contains the variable names.
- Always use double backslash \\ rather than \ in the file path definition.

The commonest cause of failure is that the number of variable names (characters strings in row 1) does not match the number of columns of information. In turn, the commonest cause of this is that you have blank spaces in your variable names:

```
state name   population   home ownership   cars   insurance
```

This is wrong because R expects seven columns of numbers when there are only five. Replace the spaces within the names by dots and it will work fine:

```
state.name   population   home.ownership   cars   insurance
```

The next most common cause of failure is that the data file contains blank spaces where there are missing values. Replace these blanks with NA in Excel (or use a different separator symbol: see below).

Finally, there can be problems when you are trying to read variables that consist of character strings containing blank spaces (as in files containing place names). You can use read.table so long as you export the file from Excel using commas to separate the fields, and you tell read.table that the separators are commas using sep=","

```
map<-read.table("c:\\temp\\bowens.csv",header=T,sep=",")
```

but it is quicker and easier to use read.csv in this case(and there is no need for header=T)

```
map<-read.csv("c:\\temp\\bowens.csv")
```

If you are tired of writing header=T in all your read.table functions, then switch to

```
read.delim("c:\\temp\\file.txt")
```

or write your own function

```
rt<-function(x) read.delim(x)
```

then use the function rt to read data-table files like this:

```
rt("c:\\temp\\regression.txt")
```

Better yet, remove the need to enter the drive and directory or the file suffix:

```
rt<-function(x) read.delim(paste("c:\\temp\\",x,".txt",sep=""))
rt("regression")
```

Browsing to Find Files

The R function for this is file.choose(). Here it is in action with read.table:

```
data<-read.table(file.choose(),header=T)
```

Once you click on your selected file this is read into the dataframe called data.

Separators and Decimal Points

The default field separator character in read.table is sep="". This separator is white space, which is produced by one or more spaces, one or more tabs \t, one or more

newlines \n, or one or more carriage returns. If you do have a different separator between the variables sharing the same line (i.e. other than a tab within a .txt
file) then there may well be a special read function for your case. Note that these
all have the sensible default that header=TRUE (the first row contains the variable names): for comma-separated fields use read.csv("c:\\temp\\file.txt"), for semicolon separated fields read.csv2("c:\\temp\\file.txt"), and for decimal points as a comma
read.delim2("c:\\temp\\file.txt"). You would use comma or semicolon separators if you had
character variables that might contain one or more blanks (e.g. country names like 'United
Kingdom' or 'United States of America').

   If you want to specify row.names then one of the columns of the dataframe must be a
vector of unique row names. This can be a single number giving the column of the table
which contains the row names, or character string giving the variable name of the table
column containing the row names (see p. 123). Otherwise if row.names is missing, the
rows are numbered.

   The default behaviour of read.table is to convert character variables into factors. If you
do *not* want this to happen (you want to keep a variable as a character vector) then use
as.is to specify the columns that should not be converted to factors:

murder<-read.table("c:\\temp\\murders.txt",header=T,as.is="region"); attach(murder)

We use the attach function so that the variables inside a dataframe can be accessed directly
by name. Technically, this means that the database is attached to the R search path, so that
the database is searched by R when evaluating a variable.

table(region)

```
region
North.Central   Northeast   South   West
          12           9      16     13
```

If we had not attached a dataframe, then we would have had to specify the name of the
dataframe first like this:

table(murder$region)

The following warning will be produced if your attach function causes a duplication of one
or more names:

```
The following object(s) are masked _by_ .GlobalEnv:
        murder
```

The reason in the present case is that we have created a dataframe called murder and attached
a variable which is also called murder. This ambiguity might cause difficulties later. The
commonest cause of this problem occurs with simple variable names like x and y. It is very
easy to end up with multiple variables of the same name within a single session that mean
totally different things. The warning after using attach should alert you to the possibility
of such problems. If the vectors sharing the same name are of different lengths, then R is
likely to stop you before you do anything too silly, but if the vectors are the same length
then you run the serious risk of fitting the wrong explanatory variable (e.g. fitting the wrong
one from two vectors both called x) or having the wrong response variable (e.g. from two
vectors both called y). The moral is:

- use longer, more self-explanatory variable names;
- do not calculate variables with the same name as a variables inside a dataframe;
- always **detach** dataframes once you are finished using them;
- **remove** calculated variables once you are finished with them (rm; see p. 8).

The best practice, however, is not to use **attach** in the first place, but to use functions like **with** instead (see p. 18). If you get into a real tangle, it is often easiest to quit R and start another R session. To check that region is not a factor, write:

is.factor(region)

```
[1]  FALSE
```

Input and Output Formats

Formatting is controlled using **escape sequences**, typically within double quotes:

\n   newline
\r   carriage return
\t   tab character
\b   backspace
\a   bell
\f   form feed
\v   vertical tab

Setting the Working Directory

You do not have to type the drive name and folder name every time you want to read or write a file, so if you use the same path frequently it is sensible to set the working directory using the **setwd** function:

setwd("c:\\temp")

. . .

. . .

read.table("daphnia.txt",header=T)

If you want to find out the name of the current working directory, use **getwd()**:

getwd()

```
[1]  "c:/temp"
```

Checking Files from the Command Line

It can be useful to check whether a given filename exists in the path where you think it should be. The function is **file.exists** and is used like this:

file.exists("c:\\temp\\Decay.txt")

```
[1]  TRUE
```

For more on file handling, see ?files.

Reading Dates and Times from Files

You need to be very careful when dealing with dates and times in any sort of computing. R has a particularly robust system for working with dates and times, which is explained in detail on p. 89. Typically, you will read dates and times as character strings, then convert them into dates and/or times within R.

Built-in Data Files

There are many built-in data sets within the base package of R. You can see their names by typing

data()

You can read the documentation for a particular data set with the usual query:

?lynx

Many of the contributed packages contain data sets, and you can view their names using the try function. This evaluates an expression and traps any errors that occur during the evaluation. The try function establishes a handler for errors that uses the default error handling protocol:

```
try(data(package="spatstat"));Sys.sleep(3)
try(data(package="spdep"));Sys.sleep(3)
try(data(package="MASS"))
```

Built-in data files can be attached in the normal way; then the variables within them accessed by their names:

```
attach(OrchardSprays)
decrease
```

Reading Data from Files with Non-standard Formats Using scan

The scan function is very flexible, but as a consequence of this, it is much harder to use than read.table. This example uses the US murder data. The filename comes first, in the usual format (enclosed in double quotes and using paired backslashes to separate the drive name from the folder name and the folder name from the file name). Then comes skip=1 because the first line of the file contains the variable names (as indicated by header=T in a read.table function). Next comes what, which is a list of length the number of variables (the number of columns to be read; 4 in this case) specifying their type (character " " in this case):

```
murders<-scan("c:\\temp\\murders.txt", skip=1, what=list("","","",""))
```

```
Read 50 records
```

The object produced by scan is a list rather than a dataframe as you can see from

```
class(murders)
```

```
[1] "list"
```

It is simple to convert the list to a dataframe using the as.data.frame function

```
murder.frame<-as.data.frame(murders)
```

You are likely to want to use the variables names from the file as variable names in the dataframe. To do this, read just the first line of the file using scan with nlines=1:

```
murder.names<-
scan("c:\\temp\\murders.txt",nlines=1,what="character",quiet=T)
murder.names
```

```
[1] "state" "population" "murder" "region"
```

Note the use of quiet=T to switch off the report of how many records were read. Now give these names to the columns of the dataframe

```
names(murder.frame)<-murder.names
```

Finally, convert columns 2 and 3 of the dataframe from factors to numbers:

```
murder.frame[,2]<-as.numeric(murder.frame[,2])
murder.frame[,3]<-as.numeric(murder.frame[,3])
summary(murder.frame)
```

```
     state              population        murder              region
Alabama     : 1   Min.     : 1.00   Min.    : 1.00   North.Central :12
Alaska      : 1   1st Qu.  :13.25   1st Qu. :11.25   Northeast     : 9
Arizona     : 1   Median   :25.50   Median  :22.50   South         :16
Arkansas    : 1   Mean     :25.50   Mean    :22.10   West          :13
California  : 1   3rd Qu.  :37.75   3rd Qu. :32.75
Colorado    : 1   Max.     :50.00   Max.    :44.00
(Other)     :44
```

You can see why people prefer to use read.table for this sort of data file:

```
murders<-read.table("c:\\temp\\murders.txt",header=T)
summary(murders)
```

```
     state              population        murder               region
Alabama     : 1   Min.     :  365   Min.    : 1.400   North.Central :12
Alaska      : 1   1st Qu.  : 1080   1st Qu. : 4.350   Northeast     : 9
Arizona     : 1   Median   : 2839   Median  : 6.850   South         :16
Arkansas    : 1   Mean     : 4246   Mean    : 7.378   West          :13
California  : 1   3rd Qu.  : 4969   3rd Qu. :10.675
Colorado    : 1   Max.     :21198   Max.    :15.100
(Other)     :44
```

Note, however, that the scan function is quicker than read.table for input of large (numeric only) matrices.

Reading Files with Different Numbers of Values per Line

Here is a case where you might want to use scan because the data are not configured like a dataframe. The file rt.txt has different numbers of values per line (a neighbours file in spatial analysis, for example; see p. 769). In this example, the file contains five lines with 1, 2, 4, 2 and 1 numbers respectively: in general, you will need to find out the number of lines of data in the file by counting the number of end-of-line control character "\n" using the length function like this:

```
line.number<-length(scan("c:\\temp\\rt.txt",sep="\n"))
```

The trick is to combine the skip and nlines options within scan to read one line at a time, skipping no lines to read the first row, skipping one row to read the second line, and so on. Note that since the values are numbers we do not need to specify what:

```
(my.list<-sapply(0:(line.number-1),
    function(x) scan("c:\\temp\\rt.txt",skip=x,nlines=1,quiet=T)))

[[1]]
[1]  138

[[2]]
[1]  27  44

[[3]]
[1]  19  20  345  48

[[4]]
[1]  115  23  66

[[5]]
[1]  59
```

The scan function has produced a list of vectors, each of a different length. You might want to know the number of numbers in each row, using length with lapply like this:

```
unlist(lapply(my.list,length))

[1]  1  2  4  2  1
```

Alternatively, you might want to create a vector containing the *last element* from each row:

```
unlist(lapply(1:length(my.list), function(i) my.list[[i]][length(my.list[[i]])]))

[1]  138  44  48  2366  59
```

The readLines Function

In some cases you might want to read each line from a file separately. The argument n=-1 means read to the end of the file. Let's try it out with the murders data (p. 100):

```
readLines("c:\\temp\\murders.txt",n=-1)
```

This produces the rather curious object of class = "character":

```
[1]  "state\tpopulation\tmurder\tregion"   "Alabama\t3615\t15.1\tSouth"
[3]  "Alaska\t365\t11.3\tWest"             "Arizona\t2212\t7.8\tWest"
```

. . . .
. . . .
[49] "West.Virginia\t1799\t6.7\tSouth"   "Wisconsin\t4589\t3\tNorth.Central"
[51] "Wyoming\t376\t6.9\tWest"

Each line has been converted into a single character string. Line [1] contains the four variable names (see above) separated by tab characters \t. Line [2] contains the first row of data for murders in Alabama, while row [51] contains the last row of data for murders in Wyoming. You can use the string-splitting function strsplit to tease apart the elements of the string (say, for the Wyoming data [51]):

```
mo<-readLines("c:\\temp\\murders.txt",n=-1)
strsplit(mo[51],"\t")
```

```
[[1]]
[1]  "Wyoming""376"  "6.9"  "West"
```

You would probably want 376 and 6.9 as numeric rather than character objects:

```
as.numeric(unlist(strsplit(mo[51],"\t")))
```

```
[1]  NA  376.0  6.9  NA
Warning message:
NAs  introduced by coercion
```

where the two names Wyoming and West have been coerced to NA, or

```
as.vector(na.omit(as.numeric(unlist(strsplit(mo[51],"\t")))))
```

```
[1]  376.0  6.9
Warning message:
NAs introduced by coercion
```

to get the numbers on their own. Here is how to extract the two numeric variables (murder = mur and population = pop) from this object using sapply:

```
mv<-sapply(2:51,function(i)
            as.vector(na.omit(as.numeric(unlist(strsplit(mo[i],"\t"))))))
pop<-mv[1,]
mur<-mv[2,]
```

and here is how to get character vectors of the state names and regions (the first and fourth elements of each row of the list called ms, called sta and reg, respectively):

```
ms<-sapply(2:51,function(i) strsplit(mo[i],"\t"))
texts<-unlist(lapply(1:50,function(i) ms[[i]][c(1,4)]))
sta<-texts[seq(1,99,2)]
reg<- texts[seq(2,100,2)]
```

Finally, we can convert all the information from readLines into a data.frame

```
data.frame(sta,pop,mur,reg)
```

	sta	pop	mur	reg
1	Alabama	3615	15.1	South
2	Alaska	365	11.3	West
...				
49	Wisconsin	4589	3.0	North.Central
50	Wyoming	376	6.9	West

This could all have been achieved in a single line with **read.table** (see above), and the **readLines** function is much more useful when the rows in the file contain different numbers of entries. Here is the simple example from p. 104 using **readLines** instead of **scan**:

```
rlines<-readLines("c:\\temp\\rt.txt")
split.lines<-strsplit(rlines,"\t")
new<-sapply(1:5,function(i) as.vector(na.omit(as.numeric(split.lines[[i]]))))
new

[[1]]
[1]  138

[[2]]
[1]  27  44

[[3]]
[1]  19  20  345  48

[[4]]
[1]  115  2366

[[5]]
[1]  59
```

The key features of this procedure are the removal of the tabs (\t) and the separation of the values with each row with the **strsplit** function, the conversion of the characters to numbers, and the removal of the NAs which are introduced by default by the **as.numeric** function. I think that **scan** (p. 104) is more intuitive in such a case.

4

Dataframes

Learning how to handle your data, how to enter it into the computer, and how to read the data into R are amongst the most important topics you will need to master. R handles data in objects known as dataframes. A **dataframe** is an object with rows and columns (a bit like a matrix). The rows contain different observations from your study, or measurements from your experiment. The columns contain the values of different variables. The values in the body of a matrix can only be numbers; those in a dataframe can also be numbers, but they could also be text (e.g. the names of factor levels for categorical variables, like male or female in a variable called gender), they could be calendar dates (e.g. 23/5/04), or they could be logical variables (TRUE or FALSE). Here is a spreadsheet in the form of a dataframe with seven variables, the leftmost of which comprises the row names, and other variables are numeric (Area, Slope, Soil pH and Worm density), categorical (Field Name and Vegetation) or logical (Damp is either true = T or false = F).

Field Name	Area	Slope	Vegetation	Soil pH	Damp	Worm density
Nash's Field	3.6	11	Grassland	4.1	F	4
Silwood Bottom	5.1	2	Arable	5.2	F	7
Nursery Field	2.8	3	Grassland	4.3	F	2
Rush Meadow	2.4	5	Meadow	4.9	T	5
Gunness' Thicket	3.8	0	Scrub	4.2	F	6
Oak Mead	3.1	2	Grassland	3.9	F	2
Church Field	3.5	3	Grassland	4.2	F	3
Ashurst	2.1	0	Arable	4.8	F	4
The Orchard	1.9	0	Orchard	5.7	F	9
Rookery Slope	1.5	4	Grassland	5	T	7
Garden Wood	2.9	10	Scrub	5.2	F	8
North Gravel	3.3	1	Grassland	4.1	F	1
South Gravel	3.7	2	Grassland	4	F	2
Observatory Ridge	1.8	6	Grassland	3.8	F	0
Pond Field	4.1	0	Meadow	5	T	6
Water Meadow	3.9	0	Meadow	4.9	T	8
Cheapside	2.2	8	Scrub	4.7	T	4
Pound Hill	4.4	2	Arable	4.5	F	5
Gravel Pit	2.9	1	Grassland	3.5	F	1
Farm Wood	0.8	10	Scrub	5.1	T	3

The R Book   Michael J. Crawley
© 2007 John Wiley & Sons, Ltd

Perhaps the most important thing about analysing your own data properly is getting your dataframe absolutely right. The expectation is that you will have used a spreadsheet such as Excel to enter and edit the data, and that you will have used plots to check for errors. The thing that takes some practice is learning exactly how to put your numbers into the spreadsheet. There are countless ways of doing it wrong, but only one way of doing it right. And this way is *not* the way that most people find intuitively to be the most obvious.

The key thing is this: all the values of the same variable must go in the same column. It does not sound like much, but this is what people tend to get wrong. If you had an experiment with three treatments (control, pre-heated and pre-chilled), and four measurements per treatment, it might seem like a good idea to create the spreadsheet like this:

control	preheated	prechilled			
6.1	6.3	7.1			
5.9	6.2	8.2			
5.8	5.8	7.3			
5.4	6.3	6.9			

However, this is not a dataframe, because values of the response variable appear in three different columns, rather than all in the same column. The correct way to enter these data is to have two columns: one for the response variable and one for the levels of the experimental factor called Treatment (control, preheated and prechilled). Here are the same data, entered correctly as a dataframe:

Response	Treatment			
6.1	control			
5.9	control			
5.8	control			
5.4	control			
6.3	preheated			
6.2	preheated			
5.8	preheated			
6.3	preheated			
7.1	prechilled			
8.2	prechilled			
7.3	prechilled			
6.9	prechilled			

A good way to practise this layout is to use the Excel function called PivotTable (found under Data on the main menu bar) on your own data: it requires your spreadsheet to be in the form of a dataframe, with each of the explanatory variables in its own column.

Once you have made your dataframe in Excel and corrected all the inevitable data-entry and spelling errors, then you need to save the dataframe in a file format that can be read by R. Much the simplest way is to save all your dataframes from Excel as tab-delimited text files: File/Save As... / then from the 'Save as type' options choose 'Text (Tab delimited)'. There is no need to add a suffix, because Excel will automatically add '.txt' to your file name. This file can then be read into R directly as a dataframe, using the read.table function.

As pointed out in Chapter 3, is important to note that read.table would fail if there were any spaces in any of the variable names in row 1 of the dataframe (the header row), such as Field Name, Soil pH or Worm Density (above), or between any of the words within the same factor level (as in many of the field names). These should be replaced by dots '.' before the dataframe is saved in Excel. Also, it is good idea to remove any apostrophes, as these can sometimes cause problems because there is more than one ASCII code for single quote. Now the dataframe can be read into R. Think of a name for the dataframe (say, 'worms' in this case) and then allocate the data from the file to the dataframe name using the **gets arrow** <- like this:

```
worms<-read.table("c:\\temp\\worms.txt",header=T)
```

Once the file has been imported to R we often want to do two things:

- use **attach** to make the variables accessible by name within the R session;
- use **names** to get a list of the variable names.

Typically, the two commands are issued in sequence, whenever a new dataframe is imported from file (but see p. 18 for superior alternatives to **attach**):

```
attach(worms)
names(worms)
[1]  "Field.Name"  "Area"   "Slope"           "Vegetation"
[5]  "Soil.pH"     "Damp"   "Worm.density"
```

To see the contents of the dataframe, just type its name:

```
worms
```

	Field.Name	Area	Slope	Vegetation	Soil.pH	Damp	Worm.density
1	Nashs.Field	3.6	11	Grassland	4.1	FALSE	4
2	Silwood.Bottom	5.1	2	Arable	5.2	FALSE	7
3	Nursery.Field	2.8	3	Grassland	4.3	FALSE	2
4	Rush.Meadow	2.4	5	Meadow	4.9	TRUE	5
5	Gunness.Thicket	3.8	0	Scrub	4.2	FALSE	6
6	Oak.Mead	3.1	2	Grassland	3.9	FALSE	2
7	Church.Field	3.5	3	Grassland	4.2	FALSE	3
8	Ashurst	2.1	0	Arable	4.8	FALSE	4
9	The.Orchard	1.9	0	Orchard	5.7	FALSE	9
10	Rookery.Slope	1.5	4	Grassland	5.0	TRUE	7
11	Garden.Wood	2.9	10	Scrub	5.2	FALSE	8
12	North.Gravel	3.3	1	Grassland	4.1	FALSE	1
13	South.Gravel	3.7	2	Grassland	4.0	FALSE	2
14	Observatory.Ridge	1.8	6	Grassland	3.8	FALSE	0

15	Pond.Field	4.1	0	Meadow	5.0	TRUE	6
16	Water.Meadow	3.9	0	Meadow	4.9	TRUE	8
17	Cheapside	2.2	8	Scrub	4.7	TRUE	4
18	Pound.Hill	4.4	2	Arable	4.5	FALSE	5
19	Gravel.Pit	2.9	1	Grassland	3.5	FALSE	1
20	Farm.Wood	0.8	10	Scrub	5.1	TRUE	3

Notice that R has expanded our abbreviated T and F into TRUE and FALSE. The object called **worms** now has all the attributes of a dataframe. For example, you can summarize it, using **summary**:

summary(worms)

```
    Field.Name           Area            Slope            Vegetation
Ashurst       : 1   Min.   :0.800   Min.    : 0.00   Arable    :3
Cheapside     : 1   1st Qu. :2.175  1st Qu. : 0.75   Grassland :9
Church.Field  : 1   Median :3.000   Median  : 2.00   Meadow    :3
Farm.Wood     : 1   Mean   :2.990   Mean    : 3.50   Orchard   :1
Garden.Wood   : 1   3rd Qu. :3.725  3rd Qu. : 5.25   Scrub     :4
Gravel.Pit    : 1   Max.   :5.100   Max.    :11.00
(Other)       :14
```

```
    Soil.pH           Damp            Worm.density
Min.    :3.500   Mode   :logical   Min.    :0.00
1st Qu. :4.100   FALSE  :14        1st Qu. :2.00
Median :4.600    TRUE   :6         Median :4.00
Mean    :4.555                     Mean    :4.35
3rd Qu. :5.000                     3rd Qu. :6.25
Max.    :5.700                     Max.    :9.00
```

Values of continuous variables are summarized under six headings: one parametric (the arithmetic mean) and five non-parametric (maximum, minimum, median, 25th percentile or first quartile, and 75th percentile or third quartile). Tukey's famous five-number function (**fivenum**; see p. 281) is slightly different, with hinges rather than first and third quartiles. Levels of categorical variables are counted. Note that the field names are not listed in full because they are unique to each row; six of them are named, then R says 'plus 14 others'.

The two functions **by** and **aggregate** allow summary of the dataframe on the basis of factor levels. For instance, it might be interesting to know the means of the numeric variables for each vegetation type. The function for this is **by**:

by(worms,Vegetation,mean)

```
Vegetation: Arable
    Field.Name      Area        Slope  Vegetation   Soil.pH        Damp
          NA   3.866667    1.333333          NA   4.833333    0.000000
Worm.density
      5.333333
- - - - - - - - - - - - - - - - - -
Vegetation: Grassland
    Field.Name      Area        Slope  Vegetation    Soil.pH        Damp
          NA   2.9111111   3.6666667          NA   4.1000000   0.1111111
Worm.density
      2.4444444
- - - - - - - - - - - - - - - - - -
```

```
Vegetation: Meadow
   Field.Name        Area      Slope  Vegetation    Soil.pH      Damp
            NA    3.466667   1.666667          NA   4.933333  1.000000
Worm.density
     6.333333
- - - - - - - - - - - - - - - - -
Vegetation: Orchard
  Field.Name   Area  Slope  Vegetation  Soil.pH  Damp
          NA    1.9    0.0          NA      5.7   0.0
Worm.density
         9.0
- - - - - - - - - - - - - - - - -
Vegetation: Scrub
   Field.Name    Area  Slope  Vegetation  Soil.pH    Damp
           NA   2.425  7.000          NA    4.800   0.500
Worm.density
       5.250
```

Notice that the logical variable Damp has been coerced to numeric (TRUE $=1$, FALSE $=0$) and then averaged. Warning messages are printed for the non-numeric variables to which the function mean is not applicable, but this is a useful and quick overview of the effects of the five types of vegetation.

Subscripts and Indices

The key thing about working effectively with dataframes is to become completely at ease with using subscripts (or indices, as some people call them). In R, subscripts appear in square brackets []. A dataframe is a two-dimensional object, comprising rows and columns. The rows are referred to by the first (left-hand) subscript, the columns by the second (right-hand) subscript. Thus

worms[3,5]

```
[1]  4.3
```

is the value of Soil.pH (the variable in column 5) in row 3. To extract a range of values (say the 14th to 19th rows) from worm density (the variable in the seventh column) we use the colon operator : to generate a series of subscripts (14, 15, 16, 17, 18 and 19):

worms[14:19,7]

```
[1] 0 6 8 4 5 1
```

To extract a group of rows and a group of columns, you need to generate a series of subscripts for both the row and column subscripts. Suppose we want Area and Slope (columns 2 and 3) from rows 1 to 5:

worms[1:5,2:3]

```
  Area  Slope
1  3.6     11
2  5.1      2
3  2.8      3
4  2.4      5
5  3.8      0
```

This next point is very important, and is hard to grasp without practice. To select *all* the entries in a *row* the syntax is 'number comma blank'. Similarly, to select all the entries in a *column* the syntax is 'blank comma number'. Thus, to select all the columns in row 3

worms[3,]

```
    Field.Name  Area  Slope  Vegetation  Soil.pH   Damp  Worm.density
3  Nursery.Field  2.8      3  Grassland      4.3  FALSE             2
```

whereas to select all of the rows in column number 3 we enter

worms[,3]

```
[1] 11 2 3 5  0 2 3 0 0 4 10 1 2 6 0 0 8 2 1 1  0
```

This is a key feature of the R language, and one that causes problems for beginners. Note that these two apparently similar commands create *objects of different classes*:

class(worms[3,])

```
[1]  "data.frame"
```

class(worms[,3])

```
[1]  "integer"
```

You can create sets of rows or columns. For instance, to extract all the rows for Field Name and the Soil pH (columns 1 and 5) use the concatenate function, c, to make a vector of the required column numbers c(1,5):

worms[,c(1,5)]

```
              Field.Name  Soil.pH
1             Nashs.Field      4.1
2          Silwood.Bottom      5.2
3           Nursery.Field      4.3
4             Rush.Meadow      4.9
5         Gunness.Thicket      4.2
6                Oak.Mead      3.9
7            Church.Field      4.2
8                 Ashurst      4.8
9             The.Orchard      5.7
10          Rookery.Slope      5.0
11            Garden.Wood      5.2
12            North.Gravel      4.1
13            South.Gravel      4.0
14       Observatory.Ridge      3.8
15             Pond.Field      5.0
16            Water.Meadow      4.9
17              Cheapside      4.7
18             Pound.Hill      4.5
19             Gravel.Pit      3.5
20              Farm.Wood      5.1
```

Selecting Rows from the Dataframe at Random

In bootstrapping or cross-validation we might want to select certain rows from the dataframe at random. We use the sample function to do this: the default replace = FALSE performs shuffling (each row is selected once and only once), while the option replace = TRUE (sampling with replacement) allows for multiple copies of certain rows. Here we use replace = F to select a unique 8 of the 20 rows at random:

worms[sample(1:20,8),]

	Field.Name	Area	Slope	Vegetation	Soil.pH	Damp	Worm.density
7	Church.Field	3.5	3	Grassland	4.2	FALSE	3
17	Cheapside	2.2	8	Scrub	4.7	TRUE	4
19	Gravel.Pit	2.9	1	Grassland	3.5	FALSE	1
14	Observatory.Ridge	1.8	6	Grassland	3.8	FALSE	0
12	North.Gravel	3.3	1	Grassland	4.1	FALSE	1
9	The.Orchard	1.9	0	Orchard	5.7	FALSE	9
11	Garden.Wood	2.9	10	Scrub	5.2	FALSE	8
8	Ashurst	2.1	0	Arable	4.8	FALSE	4

Note that the row numbers are in random sequence (not sorted), so that if you want a sorted random sample you will need to order the dataframe after the randomization.

Sorting Dataframes

It is common to want to sort a dataframe by rows, but rare to want to sort by columns. Because we are sorting by rows (the first subscript) we specify the order of the row subscripts *before* the comma. Thus, to sort the dataframe on the basis of values in one of the columns (say, Slope), we write

worms[order(Slope),]

	Field.Name	Area	Slope	Vegetation	Soil.pH	Damp	Worm.density
5	Gunness.Thicket	3.8	0	Scrub	4.2	FALSE	6
8	Ashurst	2.1	0	Arable	4.8	FALSE	4
9	The.Orchard	1.9	0	Orchard	5.7	FALSE	9
15	Pond.Field	4.1	0	Meadow	5.0	TRUE	6
16	Water.Meadow	3.9	0	Meadow	4.9	TRUE	8
12	North.Gravel	3.3	1	Grassland	4.1	FALSE	1
19	Gravel.Pit	2.9	1	Grassland	3.5	FALSE	1
2	Silwood.Bottom	5.1	2	Arable	5.2	FALSE	7
6	Oak.Mead	3.1	2	Grassland	3.9	FALSE	2
13	South.Gravel	3.7	2	Grassland	4.0	FALSE	2
18	Pound.Hill	4.4	2	Arable	4.5	FALSE	5
3	Nursery.Field	2.8	3	Grassland	4.3	FALSE	2
7	Church.Field	3.5	3	Grassland	4.2	FALSE	3
10	Rookery.Slope	1.5	4	Grassland	5.0	TRUE	7
4	Rush.Meadow	2.4	5	Meadow	4.9	TRUE	5
14	Observatory.Ridge	1.8	6	Grassland	3.8	FALSE	0
17	Cheapside	2.2	8	Scrub	4.7	TRUE	4
11	Garden.Wood	2.9	10	Scrub	5.2	FALSE	8
20	Farm.Wood	0.8	10	Scrub	5.1	TRUE	3
1	Nashs.Field	3.6	11	Grassland	4.1	FALSE	4

There are some points to notice here. Because we wanted the sorting to apply to all the columns, the column subscript (after the comma) is blank: [order(Slope),]. The original row numbers are retained in the leftmost column. Where there are ties for the sorting variable (e.g. there are five ties for Slope $= 0$) then the rows are in their original order. If you want the dataframe in reverse order (ascending order) then use the rev function outside the order function like this:

worms[rev(order(Slope)),]

	Field.Name	Area	Slope	Vegetation	Soil.pH	Damp	Worm.density
1	Nashs.Field	3.6	11	Grassland	4.1	FALSE	4
20	Farm.Wood	0.8	10	Scrub	5.1	TRUE	3
11	Garden.Wood	2.9	10	Scrub	5.2	FALSE	8
17	Cheapside	2.2	8	Scrub	4.7	TRUE	4
14	Observatory.Ridge	1.8	6	Grassland	3.8	FALSE	0
4	Rush.Meadow	2.4	5	Meadow	4.9	TRUE	5
10	Rookery.Slope	1.5	4	Grassland	5.0	TRUE	7
7	Church.Field	3.5	3	Grassland	4.2	FALSE	3
3	Nursery.Field	2.8	3	Grassland	4.3	FALSE	2
18	Pound.Hill	4.4	2	Arable	4.5	FALSE	5
13	South.Gravel	3.7	2	Grassland	4.0	FALSE	2
6	Oak.Mead	3.1	2	Grassland	3.9	FALSE	2
2	Silwood.Bottom	5.1	2	Arable	5.2	FALSE	7
19	Gravel.Pit	2.9	1	Grassland	3.5	FALSE	1
12	North.Gravel	3.3	1	Grassland	4.1	FALSE	1
16	Water.Meadow	3.9	0	Meadow	4.9	TRUE	8
15	Pond.Field	4.1	0	Meadow	5.0	TRUE	6
9	The.Orchard	1.9	0	Orchard	5.7	FALSE	9
8	Ashurst	2.1	0	Arable	4.8	FALSE	4
5	Gunness.Thicket	3.8	0	Scrub	4.2	FALSE	6

Notice, now, that when there are ties (e.g. Slope $= 0$), the original rows are also in reverse order.

More complicated sorting operations might involve two or more variables. This is achieved very simply by separating a series of variable names by commas within the order function. R will sort on the basis of the left-hand variable, with ties being broken by the second variable, and so on. Suppose that we want to order the rows of the database on worm density within each vegetation type:

worms[order(Vegetation,Worm.density),]

	Field.Name	Area	Slope	Vegetation	Soil.pH	Damp	Worm.density
8	Ashurst	2.1	0	Arable	4.8	FALSE	4
18	Pound.Hill	4.4	2	Arable	4.5	FALSE	5
2	Silwood.Bottom	5.1	2	Arable	5.2	FALSE	7
14	Observatory.Ridge	1.8	6	Grassland	3.8	FALSE	0
12	North.Gravel	3.3	1	Grassland	4.1	FALSE	1
19	Gravel.Pit	2.9	1	Grassland	3.5	FALSE	1
3	Nursery.Field	2.8	3	Grassland	4.3	FALSE	2
6	Oak.Mead	3.1	2	Grassland	3.9	FALSE	2
13	South.Gravel	3.7	2	Grassland	4.0	FALSE	2
7	Church.Field	3.5	3	Grassland	4.2	FALSE	3
1	Nashs.Field	3.6	11	Grassland	4.1	FALSE	4
10	Rookery.Slope	1.5	4	Grassland	5.0	TRUE	7
4	Rush.Meadow	2.4	5	Meadow	4.9	TRUE	5
15	Pond.Field	4.1	0	Meadow	5.0	TRUE	6
16	Water.Meadow	3.9	0	Meadow	4.9	TRUE	8

9	The.Orchard	1.9	0	Orchard	5.7	FALSE 9	
20	Farm.Wood	0.8	10	Scrub	5.1	TRUE	3
17	Cheapside	2.2	8	Scrub	4.7	TRUE	4
5	Gunness.Thicket	3.8	0	Scrub	4.2	FALSE	6
11	Garden.Wood	2.9	10	Scrub	5.2	FALSE	8

Notice that as with single-condition sorts, when there are ties (as in grassland with worm density $= 2$), the rows are in their original sequence (here, 3, 6, 13). We might want to override this by specifying a third sorting condition (e.g. soil pH):

worms[order(Vegetation,Worm.density,Soil.pH),]

	Field.Name	Area	Slope	Vegetation	Soil.pH	Damp	Worm.density
8	Ashurst	2.1	0	Arable	4.8	FALSE	4
18	Pound.Hill	4.4	2	Arable	4.5	FALSE	5
2	Silwood.Bottom	5.1	2	Arable	5.2	FALSE	7
14	Observatory.Ridge	1.8	6	Grassland	3.8	FALSE	0
19	Gravel.Pit	2.9	1	Grassland	3.5	FALSE	1
12	North.Gravel	3.3	1	Grassland	4.1	FALSE	1
6	Oak.Mead	3.1	2	Grassland	3.9	FALSE	2
13	South.Gravel	3.7	2	Grassland	4.0	FALSE	2
3	Nursery.Field	2.8	3	Grassland	4.3	FALSE	2
7	Church.Field	3.5	3	Grassland	4.2	FALSE	3
1	Nashs.Field	3.6	11	Grassland	4.1	FALSE	4
10	Rookery.Slope	1.5	4	Grassland	5.0	TRUE	7
4	Rush.Meadow	2.4	5	Meadow	4.9	TRUE	5
15	Pond.Field	4.1	0	Meadow	5.0	TRUE	6
16	Water.Meadow	3.9	0	Meadow	4.9	TRUE	8
9	The.Orchard	1.9	0	Orchard	5.7	FALSE	9
20	Farm.Wood	0.8	10	Scrub	5.1	TRUE	3
17	Cheapside	2.2	8	Scrub	4.7	TRUE	4
5	Gunness.Thicket	3.8	0	Scrub	4.2	FALSE	6
11	Garden.Wood	2.9	10	Scrub	5.2	FALSE	8

The rule is this: if in doubt, sort using more variables than you think you need. That way you can be absolutely certain that the rows are in the order you expect them to be in. This is exceptionally important when you begin to make assumptions about the variables associated with a particular value of the response variable on the basis of its row number.

Perhaps you want only certain columns in the sorted dataframe? Suppose we want Vegetation, Worm.density, Soil.pH and Slope, and we want them in that order from left to right, then we specify the column numbers in the sequence we want them to appear as a vector, thus: c(4,7,5,3):

worms[order(Vegetation,Worm.density),c(4,7,5,3)]

	Vegetation	Worm.density	Soil.pH	Slope
8	Arable	4	4.8	0
18	Arable	5	4.5	2
2	Arable	7	5.2	2
14	Grassland	0	3.8	6
12	Grassland	1	4.1	1
19	Grassland	1	3.5	1
3	Grassland	2	4.3	3
6	Grassland	2	3.9	2

13	Grassland	2	4.0	2
7	Grassland	3	4.2	3
1	Grassland	4	4.1	11
10	Grassland	7	5.0	4
4	Meadow	5	4.9	5
15	Meadow	6	5.0	0
16	Meadow	8	4.9	0
9	Orchard	9	5.7	0
20	Scrub	3	5.1	10
17	Scrub	4	4.7	8
5	Scrub	6	4.2	0
11	Scrub	8	5.2	10

You can select the columns on the basis of their variables names, but this is more fiddly to type, because you need to put the variable names in quotes like this:

```
worms[order(Vegetation,Worm.density),
c("Vegetation", "Worm.density", "Soil.pH", "Slope")]
```

Using Logical Conditions to Select Rows from the Dataframe

A very common operation is selecting certain rows from the dataframe on the basis of values in one or more of the variables (the columns of the dataframe). Suppose we want to restrict the data to cases from damp fields. We want all the columns, so the syntax for the subscripts is ['which rows', blank]:

```
worms[Damp == T,]
```

	Field.Name	Area	Slope	Vegetation	Soil.pH	Damp	Worm.density
4	Rush.Meadow	2.4	5	Meadow	4.9	TRUE	5
10	Rookery.Slope	1.5	4	Grassland	5.0	TRUE	7
15	Pond.Field	4.1	0	Meadow	5.0	TRUE	6
16	Water.Meadow	3.9	0	Meadow	4.9	TRUE	8
17	Cheapside	2.2	8	Scrub	4.7	TRUE	4
20	Farm.Wood	0.8	10	Scrub	5.1	TRUE	3

Note that because Damp is a logical variable (with just two potential values, TRUE or FALSE) we can refer to true or false in abbreviated form, T or F. Also notice that the T in this case is not enclosed in quotes: the T means true, not the character string 'T'. The other important point is that the symbol for the logical condition is == (two successive equals signs with no gap between them; see p. 27).

The logic for the selection of rows can refer to values (and functions of values) in more than one column. Suppose that we wanted the data from the fields where worm density was higher than the median (>median(Worm.density)) and soil pH was less than 5.2. In R, the logical operator for AND is the & symbol:

```
worms[Worm.density > median(Worm.density) & Soil.pH < 5.2,]
```

	Field.Name	Area	Slope	Vegetation	Soil.pH	Damp	Worm.density
4	Rush.Meadow	2.4	5	Meadow	4.9	TRUE	5
5	Gunness.Thicket	3.8	0	Scrub	4.2	FALSE	6

10	Rookery.Slope	1.5	4	Grassland	5.0	TRUE	7
15	Pond.Field	4.1	0	Meadow	5.0	TRUE	6
16	Water.Meadow	3.9	0	Meadow	4.9	TRUE	8
18	Pound.Hill	4.4	2	Arable	4.5	FALSE	5

Suppose that we want to extract all the columns that contain numbers (rather than characters or logical variables) from the dataframe. The function is.numeric can be applied across all the columns of worms using sapply to create subscripts like this:

worms[,sapply(worms,is.numeric)]

	Area	Slope	Soil.pH	Worm.density
1	3.6	11	4.1	4
2	5.1	2	5.2	7
3	2.8	3	4.3	2
4	2.4	5	4.9	5
5	3.8	0	4.2	6
6	3.1	2	3.9	2
7	3.5	3	4.2	3
8	2.1	0	4.8	4
9	1.9	0	5.7	9
10	1.5	4	5.0	7
11	2.9	10	5.2	8
12	3.3	1	4.1	1
13	3.7	2	4.0	2
14	1.8	6	3.8	0
15	4.1	0	5.0	6
16	3.9	0	4.9	8
17	2.2	8	4.7	4
18	4.4	2	4.5	5
19	2.9	1	3.5	1
20	0.8	10	5.1	3

We might want to extract the columns that were factors:

worms[,sapply(worms,is.factor)]

	Field.Name	Vegetation
1	Nashs.Field	Grassland
2	Silwood.Bottom	Arable
3	Nursery.Field	Grassland
4	Rush.Meadow	Meadow
5	Gunness.Thicket	Scrub
6	Oak.Mead	Grassland
7	Church.Field	Grassland
8	Ashurst	Arable
9	The.Orchard	Orchard
10	Rookery.Slope	Grassland
11	Garden.Wood	Scrub
12	North.Gravel	Grassland
13	South.Gravel	Grassland
14	Observatory.Ridge	Grassland

```
15      Pond.Field      Meadow
16    Water.Meadow      Meadow
17      Cheapside       Scrub
18     Pound.Hill       Arable
19     Gravel.Pit    Grassland
20      Farm.Wood       Scrub
```

Because worms is a dataframe, the characters have all been coerced to factors, so worms[,sapply(worms,is.character)] produces the answer NULL.

To drop a row or rows from the dataframe, use **negative subscripts**. Thus to drop the middle 10 rows (i.e. row numbers 6 to 15 inclusive) do this:

worms[-(6:15),]

```
                Field.Name  Area  Slope  Vegetation  Soil.pH  Damp   Worm.density
1               Nashs.Field  3.6    11   Grassland     4.1   FALSE         4
2            Silwood.Bottom  5.1     2      Arable     5.2   FALSE         7
3             Nursery.Field  2.8     3   Grassland     4.3   FALSE         2
4              Rush.Meadow   2.4     5      Meadow     4.9    TRUE         5
5           Gunness.Thicket  3.8     0       Scrub     4.2   FALSE         6
16             Water.Meadow  3.9     0      Meadow     4.9    TRUE         8
17               Cheapside   2.2     8       Scrub     4.7    TRUE         4
18              Pound.Hill   4.4     2      Arable     4.5   FALSE         5
19              Gravel.Pit   2.9     1   Grassland     3.5   FALSE         1
20               Farm.Wood   0.8    10       Scrub     5.1    TRUE         3
```

Here are all the rows that are not grasslands (the logical symbol ! means NOT):

worms[!(Vegetation=="Grassland"),]

```
                Field.Name  Area  Slope  Vegetation  Soil.pH  Damp   Worm.density
2            Silwood.Bottom  5.1     2      Arable     5.2   FALSE         7
4              Rush.Meadow   2.4     5      Meadow     4.9    TRUE         5
5           Gunness.Thicket  3.8     0       Scrub     4.2   FALSE         6
8                  Ashurst   2.1     0      Arable     4.8   FALSE         4
9               The.Orchard  1.9     0     Orchard     5.7   FALSE         9
11              Garden.Wood  2.9    10       Scrub     5.2   FALSE         8
15               Pond.Field  4.1     0      Meadow     5.0    TRUE         6
16             Water.Meadow  3.9     0      Meadow     4.9    TRUE         8
17               Cheapside   2.2     8       Scrub     4.7    TRUE         4
18              Pound.Hill   4.4     2      Arable     4.5   FALSE         5
20               Farm.Wood   0.8    10       Scrub     5.1    TRUE         3
```

If you want to use minus signs rather than logical NOT to drop rows from the dataframe, the expression you use must evaluate to numbers. The which function is useful for this. Let's use this technique to drop the non-damp fields.

worms[-which(Damp==F),]

```
                Field.Name  Area  Slope  Vegetation  Soil.pH  Damp  Worm.density
4              Rush.Meadow   2.4     5      Meadow     4.9   TRUE        5
10           Rookery.Slope   1.5     4   Grassland     5.0   TRUE        7
15              Pond.Field   4.1     0      Meadow     5.0   TRUE        6
16             Water.Meadow  3.9     0      Meadow     4.9   TRUE        8
17               Cheapside   2.2     8       Scrub     4.7   TRUE        4
20               Farm.Wood   0.8    10       Scrub     5.1   TRUE        3
```

which achieves the same end as the more elegant

```
worms[!Damp==F,]
```

or even simpler,

```
worms[Damp==T,]
```

Omitting Rows Containing Missing Values, NA

In statistical modelling it is often useful to have a dataframe that contains no missing values in the response or explanatory variables. You can create a shorter dataframe using the na.omit function. Here is a sister dataframe of worms in which certain values are NA:

```
data<-read.table("c:\\temp\\worms.missing.txt",header=T)
data
```

	Field.Name	Area	Slope	Vegetation	Soil.pH	Damp	Worm.density
1	Nashs.Field	3.6	11	Grassland	4.1	FALSE	4
2	Silwood.Bottom	5.1	NA	Arable	5.2	FALSE	7
3	Nursery.Field	2.8	3	Grassland	4.3	FALSE	2
4	Rush.Meadow	2.4	5	Meadow	4.9	TRUE	5
5	Gunness.Thicket	3.8	0	Scrub	4.2	FALSE	6
6	Oak.Mead	3.1	2	Grassland	3.9	FALSE	2
7	Church.Field	3.5	3	Grassland	NA	NA	NA
8	Ashurst	2.1	0	Arable	4.8	FALSE	4
9	The.Orchard	1.9	0	Orchard	5.7	FALSE	9
10	Rookery.Slope	1.5	4	Grassland	5.0	TRUE	7
11	Garden.Wood	2.9	10	Scrub	5.2	FALSE	8
12	North.Gravel	3.3	1	Grassland	4.1	FALSE	1
13	South.Gravel	3.7	2	Grassland	4.0	FALSE	2
14	Observatory.Ridge	1.8	6	Grassland	3.8	FALSE	0
15	Pond.Field	4.1	0	Meadow	5.0	TRUE	6
16	Water.Meadow	3.9	0	Meadow	4.9	TRUE	8
17	Cheapside	2.2	8	Scrub	4.7	TRUE	4
18	Pound.Hill	4.4	2	Arable	4.5	FALSE	5
19	Gravel.Pit	NA	1	Grassland	3.5	FALSE	1
20	Farm.Wood	0.8	10	Scrub	5.1	TRUE	3

By inspection we can see that we should like to leave out row 2 (one missing value), row 7 (three missing values) and row 19 (one missing value). This could not be simpler:

```
na.omit(data)
```

	Field.Name	Area	Slope	Vegetation	Soil.pH	Damp	Worm.density
1	Nashs.Field	3.6	11	Grassland	4.1	FALSE	4
3	Nursery.Field	2.8	3	Grassland	4.3	FALSE	2
4	Rush.Meadow	2.4	5	Meadow	4.9	TRUE	5

```
5       Gunness.Thicket  3.8   0     Scrub  4.2  FALSE  6
6             Oak.Mead  3.1   2  Grassland  3.9  FALSE  2
8             Ashurst  2.1   0    Arable  4.8  FALSE  4
9         The.Orchard  1.9   0    Orchard  5.7  FALSE  9
10      Rookery.Slope  1.5   4  Grassland  5.0   TRUE  7
11       Garden.Wood  2.9  10     Scrub  5.2  FALSE  8
12       North.Gravel  3.3   1  Grassland  4.1  FALSE  1
13       South.Gravel  3.7   2  Grassland  4.0  FALSE  2
14  Observatory.Ridge  1.8   6  Grassland  3.8  FALSE  0
15         Pond.Field  4.1   0     Meadow  5.0   TRUE  6
16       Water.Meadow  3.9   0     Meadow  4.9   TRUE  8
17          Cheapside  2.2   8      Scrub  4.7   TRUE  4
18          Pound.Hill  4.4   2    Arable  4.5  FALSE  5
20           Farm.Wood  0.8  10     Scrub  5.1   TRUE  3
```

and you see that rows 2, 7 and 19 have been omitted in creating the new dataframe. Alternatively, you can use the na.exclude function. This differs from na.omit only in the class of the na.action attribute of the result, which gives different behaviour in functions making use of naresid and napredict: when na.exclude is used the residuals and predictions are padded to the correct length by inserting NAs for cases omitted by na.exclude (in this example they would be of length 20, whereas na.omit would give residuals and predictions of length 17).

new.frame<-na.exclude(data)

The function to test for the presence of missing values across a dataframe is complete.cases:

complete.cases(data)

```
 [1]  TRUE FALSE  TRUE  TRUE  TRUE  TRUE FALSE  TRUE  TRUE  TRUE  TRUE  TRUE
[13]  TRUE  TRUE  TRUE  TRUE  TRUE  TRUE FALSE  TRUE
```

You could use this as a less efficient analogue of na.omit(data), but why would you?

data[complete.cases(data),]

It is well worth checking the individual variables separately, because it is possible that one (or a few) variable(s) contributes most of the missing values, and it may be preferable to remove these variables from the modelling rather than lose the valuable information about the other explanatory variables associated with these cases. Use summary to count the missing values for each variable in the dataframe, or use apply with the function is.na to sum up the missing values in each variable:

apply(apply(data,2,is.na),2,sum)

```
Field.Name  Area  Slope  Vegetation  Soil.pH  Damp  Worm.density
         0     1      1           0        1     1             1
```

You can see that in this case no single variable contributes more missing values than any other.

Using order and unique to Eliminate Pseudoreplication

In this rather more complicated example, you are asked to extract a single record for each vegetation type, and that record is to be the case within that vegetation type that has the greatest worm density. There are two steps to this: first order all of the rows of the dataframe using rev(order(Worm.density)), then select the subset of these rows which is unique for vegetation type:

worms[rev(order(Worm.density)),][unique(Vegetation),]

	Field.Name	Area	Slope	Vegetation	Soil.pH	Damp	Worm.density
16	Water.Meadow	3.9	0	Meadow	4.9	TRUE	8
9	The.Orchard	1.9	0	Orchard	5.7	FALSE	9
11	Garden.Wood	2.9	10	Scrub	5.2	FALSE	8
2	Silwood.Bottom	5.1	2	Arable	5.2	FALSE	7
10	Rookery.Slope	1.5	4	Grassland	5.0	TRUE	7

Complex Ordering with Mixed Directions

Sometimes there are multiple sorting variables, but the variables have to be sorted in opposing directions. In this example, the task is to order the database first by vegetation type in alphabetical order (the default) and then within each vegetation type to sort by worm density in decreasing order (highest densities first). The trick here is to use order (rather than rev(order())) but to put a minus sign in front of Worm.density like this:

worms[order(Vegetation,-Worm.density),]

	Field.Name	Area	Slope	Vegetation	Soil.pH	Damp	Worm.density
2	Silwood.Bottom	5.1	2	Arable	5.2	FALSE	7
18	Pound.Hill	4.4	2	Arable	4.5	FALSE	5
8	Ashurst	2.1	0	Arable	4.8	FALSE	4
10	Rookery.Slope	1.5	4	Grassland	5.0	TRUE	7
1	Nashs.Field	3.6	11	Grassland	4.1	FALSE	4
7	Church.Field	3.5	3	Grassland	4.2	FALSE	3
3	Nursery.Field	2.8	3	Grassland	4.3	FALSE	2
6	Oak.Mead	3.1	2	Grassland	3.9	FALSE	2
13	South.Gravel	3.7	2	Grassland	4.0	FALSE	2
12	North.Gravel	3.3	1	Grassland	4.1	FALSE	1
19	Gravel.Pit	2.9	1	Grassland	3.5	FALSE	1
14	Observatory.Ridge	1.8	6	Grassland	3.8	FALSE	0
16	Water.Meadow	3.9	0	Meadow	4.9	TRUE	8
15	Pond.Field	4.1	0	Meadow	5.0	TRUE	6
4	Rush.Meadow	2.4	5	Meadow	4.9	TRUE	5
9	The.Orchard	1.9	0	Orchard	5.7	FALSE	9
11	Garden.Wood	2.9	10	Scrub	5.2	FALSE	8
5	Gunness.Thicket	3.8	0	Scrub	4.2	FALSE	6
17	Cheapside	2.2	8	Scrub	4.7	TRUE	4
20	Farm.Wood	0.8	10	Scrub	5.1	TRUE	3

Using the minus sign only works when sorting numerical variables. For factor levels you can use the rank function to make the levels numeric like this:

worms[order(-rank(Vegetation),-Worm.density),]

	Field.Name	Area	Slope	Vegetation	Soil.pH	Damp	Worm.density
11	Garden.Wood	2.9	10	Scrub	5.2	FALSE	8
5	Gunness.Thicket	3.8	0	Scrub	4.2	FALSE	6
17	Cheapside	2.2	8	Scrub	4.7	TRUE	4
20	Farm.Wood	0.8	10	Scrub	5.1	TRUE	3
9	The.Orchard	1.9	0	Orchard	5.7	FALSE	9
16	Water.Meadow	3.9	0	Meadow	4.9	TRUE	8
15	Pond.Field	4.1	0	Meadow	5.0	TRUE	6
4	Rush.Meadow	2.4	5	Meadow	4.9	TRUE	5
10	Rookery.Slope	1.5	4	Grassland	5.0	TRUE	7
1	Nashs.Field	3.6	11	Grassland	4.1	FALSE	4
7	Church.Field	3.5	3	Grassland	4.2	FALSE	3
3	Nursery.Field	2.8	3	Grassland	4.3	FALSE	2
6	Oak.Mead	3.1	2	Grassland	3.9	FALSE	2
13	South.Gravel	3.7	2	Grassland	4.0	FALSE	2
12	North.Gravel	3.3	1	Grassland	4.1	FALSE	1
19	Gravel.Pit	2.9	1	Grassland	3.5	FALSE	1
14	Observatory.Ridge	1.8	6	Grassland	3.8	FALSE	0
2	Silwood.Bottom	5.1	2	Arable	5.2	FALSE	7
18	Pound.Hill	4.4	2	Arable	4.5	FALSE	5
8	Ashurst	2.1	0	Arable	4.8	FALSE	4

It is less likely that you will want to select *columns* on the basis of logical operations, but it is perfectly possible. Suppose that for some reason you want to select the columns that contain the character 'S' (upper-case S). In R the function for this is **grep**, which returns the subscript (a number or set of numbers) indicating which character strings within a vector of character strings contained an upper-case S. The names of the variables within a dataframe are obtained by the **names** function:

names(worms)

```
[1]  "Field.Name"  "Area"   "Slope"          "Vegetation"
[5]  "Soil.pH"     "Damp"   "Worm.density"
```

so we want our function **grep** to pick out variables numbers 3 and 5 because they are the only ones containing upper-case S:

grep("S",names(worms))

```
[1]  3  5
```

Finally, we can use these numbers as subscripts [,c(3,5)] to select columns 3 and 5:

worms[,grep("S",names(worms))]

	Slope	Soil.pH
1	11	4.1
2	2	5.2
3	3	4.3
4	5	4.9
5	0	4.2
6	2	3.9
7	3	4.2
8	0	4.8
9	0	5.7

```
10    4   5.0
11   10   5.2
12    1   4.1
13    2   4.0
14    6   3.8
15    0   5.0
16    0   4.9
17    8   4.7
18    2   4.5
19    1   3.5
20   10   5.1
```

A Dataframe with Row Names instead of Row Numbers

You can suppress the creation of row numbers and allocate your own unique names to each row by altering the syntax of the read.table function. The first column of the worms database contains the names of the fields in which the other variables were measured. Up to now, we have read this column as if it was the first variable (p. 107).

```
detach(worms)
worms<-read.table("c:\\temp\\worms.txt",header=T,row.names=1)
worms
```

	Area	Slope	Vegetation	Soil.pH	Damp	Worm.density
Nashs.Field	3.6	11	Grassland	4.1	FALSE	4
Silwood.Bottom	5.1	2	Arable	5.2	FALSE	7
Nursery.Field	2.8	3	Grassland	4.3	FALSE	2
Rush.Meadow	2.4	5	Meadow	4.9	TRUE	5
Gunness.Thicket	3.8	0	Scrub	4.2	FALSE	6
Oak.Mead	3.1	2	Grassland	3.9	FALSE	2
Church.Field	3.5	3	Grassland	4.2	FALSE	3
Ashurst	2.1	0	Arable	4.8	FALSE	4
The.Orchard	1.9	0	Orchard	5.7	FALSE	9
Rookery.Slope	1.5	4	Grassland	5.0	TRUE	7
Garden.Wood	2.9	10	Scrub	5.2	FALSE	8
North.Gravel	3.3	1	Grassland	4.1	FALSE	1
South.Gravel	3.7	2	Grassland	4.0	FALSE	2
Observatory.Ridge	1.8	6	Grassland	3.8	FALSE	0
Pond.Field	4.1	0	Meadow	5.0	TRUE	6
Water.Meadow	3.9	0	Meadow	4.9	TRUE	8
Cheapside	2.2	8	Scrub	4.7	TRUE	4
Pound.Hill	4.4	2	Arable	4.5	FALSE	5
Gravel.Pit	2.9	1	Grassland	3.5	FALSE	1
Farm.Wood	0.8	10	Scrub	5.1	TRUE	3

Notice that the field names column is not now headed by a variable name, and that the row numbers, as intended, have been suppressed.

Creating a Dataframe from Another Kind of Object

We have seen that the simplest way to create a dataframe in R is to read a table of data from an external file using the read.table function. Alternatively, you can create a dataframe by using the **data.frame** function to bind together a number of objects. Here are three vectors of the same length:

```
x<-runif(10)
y<-letters[1:10]
z<-sample(c(rep(T,5),rep(F,5)))
```

To make them into a dataframe called new, just type:

```
new<-data.frame(y,z,x)
new
```

```
      y     z          x
1     a   TRUE   0.72675982
2     b   FALSE  0.83847227
3     c   FALSE  0.61765685
4     d   TRUE   0.78541650
5     e   FALSE  0.51168828
6     f   TRUE   0.53526324
7     g   TRUE   0.05552335
8     h   TRUE   0.78486234
9     i   FALSE  0.68385443
10    j   FALSE  0.89367837
```

Note that the order of the columns is controlled simply the sequence of the vector names (left to right) specified within the **data.frame** function.

In this example, we create a table of counts of random integers from a Poisson distribution, then convert the table into a dataframe. First, we make a **table** object:

```
y<-rpois(1500,1.5)
table(y)
```

```
y
  0    1    2    3   4    5  6  7
344  502  374  199  63  11  5  2
```

Now it is simple to convert this **table** object into a dataframe with two variables, the count and the frequency using the **as.data.frame** function:

```
as.data.frame(table(y))
```

```
   y  Freq
1  0   344
2  1   502
3  2   374
4  3   199
5  4    63
6  5    11
7  6     5
8  7     2
```

In some cases you might want to expand a dataframe like the one above such that it had a separate row for every distinct count (i.e. 344 rows with $y = 0$, 502 rows with $y = 1$, 374 rows with $y = 2$, and so on). You use **lapply** with **rep** for this:

```
short.frame<-as.data.frame(table(y))
long<-as.data.frame(lapply(short.frame, function(x) rep(x, short.frame$Freq)))
long[,1]
```

```
[1]  0 0 0 0 0 0 0 0 0 0 0 0 0 0 0 0 0 0 0 0 0 0 0 0 0 0 0
     0 0 0 0 0 0 0 0 0 0
[38] 0 0 0 0 0 0 0 0 0 0 0 0 0 0 0 0 0 0 0 0 0 0 0 0 0 0 0
     0 0 0 0 0 0 0 0 0 0
[75] 0 0 0 0 0 0 0 0 0 0 0 0 0 0 0 0 0 0 0 0 0 0 0 0 0 0 0
     0 0 0 0 0 0 0 0 0 0
...
...
[1444] 4 4 4 4 4 4 4 4 4 4 4 4 4 4 4 4 4 4 4 4 4 4 4 4 4 4
       4 4 4 4 4 4 4 4 4 4
[1481] 4 4 5 5 5 5 5 5 5 5 5 5 5 6 6 6 6 6 7 7
```

Note the use of the anonymous function to generate the repeats of each row by the value specified in Freq. We drop the second column of long because it is redundant (the Freq value).

Eliminating Duplicate Rows from a Dataframe

Sometimes a dataframe will contain duplicate rows where all the variables have exactly the same values in two or more rows. Here is a simple example:

```
dups<-read.table("c:\\temp\\dups.txt",header=T)
dups
```

```
  var1  var2  var3  var4
1   1     2     3     1
2   1     2     2     1
3   3     2     1     1
4   4     4     2     1
5   3     2     1     1
6   6     1     2     5
7   1     2     3     2
```

Note that row number 5 is an exact duplicate of row number 3. To create a dataframe with all the duplicate rows stripped out, use the unique function like this:

```
unique(dups)
```

```
  var1  var2  var3  var4
1   1     2     3     1
2   1     2     2     1
3   3     2     1     1
4   4     4     2     1
6   6     1     2     5
7   1     2     3     2
```

Notice that the row names in the new dataframe are the same as in the original, so that you can spot that row number 5 was removed by the operation of the function unique.

To view the rows that are duplicates in a dataframe (if any) use the duplicated function:

```
dups[duplicated(dups),]
```

```
  var1  var2  var3  var4
5   3     2     1     1
```

Dates in Dataframes

There is an introduction to the complexities of using dates and times in dataframes on pp. 89–95. Here we work with a simple example:

```
nums<-read.table("c:\\temp\\sortdata.txt",header=T)
attach(nums)
names(nums)
```

```
[1]  "name"  "date"  "response"  "treatment"
```

The idea is to order the rows by date. The ordering is to be applied to all four columns of the dataframe. Note that ordering on the basis of our variable called date does not work in the way we want it to:

```
nums[order(date),]
```

```
         name        date      response  treatment
53      rachel   01/08/2003  32.98792196        B
65      albert   02/06/2003  38.41979568        A
6          ann   02/07/2003   2.86983693        B
10      cecily   02/11/2003   6.81467571        A
4          ian   02/12/2003   2.09505949        A
29     michael   03/05/2003  15.59890900        B
...
```

This is because of the format used for depicting date in the dataframe called nums: date is a character string in which the first characters are the day, then the month, then the year. When we sort by date, we typically want 2001 to come before 2006, May 2006 before September 2006 and 12 May 2006 before 14 May 2006. In order to sort by date we need first to convert our variable into date-time format using the **strptime** function (see p. 92 for details):

```
dates<-strptime(date,format="%d/%m/%Y")
dates
```

```
[1]  "2003-08-25"  "2003-05-21"  "2003-10-12"  "2003-12-02"  "2003-10-18"
[6]  "2003-07-02"  "2003-09-27"  "2003-06-05"  "2003-06-11"  "2003-11-02"
```

Note how **strptime** has produced a date object with year first, then a hyphen, then month, then a hyphen, then day which will sort into the desired sequence. We bind the new variable to the dataframe called **nums** like this:

```
nums<-cbind(nums,dates)
```

Now that the new variable is in the correct format the dates can be sorted as characters:

```
nums[order(as.character(dates)),1:4]
```

```
         name        date      response  treatment
49      albert   21/04/2003  30.66632632        A
63       james   24/04/2003  37.04140266        A
24        john   27/04/2003  12.70257306        A
33     william   30/04/2003  18.05707279        B
29     michael   03/05/2003  15.59890900        B
71         ian   06/05/2003  39.97237868        A
50      rachel   09/05/2003  30.81807436        B
```

Note the use of subscripts to omit the new **dates** variable by selecting only columns 1 to 4 of the dataframe. Another way to extract elements of a dataframe is to use the **subset** function with **select** like this:

```
subset(nums,select=c("name","dates"))
```

```
            name       dates
1         albert  2003-08-25
2            ann  2003-05-21
3           john  2003-10-12
4            ian  2003-12-02
5        michael  2003-10-18
...
...
73      georgina  2003-05-24
74      georgina  2003-08-16
75       heather  2003-11-14
76     elizabeth  2003-06-23
```

Selecting Variables on the Basis of their Attributes

In this example, we want to extract all of the columns from **nums** (above) that are numeric. Use **sapply** to obtain a vector of logical values:

```
sapply(nums,is.numeric)
```

```
 name    date  response  treatment  dates
FALSE   FALSE      TRUE      FALSE   TRUE
```

Now use this object to form the column subscripts to extract the two numeric variables:

```
nums[,sapply(nums,is.numeric)]
```

```
     response        dates
1  0.05963704  2003-08-25
2  1.46555993  2003-05-21
3  1.59406539  2003-10-12
4  2.09505949  2003-12-02
```

Note that **dates** is numeric but **date** was not (it is a factor, having been converted from a character string by the **read.table** function).

Using the **match** Function in Dataframes

The worms dataframe (above) contains fields of five different vegetation types:

```
unique(worms$Vegetation)
```

```
[1] Grassland  Arable  Meadow  Scrub  Orchard
```

and we want to know the appropriate herbicides to use in each of the 20 fields. The herbicides are in a separate dataframe that contains the recommended herbicides for a much larger set of plant community types:

```
herbicides<-read.table("c:\\temp\\herbicides.txt",header=T)
herbicides
```

```
        Type  Herbicide
1    Woodland  Fusilade
2     Conifer  Weedwipe
3      Arable  Twinspan
4        Hill  Weedwipe
5     Bracken  Fusilade
6       Scrub  Weedwipe
7   Grassland  Allclear
8       Chalk  Vanquish
9      Meadow  Propinol
10       Lawn  Vanquish
11    Orchard  Fusilade
12      Verge  Allclear
```

The task is to create a vector of length 20 (one for every field in worms) containing the name of the appropriate herbicide. The first value needs to be Allclear because Nash's Field is grassland, and the second needs to be Twinspan because Silwood Bottom is arable, and so on. The first vector in match is worms$Vegetation and the second vector in match is herbicides$Type. The result of this match is used as a vector of subscripts to extract the relevant herbicides from herbicides$Herbicide like this:

herbicides$Herbicide[match(worms$Vegetation,herbicides$Type)]

```
 [1]  Allclear  Twinspan  Allclear  Propinol  Weedwipe  Allclear
      Allclear  Twinspan
 [9]  Fusilade  Allclear  Weedwipe  Allclear  Allclear  Allclear
      Propinol  Propinol
[17]  Weedwipe  Twinspan  Allclear  Weedwipe
```

You could add this information as a new column in the worms dataframe:

worms$hb<-herbicides$Herbicide[match(worms$Vegetation,herbicides$Type)]

or create a new dataframe called recs containing the herbicide recommendations:

recs<-data.frame(
    worms,hb=herbicides$Herbicide[match(worms$Vegetation,herbicides$Type)])
recs

```
                 Field.Name  Area  Slope  Vegetation  Soil.pH   Damp  Worm.density        hb
1               Nashs.Field   3.6     11   Grassland      4.1  FALSE             4  Allclear
2            Silwood.Bottom   5.1      2      Arable      5.2  FALSE             7  Twinspan
3             Nursery.Field   2.8      3   Grassland      4.3  FALSE             2  Allclear
4               Rush.Meadow   2.4      5      Meadow      4.9   TRUE             5  Propinol
5           Gunness.Thicket   3.8      0       Scrub      4.2  FALSE             6  Weedwipe
6                  Oak.Mead   3.1      2   Grassland      3.9  FALSE             2  Allclear
7              Church.Field   3.5      3   Grassland      4.2  FALSE             3  Allclear
8                   Ashurst   2.1      0      Arable      4.8  FALSE             4  Twinspan
9               The.Orchard   1.9      0     Orchard      5.7  FALSE             9  Fusilade
10             Rookery.Slope  1.5      4   Grassland      5.0   TRUE             7  Allclear
11              Garden.Wood   2.9     10       Scrub      5.2  FALSE             8  Weedwipe
12              North.Gravel   3.3      1   Grassland      4.1  FALSE             1  Allclear
13              South.Gravel   3.7      2   Grassland      4.0  FALSE             2  Allclear
14         Observatory.Ridge  1.8      6   Grassland      3.8  FALSE             0  Allclear
15               Pond.Field   4.1      0      Meadow      5.0   TRUE             6  Propinol
16              Water.Meadow   3.9      0      Meadow      4.9   TRUE             8  Propinol
17                Cheapside   2.2      8       Scrub      4.7   TRUE             4  Weedwipe
```

```
18   Pound.Hill   4.4    2     Arable   4.5   FALSE   5   Twinspan
19   Gravel.Pit   2.9    1   Grassland  3.5   FALSE   1   Allclear
20   Farm.Wood    0.8   10      Scrub   5.1    TRUE   3   Weedwipe
```

Merging Two Dataframes

Suppose we have two dataframes, the first containing information on plant life forms and the second containing information of time of flowering. We want to produce a single dataframe showing information on both life form and flowering time. Both dataframes contain variables for genus name and species name:

(lifeforms<-read.table("c:\\temp\\lifeforms.txt",header=T))

```
     Genus      species  lifeform
1     Acer   platanoides     tree
2     Acer     palmatum      tree
3    Ajuga     reptans       herb
4   Conyza   sumatrensis    annual
5   Lamium      album        herb
```

(flowering<-read.table("c:\\temp\\fltimes.txt",header=T))

```
       Genus        species   flowering
1       Acer     platanoides       May
2      Ajuga        reptans        June
3    Brassica        napus        April
4   Chamerion  angustifolium       July
5     Conyza      bilbaoana      August
6     Lamium        album       January
```

Because at least one of the variable names is identical in the two dataframes (in this case, two variables are identical, namely Genus and species) we can use the simplest of all merge commands:

merge(flowering,lifeforms)

```
     Genus      species  flowering  lifeform
1     Acer   platanoides     May       tree
2    Ajuga     reptans      June       herb
3   Lamium      album      January     herb
```

The important point to note is that the merged dataframe contains only those rows which had *complete* entries in both dataframes. Two rows from the lifeforms database were excluded because there were no flowering time data for them (*Acer platanoides* and *Conyza sumatrensis*), and three rows from the flowering-time database were excluded because there were no lifeform data for them (*Chamerion angustifolium, Conyza bilbaoana* and *Brassica napus*).

If you want to include all the species, with missing values (NA) inserted when flowering times or lifeforms are not known, then use the all=T option:

(both<-merge(flowering,lifeforms,all=T))

```
     Genus      species  flowering  lifeform
1     Acer   platanoides     May       tree
2     Acer     palmatum     <NA>       tree
3    Ajuga     reptans      June       herb
```

4	Brassica	napus	April	<NA>
5	Chamerion	angustifolium	July	<NA>
6	Conyza	bilbaoana	August	<NA>
7	Conyza	sumatrensis	<NA>	annual
8	Lamium	album	January	herb

One complexity that often arises is that the same variable has *different names* in the two dataframes that need to be merged. The simplest solution is often to edit the variable names in Excel before reading them into R, but failing this, you need to specify the names in the first dataframe (known conventionally as the *x* dataframe) and the second dataframe (known conventionally as the *y* dataframe) using the by.x and by.y options in merge. We have a third dataframe containing information on the seed weights of all eight species, but the variable Genus is called 'name1' and the variable species is called 'name2'.

```
(seeds<-read.table("c:\\temp\\seedwts.txt",header=T))
```

	name1	name2	seed
1	Acer	platanoides	32.0
2	Lamium	album	12.0
3	Ajuga	reptans	4.0
4	Chamerion	angustifolium	1.5
5	Conyza	bilbaoana	0.5
6	Brassica	napus	7.0
7	Acer	palmatum	21.0
8	Conyza	sumatrensis	0.6

Just using merge(both,seeds) fails miserably: you should try it, to see what happens. We need to inform the merge function that Genus and name1 are synonyms (different names for the same variable), as are species and name2.

```
merge(both,seeds,by.x=c("Genus","species"),by.y=c("name1","name2"))
```

	Genus	species	flowering	lifeform	seed
1	Acer	palmatum	<NA>	tree	21.0
2	Acer	platanoides	May	tree	32.0
3	Ajuga	reptans	June	herb	4.0
4	Brassica	napus	April	<NA>	7.0
5	Chamerion	angustifolium	July	<NA>	1.5
6	Conyza	bilbaoana	August	<NA>	0.5
7	Conyza	sumatrensis	<NA>	annual	0.6
8	Lamium	album	January	herb	12.0

Note that the variable names used in the merged dataframe are the names used in the *x* dataframe.

Adding Margins to a Dataframe

Suppose we have a dataframe showing sales by season and by person:

```
frame<-read.table("c:\\temp\\sales.txt",header=T)
frame
```

	name	spring	summer	autumn	winter
1	Jane.Smith	14	18	11	12
2	Robert.Jones	17	18	10	13

```
3        Dick.Rogers   12  16   9  14
4  William.Edwards   15  14  11  10
5      Janet.Jones   11  17  11  16
```

and we want to add margins to this dataframe showing departures of the seasonal means from the overall mean (as an extra row at the bottom) and departures of the peoples' means (as an extra column on the right). Finally, we want the sales in the body of the dataframe to be represented by departures from the overall mean.

```
people<-rowMeans(frame[,2:5])
people<-people-mean(people)
people
```

```
   1     2      3      4      5
0.30  1.05  -0.70  -0.95   0.30
```

It is very straightforward to add a new column to the dataframe using cbind:

```
(new.frame<-cbind(frame,people))
```

```
                name  spring  summer  autumn  winter  people
1        Jane.Smith      14      18      11      12    0.30
2      Robert.Jones      17      18      10      13    1.05
3       Dick.Rogers      12      16       9      14   -0.70
4   William.Edwards      15      14      11      10   -0.95
5       Janet.Jones      11      17      11      16    0.30
```

Robert Jones is the most effective sales person $(+1.05)$ and William Edwards is the least effective (-0.95). The column means are calculated in a similar way:

```
seasons<-colMeans(frame[,2:5])
seasons<-seasons-mean(seasons)
seasons
```

```
spring  summer  autumn  winter
  0.35    3.15   -3.05   -0.45
```

Sales are highest in summer $(+3.15)$ and lowest in autumn (-3.05).

Now there is a hitch, however, because there are only four column means but there are six columns in new.frame, so we can not use rbind directly. The simplest way to deal with this is to make a copy of one of the rows of the new dataframe

```
new.row<-new.frame[1,]
```

and then edit this to include the values we want: a label in the first column to say 'seasonal means' then the four column means, and then a zero for the grand mean of the effects:

```
new.row[1]<-"seasonal effects"
new.row[2:5]<-seasons
new.row[6]<-0
```

Now we can use rbind to add our new row to the bottom of the extended dataframe:

```
(new.frame<-rbind(new.frame,new.row))
```

```
              name  spring  summer  autumn  winter  people
1      Jane.Smith   14.00   18.00   11.00   12.00    0.30
2    Robert.Jones   17.00   18.00   10.00   13.00    1.05
```

3	Dick.Rogers	12.00	16.00	9.00	14.00	-0.70
4	William.Edwards	15.00	14.00	11.00	10.00	-0.95
5	Janet.Jones	11.00	17.00	11.00	16.00	0.30
11	seasonal effects	0.35	3.15	-3.05	-0.45	0.00

The last task is to replace the counts of sales in the dataframe new.frame[1:5,2:5] by departures from the overall mean sale per person per season (the grand mean, $gm = 13.45$). We need to use unlist to stop R from estimating a separate mean for each column, then create a vector of length 4 containing repeated values of the grand mean (one for each column of sales). Finally, we use sweep to subtract the grand mean from each value.

```
gm<-mean(unlist(new.frame[1:5,2:5]))
gm<-rep(gm,4)
new.frame[1:5,2:5]<-sweep(new.frame[1:5,2:5],2,gm)
new.frame
```

	name	spring	summer	autumn	winter	people
1	Jane.Smith	0.55	4.55	-2.45	-1.45	0.30
2	Robert.Jones	3.55	4.55	-3.45	-0.45	1.05
3	Dick.Rogers	-1.45	2.55	-4.45	0.55	-0.70
4	William.Edwards	1.55	0.55	-2.45	-3.45	-0.95
5	Janet.Jones	-2.45	3.55	-2.45	2.55	0.30
11	seasonal effects	0.35	3.15	-3.05	-0.45	0.00

The best per-season performance was shared by Jane Smith and Robert Jones who each sold 4.55 units more than the overall average in summer.

Summarizing the Contents of Dataframes

There are three useful functions here

- summary   summarize all the contents of all the variables
- aggregate   create a table after the fashion of tapply
- by   perform functions for each level of specified factors

Use of summary and by with the worms database on p. 110.
   The other useful function for summarizing a dataframe is aggregate. It is used like tapply (see p. 18) to apply a function (mean in this case) to the levels of a specified categorical variable (Vegetation in this case) for a specified range of variables (Area, Slope, Soil.pH and Worm.density are defined using their subscripts as a column index in worms[,c(2,3,5,7)]):

```
aggregate(worms[,c(2,3,5,7)],by=list(veg=Vegetation),mean)
```

	veg	Area	Slope	Soil.pH	Worm.density
1	Arable	3.866667	1.333333	4.833333	5.333333
2	Grassland	2.911111	3.666667	4.100000	2.444444
3	Meadow	3.466667	1.666667	4.933333	6.333333
4	Orchard	1.900000	0.000000	5.700000	9.000000
5	Scrub	2.425000	7.000000	4.800000	5.250000

The by argument needs to be a list even if, as here, we have only one classifying factor. Here are the aggregated summaries for Vegetation and Damp:

aggregate(worms[,c(2,3,5,7)],by=list(veg=Vegetation,d=Damp),mean)

```
        veg     d      Area     Slope   Soil.pH  Worm.density
1    Arable  FALSE  3.866667  1.333333  4.833333      5.333333
2 Grassland  FALSE  3.087500  3.625000  3.987500      1.875000
3   Orchard  FALSE  1.900000  0.000000  5.700000      9.000000
4     Scrub  FALSE  3.350000  5.000000  4.700000      7.000000
5 Grassland   TRUE  1.500000  4.000000  5.000000      7.000000
6    Meadow   TRUE  3.466667  1.666667  4.933333      6.333333
7     Scrub   TRUE  1.500000  9.000000  4.900000      3.500000
```

Note that this summary is unbalanced because there were no damp arable or orchard sites and no dry meadows.

Graphics

Producing high-quality graphics is one of the main reasons for doing statistical computing. The particular plot function you need will depend on the number of variables you want to plot and the pattern you wish to highlight. The plotting functions in this chapter are dealt with under four headings:

- plots with two variables;
- plots for a single sample;
- multivariate plots;
- special plots for particular purposes.

Changes to the detailed look of the graphs are dealt with in Chapter 27.

Plots with Two Variables

With two variables (typically the *response variable* on the y axis and the *explanatory variable* on the x axis), the kind of plot you should produce depends upon the nature of your explanatory variable. When the explanatory variable is a continuous variable, such as length or weight or altitude, then the appropriate plot is a **scatterplot**. In cases where the explanatory variable is categorical, such as genotype or colour or gender, then the appropriate plot is either a **box-and-whisker plot** (when you want to show the scatter in the raw data) or a **barplot** (when you want to emphasize the effect sizes).

The most frequently used plotting functions for two variables in R are the following:

- plot(x,y)   scatterplot of y against x
- plot(factor, y)   box-and-whisker plot of y at levels of factor
- barplot(y)   heights from a vector of y values

Plotting with two continuous explanatory variables: scatterplots

The plot function draws axes and adds a scatterplot of points. Two extra functions, points and lines, *add* extra points or lines to an *existing* plot. There are two ways of specifying plot, points and lines and you should choose whichever you prefer:

- Cartesian   plot(x,y)

- formula   plot(y~x)

The advantage of the formula-based plot is that the plot function and the model fit look and feel the same (response variable, tilde, explanatory variable). If you use Cartesian plots (eastings first, then northings, like the grid reference on a map) then the plot has '*x* then *y*' while the model has '*y* then *x*'.

At its most basic, the plot function needs only two arguments: first the name of the explanatory variable (*x* in this case), and second the name of the response variable (*y* in this case): plot(x,y). The data we want to plot are read into R from a file:

```
data1<-read.table("c:\\temp\\scatter1.txt",header=T)
attach(data1)
names(data1)
```

```
[1]  "xv"  "ys"
```

Producing the scatterplot could not be simpler: just type

```
plot(xv,ys,col="red")
```

with the vector of *x* values first, then the vector of *y* values (changing the colour of the points is optional). Notice that the axes are labelled with the variable names, unless you chose to override these with xlab and ylab. It is often a good idea to have longer, more explicit labels for the axes than are provided by the variable names that are used as default options (*xv* and *ys* in this case). Suppose we want to change the label '*xv*' into the longer label 'Explanatory variable' and the label on the *y* axis from '*ys*' to 'Response variable'. Then we use xlab and ylab like this:

```
plot(xv,ys,col="red,xlab="Explanatory variable",ylab="Response variable")
```

The great thing about graphics in R is that it is extremely straightforward to add things to your plots. In the present case, we might want to add a regression line through the cloud of data points. The function for this is abline which can take as its argument the linear model object lm(ys~xv) (as explained on p. 387):

```
abline(lm(ys~xv))
```

Just as it is easy to add lines to the plot, so it is straightforward to add more points. The extra points are in another file:

```
data2<-read.table("c:\\temp\\scatter2.txt",header=T)
attach(data2)
names(data2)
```

```
[1]  "xv2"  "ys2"
```

The new points (xv2,ys2) are added using the points function like this:

```
points(xv2,ys2,col="blue")
```

and we can finish by adding a regression line to the extra points:

```
abline(lm(ys2~xv2))
```

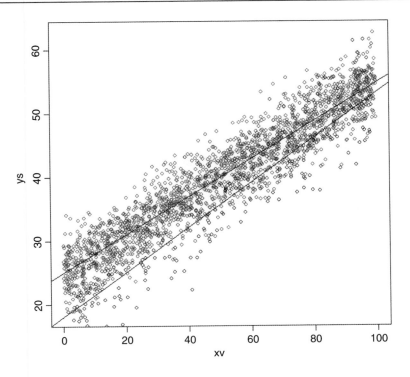

This example shows a very important feature of the plot function. Notice that several of the lower values from the second (blue) data set have *not* appeared on the graph. This is because (unless we say otherwise at the outset) R chooses 'pretty' scaling for the axes based on the data range in the *first* set of points to be drawn. If, as here, the range of subsequent data sets lies outside the scale of the x and y axes, then points are simply left off without any warning message.

One way to cure this problem is to plot all the data with type="n" so that the axes are scaled to encompass all the points form all the data sets (using the concatenation function), then to use points and lines to add the data to the blank axes, like this:

```
plot(c(xv,xv2),c(ys,ys2),xlab="x",ylab="y",type="n")
points(xv,ys,col="red")
points(xv2,ys2,col="blue")
abline(lm(ys~xv))
abline(lm(ys2~xv2))
```

Now all of the points from both data sets appear on the scattergraph. Another way to ensure that all the data are within the plotted axes is to scale the axes yourself, rather than rely on R's choice of pretty scaling, using xlim and ylim. Each of these requires a vector of length 2 to show the minimum and maximum values for each axis. These values are automatically rounded to make them pretty for axis labelling. You will want to control the scaling of the axes when you want two comparable graphs side by side, or when you want to overlay several lines or sets of points on the same axes. Remember that the initial plot function sets the axes scales: this can be a problem if subsequent lines or points are off-scale.

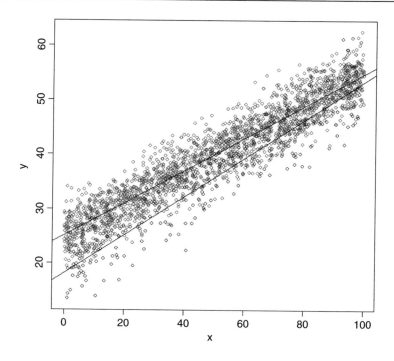

A good way to find out the axis values is to use the range function applied to the data sets in aggregate:

```
range(c(xv,xv2))
```

```
[1]  0.02849861 99.93262000
```

```
range(c(ys,ys2))
```

```
[1] 13.41794 62.59482
```

Here the x axis needs to go from 0.02 up to 99.93 (0 to 100 would be pretty) and the y axis needs to go from 13.4 up to 62.6 (0 to 80 would be pretty). This is how the axes are drawn; the points and lines are added exactly as before:

```
plot(c(xv,xv2),c(ys,ys2),xlim=c(0,100),ylim=c(0,80),xlab="x",ylab="y",type="n")
points(xv,ys,col="red")
points(xv2,ys2,col="blue")
abline(lm(ys~xv))
abline(lm(ys2~xv2))
```

Adding a legend to the plot to explain the difference between the two colours of points would be useful. The thing to understand about the legend function is that the number of lines of text inside the legend box is determined by the length of the vector containing the labels (2 in this case: c("treatment","control") The other two vectors must be of the same length as this: for the plotting symbols pch=c(1,1) and the colours col=c(2,4). The legend function can be used with locator(1) to allow you to select exactly where on the plot surface the legend box should be placed. Click the mouse button when the cursor is where you want the *top left* of the box around the legend to be. It is useful to know the first six colours (col=) used by the plot function:

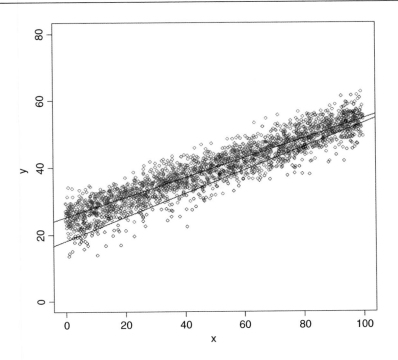

- 1 black (the default)
- 2 red
- 3 green
- 4 blue
- 5 pale blue
- 6 purple

Here, the red circles (col = 2) are the treatment values and the blue circles (col = 4) are the control values. Both are represented by open plotting symbols pch=c(1,1).

```
legend(locator(1),c("treatment","control"),pch=c(1,1),col=c(2,4))
```

This is about as complicated as you would want to make any figure. Adding more information would begin to detract from the message.

Changing the plotting characters used in the points function and in scatterplots involves the function pch: here is the full set of plotting characters:

```
plot(0:10,0:10,type="n",xlab="",ylab="")
k<- -1
for (i in c(2,5,8)) {
for (j in 0:9) {
k<-k+1
points(i,j,pch=k,cex=2)}}
```

Starting at $y = 0$ and proceeding vertically from $x = 2$, you see plotting symbols 0 (open square), 1 (the default open circle), 2 (open triangle), 3 (plus) etc., up to 25 (by which point

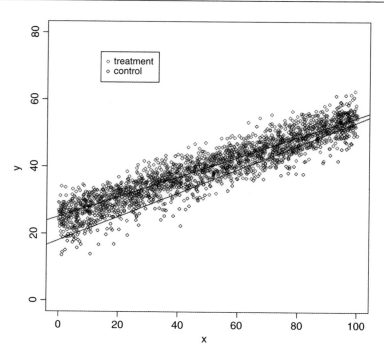

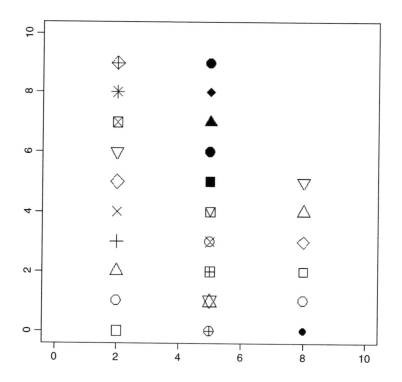

the characters are repeating). My favourite symbol for scatterplots with few data points is pch=16, the solid circle (in the central column), but with many points you might prefer to use pch="." or pch=20. Higher values for pch include the upper-case characters from the top row of the keyboard such as !, $, %, &, * (33–47), numerals (48–57), logical symbols (58–64) and letters (65 and up).

Identifying individuals in scatterplots

The best way to identify multiple individuals in scatterplots is to use a combination of colours and symbols, using as.numeric to convert the grouping factor (the variable acting as the subject identifier) into a colour and/or a symbol. Here is an example where reaction time is plotted against days of sleep deprivation for 18 subjects:

```
data<-read.table("c:\\temp\\sleep.txt",header=T)
attach(data)
Subject<-factor(Subject)
plot(Days,Reaction,col=as.numeric(Subject),pch=as.numeric(Subject))
```

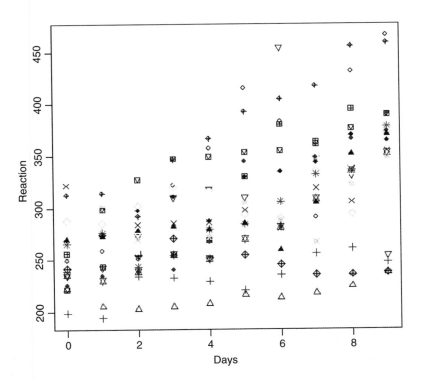

The individuals stand out more clearly from one another when both colour and symbol are used.

Using a third variable to label a scatterplot

The following example concerns the response of a grass species *Festuca rubra* as measured by its biomass in small samples (FR) to two explanatory variables, soil pH and total hay yield (the mass of all plant species combined). A scatterplot of pH against hay shows the locations of the various samples. The idea is to use the text function to label each of the points on the scatterplot with the dry mass of *F. rubra* in that particular sample, to see whether there is systematic variation in the mass of *Festuca* with changes in hay yield and soil pH.

```
data<-read.table("c:\\temp\\pgr.txt",header=T)
attach(data)
names(data)
```

```
[1]  "FR"  "hay"  "pH"
```

```
plot(hay,pH)
text(hay, pH, labels=round(FR, 2), pos=1, offset=0.5,cex=0.7)
```

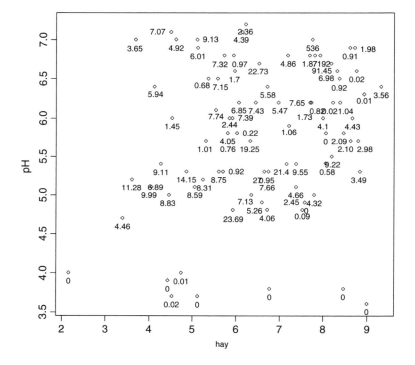

The labels are *centred* on the x value of the point (pos=1) and are *offset half a character below* the point (offset=0.5). They show the value of *FR* rounded to two significant digits (labels=round(FR, 2)) at 70% character expansion (cex=0.7). There is an obvious problem with this method when there is lots of overlap between the labels (as in the top right), but the technique works well for more widely spaced points. The plot shows that high values of *Festuca* biomass are concentrated at intermediate values of both soil pH and hay yield.

You can also use a third variable to choose the colour of the points in your scatterplot. Here the points with *FR* above median are shown in red, the others in black:

```
plot(hay,pH,pch=16,col=ifelse(FR>median(FR),"red","black"))
```

For three-dimensional plots see image, contour and wireframe on p. 843, and for more on adding text to plots see p. 143.

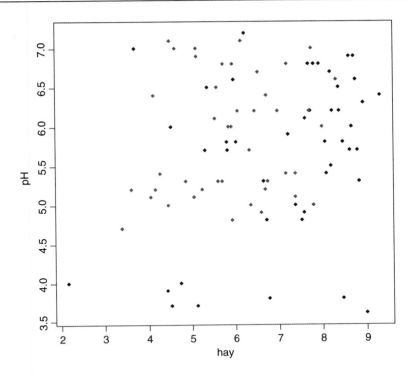

Adding text to scatterplots

It is very easy to add text to graphics. Suppose you wanted to add the text '(b)' to a plot at the location $x = 0.8$ and $y = 45$; just type

```
text(0.8,45,"(b)")
```

In this example we want to produce a map of place names and the place names are in a file called map.places.csv, but their coordinates are in another, much longer file called bowens.csv, containing many more place names than we want to plot. If you have factor level names with spaces in them (e.g. multiple words), then the best format for reading files is comma-delimited ('.csv' rather than the standard tab-delimited, '.txt' files). You read them into a dataframe in R using read.csv in place of read.table:

```
map.places <-read.csv("c:\\temp\\map.places.csv",header=T)
attach(map.places)
names(map.places)
```

```
[1] "wanted"
```

```
map.data<-read.csv("c:\\temp\\bowens.csv",header=T)
attach(map.data)
names(map.data)
```

```
[1] "place" "east" "north"
```

There is a slight complication to do with the coordinates. The northernmost places are in a different 100 km square so, for instance, a northing of 3 needs to be altered to 103. It is convenient that all of the values that need to be changed have northings < 60 in the dataframe:

```
nn<-ifelse(north<60,north+100,north)
```

This says change all of the northings for which north < 60 is TRUE to nn<-north+100, and leave unaltered all the others (FALSE) as nn<-north.

We begin by plotting a blank space (type="n") of the right size (eastings from 20 to 100 and northings from 60 to 110) with blank axis labels ":

```
plot(c(20,100),c(60,110),type="n",xlab="" ylab="")
```

The trick is to select the appropriate places in the vector called place and use text to plot each name in the correct position (east[i],nn[i]). For each place name in wanted we find the correct subscript for that name within place using the which function:

```
for (i in 1:length(wanted)){
ii <- which(place == as.character(wanted[i]))
text(east[ii], nn[ii], as.character(place[ii]), cex = 0.6) }
```

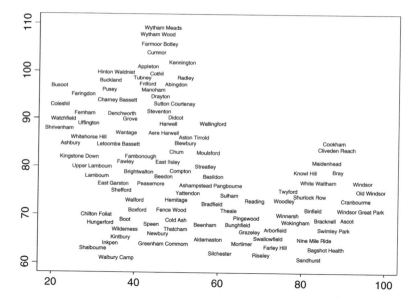

Drawing mathematical functions

The curve function is used for this. Here is a plot of $x^3 - 3x$ between $x = -2$ and $x = 2$:

```
curve(x^3-3*x, -2, 2)
```

Here is the more cumbersome code to do the same thing using plot:

```
x<-seq(-2,2,0.01)
y<-x^3-3*x
plot(x,y,type="l")
```

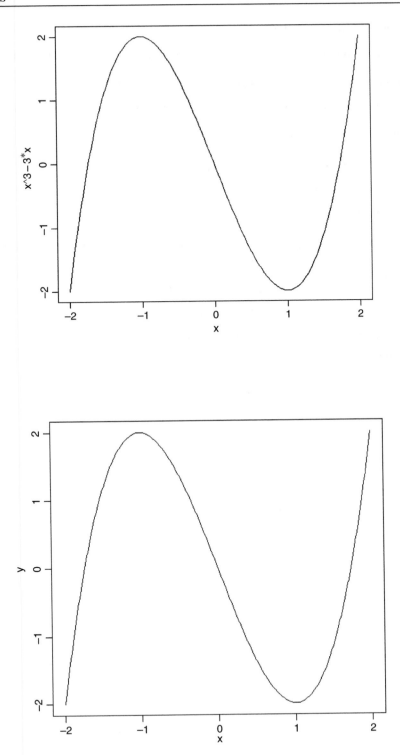

Adding other shapes to a plot

Once you have produced a set of axes using plot it is straightforward to locate and insert other kinds of things. Here are two unlabelled axes, without tick marks (xaxt="n"), both scaled 0 to 10 but without any of the 11 points drawn on the axes (type="n"):

plot(0:10,0:10,xlab="",ylab="",xaxt="n",yaxt="n",type="n")

You can easily add extra graphical objects to plots:

- rect   rectangles
- arrows   arrows and bars
- polygon   more complicated straight-sided shapes

For the purposes of demonstration we shall add a single-headed arrow, a double-headed arrow, a rectangle and a six-sided polygon to this space.

We want to put a solid square object in the top right-hand corner, and we know the precise coordinates to use. The syntax for the rect function is to provide four numbers:

rect(xleft, ybottom, xright, ytop)

so to plot the square from (6,6) to (9,9) involves

rect(6,6,9,9)

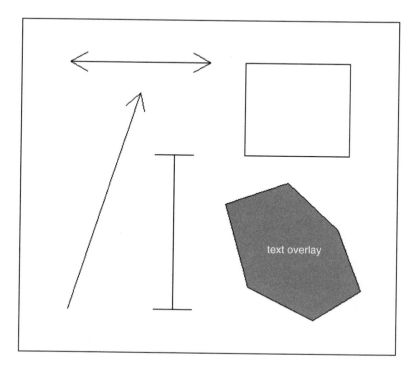

You can fill the shape with solid colour (col) or with shading lines (density, angle) as described on p. 829. You might want R to tell you the coordinates of the corners of the rectangle and you can use the locator() function for this. The rect function does not accept locator as its arguments but you can easily write a function (here called corners) to do this:

```
corners<-function(){
coos<-c(unlist(locator(1)),unlist(locator(1)))
rect(coos[1],coos[2],coos[3],coos[4])
}
```

Run the function like this:

```
corners()
```

Then click in the bottom left-hand corner and again in the top right-hand corner, and a rectangle will be drawn.

Drawing arrows as in the diagram above is straightforward. The syntax for the arrows function to draw a line from the point (x0, y0) to the point (x1, y1) with the arrowhead, by default, at the 'second' end (x1, y1) is:

```
arrows(x0, y0, x1, y1)
```

Thus, to draw an arrow from (1,1) to (3,8) with the head at (3,8) type

```
arrows(1,1,3,8)
```

A double-headed arrow from (1,9) to (5,9) is produced by adding code=3 like this:

```
arrows(1,9,5,9,code=3)
```

A vertical bar with two square ends (e.g. like an error bar) uses angle = 90 instead of the default angle = 30)

```
arrows(4,1,4,6,code=3,angle=90)
```

Here is a function that draws an arrow from the cursor position of your first click to the position of your second click:

```
click.arrows<-function(){
coos<-c(unlist(locator(1)),unlist(locator(1)))
arrows(coos[1],coos[2],coos[3],coos[4])
}
```

To run this, type

```
click.arrows()
```

We now wish to draw a polygon. To do this, it is often useful to save the values of a series of locations. Here we intend to save the coordinates of six points in a vector called locations to define a polygon for plotting

```
locations<-locator(6)
```

After you have clicked over the sixth location, control returns to the screen. What kind of object has locator produced?

```
class(locations)
```

```
[1] "list"
```

It has produced a list, and we can extract the vectors of x and y values from the list using $ to name the elements of the list (R has created the helpful names x and y):

locations$x

```
[1] 1.185406 3.086976 5.561308 6.019518 4.851083 1.506152
```

locations$y

```
[1] 8.708933 9.538905 8.060518 4.377520 1.239191 2.536021
```

Now we draw the lavender-coloured polygon like this

polygon(locations,col="lavender")

Note that the polygon function has automatically closed the shape, drawing a line from the last point to the first.

The polygon function can be used to draw more complicated shapes, even curved ones. In this example we are asked to shade the area beneath a standard normal curve for values of x that are less than or equal to -1. First draw the line for the standard normal:

```
xv<-seq(-3,3,0.01)
yv<-dnorm(xv)
plot(xv,yv,type="l")
```

Then fill the area to the left of $xv \le -1$ in red:

polygon(c(xv[xv<=-1],-1),c(yv[xv<=-1],yv[xv==-3]),col="red")

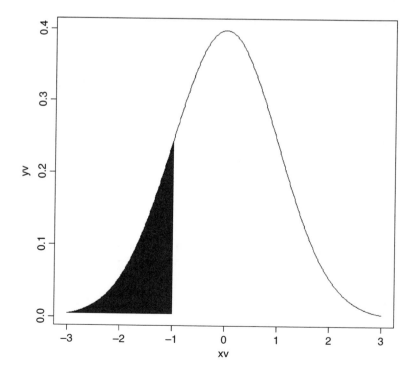

Note the insertion of the point (-1, yv[xv== -3]) to draw the polygon down to the x axis.

Smooth curves

Up to this point our response variable was shown as a scatter of data points. In many cases, however, we want to show the response as a smooth curve. The important tip is that to produce reasonably smooth-looking curves in R you should draw about 100 straight line sections between the minimum and maximum values of your x axis.

The Ricker curve is named after the famous Canadian fish biologist who introduced this two-parameter hump-shaped model for describing recruitment to a fishery y as a function of the density of the parental stock, x. We wish to compare two Ricker curves with the following parameter values:

$$y_A = 482xe^{-0.045x}, \qquad y_B = 518xe^{-0.055x}.$$

The first decision to be made is the range of x values for the plot. In our case this is easy because we know from the literature that the minimum value of x is 0 and the maximum value of x is 100. Next we need to generate about 100 values of x at which to calculate and plot the smoothed values of y:

```
xv<-0:100
```

Next, calculate vectors containing the values of y_A and y_B at each of these x values:

```
yA<-482*xv*exp(-0.045*xv)
yB<-518*xv*exp(-0.055*xv)
```

We are now ready to draw the two curves, but we do not know how to scale the y axis. We could find the maximum and minimum values of y_A and y_B then use ylim to specify the extremes of the y axis, but it is more convenient to use the option type="n" to draw the axes without any data, then use lines to add the two smooth functions later. The blank axes (not shown here) are produced like this:

```
plot(c(xv,xv),c(yA,yB),xlab="stock",ylab="recruits",type="n")
```

We want to draw the smooth curve for y_A as a solid blue line (lty = 1, col = "blue"),

```
lines(xv,yA,lty=1,col="blue")
```

and the curve for y_B as a navy blue dotted line (lty = 2, col = "navy"),

```
lines(xv,yB,lty=2,col="navy")
```

Fitting non-linear parametric curves through a scatterplot

Here is a set of data showing a response variable, y (recruits to a fishery), as a function of a continuous explanatory variable x (the size of the fish stock):

```
rm(x,y)
info<-read.table("c:\\temp\\plotfit.txt",header=T)
attach(info)
names(info)
```

```
[1] "x" "y"
```

```
plot(x,y,xlab="stock",ylab="recruits",pch=16)
```

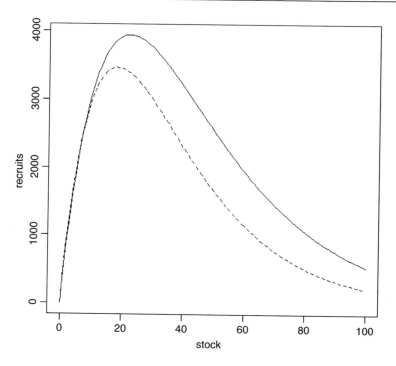

We do not know the parameter values of the best-fit curve in advance, but we can estimate them from the data using non-linear least squares nls. Note that you need to provide an initial guesstimate for the two parameter values in what is called a 'start list' (p. 663 for details):

model<-nls(y~a*x*exp(-b*x),start=list(a=500,b=0.05))

```
Formula: y ~ a * x * exp(-b * x)

Parameters:
      Estimate    Std. Error    t value    Pr(>|t|)
a     4.820e+02   1.593e+01      30.26     <2e-16    ***
b     4.461e-02   8.067e-04      55.29     <2e-16    ***
Residual standard error: 204.2 on 27 degrees of freedom
```

So the least-squares estimates of the two parameters are $a = 482.0$ and $b = 0.044\,61$, and their standard errors are 15.93 and 0.000 806 7 respectively. We already have a set of x values, xv (above), so we can use predict to add this as a dashed line to the plot:

lines(xv,predict(model,list(x=xv)),lty=2)

Next, you want to compare this regression line with a theoretical model, which was

$$y = 480xe^{-0.047x}.$$

We need to evaluate y across the xv values for the theoretical model:

yv<-480*xv*exp(-0.047*xv)

Now use the lines function to add this second curve to the plot as a solid line:

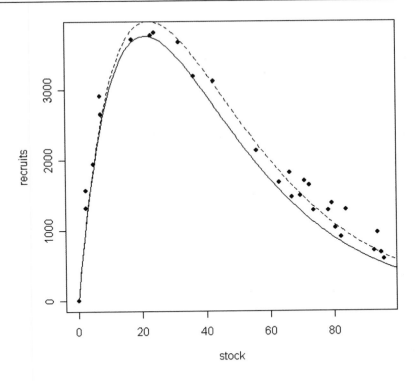

lines(xv,yv)

Notice that the regression model (dashed line) predicts the values of y for $x > 30$ much better than the theoretical model (solid line), and that both models slightly underestimate the values of y for $x < 20$. The plot is imperfect to the extent that the maximum of the dashed regression curve does not fit within the limits of the y axis. You should correct this as an exercise.

Fitting non-parametric curves through a scatterplot

It is common to want to fit a non-parametric smoothed curve through data, especially when there is no obvious candidate for a parametric function. R offers a range of options:

- lowess (a non-parametric curve fitter);
- loess (a modelling tool);
- gam (fits generalized additive models; p. 611);
- lm for polynomial regression (fit a linear model involving powers of x).

We will illustrate these options using the jaws data. First, we load the data:

```
data<-read.table("c:\\temp\\jaws.txt",header=T)
attach(data)
names(data)
```

```
[1] "age"  "bone"
```

Before we fit our curves to the data, we need to consider how best to display the results together.

Without doubt, the graphical parameter you will change most often just happens to be the least intuitive to use. This is the number of graphs per screen, called somewhat unhelpfully, mfrow. The idea is simple, but the syntax is hard to remember. You need to specify the number of rows of plots you want, and number of plots per row, in a vector of two numbers. The first number is the number of rows and the second number is the number of graphs per row. The vector is made using c in the normal way. The default single-plot screen is

par(mfrow=c(1,1))

Two plots side by side is

par(mfrow=c(1,2))

A panel of four plots in a 2 × 2 square is

par(mfrow=c(2,2))

To move from one plot to the next, you need to execute a new plot function. Control stays within the same plot frame while you execute functions like points, lines or text. Remember to return to the default single plot when you have finished your multiple plot by executing par(mfrow=c(1,1)). If you have more than two graphs per row or per column, the character expansion cex is set to 0.5 and you get half-size characters and labels.

Let us now plot our four graphs:

```
par(mfrow=c(2,2))
plot(age,bone,pch=16)
text(45,20,"lowess",pos=2)
lines(lowess(age,bone))

plot(age,bone,pch=16)
text(45,20,"loess",pos=2)
model<-loess(bone~age)
xv<-0:50
yv<-predict(model,data.frame(age=xv))
lines(xv,yv)

plot(age,bone,pch=16)
text(45,20,"gam",pos=2)
library(mgcv)
model<-gam(bone~s(age))
yv<-predict(model,list(age=xv))
lines(xv,yv)

plot(age,bone,pch=16)
text(45,20,"polynomial",pos=2)
model<-lm(bone~age+I(age^2)+I(age^3))
yv<-predict(model,list(age=xv))
lines(xv,yv)
```

The lowess function (top left) is a curve-smoothing function that returns x and y coordinates that are drawn on the plot using the lines function. The modern loess function (top right) is a modelling tool (y~x) from which the coordinates to be drawn by lines are extracted using

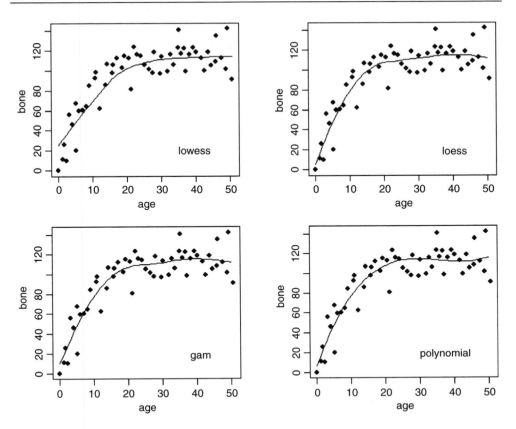

predict with a data.frame function containing the *x* values for plotting. Alternatively, you can use a generalized additive model, gam (bottom left) to generate the smoothed curve (y~s(x)) using predict. Finally, you might use a linear model with a polynomial function of *x* (here a cubic, bottom right).

Joining the dots

Sometimes you want to join the points on a scatterplot by lines. The trick is to ensure that the points on the *x* axis are ordered: if they are not ordered, the result is a mess.

```
smooth<-read.table("c:\\temp\\smoothing.txt",header=T)
attach(smooth)
names(smooth)
```

```
[1] "x" "y"
```

Begin by producing a vector of subscripts representing the ordered values of the explanatory variable. Then draw lines with the vector as subscripts to both the *x* and *y* variables:

```
sequence<-order(x)
lines(x[sequence],y[sequence])
```

If you do not order the *x* values, and just use the lines function, this is what happens:

```
plot(x,y,pch=16)
lines(x,y)
```

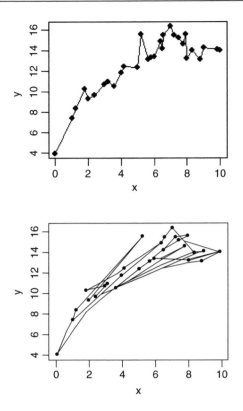

Plotting with a categorical explanatory variable

When the explanatory variable is categorical rather than continuous, we cannot produce a scatterplot. Instead, we choose between a **barplot** and a **boxplot**. I prefer box-and-whisker plots because they convey so much more information, and this is the default plot in R with a categorical explanatory variable.

Categorical variables are **factors** with two or more **levels** (see p. 26). Our first example uses the factor called month (with levels 1 to 12) to investigate weather patterns at Silwood Park:

```
weather<-read.table("c:\\temp\\SilwoodWeather.txt",header=T)
attach(weather)
names(weather)
```

```
[1] "upper" "lower" "rain" "month" "yr"
```

There is one bit of housekeeping we need to do before we can plot the data. We need to declare month to be a factor. At the moment, R just thinks it is a number:

```
month<-factor(month)
```

Now we can plot using a categorical explanatory variable (month) and, because the first variable is a factor, we get a boxplot rather than a scatterplot:

```
plot(month,upper)
```

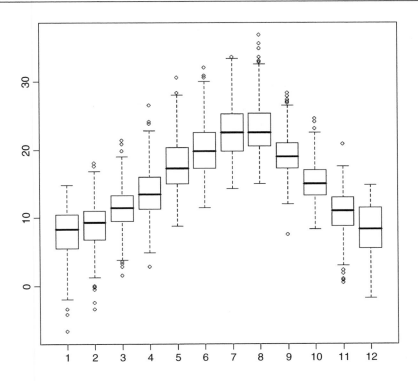

Note that there are no axis labels in the default box-and-whisker plot, and to get informative labels we should need to type

plot(month,upper,ylab="daily maximum temperature",xlab="month")

The boxplot summarizes a great deal of information very clearly. The horizontal line shows the **median** upper daily temperature for each month. The bottom and top of the box show the 25th and 75th **percentiles**, respectively (i.e. the location of the middle 50% of the data, also called the first and third **quartiles**). The vertical dashed lines are called the 'whiskers'. They show one of two things: either the maximum value or 1.5 times the **interquartile range** of the data, whichever is the smaller. The quantity '1.5 times the interquartile range of the data' is roughly 2 standard deviations, and the interquartile range is the difference in the response variable between its first and third quartiles. Points more than 1.5 times the interquartile range *above the third quartile* and points more than 1.5 times the interquartile range *below the first quartile* are defined as **outliers** and plotted individually. Thus, when there are no outliers the whiskers show the maximum and minimum values (here only in month 12). Boxplots not only show the location and spread of data but also indicate skewness (which shows up as asymmetry in the sizes of the upper and lower parts of the box). For example, in February the range of lower temperatures was much greater than the range of higher temperatures. Boxplots are also excellent for spotting errors in the data when the errors are represented by extreme outliers.

Data for the next example come from an experiment on plant competition, with five factor levels in a single categorical variable called clipping: a control (unclipped), two root clipping treatments (r5 and r10) and two shoot clipping treatments (n25 and n50) in which the leaves of neighbouring plants were reduced by 25% and 50%. The response variable is yield at maturity (a dry weight).

```
trial<-read.table("c:\\temp\\compexpt.txt",header=T)
attach(trial)
names(trial)
```

```
[1] "biomass" "clipping"
```

The boxplot is created with exactly the same syntax as the scatterplot, with the *x* variable first, then the *y* variable, the only difference being that the *x* variable is categorical rather than continuous:

```
plot(clipping,biomass,xlab="treatment",ylab="yield")
```

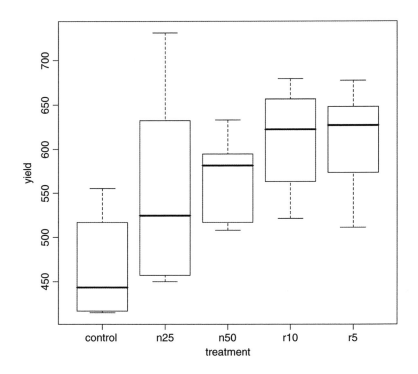

In this case there are no outliers, so the top of the upper bar represents the maximum value of the response for that factor level, and the bottom of the lower bar represents the minimum value. Because there are no outliers, the upper and lower bars are of different lengths.

Boxplots with notches to indicate significant differences

Boxplots are very good at showing the distribution of the data points around the median: for instance, the range of values for n25 is much greater than for any other treatment, and the control data are skew, with a bigger range of values for the third quartile than for the second (the upper box is bigger than the lower box). On the other hand, boxplots are not so good at indicating whether or not the median values are significantly different from one another. Tukey invented **notches** to get the best of both worlds. The notches are drawn as a 'waist' on either side of the median and are intended to give a rough impression of the significance of the differences between two medians. Boxes in which *the notches do not overlap* are likely to prove to have significantly different medians under an appropriate test. Boxes with overlapping notches probably do not have significantly different medians. The size of the notch increases with the magnitude of the interquartile range and declines with the square root of the replication, like this:

$$\text{notch} = \mp 1.58 \frac{IQR}{\sqrt{n}},$$

where *IQR* is the interquartile range and *n* is the replication per sample. Notches are based on assumptions of asymptotic normality of the median and roughly equal sample sizes for the two medians being compared, and are said to be rather insensitive to the underlying distributions of the samples. The idea is to give roughly a 95% confidence interval for the difference in two medians, but the theory behind this is somewhat vague.

When the sample sizes are small and/or the within-sample variance is high, the notches are not drawn as you might expect them (i.e. as a waist within the box). Instead, the notches are extended *above* the 75th percentile and/or *below* the 25th percentile. This looks odd, but it is an intentional feature, intended to act as a warning of the likely invalidity of the test. You can see this in action for the biomass data:

```
par(mfrow=c(1,2))
boxplot(biomass~clipping)
boxplot(biomass~clipping,notch=T)
```

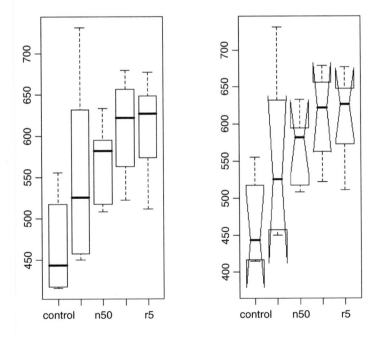

Note the different scales on the two *y* axes: our original boxplot on the left and the new boxplot with notches on the right. The notches for the three rightmost boxes do not overlap

the notches for the controls, but the notches for n25 overlap those of both the controls
and the three rightmost boxes. For each factor level, however, one or both of the notches
extends beyond the limits of the box, indicating that we should not have any confidence in
the significance of the pairwise comparisons. In this example, replication is too low, and
within-level variance is too great, to justify such a test.

Rather than use plot to produce a boxplot, an alternative is to use a barplot to show
the heights of the five mean values from the different treatments. We need to calculate the
means using the function tapply like this:

```
means<-tapply(biomass,clipping,mean)
```

Then the barplot is produced very simply:

```
par(mfrow=c(1,1))
barplot(means,xlab="treatment",ylab="yield")
```

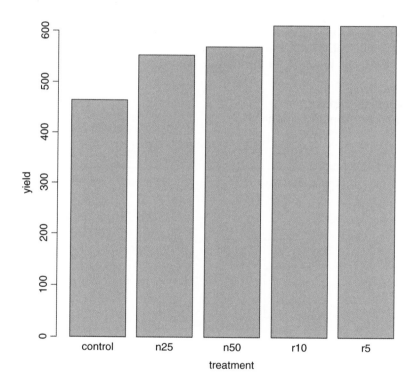

Unless we add error bars to such a barplot, the graphic gives no indication of the extent
of the uncertainty associated with each of the estimated treatment means. A function to
produce barplots with error bars is described on p. 462, where it is applied to these plant
competition data.

Plots for multiple comparisons

When there are many levels of a categorical explanatory variable, we need to be cautious
about the statistical issues involved with multiple comparisons (see p. 483). Here we contrast

two graphical techniques for displaying multiple comparisons: boxplots with notches, and Tukey's 'honest significant difference'.

The data show the response of yield to a categorical variable (fact) with eight levels representing eight different genotypes of seed (cultivars) used in the trial:

```
data<-read.table("c:\\temp\\box.txt",header=T)
attach(data)
names(data)
```

```
[1] "fact" "response"
```

```
plot(response~factor(fact))
```

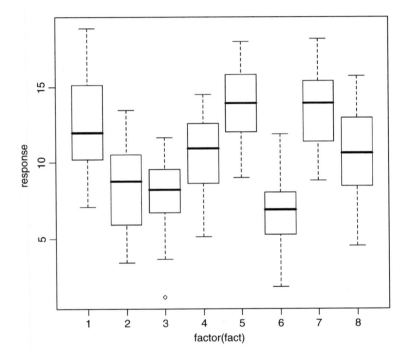

Because the genotypes (factor levels) are unordered, it is hard to judge from the plot which levels might be significantly different from which others. We start, therefore, by calculating an index which will rank the mean values of response across the different factor levels:

```
index<-order(tapply(response,fact,mean))
ordered<-factor(rep(index,rep(20,8)))
boxplot(response~ordered,notch=T,names=as.character(index),
                    xlab="ranked treatments",ylab="response")
```

There are several points to clarify here. We plot the response as a function of the factor called ordered (rather than fact) so that the boxes are ranked from lowest mean yield on the left (cultivar 6) to greatest mean on the right (cultivar 5). We change the names of the boxes to reflect the values of index (i.e. the original values of fact: otherwise they would read 1

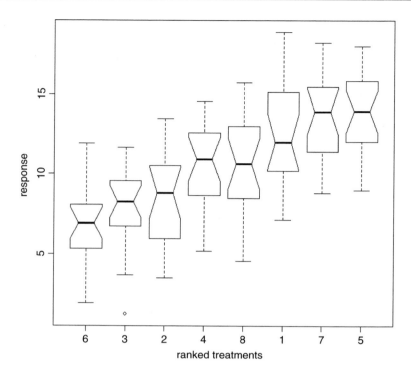

to 8). Note that the vector called index is of length 8 (the number of boxes on the plot), but ordered is of length 160 (the number of values of response). Looking at the notches, no two adjacent pairs of medians appear to be significantly different, but the median of treatment 4 appears to be significantly greater than the median of treatment 6, and the median of treatment 5 appears to be significantly greater than the median of treatment 8 (but only just).

The statistical analysis of these data might involve user-specified contrasts (p. 370), once it is established that there are significant differences to be explained. This we assess with a one-way analysis of variance to test the hypothesis that at least one of the means is significantly different from the others (see p. 457):

```
model<-aov(response~factor(fact))
summary(model)
```

	Df	Sum Sq	Mean Sq	F value	Pr(>F)	
factor(fact)	7	925.70	132.24	17.477	< 2.2e-16	***
Residuals	152	1150.12	7.57			

Indeed, there is very compelling evidence ($p < 0.0001$) for accepting that there are significant differences between the mean yields of the eight different crop cultivars.

Alternatively, if you want to do multiple comparisons, then because there is no *a priori* way of specifying contrasts between the eight treatments, you might use Tukey's honest significant difference (see p. 483):

```
plot(TukeyHSD(model))
```

Comparisons having intervals that do not overlap the vertical dashed line are significantly different. The vertical dashed line indicates no difference between the mean values for the

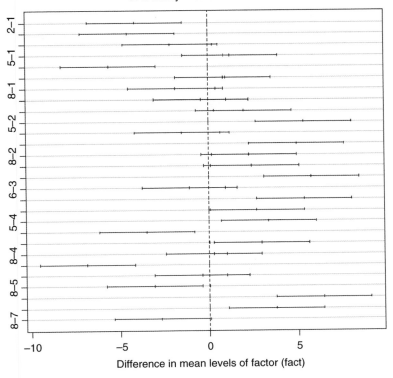

95% family-wise confidence level

Difference in mean levels of factor (fact)

factor-level comparisons indicated on the *y* axis. Thus, we can say that the contrast between cultivars 8 and 7 (8-7) falls just short of significance (despite the fact that their notches do not overlap; see above), but the comparisons 7-6 and 8-6 are both significant (their *boxes* do not overlap, let alone their notches). The missing comparison labels on the *y* axis have to be inferred from a knowledge of the number of factor levels (8 in this example). So, since 8 vs. 7 is labelled, the next one up must be 8-6 and the one above that is 7-6, then we find the labelled 8-5, so it must be 7-5 above that and 6-5 above that, then the labelled 8-4, and so on.

Plots for Single Samples

When we have a just one variable, the choice of plots is more restricted:

- histograms to show a frequency distribution;
- index plots to show the values of *y* in sequence;
- time-series plots;
- compositional plots like pie diagrams

Histograms

The commonest plots for a single sample are histograms and index plots. Histograms are excellent for showing the mode, the spread and the symmetry (skew) of a set of data. The R function hist is deceptively simple. Here is a histogram of 1000 random points drawn from a Poisson distribution with a mean of 1.7:

```
hist(rpois(1000,1.7),
          main="",xlab="random numbers from a Poisson with mean 1.7")
```

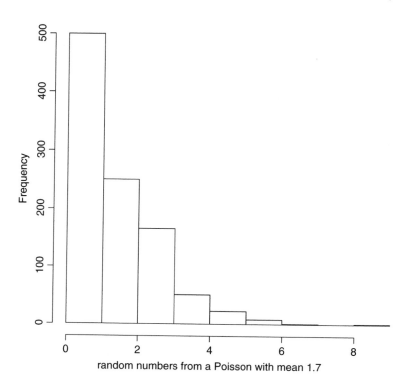

This illustrates perfectly one of the big problems with histograms: it is not clear what the bar of height 500 is showing. Is it the frequency of zeros, or the frequency of zeros and ones lumped together? What we really want in a case like this is a separate histogram bar for each integer value from 0 to 8. We achieve this by specifying the breaks on the x axis to be at -0.5, 0.5, 1.5, ..., like this:

```
hist(rpois(1000,1.7),breaks=seq(-0.5,9.5,1),
          main="",xlab="random numbers from a Poisson with mean 1.7")
```

That's more like it. Now we can see that the mode is 1 (not 0), and that 2s are substantially more frequent than 0s. The distribution is said to be 'skew to the right' (or 'positively skew') because the long tail is on the right-hand side of the histogram.

Overlaying histograms with smooth density functions

If it is in any way important, then you should always specify the break points yourself. Unless you do this, the hist function may not take your advice about the number of bars

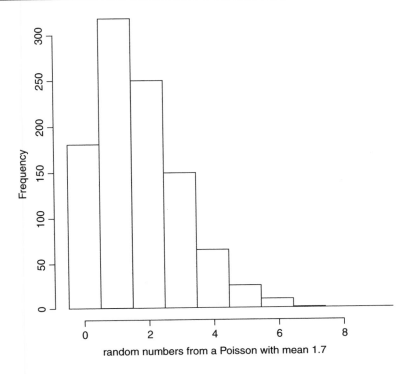

random numbers from a Poisson with mean 1.7

or the width of bars. For small-integer data (less than 20, say), the best plan is to have one bin for each value. You create the breaks by starting at −0.5 to accommodate the zeros and going up to max(y) + 0.5 to accommodate the biggest count. Here are 158 random integers from a negative binomial distribution with $\mu = 1.5$ and $k = 1.0$:

```
y<-rnbinom(158,mu=1.5,size=1)
bks<- -0.5:(max(y)+0.5)
hist(y,bks,main="")
```

To get the best fit of a density function for this histogram we should estimate the parameters of our particular sample of negative binomially distributed counts:

```
mean(y)
```

```
[1] 1.772152
```

```
var(y)
```

```
[1] 4.228009
```

```
mean(y)^2/(var(y)-mean(y))
```

```
[1] 1.278789
```

In R, the parameter k of the negative binomial distribution is known as size and the mean is known as mu. We want to generate the probability density for each count between 0 and 11, for which the R function is dnbinom:

```
xs<-0:11
ys<-dnbinom(xs,size=1.2788,mu=1.772)
lines(xs,ys*158)
```

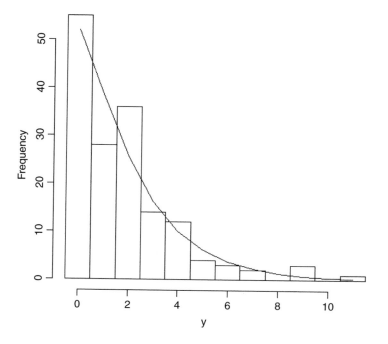

Not surprisingly, since we generated the data, the negative binomial distribution is a very good description of the frequency distribution. The frequency of 1s is a bit low and of 2s is a bit high, but the other frequencies are very well described.

Density estimation for continuous variables

The problems associated with drawing histograms of continuous variables are much more challenging. The subject of density estimation is an important issue for statisticians, and whole books have been written about it (Silverman 1986; Scott 1992). You can get a feel for what is involved by browsing the ?density help window. The algorithm used in density.default disperses the mass of the empirical distribution function over a regular grid of at least 512 points, uses the fast Fourier transform to convolve this approximation with a discretized version of the kernel, and then uses linear approximation to evaluate the density at the specified points. The choice of bandwidth is a compromise between smoothing enough to rub out insignificant bumps and smoothing too much so that real peaks are eliminated. The rule of thumb for bandwidth is

$$b = \frac{\max(x) - \min(x)}{2(1 + \log_2 n)}$$

(where n is the number of data points). For details see Venables and Ripley (2002). We can compare hist with Venables and Ripley's truehist for the Old Faithful eruptions data. The rule of thumb for bandwidth gives:

```
library(MASS)
attach(faithful)
(max(eruptions)-min(eruptions))/(2*(1+log(length(eruptions),base=2)))
```

[1] 0.192573

but this produces much too bumpy a fit. A bandwidth of 0.6 looks much better:

```
par(mfrow=c(1,2))
hist(eruptions,15,freq=FALSE,main="",col=27)
lines(density(eruptions,width=0.6,n=200))
truehist(eruptions,nbins=15,col=27)
lines(density(eruptions,n=200))
```

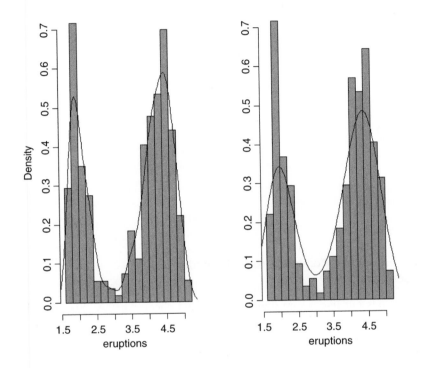

Note that although we asked for 15 bins, we actually got 18. Note also, that although both histograms have 18 bins, they differ substantially in the heights of several of the bars. The left hist has two peaks above density $=0.5$ while truehist on the right has three. There is a sub-peak in the trough of hist at about 3.5 but not of truehist. And so on. Such are the problems with histograms. Note, also, that the default probability density curve (on the right) picks out the heights of the peaks and troughs much less well than our bandwidth of 0.6 (on the left).

Index plots

The other plot that is useful for single samples is the index plot. Here, plot takes a single argument which is a continuous variable and plots the values on the y axis, with the x coordinate determined by the position of the number in the vector (its 'index', which is 1 for the first number, 2 for the second, and so on up to length(y) for the last value). This kind of plot is especially useful for error checking. Here is a data set that has not yet been quality checked, with an index plot of response$y:

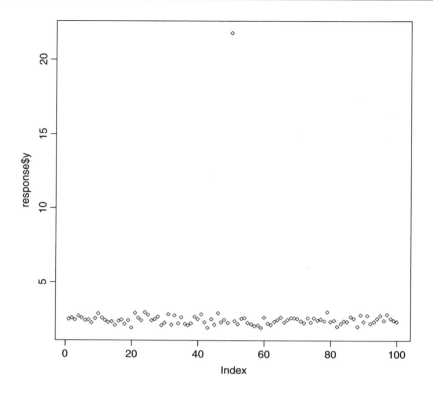

```
response<-read.table("c:\\temp\\das.txt",header=T)
plot(response$y)
```

The error stands out like a sore thumb. We should check whether this might have been a data-entry error, such as a decimal point in the wrong place. But which value is it, precisely, that is wrong? What is clear is that it is the only point for which $y > 15$, so we can use the which function to find out its index (the subscript within y):

```
which(response$y > 15)
```

```
[1] 50
```

We can then use this value as the subscript to see the precise value of the erroneous y:

```
response$y[50]
```

```
[1] 21.79386
```

Having checked in the lab notebook, it is obvious that this number should be 2.179 rather than 21.79, so we replace the 50th value of y with the correct value:

```
response$y[50]<-2.179386
```

Now we can repeat the index plot to see if there are any other obvious mistakes

```
plot(response$y)
```

That's more like it.

Time series plots

When a time series is complete, the time series plot is straightforward, because it just amounts to joining the dots in an ordered set of *y* values. The issues arise when there are missing values in the time series, particularly groups of missing values for which periods we typically know nothing about the behaviour of the time series.

There are two functions in R for plotting time series data: ts.plot and plot.ts. Here is ts.plot in action, producing three time series on the same axes using different line types:

```
data(UKLungDeaths)
ts.plot(ldeaths, mdeaths, fdeaths, xlab="year", ylab="deaths", lty=c(1:3))
```

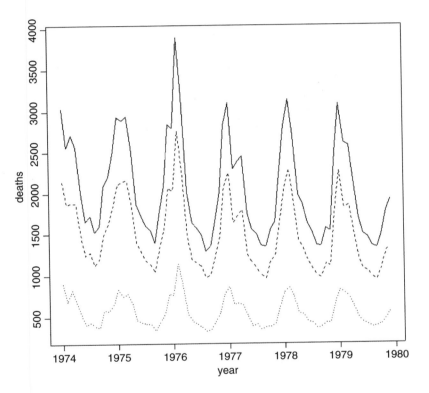

The upper, solid line shows total deaths, the heavier dashed line shows male deaths and the faint dotted line shows female deaths. The difference between the sexes is clear, as is the pronounced seasonality, with deaths peaking in midwinter.

The alternative function plot.ts works for plotting objects inheriting from class ts (rather than simple vectors of numbers in the case of ts.plot).

```
data(sunspots)
plot(sunspots)
```

The simple statement plot(sunspots) works because sunspots inherits from the time series class:

```
class(sunspots)
```

```
[1] "ts"
```

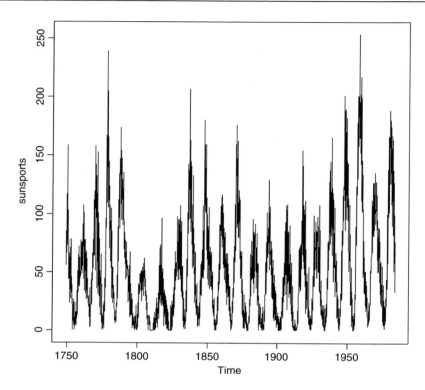

```
is.ts(sunspots)
```

```
[1] TRUE
```

Pie charts

Statisticians don't like pie charts because they think that people should know what 50% looks like. Pie charts, however, can sometimes be useful to illustrate the proportional make-up of a sample in presentations. The function **pie** takes a vector of numbers, turns them into proportions, and divides up the circle on the basis of those proportions. It is essential to use a label to indicate which pie segment is which. The label is provided as a vector of character strings, here called **data$names**: Because there are blank spaces in some of the names ('oil shales' and 'methyl clathrates') we cannot use **read.table** with a tab-delimited text file to enter the data. Instead, we save the file called piedata from Excel as a comma-delimited file, with a '.csv' extention, and input the data to R using **read.csv** in place of **read.table**, like this:

```
data<-read.csv("c:\\temp\\piedata.csv",header=T)
data
```

```
                 names    amounts
1                 coal          4
2                  oil          2
3                  gas          1
4           oil shales         3
5    methyl clathrates         6
```

The pie chart is created like this:

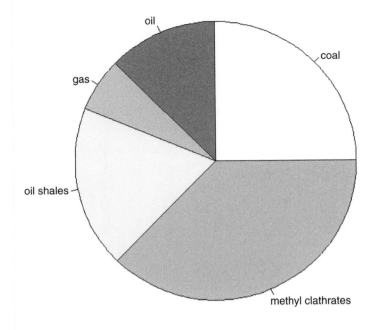

```
pie(data$amounts,labels=as.character(data$names))
```

You can change the colours of the segments if you want to (p. 855).

The stripchart function

For sample sizes that are too small to use box-and-whisker plots, an alternative plotting method is to use the **stripchart** function. The point of using **stripchart** is to look carefully at the location of individual values within the small sample, and to compare values across cases. The **stripchart** plot can be specified by a model formula y~factor and the strips can be specified to run vertically rather than horizontally. Here is an example from the built-in OrchardSprays data set where the response variable is called **decrease** and there is a single categorical variable called **treatment** (with eight levels A–H). Note the use of with instead of **attach**:

```
data(OrchardSprays)
with(OrchardSprays,
    stripchart(decrease ~ treatment,
    ylab = "decrease", vertical = TRUE, log = "y"))
```

This has the layout of the box-and-whisker plot, but shows all the raw data values. Note the logarithmic y axis and the vertical alignment of the eight strip charts.

Plots with multiple variables

Initial data inspection using plots is even more important when there are many variables, any one of which might contain mistakes or omissions. The principal plot functions when there are multiple variables are:

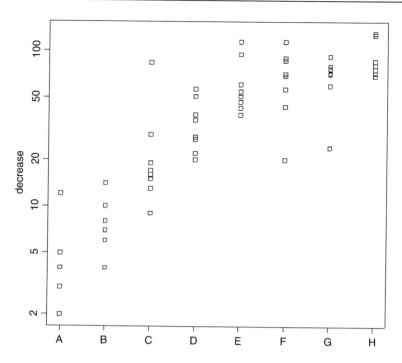

- **pairs** for a matrix of scatterplots of every variable against every other;
- **coplot** for conditioning plots where *y* is plotted against *x* for different values of *z*;
- **xyplot** where a set of panel plots is produced.

We illustrate these functions with the ozone data.

The **pairs** function

With two or more continuous explanatory variables (i.e. in a multiple regression; see p. 433) it is valuable to be able to check for subtle dependencies between the explanatory variables. The **pairs** function plots every variable in the dataframe on the *y* axis against every other variable on the *x* axis: you will see at once what this means from the following example:

```
ozonedata<-read.table("c:\\temp\\ozone.data.txt",header=T)
attach(ozonedata)
names(ozonedata)
```

```
[1] "rad"  "temp"  "wind"  "ozone"
```

The **pairs** function needs only the name of the whole dataframe as its first argument. We exercise the option to add a non-parametric smoother to the scatterplots:

```
pairs(ozonedata,panel=panel.smooth)
```

The response variables are named in the rows and the explanatory variables are named in the columns. In the upper row, labelled rad, the response variable (on the *y* axis) is solar radiation. In the bottom row the response variable, ozone, is on the *y* axis of all three panels. Thus, there appears to be a strong negative non-linear relationship between ozone and wind

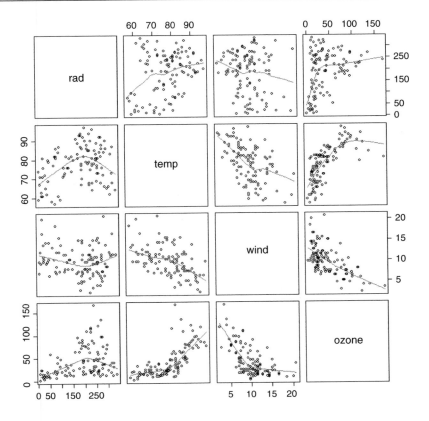

speed, a positive non-linear relationship between air temperature and ozone (middle panel in the bottom row) and an indistinct, perhaps humped, relationship between ozone and solar radiation (left-most panel in the bottom row). As to the explanatory variables, there appears to be a negative correlation between wind speed and temperature.

The coplot function

A real difficulty with multivariate data is that the relationship between two variables may be obscured by the effects of other processes. When you draw a two-dimensional plot of y against x, then all of the effects of the other explanatory variables are squashed flat onto the plane of the paper. In the simplest case, we have one response variable (ozone) and just two explanatory variables (wind speed and air temperature). The function is written like this:

```
coplot(ozone~wind|temp,panel = panel.smooth)
```

With the response (ozone) on the left of the tilde and the explanatory variable on the x axis (wind) on the right, with the conditioning variable after the conditioning operator | (here read as 'given temp'). An option employed here is to fit a non-parametric smoother through the scatterplot in each of the panels.

The coplot panels are ordered from lower-left to upper right, associated with the values of the conditioning variable in the upper panel (temp) from left to right. Thus, the lower-left plot is for the lowest temperatures (56–72 degrees F) and the upper right plot is for the highest temperatures (82–96 degrees F). This coplot highlights an interesting interaction. At the two lowest levels of the conditioning variable, temp, there is little or no relationship

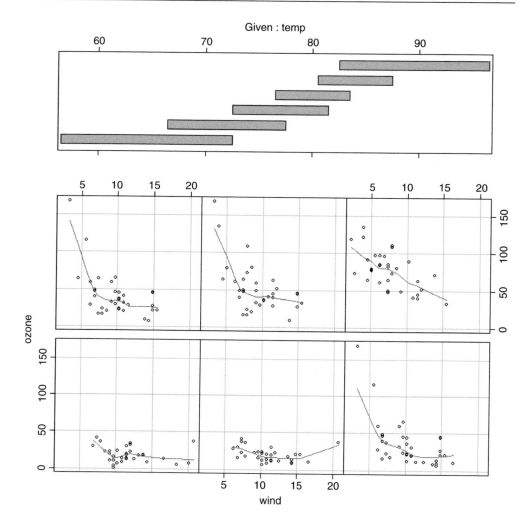

between ozone and wind speed, but in the four remaining panels (at higher temperatures) there is a distinct negative relationship between wind speed and ozone concentration. The hard thing to understand about **coplot** involves the 'shingles' that are shown in the upper margin (given temp in this case). The overlap between the shingles is intended to show how much overlap there is between one panel and the next in terms of the data points they have in common. In this default configuration, half of the data in a panel is shared with the panel to the left, and half of the data is shared with the panel to the right (**overlap = 0.5**). You can alter the shingle as far as the other extreme, when all the data points in a panel are unique to that panel (there is no overlap between adjacent shingles; **overlap = -0.05**).

Interaction plots

These are useful when the response to one factor depends upon the level of another factor. They are a particularly effective graphical means of interpreting the results of factorial experiments (p. 466). Here is an experiment with grain yields in response to irrigation and fertilizer application:

```
yields<-read.table("c:\\temp\\splityield.txt",header=T)
attach(yields)
names(yields)
```

```
[1]  "yield"  "block"  "irrigation"  "density"  "fertilizer"
```

The interaction plot has a rather curious syntax, because the response variable (yield) comes *last* in the list of arguments. The factor listed first forms the *x* axis of the plot (three levels of fertilizer), and the factor listed second produces the family of lines (two levels of irrigation). The lines join the mean values of the response for each combination of factor levels:

interaction.plot(fertilizer,irrigation, yield)

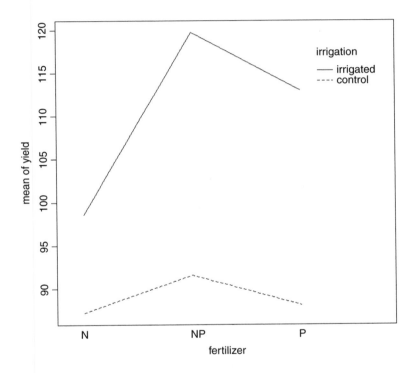

The interaction plot shows that the mean response to fertilizer depends upon the level of irrigation, as evidenced by the fact that the lines are not parallel.

Special Plots

Trellis graphics

The main purpose of trellis graphics is to produce multiple plots per page and multi-page plots. The plots are produced in adjacent panels, typically with one plot for each level of a categorical variable (called the **conditioning variable**). For instance, you might plot weight against age for each of two genders (males and females). The response variable is weight, the continuous explanatory variable is age (also called the **primary covariate** in documentation on trellis graphics) and the categorical explanatory variable is gender (a factor with two levels). In a case like this, the default would produce two panels side by side in one row, with the panel for females on the left (simply because 'f' comes before 'm' in the alphabet). In the jargon of trellis graphics, gender is a **grouping factor** that divides the observations into distinct groups. Here are the data:

```
data<-read.table("c:\\temp\\panels.txt",header=T)
attach(data)
names(data)
```

```
[1]  "age"  "weight"  "gender"
```

The package for producing trellis graphics in R is called lattice (not trellis as you might have guessed, because that name was pre-empted by a commercial package):

```
library(lattice)
```

The panel plots are created by the xyplot function, using a formula to indicate the grouping structure: weight ~ age | gender. This is read as 'weight is plotted as a function of age, given gender' (the vertical bar | is the 'given' symbol).

```
xyplot(weight ~ age | gender)
```

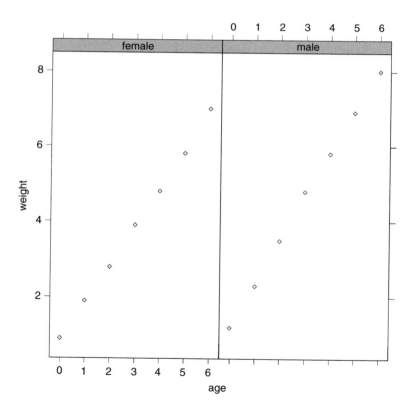

Trellis graphics is a framework for data visualization developed at Bell Laboratories by Rick Becker, Bill Cleveland and others, extending the ideas about what makes for an effective graph (layout, colour, style, symbol sizes and so forth) presented in Cleveland (1993). The interface is based on the implementation in S-PLUS, but there are several differences, and code produced for S-PLUS might not work in R.

Most of the high-level trellis functions in S-PLUS are implemented in R, with the exception of the pie chart:

- barchart for barplots;

- bwplot for box-and-whisker plots;

- densityplot for kernel density plots;

- dotplot for dot plots;

- histogram for panels of histograms;

- qqmath for quantile plots against mathematical distributions;

- stripplot for a one-dimensional scatterplot;

- qq for a QQ plot for comparing two distributions;

- xyplot for a scatterplot;

- levelplot for creating level plots (similar to image plots);

- contourplot for contour plots;

- cloud for three-dimensional scatterplots;

- wireframe for 3D surfaces (similar to persp plots);

- splom for a scatterplot matrix;

- parallel for creating parallel coordinate plots;

- rfs to produce a residual and fitted value plot (see also oneway);

- tmd for a Tukey mean–difference plot.

The lattice package has been developed by Deepayan Sarkar, and the plots created by lattice are rendered by the Grid Graphics engine for R (developed by Paul Murrell). Lattice plots are highly customizable via user-modifiable settings, but these are completely unrelated to base graphics settings. In particular, changing par() settings usually has no effect on lattice plots. To read more about the background and capabilities of the lattice package, type

```
help(package = lattice)
```

Here is an example trellis plot for the interpretation of a designed experiment where all the explanatory variables are categorical. It uses bwplot to illustrate the results of a three-way analysis of variance (p. 479).

```
data<-read.table("c:\\temp\\daphnia.txt",header=T)
attach(data)
names(data)
```

```
[1]  "Growth.rate"  "Water"  "Detergent"  "Daphnia"
```

```
library(lattice)
trellis.par.set(col.whitebg())
bwplot(Growth.rate~Water+Daphnia|Detergent)
```

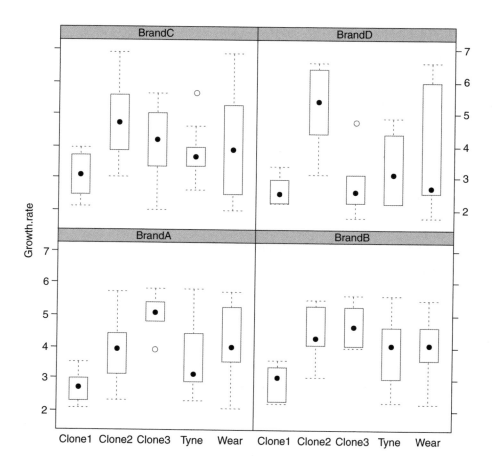

Design plots

An effective way of visualizing effect sizes in designed experiments is the plot.design function which is used just like a model formula:

plot.design(Growth.rate~Water*Detergent*Daphnia)

This shows the main effects of the three factors, drawing attention to the major differences between the daphnia clones and the small differences between the detergent brands A, B and C. The default (as here) is to plot means, but other functions can be specified such as median, var or sd. Here are the standard deviations for the different factor levels

plot.design(Growth.rate~Water*Detergent*Daphnia,fun="sd")

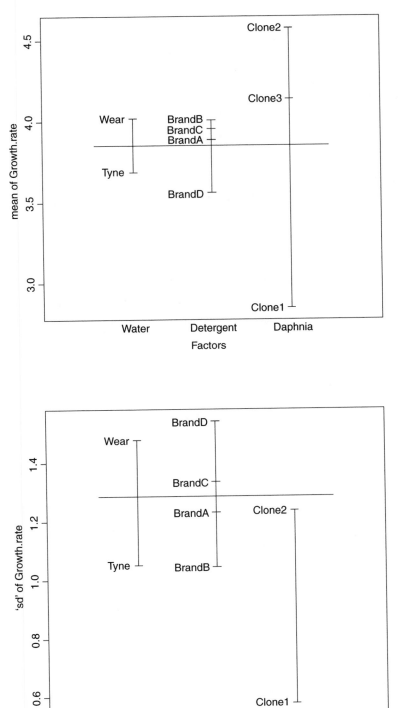

Effect sizes

An alternative is to use the **effects** package which takes a model object (a linear model or a generalized linear model) and provides trellis plots of specified effects

```
install.packages("effects")
library(effects)
model<-lm(Growth.rate~Water*Detergent*Daphnia)
```

First calculate all the effects using the **all.effects** function, then plot this object, specifying the interaction you want to see, using double quotes:

```
daph.effects<-all.effects(model)
plot(daph.effects,"Water:Detergent:Daphnia")
```

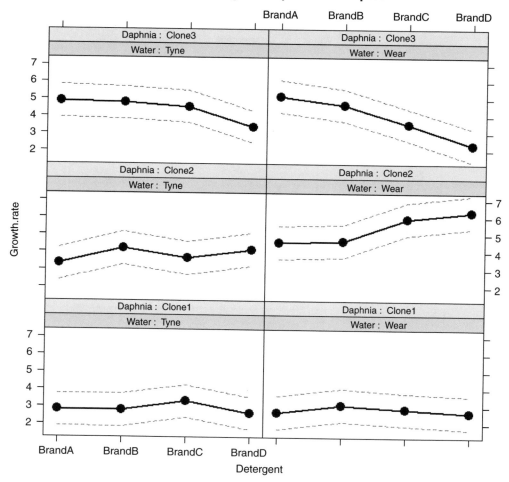

Bubble plots

The bubble plot is useful for illustrating variation in a third variable across different locations in the x–y plane. Here are data on grass yields ate different combinations of biomass and soil pH:

```
ddd<-read.table("c:\\temp\\pgr.txt",header=T)
attach(ddd)
names(ddd)
```

```
[1]  "FR"  "hay"  "pH"
```

```
bubble.plot(hay,pH,FR)
```

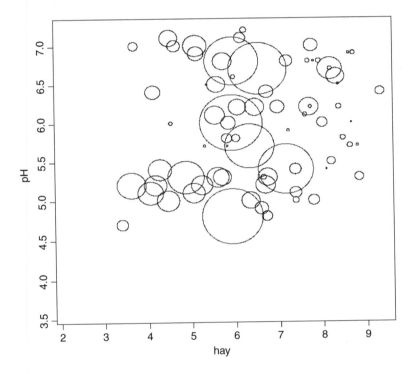

In the vicinity of hay = 6 and pH = 6 *Festuca rubra* shows one very high value, four intermediate values, two low values and one very low value. Evidently, hay crop and soil pH are not the only factors determining the abundance of *F. rubra* in this experiment. Here is a simple function for drawing bubble plots (see also p. 853):

```
bubble.plot<-function(xv,yv,rv,bs=0.1){
r<-rv/max(rv)
yscale<-max(yv)-min(yv)
xscale<-max(xv)-min(xv)

plot(xv,yv,type="n", xlab=deparse(substitute(xv)),
          ylab=deparse(substitute(yv)))
```

```
for (i in 1:length(xv)) bubble(xv[i],yv[i],r[i],bs,xscale,yscale) }

bubble<-function (x,y,r,bubble.size,xscale,yscale) {
theta<-seq(0,2*pi,pi/200)
yv<-r*sin(theta)*bubble.size*yscale
xv<-r*cos(theta)* bubble.size*xscale
lines(x+xv,y+yv) }
```

Plots with many identical values

Sometimes, especially with count data, it happens that two or more points fall in exactly
the same location in a scatterplot. In such a case, the repeated values of *y* are hidden, one
buried beneath the other, and you might want to indicate the number of cases represented
at each point on the scatterplot. The function to do this is called **sunflowerplot**, so-called
because it produces one 'petal' of a flower for each value of *y* (if there is more than one)
that is located at that particular point. Here it is in action:

```
numbers<-read.table("c:\\temp\\longdata.txt",header=T)
attach(numbers)
names(numbers)
```

```
[1]  "xlong"  "ylong"
```

```
sunflowerplot(xlong,ylong)
```

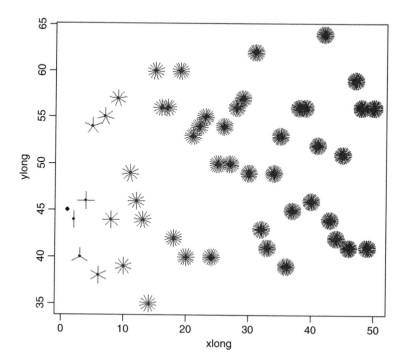

As you can see, the replication at each point increases as *x* increases from 1 on the left
to 50 on the right. The petals stop being particularly informative once there are more than

about 20 of them (about half way along the x axis). Single values (as on the extreme left) are shown without any petals, while two points in the same place have two petals. As an option, you can specify two vectors containing the unique values of x and y with a third vector containing the frequency of each combination (the number of repeats of each value).

Summary

It is worth restating the really important things about plotting:

- **Plots**: plot(x,y) gives a scatterplot if x is continuous, and a box-and-whisker plot if x is a factor. Some people prefer the alternative syntax plot(y~x) using 'tilde' as in a model formula.

- **Type** of plot: Options include lines type="l" or null (axes only) type="n".

- **Lines**: lines(x,y) plots a smooth function of y against x using the x and y values provided. You might prefer lines(y~x).

- **Line types**: Useful with multiple line plots, lty=2 (an option in plot or lines).

- **Points**: points(x,y) adds another set of data points to a plot. You might prefer points(y~x).

- **Plotting characters** for different data sets: pch=2 or pch="*" (an option in points or plot)

- Setting non-default limits to the x or y axis scales uses xlim=c(0,25) and/or ylim=c(0,1) as an option in plot.

6

Tables

The alternative to using graphics is to summarize your data in tabular form. Broadly speaking, if you want to convey *detail* use a table, and if you want to show *effects* then use graphics. You are more likely to want to use a table to summarize data when your explanatory variables are categorical (such as people's names, or different commodities) than when they are continuous (in which case a scatterplot is likely to be more informative; see p. 135).

Summary Tables

The most important function in R for generating summary tables is the somewhat obscurely named tapply function. It is called tapply because it applies a named function (such as mean or variance) across specified margins (factor levels) to create a table. If you have used the PivotTable function in Excel you will be familiar with the concept.

Here is tapply in action:

```
data<-read.table("c:\\temp\\Daphnia.txt",header=T)
attach(data)
names(data)
```

```
[1] "Growth.rate" "Water" "Detergent"  "Daphnia"
```

The response variable is growth rate of the animals, and there are three categorical explanatory variables: the river from which the water was sampled, the kind of detergent experimentally added, and the clone of daphnia employed in the experiment. In the simplest case we might want to tabulate the mean growth rates for the four brands of detergent tested,

```
tapply(Growth.rate,Detergent,mean)
```

```
  BrandA     BrandB     BrandC     BrandD
3.884832   4.010044   3.954512   3.558231
```

or for the two rivers,

```
tapply(Growth.rate,Water,mean)
```

```
    Tyne       Wear
3.685862   4.017948
```

or for the three daphnia clones,

tapply(Growth.rate,Daphnia,mean)

```
   Clone1      Clone2      Clone3
2.839875   4.577121   4.138719
```

Two-dimension summary tables are created by replacing the single explanatory variable (the second argument in the function call) by a list indicating which variable is to be used for the rows of the summary table and which variable is to be used for creating the columns of the summary table. To get the daphnia clones as the rows and detergents as the columns, we write list(Daphnia,Detergent) – rows first then columns – and use tapply to create the summary table as follows:

tapply(Growth.rate,list(Daphnia,Detergent),mean)

```
         BrandA      BrandB     BrandC      BrandD
Clone1  2.732227  2.929140  3.071335  2.626797
Clone2  3.919002  4.402931  4.772805  5.213745
Clone3  5.003268  4.698062  4.019397  2.834151
```

If we wanted the median values (rather than the means), then we would just alter the third argument of the tapply function like this:

tapply(Growth.rate,list(Daphnia,Detergent),median)

```
         BrandA      BrandB     BrandC      BrandD
Clone1  2.705995  3.012495  3.073964  2.503468
Clone2  3.924411  4.282181  4.612801  5.416785
Clone3  5.057594  4.627812  4.040108  2.573003
```

To obtain a table of the standard errors of the means (where each mean is based on 6 numbers -2 replicates and 3 rivers) the function we want to apply is $\sqrt{s^2/n}$. There is no built-in function for the standard error of a mean, so we create what is known as an **anonymous function** inside the tapply function with function(x)sqrt(var(x)/length(x)):

tapply(Growth.rate,list(Daphnia,Detergent), function(x)sqrt(var(x)/length(x)))

```
          BrandA        BrandB       BrandC        BrandD
Clone1  0.2163448  0.2319320  0.3055929  0.1905771
Clone2  0.4702855  0.3639819  0.5773096  0.5520220
Clone3  0.2688604  0.2683660  0.5395750  0.4260212
```

When tapply is asked to produce a three-dimensional table, it produces a stack of two-dimensional tables, the number of stacked tables being determined by the number of levels of the categorical variable that comes third in the list (Water in this case):

tapply(Growth.rate,list(Daphnia,Detergent,Water),mean)

```
, ,Tyne
         BrandA      BrandB     BrandC      BrandD
Clone1  2.811265  2.775903  3.287529  2.597192
Clone2  3.307634  4.191188  3.620532  4.105651
Clone3  4.866524  4.766258  4.534902  3.365766
```

```
, ,Wear
          BrandA      BrandB      BrandC      BrandD
Clone1  2.653189   3.082377   2.855142   2.656403
Clone2  4.530371   4.614673   5.925078   6.321838
Clone3  5.140011   4.629867   3.503892   2.302537
```

In cases like this, the function ftable (which stands for 'flat table') often produces more pleasing output:

ftable(tapply(Growth.rate,list(Daphnia,Detergent,Water),mean))

```
                     Tyne        Wear
Clone1   BrandA   2.811265   2.653189
         BrandB   2.775903   3.082377
         BrandC   3.287529   2.855142
         BrandD   2.597192   2.656403

Clone2   BrandA   3.307634   4.530371
         BrandB   4.191188   4.614673
         BrandC   3.620532   5.925078
         BrandD   4.105651   6.321838

Clone3   BrandA   4.866524   5.140011
         BrandB   4.766258   4.629867
         BrandC   4.534902   3.503892
         BrandD   3.365766   2.302537
```

Notice that the order of the rows, columns or tables is determined by the alphabetical sequence of the factor levels (e.g. Tyne comes before Wear in the alphabet). If you want to override this, you must specify that the factor is.ordered in a non-standard way:

water<-factor(Water,levels=c("Wear","Tyne"),ordered=is.ordered(Water))

Now the summary statistics for the Wear appear in the left-hand column of output:

ftable(tapply(Growth.rate,list(Daphnia,Detergent,water),mean))

```
                     Wear        Tyne
Clone1   BrandA   2.653189   2.811265
         BrandB   3.082377   2.775903
         BrandC   2.855142   3.287529
         BrandD   2.656403   2.597192

Clone2   BrandA   4.530371   3.307634
         BrandB   4.614673   4.191188
         BrandC   5.925078   3.620532
         BrandD   6.321838   4.105651

Clone3   BrandA   5.140011   4.866524
         BrandB   4.629867   4.766258
         BrandC   3.503892   4.534902
         BrandD   2.302537   3.365766
```

The function to be applied in generating the table can be supplied with extra arguments:

tapply(Growth.rate,Detergent,mean,trim=0.1)

```
  BrandA      BrandB      BrandC      BrandD
3.874869   4.019206   3.890448   3.482322
```

An extra argument is essential if you want means when there are missing values:

tapply(Growth.rate,Detergent,mean,na.rm=T)

You can use tapply to create new, abbreviated dataframes comprising summary parameters estimated from larger dataframe. Here, for instance, is a dataframe of mean growth rate classified by detergent and daphina clone (i.e. averaged over river water and replicates). The trick is to convert the factors to numbers before using tapply, then using these numbers to extract the relevant levels from the original factors:

dets<-as.vector(tapply(as.numeric(Detergent),list(Detergent,Daphnia),mean))
levels(Detergent)[dets]

```
[1]  "BrandA"  "BrandB"  "BrandC"  "BrandD"  "BrandA"  "BrandB"
     "BrandC"  "BrandD"
[9]  "BrandA"  "BrandB"  "BrandC"  "BrandD"
```

clones<-as.vector(tapply(as.numeric(Daphnia),list(Detergent,Daphnia),mean))
levels(Daphnia)[clones]

```
[1]  "Clone1"  "Clone1"  "Clone1"  "Clone1"  "Clone2"  "Clone2"
     "Clone2"  "Clone2"
[9]  "Clone3"  "Clone3"  "Clone3"  "Clone3"
```

You will see that these vectors of factor levels are the correct length for the new reduced dataframe (12, rather than the original length 72). The 12 mean values

tapply(Growth.rate,list(Detergent,Daphnia),mean)

```
              Clone1      Clone2      Clone3
BrandA   2.732227    3.919002    5.003268
BrandB   2.929140    4.402931    4.698062
BrandC   3.071335    4.772805    4.019397
BrandD   2.626797    5.213745    2.834151
```

can now be converted into a vector called means, and the three new vectors combined into a dataframe:

means<-as.vector(tapply(Growth.rate,list(Detergent,Daphnia),mean))
detergent<-levels(Detergent)[dets]
daphnia<-levels(Daphnia)[clones]
data.frame(means,detergent,daphnia)

```
        means    detergent   daphnia
1    2.732227       BrandA    Clone1
2    2.929140       BrandB    Clone1
3    3.071335       BrandC    Clone1
4    2.626797       BrandD    Clone1
5    3.919002       BrandA    Clone2
6    4.402931       BrandB    Clone2
7    4.772805       BrandC    Clone2
8    5.213745       BrandD    Clone2
9    5.003268       BrandA    Clone3
10   4.698062       BrandB    Clone3
11   4.019397       BrandC    Clone3
12   2.834151       BrandD    Clone3
```

The same result can be obtained using the as.data.frame.table function

```
as.data.frame.table(tapply(Growth.rate,list(Detergent,Daphnia),mean))
```

	Var1	Var2	Freq
1	BrandA	Clone1	2.732227
2	BrandB	Clone1	2.929140
3	BrandC	Clone1	3.071335
4	BrandD	Clone1	2.626797
5	BrandA	Clone2	3.919002
6	BrandB	Clone2	4.402931
7	BrandC	Clone2	4.772805
8	BrandD	Clone2	5.213745
9	BrandA	Clone3	5.003268
10	BrandB	Clone3	4.698062
11	BrandC	Clone3	4.019397
12	BrandD	Clone3	2.834151

but you would need to edit the names like this:

```
new<-as.data.frame.table(tapply(Growth.rate,list(Detergent,Daphnia),mean))
names(new)<-c("detergents","daphina","means")
```

Tables of Counts

Here are simulated data from a trial in which red blood cells were counted on 10 000 slides. The mean number of cells per slide (μ) was 1.2 and the distribution had an aggregation parameter $k = 0.63$ (known in R as size). The probability for a negative binomial distribution (prob in R) is given by $k/(\mu + k) = 0.63/1.83$ so

```
cells<-rnbinom(10000,size=0.63,prob=0.63/1.83)
```

We want to count how many times we got no red blood cells on the slide, and how often we got 1, 2, 3, ... cells. The R function for this is table:

```
table(cells)
```

```
cells
   0      1      2      3      4      5      6     7     8     9    10    11    12    13    14    15
5149   2103   1136    629    364    226    158    81    52    33    22    11    11     6     9     5
  16     17     24
   3      1      1
```

That's all there is to it. You will get slightly different values because of the randomization. We found 5149 slides with no red blood cells and one slide with a massive 24 red blood cells.

We often want to count separately for each level of a factor. Here we know that the first 5000 samples came from male patients and the second 5000 from females:

```
gender<-rep(c("male","female"),c(5000,5000))
```

To tabulate the counts separately for the two sexes we just write

```
table(cells,gender)
```

	gender	
cells	female	male
0	2646	2503
1	1039	1064
2	537	599
3	298	331
4	165	199
5	114	112
6	82	76
7	43	38
8	30	22
9	16	17
10	7	15
11	5	6
12	5	6
13	6	0
14	3	6
15	3	2
16	0	3
17	1	0
24	0	1

Evidently there are no major differences between the sexes in these red blood counts. A statistical comparison of the two sets of counts involves the use of log-linear models (see p. 556). Here, we need only note that the slightly higher mean count for males is not statistically significant ($p = 0.061$ in a GLM with quasi-Poisson errors):

tapply(cells,gender,mean)

```
female    male
1.1562  1.2272
```

Expanding a Table into a Dataframe

For the purposes of model-fitting, we often want to expand a table of explanatory variables to create a dataframe with as many repeated rows as specified by a count. Here are the data:

```
count.table<-read.table("c:\\temp \\tabledata.txt",header=T)
attach(count.table)
names(count.table)
```

```
[1]  "count"  "sex"  "age"  "condition"
```

count.table

	count	sex	age	condition
1	12	male	young	healthy
2	7	male	old	healthy
3	9	female	young	healthy
4	8	female	old	healthy
5	6	male	young	parasitized
6	7	male	old	parasitized
7	8	female	young	parasitized
8	5	female	old	parasitized

The idea is to create a new dataframe with a separate row for each case. That is to say, we want 12 copies of the first row (for healthy young males), seven copies of the second row (for healthy old males), and so on. The trick is to use lapply to apply the repeat function rep to each variable in count.table such that each row is repeated by the number of times specified in the vector called count:

lapply(count.table,function(x)rep(x, count.table$count))

Then we convert this object from a list to a dataframe using as.data.frame like this:

dbtable<-as.data.frame(lapply(count.table,function(x) rep(x, count.table$count)))

To tidy up, we probably want to remove the redundant vector of counts:

dbtable<-dbtable[,-1]
dbtable

```
        sex      age     condition
1      male    young      healthy
2      male    young      healthy
3      male    young      healthy
4      male    young      healthy
5      male    young      healthy
6      male    young      healthy
7      male    young      healthy
8      male    young      healthy
9      male    young      healthy
10     male    young      healthy
11     male    young      healthy
12     male    young      healthy
13     male      old      healthy
14     male      old      healthy
15     male      old      healthy
16     male      old      healthy
...
60   female      old   parasitized
61   female      old   parasitized
62   female      old   parasitized
```

Now we can use the contents of dbtable as explanatory variables in modelling other responses of each of the 62 cases (e.g. the animals' body weights).

Converting from a Dataframe to a Table

The reverse procedure of creating a table from a dataframe is much more straightforward, and involves nothing more than the table function:

table(dbtable)

```
, , condition = healthy

        age
sex        old   young
  female    8       9
    male    7      12
```

```
,  , condition = parasitized
      age
sex          old   young
  female      5       8
  male        7       6
```

You might want this tabulated object itself to be another dataframe, in which case use

as.data.frame(table(dbtable))

```
         sex      age     condition   Freq
1    female     old       healthy      8
2      male     old       healthy      7
3    female   young       healthy      9
4      male   young       healthy     12
5    female     old   parasitized      5
6      male     old   parasitized      7
7    female   young   parasitized      8
8      male   young   parasitized      6
```

You will see that R has invented the variable name `Freq` for the counts of the various contingencies. To change this to 'count' use names with the appropriate subscript [4]:

frame<-as.data.frame(table(dbtable))
names(frame)[4]<-"count"
frame

```
         sex      age     condition   count
1    female     old       healthy      8
2      male     old       healthy      7
3    female   young       healthy      9
4      male   young       healthy     12
5    female     old   parasitized      5
6      male     old   parasitized      7
7    female   young   parasitized      8
8      male   young   parasitized      6
```

Calculating tables of proportions

The *margins* of a table (the row totals or the column totals) are often useful for calculating proportions instead of counts. Here is a data matrix called counts:

counts<-matrix(c(2,2,4,3,1,4,2,0,1,5,3,3),nrow=4)
counts

```
        [,1]  [,2]  [,3]
[1,]     2     1     1
[2,]     2     4     5
[3,]     4     2     3
[4,]     3     0     3
```

The proportions will be different when they are expressed as a fraction of the row totals or as a fraction of the column totals. You need to remember that the row subscripts come first, which is why margin number 1 refers to the row totals:

prop.table(counts,1)

```
              [,1]          [,2]          [,3]
[1,]    0.5000000    0.2500000    0.2500000
[2,]    0.1818182    0.3636364    0.4545455
[3,]    0.4444444    0.2222222    0.3333333
[4,]    0.5000000    0.0000000    0.5000000
```

The column totals are the second margin, so to express the counts as proportions of the relevant column total use:

prop.table(counts,2)

```
              [,1]          [,2]             [,3]
[1,]    0.1818182    0.1428571    0.08333333
[2,]    0.1818182    0.5714286    0.41666667
[3,]    0.3636364    0.2857143    0.25000000
[4,]    0.2727273    0.0000000    0.25000000
```

To check that the column proportions sum to one, use colSums like this:

colSums(prop.table(counts,2))

```
[1] 1 1 1
```

If you want the proportions expressed as a fraction of the grand total sum(counts), then simply omit the margin number:

prop.table(counts)

```
              [,1]            [,2]             [,3]
[1,]    0.06666667    0.03333333    0.03333333
[2,]    0.06666667    0.13333333    0.16666667
[3,]    0.13333333    0.06666667    0.10000000
[4,]    0.10000000    0.00000000    0.10000000
```

sum(prop.table(counts))

```
[1] 1
```

In any particular case, you need to choose carefully whether it makes sense to express your counts as proportions of the row totals, column totals or grand total.

The scale function

For a numeric matrix, you might want to scale the values within a column so that they have a mean of 0. You might also want to know the standard deviation of the values within each column. These two actions are carried out simultaneously with the scale function:

scale(counts)

```
              [,1]          [,2]             [,3]
[1,]    -0.7833495    -0.439155    -1.224745
[2,]    -0.7833495     1.317465     1.224745
[3,]     1.3055824     0.146385     0.000000
[4,]     0.2611165    -1.024695     0.000000
```

attr(,"scaled:center")
```
[1]   2.75   1.75   3.00
```

```
attr(,"scaled:scale")
[1]   0.9574271   1.7078251   1.6329932
```

The values in the table are the counts minus the column means of the counts. The means of the columns – `attr(,"scaled:center")` – are 2.75, 1.75 and 3.0, while the standard deviations of the columns – `attr(,"scaled:scale")` – are 0.96, 1.71 and 1.63. To check that the scales are the standard deviations (sd) of the counts within a column, you could use apply to the columns (margin = 2) like this:

```
apply(counts,2,sd)
```

```
[1]   0.9574271   1.7078251   1.6329932
```

The expand.grid function

This is a useful function for generating tables of combinations of factor levels. Suppose we have three variables: height with five levels between 60 and 80 in steps of 5, weight with five levels between 100 and 300 in steps of 50, and two sexes.

```
expand.grid(height = seq(60, 80, 5), weight = seq(100, 300, 50),
sex = c("Male","Female"))
```

```
     height   weight      sex
1        60      100     Male
2        65      100     Male
3        70      100     Male
4        75      100     Male
5        80      100     Male
6        60      150     Male
7        65      150     Male
8        70      150     Male
9        75      150     Male
10       80      150     Male
11       60      200     Male
...
47       65      300   Female
48       70      300   Female
49       75      300   Female
50       80      300   Female
```

The model.matrix function

Creating tables of dummy variables for use in statistical modelling is extremely easy with the model.matrix function. You will see what the function does with a simple example. Suppose that our dataframe contains a factor called parasite indicating the identity of a gut parasite. The variable called parasite has five levels: *vulgaris, kochii, splendens, viridis* and *knowlesii*. Note that there was no header row in the data file, so the variable name parasite had to be added subsequently, using names:

```
data<-read.table("c:\\temp \\parasites.txt")
names(data)<-"parasite"
attach(data)
```

In our modelling we want to create a two-level dummy variable (present/absent) for each parasite species, so that we can ask questions such as whether the mean value of the response variable is significantly different in cases where *vulgaris* is present and when it is absent. The long-winded way of doing this is to create a new factor for each species:

```
vulgaris<-factor(1*(parasite=="vulgaris"))
kochii<-factor(1*(parasite=="kochii"))
```

and so on, with 1 for TRUE (present) and 0 for FALSE (absent). This is how easy it is to do with model.matrix:

```
model.matrix(~parasite-1)
```

	parasite kochii	parasiteknowlesii	parasitesplendens	parasiteviridis
1	0	0	0	0
2	0	0	1	0
3	0	1	0	0
4	0	0	0	0
5	0	1	0	0
6	0	0	0	1
7	0	0	1	0
8	0	0	1	0
9	0	0	0	1
10	0	0	0	0
11	0	0	1	0
12	0	0	0	1
13	0	0	1	0

The -1 in the model formula ensures that we create a dummy variable for each of the five parasite species (technically, it suppresses the creation of an intercept). Now we can join these five columns of dummy variables to the dataframe containing the response variable and the other explanatory variables,

```
new.frame<-data.frame(original.frame, model.matrix(~parasite-1))
attach(new.frame)
```

after which we can use variable names like parasiteknowlesii in statistical modelling.

7

Mathematics

You can do a lot of maths in R. Here we concentrate on the kinds of mathematics that find most frequent application in scientific work and statistical modelling:

- functions;
- continuous distributions;
- discrete distributions;
- matrix algebra;
- calculus;
- differential equations.

Mathematical Functions

For the kinds of functions you will meet in statistical computing there are only three mathematical rules that you need to learn: these are concerned with powers, exponents and logarithms. In the expression x^b the explanatory variable is raised to the **power** b. In e^x the explanatory variable appears as a power – in this special case, of $e = 2.71828$, of which x is the **exponent**. The inverse of e^x is the **logarithm** of x, denoted by $\log(x)$ – note that all our logs are to the base e and that, for us, writing $\log(x)$ is the same as $\ln(x)$.

It is also useful to remember a handful of mathematical facts that are useful for working out **behaviour at the limits**. We would like to know what happens to y when x gets very large (e.g. $x \to \infty$) and what happens to y when x goes to 0 (i.e. what the intercept is, if there is one). These are the most important rules:

- Anything to the power zero is 1: $x^0 = 1$.
- One raised to any power is still 1: $1^x = 1$.
- Infinity plus 1 is infinity: $\infty + 1 = \infty$.
- One over infinity (the reciprocal of infinity, ∞^{-1}) is zero: $\frac{1}{\infty} = 0$.
- A number bigger than 1 raised to the power infinity is infinity: $1.2^\infty = \infty$.

The R Book   Michael J. Crawley
© 2007 John Wiley & Sons, Ltd

- A fraction (e.g. 0.99) raised to the power infinity is zero: $0.99^\infty = 0$.

- Negative powers are reciprocals: $x^{-b} = \dfrac{1}{x^b}$.

- Fractional powers are roots: $x^{1/3} = \sqrt[3]{x}$.

- The base of natural logarithms, e, is 2.718 28, so $e^\infty = \infty$.

- Last, but perhaps most usefully: $e^{-\infty} = \dfrac{1}{e^\infty} = \dfrac{1}{\infty} = 0$.

There are built-in functions in R for logarithmic, probability and trigonometric functions (p. 11).

Logarithmic functions

The logarithmic function is given by

$$y = a \ln(bx).$$

Here the logarithm is to base e. The exponential function, in which the response y is the antilogarithm of the continuous explanatory variable x, is given by

$$y = ae^{bx}.$$

Both these functions are smooth functions, and to draw smooth functions in R you need to generate a series of 100 or more regularly spaced x values between min(x) and max(x):

x<-seq(0,10,0.1)

In R the exponential function is exp and the natural log function (ln) is log. Let $a = b = 1$. To plot the exponential and logarithmic functions with these values together in a row, write

y<-exp(x)
plot(y~x,type="l",main="Exponential")

y<-log(x)
plot(y~x,type="l",main="Logarithmic")

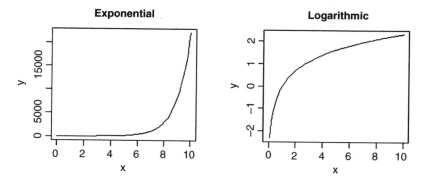

Note that the plot function can be used in an alternative way, specifying the Cartesian coordinates of the line using plot(x,y) rather than the formula plot(y~x) (see p. 181).

These functions are most useful in modelling process of exponential growth and decay.

Trigonometric functions

Here are the cosine (base/hypotenuse), sine (perpendicular/hypotenuse) and tangent (per-pendicular/base) functions of x (measured in radians) over the range 0 to 2π. Recall that the full circle is 2π radians, so 1 radian $= 360/2\pi = 57.295\,78$ degrees.

```
x<-seq(0,2*pi,2*pi/100)
y1<-cos(x)
y2<-sin(x)
plot(y1~x,type="l",main="cosine")
plot(y2~x,type="l",main="sine")
y3<-tan(x)
plot(y3~x,type="l",ylim=c(-3,3),main="tangent")
```

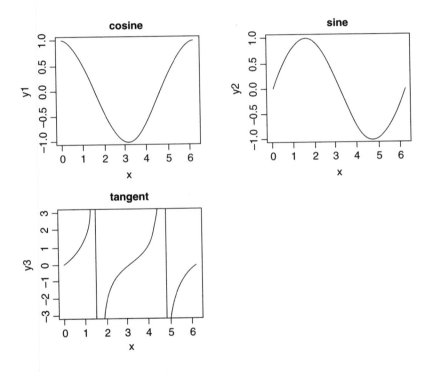

The tangent of x has discontinuities, shooting off to positive infinity at $x = \pi/2$ and again at $x = 3\pi/2$. Restricting the range of values plotted on the y axis (here from -3 to $+3$) therefore gives a better picture of the shape of the tan function. Note that R joins the plus infinity and minus infinity 'points' with a straight line at $x = \pi/2$ and at $x = 3\pi/2$ within the frame of the graph defined by ylim.

Power laws

There is an important family of two-parameter mathematical functions of the form

$$y = ax^b$$

known as **power laws**. Depending on the value of the power, b, the relationship can take one of five forms. In the trivial case of $b = 0$ the function is $y = a$ (a horizontal straight line). The four more interesting shapes are as follows:

```
x<-seq(0,1,0.01)
y<-x^0.5
plot(x,y,type="l",main="0<b<1")
y<-x
plot(x,y,type="l",main="b=1")
y<-x^2
plot(x,y,type="l",main="b>1")
y<-1/x
plot(x,y,type="l",main="b<0")
```

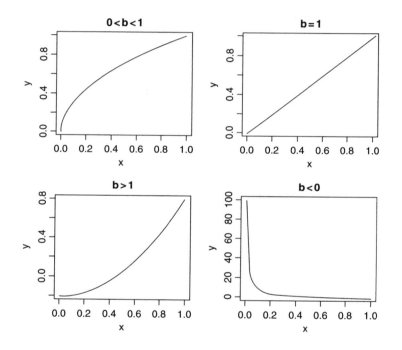

These functions are useful in a wide range of disciplines. The parameters a and b are easy to estimate from data because the function is linearized by a log-log transformation,

$$\log(y) = \log(ax^b) = \log(a) + b\log(x),$$

so that on log-log axes the intercept is $\log(a)$ and the slope is b. These are often called **allometric** relationships because when $b \neq 1$ the proportion of x that becomes y varies with x.

An important empirical relationship from ecological entomology that has applications in a wide range of statistical analysis is known as **Taylor's power law**. It has to do with the relationship between the variance and the mean of a sample. In elementary statistical models, the variance is assumed to be constant (i.e. the variance does not depend upon the mean). In field data, however, Taylor found that variance increased with the mean according to a power law, such that on log-log axes the data from most systems fell above a line through the origin with slope $= 1$ (the pattern shown by data that are Poisson distributed, where the variance is equal to the mean) and below a line through the origin with a slope of 2. Taylor's power law states that, for a particular system:

- log(variance) is a linear function of log(mean);

- the scatter about this straight line is small;

- the slope of the regression of log(variance) against log(mean) is greater than 1 and less than 2;

- the parameter values of the log-log regression are fundamental characteristics of the system.

Polynomial functions

Polynomial functions are functions in which x appears several times, each time raised to a different power. They are useful for describing curves with humps, inflections or local maxima like these:

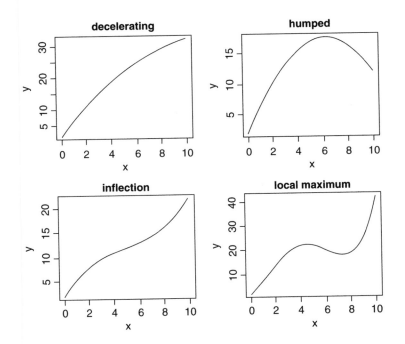

The top left-hand panel shows a decelerating positive function, modelled by the quadratic

```
x<-seq(0,10,0.1)
y1<-2+5*x-0.2*x^2
```

Making the negative coefficient of the x^2 term larger produces a curve with a hump as in the top right-hand panel:

```
y2<-2+5*x-0.4*x^2
```

Cubic polynomials can show points of inflection, as in the lower left-hand panel:

```
y3<-2+4*x-0.6*x^2+0.04*x^3
```

Finally, polynomials containing powers of 4 are capable of producing curves with local maxima, as in the lower right-hand panel:

```
y4<-2+4*x+2*x^2-0.6*x^3+0.04*x^4
par(mfrow=c(2,2)
plot(x,y1,type="l",ylab="y",main="decelerating")
plot(x,y2,type="l",ylab="y",main="humped")
plot(x,y3,type="l",ylab="y",main="inflection")
plot(x,y4,type="l",ylab="y",main="local maximum")
```

Inverse polynomials are an important class of functions which are suitable for setting up generalized linear models with gamma errors and inverse link functions:

$$\frac{1}{y} = a + bx + cx^2 + dx^3 + \cdots + zx^n.$$

Various shapes of function are produced, depending on the order of the polynomial (the maximum power) and the signs of the parameters:

```
par(mfrow=c(2,2))
y1<-x/(2+5*x)
y2<-1/(x-2+4/x)
y3<-1/(x^2-2+4/x)
plot(x,y1,type="l",ylab="y",main="Michaelis-Menten")
plot(x,y2,type="l",ylab="y",main="shallow hump")
plot(x,y3,type="l",ylab="y",main="steep hump")
```

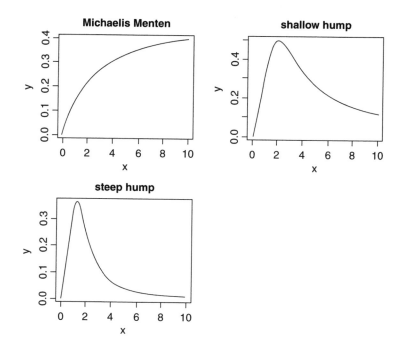

There are two ways of parameterizing the Michaelis–Menten equation:

$$y = \frac{ax}{1+bx} \quad \text{and} \quad y = \frac{x}{c+dx}.$$

In the first case, the asymptotic value of y is a/b and in the second it is $1/d$.

Gamma function

The gamma function $\Gamma(t)$ is an extension of the factorial function, $t!$, to positive real numbers:

$$\Gamma(t) = \int_0^\infty x^{t-1}e^{-x}dx.$$

It looks like this:

```
t<-seq(0.2,4,0.01)
plot(t,gamma(t),type="l")
abline(h=1,lty=2)
```

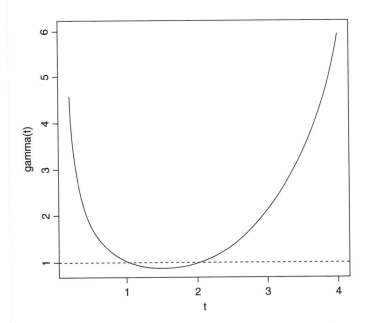

Note that $\Gamma(t)$ is equal to 1 at both $t=1$ and $t=2$. For integer values of t, $\Gamma(t+1)=t!$, and.

Asymptotic functions

Much the most commonly used asymptotic function is

$$y = \frac{ax}{1+bx},$$

which has a different name in almost every scientific discipline. For example, in biochemistry it is called Michaelis–Menten, and shows reaction rate as a function of enzyme concentration; in ecology it is called Holling's disc equation and shows predator feeding rate as a function of prey density. The graph passes through the origin and rises with diminishing returns to an asymptotic value at which increasing the value of x does not lead to any further increase in y.

The other common function is the asymptotic exponential

$$y = a(1 - e^{-bx}).$$

This, too, is a two-parameter model, and in many cases the two functions would describe data equally well (see p. 664 for an example of this comparison).

Let's work out the behaviour at the limits of our two asymptotic functions, starting with the asymptotic exponential. For $x = 0$ we have

$$y = a(1 - e^{-b \times 0}) = a(1 - e^0) = a(1 - 1) = a \times 0 = 0,$$

so the graph goes through the origin. At the other extreme, for $x = \infty$, we have

$$y = a(1 - e^{-b \times \infty}) = a(1 - e^{-\infty}) = a(1 - 0) = a(1) = a,$$

which demonstrates that the relationship is asymptotic, and that the asymptotic value of y is a.

For the Michaelis–Menten equation, determining the behaviour at the limits is somewhat more difficult, because for $x = \infty$ we end up with $y = \infty/\infty$ which you might imagine is always going to be 1 no matter what the values of a and b. In fact, there is a special mathematical rule for this case, called l'Hospital's rule: when you get a ratio of infinity to infinity, you work out the ratio of the derivatives to obtain the behaviour at the limit. For $x = 0$ the limit is easy:

$$y = \frac{a \times 0}{1 + b \times 0} = \frac{0}{1 + 0} = \frac{0}{1} = 0.$$

For $x = \infty$ we get $y = \infty/(1 + \infty) = \infty/\infty$. The numerator is ax so its derivative with respect to x is a. The denominator is $1 + bx$ so its derivative with respect to x is $0 + b = b$. So the ratio of the derivatives is a/b, and this is the asymptotic value of the Michaelis–Menten equation.

Parameter estimation in asymptotic functions

There is no way of linearizing the asymptotic exponential model, so we must resort to non-linear least squares (nls) to estimate parameter values for it (p. 662). One of the advantages of the Michaelis–Menten function is that it is easy to linearize. We use the **reciprocal transformation**

$$\frac{1}{y} = \frac{1 + bx}{ax},$$

which, at first glance, isn't a big help. But we can separate the terms on the right because they have a common denominator. Then we can cancel the xs, like this:

$$\frac{1}{y} = \frac{1}{ax} + \frac{bx}{ax} = \frac{1}{ax} + \frac{b}{a}$$

so if we put $y = 1/y$, $x = 1/x$, $A = 1/a$, and $C = b/a$, we see that

$$Y = AX + C$$

which is linear: C is the intercept and A is the slope. So to estimate the values of a and b from data, we would transform both x and y to reciprocals, plot a graph of $1/y$ against $1/x$, carry out a linear regression, then back-transform, to get:

$$a = \frac{1}{A},$$
$$b = aC.$$

Suppose that we knew that the graph passed through the two points (0.2, 44.44) and (0.6, 70.59). How do we work out the values of the parameters a and b? First, we calculate the four reciprocals. The slope of the linearized function, A, is the change in $1/y$ divided by the change in $1/x$:

(1/44.44 - 1/70.59)/(1/0.2 - 1/0.6)

[1] 0.002500781

so $a = 1/A = 1/0.0025 = 400$. Now we rearrange the equation and use one of the points (say $x = 0.2$, $y = 44.44$) to get the value of b:

$$b = \frac{1}{x}\left(\frac{ax}{y} - 1\right) = \frac{1}{0.2}\left(\frac{400 \times 0.2}{44.44} - 1\right) = 4.$$

Sigmoid (S-shaped) functions

The simplest S-shaped function is the **two-parameter logistic** where, for $0 \le y \le 1$,

$$y = \frac{e^{a+bx}}{1 + e^{a+bx}}$$

which is central to the fitting of generalized linear models for proportion data (Chapter 16).
   The **three-parameter logistic** function allows y to vary on any scale:

$$y = \frac{a}{1 + be^{-cx}}.$$

The intercept is $a/(1 + b)$, the asymptotic value is a and the initial slope is measured by c. Here is the curve with parameters 100, 90 and 1.0:

```
par(mfrow=c(2,2))
x<-seq(0,10,0.1)
y<-100/(1+90*exp(-1*x))
plot(x,y,type="l",main="three-parameter logistic")
```

The **four-parameter logistic** function has asymptotes at the left-(a) and right-hand (b) ends of the x axis and scales (c) the response to x about the midpoint (d) where the curve has its inflexion:

$$y = a + \frac{b - a}{1 + e^{c(d-x)}}.$$

Letting $a = 20$, $b = 120$, $c = 0.8$ and $d = 3$, the function

$$y = 20 + \frac{100}{1 + e^{0.8 \times (3-x)}}$$

looks like this

```
y<-20+100/(1+exp(0.8*(3-x)))
plot(x,y,ylim=c(0,140),type="l",main="four-parameter logistic")
```

Negative sigmoid curves have the parameter $c < 0$ as for the function

$$y = 20 + \frac{100}{1 + e^{-0.8 \times (3-x)}}.$$

An asymmetric S-shaped curve much used in demography and life insurance work is the **Gompertz growth model**,

$$y = ae^{be^{cx}}.$$

The shape of the function depends on the signs of the parameters b and c. For a negative sigmoid, b is negative (here -1) and c is positive (here $+0.02$):

```
x<- -200:100
y<-100*exp(-exp(0.02*x))
plot(x,y,type="l",main="negative Gompertz")
```

For a positive sigmoid both parameters are negative:

```
x<- 0:100
y<- 50*exp(-5*exp(-0.08*x))
plot(x,y,type="l",main="positive Gompertz")
```

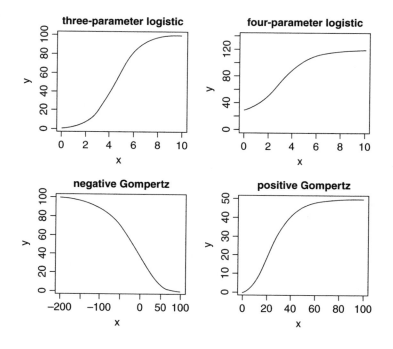

Biexponential model

This is a useful four-parameter non-linear function, which is the sum of two exponential functions of x:

$$y = ae^{bx} + ce^{dx}.$$

Various shapes depend upon the signs of the parameters b, c and d:

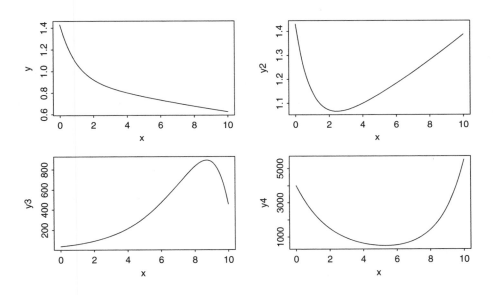

the upper left-hand panel shows c positive, b and d negative; the upper right-hand panel shows c and d positive, b negative; the lower left-hand panel shows c and d negative, b positive; and the lower right panel shows c and b negative, d positive. When b, c and d are all negative, this function is known as the **first-order compartment model** in which a drug administered at time 0 passes through the system with its dynamics affected by elimination, absorption and clearance.

Transformations of the response and explanatory variables

We have seen the use of transformation to linearize the relationship between the response and the explanatory variables:

- $\log(y)$ against x for exponential relationships;
- $\log(y)$ against $\log(x)$ for power functions;
- $\exp(y)$ against x for logarithmic relationships;
- $1/y$ against $1/x$ for asymptotic relationships;
- $\log(p/(1-p))$ against x for proportion data.

Other transformations are useful for variance stabilization:

- $\sqrt{y}$ to stabilize the variance for count data;
- arcsin(y) to stabilize the variance of percentage data.

Probability functions

There are many specific probability distributions in R (normal, Poisson, binomial, etc.), and these are discussed in detail later. Here we look at the base mathematical functions that deal with elementary probability. The **factorial** function gives the number of permutations of n items. How many ways can 4 items be arranged? The first position could have any one of the 4 items in it, but by the time we get to choosing the second item we shall already have specified the first item so there are just $4 - 1 = 3$ ways of choosing the second item. There are only $4 - 2 = 2$ ways of choosing the third item, and by the time we get to the last item we have no degrees of freedom at all: the last number must be the one item out of four that we have not used in positions 1, 2 or 3. So with 4 items the answer is $4 \times (4 - 1) \times (4 - 2) \times (4 - 3)$ which is $4 \times 3 \times 2 \times 1 = 24$. In general, factorial($n$) is given by

$$n! = n(n - 1)(n - 2) \ldots \times 3 \times 2.$$

The R function is **factorial** and we can plot it for values of x from 0 to 10 using the step option **type="s"**, in plot with a logarithmic scale on the y axis **log="y"**,

```
x<-0:6
plot(x,factorial(x),type="s",main="factorial x",log="y")
```

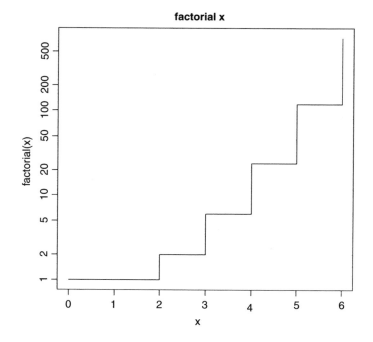

The other important base function for probability calculations in R is the choose function which calculates **binomial coefficients**. These show the number of ways there are of selecting x items out of n items when the item can be one of just two types (e.g. either male or female, black or white, solvent or insolvent). Suppose we have 8 individuals and we want to know how many ways there are that 3 of them could be males (and hence 5 of them females). The answer is given by

$$\binom{n}{x} = \frac{n!}{x!(n-x)!},$$

so with $n = 8$ and $x = 3$ we get

$$\binom{n}{x} = \frac{8!}{3!(8-3)!} = \frac{8 \times 7 \times 6}{3 \times 2} = 56$$

and in R

choose(8,3)

[1] 56

Obviously there is only one way that all 8 individuals could be male or female, so there is only one way of getting 0 or 8 'successes'. One male could be the first individual you select, or the second, or the third, and so on. So there are 8 ways of selecting 1 out of 8. By the same reasoning, there must be 8 ways of selecting 7 males out of 8 individuals (the lone female could be in any one of the 8 positions). The following is a graph of the number of ways of selecting from 0 to 8 males out of 8 individuals:

plot(0:8,choose(8,0:8),type="s",main="binomial coefficients")

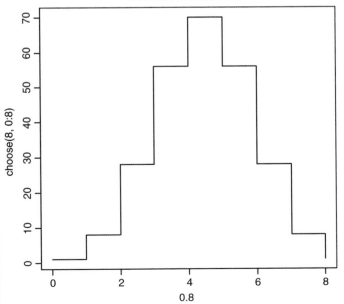

Continuous Probability Distributions

R has a wide range of built-in probability distributions, for each of which four functions are available: the probability density function (which has a **d** prefix); the cumulative probability (**p**); the quantiles of the distribution (**q**); and random numbers generated from the distribution (**r**). Each letter can be prefixed to the R function names in Table 7.1 (e.g. **dbeta**).

Table 7.1. The probability distributions supported by R. The meanings of the parameters are explained in the text.

R function	Distribution	Parameters
beta	beta	shape1, shape2
binom	binomial	sample size, probability
cauchy	Cauchy	location, scale
exp	exponential	rate (optional)
chisq	chi-squared	degrees of freedom
f	Fisher's F	df1, df2
gamma	gamma	shape
geom	geometric	probability
hyper	hypergeometric	m, n, k
lnorm	lognormal	mean, standard deviation
logis	logistic	location, scale
nbinom	negative binomial	size, probability
norm	normal	mean, standard deviation
pois	Poisson	mean
signrank	Wilcoxon signed rank statistic	sample size n
t	Student's t	degrees of freedom
unif	uniform	minimum, maximum (opt.)
weibull	Weibull	shape
wilcox	Wilcoxon rank sum	m, n

The **cumulative probability** function is a straightforward notion: it is an S-shaped curve showing, for any value of x, the probability of obtaining a sample value that is less than or equal to x. Here is what it looks like for the normal distribution:

```
curve(pnorm(x),-3,3)
arrows(-1,0,-1,pnorm(-1),col="red")
arrows(-1,pnorm(-1),-3,pnorm(-1),col="green")
```

The value of $x(-1)$ leads up to the cumulative probability (red arrow) and the probability associated with obtaining a value of this size (-1) or smaller is on the y axis (green arrow). The value on the y axis is 0.158 655 3:

```
pnorm(-1)
```

```
[1]  0.1586553
```

The **probability density** is the slope of this curve (its 'derivative'). You can see at once that the slope is never negative. The slope starts out very shallow up to about $x = -2$, increases up to a peak (at $x = 0$ in this example) then gets shallower, and becomes very small indeed above about $x = 2$. Here is what the density function of the normal (**dnorm**) looks like:

```
curve(dnorm(x),-3,3)
```

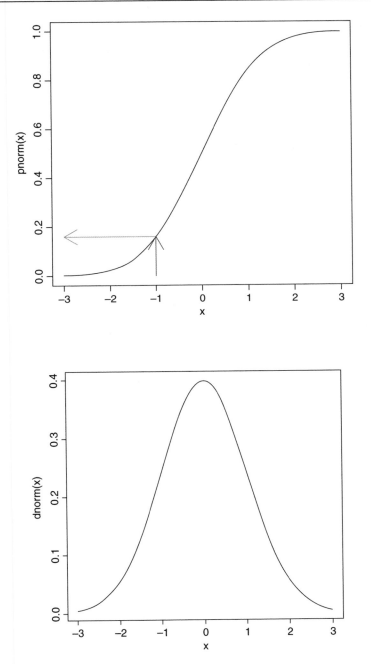

For a discrete random variable, like the Poisson or the binomial, the probability density function is straightforward: it is simply a histogram with the y axis scaled as probabilities rather than counts, and the discrete values of x (0, 1, 2, 3, ...) on the horizontal axis. But for a continuous random variable, the definition of the probability density function is more subtle: it does not have probabilities on the y axis, but rather the derivative (the slope) of the cumulative probability function at a given value of x.

Normal distribution

This distribution is central to the theory of parametric statistics. Consider the following simple exponential function:

$$y = \exp(-|x|^m).$$

As the power (m) in the exponent increases, the function becomes more and more like a step function. The following panels show the relationship between y and x for $m = 1, 2, 3$ and 8, respectively:

```
par(mfrow=c(2,2))
x<-seq(-3,3,0.01)
y<-exp(-abs(x))
plot(x,y,type="l")
y<-exp(-abs(x)^2)
plot(x,y,type="l")
y<-exp(-abs(x)^3)
plot(x,y,type="l")
y<-exp(-abs(x)^8)
plot(x,y,type="l")
```

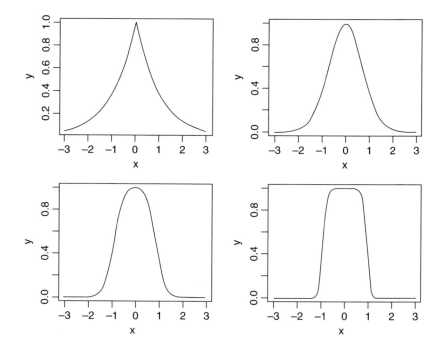

The second of these panels (top right), where $y = \exp(-x^2)$, is the basis of an extremely important and famous probability density function. Once it has been scaled, so that the integral (the area under the curve from $-\infty$ to $+\infty$) is unity, this is the normal distribution.

Unfortunately, the scaling constants are rather cumbersome. When the distribution has mean 0 and standard deviation 1 (the **standard normal** distribution) the equation becomes:

$$f(z) = \frac{1}{\sqrt{2\pi}} e^{-z^2/2}.$$

Suppose we have measured the heights of 100 people. The mean height was 170 cm and the standard deviation was 8 cm (top left panel, below). We can ask three sorts of questions about data like these: what is the probability that a randomly selected individual will be:

- shorter than a particular height?

- taller than a particular height?

- between one specified height and another?

The area under the whole curve is exactly 1; everybody has a height between minus infinity and plus infinity. True, but not particularly helpful. Suppose we want to know the probability that one of our people, selected at random from the group, will be less than 160 cm tall. We need to convert this height into a value of z; that is to say, we need to convert 160 cm into *a number of standard deviations from the mean*. What do we know about the standard normal distribution? It has a mean of 0 and a standard deviation of 1. So we can convert any value y, from a distribution with mean $\bar{y}$ and standard deviation s very simply by calculating:

$$z = \frac{y - \bar{y}}{s}.$$

So we convert 160 cm into a number of standard deviations. It is less than the mean height (170 cm) so its value will be negative:

$$z = \frac{160 - 170}{8} = -1.25.$$

Now we need to find the probability of a value of the standard normal taking a value of -1.25 or smaller. This is *the area under the left hand tail* (the integral) of the density function. The function we need for this is pnorm: we provide it with a value of z (or, more generally, with a quantile) and it provides us with the probability we want:

pnorm(-1.25)

[1] 0.1056498

So the answer to our first question (the red area, top right) is just over 10%.

Next, what is the probability of selecting one of our people and finding that they are taller than 185 cm (bottom left)? The first two parts of the exercise are exactly the same as before. First we convert our value of 185 cm into a number of standard deviations:

$$z = \frac{185 - 170}{8} = 1.875.$$

Then we ask what probability is associated with this, using pnorm:

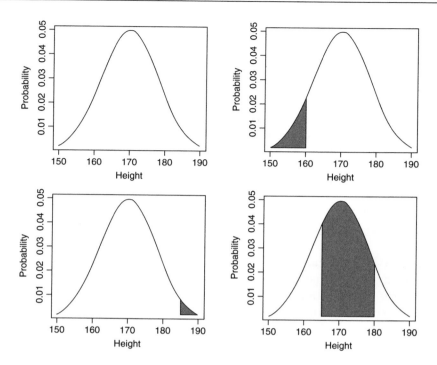

pnorm(1.875)

```
[1]  0.9696036
```

But this is the answer to a different question. This is the probability that someone will be *less* than or equal to 185 cm tall (that is what the function pnorm has been written to provide). All we need to do is to work out the complement of this:

1-pnorm(1.875)

```
[1]  0.03039636
```

So the answer to the second question is about 3%.

Finally, we might want to know the probability of selecting a person between 165 cm and 180 cm. We have a bit more work to do here, because we need to calculate two z values:

$$z_1 = \frac{165 - 170}{8} = -0.625 \quad \text{and} \quad z_2 = \frac{(180 - 170)}{8} = 1.25.$$

The important point to grasp is this: we want the probability of selecting a person between these two z values, so we *subtract the smaller probability from the larger probability*:

pnorm(1.25)-pnorm(-0.625)

```
[1]  0.6283647
```

Thus we have a 63% chance of selecting a medium-sized person (taller than 165 cm and shorter than 180 cm) from this sample with a mean height of 170 cm and a standard deviation of 8 cm (bottom right, above).

The central limit theorem

If you take repeated samples from a population with finite variance and calculate their averages, then the averages will be normally distributed. This is called the **central limit theorem**. Let's demonstrate it for ourselves. We can take five uniformly distributed random numbers between 0 and 10 and work out the average. The average will be low when we get, say, 2,3,1,2,1 and big when we get 9,8,9,6,8. Typically, of course, the average will be close to 5. Let's do this 10 000 times and look at the distribution of the 10 000 means. The data are rectangularly (uniformly) distributed on the interval 0 to 10, so the distribution of the raw data should be flat-topped:

```
hist(runif(10000)*10,main="")
```

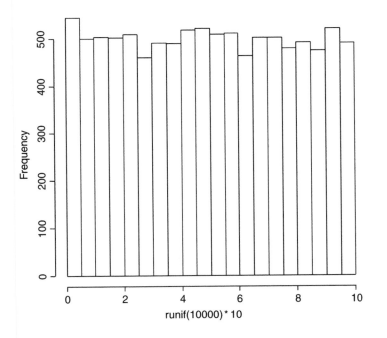

What about the distribution of sample means, based on taking just 5 uniformly distributed random numbers?

```
means<-numeric(10000)
for (i in 1:10000){
means[i]<-mean(runif(5)*10)
}
hist(means,ylim=c(0,1600))
```

Nice, but how close is this to a normal distribution? One test is to draw a normal distribution with the same parameters on top of the histogram. But what are these parameters? The normal is a two-parameter distribution that is characterized by its mean and its standard deviation. We can estimate these two parameters from our sample of 10 000 means (your values will be slightly different because of the randomization):

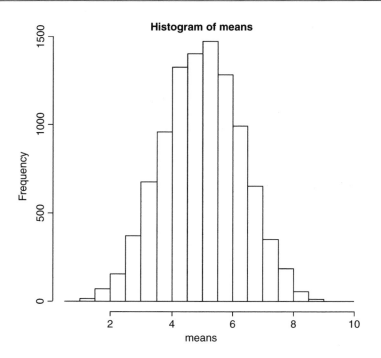

mean(means)

```
[1] 4.998581
```

sd(means)

```
[1] 1.289960
```

Now we use these two parameters in the probability density function of the normal distribution (dnorm) to create a normal curve with our particular mean and standard deviation. To draw the smooth line of the normal curve, we need to generate a series of values for the x axis; inspection of the histograms suggest that sensible limits would be from 0 to 10 (the limits we chose for our uniformly distributed random numbers). A good rule of thumb is that for a smooth curve you need at least 100 values, so let's try this:

xv<-seq(0,10,0.1)

There is just one thing left to do. The probability density function has an integral of 1.0 (that's the area beneath the normal curve), but we had 10 000 samples. To scale the normal probability density function to our particular case, however, depends on the height of the highest bar (about 1500 in this case). The height, in turn, depends on the chosen bin widths; if we doubled with width of the bin there would be roughly twice as many numbers in the bin and the bar would be twice as high on the y axis. To get the height of the bars on our frequency scale, therefore, we multiply the total frequency, 10 000 by the bin width, 0.5 to get 5000. We multiply 5000 by the probability density to get the height of the curve. Finally, we use lines to overlay the smooth curve on our histogram:

yv<-dnorm(xv,mean=4.998581,sd=1.28996)*5000
lines(xv,yv)

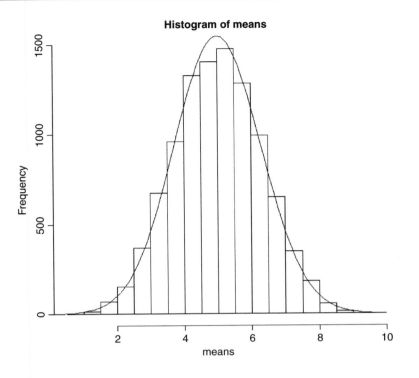

Histogram of means

The fit is excellent. The central limit theorem really works. Almost any distribution, even a 'badly behaved' one like the uniform distribution we worked with here, will produce a normal distribution of sample means taken from it.

A simple example of the operation of the central limit theorem involves the use of dice. Throw one die lots of times and each of the six numbers should come up equally often: this is an example of a **uniform** distribution:

```
par(mfrow=c(2,2))
hist(sample(1:6,replace=T,10000),breaks=0.5:6.5,main="",xlab="one die")
```

Now throw two dice and add the scores together: this is the ancient game of craps. There are 11 possible scores from a minimum of 2 to a maximum of 12. The most likely score is 7 because there are 6 ways that this could come about:

$$1,6 \quad 6,1 \quad 2,5 \quad 5,2 \quad 3,4 \quad 4,3$$

For many throws of craps we get a **triangular** distribution of scores, centred on 7:

```
a<-sample(1:6,replace=T,10000)
b<-sample(1:6,replace=T,10000)
hist(a+b,breaks=1.5:12.5,main="", xlab="two dice")
```

There is already a clear indication of central tendency and spread. For three dice we get

```
c<-sample(1:6,replace=T,10000)
hist(a+b+c,breaks=2.5:18.5,main="", xlab="three dice")
```

and the bell shape of the normal distribution is starting to emerge. By the time we get to five dice, the **binomial** distribution is virtually indistinguishable from the normal:

```
d<-sample(1:6,replace=T,10000)
e<-sample(1:6,replace=T,10000)
hist(a+b+c+d+e,breaks=4.5:30.5,main="", xlab="five dice")
```

The smooth curve is given by a normal distribution with the same mean and standard deviation:

```
mean(a+b+c+d+e)
```

```
[1] 17.5937
```

```
sd(a+b+c+d+e)
```

```
[1] 3.837668
```

```
lines(seq(1,30,0.1),dnorm(seq(1,30,0.1),17.5937,3.837668)*10000)
```

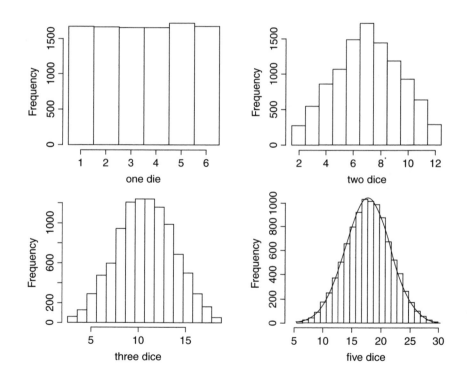

Maximum likelihood with the normal distribution

The probability density of the normal is

$$f(y|\mu, \sigma) = \frac{1}{\sigma\sqrt{2\pi}} \exp\left[-\frac{(y-\mu)^2}{2\sigma^2}\right],$$

which is read as saying the probability of getting a data value y, given ($|$) a mean of μ and a variance of σ^2, is calculated from this rather complicated-looking two-parameter exponential function. For any given combination of μ and σ^2, it gives a value between 0 and 1. Recall that likelihood is the product of the probability densities, for each of the values of the response variable, y. So if we have n values of y in our experiment, the likelihood function is

$$L(\mu, \sigma) = \prod_{i=1}^{n} \left(\frac{1}{\sigma\sqrt{2\pi}} \exp\left[-\frac{(y_i - \mu)^2}{2\sigma^2} \right] \right),$$

where the only change is that y has been replaced by y_i and we multiply together the probabilities for each of the n data points. There is a little bit of algebra we can do to simplify this: we can get rid of the product operator, Π, in two steps. First, for the constant term: that, multiplied by itself n times, can just be written as $1/(\sigma\sqrt{2\pi})^n$. Second, remember that the product of a set of antilogs (exp) can be written as the antilog of a sum of the values of x_i like this: $\prod \exp(x_i) = \exp(\sum x_i)$. This means that the product of the right-hand part of the expression can be written as

$$\exp\left[-\frac{\sum_{i=1}^{n} (y_i - \mu)^2}{2\sigma^2} \right]$$

so we can rewrite the likelihood of the normal distribution as

$$L(\mu, \sigma) = \frac{1}{\left(\sigma\sqrt{2\pi}\right)^n} \exp\left[-\frac{1}{2\sigma^2} \sum_{i=1}^{n} (y_i - \mu)^2 \right].$$

The two parameters μ and σ are unknown, and the purpose of the exercise is to use statistical modelling to determine their maximum likelihood values from the data (the n different values of y). So how do we find the values of μ and σ that maximize this likelihood? The answer involves calculus: first we find the derivative of the function with respect to the parameters, then set it to zero, and solve.

It turns out that because of the exp function in the equation, it is easier to work out the log of the likelihood,

$$l(\mu, \sigma) = -\frac{n}{2} \log(2\pi) - n \log(\sigma) - \sum (y_i - \mu)^2 / 2\sigma^2,$$

and maximize this instead. Obviously, the values of the parameters that maximize the log-likelihood $l(\mu, \sigma) = \log(L(\mu, \sigma))$ will be the same as those that maximize the likelihood. From now on, we shall assume that summation is over the index i from 1 to n.

Now for the calculus. We start with the mean, μ. The derivative of the log-likelihood with respect to μ is

$$\frac{dl}{d\mu} = \sum (y_i - \mu)/\sigma^2.$$

Set the derivative to zero and solve for μ:

$$\sum (y_i - \mu)/\sigma^2 = 0 \quad \text{so} \quad \sum (y_i - \mu) = 0.$$

Taking the summation through the bracket, and noting that $\sum \mu = n\mu$,

$$\sum y_i - n\mu = 0 \quad \text{so} \quad \sum y_i = n\mu \quad \text{and} \quad \mu = \frac{\sum y_i}{n}.$$

The maximum likelihood estimate of μ is the arithmetic mean.

Next we find the derivative of the log-likelihood with respect to σ:

$$\frac{\mathrm{d}l}{\mathrm{d}\sigma} = -\frac{n}{\sigma} + \frac{\sum (y_i - \mu)^2}{\sigma^3},$$

recalling that the derivative of $\log(x)$ is $1/x$ and the derivative of $-1/x^2$ is $2/x^3$. Solving, we get

$$-\frac{n}{\sigma} + \frac{\sum (y_i - \mu)^2}{\sigma^3} = 0 \quad \text{so} \quad \sum (y_i - \mu)^2 = \sigma^3 \left(\frac{n}{\sigma}\right) = \sigma^2 n$$

$$\sigma^2 = \frac{\sum (y_i - \mu)^2}{n}.$$

The maximum likelihood estimate of the variance σ^2 is the mean squared deviation of the y values from the mean. This is a biased estimate of the variance, however, because it does not take account of the fact that we estimated the value of μ from the data. To unbias the estimate, we need to lose 1 degree of freedom to reflect this fact, and divide the sum of squares by $n - 1$ rather than by n (see p. 52 and restricted maximum likelihood estimators in Chapter 19).

Here, we illustrate R's built-in probability functions in the context of the normal distribution. The density function dnorm has a value of z (a quantile) as its argument. Optional arguments specify the mean and standard deviation (default is the standard normal with mean 0 and standard deviation 1). Values of z outside the range -3.5 to $+3.5$ are very unlikely.

```
par(mfrow=c(2,2))
curve(dnorm,-3,3,xlab="z",ylab="Probability density",main="Density")
```

The probability function pnorm also has a value of z (a quantile) as its argument. Optional arguments specify the mean and standard deviation (default is the standard normal with mean 0 and standard deviation 1). It shows the cumulative probability of a value of z less than or equal to the value specified, and is an S-shaped curve:

```
curve(pnorm,-3,3,xlab="z",ylab="Probability",main="Probability")
```

Quantiles of the normal distribution qnorm have a cumulative probability as their argument. They perform the opposite function of pnorm, returning a value of z when provided with a probability.

```
curve(qnorm,0,1,xlab="p",ylab="Quantile (z)",main="Quantiles")
```

The normal distribution random number generator rnorm produces random real numbers from a distribution with specified mean and standard deviation. The first argument is the number of numbers that you want to be generated: here are 1000 random numbers with mean 0 and standard deviation 1:

```
y<-rnorm(1000)
hist(y,xlab="z",ylab="frequency",main="Random numbers")
```

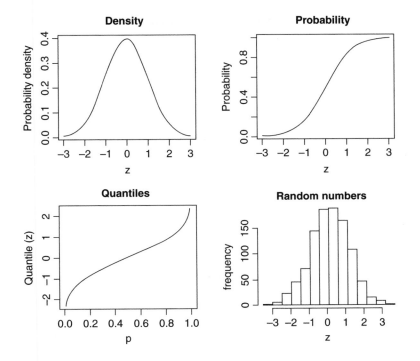

The four functions (d, p, q and r) work in similar ways with all the other probability distributions.

Generating random numbers with exact mean standard deviation

If you use a random number generator like rnorm then, naturally, the sample you generate will not have exactly the mean and standard deviation that you specify, and two runs will produce vectors with different means and standard deviations. Suppose we want 100 normal random numbers with a mean of exactly 24 and a standard deviation of precisely 4:

```
yvals<-rnorm(100,24,4)
mean(yvals)
```

```
[1] 24.2958
```

```
sd(yvals)
```

```
[1] 3.5725
```

Close, but not spot on. If you want to generate random numbers with an exact mean and standard deviation, then do the following:

```
ydevs<-rnorm(100,0,1)
```

Now compensate for the fact that the mean is not exactly 0 and the standard deviation is not exactly 1 by expressing all the values as departures from the sample mean scaled in units of the sample's standard deviations:

```
ydevs<-(ydevs-mean(ydevs))/sd(ydevs)
```

Check that the mean is zero and the standard deviation is exactly 1:

```
mean(ydevs)
```

```
[1] -2.449430e-17
```

```
sd(ydevs)
```

```
[1]  1
```

The mean is as close to zero as makes no difference, and the standard deviation is one. Now multiply this vector by your desired standard deviation and add to your desired mean value to get a sample with exactly the means and standard deviation required:

```
yvals<-24 + ydevs*4
mean(yvals)
```

```
[1] 24
```

```
sd(yvals)
```

```
[1]  4
```

Comparing data with a normal distribution

Various tests for normality are described on p. 281. Here we are concerned with the task of comparing a histogram of real data with a smooth normal distribution with the same mean and standard deviation, in order to look for evidence of non-normality (e.g. skew or kurtosis).

```
par(mfrow=c(1,1))
fishes<-read.table("c:\\temp\\fishes.txt",header=T)
attach(fishes)
names(fishes)
```

```
[1] "mass"
```

```
mean(mass)
```

```
[1]  4.194275
```

```
max(mass)
```

```
[1]  15.53216
```

Now the histogram of the mass of the fish is produced, specifying integer bins that are 1 gram in width, up to a maximum of 16.5 g:

```
hist(mass,breaks=-0.5:16.5,col="green",main="")
```

For the purposes of demonstration, we generate everything we need *inside* the lines function: the sequence of x values for plotting (0 to 16), and the height of the density function (the number of fish (length(mass)) times the probability density for each member of this sequence, for a normal distribution with mean(mass) and standard deviation sqrt(var(mass)) as its parameters, like this:

```
lines(seq(0,16,0.1),length(mass)*dnorm(seq(0,16,0.1),mean(mass),sqrt(var(mass))))
```

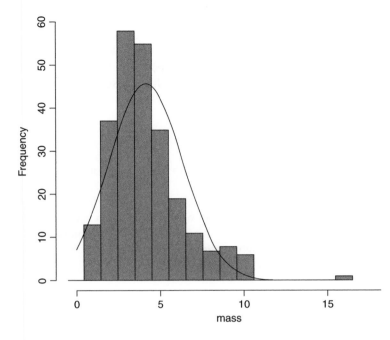

The distribution of fish sizes is clearly *not* normal. There are far too many fishes of 3 and 4 grams, too few of 6 or 7 grams, and too many really big fish (more than 8 grams). This kind of skewed distribution is probably better described by a gamma distribution than a normal distribution (see p. 231).

Other distributions used in hypothesis testing

The main distributions used in hypothesis testing are: the **chi-squared**, for testing hypotheses involving count data; **Fisher's F**, in analysis of variance (ANOVA) for comparing two variances; and **Student's t**, in small-sample work for comparing two parameter estimates. These distributions tell us the size of the test statistic that could be expected by chance alone when nothing was happening (i.e. when the null hypothesis was true). Given the rule that a big value of the test statistic tells us that something *is* happening, and hence that the null hypothesis is false, these distributions define what constitutes a big value of the test statistic.

For instance, if we are doing a chi-squared test, and our test statistic is 14.3 on 9 d.f., we need to know whether this is a large value (meaning the null hypothesis is false) or a small value (meaning the null hypothesis is accepted, or at least cannot be rejected). In the old days we would have looked up the value in chi-squared tables. We would have looked in the row labelled 9 (the degrees of freedom row) and the column headed by $\alpha = 0.05$. This is the conventional value for the acceptable probability of committing a Type I error: that is to say we allow a 1 in 20 chance of rejecting the null hypothesis when it is actually true; see p. 317). Nowadays, we just type:

1-pchisq(14.3,9)

[1] 0.1120467

This indicates that 14.3 is actually a relatively small number when we have 9 d.f. We would conclude that nothing is happening, because a value of chi-squared as large as 14.3 has a greater than an 11% probability of arising by chance alone. We would want the probability to be less than 5% before we rejected the null hypothesis. So how large would the test statistic need to be, before we would reject the null hypothesis? We use qchisq to answer this. Its two arguments are $1 - \alpha$ and the number of degrees of freedom:

qchisq(0.95,9)

[1] 16.91898

So the test statistic would need to be larger than 16.92 in order for us to reject the null hypothesis when there were 9 d.f.

We could use pf and qf in an exactly analogous manner for Fisher's F. Thus, the probability of getting a variance ratio of 2.85 by chance alone when the null hypothesis is true, given that we have 8 d.f. in the numerator and 12 d.f. in the denominator, is just under 5% (i.e. the value is just large enough to allow us to reject the null hypothesis):

1-pf(2.85,8,12)

[1] 0.04992133

Note that with pf, degrees of freedom in the numerator (8) come first in the list of arguments, followed by d.f. in the denominator (12).

Similarly, with Student's t statistics and pt and qt. For instance, the value of t in tables for a two-tailed test at $\alpha/2 = 0.025$ with d.f. $= 10$ is

qt(0.975,10)

[1] 2.228139

chi-squared

Perhaps the best known of all the statistical distributions, introduced to generations of school children in their geography lessons, and comprehensively misunderstood thereafter. It is a special case of the gamma distribution (p. 229) characterized by a single parameter, the number of degrees of freedom. The mean is equal to the degrees of freedom ν ('nu', pronounced 'new'), and the variance is equal to 2ν. The density function looks like this:

$$f(x) = \frac{1}{2^{\nu/2}\Gamma(\nu/2)} x^{\nu/2-1} e^{-x/2},$$

where Γ is the gamma function (see p. 201). The chi-squared is important because many quadratic forms follow the chi-squared distribution under the assumption that the data follow the normal distribution. In particular, the sample variance is a scaled chi-squared variable. Likelihood ratio statistics are also approximately distributed as a chi-squared (see the F distribution, below).

When the cumulative probability is used, an optional third argument can be provided to describe non-centrality. If the non-central chi-squared is the sum of ν independent normal random variables, then the non-centrality parameter is equal to the sum of the squared means of the normal variables. Here are the cumulative probability plots for a non-centrality parameter (ncp) based on three normal means (of 1, 1.5 and 2) and another with 4 means and ncp $= 10$:

```
par(mfrow=c(1,2))
x<-seq(0,30,.25)
plot(x,pchisq(x,3,7.25),type="l",ylab="p(x)",xlab="x")
plot(x,pchisq(x,5,10),type="l",ylab="p(x)",xlab="x")
```

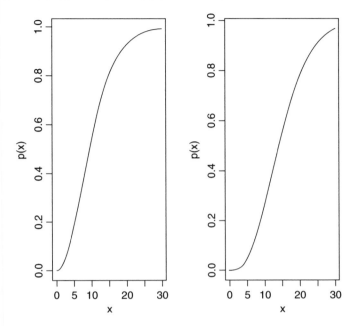

The cumulative probability on the left has 3 d.f. and non-centrality parameter $(1^2 + 1.5^2 + 2^2 = 7.25)$, while the distribution on the right has 4 d.f. and non-centrality 10 (note the longer left-hand tail at low probabilities).

chi-squared is also used to establish confidence intervals for sample variances. The quantity

$$\frac{(n-1)s^2}{\sigma^2}$$

is the degrees of freedom $(n-1)$ multiplied by the ratio of the sample variance s^2 to the unknown population variance σ^2. This follows a chi-squared distribution, so we can establish a 95% confidence interval for σ^2 as follows:

$$\frac{(n-1)s^2}{\chi^2_{1-\alpha/2}} \leq \sigma^2 \leq \frac{(n-1)s^2}{\chi^2_{\alpha/2}}$$

Suppose the sample variance $s^2 = 10.2$ on 8 d.f. Then the interval on σ^2 is given by

`8*10.2/qchisq(.975,8)`

`[1] 4.65367`

`8*10.2/qchisq(.025,8)`

`[1] 37.43582`

which means that we can be 95% confident that the population variance lies in the range $4.65 \leq \sigma^2 \leq 37.44$

Fisher's F

This is the famous variance ratio test that occupies the penultimate column of every ANOVA table. The ratio of treatment variance to error variance follows the F distribution, and you will often want to use the quantile qf to look up critical values of F. You specify, in order, the probability of your one-tailed test (this will usually be 0.95), then the two degrees of freedom: numerator first, then denominator. So the 95% value of F with 2 and 18 d.f. is

```
qf(.95,2,18)
```

```
[1] 3.554557
```

This is what the density function of F looks like for 2 and 18 d.f. (left) and 6 and 18 d.f. (right):

```
x<-seq(0.05,4,0.05)
plot(x,df(x,2,18),type="l",ylab="f(x)",xlab="x")
plot(x,df(x,6,18),type="l",ylab="f(x)",xlab="x")
```

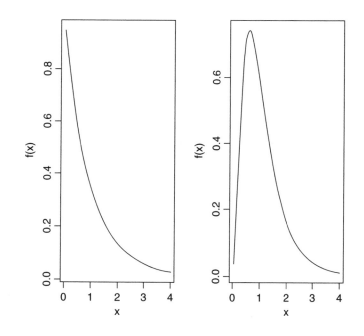

The F distribution is a two-parameter distribution defined by the density function

$$f(x) = \frac{r\Gamma(1/2(r+s))}{s\Gamma(1/2r)\Gamma(1/2s)} \frac{(rx/s)^{(r-1)/2}}{[1+(rx/s)]^{(r+s)/2}}$$

where r is the degrees of freedom in the numerator and s is the degrees of freedom in the denominator. The distribution is named after R.A. Fisher, the father of analysis of variance, and principal developer of quantitative genetics. It is central to hypothesis testing, because of its use in *assessing the significance of the differences between two variances*. The test statistic is calculated by dividing the larger variance by the smaller variance. The two variances are significantly different when this ratio is larger than the critical value of Fisher's F. The degrees of freedom in the numerator and in the denominator allow the calculation of the critical value of the test statistic. When there is a single degree of freedom in the numerator, the distribution is equal to the square of Student's t: $F = t^2$. Thus, while the rule of thumb for the critical value of t is 2, so the rule of thumb for $F = t^2 = 4$. To see how well the rule of thumb works, we can plot critical F against d.f. in the numerator:

```
df<-seq(1,30,.1)
plot(df,qf(.95,df,30),type="l",ylab="Critical F")
lines(df,qf(.95,df,10),lty=2)
```

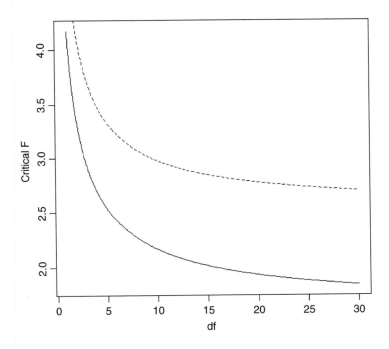

You see that the rule of thumb (critical $F = 4$) quickly becomes much too large once the d.f. in the numerator (on the x axis) is larger than 2. The lower (solid) line shows the critical values of F when the denominator has 30 d.f. and the upper (dashed) line shows the case in which the denominator has 10 d.f.

The shape of the density function of the F distribution depends on the degrees of freedom in the numerator.

```
x<-seq(0.01,3,0.01)
plot(x,df(x,1,10),type="l",ylim=c(0,1),ylab="f(x)")
lines(x,df(x,2,10),lty=6)
lines(x,df(x,5,10),lty=2)
lines(x,df(x,30,10),lty=3)
```

The probability density $f(x)$ declines monotonically when the numerator has 1 d.f. or 2 d.f., but rises to a maximum for d.f. of 3 or more (5 and 30 are shown here): all the graphs have 10 d.f. in the denominator.

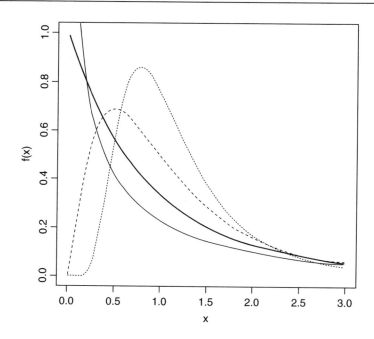

Student's t

This famous distribution was first published by W.S. Gossett in 1908 under the pseudonym of 'Student' because his then employer, the Guinness brewing company in Dublin, would not permit employees to publish under their own names. It is a model with one parameter, r, with density function:

$$f(x) = \frac{\Gamma\left(1/2(r+1)\right)}{(\pi r)^{1/2}\Gamma\left(1/2r\right)} \left(1 + \frac{x^2}{r}\right)^{-(r+1)/2}$$

where $-\infty < x < +\infty$. This looks very complicated, but if all the constants are stripped away, you can see just how simple the underlying structure really is

$$f(x) = \left(1 + x^2\right)^{-1/2}.$$

We can plot this for values of x from -3 to $+3$ as follows:

```
x<-seq(-3,3,0.01)
fx<-(1+x^2)^(-0.5)
plot(x,fx,type="l")
```

The main thing to notice is how fat the tails of the distribution are, compared with the normal distribution. The plethora of constants is necessary to scale the density function so that its integral is 1. If we define U as

$$U = \frac{(n-1)}{\sigma^2}s^2,$$

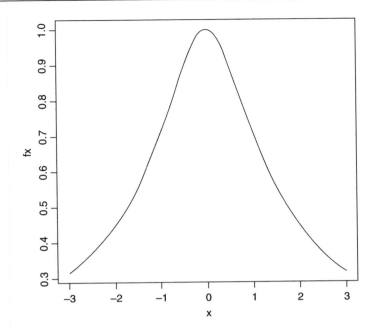

then this is chi-squared distributed on $n-1$ d.f. (see above). Now define V as

$$V = \frac{n^{1/2}}{\sigma}(\bar{y} - \mu)$$

and note that this is normally distributed with mean 0 and standard deviation 1 (the standard normal distribution), so

$$\frac{V}{(U/(n-1))^{1/2}}$$

is the ratio of a normal distribution and a chi-squared distribution. You might like to compare this with the F distribution (above), which is the ratio of two chi-squared distributed random variables.

At what point does the rule of thumb for Student's $t=2$ break down so seriously that it is actually misleading? To find this out, we need to plot the value of Student's t against sample size (actually against degrees of freedom) for small samples. We use qt (quantile of t) and fix the probability at the two-tailed value of 0.975:

```
plot(1:30,qt(0.975,1:30), ylim=c(0,12),type="l",ylab="Student"s t
value",xlab="d.f.")
```

As you see, the rule of thumb only becomes really hopeless for degrees of freedom less than about 5 or so. For most practical purposes $t \approx 2$ really is a good working rule of thumb. So what does the t distribution look like, compared to a normal? Let's redraw the standard normal as a dotted line (lty=2):

```
xvs<-seq(-4,4,0.01)
plot(xvs,dnorm(xvs),type="l",lty=2,ylab="Probability density",xlab="Deviates")
```

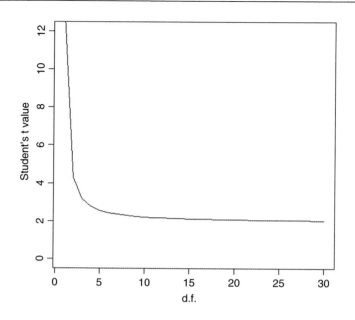

Now we can overlay Student's t with d.f. $= 5$ as a solid line to see the difference:

lines(xvs,dt(xvs,df=5))

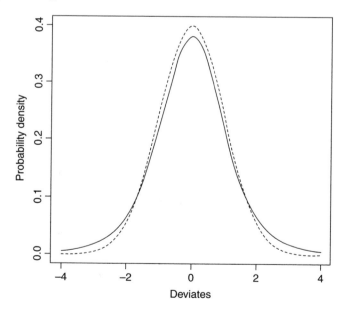

The difference between the normal (dotted) and Student's t distributions (solid line) is that the t distribution has 'fatter tails'. This means that extreme values are more likely with a t distribution than with a normal, and the confidence intervals are correspondingly broader. So instead of a 95% interval of ± 1.96 with a normal distribution we should have a 95% interval of ± 2.57 for a Student's t distribution with 5 degrees of freedom:

```
qt(0.975,5)
```

```
[1] 2.570582
```

The gamma distribution

The gamma distribution is useful for describing a wide range of processes where the data are positively skew (i.e. non-normal, with a long tail on the right). It is a two-parameter distribution, where the parameters are traditionally known as shape and rate. Its density function is:

$$f(x) = \frac{1}{\beta^\alpha \Gamma(\alpha)} x^{\alpha-1} e^{-x/\beta}$$

where α is the shape parameter and β^{-1} is the rate parameter (alternatively, β is known as the scale parameter). Special cases of the gamma distribution are the **exponential** ($\alpha = 1$) and **chi-squared** ($\alpha = v/2$, $\beta = 2$).

To see the effect of the shape parameter on the probability density, we can plot the gamma distribution for different values of shape and rate over the range 0.01 to 4:

```
x<-seq(0.01,4,.01)
par(mfrow=c(2,2))
y<-dgamma(x,.5,.5)
plot(x,y,type="l")
y<-dgamma(x,.8,.8)
plot(x,y,type="l")
y<-dgamma(x,2,2)
plot(x,y,type="l")
y<-dgamma(x,10,10)
plot(x,y,type="l")
```

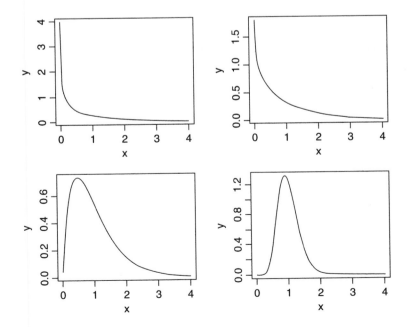

The graphs from top left to bottom right show different values of α: 0.5, 0.8, 2 and 10. Note how $\alpha < 1$ produces monotonic declining functions and $\alpha > 1$ produces humped curves that pass through the origin, with the degree of skew declining as α increases.

The mean of the distribution is $\alpha\beta$, the variance is $\alpha\beta^2$, the skewness is $2/\sqrt{\alpha}$ and the kurtosis is $6/\alpha$. Thus, for the exponential distribution we have a mean of β, a variance of β^2, a skewness of 2 and a kurtosis of 6, while for the chi-squared distribution we have a mean of ν, a variance of 2ν a skewness of $2\sqrt{2/\nu}$ and a kurtosis of $12/\nu$. Observe also that

$$\frac{1}{\beta} = \frac{\text{mean}}{\text{variance}},$$

$$shape = \frac{1}{\beta} \times \text{mean}.$$

We can now answer questions like this: what value is 95% quantile expected from a gamma distribution with mean $=2$ and variance $=3$? This implies that rate is 2/3 and shape is 4/3 so:

```
qgamma(0.95,2/3,4/3)
```

```
[1] 1.732096
```

An important use of the gamma distribution is in describing continuous measurement data that are *not* normally distributed. Here is an example where body mass data for 200 fishes are plotted as a histogram and a gamma distribution with the same mean and variance is overlaid as a smooth curve:

```
fishes<-read.table("c:\\temp\\fishes.txt",header=T)
attach(fishes)
names(fishes)
```

```
[1] "mass"
```

First, we calculate the two parameter values for the gamma distribution:

```
rate<-mean(mass)/var(mass)
shape<-rate*mean(mass)
rate
```

```
[1] 0.8775119
```

```
shape
```

```
[1] 3.680526
```

We need to know the largest value of mass, in order to make the bins for the histogram:

```
max(mass)
```

```
[1] 15.53216
```

Now we can plot the histogram, using break points at 0.5 to get integer-centred bars up to a maximum of 16.5 to accommodate our biggest fish:

```
hist(mass,breaks=-0.5:16.5,col="green",main="")
```

The density function of the gamma distribution is overlaid using lines like this:

```
lines(seq(0.01,15,0.01),length(mass)*dgamma(seq(0.01,15,0.01),shape,rate))
```

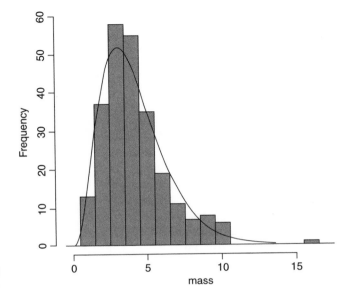

The fit is much better than when we tried to fit a normal distribution to these same data earlier (see p. 221).

The exponential distribution

This is a one-parameter distribution that is a special case of the gamma distribution. Much used in survival analysis, its density function is given on p. 792 and its use in survival analysis explained on p. 802. The random number generator of the exponential is useful for Monte Carlo simulations of time to death when the hazard (the instantaneous risk of death) is constant with age. You specify the hazard, which is the reciprocal of the mean age at death:

```
rexp(15,0.1)
 [1]   9.811752   5.738169  16.261665  13.170321   1.114943
 [6]   1.986883   5.019848   9.399658  11.382526   2.121905
[11]  10.941043   5.868017   1.019131  13.040792  38.023316
```

These are 15 random lifetimes with an expected value of $1/0.1 = 10$ years; they give a sample mean of 9.66 years.

The beta distribution

This has two positive constants, a and b, and x is bounded $0 \le x \le 1$:

$$f(x) = \frac{\Gamma(a+b)}{\Gamma(a)\Gamma(b)} x^{a-1}(1-x)^{b-1}.$$

In R we generate a family of density functions like this:

```
x<-seq(0,1,0.01)
fx<-dbeta(x,2,3)
plot(x,fx,type="l")
fx<-dbeta(x,0.5,2)
plot(x,fx,type="l")
fx<-dbeta(x,2,0.5)
plot(x,fx,type="l")
fx<-dbeta(x,0.5,0.5)
plot(x,fx,type="l")
```

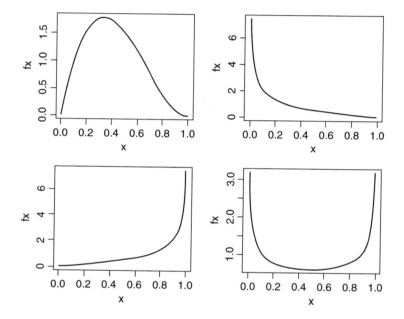

The important point is whether the parameters are greater or less than 1. When both are greater than 1 we get an n-shaped curve which becomes more skew as $b > a$ (top left). If $0 < a < 1$ and $b > 1$ then the density is negative (top right), while for $a > 1$ and $0 < b < 1$ the density is positive (bottom left). The function is U-shaped when both a and b are positive fractions. If $a = b = 1$, then we obtain the uniform distribution on [0,1].

Here are 20 random numbers from the beta distribution with shape parameters 2 and 3:

```
rbeta(20,2,3)
```

```
 [1]  0.5820844  0.5150638  0.5420181  0.1110348  0.5012057  0.3641780
 [7]  0.1133799  0.3340035  0.2802908  0.3852897  0.6496373  0.3377459
[13]  0.1743189  0.4568897  0.7343201  0.3040988  0.5670311  0.2241543
[19]  0.6358050  0.5932503
```

Cauchy

This is a long-tailed two-parameter distribution, characterized by a location parameter a and a scale parameter b. It is real-valued, symmetric about a (which is also its median), and is a curiosity in that it has long enough tails that the expectation does not exist – indeed, it has no moments at all (it often appears in counter-examples in maths books). The harmonic mean of a variable with positive density at 0 is typically distributed as Cauchy, and the Cauchy distribution also appears in the theory of Brownian motion (e.g. random walks). The general form of the distribution is

$$f(x) = \frac{1}{\pi b(1 + ((x-a)/b)^2)},$$

for $-\infty < x < \infty$. There is also a one-parameter version, with $a = 0$ and $b = 1$, which is known as the standard Cauchy distribution and is the same as Student's t distribution with one degree of freedom:

$$f(x) = \frac{1}{\pi(1 + x^2)},$$

for $-\infty < x < \infty$.

```
par(mfrow=c(1,2))
plot(-200:200,dcauchy(-200:200,0,10),type="l",ylab="p(x)",xlab="x")
plot(-200:200,dcauchy(-200:200,0,50),type="l",ylab="p(x)",xlab="x")
```

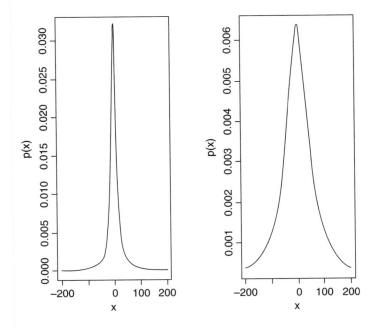

Note the very long, fat tail of the Cauchy distribution. The left-hand density function has scale $= 10$ and the right hand plot has scale $= 50$; both have location $= 0$.

The lognormal distribution

The lognormal distribution takes values on the positive real line. If the logarithm of a lognormal deviate is taken, the result is a normal deviate, hence the name. Applications for the lognormal include the distribution of particle sizes in aggregates, flood flows, concentrations of air contaminants, and failure times. The hazard function of the lognormal is increasing for small values and then decreasing. A mixture of heterogeneous items that individually have monotone hazards can create such a hazard function.

Density, cumulative probability, quantiles and random generation for the lognormal distribution employ the function dlnorm like this:

dlnorm(x, meanlog=0, sdlog=1)

The mean and standard deviation are optional, with default meanlog $=0$ and sdlog $=1$. Note that these are *not* the mean and standard deviation; the lognormal distribution has mean $e^{\mu+\sigma^2/2}$, variance $(e^{\sigma^2}-1)e^{2\mu+\sigma^2}$, skewness $(e^{\sigma^2}+2)\sqrt{e^{\sigma^2}-1}$ and kurtosis $e^{4\sigma^2}+2e^{3\sigma^2}+3e^{2\sigma^2}-6$.

```
par(mfrow=c(1,1))
plot(seq(0,10,0.05),dlnorm(seq(0,10,0.05)), type="l",xlab="x",ylab="LogNormal f(x)")
```

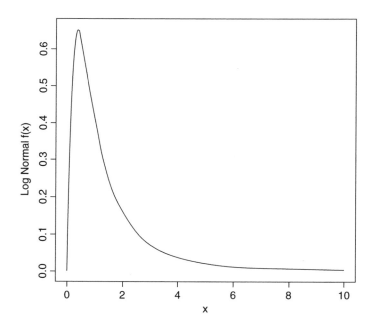

The extremely long tail and exaggerated positive skew are characteristic of the lognormal distribution. Logarithmic transformation followed by analysis with normal errors is often appropriate for data such as these.

The logistic distribution

The logistic is the canonical link function in generalized linear models with binomial errors and is described in detail in Chapter 16 on the analysis of proportion data. The cumulative probability is a symmetrical S-shaped distribution that is bounded above by 1 and below by 0. There are two ways of writing the cumulative probability equation:

$$p(x) = \frac{e^{a+bx}}{1 + e^{a+bx}}$$

and

$$p(x) = \frac{1}{1 + \beta e^{-\alpha x}}$$

The great advantage of the first form is that it linearizes under the log-odds transformation (see p. 572) so that

$$\ln\left(\frac{p}{q}\right) = a + bx,$$

where p is the probability of success and $q = (1 - p)$ is the probability of failure. The logistic is a unimodal, symmetric distribution on the real line with tails that are longer than the normal distribution. It is often used to model growth curves, but has also been used in bioassay studies and other applications. A motivation for using the logistic with growth curves is that the logistic distribution function $f(x)$ has the property that the derivative of $f(x)$ with respect to x is proportional to $[f(x) - A][B - f(x)]$ with $A < B$. The interpretation is that the rate of growth is proportional to the amount already grown, times the amount of growth that is still expected.

```
par(mfrow=c(1,2))
plot(seq(-5,5,0.02),dlogis(seq(-5,5,.02)), type="l",ylab="Logistic f(x)")
plot(seq(-5,5,0.02),dnorm(seq(-5,5,.02)), type="l",ylab="Normal f(x)")
```

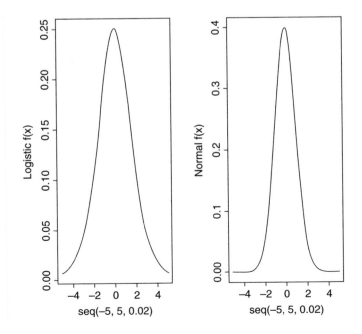

Here, the logistic density function dlogis (left) is compared with an equivalent normal density function dnorm (right) using the default mean 0 and standard deviation 1 in both cases. Note the much fatter tails of the logistic (still substantial probability at ±4 standard deviations. Note also the difference in the scales of the two y axes (0.25 for the logistic, 0.4 for the normal).

The log-logistic distribution

The log-logistic is a very flexible four-parameter model for describing growth or decay processes:

$$y = a + b \left[\frac{\exp(c(\log(x) - d))}{1 + \exp(c(\log(x) - d))} \right].$$

Here are two cases. The first is a negative sigmoid with $c = -1.59$ and $a = -1.4$:

```
x<-seq(0.1,1,0.01)
y<--1.4+2.1*(exp(-1.59*log(x)-1.53)/(1+exp(-1.59*log(x)-1.53)))
plot(log(x),y,type="l", main="c = -1.59")
```

For the second we have $c = 1.59$ and $a = 0.1$:

```
y<-0.1+2.1*(exp(1.59*log(x)-1.53)/(1+exp(1.59*log(x)-1.53)))
plot(log(x),y,type="l",main="c = -1.59")
```

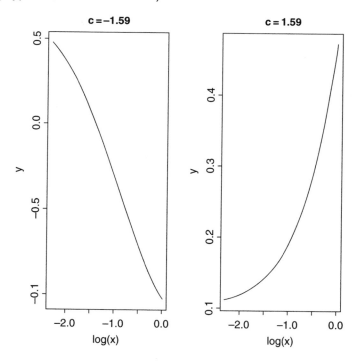

The Weibull distribution

The origin of the Weibull distribution is in *weakest link analysis*. If there are r links in a chain, and the strengths of each link Z_i are independently distributed on $(0, \infty)$, then the distribution of weakest links $V = \min(Z_j)$ approaches the Weibull distribution as the number of links increases.

The Weibull is a two-paramter model that has the exponential distribution as a special case. Its value in demographic studies and survival analysis is that it allows for the death rate to increase or to decrease with age, so that all three types of survivorship curve can be analysed (as explained on p. 802). The density, survival and hazard functions with $\lambda = \mu^{-\alpha}$ are:

$$f(t) = \alpha\lambda t^{\alpha-1}e^{-\lambda t^{\alpha}},$$

$$S(t) = e^{-\lambda t^{\alpha}},$$

$$h(t) = \frac{f(t)}{S(t)} = \alpha\lambda t^{\alpha-1}.$$

The mean of the Weibull distribution is $\Gamma(1 + \alpha^{-1})\mu$ and the variance is $\mu^2(\Gamma(1 + 2/\alpha) - (\Gamma(1 + 1/\alpha))^2)$, and the parameter α describes the shape of the hazard function (the background to determining the likelihood equations is given by Aitkin $et\ al.$ (1989, pp. 281–283). For $\alpha = 1$ (the exponential distribution) the hazard is constant, while for $\alpha > 1$ the hazard increases with age and for $\alpha < 1$ the hazard decreases with age.

Because the Weibull, lognormal and log-logistic all have positive skewness, it is difficult to discriminate between them with small samples. This is an important problem, because each distribution has differently shaped hazard functions, and it will be hard, therefore, to discriminate between different assumptions about the age-specificity of death rates. In survival studies, parsimony requires that we fit the exponential rather than the Weibull unless the shape parameter α is significantly different from 1.

Here is a family of three Weibull distributions with $\alpha = 1$, 2 and 3 (dotted, dashed and solid lines, respectively). Note that for large values of α the distribution becomes symmetrical, while for $\alpha \leq 1$ the distribution has its mode at $t = 0$.

```
a<-3
l<-1
t<-seq(0,1.8,.05)
ft<-a*l*t^(a-1)*exp(-l*t^a)
plot(t,ft,type="l")
a<-1
ft<-a*l*t^(a-1)*exp(-l*t^a)
lines(t,ft,type="l",lty=2)
a<-2
ft<-a*l*t^(a-1)*exp(-l*t^a)
lines(t,ft,type="l",lty=3)
```

Multivariate normal distribution

If you want to generate two (or more) vectors or normally distributed random numbers that are correlated with one another to a specified degree, then you need the mvrnorm function from the MASS library:

```
library(MASS)
```

Suppose we want two vectors of 1000 random numbers each. The first vector has a mean of 50 and the second has a mean of 60. The difference from rnorm is that we need to specify their covariance as well as the standard deviations of each separate variable. This is achieved with a positive-definite symmetric matrix specifying the covariance matrix of the variables.

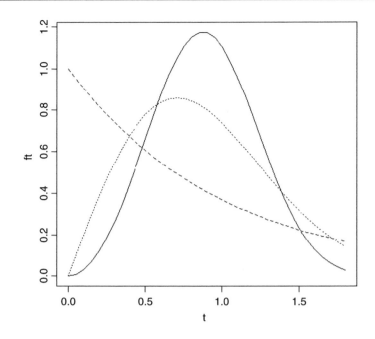

```
xy<-mvrnorm(1000,mu=c(50,60),matrix(c(4,3.7,3.7,9),2))
```

We can check how close the variances are to our specified values:

```
var(xy)
```

```
            [,1]        [,2]
[1,]   3.983063   3.831880
[2,]   3.831880   8.922865
```

Not bad: we said the covariance should be 3.70 and the simulated data are 3.83. We extract the two separate vectors x and y and plot them to look at the correlation

```
x<-xy[,1]
y<-xy[,2]
plot(x,y,pch=16,ylab="y",xlab="x")
```

It is worth looking at the variances of x and y in more detail:

```
var(x)
```

```
[1]  3.983063
```

```
var(y)
```

```
[1]  8.922865
```

If the two samples were *independent*, then the variance of the sum of the two variables would be equal to the sum of the two variances. Is this the case here?

```
var(x+y)
```

```
[1]   20.56969
```

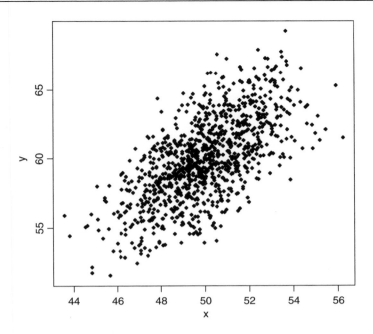

var(x)+var(y)

```
[1]  12.90593
```

No it isn't. The variance of the sum (20.57) is much greater than the sum of the variances (12.91). This is because x and y are positively correlated; big values of x tend to be associated with big values of y and vice versa. This being so, we would expect the variance of the difference between x and y to be less than the sum of the two variances:

var(x-y)

```
[1]  5.242167
```

As predicted, the variance of the difference (5.24) is much less than the sum of the variances (12.91). We conclude that *the variance of a sum of two variables is only equal to the variance of the difference of two variables when the two variables are independent.* What about the covariance of x and y? We found this already by applying the var function to the matrix xy (above). We specified that the covariance should be 3.70 in calling the multivariate normal distribution, and the difference between 3.70 and 3.831 880 is simply due to the random selection of points. The covariance is related to the separate variances through the correlation coefficient ρ as follows (see p. 310):

$$\mathrm{cov}(x, y) = \rho\sqrt{s_x^2 s_y^2}.$$

For our example, this checks out as follows, where the sample value of ρ is cor(x,y)

cor(x,y)*sqrt(var(x)*var(y))

```
[1]  3.83188
```

which is our observed covariance between x and y with $\rho = 0.642\,763\,5$.

The uniform distribution

This is the distribution that the random number generator in your calculator hopes to emulate. The idea is to generate numbers between 0 and 1 where every possible real number on this interval has exactly the same probability of being produced. If you have thought about this, it will have occurred to you that there is something wrong here. Computers produce numbers by following recipes. If you are following a recipe then the outcome is predictable. If the outcome is predictable, then how can it be random? As John von Neumann once said: 'Anyone who uses arithmetic methods to produce random numbers is in a state of sin.' This raises the question as to what, exactly, a computer-generated random number is. The answer turns out to be scientifically very interesting and very important to the study of encryption (for instance, any pseudorandom number sequence generated by a linear recursion is insecure, since from a sufficiently long sub-sequence of the outputs, one can predict the rest of the outputs). If you are interested, look up the Mersenne twister online. Here we are only concerned with how well the modern pseudo-random number generator performs. Here is the outcome of the R function runif simulating the throwing a 6-sided die 10 000 times: the histogram ought to be flat:

```
x<-ceiling(runif(10000)*6)
table(x)
```

```
x
   1     2     3     4     5     6
1620  1748  1607  1672  1691  1662
```

```
hist(x,breaks=0.5:6.5,main="")
```

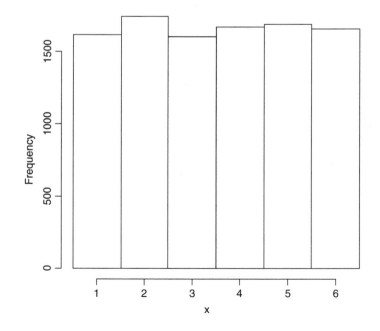

This is remarkably close to theoretical expectation, reflecting the very high efficiency of R's random-number generator. Try mapping 1 000 000 points to look for gaps:

```
x<-runif(1000000)
y<-runif(1000000)
plot(x,y,pch=16)
```

This produced an absolutely solid black map for me: there were no holes in the pattern, so there were no pairs of numbers that were not generated at this scale of resolution (pch=16). For a more thorough check we can count the frequency of combinations of numbers: with 36 cells, the expected frequency is $1\,000\,000/36 = 27\,777.78$: numbers per cell. We use the cut function to produce 36 bins:

```
table(cut(x,6),cut(y,6))
```

	(-0.001,0.166]	(0.166,0.333]	(0.333,0.5]	(0.5,0.667]	(0.667,0.834]	(0.834,1]
(-0.001,0.166]	27541	27795	27875	27851	27664	27506
(0.166,0.333]	27908	28033	27975	27859	27862	27600
(0.333,0.5]	27509	27827	27991	27689	27878	27733
(0.5,0.667]	27718	28074	27548	28062	27777	27760
(0.667,0.834]	27820	28084	27466	27753	27784	27454
(0.834,1]	27463	27997	27982	27685	27571	27906

As you can see the observed frequencies are remarkably close to expectation.

Plotting empirical cumulative distribution functions

The function ecdf is used to compute or plot an empirical cumulative distribution function. Here it is in action for the fishes data (p. 220 and 230):

```
fishes<-read.table("c:\\temp\\fishes.txt",header=T)
attach(fishes)
names(fishes)

[1]  "mass"

plot(ecdf(mass))
```

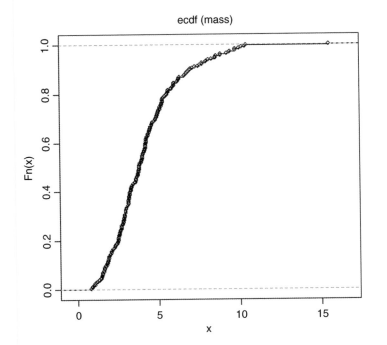

The pronounced positive skew in the data is evident from the fact that the left-hand side of the cumulative distribution is much steeper than the right-hand side (and see p. 230).

Discrete probability distributions

The Bernoulli distribution

This is the distribution underlying tests with a binary response variable. The response takes one of only two values: it is 1 with probability p (a 'success') and is 0 with probability $1 - p$ (a 'failure'). The density function is given by:

$$p(X) = p^x(1-p)^{1-x}$$

The statistician's definition of **variance** is the expectation of x^2 minus the square of the expectation of x: $\sigma^2 = E(X^2) - [E(X)]^2$. We can see how this works with a simple distribution like the Bernoulli. There are just two outcomes in $f(x)$: a success, where $x = 1$ with probability p and a failure, where $x = 0$ with probability $1 - p$. Thus, the expectation of x is

$$E(X) = \sum xf(x) = 0 \times (1-p) + 1 \times p = 0 + p = p$$

and expectation of x^2 is

$$E(X^2) = \sum x^2 f(x) = 0^2 \times (1-p) + 1^2 \times p = 0 + p = p,$$

so the variance of the Bernoulli is

$$\text{var}(X) = E(X^2) - [E(X)]^2 = p - p^2 = p(1-p) = pq.$$

The binomial distribution

This is a one-parameter distribution in which p describes the probability of success in a binary trial. The probability of x successes out of n attempts is given by multiplying together the probability of obtaining one specific realization and the number of ways of getting that realization.

We need a way of generalizing the number of ways of getting x items out of n items. The answer is the combinatorial formula

$$\binom{n}{x} = \frac{n!}{x!(n-x)!},$$

where the 'exclamation mark' means 'factorial'. For instance, $5! = 5 \times 4 \times 3 \times 2 = 120$. This formula has immense practical utility. It shows you at once, for example, how unlikely you are to win the National Lottery in which you are invited to select six numbers between 1 and 49. We can use the built-in factorial function for this

```
factorial(49)/(factorial(6)*factorial(49-6))
```

```
[1]  13983816
```

which is roughly a 1 in 14 million chance of winning the jackpot. You are more likely to die between buying your ticket and hearing the outcome of the draw. As we have seen (p. 11), there is a built-in R function for the combinatorial function

choose(49,6)

[1] 13983816

and we use the choose function from here on.

The general form of the binomial distribution is given by

$$p(x) = \binom{n}{x} p^x (1-p)^{n-x},$$

using the combinatorial formula above. The mean of the binomial distribution is np and the variance is $np(1-p)$.

Since $1 - p$ is less than 1 it is obvious that **the variance is less than the mean** for the binomial distribution (except, of course, in the trivial case when $p = 0$ and the variance is 0). It is easy to visualize the distribution for particular values of n and p.

```
p<-0.1
n<-4
x<-0:n
px<-choose(n,x)*p^x*(1-p)^(n-x)
barplot(px,names=as.character(x))
```

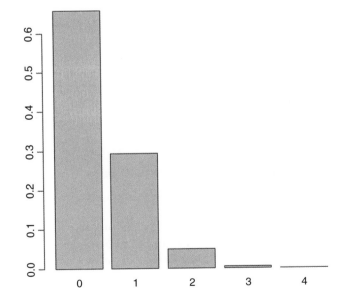

The four distribution functions available for the binomial in R (density, cumulative probability, quantiles and random generation) are used like this:

dbinom(x, size, prob)

The density function shows the probability for the specified count x (e.g. the number of parasitized fish) out of a sample of $n = $ size, with probability of success $= $ prob. So if we catch four fish when 10% are parasitized in the parent population, we have size $= 4$ and prob $= 0.1$, so a graph of probability density against number of parasitized fish can be obtained like this:

barplot(dbinom(0:4,4,0.1),names=as.character(0:4),xlab="x",ylab="f(x)")

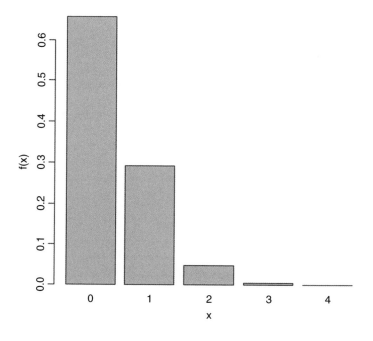

The most likely number of parasitized fish is 0. Note that we can generate the sequence of x values we want to plot (0:4 in this case) *inside* the density function.

The cumulative probability shows the sum of the probability densities up to and including $p(x)$, plotting cumulative probability against the number of successes, for a sample of $n = $ size and probability $= $ prob. Our fishy plot looks like this:

barplot(pbinom(0:4,4,0.1),names=as.character(0:4),xlab="x",ylab="p(x)")

This says that the probability of getting 2 or fewer parasitized fish out of a sample of 4 is very close to 1.

The quantiles function asks 'with specified probability p (often 0.025 and 0.975 for two-tailed 95% tests), what is the expected number of fish to be caught in a sample of $n = $ size and a probability $= $ prob?'. So for our example, the two-tailed (no pun intended) lower and upper 95% expected catches of parasitized fish are

qbinom(.025,4,0.1)

[1] 0

qbinom(.975,4,0.1)

[1]   2

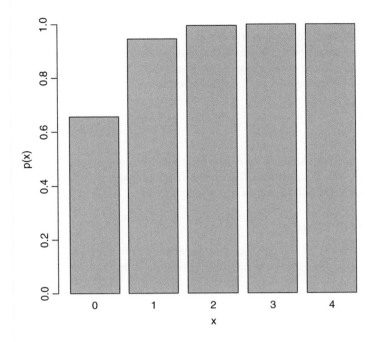

which means that with 95% certainty we shall catch between 0 and 2 parasitized fish out of 4 if we repeat the sampling exercise. We are very unlikely to get 3 or more parasitized fish out of a sample of 4 if the proportion parasitized really is 0.1.

This kind of calculation is very important in power calculations in which we are interested in determining whether or not our chosen sample size ($n = 4$ in this case) is capable of doing the job we ask of it. Suppose that the fundamental question of our survey is whether or not the parasite is present in a given lake. If we find one or more parasitized fish then the answer is clearly 'yes'. But how likely are we to miss out on catching any parasitized fish and hence of concluding, wrongly, that the parasites are not present in the lake? With out sample size of $n = 4$ and $p = 0.1$ we have a probability of missing the parasite of 0.9 for each fish caught and hence a probability of $0.9^4 = 0.6561$ of missing out altogether on finding the parasite. This is obviously unsatisfactory. We need to think again about the sample size. What is the smallest sample, n, that makes the probability of missing the parasite altogether less than 0.05?

We need to solve

Taking logs, $$0.05 = 0.9^n$$

$$\log(0.05) = n \log(0.9)$$

so

$$n = \frac{\log(0.05)}{\log(0.9)} = 28.433\,16$$

which means that to make our journey worthwhile we should keep fishing until we have found more than 28 unparasitized fishes, before we reject the hypothesis that parasitism is present at a rate of 10%. Of course, it would take a much bigger sample to reject a hypothesis of presence at a much lower rate.

Random numbers are generated from the binomial distribution like this. The first argument is the number of random numbers we want. The second argument is the sample size ($n=4$) and the third is the probability of success ($p=0.1$).

```
rbinom(10,4,0.1)
```

```
[1] 0 0 0 0 0 1 0 1 0 1
```

Here we repeated the sampling of 4 fish ten times. We got 1 parasitized fish out of 4 on three occasions, and 0 parasitized fish on the remaining seven occasions. We never caught 2 or more parasitized fish in any of these samples of 4.

The geometric distribution

Suppose that a series of independent Bernoulli trails with probability p are carried out at times 1, 2, 3, ... Now let W be the waiting time until the first success occurs. So

$$P(W > x) = (1 - p)^x,$$

which means that

$$P(W = x) = P(W > x - 1) - P(W > x)$$

The density function, therefore, is

$$f(x) = p(1 - p)^{x-1}$$

```
fx<-dgeom(0:20,0.2)
barplot(fx,names=as.character(0:20),xlab="x",ylab="f(x)")
```

For the geometric distribution,

- the mean is $\dfrac{1 - p}{p}$,

- the variance is $\dfrac{1 - p}{p^2}$.

The geometric has a very long tail. Here are 100 random numbers from a geometric distribution with $p=0.1$: the mode is 0 but outlying values as large as 43 and 51 have been generated.

```
table(rgeom(100,0.1)
```

0	1	2	3	4	5	7	8	9	10	11	12	13	14	15	16	17	18	19	21
13	8	9	6	4	12	5	4	3	4	3	6	1	1	3	1	1	3	1	1

22	23	25	26	30	43	51
1	2	1	3	2	1	1

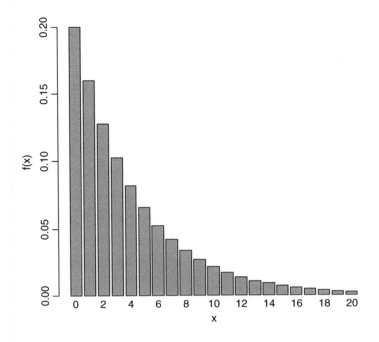

The hypergeometric distribution

'Balls in urns' are the classic sort of problem solved by this distribution. The density function of the hypergeometric is

$$f(x) = \frac{\binom{b}{x}\binom{N-b}{n-x}}{\binom{N}{n}}.$$

Suppose that there are N coloured balls in the statistician's famous urn: b of them are blue and $r = N - b$ of them are red. Now a sample of n balls is removed from the urn; this is sampling *without replacement*. Now $f(x)$ gives the probability that x of these n balls are blue.

The built-in functions for the hypergeometric are used like this:

```
dhyper(q, b,r,n),
rhyper(m, b,r,n).
```

Here

- q is a vector of values of a random variable representing the number of blue balls out of a sample of size n drawn from an urn containing b blue balls and r red ones.

- b is the number of blue balls in the urn. This could be a vector with non-negative integer elements

- r is the number of red balls in the urn $= N - b$. This could also be a vector with non-negative integer elements

- n is the number of balls drawn from an urn with b blue and r red balls. This can be a vector like b and r.

- p is a vector of probabilities with values between 0 and 1.

- m is the number of hypergeometrically distributed random numbers to be generated.

Let the urn contain $N = 20$ balls of which 6 are blue and 14 are red. We take a sample of $n = 5$ balls so x could be 0, 1, 2, 3, 4 or 5 of them blue, but since the proportion blue is only 6/20 the higher frequencies are most unlikely. Our example is evaluated like this:

```
ph<-numeric(6)
for(i in 0:5) ph[i]<-dhyper(i,6,14,5)
barplot(ph,names=as.character(0:5))
```

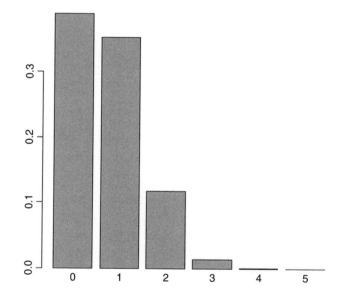

We are very unlikely to get more than 2 red balls out of 5. The most likely outcome is that we get 0 or 1 red ball out of 5. We can simulate a set of Monte Carlo trials of size 5. Here are the numbers of red balls obtained in 20 realizations of our example

```
rhyper(20,6,14,5)
```

```
[1]  1  1  1  2  1  2  0  1  3  2  3  0  2  0  1  1  2  1  1  2
```

The binomial distribution is a limiting case of the hypergeometric which arises as N, b and r approach infinity in such a way that b/N approaches p, and r/N approaches $1 - p$ (see p. 242). This is because as the numbers get large, the fact that we are *sampling without replacement* becomes irrelevant. The binomial distribution assumes sampling with replacement from a finite population, or sampling without replacement from an infinite population.

The multinomial distribution

Suppose that there are t possible outcomes from an experimental trial, and the outcome i has probability p_i. Now allow n independent trials where $n = n_1 + n_2 + \ldots + n_t$ and ask what is the probability of obtaining the vector of N_i occurrences of the ith outcome:

$$P(N_i = n_i) = \frac{n!}{n_1! n_2! n_3! \ldots n_t!} p_1^{n_1} p_2^{n_2} p_3^{n_3} \ldots p_t^{n_t},$$

where i goes from 1 to t. Take an example with three outcomes, where the first outcome is twice as likely as the other two ($p_1 = 0.5$, $p_2 = 0.25$ and $p_3 = 0.25$). We do 4 trials with $n_1 = 6$, $n_2 = 5$, $n_3 = 7$ and $n_4 = 6$, so $n = 24$. We need to evaluate the formula for $i = 1$, 2 and 3 (because there are three possible outcomes). It is sensible to start by writing a function called multi to carry out the calculations for any numbers of successes a, b and c for the three outcomes given our three probabilities 0.5, 0.25 and 0.25:

```
multi<-function(a,b,c) {
      factorial(a+b+c)/(factorial(a)*factorial(b)*factorial(c))*.5^a*.25^b*.25^c}
```

Now put the function in a loop to work out the probability of getting the required patterns of success, psuc, for the three outcomes. We illustrate just one case, in which the third outcome is fixed at four successes. This means that the first and second cases vary stepwise between 19 and 1 and 9 and 11 respectively:

```
psuc<-numeric(11)
for (i in 0:10) psuc[i]<-multi(19-i,1+i,4)
barplot(psuc,names=as.character(0:10))
```

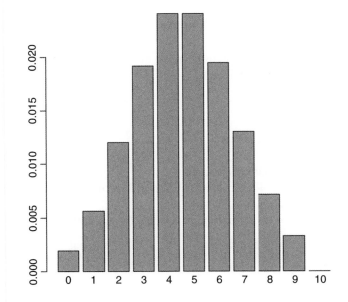

The most likely outcome here is that we would get $19 - 4 = 15$ or $19 - 5 = 14$ successes of type 1 in a trial of size 24 with probabilities 0.5, 0.25 and 0.25 when the number of successes of the third case was 4 out of 24. You can easily modify the function to deal with other probabilities and other number of outcomes.

The Poisson distribution

This is one of the most useful and important of the discrete probability distributions for describing count data. We know how many times something happened (e.g. kicks from cavalry horses, lightening strikes, bomb hits), but we have no way of knowing how many times it did not happen. The Poisson is a one-parameter distribution with the interesting property that its variance is equal to its mean. A great many processes show variance increasing with the mean, often faster than linearly (see the negative binomial distribution below). The density function of the Poisson shows the probability of obtaining a count of x when the mean count per unit is λ:

$$p(x) = \frac{e^{-\lambda} \lambda^x}{x!}.$$

The zero term of the Poisson (the probability of obtaining a count of zero) is obtained by setting $x = 0$:

$$p(0) = e^{-\lambda},$$

which is simply the antilog of minus the mean. Given $p(0)$, it is clear that $p(1)$ is just

$$p(1) = p(0)\lambda = \lambda e^{-\lambda},$$

and any subsequent probability is readily obtained by multiplying the previous probability by the mean and dividing by the count, thus:

$$p(x) = p(x-1)\frac{\lambda}{x}.$$

Functions for the density, cumulative distribution, quantiles and random number generation of the Poisson distribution are obtained by

```
dpois(x, lambda)
ppois(q, lambda)
qpois(p, lambda)
rpois(n, lambda)
```

where lambda is the mean count per sample.

The Poisson distribution holds a central position in three quite separate areas of statistics:

- in the description of random spatial point patterns (see p. 749);
- as the frequency distribution of counts of rare but independent events (see p. 208);
- as the error distribution in GLMs for count data (see p. 527).

If we wanted 600 simulated counts from a Poisson distribution with a mean of, say, 0.90 blood cells per slide, we just type:

```
count<-rpois(600,0.9)
```

We can use table to see the frequencies of each count generated:

table(count)

```
count
   0     1     2     3    4    5
 244   212   104    33    6    1
```

or hist to see a histogram of the counts:

hist(count,breaks = - 0.5:6.5,main="")

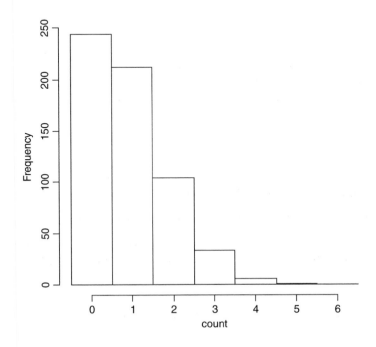

Note the use of the vector of break points on integer increments from −0.5 to create integer bins for the histogram bars.

The negative binomial distribution

This discrete, two-parameter distribution is useful for describing the distribution of count data, where the variance is often much greater than the mean. The two parameters are the mean μ and the clumping parameter k, given by

$$k = \frac{\mu^2}{\sigma^2 - \mu}$$

The smaller the value of k, the greater the degree of clumping. The density function is

$$p(x) = \left(1 + \frac{\mu}{k}\right)^{-k} \frac{(k + x - 1)!}{x!(k - 1)!} \left(\frac{\mu}{\mu + k}\right)^x.$$

The zero term is found by setting $x = 0$ and simplifying:

$$p(0) = \left(1 + \frac{\mu}{k}\right)^{-k}$$

Successive terms in the distribution can then be computed iteratively from

$$p(x) = p(x-1)\left(\frac{k+x-1}{x}\right)\left(\frac{\mu}{\mu+k}\right).$$

An initial estimate of the value of k can be obtained from the sample mean and variance

$$k \approx \frac{\bar{x}^2}{s^2 - \bar{x}}.$$

Since k cannot be negative, it is clear that the negative binomial distribution should not be fitted to data where the variance is less than the mean.

The maximum likelihood estimate of k is found numerically, by iterating progressively more fine-tuned values of k until the left- and right-hand sides of the following equation are equal:

$$n \ln\left(1 + \frac{\mu}{k}\right) = \sum_{x=0}^{max}\left(\frac{A(x)}{k+x}\right)$$

where the vector $A(x)$ contains the total frequency of values *greater* than x. You could write a function to work out the probability densities like this:

```
negbin<-function(x,u,k) (1+u/k)^(-k)*(u/(u+k))^x*gamma(k+x)/(factorial(x)*gamma(k))
```

then use the function to produce a barplot of probability densities for a range of x values (say 0 to 10, for a distribution with specified mean and aggregation parameter (say $\mu = 0.8$, $k = 0.2$) like this

```
xf<-sapply(0:10, function(i) negbin(i,0.8,0.2))
barplot(xf,names=as.character(0:10),xlab="count",ylab="probability density")
```

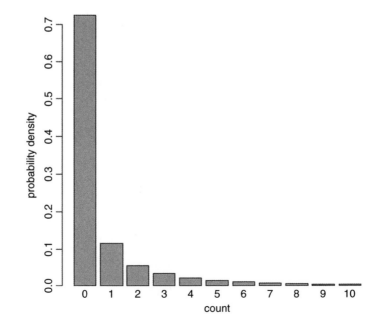

There is another, quite different way of looking at the negative binomial distribution. Here, the response variable is the waiting time W_r for the rth success:

$$f(x) = \binom{x-1}{r-1} p^r (1-p)^{x-r}$$

It is important to realize that x starts at r and increases from there (obviously, the rth success cannot occur before the rth attempt). The function dnbinom represents the number of failures x (e.g. tails in coin tossing) before size successes (or heads in coin tossing) are achieved, when the probability of a success (or of a head) is prob:

dnbinom(x, size, prob)

Suppose we are interested in the distribution of waiting times until the 5th success occurs in a negative binomial process with $p = 0.1$. We start the sequence of x values at 5

plot(5:100,dnbinom(5:100,5,0.1),type="s",xlab="x",ylab="f(x)")

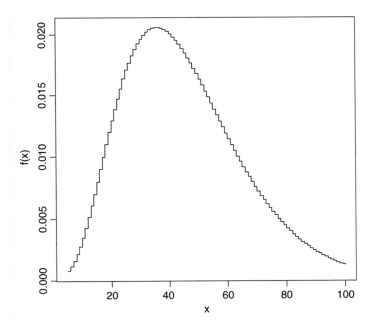

This shows that the most likely waiting time for the 5th success, when probability of a success is 1/10, is about 31 trials after the 5th trial. Note that the negative binomial distribution is quite strongly skew to the right.

It is easy to generate negative binomial data using the random number generator:

rnbinom(n, size, prob)

The number of random numbers required is n. When the second parameter, size, is set to 1 the distribution becomes the geometric (see above). The final parameter, prob, is the probability of success per trial, p. Here we generate 100 counts with a mean of 0.6:

count<-rnbinom(100,1,0.6)

We can use table to see the frequency of the different counts:

table(count)

```
  0   1   2   3   5   6
 65  18  13   2   1   1
```

It is sensible to check that the mean really is 0.6 (or very close to it)

mean(count)

[1] 0.61

The variance will be substantially greater than the mean

var(count)

[1] 1.129192

and this gives an estimate of k of

$$\frac{0.61^2}{1.129 - 0.61} = 0.717.$$

The following data show the number of spores counted on 238 buried glass slides. We are interested in whether these data are well described by a negative binomial distribution. If they are we would like to find the maximum likelihood estimate of the aggregation parameter k.

```
x<-0:12
freq<-c(131,55,21,14,6,6,2,0,0,0,0,2,1)
barplot(freq,names=as.character(x),ylab="frequency",xlab="spores")
```

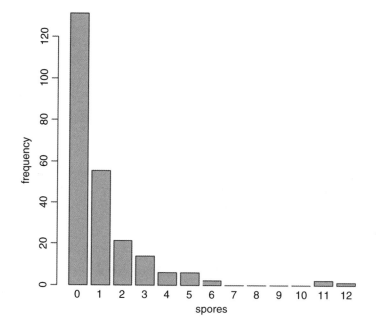

We start by looking at the variance – mean ratio of the counts. We cannot use mean and variance directly, because our data are *frequencies* of counts, rather than counts themselves. This is easy to rectify: we use rep to create a vector of counts y in which each count (x) is repeated the relevant number of times (freq). Now we can use mean and var directly:

```
y<-rep(x,freq)
mean(y)
```

```
[1]  1.004202
```

```
var(y)
```

```
[1]  3.075932
```

This shows that the data are highly aggregated (the variance mean ratio is roughly 3, recalling that it would be 1 if the data were Poisson distributed). Our rough estimate of k is therefore

```
mean(y)^2/(var(y)-mean(y))
```

```
[1]  0.4867531
```

Here is a function that takes a vector of frequencies of counts x (between 0 and length(x) – 1) and computes the maximum likelihood estimate of the aggregation parameter, k:

```
kfit <-function(x)
{
    lhs<-numeric()
    rhs<-numeric()
    y <-0:(length(x) - 1)
    j<-0:(length(x)-2)
    m <-sum(x * y)/(sum(x))
    s2 <-(sum(x * y^2) - sum(x * y)^2/sum(x))/(sum(x)- 1)
    k1 <-m^2/(s2 - m)
    a<-numeric(length(x)-1)
    for(i in 1:(length(x) - 1)) a[i] <-sum(x [- c(1:i)])
    i<-0
    for (k in seq(k1/1.2,2*k1,0.001)) {
    i<-i+1
    lhs[i] <-sum(x) * log(1 + m/k)
    rhs[i] <-sum(a/(k + j))
        }
    k<-seq(k1/1.2,2*k1,0.001)
    plot(k, abs(lhs-rhs),xlab="k",ylab="Difference",type="l")

    d<-min(abs(lhs-rhs))
    sdd<-which(abs(lhs-rhs)==d)
    k[sdd]

}
```

We can try it out with our spore count data.

kfit(freq)

```
[1]  0.5826276
```

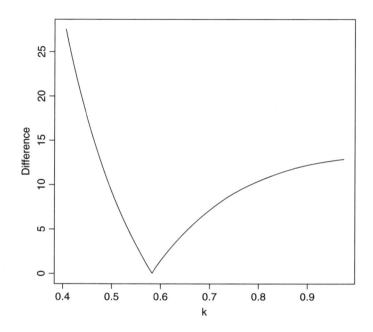

The minimum difference is close to zero and occurs at about $k = 0.55$. The printout shows that the maximum likelihood estimate of k is 0.582 (to the 3 decimal places we simulated; the last 4 decimals (6276) are meaningless and would not be printed in a more polished function).

How would a negative binomial distribution with a mean of 1.0042 and a k value of 0.583 describe our count data? The expected frequencies are obtained by multiplying the probability density (above) by the total sample size (238 slides in this case).

```
nb<-238*(1+1.0042/0.582)^(-0.582)*factorial(.582+(0:12)-1)/
    (factorial(0:12)*factorial(0.582-1))*(1.0042/(1.0042+0.582))^(0:12)
```

We shall compare the observed and expected frequencies using barplot. We need to alternate the observed and expected frequencies. There are three steps to the procedure:

- Concatenate the observed and expected frequencies in an alternating sequence.

- Create list of labels to name the bars (alternating blanks and counts).

- Produce a legend to describe the different bar colours.

The concatenated list of frequencies (called both) is made like this, putting the 13 observed counts (freq) in the odd-numbered bars and the 13 expected counts (nb) in the even-numbered bars (note the use of modulo %% to do this):

```
both<-numeric(26)
both[1:26 %% 2 != 0]<-freq
both[1:26 %% 2 == 0]<-nb
```

Now we can draw the combined barplot:

```
barplot(both,col=rep(c(1,0),13),ylab="frequency")
```

Because two adjacent bars refer to the same count (the observed and expected frequencies) we do not want to use barplot's built-in names argument for labelling the bars (it would want to write a label on every bar, 26 labels in all). Instead, we want to write the count just once for each pair of bars, located between the observed and expected bars, using as.character(0:12). This is a job for the mtext function, which writes text in the margins of plot. We need to specify the margin in which we want to write. The bottom margin in side = 1. The only slightly tricky thing is to work out the x coordinates for the 13 labels along the axis. To see how the x axis has been scaled by the barplot function, allocate the barplot function (as above) to a vector name, then inspect its contents:

```
xscale<-barplot(both,col=rep(c(1,0),13),ylab="frequency")
as.vector(xscale)
```

```
 [1]   0.7   1.9   3.1   4.3   5.5   6.7   7.9   9.1  10.3  11.5  12.7  13.9  15.1
[14]  16.3  17.5  18.7  19.9  21.1  22.3  23.5  24.7  25.9  27.1  28.3  29.5  30.7
```

Here you can see that the left-hand side of the first bar is at $x = 0.7$ and the 26th bar is at $x = 30.7$. A little experimentation will show that we want to put out first label at $x = 1.4$ and then at intervals of 2.4 (two bars are separated by, for instance, $11.5 - 9.1 = 2.4$). We specify the sequence seq(1.4,30.2,2.4) as the argument to at within mtext:

```
mtext(as.character(0:12),side=1,at=seq(1.4,30.2,2.4))
```

The default used here is for mtext to write the labels in line number 0 away from the side in question: if you want to change this, add the argument line=1 to mtext.

The legend function creates a legend to show which bars represent the observed frequencies (black in this case) and which represent the expected, negative binomial frequencies (open bars). Just click when the cursor is in the position where you want the *top left-hand corner* of the legend box to be:

```
legend(locator(1),c("observed","expected"),fill=c("black","white"))
```

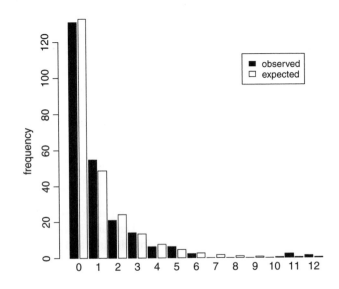

The fit is very close, so we can be reasonably confident in describing the observed counts as negative binomially distributed. The tail of the observed distribution is rather fatter than the expected negative binomial tail, so we might want to measure the lack of fit between observed and expected distributions. A simple way to do this is to use Pearson's chi-squared taking care to use only those cases where the expected frequency nb is greater than 5:

sum(((freq-nb)^2/nb)[nb > 5])

[1] 1.634975

This is based on five legitimate comparisons

sum(nb>5)

[1] 5

and hence on $5 - p - 1 = 2$ d.f. because we have estimated $p = 2$ parameters from the data in estimating the expected distribution (the mean and k of the negative binomial) and lost one degree of freedom for contingency (the total number of counts must add up to 238). Our calculated value of chi-squared $= 1.63$ is much less than the value in tables

qchisq(0.95,2)

[1] 5.991465

so we accept the hypothesis that our data are not significantly different from a negative binomial with mean $= 1.0042$ and $k = 0.582$.

The Wilcoxon rank-sum statistic

This function calculates the distribution of the Wilcoxon rank-sum statistic (also known as Mann–Whitney), and returns values for the exact probability at discrete values of q:

dwilcox(q, m, n)

Here q is a vector of quantiles, m is the number of observations in sample x (a positive integer not greater than 50), and n is the number of observations in sample y (also a positive integer not greater than 50). The Wilcoxon rank-sum statistic is the sum of the ranks of x in the combined sample c(x,y). The Wilcoxon rank-sum statistic takes on values W between the limits:

$$\frac{m(m+1)}{2} \le W \le \frac{m(m+2n+1)}{2}.$$

This statistic can be used for a non-parametric test of location shift between the parent populations x and y.

Matrix Algebra

There is a comprehensive set of functions for handling matrices in R. We begin with a matrix called a that has three rows and two columns. Data are typically entered into matrices columnwise, so the first three numbers (1, 0, 4) go in column 1 and the second three numbers (2, −1, 1) go in column 2:

```
a<-matrix(c(1,0,4,2,-1,1),nrow=3)
a
```

```
       [,1]   [,2]
[1,]      1      2
[2,]      0     -1
[3,]      4      1
```

Our second matrix, called *b*, has the same number of columns as *A* has rows (i.e. three in this case). Entered columnwise, the first two numbers (1, −1) go in column 1, the second two numbers (2, 1) go in column 2, and the last two numbers (1, 0) go in column 3:

```
b<-matrix(c(1,-1,2,1,1,0),nrow=2)
b
```

```
       [,1]   [,2]   [,3]
[1,]      1      2      1
[2,]     -1      1      0
```

Matrix multiplication

To multiply one matrix by another matrix you take the rows of the first matrix and the columns of the second matrix. Put the first row of *a* side by side with the first column of *b*:

```
a[1,]
```

```
[1]  1  2
```

```
b[,1]
```

```
[1]  1  -1
```

and work out the point products:

```
a[1,]*b[,1]
```

```
[1]  1  -2
```

then add up the point products

```
sum(a[1,]*b[,1])
```

```
[1]  -1
```

The sum of the point products is −1 and this is the first element of the product matrix. Next, put the first row of *a* with the second column of *b*:

```
a[1,]
```

```
[1]  1  2
```

```
b[,2]
```

```
[1]  2  1
```

```
a[1,]*b[,2]
```

```
[1]  2  2
```

sum(a[1,]*b[,2])

```
[1]   4
```

so the point products are 2, 2 and the sum of the point products is $2+2=4$. So 4 goes in row 1 and column 2 of the answer. Then take the last column of b and match it against the first row of a:

a[1,]*b[,3]

```
[1]   1   0
```

sum(a[1,]*b[,3])

```
[1]   1
```

so the sum of the point products is $1+0=1$. This goes in row 1, column 3 of the answer. And so on. We repeat these steps for row 2 of matrix a $(0, -1)$ and then again for row 3 of matrix a $(4, 1)$ to obtain the complete matrix of the answer. In R, the symbol for matrix multiplication is %*%. Here is the full answer:

a%*%b

```
     [,1] [,2] [,3]
[1,]   -1    4    1
[2,]    1   -1    0
[3,]    3    9    4
```

where you see the values we calculated by hand $(-1, 4, 1)$ in the first row.

It is important to understand that with matrices a times b is not the same as b times a. The matrix resulting from a multiplication has the number of rows of the matrix on the left (a has 3 rows in the case above). But b has just two rows, so multiplication

b%*%a

```
     [,1] [,2]
[1,]    5    1
[2,]   -1   -3
```

produces a matrix with 2 rows. The value 5 in row 1 column 1 of the answer is the sum of the point products $(1 \times 1) + (2 \times 0) + (1 \times 4) = 1+0+4=5$.

Diagonals of matrices

To create a diagonal matrix of 3 rows and 3 columns, with 1s on the diagonal use the diag function like this:

(ym<-diag(1,3,3))

```
     [,1] [,2] [,3]
[1,]    1    0    0
[2,]    0    1    0
[3,]    0    0    1
```

You can alter the values of the diagonal elements of a matrix like this:

diag(ym)<-1:3
ym

```
        [,1] [,2] [,3]
[1,]      1    0    0
[2,]      0    2    0
[3,]      0    0    3
```

or extract a vector containing the diagonal elements of a matrix like this:

diag(ym)

```
[1]  1  2  3
```

You might want to extract the diagonal of a variance – covariance matrix:

M <-cbind(X=1:5, Y=rnorm(5))
var(M)

```
              X             Y
X   2.50000000   0.04346324
Y   0.04346324   0.88056034
```

diag(var(M))

```
        X             Y
2.5000000   0.8805603
```

Determinant

The determinant of the square (2×2) array

$$\begin{bmatrix} a & b \\ c & d \end{bmatrix}$$

is defined for any numbers a, b, c and d as

$$\begin{vmatrix} a & b \\ c & d \end{vmatrix} \equiv ad - bc.$$

Suppose that A is a square matrix of order (3×3):

$$A = \begin{bmatrix} a_{11} & a_{12} & a_{13} \\ a_{21} & a_{22} & a_{23} \\ a_{31} & a_{32} & a_{23} \end{bmatrix}.$$

Then the third-order determinant of A is defined to be the number

$$\det A = a_{11} \begin{vmatrix} a_{22} & a_{23} \\ a_{32} & a_{33} \end{vmatrix} - a_{12} \begin{vmatrix} a_{21} & a_{23} \\ a_{31} & a_{33} \end{vmatrix} + a_{13} \begin{vmatrix} a_{21} & a_{22} \\ a_{31} & a_{32} \end{vmatrix}.$$

Applying the rule $\begin{vmatrix} a & b \\ c & d \end{vmatrix} \equiv ad - bc$ to this equation gives

$$\det A = a_{11}a_{22}a_{a33} - a_{11}a_{23}a_{32} + a_{12}a_{23}a_{31} - a_{12}a_{21}a_{33} + a_{13}a_{21}a_{32} - a_{13}a_{22}a_{31}.$$

Take a numerical example:

$$A = \begin{bmatrix} 1 & 2 & 3 \\ 2 & 1 & 1 \\ 4 & 1 & 2 \end{bmatrix}.$$

This has determinant

$$\det A = (1 \times 1 \times 2) - (1 \times 1 \times 1) + (2 \times 1 \times 4) - (2 \times 2 \times 2) + (3 \times 2 \times 1) - (3 \times 1 \times 4)$$
$$= 2 - 1 + 8 - 8 + 6 - 12 = -5.$$

Here is the example in R using the determinant function det:

```
A<-matrix(c(1,2,4,2,1,1,3,1,2),nrow=3)
A
```

```
     [,1] [,2] [,3]
[1,]    1    2    3
[2,]    2    1    1
[3,]    4    1    2
```

```
det(A)
```

```
[1] -5
```

The great thing about determinants is that if any row or column of a determinant is multiplied by a scalar λ, then the value of the determinant is multiplied by λ (since a factor λ will appear in each of the products). Also, if all the elements of a row or a column are zero then the determinant $|A| = 0$. Again, if all the corresponding elements of two rows or columns of $|A|$ are equal then $|A| = 0$.

For instance, here is the bottom row of A multiplied by 3:

```
B<-A
B[3,]<-3*B[3,]
B
```

```
     [,1] [,2] [,3]
[1,]    1    2    3
[2,]    2    1    1
[3,]   12    3    6
```

and here is the determinant:

```
det(B)
```

```
[1] -15
```

Here is an example when all the elements of column 2 are zero, so det $C = 0$:

```
C<-A
C[,2]<-0
C
```

```
     [,1] [,2] [,3]
[1,]    1    0    3
[2,]    2    0    1
[3,]    4    0    2
```

det(C)

[1] 0

If $\det A \neq 0$ then the rows and columns of A must be linearly independent. This important concept is expanded in terms of contrast coefficients on p. 372.

Inverse of a matrix

The operation of division is not defined for matrices. However, for a square matrix that has $|A| \neq 0$ a multiplicative inverse matrix denoted by A^{-1} can be defined. This multiplicative inverse A^{-1} is unique and has the property that

$$A^{-1}A = AA^{-1} = I,$$

where I is the unit matrix. So if A is a square matrix for which $|A| \neq 0$ the matrix inverse is defined by the relationship

$$A^{-1} = \frac{\text{adj}A}{|A|},$$

where the adjoint matrix of A (adj A) is the matrix of cofactors of A. The cofactors of A are computed as $A_{ij} = (-1)^{i+j}M_{ij}$, where M_{ij} are the 'minors' of the elements a_{ij} (these are the determinants of the matrices of A from which row i and column j have been deleted). The properties of the inverse matrix can be laid out for two non-singular square matrices, A and B, of the same order as follows:

$$AA^{-1} = A^{-1}A = I$$
$$(AB)^{-1} = B^{-1}A^{-1}$$
$$(A^{-1})' = (A')^{-1}$$
$$(A^{-1})^{-1} = A$$
$$|A| = \frac{1}{|A^{-1}|}$$

Here is R's version of the inverse of the 3×3 matrix A (above) using the ginv function from the MASS library

library(MASS)
ginv(A)

```
              [,1]  [,2]  [,3]
[1,] -2.000000e-01   0.2   0.2
[2,] -2.224918e-16   2.0  -1.0
[3,]  4.000000e-01  -1.4   0.6
```

where the number in row 2 column 1 is a zero (except for rounding error).

Here is the penultimate rule $(A^{-1})^{-1} = A$ evaluated by R:

ginv(ginv(A))

```
      [,1] [,2] [,3]
[1,]    1    2    3
[2,]    2    1    1
[3,]    4    1    2
```

and here is the last rule $|A| = 1/|A^{-1}|$:

1/det(ginv(A))

[1] -5

Eigenvalues and eigenvectors

We have a square matrix A and two column vectors X and K, where

$$AX = K,$$

and we want to discover the scalar multiplier λ such that

$$AX = \lambda X.$$

This is equivalent to $(A - \lambda I)X = 0$, where I is the unit matrix. This can only have one non-trivial solution when the determinant associated with the coefficient matrix A vanishes, so we must have

$$|A - \lambda I| = 0.$$

When expanded, this determinant gives rise to an algebraic equation of degree n in λ called the **characteristic equation**. It has n roots $\lambda_1, \lambda_2, \ldots, \lambda_n$, each of which is called an **eigenvalue**. The corresponding solution vector X_i is called an **eigenvector** of A corresponding to λ_i.

Here is an example from population ecology. The matrix A shows the demography of different age classes: the top row shows fecundity (the number of females born per female of each age) and the sub-diagonals show survival rates (the fraction of one age class that survives to the next age class). When these numbers are constants the matrix is known as the **Leslie matrix**. In the absence of density dependence the constant parameter values in A will lead either to exponential increase in total population size (if $\lambda_1 > 1$) or exponential decline (if $\lambda_1 < 1$) once the initial transients in age structure have damped away. Once exponential growth has been achieved, then the age structure, as reflected by the proportion of individuals in each age class, will be a constant. This is known as the first eigenvector.

Consider the Leslie matrix, L, which is to be multiplied by a column matrix of age-structured population sizes, n:

```
L<-c(0,0.7,0,0,6,0,0.5,0,3,0,0,0.3,1,0,0,0)
L<-matrix(L,nrow=4)
```

Note that the elements of the matrix are entered in columnwise, not row-wise sequence. We make sure that the Leslie matrix is properly conformed:

```
L
        [,1]  [,2]  [,3]  [,4]
[1,]    0.0   6.0   3.0    1
[2,]    0.7   0.0   0.0    0
[3,]    0.0   0.5   0.0    0
[4,]    0.0   0.0   0.3    0
```

The top row contains the age-specific fecundities (e.g. 2-year-olds produce 6 female off-spring per year), and the sub-diagonal contains the survivorships (70% of 1-year-olds become 2-year-olds, etc.). Now the population sizes at each age go in a column vector, n:

```
n<-c(45,20,17,3)
n<-matrix(n,ncol=1)
n
```

```
        [,1]
[1,]     45
[2,]     20
[3,]     17
[4,]      3
```

Population sizes next year in each of the four age classes are obtained by matrix multiplication, $\%*\%$

```
L %*% n
```

```
        [,1]
[1,]   174.0
[2,]    31.5
[3,]    10.0
[4,]     5.1
```

We can check this the long way. The number of juveniles next year (the first element of n) is the sum of all the babies born last year:

```
45*0+20*6+17*3+3*1
```

```
[1] 174
```

We write a function to carry out the matrix multiplication, giving next year's population vector as a function of this year's:

```
fun<-function(x) L %*% x
```

Now we can simulate the population dynamics over a period long enough (say, 40 generations) for the age structure to approach stability. So long as the population growth rate $\lambda > 1$ the population will increase exponentially, once the age structure has stabilized:

```
n<-c(45,20,17,3)
n<-matrix(n,ncol=1)
structure<-numeric(160)
dim(structure)<-c(40,4)
for (i in 1:40) {
n<-fun(n)
structure[i,]<-n
}
matplot(1:40,log(structure),type="l")
```

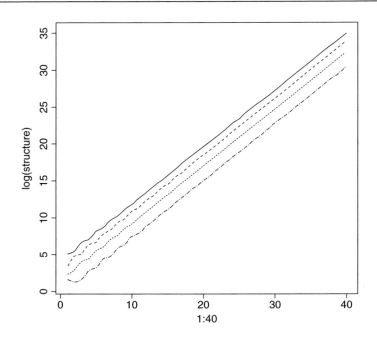

You can see that after some initial transient fluctuations, the age structure has more or less stabilized by year 20 (the lines for log population size of juveniles (top line), 1-, 2- and 3-year-olds are parallel). By year 40 the population is growing exponentially in size, multiplying by a constant of λ each year.

The population growth rate (the per-year multiplication rate, λ) is approximated by the ratio of total population sizes in the 40th and 39th years:

```
sum(structure[40,])/sum(structure[39,])
```

```
[1]  2.164035
```

and the approximate stable age structure is obtained from the 40th value of *n*:

```
structure[40,]/sum(structure[40,])
```

```
[1]  0.709769309  0.230139847  0.052750539  0.007340305
```

The exact values of the population growth rate and the stable age distribution are obtained by matrix algebra: they are the dominant eigenvalue and dominant eigenvector, respectively. Use the function **eigen** applied to the Leslie matrix, *L*, like this:

```
eigen(L)
```

```
$values
[1]  2.1694041+0.0000000i  -1.9186627+0.0000000i  -0.1253707+0.0975105i
[4]  -0.1253707-0.0975105i
$vectors
                    [,1]            [,2]                    [,3]
[1,]  -0.949264118+0i  -0.93561508+0i  -0.01336028-0.03054433i
[2,]  -0.306298338+0i   0.34134741+0i  -0.03616819+0.14241169i
[3,]  -0.070595039+0i  -0.08895451+0i   0.36511901-0.28398118i
[4,]  -0.009762363+0i   0.01390883+0i  -0.87369452+0.00000000i
```

```
                              [,4]
[1,]  -0.01336028+0.03054433i
[2,]  -0.03616819-0.14241169i
[3,]   0.36511901+0.28398118i
[4,]  -0.87369452+0.00000000i
```

The dominant eigenvalue is 2.1694 (compared with our empirical approximation of 2.1640 after 40 years). The stable age distribution is given by the first eigenvector, which we need to turn into proportions

eigen(L)$vectors[,1]/sum(eigen(L)$vectors[,1])

```
[1]  0.710569659+0i  0.229278977+0i  0.052843768+0i  0.007307597+0i
```

This compares with our approximation (above) in which the proportion in the first age class was 0.70977 after 40 years (rather than 0.71057).

Matrices in statistical models

Perhaps the main use of matrices in R is in statistical calculations, in generalizing the calculation of sums of squares and sums of products (see p. 388 for background). Here are the data used in Chapter 10 to introduce the calculation of sums of squares in linear regression:

numbers<-read.table("c:\\temp\\tannin.txt",header=T)
attach(numbers)
names(numbers)

```
[1]  "growth"  "tannin"
```

The response variable is growth (y) and the explanatory variable is tannin concentration (x) in the diet of a group of insect larvae. We need the famous five (see p. 270): the sum of the y values,

growth

```
[1]  12  10  8  11  6  7  2  3  3
```

sum(growth)

```
[1]  62
```

the sum of the squares of the y values,

growth^2

```
[1]  144  100  64  121  36  49  4  9  9
```

sum(growth^2)

```
[1]  536
```

the sum of the x values,

tannin

```
[1]  0  1  2  3  4  5  6  7  8
```

sum(tannin)

```
[1]  36
```

the sum of the squares of the x values,

tannin^2

```
[1]  0  1  4  9  16  25  36  49  64
```

sum(tannin^2)

```
[1]  204
```

and finally, to measure the covariation between x and y, we need the sum of the products,

growth*tannin

```
[1]  0  10  16  33  24  35  12  21  24
```

sum(growth*tannin)

```
[1]  175
```

You can see at once that for more complicated models (such as multiple regression) it is essential to be able to generalize and streamline this procedure. This is where matrices come in. Matrix multiplication involves the calculation of sums of products where a row vector is multiplied by a column vector of the same length to obtain a single value. Thus, we should be able to obtain the required sum of products, 175, by using matrix multiplication symbol %*% in place of the regular multiplication symbol:

growth%*%tannin

```
       [,1]
[1,]   175
```

That works fine. But what about sums of squares? Surely if we use matrix multiplication on the same vector we will get an object with many rows (9 in this case). Not so.

growth%*%growth

```
       [,1]
[1,]   536
```

R has coerced the left-hand vector of growth into a row vector in order to obtain the desired result. You can override this, if for some reason you wanted the answer to have 9 rows, by specifying the transpose t() of the right-hand growth vector,

growth%*%t(growth)

```
      [,1] [,2] [,3] [,4] [,5] [,6] [,7] [,8] [,9]
[1,]  144  120   96  132   72   84   24   36   36
[2,]  120  100   80  110   60   70   20   30   30
[3,]   96   80   64   88   48   56   16   24   24
[4,]  132  110   88  121   66   77   22   33   33
[5,]   72   60   48   66   36   42   12   18   18
[6,]   84   70   56   77   42   49   14   21   21
[7,]   24   20   16   22   12   14    4    6    6
[8,]   36   30   24   33   18   21    6    9    9
[9,]   36   30   24   33   18   21    6    9    9
```

but, of course, that is not what we want. R's default is what we need. So this should also work in obtaining the sum of squares of the explanatory variable:

tannin%*%tannin

```
      [,1]
[1,]   204
```

So far, so good. But how do we obtain the sums using matrix multiplication? The trick here is to matrix multiply the vector by a vector of 1s: here are the sum of the *y* values,

growth%*%rep(1,9)

```
      [,1]
[1,]    62
```

and the sum of the *x* values,

tannin%*%rep(1,9)

```
      [,1]
[1,]    36
```

Finally, can we use matrix multiplication to arrive at the sample size, *n*? We do this by matrix multiplying a row vector of 1s by a column vector of 1s. This rather curious operation produces the right result, by adding up the nine 1s that result from the nine repeats of the calculation 1×1:

rep(1,9)%*%rep(1,9)

```
      [,1]
[1,]     9
```

But how do we get all of the famous five in a single matrix? The thing to understand is the dimensionality of such a matrix. It needs to contain sums as well as sums of products. We have two variables (growth and tannin) and their matrix multiplication produces a single scalar value (see above). In order to get to the sums of squares as well as the sums of products we use cbind to create a 9×2 matrix like this:

a<-cbind(growth,tannin)
a

```
       growth  tannin
[1,]       12       0
[2,]       10       1
[3,]        8       2
[4,]       11       3
[5,]        6       4
[6,]        7       5
[7,]        2       6
[8,]        3       7
[9,]        3       8
```

To obtain a results table with 2 rows rather than 9 rows we need to multiply the transpose of matrix *a* by matrix *a*:

t(a)%*%a

```
         growth  tannin
growth      536     175
tannin      175     204
```

That's OK as far as it goes, but it has only given us the sums of squares (536 and 204) and the sum of products (175). How do we get the sums as well? The trick is to bind a column of 1s onto the left of matrix a:

```
b<-cbind(1,growth,tannin)
b
```

```
      growth  tannin
[1,]  1    12       0
[2,]  1    10       1
[3,]  1     8       2
[4,]  1    11       3
[5,]  1     6       4
[6,]  1     7       5
[7,]  1     2       6
[8,]  1     3       7
[9,]  1     3       8
```

It would look better if the first column had a variable name: let's call it 'sample':

```
dimnames(b)[[2]] [1]<-"sample"
```

Now to get a summary table of sums as well as sums of products, we matrix multiply b by itself. We want the answer to have three rows (rather than nine) so we matrix multiply the transpose of b (which has three rows) by b (which has nine rows):

```
t(b)%*%b
```

```
        sample  growth  tannin
sample       9      62      36
growth      62     536     175
tannin      36     175     204
```

So there you have it. All of the famous five, plus the sample size, in a single matrix multiplication.

Statistical models in matrix notation

We continue this example to show how matrix algebra is used to generalize the procedures used in linear modelling (such as regression or analysis of variance) based on the values of the famous five. We want to be able to determine the parameter estimates (such as the intercept and slope of a linear regression) and to apportion the total sum of squares between variation explained by the model (SSR) and unexplained variation (SSE). Expressed in matrix terms, the linear regression model is

$$Y = Xb + e,$$

and we want to determine the least-squares estimate of b, given by

$$b = (X'X) - 1X'Y,$$

and then carry out the analysis of variance

$$b'X'Y.$$

We look at each of these in turn.

The response variable Y, 1 and the errors e are simple $n \times 1$ column vectors, X is an $n \times 2$ matrix and β is a 2×1 vector of coefficients, as follows:

$$Y = \begin{bmatrix} 12 \\ 10 \\ 8 \\ 11 \\ 6 \\ 7 \\ 2 \\ 3 \\ 3 \end{bmatrix}, X = \begin{bmatrix} 1 & 0 \\ 1 & 1 \\ 1 & 2 \\ 1 & 3 \\ 1 & 4 \\ 1 & 5 \\ 1 & 6 \\ 1 & 7 \\ 1 & 8 \end{bmatrix}, e = \begin{bmatrix} e_1 \\ e_2 \\ e_3 \\ e_4 \\ e_5 \\ e_6 \\ e_7 \\ e_8 \\ e_9 \end{bmatrix}, 1 = \begin{bmatrix} 1 \\ 1 \\ 1 \\ 1 \\ 1 \\ 1 \\ 1 \\ 1 \\ 1 \end{bmatrix}, \beta = \begin{bmatrix} \beta_0 \\ \beta_1 \end{bmatrix}.$$

The y vector and the 1 vector are created like this:

```
Y<-growth
one<-rep(1,9)
```

The sample size is given by $1'1$ (transpose of vector 1 times vector 1):

```
t(one) %*% one
```

```
        [,1]
[,1]     9
```

The vector of explanatory variable(s) X is created by binding a column of ones to the left

```
X<-cbind(1,tannin)
X
```

```
        tannin
[1,]  1      0
[2,]  1      1
[3,]  1      2
[4,]  1      3
[5,]  1      4
[6,]  1      5
[7,]  1      6
[8,]  1      7
[9,]  1      8
```

In this notation

$$\sum y^2 = y_1^2 + y_2^2 + \cdots + y_n^2 = Y'Y,$$

```
t(Y)%*%Y
```

```
        [,1]
[,1]    536
```

$$\sum y = n\bar{y} = y_1 + y_2 + \cdots + y_n = 1'Y$$

```
t(one)%*%Y
```

```
        [,1]
[,1]     62
```

$$\left(\sum y\right)^2 = Y'11'Y$$

t(Y) %*% one %*% t(one) %*% Y

```
      [1,]
[1,] 3844
```

For the matrix of explanatory variables, we see that X'X gives a 2×2 matrix containing n, $\sum x$ and $\sum x^2$. The numerical values are easy to find using matrix multiplication %*%

t(X)%*%X

```
            tannin
         9      36
tannin  36     204
```

Note that X'X (a 2×2 matrix) is completely different from XX' (a 9×9 matrix). The matrix X'Y gives a 2×1 matrix containing $\sum y$ and the sum of products $\sum xy$:

t(X)%*%Y

```
         [,1]
          62
tannin   175
```

Now, using the beautiful symmetry of the normal equations

$$b_0 n + b_1 \sum x = \sum y,$$
$$b_0 \sum x + b_1 \sum x^2 = \sum xy,$$

we can write the regression directly in matrix form as

$$X'Xb = X'Y$$

because we already have the necessary matrices to form the left- and right-hand sides. To find the least-squares parameter values b we need to divide both sides by X'X. This involves calculating the inverse of the X'X matrix. The inverse exists only when the matrix is square and when its determinant is non-singular. The inverse contains $-\bar{x}$ and $\sum x^2$ as its terms, with $SSX = \sum (x - \bar{x})^2$, the sum of the squared differences between the x values and mean x, or $n.SSX$ as the denominator:

$$(X'X)^{-1} = \begin{bmatrix} \dfrac{\sum x^2}{n \sum (x - \bar{x})^2} & \dfrac{-\bar{x}}{\sum (x - \bar{x})^2} \\ \dfrac{-\bar{x}}{\sum (x - \bar{x})^2} & \dfrac{1}{\sum (x - \bar{x})^2} \end{bmatrix}.$$

When every element of a matrix has a common factor, it can be taken outside the matrix. Here, the term $1/(n.SSX)$ can be taken outside to give

$$(X'X)^{-1} = \frac{1}{n \sum (x - \bar{x})^2} \begin{bmatrix} \sum x^2 & -\sum x \\ -\sum x & n \end{bmatrix}.$$

Computing the numerical value of this is easy using the matrix function ginv:

```
library(MASS)
ginv(t(X)%*%X)
```

```
              [,1]           [,2]
[1,]   0.37777778  -0.06666667
[2,]  -0.06666667   0.01666667
```

Now we can solve the normal equations

$$(X'X) - 1(X'X)b = (X'X) - 1X'Y,$$

using the fact that $(X'X) - 1(X'X) = I$ to obtain the important general result:

$$b = (X'X) - 1X'Y,$$

```
ginv(t(X)%*%X)%*%t(X)%*%Y
```

```
              [,1]
[1,]   11.755556
[2,]   -1.216667
```

which you will recognize from our hand calculations as the intercept and slope respectively (see p. 392). The ANOVA computations are as follows. The correction factor is

$$CF = Y'11'Y/n$$

```
CF<-t(Y) %*% one %*% t(one)%*% Y/9
CF
```

```
              [,1]
[1,]   427.1111
```

The total sum of squares, SSY, is $Y'Y - CF$:

```
t(Y)%*%Y-CF
```

```
              [,1]
[1,]   108.8889
```

The regression sum of squares, SSR, is $b'X'Y - CF$:

```
b<-ginv(t(X)%*%X)%*%t(X)%*%Y
t(b)%*%t(X)%*%Y-CF
```

```
              [,1]
[1,]   88.81667
```

and the error sum of squares, SSE, is $Y'Y - b'X'Y$

```
t(Y) %*% Y - t(b) %*% t(X) %*% Y
```

```
              [,1]
[1,]   20.07222
```

You should check these figures against the hand calculations on p. 396. Obviously, this is not a sensible way to carry out a single linear regression, but it demonstrates how to generalize the calculations for cases that have two or more continuous explanatory variables.

Solving systems of linear equations using matrices

Suppose we have two equations containing two unknown variables:

$$3x + 4y = 12,$$

$$x + 2y = 8.$$

We can use the function solve to find the values of the variables if we provide it with two matrices:

- a square matrix A containing the *coefficients* (3, 1, 4 and 2, columnwise);

- a column vector kv containing the *known values* (12 and 8).

We set the two matrices up like this (columnwise, as usual)

```
A<-matrix(c(3,1,4,2),nrow=2)
A
```

```
     [,1]  [,2]
[1,]    3     4
[2,]    1     2
```

```
kv<-matrix(c(12,8),nrow=2)
kv
```

```
     [,1]
[1,]   12
[2,]    8
```

Now we can solve the simultaneous equations

```
solve(A,kv)
```

```
     [,1]
[1,]   -4
[2,]    6
```

to give $x = -4$ and $y = 6$ (which you can easily verify by hand). The function is most useful when there are many simultaneous equations to be solved.

Calculus

The rules of differentiation and integration are known to R. You will use them in modelling (e.g. in calculating starting values in non-linear regression) and for numeric minimization using optim. Read the help files for D and integrate to understand the limitations of these functions.

Derivatives

The R function for symbolic and algorithmic derivatives of simple expressions is D. Here are some simple examples to give you the idea. See also ?deriv.

```
D(expression(2*x^3),"x")
```

```
2 * (3 * x^2)
```

```
D(expression(log(x)),"x")
```

```
1/x
```

```
D(expression(a*exp(-b * x)),"x")
```

```
-(a * (exp(-b * x) * b))
```

```
D(expression(a/(1+b*exp(-c * x))),"x")
```

```
a * (b * (exp(-c * x) * c))/(1 + b * exp(-c * x))^2
```

```
trig.exp <-expression(sin(cos(x + y^2)))
D(trig.exp, "x")
```

```
-(cos(cos(x + y^2)) * sin(x + y^2))
```

Integrals

The R function is integrate. Here are some simple examples to give you the idea:

```
integrate(dnorm,0,Inf)
```

```
0.5 with absolute error < 4.7e-05
```

```
integrate(dnorm,-Inf,Inf)
```

```
1 with absolute error < 9.4e-05
```

```
integrate(function(x) rep(2, length(x)), 0, 1)
```

```
2 with absolute error < 2.2e-14
```

```
integrand <-function(x) {1/((x+1)*sqrt(x))}
integrate(integrand, lower = 0, upper = Inf)
```

```
3.141593 with absolute error < 2.7e-05
```

```
xv<-seq(0,10,0.1)
plot(xv,integrand(xv),type="l")
```

The area under the curve is $\pi = 3.141\,593$.

Differential equations

We need to solve a system of ordinary differential equations (ODEs) and choose to use the classical Runge–Kutta fourth-order integration function rk4 from the odesolve package:

```
install.packages("odesolve")
library(odesolve)
```

The example involves a simple resource-limited plant herbivore where $V =$ vegetation and $N =$ herbivore population. We need to specify two differential equations: one for the vegetation (dV/dt) and one for the herbivore population (dN/dt):

$$\frac{dV}{dt} = rV\left(\frac{K - V}{K}\right) - bVN,$$

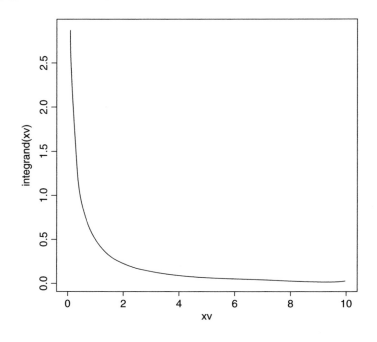

$$\frac{dN}{dt} = cVN - dN.$$

The steps involved in solving these ODEs in R are as follows:

- Define a function (called **phmodel** in this case).
- Name the response variables V and N from x[1] and x[2].
- Write the vegetation equation as dv using **with**.
- Write the herbivore equation as dn using **with**.
- Combine these vectors into a list called res.
- Generate a time series over which to solve the equations.
- Here, t is from 0 to 500 in steps of 1.
- Set the parameter values in **parms**.
- Set the starting values for V and N in y and xstart.
- Use **rk4** to create a dataframe with the V and N time series.

```
phmodel <-function(t, x, parms) {
v<-x[1]
n<-x[2]
with(as.list(parms), {
dv<-r*v*(K-v)/K - b*v*n
dn<-c*v*n – d*n
res<-c(dv, dn)
```

```
list(res)
})}
```

```
times <-seq(0, 500, length=501)
```

```
parms <-c(r=0.4, K=1000, b=0.02, c=0.01, d=0.3)
```

```
y<-xstart <-c(v=50, n=10)
output <-as.data.frame(rk4(xstart, times, phmodel, parms))
```

```
plot (output$time, output$v,
ylim=c(0,60),type="n",ylab="abundance",xlab="time")
lines (output$time, output$v)
lines (output$time, output$n,lty=2)
```

The output shows plants abundance as a solid line against time and herbivore abundance as a dotted line:

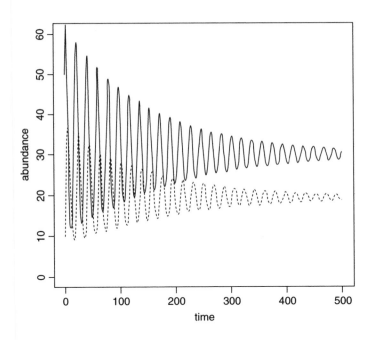

The system exhibits damped oscillations to a stable point equilibrium at which dV/dt and dN/dt are both equal to zero, so equilibrium plant abundance $= d/c = 0.3/0.01 = 30$ and equilibrium herbivore abundance $= r(K - V^*)/bK = 19.4$.

8

Classical Tests

There is absolutely no point in carrying out an analysis that is more complicated than it needs to be. Occam's razor applies to the choice of statistical model just as strongly as to anything else: simplest is best. The so-called classical tests deal with some of the most frequently used kinds of analysis for single-sample and two-sample problems.

Single Samples

Suppose we have a single sample. The questions we might want to answer are these:

- What is the mean value?
- Is the mean value significantly different from current expectation or theory?
- What is the level of uncertainty associated with our estimate of the mean value?

In order to be reasonably confident that our inferences are correct, we need to establish some facts about the distribution of the data:

- Are the values normally distributed or not?
- Are there outliers in the data?
- If data were collected over a period of time, is there evidence for serial correlation?

Non-normality, outliers and serial correlation can all invalidate inferences made by standard parametric tests like Student's *t* test. It is much better in cases with non-normality and/or outliers to use a non-parametric technique such as Wilcoxon's signed-rank test. If there is serial correlation in the data, then you need to use time series analysis or mixed-effects models.

Data summary

To see what is involved in summarizing a single sample, read the data called *y* from the file called das.txt:

```
data<-read.table("c:\\temp\\das.txt",header=T)
```

The R Book   Michael J. Crawley
© 2007 John Wiley & Sons, Ltd

names(data)

```
[1]  "y"
```

attach(data)

As usual, we begin with a set of single sample plots: an index plot (scatterplot with a single argument, in which data are plotted in the order in which they appear in the dataframe), a box-and-whisker plot (see p. 155) and a frequency plot (a histogram with bin-widths chosen by R):

```
par(mfrow=c(2,2))
plot(y)
boxplot(y)
hist(y,main="")
y2<-y
y2[52]<-21.75
plot(y2)
```

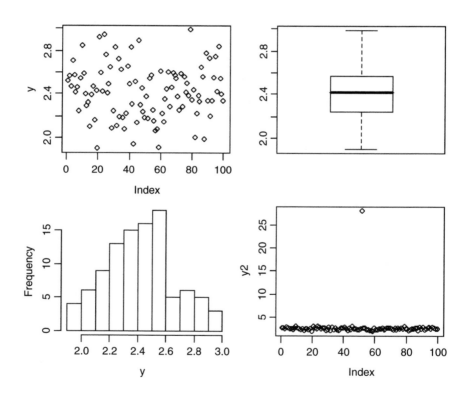

The index plot (bottom right) is particularly valuable for drawing attention to mistakes in the dataframe. Suppose that the 52nd value had been entered as 21.75 instead of 2.175: the mistake stands out like a sore thumb in the plot (bottom right).

Summarizing the data could not be simpler. We use the built-in function called summary like this:

summary(y)

```
 Min.  1st Qu.  Median   Mean  3rd Qu.    Max.
1.904    2.241   2.414  2.419    2.568   2.984
```

This gives us six pieces of information about the vector called y. The smallest value is 1.904 (labelled Min. for minimum) and the largest value is 2.984 (labelled Max. for maximum). There are two measures of central tendency: the median is 2.414 and the arithmetic mean in 2.419. The other two figures (labelled 1st Qu. and 3rd Qu.) are the first and third quartiles (see p. 155).

An alternative is Tukey's 'five-number summary' which comprises minimum, lower-hinge, median, upper-hinge and maximum for the input data. Hinges are close to the first and third quartiles (compare with summary, above), but different for small samples (see below):

fivenum(y)

```
[1]  1.903978  2.240931  2.414137  2.569583  2.984053
```

This is how the fivenum summary is produced: x takes the sorted values of y, and n is the length of y. Then five numbers, d, are calculated to use as subscripts to extract five averaged values from x like this:

x<-sort(y)
n<-length(y)
d <- c(1, 0.5 * floor(0.5 * (n + 3)), 0.5 * (n + 1),
    n + 1 - 0.5 * floor(0.5 * (n + 3)), n)
0.5 * (x[floor(d)] + x[ceiling(d)])

```
[1]  1.903978  2.240931  2.414137  2.569583  2.984053
```

where the d values are

```
[1]   1.0   25.5   50.5   75.5   100.0
```

with floor and ceiling providing the lower and upper subscripts for averaging.

Plots for testing normality

The simplest test of normality (and in many ways the best) is the 'quantile–quantile plot'. This plots the ranked samples from our distribution against a similar number of ranked quantiles taken from a normal distribution. If our sample is normally distributed then the line will be straight. Departures from normality show up as various sorts of non-linearity (e.g. S-shapes or banana shapes). The functions you need are qqnorm and qqline (quantile–quantile plot against a normal distribution):

qqnorm(y)
qqline(y,lty=2)

This shows a slight S-shape, but there is no compelling evidence of non-normality (our distribution is somewhat skew to the left; see the histogram, above).

Test for normality

We might use shapiro.test for testing whether the data in a vector come from a normal distribution. Let's generate some data that are lognormally distributed, so we should want them to *fail* the normality test:

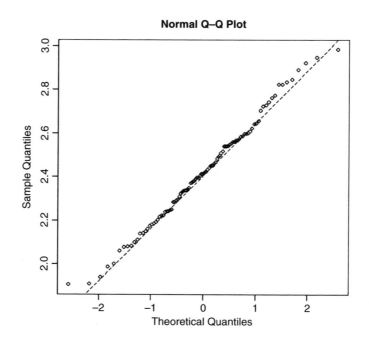

x<-exp(rnorm(30))
shapiro.test(x)

```
    Shapiro-Wilk normality test

data: x
W = 0.5701, p-value = 3.215e-08
```

They do: $p < 0.000\,001$. A p value is an estimate of the probability that a particular result, or a result more extreme than the result observed, could have occurred by chance, *if the null hypothesis were true*. In short, the p value is a measure of the credibility of the null hypothesis. If something is sufficiently unlikely to have occurred by chance (say, $p < 0.05$), we say that it is statistically significant. For example, in comparing two sample means, where the null hypothesis is that the means are the same, a low p value means that the hypothesis is unlikely to be true and the difference is statistically significant. A large p value (say $p = 0.23$) means that there is no compelling evidence on which to reject the null hypothesis. Of course, saying 'we do not reject the null hypothesis' and 'the null hypothesis is true' are two quite different things. For instance, we may have failed to reject a false null hypothesis because our sample size was too low, or because our measurement error was too large. Thus, p values are interesting, but they don't tell the whole story: effect sizes and sample sizes are equally important in drawing conclusions.

An example of single-sample data

We can investigate the issues involved by examining the data from Michelson's famous experiment in 1879 to measure the speed of light (see Michelson, 1880). The dataframe called light contains his results (km s^{-1}), but with 299 000 subtracted.

light<-read.table("c:\\temp\\light.txt",header=T)

```
attach(light)
names(light)
```

```
[1]  "speed"
```

```
hist(speed,main="")
```

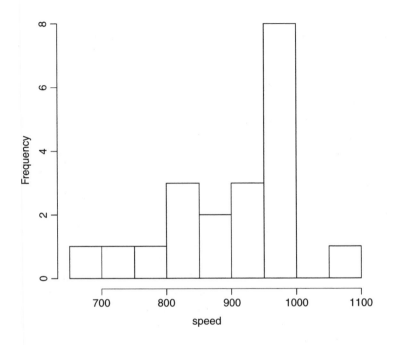

We get a summary of the non-parametric descriptors of the sample like this:

```
summary(speed)
```

```
Min.   1st Qu.   Median   Mean   3rd Qu.   Max.
 650       850      940    909       980   1070
```

From this, you see at once that the median (940) is substantially bigger than the mean (909), as a consequence of the strong negative skew in the data seen in the histogram. The **interquartile range** is the difference between the first and third quartiles: $980 - 850 = 130$. This is useful in the detection of outliers: a good rule of thumb is this: an **outlier** is a value more than 1.5 times the interquartile range above the third quartile or below the first quartile ($130 \times 1.5 = 195$). In this case, therefore, outliers would be measurements of speed that were less than $850 - 195 = 655$ or greater than $980 + 195 = 1175$. You will see that there are no large outliers in this data set, but one or more small outliers (the minimum is 650).

   We want to test the hypothesis that Michelson's estimate of the speed of light is significantly different from the value of 299 990 thought to prevail at the time. Since the data have all had 299 000 subtracted from them, the test value is 990. Because of the non-normality, the use of Student's t test in this case is ill advised. The correct test is Wilcoxon's signed-rank test.

```
wilcox.test(speed,mu=990)
```

```
 Wilcoxon signed rank test with continuity correction
data: speed
V = 22.5,  p-value = 0.00213
alternative hypothesis: true mu is not equal to 990

Warning message:
Cannot compute exact p-value with ties in: wilcox.test.default(speed,
mu = 990)
```

We reject the null hypothesis and accept the alternative hypothesis because $p = 0.002\,13$ (i.e. much less than 0.05). The speed of light is significantly less than 299 990.

Bootstrap in hypothesis testing

You have probably heard the old phrase about 'pulling yourself up by your own bootlaces'. That is where the term 'bootstrap' comes from. It is used in the sense of getting 'something for nothing'. The idea is very simple. You have a single sample of n measurements, but you can sample from this in very many ways, so long as you allow some values to appear more than once, and other samples to be left out (i.e. **sampling with replacement**). All you do is calculate the sample mean lots of times, once for each sampling from your data, then obtain the confidence interval by looking at the extreme highs and lows of the estimated means using a function called quantile to extract the interval you want (e.g. a 95% interval is specified using c(0.0275, 0.975) to locate the lower and upper bounds).

Our sample mean value of y is 909. The question we have been asked to address is this: how likely is it that the population mean that we are trying to estimate with our random sample of 100 values is as big as 990? We take 10 000 random samples with replacement using $n = 100$ from the 100 values of light and calculate 10 000 values of the mean. Then we ask: what is the probability of obtaining a mean as large as 990 by inspecting the right-hand tail of the cumulative probability distribution of our 10 000 bootstrapped mean values? This is not as hard as it sounds:

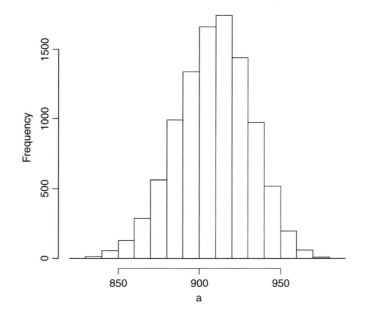

```
a<-numeric(10000)
for(i in 1:10000) a[i]<-mean(sample(speed,replace=T))
hist(a,main="")
```

The test value of 990 is way off the scale to the right, so a mean of 990 is clearly most unlikely, given the data with $\max(a) = 979$. In our 10 000 samples of the data, we never obtained a mean value greater than 979, so the probability that the mean is 990 is clearly $p < 0.0001$.

Higher-order moments of a distribution: quantifying non-normality

So far, and without saying so explicitly, we have encountered the first two moments of a sample distribution. The quantity $\sum y$ was used in the context of defining the arithmetic mean of a single sample: this is the first moment $\bar{y} = \sum y/n$. The quantity $\sum(y-\bar{y})^2$, the sum of squares, was used in calculating sample variance, and this is the second moment of the distribution $s^2 = \sum(y-\bar{y})^2/(n-1)$. Higher-order moments involve powers of the difference greater than 2 such as $\sum(y-\bar{y})^3$ and $\sum(y-\bar{y})^4$.

Skew

Skew (or skewness) is the dimensionless version of the third moment about the mean,

$$m_3 = \frac{\sum(y-\bar{y})^3}{n},$$

which is rendered dimensionless by dividing by the cube of the standard deviation of y (because this is also measured in units of y^3):

$$s_3 = \mathrm{sd}(y)^3 = (\sqrt{s^2})^3$$

The skew is then given by

$$\mathrm{skew} =_{\gamma_1} = \frac{m_3}{s_3}.$$

It measures the extent to which a distribution has long, drawn-out *tails* on one side or the other. A normal distribution is symmetrical and has $\gamma_1 = 0$. Negative values of γ_1 mean skew to the left (negative skew) and positive values mean skew to the right.

To test whether a particular value of skew is significantly different from 0 (and hence the distribution from which it was calculated is significantly non-normal) we divide the estimate of skew by its approximate standard error:

$$se_{\gamma_1} = \sqrt{\frac{6}{n}}$$

It is straightforward to write an R function to calculate the degree of skew for any vector of numbers, x, like this:

```
skew<-function(x){
m3<-sum((x-mean(x))^3)/length(x)
s3<-sqrt(var(x))^3
m3/s3 }
```

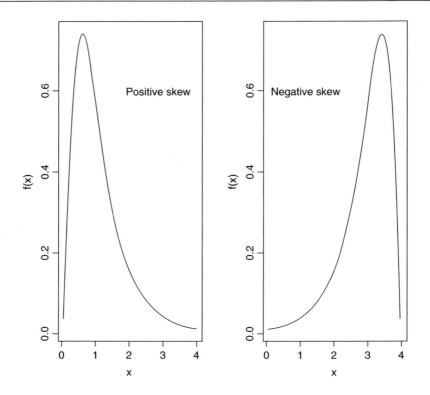

Note the use of the length(x) function to work out the sample size, n, whatever the size of the vector x. The last expression inside a function is not assigned to a variable name, and is returned as the value of skew(x) when this is executed from the command line.

```
data<-read.table("c:\\temp\\skewdata.txt",header=T)
attach(data)
names(data)
```

```
[1]  "values"
```

hist(values)

The data appear to be positively skew (i.e. to have a longer tail on the right than on the left). We use the new function skew to quantify the degree of skewness:

skew(values)

```
[1]  1.318905
```

Now we need to know whether a skew of 1.319 is significantly different from zero. We do a t test, dividing the observed value of skew by its standard error $\sqrt{6/n}$:

skew(values)/sqrt(6/length(values))

```
[1]  2.949161
```

Finally, we ask what is the probability of getting a t value of 2.949 by chance alone, when the skew value really is zero.

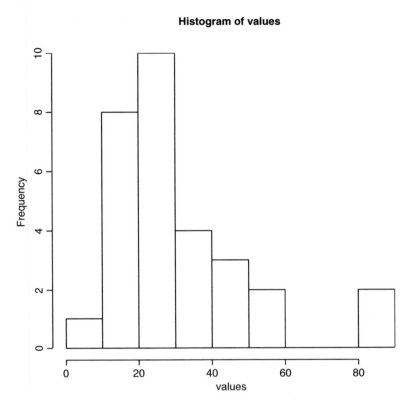

Histogram of values

1-pt(2.949,28)

```
[1]  0.003185136
```

We conclude that these data show significant non-normality ($p < 0.0032$).

The next step might be to look for a transformation that normalizes the data by reducing the skewness. One way of drawing in the larger values is to take square roots, so let's try this to begin with:

skew(sqrt(values))/sqrt(6/length(values))

```
[1]  1.474851
```

This is not significantly skew. Alternatively, we might take the logs of the values:

skew(log(values))/sqrt(6/length(values))

```
[1]  -0.6600605
```

This is now slightly skew to the left (negative skew), but the value of Student's t is smaller than with a square root transformation, so we might prefer a log transformation in this case.

Kurtosis

This is a measure of non-normality that has to do with the peakyness, or flat-toppedness, of a distribution. The normal distribution is bell-shaped, whereas a kurtotic distribution is other than bell-shaped. In particular, a more flat-topped distribution is said to be **platykurtic**, and a more pointy distribution is said to be **leptokurtic**.

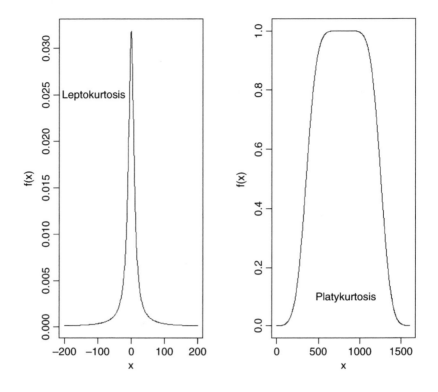

Kurtosis is the dimensionless version of the fourth moment about the mean,

$$m_4 = \frac{\sum(y - \bar{y})^4}{n},$$

which is rendered dimensionless by dividing by the square of the variance of y (because this is also measured in units of y^4):

$$s_4 = (\mathrm{var}(y))^2 = (s^2)^2$$

Kurtosis is then given by

$$\mathrm{kurtosis} = \gamma_2 = \frac{m_4}{s_4} - 3$$

The minus 3 is included because a normal distribution has $m_4/s_4 = 3$. This formulation therefore has the desirable property of giving zero kurtosis for a normal distribution, while a flat-topped (platykurtic) distribution has a negative value of kurtosis, and a pointy (leptokurtic) distribution has a positive value of kurtosis. The approximate standard error of kurtosis is

$$se_{\gamma_2} = \sqrt{\frac{24}{n}}.$$

An R function to calculate kurtosis might look like this:

```
kurtosis<-function(x) {
m4<-sum((x-mean(x))^4)/length(x)
s4<-var(x)^2
m4/s4 - 3 }
```

For our present data, we find that kurtosis is not significantly different from normal:

```
kurtosis(values)
```

```
[1]  1.297751
```

```
kurtosis(values)/sqrt(24/length(values))
```

```
[1]  1.45093
```

Two samples

The classical tests for two samples include:

- comparing two variances (Fisher's F test, var.test)
- comparing two sample means with normal errors (Student's t test, t.test)
- comparing two means with non-normal errors (Wilcoxon's rank test, wilcox.test)
- comparing two proportions (the binomial test, prop.test)
- correlating two variables (Pearson's or Spearman's rank correlation, cor.test)
- testing for independence of two variables in a contingency table (chi-squared, chisq.test, or Fisher's exact test, fisher.test).

Comparing two variances

Before we can carry out a test to compare two sample means (see below), we need to test whether the sample variances are significantly different (see p. 294). The test could not be simpler. It is called Fisher's F test after the famous statistician and geneticist R.A. Fisher, who worked at Rothamsted in south-east England. To compare two variances, all you do is divide the larger variance by the smaller variance. Obviously, if the variances are the same, the ratio will be 1. In order to be significantly different, the ratio will need to be significantly bigger than 1 (because the larger variance goes on top, in the numerator). How will we know a significant value of the variance ratio from a non-significant one? The answer, as always, is to look up the *critical value* of the variance ratio. In this case, we want critical values of Fisher's F. The R function for this is qf, which stands for 'quantiles of the F distribution'.

For our example of ozone levels in market gardens (see p. 51) there were 10 replicates in each garden, so there were $10 - 1 = 9$ degrees of freedom for each garden. In comparing two gardens, therefore, we have 9 d.f. in the numerator and 9 d.f. in the denominator. Although F tests in analysis of variance are typically one-tailed (the treatment variance is expected to be larger than the error variance if the means are significantly different; see p. 451), in this case, we had no expectation as to which garden was likely to have the higher variance, so we carry out a two-tailed test ($p = 1 - \alpha/2$). Suppose we work at the traditional $\alpha = 0.05$, then we find the critical value of F like this:

```
qf(0.975,9,9)
```

```
4.025994
```

This means that a calculated variance ratio will need to be greater than or equal to 4.02 in order for us to conclude that the two variances are significantly different at $\alpha = 0.05$.

To see the test in action, we can compare the variances in ozone concentration for market gardens B and C:

```
f.test.data<-read.table("c:\\temp\\f.test.data.txt",header=T)
attach(f.test.data)
names(f.test.data)
```

```
[1]  "gardenB"  "gardenC"
```

First, we compute the two variances:

```
var(gardenB)
```

```
[1]  1.333333
```

```
var(gardenC)
```

```
[1]  14.22222
```

The larger variance is clearly in garden C, so we compute the F ratio like this:

```
F.ratio<-var(gardenC)/var(gardenB)
F.ratio
```

```
[1]  10.66667
```

The variance in garden C is more than 10 times as big as the variance in garden B. The critical value of F for this test (with 9 d.f. in both the numerator and the denominator) is 4.026 (see qf, above), so, since the calculated value is larger than the critical value we reject the null hypothesis. The null hypothesis was that the two variances were not significantly different, so we accept the alternative hypothesis that the two variances are significantly different. In fact, it is better practice to present the p value associated with the calculated F ratio rather than just to reject the null hypothesis; to do this we use pf rather than qf. We double the resulting probability to allow for the two-tailed nature of the test:

```
2*(1-pf(F.ratio,9,9))
```

```
[1]  0.001624199
```

so the probability that the variances are the same is $p < 0.002$. Because the variances are significantly different, it would be wrong to compare the two sample means using Student's t-test.

There is a built-in function called var.test for speeding up the procedure. All we provide are the names of the two variables containing the raw data whose variances are to be compared (we don't need to work out the variances first):

var.test(gardenB,gardenC)

```
        F test to compare two variances
data:  gardenB  and  gardenC
F = 0.0938,  num  df = 9,  denom  df = 9,  p-value = 0.001624
alternative hypothesis: true ratio of variances is not equal to 1
95  percent  confidence  interval:
0.02328617  0.37743695
sample estimates:
ratio  of  variances
          0.09375
```

Note that the variance ratio, F, is given as roughly 1/10 rather than roughly 10 because var.test put the variable name that came first in the alphabet (gardenB) on top (i.e. in the numerator) instead of the bigger of the two variances. But the p value of 0.0016 is correct, and we reject the null hypothesis. These two variances are highly significantly different. This test is highly sensitive to outliers, so use it with care.

It is important to know whether variance differs significantly from sample to sample. Constancy of variance (**homoscedasticity**) is the most important assumption underlying regression and analysis of variance (p. 389). For comparing the variances of two samples, Fisher's F test is appropriate (p. 225). For multiple samples you can choose between the Bartlett test and the Fligner–Killeen test. Here are both tests in action:

refs<-read.table("c:\\temp\\refuge.txt",header=T)
attach(refs)
names(refs)

```
[1]  "B"  "T"
```

T is an ordered factor with 9 levels. Each level produces 30 estimates of yields except for level 9 which is a single zero. We begin by looking at the variances:

tapply(B,T,var)

```
       1          2          3          4          5          6          7          8
1354.024  2025.431  3125.292  1077.030  2542.599  2221.982  1445.490  1459.955
       9
      NA
```

When it comes to the variance tests we shall have to leave out level 9 of T because the tests require at least two replicates at each factor level. We need to know which data point refers to treatment $T = 9$:

which(T==9)

```
[1]  31
```

So we shall omit the 31st data point using negative subscripts. First Bartlett:

bartlett.test(B[-31],T[-31])

```
        Bartlett test of homogeneity of variances

data:  B[-31]  and  T[-31]
Bartlett's  K-squared = 13.1986,  df = 7,  p-value = 0.06741
```

So there is no significant difference between the eight variances ($p = 0.067$). Now Fligner:

```
fligner.test(B[-31],T[-31])
```

```
          Fligner-Killeen test of homogeneity of variances

data:  B[-31]  and  T[-31]
Fligner-Killeen:med  chi-squared = 14.3863,  df = 7,  p-value = 0.04472
```

Hmm. This test says that there *are* significant differences between the variances ($p < 0.05$). What you do next depends on your outlook. There are obviously some close-to-significant differences between these eight variances, but if you simply look at a plot of the data, plot(T,B), the variances appear to be very well behaved. A linear model shows some slight pattern in the residuals

```
model<-lm(B~T)
plot(model)
```

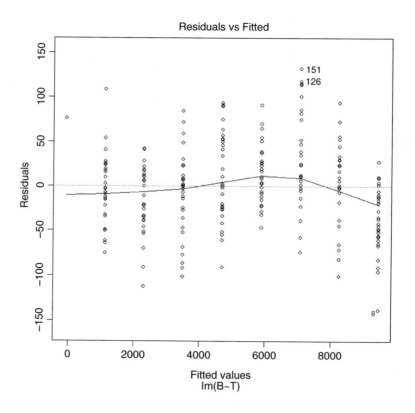

The various tests can give wildly different interpretations. Here are the ozone data from three market gardens:

```
ozone<-read.table("c:\\temp\\gardens.txt",header=T)
attach(ozone)
names(ozone)
```

```
[1]  "gardenA"  "gardenB"  "gardenC"
```

```
y<-c(gardenA,gardenB,gardenC)
garden<-factor(rep(c("A","B","C"),c(10,10,10)))
```

The question is whether the variance in ozone concentration differs from garden to garden or not. Fisher's F test comparing gardens B and C says that variance is significantly greater in garden C:

```
var.test(gardenB,gardenC)
```

```
        F test to compare two variances

data: gardenB and gardenC
F = 0.0938, num df = 9, denom df = 9, p-value = 0.001624
alternative hypothesis: true ratio of variances is not equal to 1
95 percent confidence interval:
0.02328617  0.37743695
sample estimates:
ratio of variances
        0.09375
```

Bartlett's test, likewise, says there is a highly significant difference in variance:

```
bartlett.test(y~garden)
```

```
        Bartlett test of homogeneity of variances

data: y by garden
Bartlett's K-squared = 16.7581, df = 2, p-value = 0.0002296
```

In contrast, the Fligner–Killeen test (preferred over Bartlett's test by many statisticians) says there is no compelling evidence for non-constancy of variance (**heteroscedasticity**) in these data:

```
fligner.test(y~garden)
```

```
        Fligner-Killeen test of homogeneity of variances

data: y by garden
Fligner-Killeen: med chi-squared = 1.8061, df = 2, p-value = 0.4053
```

The reason for the difference is that Fisher and Bartlett are very sensitive to outliers, whereas Fligner–Killeen is not (it is a non-parametric test which uses the ranks of the absolute values of the centred samples and weights $a(i) = qnorm((1 + i/(n+1))/2)$. Of the many tests for homogeneity of variances, this is the most robust against departures from normality (Conover et al., 1981).

You can use either a formula (as above) or a list (as below) to run these tests:

```
fligner.test(list(gardenA,gardenB,gardenC))
```

Comparing two means

Given what we know about the variation from replicate to replicate within each sample (the within-sample variance), how likely is it that our two sample means were drawn from populations with the same average? If it is highly likely, then we shall say that our two sample means are not significantly different. If it is rather unlikely, then we shall say that our sample means are significantly different. But perhaps a better way to proceed is to work out the probability that the two samples were indeed drawn from populations with the same mean. If this probability is very low (say, less than 5% or less than 1%) then we can be reasonably certain (95% or 99% in these two examples) than the means really are different from one another. Note, however, that we can never be 100% certain; the apparent difference might just be due to random sampling – we just happened to get a lot of low values in one sample, and a lot of high values in the other.

There are two simple tests for comparing two sample means:

- Student's t test when the samples are independent, the variances constant, and the errors are normally distributed;

- Wilcoxon's rank-sum test when the samples are independent but the errors are *not* normally distributed (e.g. they are ranks or scores or some sort).

What to do when these assumptions are violated (e.g. when the variances are different) is discussed later on.

Student's t test

Student was the pseudonym of W.S. Gossett who published his influential paper in *Biometrika* in 1908. He was prevented from publishing under his own name by dint of the archaic employment laws in place at the time, which allowed his employer, the Guinness Brewing Company, to prevent him publishing independent work. Student's t distribution, later perfected by R.A. Fisher, revolutionized the study of small-sample statistics where inferences need to be made on the basis of the sample variance s^2 with the population variance σ^2 unknown (indeed, usually unknowable). The test statistic is the number of standard errors by which the two sample means are separated:

$$t = \frac{\text{difference between the two means}}{\text{standard error of the difference}} = \frac{\bar{y}_A - \bar{y}_B}{se_{\text{diff}}}.$$

We know the standard error of the mean (see p. 54) but we have not yet met the standard error of the difference between two means. For two independent (i.e. non-correlated) variables, *the variance of a difference is the sum of the separate variances.* This important result allows us to write down the formula for the standard error of the difference between two sample means:

For 2 independent samples

$\sigma^2_{\bar{y}_A - \bar{y}_B} = \sigma^2_A + \sigma^2_B$

$$se_{\text{diff}} = \sqrt{\frac{s_A^2}{n_A} + \frac{s_B^2}{n_B}}$$

We now have everything we need to carry out Student's t test. Our null hypothesis is that the two sample means are the same, and we shall accept this unless the value of Student's t is so large that it is unlikely that such a difference could have arisen by chance alone. For the ozone example introduced on p. 289, each sample has 9 degrees of freedom, so we have 18 d.f. in total. Another way of thinking of this is to reason that the complete sample size as 20, and we have estimated two parameters from the data, $\bar{y}_A$ and $\bar{y}_B$, so we have $20 - 2 = 18$ d.f. We typically use 5% as the chance of rejecting the null hypothesis when it is true (this is the Type I error rate). Because we didn't know in advance which of the two gardens was going to have the higher mean ozone concentration (and we usually don't), this is a two-tailed test, so the *critical value* of Student's t is:

qt(0.975,18)

[1] 2.100922

This means that our test statistic needs to be bigger than 2.1 in order to reject the null hypothesis, and hence to conclude that the two means are significantly different at $\alpha = 0.05$.
 The dataframe is attached like this:

t.test.data<-read.table("c:\\temp\\t.test.data.txt",header=T)
attach(t.test.data)
names(t.test.data)

[1] "gardenA" "gardenB"

A useful graphical test for two samples employs the notch option of boxplot:

ozone<-c(gardenA,gardenB)
label<-factor(c(rep("A",10),rep("B",10)))
boxplot(ozone~label,notch=T,xlab="Garden",ylab="Ozone")

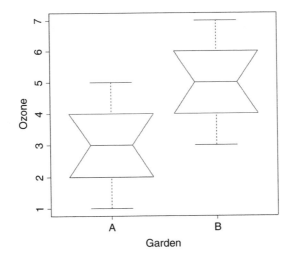

Because the notches of two plots do not overlap, we conclude that the medians are significantly different at the 5% level. Note that the variability is similar in both gardens, both in terms of the range + (the whiskers) + and the interquartile range + (the boxes).
 To carry out a t test long-hand, we begin by calculating the variances of the two samples:

s2A<-var(gardenA)
s2B<-var(gardenB)

The value of the test statistic for Student's t is the difference divided by the standard error of the difference. The numerator is the difference between the two means, and the denominator is the square root of the sum of the two variances divided by their sample sizes:

(mean(gardenA)-mean(gardenB))/sqrt(s2A/10+s2B/10)

which gives the value of Student's t as

[1] -3.872983

With *t*-tests you can ignore the minus sign; it is only the absolute value of the difference between the two sample means that concerns us. So the calculated value of the test statistic is 3.87 and the critical value is 2.10 (qt(0.975,18), above). We say that, since the calculated value is larger than the critical value, we reject the null hypothesis. Notice that the wording is exactly the same as it was for the *F* test (above). Indeed, the wording is always the same for all kinds of tests, and you should try to memorize it. The abbreviated form is easier to remember: 'larger reject, smaller accept'. The null hypothesis was that the two means are not significantly different, so we reject this and accept the alternative hypothesis that the two means are significantly different. Again, rather than merely rejecting the null hypothesis, it is better to state the probability that data as extreme as this (or more extreme) would be observed if the mean values were the same. For this we use pt rather than qt, and $2 \times$ pt because we are doing a two-tailed test:

2*pt(-3.872983,18)

[1] 0.001114540

so $p < 0.0015$.

You won't be surprised to learn that there is a built-in function to do all the work for us. It is called, helpfully, t.test and is used simply by providing the names of the two vectors containing the samples on which the test is to be carried out (gardenA and gardenB in our case).

t.test(gardenA,gardenB)

There is rather a lot of output. You often find this: the simpler the statistical test, the more voluminous the output.

```
        Welch Two Sample t-test

data:  gardenA  and  gardenB
t = -3.873, df = 18, p-value = 0.001115
alternative hypothesis: true difference in means is not equal to 0
95 percent confidence interval:
-3.0849115  -0.9150885
sample estimates:
mean of x mean of y
        3         5
```

The result is exactly the same as we obtained the long way. The value of *t* is -3.873 and since *the sign is irrelevant in a t test* we reject the null hypothesis because the test statistic is larger than the critical value of 2.1. The mean ozone concentration is significantly higher in garden B than in garden A. The computer output also gives a *p* value and a confidence interval. Note that, because the means are significantly different, *the confidence interval on the difference does not include zero* (in fact, it goes from -3.085 up to -0.915). You might present the result like this: 'Ozone concentration was significantly higher in garden B (mean $= 5.0$ pphm) than in garden A (mean $= 3.0$ pphm; $t = 3.873$, $p = 0.0011$ (2-tailed), d.f. $= 18$).'

There is a formula-based version of t.test that you can use when your explanatory variable consists of a two-level factor (see p. 492).

Wilcoxon rank-sum test

This is a non-parametric alternative to Student's *t* test, which we could use if the errors were non-normal. The Wilcoxon rank-sum test statistic, *W*, is calculated as follows. Both samples are put into a single array with their sample names clearly attached (*A* and *B* in this case, as explained below). Then the aggregate list is sorted, taking care to keep the sample labels with their respective values. A rank is assigned to each value, with ties getting the appropriate average rank (two-way ties get (rank $i+$ (rank $i+1$))/2, three-way ties get (rank $i+$ (rank $i+1$) + (rank $i+2$))/3, and so on). Finally the ranks are added up for each of the two samples, and significance is assessed on size of the smaller sum of ranks.

First we make a combined vector of the samples:

```
ozone<-c(gardenA,gardenB)
ozone
```

```
[1]  3  4  4  3  2  3  1  3  5  2  5  5  6  7  4  4  3  5  6  5
```

Then we make a list of the sample names:

```
label<-c(rep("A",10),rep("B",10))
label
```

```
[1]  "A"  "A"  "A"  "A"  "A"  "A"  "A"  "A"  "A"  "A"  "B"  "B"  "B"  "B"
     "B"  "B"  "B"  "B"  "B"  "B"
```

Now use the built-in function rank to get a vector containing the ranks, smallest to largest, within the combined vector:

```
combined.ranks<-rank(ozone)
combined.ranks
```

```
 [1]   6.0  10.5  10.5   6.0   2.5   6.0   1.0   6.0  15.0   2.5  15.0  15.0
      18.5  20.0  10.5
[16]  10.5   6.0  15.0  18.5  15.0
```

Notice that the ties have been dealt with by averaging the appropriate ranks. Now all we need to do is calculate the sum of the ranks for each garden. We use tapply with sum as the required operation

```
tapply(combined.ranks,label,sum)
```

```
 A    B
66   144
```

Finally, we compare the smaller of the two values (66) with values in tables of Wilcoxon rank sums (e.g. Snedecor and Cochran, 1980, p. 555), and reject the null hypothesis if our value of 66 is *smaller* than the value in tables. For samples of size 10 and 10 like ours, the 5% value in tables is 78. Our value is smaller than this, so we reject the null hypothesis. The two sample means are significantly different (in agreement with our earlier *t* test).

We can carry out the whole procedure automatically, and avoid the need to use tables of critical values of Wilcoxon rank sums, by using the built-in function wilcox.test:

```
wilcox.test(gardenA,gardenB)
```

```
        Wilcoxon rank sum test with continuity correction

data:  gardenA and gardenB
```

```
W = 11,  p-value = 0.002988
alternative hypothesis:  true  mu  is  not  equal  to  0

Warning  message:
Cannot  compute  exact  p-value  with  ties  in:
wilcox.test.default(gardenA,  gardenB)
```

The function uses a normal approximation algorithm to work out a z value, and from this a p value to assess the hypothesis that the two means are the same. This p value of 0.002 988 is much less than 0.05, so we reject the null hypothesis, and conclude that the mean ozone concentrations in gardens A and B are significantly different. The warning message at the end draws attention to the fact that there are ties in the data (repeats of the same ozone measurement), and this means that the p value cannot be calculated exactly (this is seldom a real worry).

It is interesting to compare the p values of the t test and the Wilcoxon test with the same data: $p = 0.001\,115$ and 0.002 988, respectively. The non-parametric test is much more appropriate than the t test when the errors are not normal, and the non-parametric test is about 95% as powerful with normal errors, and can be *more* powerful than the t test if the distribution is strongly skewed by the presence of outliers. Typically, as here, the t test will give the lower p value, so the Wilcoxon test is said to be conservative: if a difference is significant under a Wilcoxon test it would have been even more significant under a t test.

Tests on paired samples

Sometimes, two-sample data come from paired observations. In this case, we might expect a correlation between the two measurements, either because they were made on the same individual, or were taken from the same location. You might recall that the variance of a difference is the average of

$$(y_A - \mu_A)^2 + (y_B - \mu_B)^2 - 2(y_A - \mu_A)(y_B - \mu_B)$$

which is the variance of sample A, plus the variance of sample B, minus twice the covariance of A and B. When the covariance of A and B is *positive*, this is a great help because it reduces the variance of the difference, which makes it easier to detect significant differences between the means. Pairing is not always effective, because the correlation between y_A and y_B may be weak.

The following data are a composite biodiversity score based on a kick sample of aquatic invertebrates.

```
streams<-read.table("c:\\temp\\streams.txt",header=T)
attach(streams)
names(streams)
```

```
[1]  "down"  "up"
```

The elements are paired because the two samples were taken on the same river, one upstream and one downstream from the same sewage outfall.

If we ignore the fact that the samples are paired, it appears that the sewage outfall has no impact on biodiversity score ($p = 0.6856$):

t.test(down,up)

```
        Welch Two Sample t-test
data: down and up
t = -0.4088,  df = 29.755,  p-value = 0.6856
alternative hypothesis: true difference in means is not equal to 0
95 percent confidence interval:
-5.248256  3.498256
sample estimates:
mean of x  mean of y
  12.500     13.375
```

However, if we allow that the samples are paired (simply by specifying the option paired=T), the picture is completely different:

t.test(down,up,paired=T)

```
        Paired t-test
data: down and up
t = -3.0502,  df = 15,  p-value = 0.0081
alternative hypothesis: true difference in means is not equal to 0
95 percent confidence interval:
-1.4864388  -0.2635612
sample estimates:
mean of the differences
              -0.875
```

Now the difference between the means is highly significant ($p = 0.0081$). The moral is clear. If you can do a paired t test, then you should always do the paired test. It can never do any harm, and sometimes (as here) it can do a huge amount of good. In general, if you have information on *blocking* or *spatial correlation* (in this case, the fact that the two samples came from the same river), then you should always use it in the analysis.

Here is the same paired test carried out as a one-sample t test on the differences between the pairs:

d <- up-down
t.test(d)

```
        One Sample t-test
data:  d
t = 3.0502,  df = 15,  p-value = 0.0081
alternative hypothesis:  true mean is not equal to 0
95 percent confidence interval:
0.2635612  1.4864388
sample estimates:
mean of x
   0.875
```

As you see, the result is identical to the two-sample t test with paired=T ($p = 0.0081$). The upstream values of the biodiversity score were greater by 0.875 on average, and this difference is highly significant. Working with the differences has halved the number of degrees of freedom (from 30 to 15), but it has more than compensated for this by reducing the error variance, because there is such a strong positive correlation between y_A and y_B.

The sign test

This is one of the simplest of all statistical tests. Suppose that you cannot *measure* a difference, but you can *see* it (e.g. in judging a diving contest). For example, nine springboard divers were scored as better or worse, having trained under a new regime and under the conventional regime (the regimes were allocated in a randomized sequence to each athlete: new then conventional, or conventional then new). Divers were judged twice: one diver was worse on the new regime, and 8 were better. What is the evidence that the new regime produces significantly better scores in competition? The answer comes from a two-tailed binomial test. How likely is a response of 1/9 (or 8/9 or more extreme than this, i.e. 0/9 or 9/9) if the populations are actually the same (i.e. $p = 0.5$)? We use a binomial test for this, specifying the number of 'failures' (1) and the total sample size (9):

binom.test(1,9)

```
        Exact binomial test
data: 1 out of 9
number of successes = 1,  n = 9,  p-value = 0.0391
alternative hypothesis:  true p is not equal to 0.5
```

We would conclude that the new training regime is significantly better than the traditional method, because $p < 0.05$.

It is easy to write a function to carry out a sign test to compare two samples, x and y

```
sign.test <- function(x, y)
{
if(length(x) != length(y)) stop("The two variables must be the same length")
d <- x - y
binom.test(sum(d > 0), length(d))
}
```

The function starts by checking that the two vectors are the same length, then works out the vector of the differences, d. The binomial test is then applied to the number of positive differences (sum(d > 0)) and the total number of numbers (length(d)). If there was no difference between the samples, then on average, the sum would be about half of length(d). Here is the sign test used to compare the ozone levels in gardens A and B (see above):

sign.test(gardenA,gardenB)

```
        Exact binomial test

data: sum(d > 0) and length(d)
number of successes = 0,  number of trials = 10,  p-value = 0.001953
alternative hypothesis: true probability of success is not equal to 0.5
95 percent confidence interval:
0.0000000  0.3084971
sample estimates:
probability of success
                0
```

Note that the p value (0.002) from the sign test is larger than in the equivalent t test ($p = 0.0011$) that we carried out earlier. This will generally be the case: other things being equal, the parametric test will be more powerful than the non-parametric equivalent.

Binomial test to compare two proportions

Suppose that only four females were promoted, compared to 196 men. Is this an example of blatant sexism, as it might appear at first glance? Before we can judge, of course, we need to know the number of male and female candidates. It turns out that 196 men were promoted out of 3270 candidates, compared with 4 promotions out of only 40 candidates for the women. Now, if anything, it looks like the females did better than males in the promotion round (10% success for women versus 6% success for men).

The question then arises as to whether the apparent positive discrimination in favour of women is statistically significant, or whether this sort of difference could arise through chance alone. This is easy in R using the built-in binomial proportions test prop.test in which we specify two vectors, the first containing the number of successes for females and males c(4,196) and second containing the total number of female and male candidates c(40,3270):

```
prop.test(c(4,196),c(40,3270))
```

```
2-sample test for equality of proportions with continuity correction

data: c(4, 196) out of c(40, 3270)
X-squared = 0.5229,  df = 1,  p-value = 0.4696
alternative hypothesis: two.sided
95 percent confidence interval:
-0.06591631  0.14603864
sample estimates:
    prop 1       prop 2
0.10000000  0.05993884
```

```
Warning message:
Chi-squared approximation may be incorrect in: prop.test(c(4, 196),
c(40, 3270))
```

There is no evidence in favour of positive discrimination ($p = 0.4696$). A result like this will occur more than 45% of the time by chance alone. Just think what would have happened if one of the successful female candidates had not applied. Then the same promotion system would have produced a female success rate of 3/39 instead of 4/40 (7.7% instead of 10%). In small samples, small changes have big effects.

Chi-squared contingency tables

A great deal of statistical information comes in the form of *counts* (whole numbers or integers); the number of animals that died, the number of branches on a tree, the number of days of frost, the number of companies that failed, the number of patients who died. With count data, the number 0 is often the value of a response variable (consider, for example, what a 0 would mean in the context of the examples just listed). The analysis of count data in tables is discussed in more detail in Chapters 14 and 15.

The dictionary definition of contingency is 'a thing dependent on an uncertain event' (OED, 2004). In statistics, however, the contingencies are *all the events that could possibly happen.* A contingency table shows the counts of how many times each of the contingencies actually happened in a particular sample. Consider the following example that has to do with the relationship between hair colour and eye colour in white people. For simplicity, we just chose two contingencies for hair colour: 'fair' and 'dark'. Likewise we just chose two contingencies for eye colour: 'blue' and 'brown'. Each of these two categorical variables, eye colour and hair colour, has two levels ('blue' and 'brown', and 'fair' and 'dark', respectively). Between them, they define four possible outcomes (the contingencies): fair hair and blue eyes, fair hair and brown eyes, dark hair and blue eyes, and dark hair and brown eyes. We take a sample of people and count how many of them fall into each of these four categories. Then we fill in the 2×2 contingency table like this:

	Blue eyes	Brown eyes
Fair hair	38	11
Dark hair	14	51

These are our observed frequencies (or counts). The next step is very important. In order to make any progress in the analysis of these data we need a *model* which predicts the expected frequencies. What would be a sensible model in a case like this? There are all sorts of complicated models that you might select, but the simplest model (Occam's razor, or the principle of parsimony) is that hair colour and eye colour are *independent*. We may not believe that this is actually true, but the hypothesis has the great virtue of being falsifiable. It is also a very sensible model to choose because it makes it easy to predict the expected frequencies based on the assumption that the model is true. We need to do some simple probability work. What is the probability of getting a random individual from this sample whose hair was fair? A total of 49 people $(38 + 11)$ had fair hair out of a total sample of 114 people. So the probability of fair hair is 49/114 and the probability of dark hair is 65/114. Notice that because we have only two levels of hair colour, these two probabilities add up to 1 $((49 + 65)/114)$. What about eye colour? What is the probability of selecting someone at random from this sample with blue eyes? A total of 52 people had blue eyes $(38 + 14)$ out of the sample of 114, so the probability of blue eyes is 52/114 and the probability of brown eyes is 62/114. As before, these sum to 1 $((52 + 62)/114)$. It helps to add the subtotals to the margins of the contingency table like this:

	Blue eyes	Brown eyes	Row totals
Fair hair	38	11	49
Dark hair	14	51	65
Column totals	52	62	114

Now comes the important bit. We want to know the expected frequency of people with fair hair *and* blue eyes, to compare with our observed frequency of 38. Our model says that the two are independent. This is essential information, because it allows us to calculate the expected probability of fair hair and blue eyes. If, and only if, the two traits are independent, then the probability of having fair hair and blue eyes is the product of the two probabilities. So, following our earlier calculations, the probability of fair hair and blue eyes is $49/114 \times 52/114$. We can do exactly equivalent things for the other three cells of the contingency table:

	Blue eyes	Brown eyes	Row totals
Fair hair	$49/114 \times 52/114$	$49/114 \times 62/114$	49
Dark hair	$65/114 \times 52/114$	$65/114 \times 62/114$	65
Column totals	52	62	114

Now we need to know how to calculate the expected frequency. It couldn't be simpler. It is just the probability multiplied by the total sample ($n = 114$). So the expected frequency of blue eyes and fair hair is $49/114 \times 52/114 \times 114 = 22.35$, which is much less than our observed frequency of 38. It is beginning to look as if our hypothesis of independence of hair and eye colour is false.

You might have noticed something useful in the last calculation: two of the sample sizes cancel out. Therefore, the expected frequency in each cell is just the row total (R) times the column total (C) divided by the grand total (G) like this:

$$E = \frac{R \times C}{G}.$$

We can now work out the four expected frequencies:

	Blue eyes	Brown eyes	Row totals
Fair hair	22.35	26.65	49
Dark hair	29.65	35.35	65
Column totals	52	62	114

Notice that the row and column totals (the so-called 'marginal totals') are retained under the model. It is clear that the observed frequencies and the expected frequencies are different. But in sampling, everything always varies, so this is no surprise. The important question is whether the expected frequencies are *significantly* different from the observed frequencies.

We can assess the significance of the differences between observed and expected frequencies in a variety of ways:

- Pearson's chi-squared;
- *G* test;
- Fisher's exact test.

Pearson's chi-squared

We begin with Pearson's chi-squared test. The test statistic χ^2 is

$$\chi^2 = \sum \frac{(O - E)^2}{E},$$

where O is the observed frequency and E is the expected frequency. It makes the calculations easier if we write the observed and expected frequencies in parallel columns, so that we can work out the corrected squared differences more easily.

	O	E	$(O-E)^2$	$\dfrac{(O-E)^2}{E}$
Fair hair and blue eyes	38	22.35	244.92	10.96
Fair hair and brown eyes	11	26.65	244.92	9.19
Dark hair and blue eyes	14	29.65	244.92	8.26
Dark hair and brown eyes	51	35.35	244.92	6.93

All we need to do now is to add up the four components of chi-squared to get $\chi^2 = 35.33$.

The question now arises: is this a big value of chi-squared or not? This is important, because if it *is* a bigger value of chi-squared than we would expect by chance, then we should reject the null hypothesis. If, on the other hand, it is within the range of values that we would expect by chance alone, then we should accept the null hypothesis.

We always proceed in the same way at this stage. We have a calculated value of the test statistic: $\chi^2 = 35.33$. We compare this value of the test statistic with the relevant critical value. To work out the critical value of chi-squared we need two things:

- the number of degrees of freedom, and

- the degree of certainty with which to work.

In general, a contingency table has a number of rows (r) and a number of columns (c), and the degrees of freedom is given by

$$\text{d.f.} = (r-1) \times (c-1).$$

So we have $(2-1) \times (2-1) = 1$ degree of freedom for a 2×2 contingency table. You can see why there is only one degree of freedom by working through our example. Take the 'fair hair, brown eyes' box (the top right in the table) and ask how many values this could possibly take. The first thing to note is that the count could not be more than 49, otherwise the row total would be wrong. But in principle, the number in this box is free to take any value between 0 and 49. We have one degree of freedom for this box. But when we have fixed this box to be 11

	Blue eyes	Brown eyes	Row totals
Fair hair		11	49
Dark hair			65
Column totals	52	62	114

you will see that we have no freedom at all for any of the other three boxes. The top left box has to be $49 - 11 = 38$ because the row total is fixed at 49. Once the top left box is defined as 38 then the bottom left box has to be $52 - 38 = 14$ because the column total is fixed (the total number of people with blue eyes was 52). This means that the bottom right box has to be $65 - 14 = 51$. Thus, *because the marginal totals are constrained*, a 2×2 contingency table has just one degree of freedom.

The next thing we need to do is say how certain we want to be about the falseness of the null hypothesis. The more certain we want to be, the larger the value of chi-squared we would need to reject the null hypothesis. It is conventional to work at the 95% level. That is our certainty level, so our uncertainty level is $100 - 95 = 5\%$. Expressed as a fraction, this is called alpha ($\alpha = 0.05$). Technically, alpha is the probability of *rejecting* the null hypothesis when it is *true*. This is called a Type I error. A Type II error is *accepting* the null hypothesis when it is *false*.

Critical values in R are obtained by use of *quantiles* (q) of the appropriate statistical distribution. For the chi-squared distribution, this function is called qchisq. The function has two arguments: the certainty level ($p = 0.95$), and the degrees of freedom (d.f. $= 1$):

```
qchisq(0.95,1)
```

```
[1] 3.841459
```

The critical value of chi-squared is 3.841. Since the calculated value of the test statistic is *greater* than the critical value we *reject* the null hypothesis.

What have we learned so far? We have rejected the null hypothesis that eye colour and hair colour are independent. But that's not the end of the story, because we have not established the *way* in which they are related (e.g. is the correlation between them positive or negative?). To do this we need to look carefully at the data, and compare the observed and expected frequencies. If fair hair and blue eyes were positively correlated, would the observed frequency be greater or less than the expected frequency? A moment's thought should convince you that the observed frequency will be greater than the expected frequency when the traits are positively correlated (and less when they are negatively correlated). In our case we expected only 22.35 but we observed 38 people (nearly twice as many) to have both fair hair and blue eyes. So it is clear that fair hair and blue eyes are *positively* associated.

In R the procedure is very straightforward. We start by defining the counts as a 2×2 matrix like this:

```
count<-matrix(c(38,14,11,51),nrow=2)
count
```

```
      [,1]  [,2]
[1,]    38    11
[2,]    14    51
```

Notice that you enter the data *columnwise* (not row-wise) into the matrix. Then the test uses the chisq.test function, with the matrix of counts as its only argument.

```
chisq.test(count)
```

```
        Pearson's Chi-squared test with Yates' continuity correction
data: count
X-squared = 33.112, df = 1, p-value = 8.7e-09
```

The calculated value of chi-squared is slightly different from ours, because Yates' correction has been applied as the default (see Sokal and Rohlf, 1995, p. 736). If you switch the correction off (correct=F), you get the value we calculated by hand:

chisq.test(count,correct=F)

```
        Pearson's Chi-squared test

data: count
X-squared = 35.3338, df = 1, p-value = 2.778e-09
```

It makes no difference at all to the interpretation: there is a highly significant positive association between fair hair and blue eyes for this group of people. If you need to extract the frequencies expected under the null hypothesis of independence then use

chisq.test(count,correct=F)$expected

```
        [,1]        [,2]
[1,]  22.35088   26.64912
[2,]  29.64912   35.35088
```

G test of contingency

The idea is exactly the same – we are looking for evidence of non-independence of hair colour and eye colour. Even the distribution of the test statistic is the same: chi-squared. The difference is in the test statistic. Instead of computing $\sum (O-E)^2/E$ we compute the deviance from a log-linear model (see p. 552):

$$G = 2\sum O \ln\left(\frac{O}{E}\right).$$

Here are the calculations:

	O	E	$\ln\left(\dfrac{O}{E}\right)$	$O\ln\left(\dfrac{O}{E}\right)$
Fair hair and blue eyes	38	22.35	0.5307598	20.168874
Fair hair and brown eyes	11	26.65	−0.8848939	−9.733833
Dark hair and blue eyes	14	29.65	−0.7504048	−10.505667
Dark hair and brown eyes	51	35.35	0.3665272	18.692889

The test statistic G is twice the sum of the right-hand column: $2\times 18.62226 = 37.24453$. This value is compared with chi-squared in tables with 1 d.f. The calculated value of the test statistic is much greater than the critical value (3.841) so we reject the null hypothesis of independence. Hair colour and eye colour are correlated in this group of people. We need to look at the data to see which way the correlation goes. We see far more people with fair hairs and blue eyes (38) than expected under the null hypothesis of independence (22.35) so the correlation is *positive*. Pearson's chi-squared was $\chi^2 = 35.33$ (above) so the test statistics are slightly different ($\chi^2 = 37.24$ in the G test) but the interpretation is identical.

So far we have assumed equal probabilities but chisq.test can deal with cases with unequal probabilities. This example has 21 individuals distributed over four categories:

chisq.test(c(10,3,2,6))

```
        Chi-squared test for given probabilities
data: c(10, 3, 2, 6)
X-squared = 7.381, df = 3, p-value = 0.0607
```

The four counts are not significantly different from null expectation if $p = 0.25$ in each cell. However, if the null hypothesis was that the third and fourth cells had 1.5 times the probability of the first two cells, then these counts *are* highly significant.

chisq.test(c(10,3,2,6),p=c(0.2,0.2,0.3,0.3))

```
        Chi-squared test for given probabilities
data: c(10,  3,  2,  6)
X-squared = 11.3016,  df = 3,  p-value = 0.01020

Warning message:
Chi-squared approximation may be incorrect in:
chisq.test(c(10,  3,  2,  6),  p = c(0.2,  0.2,  0.3,  0.3))
```

Note the warning message associated with the low expected frequencies in cells 1 and 2.

You can use the chisq.test function with table objects as well as vectors. To test the random number generator as a simulator of the throws of a six-sided die we could simulate 100 throws like this, then use table to count the number of times each number appeared:

die<-ceiling(runif(100,0,6))
table(die)

```
die
 1   2   3   4   5   6
18  17  16  13  20  16
```

So we observed only 13 fours in this trail and 20 fives. But is this a significant departure from fairness of the die? chisq.test will answer this:

chisq.test(table(die))

```
        Chi-squared test for given probabilities
data: table(die)
X-squared = 1.64, df = 5, p-value = 0.8964
```

No. This is a fair die ($p = 0.896$). Note that the syntax is chisq.test(table(die)) not chisq.test(die) and that there are 5 degrees of freedom in this case.

Contingency tables with small expected frequencies: Fisher's exact test

When one or more of the expected frequencies is less than 4 (or 5 depending on the rule of thumb you follow) then it is wrong to use Pearson's chi-squared or log-linear models (G tests) for your contingency table. This is because small expected values inflate the value of the test statistic, and it no longer can be assumed to follow the chi-squared distribution. The individual counts are *a, b, c* and *d* like this:

	Column 1	Column 2	Row totals
Row 1	a	b	$a+b$
Row 2	c	d	$c+d$
Column totals	$a+c$	$b+d$	n

The probability of any one particular outcome is given by

$$p = \frac{(a+b)!(c+d)!(a+c)!(b+d)!}{a!b!c!d!n!}$$

where *n* is the grand total.

Our data concern the distribution of 8 ants' nests over 10 trees of each of two species (A and B). There are two categorical explanatory variables (ants and trees), and four contingencies, ants (present or absent) and trees (A or B). The response variable is the vector of four counts c(6,4,2,8) entered columnwise:

	Tree A	Tree B	Row totals
With ants	6	2	8
Without ants	4	8	12
Column totals	10	10	20

We can calculate the probability for this particular outcome:

```
factorial(8)*factorial(12)*factorial(10)*factorial(10)/
  (factorial(6)*factorial(2)*factorial(4)*factorial(8)*factorial(20))
```

```
[1]  0.07501786
```

But this is only part of the story. We need to compute the probability of outcomes that are *more extreme* than this. There are two of them. Suppose only 1 ant colony was found on Tree B. Then the table values would be 7, 1, 3, 9 but the row and column totals would be exactly the same (*the marginal totals are constrained*). The numerator always stays the same, so this case has probability

```
factorial(8)*factorial(12)*factorial(10)*factorial(10)/
  (factorial(7)*factorial(3)*factorial(1)*factorial(9)*factorial(20))
```

```
[1]  0.009526078
```

There is an even more extreme case if no ant colonies at all were found on Tree B. Now the table elements become 8, 0, 2, 10 with probability

```
factorial(8)*factorial(12)*factorial(10)*factorial(10)/
  (factorial(8)*factorial(2)*factorial(0)*factorial(10)*factorial(20))
```

```
[1]  0.0003572279
```

and we need to add these three probabilities together:

0.07501786+0.009526078+0.000352279

```
[1]  0.08489622
```

But there was no *a priori* reason for expecting the result to be in this direction. It might have been tree A that had relatively few ant colonies. We need to allow for extreme counts in the opposite direction by doubling this probability (all Fisher's exact tests are two-tailed):

2*(0.07501786+0.009526078+0.000352279)

```
[1]  0.1697924
```

This shows that there is no evidence of a correlation between tree and ant colonies. The observed pattern, or a more extreme one, could have arisen by chance alone with probability $p = 0.17$.

There is a built-in function called fisher.test, which saves us all this tedious computation. It takes as its argument a 2×2 matrix containing the counts of the four contingencies. We make the matrix like this (compare with the alternative method of making a matrix, above):

x<-as.matrix(c(6,4,2,8))
dim(x)<-c(2,2)
x

```
          [,1]   [,2]
  [1,]      6      2
  [2,]      4      8
```

we then run the test like this:

fisher.test(x)

```
        Fisher's Exact Test for Count Data

data: x
p-value = 0.1698
alternative hypothesis: true odds ratio is not equal to 1
95 percent confidence interval:
0.6026805  79.8309210
sample estimates:
odds ratio
  5.430473
```

Another way of using the function is to provide it with two vectors containing factor levels, instead of a two-dimensional matrix of counts. This saves you the trouble of counting up how many combinations of each factor level there are:

table<-read.table("c:\\temp\\fisher.txt",header=T)
attach(table)
fisher.test(tree,nests)

```
        Fisher's Exact Test for Count Data

data: tree and nests
p-value = 0.1698
```

```
alternative hypothesis: true odds ratio is not equal to 1
95 percent confidence interval:
0.6026805  79.8309210
sample estimates:
odds ratio
  5.430473
```

The fisher.test can be used with matrices much bigger than 2×2.

Correlation and covariance

With two continuous variables, x and y, the question naturally arises as to whether their values are correlated with each other. Correlation is defined in terms of the variance of x, the variance of y, and the covariance of x and y (the way the two vary together, which is to say the way they covary) on the assumption that both variables are normally distributed. We have symbols already for the two variances: s_x^2 and s_y^2. We denote the covariance of x and y by $\mathrm{cov}(x, y)$, so the correlation coefficient r is defined as

$$r = \frac{\mathrm{cov}(x, y)}{\sqrt{s_x^2 s_y^2}}.$$

We know how to calculate variances, so it remains only to work out the value of the covariance of x and y. Covariance is defined as *the expectation of the vector product x^*y*. The covariance of x and y is the expectation of the product minus the product of the two expectations. Note that when x and y are independent (i.e. they are not correlated) then the covariance between x and y is 0, so $E[xy] = E[x].E[y]$ (i.e. the product of their mean values).

Let's do a numerical example.

```
data<-read.table("c:\\temp\\twosample.txt",header=T)
attach(data)
plot(x,y)
```

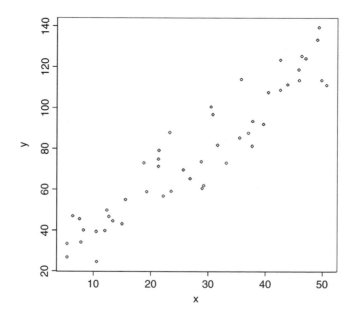

First, we need the variance of x and the variance of y:

var(x)

[1] 199.9837

var(y)

[1] 977.0153

The covariance of x and y, $\text{cov}(x, y)$, is given by the var function when we supply it with two vectors like this:

var(x,y)

[1] 414.9603

Thus, the correlation coefficient should be $414.96/\sqrt{199.98 \times 977.02}$:

var(x,y)/sqrt(var(x)*var(y))

[1] 0.9387684

Let's see if this checks out:

cor(x,y)

[1] 0.9387684

So now you know the definition of the correlation coefficient: it is the covariance divided by the geometric mean of the two variances.

Data dredging

The R function cor returns the correlation matrix of a data matrix, or a single value showing the correlation between one vector and another:

```
pollute<-read.table("c:\\temp\\Pollute.txt",header=T)
attach(pollute)
names(pollute)
```

```
[1]  "Pollution"      "Temp"   "Industry"   "Population"   "Wind"
[6]       "Rain"   "Wet.days"
```

cor(pollute)

	Pollution	Temp	Industry	Population	Wind
Pollution	1.00000000	-0.43360020	0.64516550	0.49377958	0.09509921
Temp	-0.43360020	1.00000000	-0.18788200	-0.06267813	-0.35112340
Industry	0.64516550	-0.18788200	1.00000000	0.95545769	0.23650590
Population	0.49377958	-0.06267813	0.95545769	1.00000000	0.21177156
Wind	0.09509921	-0.35112340	0.23650590	0.21177156	1.00000000
Rain	0.05428389	0.38628047	-0.03121727	-0.02606884	-0.01246601
Wet.days	0.36956363	-0.43024212	0.13073780	0.04208319	0.16694974

	Rain	Wet.days
Pollution	0.05428389	0.36956363
Temp	0.38628047	-0.43024212
Industry	-0.03121727	0.13073780
Population	-0.02606884	0.04208319
Wind	-0.01246601	0.16694974
Rain	1.00000000	0.49605834
Wet.days	0.49605834	1.00000000

The phrase 'data dredging' is used disparagingly to describe the act of trawling through a table like this, desperately looking for big values which might suggest relationships that you can publish. This behaviour is not to be encouraged. The correct approach is model simplification (see p. 327). Note that the correlations are identical in opposite halves of the matrix (in contrast to regression, where regression of y on x would give different parameter values and standard errors than regression of x on y). The correlation between two vectors produces a single value:

```
cor(Pollution,Wet.days)
```

```
[1]  0.3695636
```

Correlations with single explanatory variables can be highly misleading if (as is typical) there is substantial correlation amongst the explanatory variables (see p. 434).

Partial correlation

With more than two variables, you often want to know the correlation between x and y when a third variable, say, z, is held constant. The *partial correlation coefficient* measures this. It enables correlation due to a shared common cause to be distinguished from direct correlation. It is given by

$$r_{xy.z} = \frac{r_{xy} - r_{xz} \cdot r_{yz}}{\sqrt{(1 - r_{xz}^2)(1 - r_{yz}^2)}}.$$

Suppose we had four variables and we wanted to look at the correlation between x and y holding the other two, z and w, constant. Then

$$r_{xy.zw} = \frac{r_{xy.z} - r_{xw.z} \cdot r_{yw.z}}{\sqrt{(1 - r_{xw.z}^2)(1 - r_{yw.z}^2)}}.$$

You will need partial correlation coefficients if you want to do path analysis. R has a package called sem for carrying out structural equation modelling (including the production of path.diagram) and another called corpcor for converting correlations into partial correlations using the cor2pcor function (or vice versa with pcor2cor).

Correlation and the variance of differences between variables

Samples often exhibit positive correlations that result from pairing, as in the upstream and downstream invertebrate biodiversity data that we investigated earlier. There is an important general question about the effect of correlation on the variance of differences between variables. In the extreme, when two variables are so perfectly correlated that they are identical, then the difference between one variable and the other is zero. So it is clear that the variance of a difference will decline as the strength of positive correlation increases.

The following data show the depth of the water table (m below the surface) in winter and summer at 9 locations:

```
paired<-read.table("c:\\temp\\paired.txt",header=T)
attach(paired)
names(paired)
```

```
[1]  "Location"  "Summer"  "Winter"
```

We begin by asking whether there is a correlation between summer and winter water table depths across locations:

cor(Summer, Winter)

[1] 0.8820102

There is a strong positive correlation. Not surprisingly, places where the water table is high in summer tend to have a high water table in winter as well. If you want to determine the significance of a correlation (i.e. the p value associated with the calculated value of r) then use cor.test rather than cor. This test has non-parametric options for Kendall's *tau* or Spearman's rank, depending on the method you specify (method="k" or method="s"), but the default method is Pearson's product-moment correlation (method="p"):

cor.test(Summer, Winter)

```
        Pearson's product-moment correlation

data: Summer and Winter
t = 4.9521,  df = 7,  p-value = 0.001652
alternative hypothesis: true correlation is not equal to 0
95 percent confidence interval:
0.5259984  0.9750087
sample estimates:
      cor
0.8820102
```

The correlation is highly significant ($p = 0.00165$).

Now, let's investigate the relationship between the correlation coefficient and the three variances: the summer variance, the winter variance, and the *variance of the differences* (Summer−Winter)

varS=var(Summer)
varW=var(Winter)
varD=var(Summer - Winter)

The correlation coefficient ρ is related to these three variances by:

$$\rho = \frac{\sigma_y^2 + \sigma_z^2 - \sigma_{y-z}^2}{2\sigma_y\sigma_z}$$

So, using the values we have just calculated, we get the correlation coefficient to be

(varS+varW-varD)/(2*sqrt(varS)*sqrt(varW))

[1] 0.8820102

which checks out. We can also see whether the variance of the difference is equal to the sum of the component variances (see p. 298):

varD

[1] 0.01015

varS+varW

[1] 0.07821389

No, it is not. They would be equal only if the two samples were independent. In fact, we know that the two variables are positively correlated, so the variance of the difference should be *less* than the sum of the variances by an amount equal to $2 \times r \times s_1 \times s_2$

```
varS + varW − 2 * 0.8820102 * sqrt(varS) * sqrt(varW)
```

```
[1]  0.01015
```

That's more like it.

Scale-dependent correlations

Another major difficulty with correlations is that scatterplots can give a highly misleading impression of what is going on. The moral of this exercise is very important: things are not always as they seem. The following data show the number of species of mammals in forests of differing productivity:

```
par(mfrow=c(1,1))
rm(x,y)
productivity<-read.table("c:\\temp\\productivity.txt",header=T)
attach(productivity)
names(productivity)
```

```
[1]  "x"  "y"  "f"
```

```
plot(x,y,ylab="Mammal species",xlab="Productivity")
```

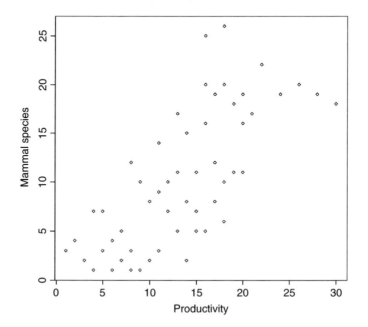

There is a very clear positive correlation: increasing productivity is associated with increasing species richness. The correlation is highly significant:

cor.test(x,y,method="spearman")

```
        Spearman's rank correlation rho

data: x and y
S = 6515, p-value = < 2.2e-16
alternative hypothesis: true rho is not equal to 0
sample estimates:
        rho
0.7516389
```

However, what if we look at the relationship for each region (*f*) separately, using xyplot from the library of lattice plots?

```
xyplot(y~x|f,
panel=function(x,y) {
panel.xyplot(x,y,pch=16)
panel.abline(lm(y~x)) })
```

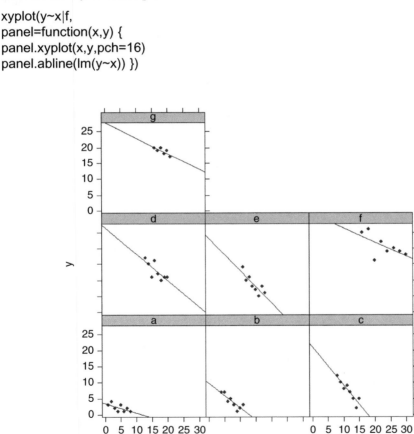

I've added the regression lines for emphasis, but the pattern is obvious. In every single case, increasing productivity is associated with *reduced* mammal species richness within each region (labelled a–g). The lesson is clear: you need to be extremely careful when looking at *correlations across different scales*. Things that are positively correlated over short time scales may turn out to be negatively correlated in the long term. Things that appear to be positively correlated at large spatial scales may turn out (as in this example) to be negatively correlated at small scales.

Kolmogorov–Smirnov test

People know this test for its wonderful name, rather than for what it actually does. It is an extremely simple test for asking one of two different questions:

- Are two sample distributions the same, or are they significantly different from one another in one or more (unspecified) ways?

- Does a particular sample distribution arise from a particular hypothesized distribution?

The two-sample problem is the one most often used. The apparently simple question is actually very broad. It is obvious that two distributions could be different because their means were different. But two distributions with exactly the same mean could be significantly different if they differed in variance, or in skew or kurtosis (see p. 287).

The Kolmogorov–Smirnov test works on **cumulative distribution functions**. These give the probability that a randomly selected value of X is less than or equal to x

$$F(x) = P[X \le x].$$

This sounds somewhat abstract. Suppose we had insect wing sizes for two geographically separated populations and we wanted to test whether the distribution of wing lengths was the same in the two places.

```
wings<-read.table("c:\\temp\\wings.txt",header=T)
attach(wings)
names(wings)
```

```
[1] "size"  "location"
```

We need to find out how many specimens there are from each location:

```
table(location)
location
 A   B
50  70
```

So the samples are of unequal size (50 insects from location A, 70 from B). It will be useful, therefore, to create two separate variables containing the wing lengths from sites A and B:

```
A<-size[location=="A"]
B<-size[location=="B"]
```

We could begin by comparing mean wing length in the two locations with a t test:

```
t.test(A,B)
```

```
        Welch Two Sample t-test
```

```
data: A and B
t = -1.6073,  df = 117.996,  p-value = 0.1107
alternative hypothesis: true difference in means is not equal to 0
95 percent confidence interval:
-2.494476  0.259348
sample estimates:
```

```
mean of x mean of y
24.11748  25.23504
```

which shows than mean wing length is not significantly different in the two locations $(p = 0.11)$.

But what about other attributes of the distribution? This is where Kolmogorov–Smirnov is really useful:

ks.test(A,B)

```
        Two-sample Kolmogorov—Smirnov test

data: A and B
D = 0.2629,  p-value = 0.02911
alternative hypothesis: two.sided
```

The two distributions are, indeed, significantly different from one another $(p < 0.05)$. But if not in their means, then in what respect do they differ? Perhaps they have different variances?

var.test(A,B)

```
        F test to compare two variances

data: A and B
F = 0.5014,  num df = 49,  denom df = 69,  p-value = 0.01192
alternative hypothesis: true ratio of variances is not equal to 1
95 percent confidence interval:
0.3006728  0.8559914
sample estimates:
ratio of variances
         0.5014108
```

Indeed they do: the variance of wing length from location B is double that from location A $(p < 0.02)$. We can finish by drawing the two histograms side by side to get a visual impression of the difference in the shape of the two distributions; open bars show the data from location B, solid bars show location A (see p. 538 for the code to draw a figure like this):

The spread of wing lengths is much greater at location B despite the fact that the mean wing length is similar in the two places. Also, the distribution is skew to the left in location B, with the result that modal wing length is greater in location B (26 mm, compared with 22 mm at A).

Power analysis

The power of a test is the probability of rejecting the null hypothesis when it is false. It has to do with Type II errors: β is the probability of accepting the null hypothesis when it is false. In an ideal world, we would obviously make β as small as possible. But there is a snag. The smaller we make the probability of committing a Type II error, the greater we make the probability of committing a Type I error, and rejecting the null hypothesis when, in fact, it is correct. A compromise is called for. Most statisticians work with $\alpha = 0.05$ and $\beta = 0.2$. Now the power of a test is defined as $1 - \beta = 0.8$ under the standard assumptions. This is used to calculate the sample sizes necessary to detect a specified difference when

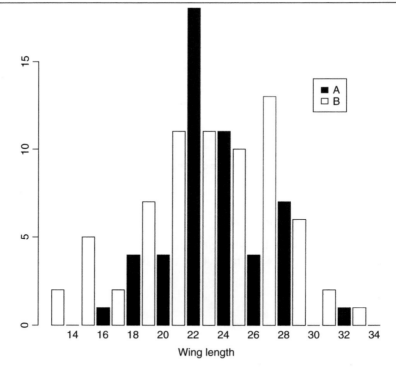

the error variance is known (or can be guessed at). Suppose that for a single sample the size of the difference you want to detect is δ and the variance in the response is s^2 (e.g. known from a pilot study or extracted from the literature). Then you will need n replicates to reject the null hypothesis with power $= 80\%$:

$$n \approx \frac{8 \times s^2}{\delta^2}$$

This is a reasonable rule of thumb, but you should err on the side of caution by having larger, not smaller, samples than these. Suppose that the mean is close to 20, and the variance is 10, but we want to detect a 10% change (i.e. $\partial = \pm 2$) with probability 0.8, then $n = 8 \times 10/2^2 = 20$. There are built-in functions in R for carrying out power analyses for ANOVA, proportion data and t tests:

power.t.test   power calculations for one- and two-sample t tests;

power.prop.test   power calculations two-sample test for proportions;

power.anova.test   power calculations for balanced one-way ANOVA tests.

The arguments to the **power.t.test** function are n (the number of observations per group), delta (the difference in means we want to be able to detect; you will need to think hard about this value), sd (the standard deviation of the sample), sig.level (the significance level (Type I error probability) where you will often accept the default value of 5%), power (the power you want the test to have where you will often accept the default value of 80%), type (the type of t test you want to carry out: two-sample, one-sample or paired) and alternative (whether you want to do a one- or a two-tailed test, where you will typically want to do the default, two-tailed test). One of the parameters n, delta, power, sd and sig.level must be passed as NULL, and that parameter is determined from the others. This sounds like a lot of work, but you will typically use all of the defaults so you only need to specify the difference, delta, and the standard deviation, sd, to work out the sample size n that will give you the power you want.

So how many replicates do we need in each of two samples to detect a difference of 10% with power =80% when the mean is 20 (i.e. delta $=2.0$) and the standard deviation is about 3.5?

```
power.t.test(delta=2,sd=3.5,power=0.8)
```

```
     Two-sample t test power calculation

            n = 49.05349
        delta = 2
           sd = 3.5
    sig.level = 0.05
        power = 0.8
  alternative = two.sided
```

If you had been working with a rule of thumb like '30 is a big enough sample' then you would be severely disappointed in this case. You simply could not have detected a difference of 10% with this experimental design. You need 50 replicates in each sample (100 replicates in all) to achieve a power of 80%. You can work out what size of difference your sample of 30 would allow you to detect, by specifying n and omitting *delta*:

```
power.t.test(n=30,sd=3.5,power=0.8)
```

```
     Two-sample t test power calculation

            n = 30
        delta = 2.574701
           sd = 3.5
    sig.level = 0.05
        power = 0.8
  alternative = two.sided
```

which shows that you could have detected a 13% change ($100 \times (22.575 - 20)/20$). The work you need to do before carrying out a power analysis before designing your experiment is to find values for the standard deviation (from the literature or by carrying out a pilot experiment) and the size of the difference your want to detect (from discussions with your sponsor or your colleagues).

Bootstrap

We want to use bootstrapping to obtain a 95% confidence interval for the mean of a vector of numbers called **values**:

```
data<-read.table("c:\\temp\\skewdata.txt",header=T)
attach(data)
names(data)
```

```
[1]  "values"
```

We shall sample with replacement from values using sample(values,replace=T), then work out the mean, repeating this operation 10 000 times, and storing the 10 000 different mean values in a vector called *ms*:

```
ms<-numeric(10000)
for (i in 1:10000){
ms[i]<-mean(sample(values,replace=T)) }
```

The answer to our problem is provided by the quantile function applied to ms: we want to know the values of ms associated with the 0.025 and the 0.975 tails of ms

```
quantile(ms,c(0.025,0.975))
```

```
    2.5%      97.5%
24.97918  37.62932
```

so the intervals below and above the mean are

```
mean(values)-quantile(ms,c(0.025,0.975))
```

```
    2.5%       97.5%
5.989472  -6.660659
```

How does this compare with the parametric confidence interval, $CI = 1.96 \times \sqrt{s^2/n}$?

```
1.96*sqrt(var(values)/length(values))
```

```
[1]  6.569802
```

Close, but not identical. Our bootstrapped intervals are skew because the data are skewed, but the parametric interval, of course, is symmetric.

Now let's see how to do the same thing using the boot function from the library called boot:

```
install.packages("boot")
library(boot)
```

The syntax of boot is very simple:

```
boot(data, statistic, R)
```

The trick to using boot lies in understanding how to write the statistic function. R is the number of resamplings you want to do (R = 10000 in this example), and data is the name of the data object to be resampled (values in this case). The attribute we want to estimate repeatedly is the mean value of values. Thus, the first argument to our function must be values. The second argument is an index (a vector of subscripts) that is used within boot to select random assortments of values. Our statistic function can use the built-in function mean to calculate the mean value of the sample of values.

```
mymean<-function(values,i) mean(values[i])
```

The key point is that we write mean(values[i]) not mean(values). Now we can run the bootstrap for 10 000 iterations

```
myboot<-boot(values,mymean,R=10000)
myboot
```

```
ORDINARY NONPARAMETRIC BOOTSTRAP
```

```
Call:
boot(data = values, statistic = mymean, R = 10000)

Bootstrap Statistics :
    original        bias   std. error
t1* 30.96866  -0.08155796   3.266455
```

The output is interpreted as follows. The `original` is the mean of the whole sample

mean(values)

```
[1]  30.96866
```

while `bias` is the difference between the arithmetic mean and the mean of the bootstrapped samples which are in the variable called myboot$t

mean(myboot$t)-mean(values)

```
[1]  -0.08155796
```

and `std. error` is the standard deviation of the simulated values in myboot$t

sqrt(var(myboot$t))

```
           [,1]
[1,]  3.266455
```

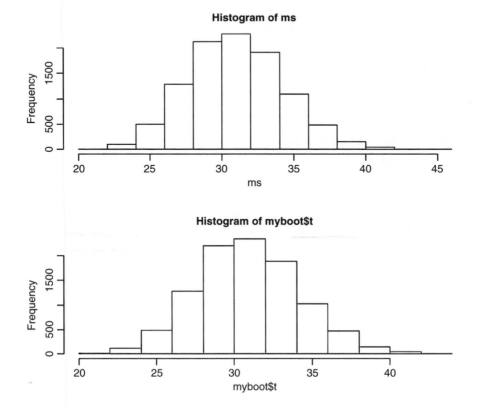

The components of **myboot** can be used to do other things. For instance, we can compare our homemade vector (*ms*, above) with a histogram of **myboot$t**:

```
par(mfrow=c(2,1))
hist(ms)
hist(myboot$t)
```

They differ in detail because they were generated with different series of random numbers. Here are the 95% intervals for comparison with ours, calculated form the quantiles of **myboot$t**:

```
mean(values)-quantile(myboot$t,c(0.025,0.975))
```

```
     2.5%       97.5%
 6.126120  -6.599232
```

There is a function **boot.ci** for calculating confidence intervals from the boot object:

```
boot.ci(myboot)
```

```
BOOTSTRAP CONFIDENCE INTERVAL CALCULATIONS
Based on 10000 bootstrap replicates
CALL :
boot.ci(boot.out = myboot)
Intervals :
Level       Normal              Basic
95%      (24.65, 37.45)    (24.37, 37.10)
Level     Percentile            BCa
95%      (24.84, 37.57)    (25.63, 38.91)
Calculations and Intervals on Original Scale
Warning message:
bootstrap variances needed for studentized intervals in:
boot.ci(myboot)
```

Normal is the parametric CI based on the standard error of the mean and the sample size (p. 54). The **Percentile** interval is the quantile from the bootstrapped estimates:

```
quantile(myboot$t,c(0.025,0.975))
```

```
     2.5%       97.5%
 24.84254   37.56789
```

which, as we saw earlier, was close to our home-made values (above). The BCa interval is the bias-corrected accelerated percentile. It is not greatly different in this case, but is the interval preferred by statisticians. A more complex example of the use of bootstrapping involving a generalized linear model is explained on p. 523. For other examples see ?boot, and for more depth read the Davison and Hinkley (1997) book from which the boot package is developed (as programmed by A.J. Canty).

9

Statistical Modelling

The hardest part of any statistical work is getting started. And one of the hardest things about getting started is choosing the right kind of statistical analysis. The choice depends on the nature of your data and on the particular question you are trying to answer. The key is to understand what kind of *response* variable you have, and to know the nature of your *explanatory* variables. The response variable is the thing you are working on: it is the variable whose variation you are attempting to understand. This is the variable that goes on the *y* axis of the graph. The explanatory variable goes on the *x* axis of the graph; you are interested in the extent to which variation in the response variable is associated with variation in the explanatory variable. You also need to consider the *way* that the variables in your analysis measure what they purport to measure. A continuous measurement is a variable such as height or weight that can take any real numbered value. A categorical variable is a factor with two or more levels: sex is a factor with two levels (male and female), and colour might be a factor with seven levels (red, orange, yellow, green, blue, indigo, violet).

It is essential, therefore, that you can answer the following questions:

- Which of your variables is the response variable?

- Which are the explanatory variables?

- Are the explanatory variables continuous or categorical, or a mixture of both?

- What kind of response variable do you have: is it a continuous measurement, a count, a proportion, a time at death, or a category?

These simple keys will lead you to the appropriate statistical method:

The explanatory variables

(a) All explanatory variables continuous	**Regression**
(b) All explanatory variables categorical	**Analysis of variance (ANOVA)**
(c) Explanatory variables both continuous and categorical	**Analysis of covariance (ANCOVA)**

The R Book Michael J. Crawley
© 2007 John Wiley & Sons, Ltd

The response variable

(a) Continuous   **Normal regression, ANOVA or ANCOVA**

(b) Proportion   **Logistic regression**

(c) Count   **Log-linear models**

(d) Binary   **Binary logistic analysis**

(e) Time at death   **Survival analysis**

The object is to determine the values of the parameters in a specific model that lead to *the best fit of the model to the data*. The data are sacrosanct, and they tell us what actually happened under a given set of circumstances. It is a common mistake to say 'the data were fitted to the model' as if the data were something flexible, and we had a clear picture of the structure of the model. On the contrary, what we are looking for is the minimal adequate model to describe the data. The model is fitted to data, not the other way around. The best model is the model that produces the least unexplained variation (the *minimal residual deviance*), subject to the constraint that all the parameters in the model should be statistically significant.

You have to specify the model. It embodies your mechanistic understanding of the explanatory variables involved, and of the way that they are related to the response variable. You want the model to be *minimal* because of the principle of parsimony, and *adequate* because there is no point in retaining an inadequate model that does not describe a significant fraction of the variation in the data. It is very important to understand that *there is not one model*; this is one of the common implicit errors involved in traditional regression and ANOVA, where the same models are used, often uncritically, over and over again. In most circumstances, there will be a large number of different, more or less plausible models that might be fitted to any given set of data. Part of the job of data analysis is to determine which, if any, of the possible models are adequate, and then, out of the set of adequate models, which is the minimal adequate model. In some cases there may be no single best model and a set of different models may all describe the data equally well (or equally poorly if the variability is great).

Maximum Likelihood

What, exactly, do we mean when we say that the parameter values should afford the 'best fit of the model to the data'? The convention we adopt is that our techniques should lead to **unbiased, variance-minimizing estimators**. We define 'best' in terms of **maximum likelihood**. This notion may be unfamiliar, so it is worth investing some time to get a feel for it. This is how it works:

- given the data,

- and given our choice of model,

- what values of the parameters of that model make the observed data most likely?

We judge the model on the basis how likely the data would be if the model were correct.

The Principle of Parsimony (Occam's Razor)

One of the most important themes running through this book concerns model simplification. The principle of parsimony is attributed to the early 14th-century English nominalist philosopher, William of Occam, who insisted that, given a set of equally good explanations for a given phenomenon, *the correct explanation is the simplest explanation*. It is called Occam's razor because he 'shaved' his explanations down to the bare minimum: his point was that in explaining something, assumptions must not be needlessly multiplied. In particular, for the purposes of explanation, things not *known* to exist should not, unless it is absolutely necessary, be postulated as existing. For statistical modelling, the principle of parsimony means that:

- models should have as few parameters as possible;

- linear models should be preferred to non-linear models;

- experiments relying on few assumptions should be preferred to those relying on many;

- models should be pared down until they are *minimal adequate*;

- simple explanations should be preferred to complex explanations.

The process of model simplification is an integral part of hypothesis testing in R. In general, a variable is retained in the model only *if it causes a significant increase in deviance when it is removed from the current model*. Seek simplicity, then distrust it.

In our zeal for model simplification, however, we must be careful not to throw the baby out with the bathwater. Einstein made a characteristically subtle modification to Occam's razor. He said: 'A model should be as simple as possible. But no simpler.' Remember, too, what Oscar Wilde said: 'Truth is rarely pure, and never simple.'

Types of Statistical Model

Fitting models to data is the central function of R. The process is essentially one of exploration; there are no fixed rules and no absolutes. The object is to determine a minimal adequate model (see Table 9.1) from the large set of potential models that might be used to describe the given set of data. In this book we discuss five types of model:

- the null model;

- the minimal adequate model;

- the current model;

- the maximal model; and

- the saturated model.

The stepwise progression from the saturated model (or the maximal model, whichever is appropriate) through a series of simplifications to the minimal adequate model is made on the basis of **deletion tests**. These are F-tests or chi-squared tests that assess the significance of the increase in deviance that results when a given term is removed from the current model.

Table 9.1. Statistical modelling involves the selection of a minimal adequate model from a potentially large set of more complex models, using stepwise model simplification.

Model	Interpretation
Saturated model	One parameter for every data point Fit: perfect Degrees of freedom: none Explanatory power of the model: none
Maximal model	Contains all (p) factors, interactions and covariates that might be of any interest. Many of the model's terms are likely to be insignificant Degrees of freedom: $n - p - 1$ Explanatory power of the model: it depends
Minimal adequate model	A simplified model with $0 \leq p' \leq p$ parameters Fit: less than the maximal model, but not significantly so Degrees of freedom: $n - p' - 1$ Explanatory power of the model: $r^2 = SSR/SSY$
Null model	Just one parameter, the overall mean $\bar{y}$ Fit: none; $SSE = SSY$ Degrees of freedom: $n - 1$ Explanatory power of the model: none

Models are representations of reality that should be both accurate and convenient. However, it is impossible to maximize a model's realism, generality and holism simultaneously, and the principle of parsimony is a vital tool in helping to choose one model over another. Thus, we would only include an explanatory variable in a model if it significantly improved the fit of the model. The fact that we went to the trouble of measuring something does not mean we have to have it in our model. Parsimony says that, other things being equal, we prefer:

- a model with $n - 1$ parameters to a model with n parameters;
- a model with $k - 1$ explanatory variables to a model with k explanatory variables;
- a linear model to a model which is curved;
- a model without a hump to a model with a hump;
- a model without interactions to a model containing interactions between factors.

Other considerations include a preference for models containing explanatory variables that are easy to measure over variables that are difficult or expensive to measure. Also, we prefer models that are based on a sound mechanistic understanding of the process over purely empirical functions.

Parsimony requires that the model should be as simple as possible. This means that the model should not contain any redundant parameters or factor levels. We achieve this by fitting a maximal model and then simplifying it by following one or more of these steps:

- remove non-significant interaction terms;
- remove non-significant quadratic or other non-linear terms;

- remove non-significant explanatory variables;

- group together factor levels that do not differ from one another;

- in ANCOVA, set non-significant slopes of continuous explanatory variables to zero.

All the above are subject, of course, to the caveats that the simplifications make good scientific sense and do not lead to significant reductions in explanatory power.

Just as there is no perfect model, so there may be no optimal scale of measurement for a model. Suppose, for example, we had a process that had Poisson errors with multiplicative effects amongst the explanatory variables. Then, one must chose between three different scales, each of which optimizes one of three different properties:

- the scale of $\sqrt{y}$ would give constancy of variance;

- the scale of $y^{2/3}$ would give approximately normal errors;

- the scale of $\ln(y)$ would give additivity.

Thus, any measurement scale is always going to be a compromise, and you should choose the scale that gives the best overall performance of the model.

Steps Involved in Model Simplification

There are no hard and fast rules, but the procedure laid out in Table 9.2 works well in practice. With large numbers of explanatory variables, and many interactions and non-linear terms, the process of model simplification can take a very long time. But this is time

Table 9.2. Model simplication process.

Step	Procedure	Explanation
1	Fit the maximal model	Fit all the factors, interactions and covariates of interest. Note the residual deviance. If you are using Poisson or binomial errors, check for overdispersion and rescale if necessary.
2	Begin model simplification	Inspect the parameter estimates using the R function summary. Remove the least significant terms first, using update -, starting with the highest-order interactions.
3	If the deletion causes an insignificant increase in deviance	Leave that term out of the model. Inspect the parameter values again. Remove the least significant term remaining.
4	If the deletion causes a significant increase in deviance	Put the term back in the model using **update +**. These are the statistically significant terms as assessed by deletion from the maximal model.
5	Keep removing terms from the model	Repeat steps 3 or 4 until the model contains nothing but significant terms. This is the minimal adequate model. If none of the parameters is significant, then the minimal adequate model is the null model.

well spent because it reduces the risk of overlooking an important aspect of the data. It is important to realize that there is no guaranteed way of finding all the important structures in a complex dataframe.

Caveats

Model simplification is an important process but it should not be taken to extremes. For example, care should be taken with the interpretation of deviances and standard errors produced with fixed parameters that have been estimated from the data. Again, the search for 'nice numbers' should not be pursued uncritically. Sometimes there are good scientific reasons for using a particular number (e.g. a power of 0.66 in an allometric relationship between respiration and body mass). It is much more straightforward, for example, to say that yield increases by 2 kg per hectare for every extra unit of fertilizer, than to say that it increases by 1.947 kg. Similarly, it may be preferable to say that the odds of infection increase 10-fold under a given treatment, than to say that the logits increase by 2.321; without model simplification this is equivalent to saying that there is a 10.186-fold increase in the odds. It would be absurd, however, to fix on an estimate of 6 rather than 6.1 just because 6 is a whole number.

Order of deletion

The data in this book fall into two distinct categories. In the case of planned experiments, all of the treatment combinations are equally represented and, barring accidents, there are no missing values. Such experiments are said to be **orthogonal**. In the case of observational studies, however, we have no control over the number of individuals for which we have data, or over the combinations of circumstances that are observed. Many of the explanatory variables are likely to be correlated with one another, as well as with the response variable. Missing treatment combinations are commonplace, and the data are said to be **non-orthogonal**. This makes an important difference to our statistical modelling because, in orthogonal designs, the variation that is attributed to a given factor is constant, and does not depend upon the order in which factors are removed from the model. In contrast, with non-orthogonal data, we find that the variation attributable to a given factor *does* depend upon the order in which factors are removed from the model. We must be careful, therefore, to judge the significance of factors in non-orthogonal studies, when they are *removed from the maximal model* (i.e. from the model including all the other factors and interactions with which they might be confounded). Remember that, *for non-orthogonal data, order matters*.

Also, if your explanatory variables are correlated with each other, then the significance you attach to a given explanatory variable will depend upon whether you delete it from a maximal model or add it to the null model. If you always test by model simplification then you won't fall into this trap.

The fact that you have laboured long and hard to include a particular experimental treatment does not justify the retention of that factor in the model if the analysis shows it to have no explanatory power. ANOVA tables are often published containing a mixture of significant and non-significant effects. This is not a problem in orthogonal designs, because sums of squares can be unequivocally attributed to each factor and interaction term. But as soon as there are missing values or unequal weights, then it is impossible to tell how the parameter estimates and standard errors of the significant terms would have been altered if the non-significant terms had been deleted. The best practice is as follows:

- Say whether your data are orthogonal or not.

- Explain any correlations amongst your explanatory variables.

- Present a minimal adequate model.

- Give a list of the non-significant terms that were omitted, and the deviance changes that resulted from their deletion.

Readers can then judge for themselves the relative magnitude of the non-significant factors, and the importance of correlations between the explanatory variables.

The temptation to retain terms in the model that are 'close to significance' should be resisted. The best way to proceed is this. If a result would have been *important* if it had been statistically significant, then it is worth repeating the experiment with higher replication and/or more efficient blocking, in order to demonstrate the importance of the factor in a convincing and statistically acceptable way.

Model Formulae in R

The structure of the model is specified in the model formula like this:

$$\text{response variable} \sim \text{explanatory variable(s)}$$

where the tilde symbol ~ reads 'is modelled as a function of' (see Table 9.3 for examples).

Table 9.3. Examples of R model formulae. In a model formula, the function I case i) stands for 'as is' and is used for generating sequences I(1:10) or calculating quadratic terms I(x^2).

Model	Model formula	Comments
Null	y ~ 1	1 is the intercept in regression models, but here it is the overall mean y
Regression	y ~ x	x is a continuous explanatory variable
Regression through origin	y ~ x-1	Do not fit an intercept
One-way ANOVA	y ~ sex	sex is a two-level categorical variable
One-way ANOVA	y ~ sex-1	as above, but do not fit an intercept (gives two means rather than a mean and a difference)
Two-way ANOVA	y ~ sex + genotype	genotype is a four-level categorical variable
Factorial ANOVA	y ~ N * P * K	N, P and K are two-level factors to be fitted along with all their interactions

Table 9.3. (Continued)

Model	Model formula	Comments
Three-way ANOVA	y ~ N*P*K – N:P:K	As above, but don't fit the three-way interaction
Analysis of covariance	y ~ x + sex	A common slope for y against x but with two intercepts, one for each sex
Analysis of covariance	y ~ x * sex	Two slopes and two intercepts
Nested ANOVA	y ~ a/b/c	Factor c nested within factor b within factor a
Split-plot ANOVA	y ~ a*b*c+Error(a/b/c)	A factorial experiment but with three plot sizes and three different error variances, one for each plot size
Multiple regression	y ~ x + z	Two continuous explanatory variables, flat surface fit
Multiple regression	y ~ x * z	Fit an interaction term as well (x + z + x:z)
Multiple regression	y ~ x + I(x^2) + z + I(z^2)	Fit a quadratic term for both x and z
Multiple regression	y <- poly(x,2) + z	Fit a quadratic polynomial for x and linear z
Multiple regression	y ~ (x + z + w)^2	Fit three variables plus all their interactions up to two-way
Non-parametric model	y ~ s(x) +s(z)	y is a function of smoothed x and z in a generalized additive model
Transformed response and explanatory variables	log(y) ~ I(1/x) + sqrt(z)	All three variables are transformed in the model

So a simple linear regression of y on x would be written as

$$y \sim x$$

and a one-way ANOVA where sex is a two-level factor would be written as

$$y \sim sex$$

The right-hand side of the model formula shows:

- the number of explanatory variables and their identities – their attributes (e.g. continuous or categorical) are usually defined prior to the model fit;
- the interactions between the explanatory variables (if any);
- non-linear terms in the explanatory variables.

On the right of the tilde, one also has the option to specify offsets or error terms in some special cases. As with the response variable, the explanatory variables can appear as transformations, or as powers or polynomials.

It is very important to note that symbols are used differently in model formulae than in arithmetic expressions. In particular:

+   indicates inclusion of an explanatory variable in the model (not addition);

-   indicates deletion of an explanatory variable from the model (not subtraction);

*   indicates inclusion of explanatory variables and interactions (not multiplication);

/   indicates nesting of explanatory variables in the model (not division);

| indicates conditioning (not 'or'), so that y ~ x | z is read as 'y as a function of x given z'.

There are several other symbols that have special meaning in model formulae. A colon denotes an interaction, so that A:B means the two-way interaction between A and B, and N:P:K:Mg means the four-way interaction between N, P, K and Mg.

Some terms can be written in an expanded form. Thus:

A*B*C   is the same as A+B+C+A:B+A:C+B:C+A:B:C

A/B/C   is the same as A+B%in%A+C%in%B%in%A

(A+B+C)^3   is the same as A*B*C

(A+B+C)^2   is the same as A*B*C – A:B:C

Interactions between explanatory variables

Interactions between two two-level categorical variables of the form A*B mean that two main effect means and one interaction mean are evaluated. On the other hand, if factor A has three levels and factor B has four levels, then seven parameters are estimated for the main effects (three means for A and four means for B). The number of interaction terms is $(a-1)(b-1)$, where a and b are the numbers of levels of the factors A and B, respectively. So in this case, R would estimate $(3-1)(4-1) = 6$ parameters for the interaction.

Interactions between two continuous variables are fitted differently. If x and z are two continuous explanatory variables, then x*z means fit x+z+x:z and the interaction term x:z behaves as if a new variable had been computed that was the pointwise product of the two vectors x and z. The same effect could be obtained by calculating the product explicitly,

product.xz <- x * z

then using the model formula y ~ x + z + product.xz. Note that the representation of the interaction by the *product* of the two continuous variables is an assumption, not a fact. The real interaction might be of an altogether different functional form (e.g. x * z^2).

Interactions between a categorical variable and a continuous variable are interpreted as an analysis of covariance; a separate slope and intercept are fitted for each level of the categorical variable. So y ~ A*x would fit three regression equations if the factor A had three levels; this would estimate six parameters from the data – three slopes and three intercepts.

The slash operator is used to denote nesting. Thus, with categorical variables A and B,

y ~ A/B

means fit 'A plus B within A'. This could be written in two other equivalent ways:

y ~ A + A:B

y ~ A + B %in% A

both of which alternatives emphasize that there is no point in attempting to estimate a main effect for B (it is probably just a factor label like 'tree number 1' that is of no scientific interest; see p. 479).

Some functions for specifying non-linear terms and higher order interactions are useful. To fit a polynomial regression in x and z, we could write

y ~ poly(x,3) + poly(z,2)

to fit a cubic polynomial in x and a quadratic polynomial in z. To fit interactions, but only up to a certain level, the ^ operator is useful. The formula

y ~ (A + B + C)^2

fits all the main effects and two-way interactions (i.e. it excludes the three-way interaction that A*B*C would have included).

The I function (upper-case letter i) stands for 'as is'. It overrides the interpretation of a model symbol as a formula operator when the intention is to use it as an arithmetic operator. Suppose you wanted to fit $1/x$ as an explanatory variable in a regression. You might try

y~1/x

but this actually does something very peculiar. It fits x nested within the intercept! When it appears in a model formula, the slash operator is assumed to imply nesting. To obtain the effect we want, we use I to write

y ~ I(1/x)

We also need to use I when we want * to represent multiplication and ^ to mean 'to the power' rather than an interaction model expansion: thus to fit x and x^2 in a quadratic regression we would write

y ~ x+I(x^2)

Creating formula objects

You can speed up the creation of complicated model formulae using paste to create series of variable names and collapse to join the variable names together by symbols. Here, for instance, is a multiple regression formula with 25 continuous explanatory variables created using the as.formula function:

```
xnames <- paste("x", 1:25, sep="")
(model.formula <- as.formula(paste("y ~ ", paste(xnames, collapse= "+"))))
```

```
y ~ x1 +   x2 +   x3 +   x4 +   x5 +   x6 +   x7 +   x8 +   x9 +  x10 +  x11 +
          x12 +  x13 +  x14 +  x15 +  x16 +  x17 +  x18 +  x19 +  x20 +  x21 +
          x22 +  x23 +  x24 +  x25
```

Multiple error terms

When there is nesting (e.g. split plots in a designed experiment; see p. 470) or temporal pseudoreplication (see p. 474) you can include an Error function as part of the model formula. Suppose you had a three-factor factorial experiment with categorical variables *A, B* and *C*. The twist is that each treatment is applied to plots of different sizes: *A* is applied to replicated whole fields, *B* is applied at random to half fields and *C* is applied to smaller split–split plots within each field. This is shown in a model formula like this:

y ~ A*B*C + Error(A/B/C)

Note that the terms within the model formula are separated by asterisks to show that it is a full factorial with all interaction terms included, whereas the terms are separated by slashes in the Error statement. There are as many terms in the Error statement as there are different sizes of plots – three in this case, although the smallest plot size (*C* in this example) can be omitted from the list – and the terms are listed left to right from the largest to the smallest plots; see p. 469 for details and examples.

The intercept as parameter 1

The simple command

y~1

causes the null model to be fitted. This works out the grand mean (the overall average) of all the data and works out the total deviance (or the total sum of squares, *SSY*, in models with normal errors and the identity link). In some cases, this may be the minimal adequate model; it is possible that none of the explanatory variables we have measured contribute anything significant to our understanding of the variation in the response variable. This is normally what you don't want to happen at the end of your three-year research project.

To remove the intercept (parameter 1) from a regression model (i.e. to force the regression line through the origin) you fit '−1' like this:

y ~ x − 1

You should not do this unless you know exactly what you are doing, and exactly why you are doing it (see p. 393 for details). Removing the intercept from an ANOVA model where all the variables are categorical has a different effect:

y ~ sex - 1

This gives the mean for males and the mean for females in the summary table, rather than the mean for females and the difference in mean for males (see p. 366).

The **update** function in model simplification

In the update function used during model simplification, the dot '.' is used to specify 'what is there already' on either side of the tilde. So if your original model said

model<-lm(y ~ A*B)

then the update function to remove the interaction term A:B could be written like this:

```
model2<-update(model, ~ .- A:B)
```

Note that there is no need to repeat the name of the response variable, and the punctuation 'tilde dot' means take model as it is, and remove from it the interaction term A:B.

Model formulae for regression

The important point to grasp is that model formulae look very like equations but there are important differences. Our simplest useful equation looks like this:

$$y = a + bx.$$

It is a two-parameter model with one parameter for the intercept, a, and another for the slope, b, of the graph of the continuous response variable y against a continuous explanatory variable x. The model formula for the same relationship looks like this:

```
y ~ x
```

The equals sign is replaced by a tilde, and all of the parameters are left out. It we had a multiple regression with two continuous explanatory variables x and z, the equation would be

```
y=a+bx+cz,
```

but the model formula is

```
y ~ x + z
```

It is all wonderfully simple. But just a minute. How does R know what parameters we want to estimate from the data? We have only told it the names of the explanatory variables. We have said nothing about how to fit them, or what sort of equation we want to fit to the data. The key to this is to understand what kind of explanatory variable is being fitted to the data. If the explanatory variable x specified on the right of the tilde is a continuous variable, then R *assumes* that you want to do a regression, and hence that you want to estimate two parameters in a linear regression whose equation is $y = a + bx$.

A common misconception is that linear models involve a straight-line relationship between the response variable and the explanatory variables. This is *not* the case, as you can see from these two linear models:

```
par(mfrow=c(1,2))
x<-seq(0,10,0.1)
plot(x,1+x-x^2/15,type="l")
plot(x,3+0.1*exp(x),type="l")
```

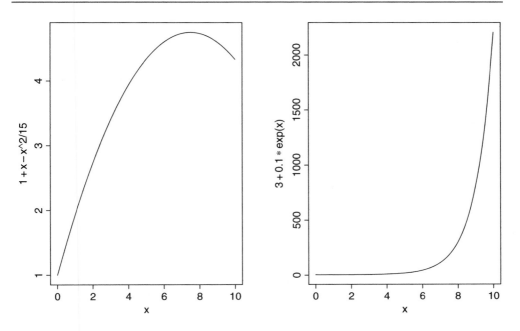

The definition of a linear model is an equation that contains mathematical variables, parameters and random variables and that is linear in the parameters and in the random variables. What this means is that if a, b and c are parameters then obviously

$$y = a + bx$$

is a linear model, but so is

$$y = a + bx - cx^2$$

because x^2 can be replaced by z which gives a linear relationship

$$y = a + bx + cz,$$

and so is

$$y = a + be^x$$

because we can create a new variable $z = \exp(x)$, so that

$$y = a + bz.$$

Some models are non-linear but can be readily linearized by transformation. For example:

$$y = \exp(a + bx)$$

is non-linear, but on taking logs of both sides, it becomes

$$\ln(y) = a + bx$$

If the equation you want to fit is more complicated than this, then you need to specify the form of the equation, and use non-linear methods (nls or nlme) to fit the model to the data (see p. 661).

Box–Cox Transformations

Sometimes it is not clear from theory what the optimal transformation of the response variable should be. In these circumstances, the Box–Cox transformation offers a simple empirical solution. The idea is to find the power transformation, λ (lambda), that maximizes the likelihood when a specified set of explanatory variables is fitted to

$$\frac{y^\lambda - 1}{\lambda}$$

as the response. The value of lambda can be positive or negative, but it cannot be zero (you would get a zero-divide error when the formula was applied to the response variable, y). For the case $\lambda = 0$ the Box–Cox transformation is defined as $\log(y)$. Suppose that $\lambda = -1$. The formula now becomes

$$\frac{y^{-1} - 1}{-1} = \frac{1/y - 1}{-1} = 1 - \frac{1}{y},$$

and this quantity is regressed against the explanatory variables and the log-likelihood computed.

In this example, we want to find the optimal transformation of the response variable, which is timber volume:

```
data<-read.delim("c:\\temp\\timber.txt") attach(data) names(data)
```

```
[1] "volume" "girth" "height"
```

We start by loading the MASS library of Venables and Ripley:

```
library(MASS)
```

The boxcox function is very easy to use: just specify the model formula, and the default options take care of everything else.

```
boxcox(volume ~ log(girth)+log(height))
```

It is clear that the optimal value of lambda is close to zero (i.e. the log transformation). We can zoom in to get a more accurate estimate by specifying our own, non-default, range of lambda values. It looks as if it would be sensible to plot from -0.5 to $+0.5$:

```
boxcox(volume ~ log(girth)+log(height),lambda=seq(-0.5,0.5,0.01))
```

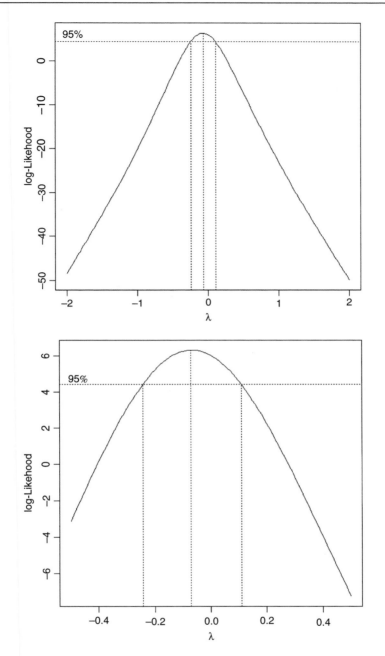

The likelihood is maximized at $\lambda \approx -0.08$, but the log-likelihood for $\lambda = 0$ is very close to the maximum. This also gives a much more straightforward interpretation, so we would go with that, and model log(volume) as a function of log(girth) and log(height) (see p. 518).

What if we had not log-transformed the explanatory variables? What would have been the optimal transformation of volume in that case? To find out, we rerun the boxcox function, simply changing the model formula like this:

```
boxcox(volume ~ girth+height)
```

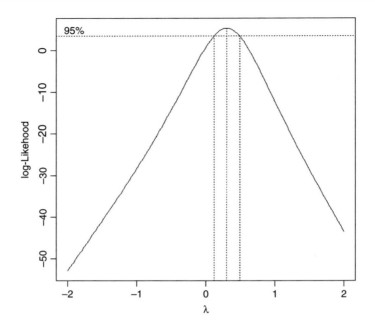

We can zoom in from 0.1 to 0.6 like this:

```
boxcox(volume ~ girth+height,lambda=seq(0.1,0.6,0.01))
```

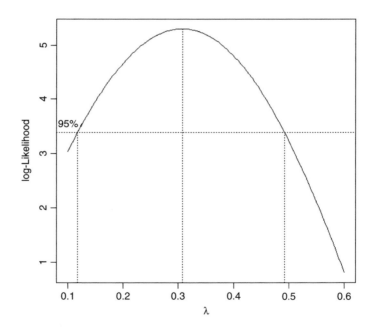

This suggests that the cube root transformation would be best ($\lambda = 1/3$). Again, this accords with dimensional arguments, since the response and explanatory variables would all have dimensions of length in this case.

Model Criticism

There is a temptation to become personally attached to a particular model. Statisticians call this 'falling in love with your model'. It is as well to remember the following truths about models:

- All models are wrong.
- Some models are better than others.
- The correct model can never be known with certainty.
- The simpler the model, the better it is.

There are several ways that we can improve things if it turns out that our present model is inadequate:

- Transform the response variable.
- Transform one or more of the explanatory variables.
- Try fitting different explanatory variables if you have any.
- Use a different error structure.
- Use non-parametric smoothers instead of parametric functions.
- Use different weights for different y values.

All of these are investigated in the coming chapters. In essence, you need a set of tools to establish whether, and how, your model is inadequate. For example, the model might:

- predict some of the y values poorly;
- show non-constant variance;
- show non-normal errors;
- be strongly influenced by a small number of influential data points;
- show some sort of systematic pattern in the residuals;
- exhibit overdispersion.

Model checking

After fitting a model to data we need to investigate how well the model describes the data. In particular, we should look to see if there are any systematic trends in the goodness of fit. For example, does the goodness of fit increase with the observation number, or is it a function of one or more of the explanatory variables? We can work with the raw residuals:

$$residuals = y - fitted \ values.$$

For instance, we should routinely plot the residuals against:

- the fitted values (to look for heteroscedasticity);
- the explanatory variables (to look for evidence of curvature);
- the sequence of data collection (to took for temporal correlation);
- standard normal deviates (to look for non-normality of errors).

Heteroscedasticity

A good model must also account for the variance–mean relationship adequately and produce additive effects on the appropriate scale (as defined by the link function). A plot of standardized residuals against fitted values should look like the sky at night (points scattered at random over the whole plotting region), with no trend in the size or degree of scatter of the residuals. A common problem is that the variance increases with the mean, so that we obtain an expanding, fan-shaped pattern of residuals (right-hand panel).

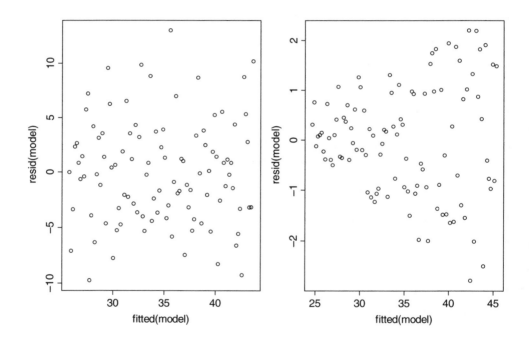

The plot on the left is what we want to see: no trend in the residuals with the fitted values. The plot on the right is a problem. There is a clear pattern of increasing residuals as the fitted values get larger. This is a picture of what heteroscedasticity looks like.

Non-normality of errors

Errors may be non-normal for several reasons. They may be skew, with long tails to the left or right. Or they may be kurtotic, with a flatter or more pointy top to their distribution. In any case, the theory is based on the assumption of normal errors, and if the errors are *not* normally distributed, then we shall not know how this affects our interpretation of the data or the inferences we make from it.

It takes considerable experience to interpret normal error plots. Here we generate a series of data sets where we introduce different but known kinds of non-normal errors. Then we plot them using a simple home-made function called mcheck (first developed by John Nelder in the original GLIM language; the name stands for model checking). The idea is to see what patterns are generated in normal plots by the different kinds of non-normality. In real applications we would use the generic plot(model) rather than mcheck (see below). First, we write the function mcheck. The idea is to produce two plots, side by side: a plot of the residuals against the fitted values on the left, and a plot of the ordered residuals against the quantiles of the normal distribution on the right.

```
mcheck <-function (obj, . . . ) {
    rs<-obj$resid
    fv<-obj$fitted
    par(mfrow=c(1,2))
    plot(fv,rs,xlab="Fitted values",ylab="Residuals")
    abline(h=0, lty=2)
    qqnorm(rs,xlab="Normal scores",ylab="Ordered residuals",main="")
    qqline(rs,lty=2)
    par(mfrow=c(1,1))
    invisible(NULL)              }
```

Note the use of $ (component selection) to extract the residuals and fitted values from the model object which is passed to the function as obj (the expression x$name is the name *component* of x). The functions qqnorm and qqline are built-in functions to produce normal probability plots. It is good programming practice to set the graphics parameters back to their default settings before leaving the function.

The aim is to create a catalogue of some of the commonest problems that arise in model checking. We need a vector of *x* values for the following regression models:

```
x<-0:30
```

Now we manufacture the response variables according to the equation

$$y = 10 + x + \varepsilon$$

where the errors, ε, have zero mean but are taken from different probability distributions in each case.

Normal errors

```
e<-rnorm(31,mean=0,sd=5)
yn<-10+x+e
mn<-lm(yn~x)
mcheck(mn)
```

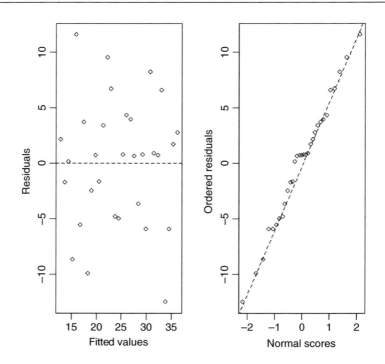

There is no suggestion of non-constant variance (left plot) and the normal plot (right) is reasonably straight. The judgement as to what constitutes an important departure from normality takes experience, and this is the reason for looking at some distinctly non-normal, but known, error structures next.

Uniform errors

```
eu<-20*(runif(31)-0.5)
yu<-10+x+eu
mu<-lm(yu ~ x)
mcheck(mu)
```

Uniform errors show up as a distinctly S-shaped pattern in the QQ plot on the right. The fit in the centre is fine, but the largest and smallest residuals are too small (they are constrained in this example to be ± 10).

Negative binomial errors

```
enb<-rnbinom(31,2,.3)
ynb<-10+x+enb
mnb<-lm(ynb ~ x)
mcheck(mnb)
```

The large negative residuals are all above the line, but the most obvious feature of the plot is the single, very large positive residual (in the top right-hand corner). In general, negative binomial errors will produce a J-shape on the QQ plot. The biggest positive residuals are much too large to have come from a normal distribution. These values may turn out to be highly influential (see below).

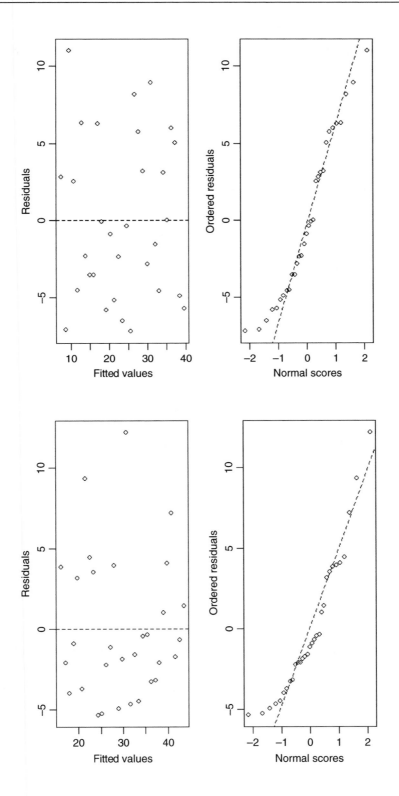

Gamma errors and increasing variance

Here the shape parameter is set to 1 and the rate parameter to $1/x$, and the variance increases with the square of the mean.

```
eg<-rgamma(31,1,1/x)
yg<-10+x+eg
mg<-lm(yg~x)
mcheck(mg)
```

The left-hand plot shows the residuals increasing steeply with the fitted values, and illustrates an asymmetry between the size of the positive and negative residuals. The right-hand plot shows the highly non-normal distribution of errors.

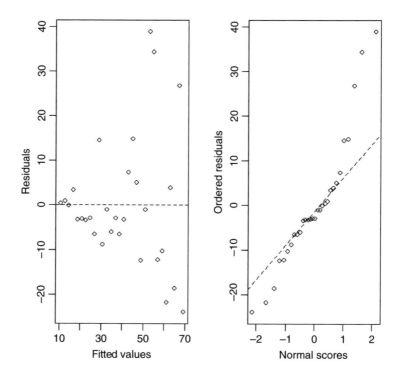

Influence

One of the commonest reasons for a lack of fit is through the existence of outliers in the data. It is important to understand, however, that a point may *appear* to be an outlier because of misspecification of the model, and not because there is anything wrong with the data. It is important to understand that analysis of residuals is a very poor way of looking for influence. Precisely because a point is highly influential, it forces the regression line close to it, and hence the influential point may have a very small residual.

Take this circle of data that shows no relationship between y and x:

```
x<-c(2,3,3,3,4)
y<-c(2,3,2,1,2)
```

We want to draw two graphs side by side, and we want them to have the same axis scales:

```
par(mfrow=c(1,2))
plot(x,y,xlim=c(0,8,),ylim=c(0,8))
```

Obviously, there is no relationship between y and x in the original data. But let's add an outlier at the point $(7,6)$ using concatenation and see what happens.

```
x1<-c(x,7)
y1<-c(y,6)
plot(x1,y1,xlim=c(0,8,),ylim=c(0,8))
abline(lm(y1~x1))
```

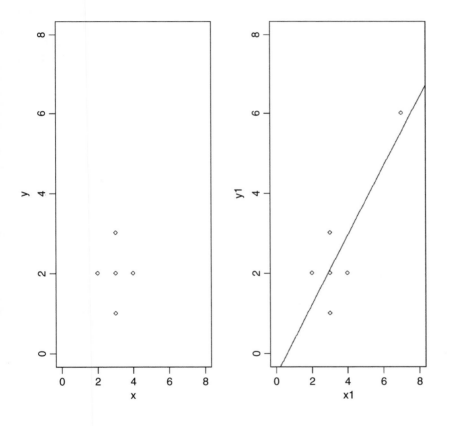

Now, there is a significant regression of y on x. The outlier is said to be highly *influential*.

To reduce the influence of outliers, there are a number of modern techniques known as **robust regression**. To see one of these in action, let's do a straightforward linear regression on these data and print the summary:

```
reg<-lm(y1~x1)
summary(reg)
```

```
Residuals:
        1          2          3          4          5          6
0.78261    0.91304   -0.08696   -1.08696   -0.95652    0.43478
```

```
Coefficients:
             Estimate   Std. Error   t value   Pr(>|t|)
(Intercept)   -0.5217       0.9876    -0.528     0.6253
x1             0.8696       0.2469     3.522     0.0244  *
```

Residual standard error: 0.9668 on 4 degrees of freedom
Multiple R-Squared: 0.7561,   Adjusted R-squared: 0.6952
F-statistic: 12.4 on 1 and 4 DF, p-value: 0.02441

The residual values make the important point that analysis of residuals is a very poor way of looking for influence. Precisely because point number 6 is so influential, it forces the regression line close to it, and hence point number 6 has a small residual (as you can see, 0.4348 is the second smallest of all the residuals). The slope of the regression line is 0.8696 with a standard error of 0.2469, and this is significantly different from 0 ($p = 0.0244$) despite the tiny sample size.

Continuing the analysis of the simple linear regression on p. 267, we investigate the function lm.influence. This produces four components: $hat, $coefficients, $sigma and $wt.res (weighted residuals):

lm.influence(lm(growth~tannin))

```
$hat
1                 2             3             4             5             6
0.3777778    0.2611111   0.1777778   0.1277778   0.1111111   0.1277778
7                 8             9
0.1777778    0.2611111   0.3777778
```

```
$coefficients
     (Intercept)           tannin
1     0.14841270   -2.619048e-02
2    -0.22690058    3.646617e-02
3    -0.39309309    5.360360e-02
4     0.58995046    5.530786e-02
5    -0.11111111   -2.794149e-18
6     0.06765747    2.537155e-02
7     0.06636637   -9.954955e-02
8     0.02873851   -1.616541e-02
9    -0.24444444    1.047619e-01
```

```
$sigma
        1           2           3           4           5           6
1.824655    1.811040    1.729448    1.320801    1.788078    1.734501
        7           8           9
1.457094    1.825513    1.757636
```

```
$wt.res
        1             2             3             4             5             6
0.2444444   -0.5388889   -1.3222222   2.8944444   -0.8888889   1.3277778
        7             8             9
-2.4555556   -0.2388889   0.9777778
```

Let's look at each of these components in turn.

The first component, $hat, is a vector containing the diagonal of the hat matrix. This is the orthogonal projector matrix onto the model space (see p. 273). The matrix X is made up of a column of 1s (corresponding to the intercept) and a column of the x values (the explanatory variable; see p. 271):

X<-cbind(1,tannin)

Then the hat matrix, H, is given by $H = X(X'X)^{-1}X'$, where X' is the transpose of X:

H<-X%*%ginv(t(X)%*%X)%*%t(X)

and we want the diagonal of this, which we could get by typing:

diag(H)

Large values of elements of this vector mean that changing y_i will have a big impact on the fitted values; i.e. the hat diagonals are measures of the leverage of y_i.

Next, $coefficients is a matrix whose ith row contains the *change* in the estimated coefficients which results when the ith case is dropped from the regression (this is different from S-PLUS, which shows the *coefficients* themselves). Data in row 9 have the biggest effect on slope and data in row 4 have the biggest effect on intercept.

The third component, $sigma, is a vector whose ith element contains the estimate of the residual standard error obtained when the ith case is dropped from the regression; thus 1.824655 is the residual standard error when point number 1 is dropped lm(growth[-1]~tannin[-1]), and the error variance in this case is $1.824655^2 = 3.329$.

Finally, $wt.res is a vector of weighted residuals (or deviance residuals in a generalized linear model) or raw residuals if weights are not set (as in this example).

A bundle of functions is available to compute some of the regression (leave-one-out deletion) diagnostics for linear and generalized linear models:

influence.measures(lm(growth ~ tannin))

```
Influence measures of
                lm(formula = growth~tannin) :

      dfb.1_    dfb.tnnn    dffit   cov.r   cook.d    hat   inf
1     0.1323   -1.11e-01    0.1323  2.167   0.01017  0.378   *
2    -0.2038    1.56e-01   -0.2058  1.771   0.02422  0.261
3    -0.3698    2.40e-01   -0.3921  1.323   0.08016  0.178
4     0.7267   -3.24e-01    0.8981  0.424   0.24536  0.128
5    -0.1011   -1.21e-17   -0.1864  1.399   0.01937  0.111
6     0.0635    1.13e-01    0.3137  1.262   0.05163  0.128
7     0.0741   -5.29e-01   -0.8642  0.667   0.27648  0.178
8     0.0256   -6.86e-02   -0.0905  1.828   0.00476  0.261
9    -0.2263    4.62e-01    0.5495  1.865   0.16267  0.378   *
```

The column names are as follows: dfb = DFBETAS, dffit = DFFITS (two terms explained in Cook and Weisberg, 1982, pp. 124–125), cov.r = covariance ratio, cook.d = Cook's distance, hat = the diagonal of the hat matrix, and inf marks influential data points with an asterisk. You can extract the influential points for plotting using $is.inf like this:

modi<-influence.measures(lm(growth ~ tannin))
which(apply(modi$is.inf, 1, any))

```
1  9
1  9
```

```
growth[which(apply(modi$is.inf, 1, any))]
```

```
[1] 12  3
```

```
tannin[which(apply(modi$is.inf, 1, any))]
```

```
[1] 0  8
```

```
summary(modi)
```

```
Potentially influential observations of
        lm(formula = growth~tannin) :

   dfb.1_ dfb.tnnn dffit cov.r  cook.d hat
1    0.13   -0.11   0.13  2.17_*   0.01  0.38
9   -0.23    0.46   0.55  1.87_*   0.16  0.38
```

```
yp<-growth[which(apply(modi$is.inf, 1, any))]
xp<-tannin[which(apply(modi$is.inf, 1, any))]
plot(tannin,growth,pch=16)
points(xp,yp,col="red",cex=1.3,pch=16)
abline(model)
```

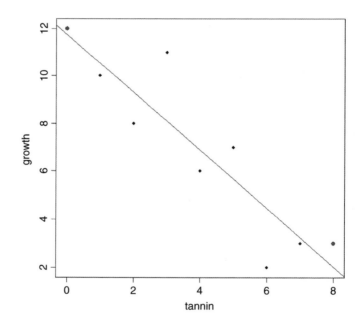

The slope would be much steeper were it not for the two points at the top left and bottom right that do not appear as black diamonds – these will be shown in red on your screen:

```
coef(lm(growth ~ tannin))
```

```
(Intercept)        tannin
  11.755556    -1.216667
```

```
coef(lm(growth ~ tannin,subset=(1:9 != 1 & 1:9 != 9)))
```

```
(Intercept)        tannin
  12.000000    -1.321429
```

Summary of Statistical Models in R

Models are fitted using one of the following model-fitting functions:

lm   fits a linear model with normal errors and constant variance; generally this is used for regression analysis using continuous explanatory variables.

aov   fits analysis of variance with normal errors, constant variance and the identity link; generally used for categorical explanatory variables or ANCOVA with a mix of categorical and continuous explanatory variables.

glm   fits generalized linear models to data using categorical or continuous explanatory variables, by specifying one of a family of **error structures** (e.g. Poisson for count data or binomial for proportion data) and a particular **link function**.

gam   fits generalized additive models to data with one of a family of error structures (e.g. Poisson for count data or binomial for proportion data) in which the continuous explanatory variables can (optionally) be fitted as arbitrary smoothed functions using non-parametric smoothers rather than specific parametric functions.

lme   and lmer fit linear mixed-effects models with specified mixtures of fixed effects and random effects and allow for the specification of correlation structure amongst the explanatory variables and autocorrelation of the response variable (e.g. time series effects with repeated measures). lmer allows for non-normal errors and non-constant variance with the same error families as a GLM.

nls   fits a non-linear regression model via least squares, estimating the parameters of a specified non-linear function.

nlme   fits a specified non-linear function in a mixed-effects model where the parameters of the non-linear function are assumed to be random effects; allows for the specification of correlation structure amongst the explanatory variables and autocorrelation of the response variable (e.g. time series effects with repeated measures).

loess   fits a local regression model with one or more continuous explanatory variables using non-parametric techniques to produce a smoothed model surface.

tree   fits a regression tree model using binary recursive partitioning whereby the data are successively split along coordinate axes of the explanatory variables so that at any node, the split is chosen that maximally distinguishes the response variable in the left and right branches. With a categorical response variable, the tree is called a classification tree, and the model used for classification assumes that the response variable follows a multinomial distribution.

For most of these models, a range of **generic functions** can be used to obtain information about the model. The most important and most frequently used are as follows:

summary   produces parameter estimates and standard errors from lm, and ANOVA tables from aov; this will often determine your choice between lm and aov. For either lm or aov you can choose summary.aov or summary.lm to get the alternative form of output (an ANOVA table or a table of parameter estimates and standard errors; see p. 364)

plot   produces diagnostic plots for model checking, including residuals against fitted values, influence tests, etc.

anova   is a wonderfully useful function for comparing different models and producing ANOVA tables.

update   is used to modify the last model fit; it saves both typing effort and computing time.

Other useful generic functions include the following:

coef   gives the coefficients (estimated parameters) from the model.

fitted   gives the fitted values, predicted by the model for the values of the explanatory variables included.

resid   gives the residuals (the differences between measured and predicted values of y).

predict   uses information from the fitted model to produce smooth functions for plotting a line through the scatterplot of your data.

Optional arguments in model-fitting functions

Unless you argue to the contrary, all of the rows in the dataframe will be used in the model fitting, there will be no offsets, and all values of the response variable will be given equal weight. Variables named in the model formula will come from the defined dataframe (data=mydata), the with function (p. 18) or from the attached dataframe (if there is one). Here we illustrate the following options:

- subset
- weights
- data
- offset
- na.action

We shall work with an example involving analysis of covariance (p. 490 for details) where we have a mix of both continuous and categorical explanatory variables:

```
data<-read.table("c:\\temp\\ipomopsis.txt",header=T)
attach(data)
names(data)
```

```
[1] "Root" "Fruit" "Grazing"
```

The response is seed production (Fruit) with a continuous explanatory variable (Root diameter) and a two-level factor Grazing (Grazed and Ungrazed).

Subsets

Perhaps the most commonly used modelling option is to fit the model to a subset of the data (e.g. fit the model to data from just the grazed plants). You could do this using subscripts on the response variable and all the explanatory variables:

```
model<-lm(Fruit[Grazing=="Grazed"] ~ Root[Grazing=="Grazed"])
```

but it is much more straightforward to use the subset argument, especially when there are lots of explanatory variables:

```
model<-lm(Fruit ~ Root,subset=(Grazing=="Grazed"))
```

The answer, of course, is the same in both cases, but the summary.lm and summary.aov tables are neater with subset. Note the round brackets used with the subset option (not the square brackets used with subscripts in the first example)

Weights

The default is for all the values of the response to have equal weights (all equal one)

$$\text{weights} = \text{rep}(1, \text{nobs})$$

There are two sorts of weights in statistical modelling, and you need to be able to distinguish between them:

- case weights give the relative importance of case, so a weight of 2 means there are two such cases;
- inverse of variances, in which the weights downplay highly variable data.

Instead of using initial root size as a covariate (as above) you could use Root as a weight in fitting a model with Grazing as the sole categorical explanatory variable:

```
model<-lm(Fruit ~ Grazing,weights=Root)
summary(model)
```

```
Call:
lm(formula = Fruit~Grazing, weights = Root)

Coefficients:
                Estimate   Std. Error   t value   Pr(>|t|)
(Intercept)       70.725        4.849     14.59     <2e-16  ***
GrazingUngrazed  -16.953        7.469     -2.27      0.029  *

Residual standard error: 62.51 on 38 degrees of freedom
Multiple R-Squared: 0.1194,   Adjusted R-squared: 0.0962
F-statistic: 5.151 on 1 and 38 DF, p-value: 0.02899
```

Needless to say, the use of weights alters the parameter estimates and their standard errors:

```
model<-lm(Fruit ~ Grazing)
summary(model)
```

```
Coefficients:
                Estimate   Std. Error   t value   Pr(>|t|)
(Intercept)       67.941        5.236    12.976   1.54e-15  ***
GrazingUngrazed  -17.060        7.404    -2.304     0.0268  *

Residual standard error: 23.41 on 38 degrees of freedom
Multiple R-Squared: 0.1226,   Adjusted R-squared: 0.09949
F-statistic: 5.309 on 1 and 38 DF, p-value: 0.02678
```

When weights (w) are specified the model is fitted using weighted least squares, in which the quantity to be minimized is $\sum w \times d^2$ (rather than $\sum d^2$), where d is the difference between the response variable and the fitted values predicted by the model.

Missing values

What to do about missing values in the dataframe is an important issue (p. 120). Ideally, of course, there are no missing values, so you don't need to worry about what action to take (na.action). If there are missing values, you have two choices:

- leave out any row of the dataframe in which one or more variables are missing, then na.action = na.omit

- fail the fitting process, so na.action = na.fail

If in doubt, you should specify na.action = na.fail because you will not get nasty surprises if unsuspected NAs in the dataframe cause strange (but unwarned) behaviour in the model. Let's introduce a missing value into the initial root weights:

```
Root[37]<-NA
model<-lm(Fruit ~ Grazing*Root)
```

The model is fitted without comment, and the only thing you might notice is that the residual degrees of freedom is reduced from 36 to 35. If you want to be warned about missing values, then use the na.action option.

```
model<-lm(Fruit ~ Grazing*Root,na.action=na.fail)
```

```
Error in na.fail.default(list(Fruit = c(59.77, 60.98, 14.73, 19.28,
34.25, :   missing values in object
```

If you are carrying out regression with time series data that include missing values then you should use na.action = NULL so that residuals and fitted values are time series as well (if the missing values were omitted, then the resulting vector would not be a time series of the correct length).

Offsets

You would not use offsets with a linear model (you could simply subtract the offset from the value of the response variable, and work with the transformed values). But with generalized linear models you may want to specify part of the variation in the response using an offset (see p. 518 for details and examples).

Dataframes containing the same variable names

If you have several different dataframes containing the same variable names (say x and y) then the simplest way to ensure that the correct variables are used in the modelling is to name the dataframe in the function call:

```
model<-lm(y ~ x,data=correct.frame)
```

The alternative is much more cumbersome to type:

```
model<-lm(correct.frame$y ~ correct.frame$x)
```

Akaike's Information Criterion

Akaike's information criterion (AIC) is known in the statistics trade as a **penalized log-likelihood**. If you have a model for which a log-likelihood value can be obtained, then

$$AIC = -2 \times \log\text{-likelihood} + 2(p+1),$$

where p is the number of parameters in the model, and 1 is added for the estimated variance (you could call this another parameter if you wanted to). To demystify AIC let's calculate it by hand. We revisit the regression data for which we calculated the log-likelihood by hand on p. 217.

```
attach(regression)
names(regression)
```

```
[1]  "growth"  "tannin"
```

```
growth
```

```
[1]  12  10  8  11  6  7  2  3  3
```

The are nine values of the response variable, growth, and we calculated the log-likelihood as -23.98941 earlier. There was only one parameter estimated from the data for these calculations (the mean value of y), so $p = 1$. This means that AIC should be

$$AIC = -2 \times -23.98941 + 2 \times (1+1) = 51.97882.$$

Fortunately, we do not need to carry out these calculations, because there is a built-in function for calculating AIC. It takes a model object as its argument, so we need to fit a one-parameter model to the growth data like this:

```
model<-lm(growth~1)
```

Then we can get the AIC directly:

```
AIC(model)
```

```
[1]  51.97882
```

AIC as a measure of the fit of a model

The more parameters that there are in the model, the better the fit. You could obtain a perfect fit if you had a separate parameter for every data point, but this model would have absolutely no explanatory power. There is always going to be a trade-off between the goodness of fit and the number of parameters required by parsimony. AIC is useful because it explicitly penalizes any superfluous parameters in the model, by adding $2(p+1)$ to the deviance.

When comparing two models, the smaller the AIC, the better the fit. This is the basis of automated model simplification using step.

You can use the function AIC to compare two models, in exactly the same way as you can use anova (as explained on p. 325). Here we develop an analysis of covariance that is introduced on p. 490.

```
model.1<-lm(Fruit~Grazing*Root)
model.2<-lm(Fruit~Grazing+Root)
```

AIC(model.1, model.2)

```
            df          AIC
model.1      5     273.0135
model.2      4     271.1279
```

Because model.2 has the *lower* AIC, we prefer it to model.l. The log-likelihood was penalized by $2 \times (4+1) = 10$ in model 1 because that model contained 4 parameters (2 slopes and 2 intercepts) and by $2 \times (3+1) = 8$ in model.2 because that model had 3 parameters (two intercepts and a common slope). You can see where the two values of AIC come from by calculation:

-2*logLik(model.1)+2*(4+1)

[1] 273.0135

-2*logLik(model.2)+2*(3+1)

[1] 271.1279

Leverage

Points increase in influence to the extent that they lie on their own, a long way from the mean value of x (to either the left or right). To account for this, measures of leverage for a given data point y are proportional to $(x - \bar{x})^2$. The commonest measure of leverage is

$$h_i = \frac{1}{n} + \frac{(x_i - \bar{x})^2}{\sum (x_j - \bar{x})^2}$$

where the denominator is *SSX*. A good rule of thumb is that a point is highly influential if its

$$h_i > \frac{2p}{n},$$

where p is the number of parameters in the model. We could easily calculate the leverage value of each point in our vector x_1 created on p. 345. It is more efficient, perhaps, to write a general function that could carry out the calculation of the h values for any vector of x values,

leverage<-function(x){1/length(x)+(x-mean(x))^2/sum((x-mean(x))^2)}

Then use the function called leverage on x_1:

leverage(x1)

```
[1]    0.3478261    0.1956522    0.1956522    0.1956522    0.1739130
[6]    0.8913043
```

This draws attention immediately to the sixth x value: its h value is more than double the next largest. The result is even clearer if we plot the leverage values

plot(leverage(x1),type="h")
abline(0.66,0,lty=2)
points(leverage(x1))

Note that if the plot directive has a single argument (as here), then the x values for the plot are taken as the *order* of the numbers in the vector to be plotted (called Index and taking the sequence 1:6 in this case). It would be useful to plot the rule-of-thumb value of what constitutes an influential point. In this case, $p = 2$ and n (the number of points on the graph) $= 6$, so a point is influential if $h_i > 0.66$.

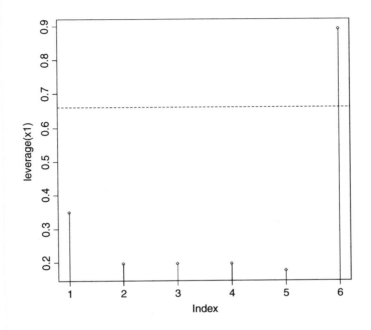

This is enough to warn us that the point (7,6) could be having a marked effect on the parameter estimates of our model. We can see if this is true by repeating the regression without the point (7,6). There are several ways of doing this. If we know the subscript of the point, [6] in this example, we can drop that point explicitly using the negative subscript convention (see p. 24).

```
reg2<-lm(y1[-6]~x1[-6])
summary(reg2)
```

```
Residuals:
          1            2            3            4            5
1.955e-16    1.000e+00    4.572e-18   -1.000e+00   -9.890e-17
Coefficients:
                Estimate    Std. Error    t value    Pr(>|t|)
(Intercept)     2.000e+00   1.770e+00        1.13       0.341
x1[-6]         -2.587e-17   5.774e-01    -4.48e-17       1.000

Residual standard error: 0.8165 on 3 degrees of freedom
Multiple R-Squared: 2.465e-32, Adjusted R-squared: -0.3333
F-statistic: 7.396e-32 on 1 and 3 DF, p-value: 1
```

The point (7,6) was indeed highly influential because without it, the slope of the graph is zero. Notice that the residuals of the points we created are not exactly zero; they are various numbers times 10^{-17}.

Alternatively, we could use **weights** to 'weight out' the point (7,6) whose influence we want to test. We need to create a vector of weights: 1s for the data we want to include and 0s for the data we want to leave out. In this simple case we could type in the weights directly like this:

```
w<-c(1,1,1,1,1,0)
```

but in general, we will want to calculate them on the basis of some logical criterion. A suitable condition for inclusion here would be $x_1 < 6$:

```
(w<-(x1<6))
```

```
[1]  TRUE  TRUE  TRUE  TRUE  TRUE  FALSE
```

Note that when we calculate the weight vector in this way, we get TRUE and FALSE rather than 1 and 0, but this works equally well. The new model looks like this:

```
reg3<-lm(y1~x1,weights=w)
summary(reg3)
```

Finally, we could use **subset** to leave out the point(s) we wanted to exclude from the model fit. Of all the options, this is the most general, and the easiest to use. As with weights, the subset is stated as part of the model specification. It says which points to include, rather than to exclude, so the logic to include any points for which $x_1 < 6$ (say):

```
reg4<-lm(y1~x1,subset=(x1<6))
summary(reg4)
```

The outputs of reg4 and reg3 are exactly the same as in reg2 using subscripts.

Misspecified Model

The model may have the wrong terms in it, or the terms may be included in the model in the wrong way. We deal with the selection of terms for inclusion in the minimal adequate model in Chapter 9. Here we simply note that **transformation of the explanatory variables** often produces improvements in model performance. The most frequently used transformations are logs, powers and reciprocals.

When both the error distribution and functional form of the relationship are unknown, there is no single specific rationale for choosing any given transformation in preference to another. The aim is pragmatic, namely to find a transformation that gives:

- constant error variance;

- approximately normal errors;

- additivity;

- a linear relationship between the response variables and the explanatory variables;

- straightforward scientific interpretation.

The choice is bound to be a compromise and, as such, is best resolved by quantitative comparison of the deviance produced under different model forms. Again, in testing for non-linearity in the relationship between y and x we might add a term in x^2 to the model; a significant parameter in the x^2 term indicates curvilinearity in the relationship between y and x.

A further element of misspecification can occur because of **structural non-linearity**. Suppose, for example, that we were fitting a model of the form

$$y = a + \frac{b}{x},$$

but the underlying process was really of the form

$$y = a + \frac{b}{c + x};$$

then the fit is going to be poor. Of course if we *knew* that the model structure was of this form, then we could fit it as a non-linear model (p. 663) or as a non-linear mixed-effects model (p. 671), but in practice this is seldom the case.

Model checking in R

The data we examine in this section are on the decay of a biodegradable plastic in soil: the response, y, is the mass of plastic remaining and the explanatory variable, x, is duration of burial:

```
Decay<-read.table("c:\\temp\\Decay.txt",header=T)
attach(Decay)
names(Decay)
```

```
[1] "time"  "amount"
```

For the purposes of illustration we shall fit a linear regression to these data and then use model-checking plots to investigate the adequacy of that model:

```
model<-lm(amount ~ time)
```

The basic model checking could not be simpler:

```
plot(model)
```

This one command produces a series of graphs, spread over four pages. The first two graphs are the most important. First, you get a plot of the residuals against the fitted values (left plot) which shows very pronounced curvature; most of the residuals for intermediate fitted values are negative, and the positive residuals are concentrated at the smallest and largest fitted values. Remember, this plot should look like the sky at night, with no pattern of any sort. This suggests systematic inadequacy in the structure of the model. Perhaps the relationship between y and x is non-linear rather than linear as we assumed here? Second, you get a QQ plot (p. 341) which indicates pronounced non-normality in the residuals (the line should be straight, not banana-shaped as here).

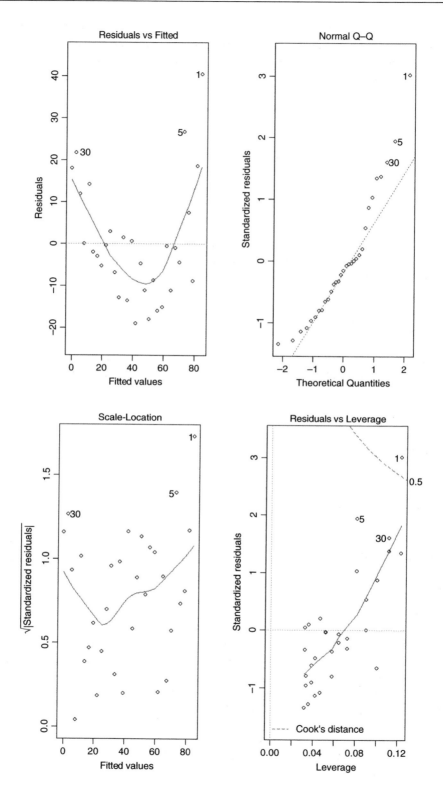

The third graph is like a positive-valued version of the first graph; it is good for detecting non-constancy of variance (heteroscedasticity), which shows up as a triangular scatter (like a wedge of cheese). The fourth graph shows a pronounced pattern in the standardized residuals as a function of the leverage. The graph also shows Cook's distance, highlighting the identity of particularly influential data points.

Cook's distance is an attempt to combine leverage and residuals in a single measure. The absolute values of the deletion residuals $|r_i^*|$ are weighted as follows:

$$C_i = |r_i^*| \left(\frac{n-p}{p} \cdot \frac{h_i}{1-h_i} \right)^{1/2}.$$

Data points 1, 5 and 30 are singled out as being influential, with point number 1 especially so. When we were happier with other aspects of the model, we would repeat the modelling, leaving out each of these points in turn. Alternatively, we could jackknife the data (see p. 422), which involves leaving every data point out, one at a time, in turn. In any event, this is clearly *not* a good model for these data. The analysis is completed on p. 407.

Extracting information from model objects

You often want to extract material from fitted models (e.g. slopes, residuals or p values) and there are three different ways of doing this:

- by name, e.g. coef(model)

- with list subscripts, e.g. summary(model)[[3]]

- using $ to name the component, e.g. model$resid

The model object we use to demonstrate these techniques is the simple linear regression that was analysed in full by hand on p. 388.

```
data<-read.table("c:\\temp\\regression.txt",header=T)
attach(data)
names(data)
```

```
[1] "growth" "tannin"
```

```
model<-lm(growth ~ tannin)
summary(model)
```

```
Call:
lm(formula = growth~tannin)

Residuals:
    Min      1Q    Median      3Q      Max
 -2.4556  -0.8889  -0.2389   0.9778   2.8944

Coefficients:
              Estimate   Std. Error   t value   Pr(>|t|)
(Intercept)    11.7556      1.0408     11.295   9.54e-06  ***
tannin         -1.2167      0.2186     -5.565   0.000846  ***

Residual standard error: 1.693 on 7 degrees of freedom
Multiple R-Squared: 0.8157,   Adjusted R-squared: 0.7893
F-statistic: 30.97 on 1 and 7 DF,  p-value: 0.000846
```

By name

You can extract the coefficients of the model, the fitted values, the residuals, the effect sizes and the variance–covariance matrix by name, as follows:

coef(model)

```
(Intercept)        tannin
  11.755556     -1.216667
```

gives the parameter estimates ('coefficients') for the intercept (a) and slope (b);

fitted(model)

```
        1          2          3          4          5          6
11.755556   10.538889   9.322222   8.105556   6.888889   5.672222
        7          8          9
 4.455556    3.238889   2.022222
```

gives the nine fitted values ($\hat{y} = a + bx$) used in calculating the residuals;

resid(model)

```
        1           2            3           4            5
0.2444444   -0.5388889   -1.3222222   2.8944444   -0.8888889
        6           7            8           9
1.3277778   -2.4555556   -0.2388889   0.9777778
```

gives the residuals (y – fitted values) for the nine data points;

effects(model)

```
(Intercept)          tannin
-20.6666667   -9.4242595    -1.3217694    2.8333333   -1.0115639   1.1435388
 -2.7013585   -0.5462557     0.6088470
attr(,"assign")
[1] 0 1
attr(,"class")
[1] "coef"
```

For a linear model fitted by lm or aov the effects are the uncorrelated single-degree-of-freedom values obtained by projecting the data onto the successive orthogonal subspaces generated by the QR decomposition during the fitting process. The first r ($=2$ in this case; the rank of the model) are associated with coefficients and the remainder span the space of residuals but are not associated with particular residuals. It produces a numeric vector of the same length as residuals of class coef. The first two rows are labelled by the corresponding coefficients (intercept and slope), and the remaining 7 rows are unlabelled.

vcov(model)

```
              (Intercept)          tannin
(Intercept)     1.0832628     -0.19116402
     tannin    -0.1911640      0.04779101
```

This extracts the variance–covariance matrix of the model's parameters.

With list subscripts

The model object is a list with many components. Here each of them is explained in turn. The first is the model formula (or 'Call') showing the response variable (growth) and the explanatory variable(s) (tannin):

summary(model)[[1]]

```
lm(formula = growth~tannin)
```

The second describes the attributes of the object called summary(model):

summary(model)[[2]]

```
growth~tannin
attr(,"variables")
list(growth, tannin)
attr(,"factors")
        tannin
growth        0
tannin        1

attr(,"term.labels")
[1] "tannin"
attr(,"order")
[1] 1 attr(,"intercept")
[1] 1
attr(,"response")
[1] 1
attr(,".Environment")
<environment: R_GlobalEnv>
attr(,"predvars")
list(growth, tannin)
attr(,"dataClasses")
   growth     tannin
"numeric"  "numeric"
```

The third gives the residuals for the nine data points:

summary(model)[[3]]

```
         1           2           3           4           5
0.2444444  -0.5388889  -1.3222222   2.8944444  -0.8888889
         6           7           8           9
1.3277778  -2.4555556  -0.2388889   0.9777778
```

The fourth gives the parameter table, including standard errors of the parameters, t values and p values:

summary(model)[[4]]

```
               Estimate   Std. Error     t value       Pr(>|t|)
(Intercept)   11.755556    1.0407991   11.294740    9.537315e-06
tannin        -1.216667    0.2186115   -5.565427    8.460738e-04
```

The fifth is concerned with whether the corresponding components of the fit (the model frame, the model matrix, the response or the QR decomposition) should be returned. The default is FALSE:

summary(model)[[5]]

```
(Intercept)    tannin
      FALSE     FALSE
```

The sixth is the residual standard error: the square root of the error variance from the summary.aov table (which is not shown here: $s^2 = 2.867$; see p. 396):

summary(model)[[6]]

```
[1] 1.693358
```

The seventh shows the number of rows in the summary.lm table (showing two parameters to have been estimated from the data with this model, and the residual degrees of freedom (d.f.= 7):

summary(model)[[7]]

```
[1] 2 7 2
```

The eighth is $r^2 = SSR/SST$, the fraction of the total variation in the response variable that is explained by the model (see p. 399 for details):

summary(model)[[8]]

```
[1] 0.8156633
```

The ninth is the adjusted R^2, explained on p. 399 but seldom used in practice:

summary(model)[[9]]

```
[1] 0.7893294
```

The tenth gives F ratio information: the three values given here are the F ratio (30.97398), the number of degrees of freedom in the model (i.e. in the numerator: numdf) and the residual degrees of freedom (i.e. in the denominator: dendf):

summary(model)[[10]]

```
    value      numdf      dendf
30.97398    1.00000    7.00000
```

The eleventh component is the correlation matrix of the parameter estimates:

summary(model)[[11]]

```
                 (Intercept)           tannin
(Intercept)      0.37777778    -0.06666667
tannin          -0.06666667     0.01666667
```

You will often want to extract elements from the parameter table that was the fourth object above. The first of these is the intercept (a, the value of growth at tannin $= 0$):

summary(model)[[4]][[1]]

```
[1] 11.75556
```

The second is the slope (b, the change in growth per unit change in tannin):

summary(model)[[4]][[2]]

```
[1] -1.216667
```

The third is the standard error of the intercept, se_a:

summary(model)[[4]][[3]]

```
[1] 1.040799
```

The fourth is the standard error of the slope, se_b:

summary(model)[[4]][[4]]

```
[1] 0.2186115
```

The fifth is the value of Student's t for the intercept $= a/se_a$:

summary(model)[[4]][[5]]

```
[1] 11.29474
```

The sixth is the value of Student's t for the slope $= b/se_b$:

summary(model)[[4]][[6]]
```
[1] -5.565427
```

The seventh is the p value for the intercept: the probability of observing a t value this large or larger, if the null hypothesis (H_0: intercept$= 0$) is true:

summary(model)[[4]][[7]]

```
[1] 9.537315e-06
```

We reject H_0 because $p < 0.05$. The eighth is the p value for the slope: the probability of observing a t value this big or larger, if the null hypothesis (H_0: slope$= 0$) is true:

summary(model)[[4]][[8]]

```
[1] 0.0008460738
```

We reject H_0 because $p < 0.05$. To save the two standard errors (1.040 799 1 and 0.218 611 5) write

sea<-summary(model)[[4]][[3]]
seb<-summary(model)[[4]][[4]]

Extracting components of the model using $

Another way to extract model components is to use the $ symbol. To get the intercept (a) and the slope (b) of the regression, type

model$coef

```
(Intercept)        tannin
  11.755556    -1.216667
```

To get the fitted values ($\hat{y} = a + bx$) used in calculating the residuals, type

model$fitted

```
          1             2             3             4             5             6
11.755556    10.538889    9.322222    8.105556    6.888889    5.672222
          7             8             9
 4.455556     3.238889    2.022222
```

To get the residuals themselves, type

model$resid

```
        1            2            3            4            5
0.2444444   -0.5388889   -1.3222222    2.8944444   -0.8888889
        6            7            8            9
1.3277778   -2.4555556   -0.2388889    0.9777778
```

Finally, the residual degrees of freedom (9 points – 2 estimated parameters $= 7$ d.f.) are

model$df

```
[1]  7
```

Extracting components from the summary.aov table

summary.aov(model)

```
            Df    Sum Sq   Mean Sq   F value     Pr(>F)
tannin       1    88.817    88.817    30.974   0.000846  ***
Residuals    7    20.072     2.867
```

You can get the degrees of freedom, sums of squares, mean squares, F ratios and p values out of the ANOVA table for a model like this:

summary.aov(model)[[1]][[1]]

```
[1]  1  7
```

summary.aov(model)[[1]][[2]]

```
[1]  88.81667  20.07222
```

summary.aov(model)[[1]][[3]]

```
[1]  88.816667  2.867460
```

summary.aov(model)[[1]][[4]]

```
[1]   30.97398 NA
```

summary.aov(model)[[1]][[5]]

```
[1]  0.0008460738  NA
```

You should experiment to see the components of the model object itself (e.g. model[[3]]).

The summary.lm table for continuous and categorical explanatory variables

It is important to understand the difference between summary.lm and summary.aov for the same model. Here is a one-way analysis of variance of the plant competition experiment (p. 155):

comp<-read.table("c:\\temp\\competition.txt",header=T)
attach(comp)
names(comp)

```
[1]  "biomass"  "clipping"
```

The categorical explanatory variable is clipping and it has five levels as follows:

levels(clipping)

```
[1] "control" "n25" "n50" "r10" "r5"
```

The analysis of variance model is fitted like this:

model<-lm(biomass ~ clipping)

and the two different summaries of it are:

summary.aov(model)

```
            Df   Sum Sq   Mean Sq   F value    Pr(>F)
clipping    4    85356    21339     4.3015     0.008752  **
Residuals   25   124020   4961
```

summary.lm(model)

```
Coefficients:
               Estimate   Std. Error   t value   Pr(>|t|)
(Intercept)    465.17     28.75        16.177    9.4e-15   ***
clippingn25    88.17      40.66        2.168     0.03987   *
clippingn50    104.17     40.66        2.562     0.01683   *
clippingr10    145.50     40.66        3.578     0.00145   **
clippingr5     145.33     40.66        3.574     0.00147   **
Residual standard error: 70.43 on 25 degrees of freedom
Multiple R-Squared: 0.4077, Adjusted R-squared: 0.3129
F-statistic: 4.302 on 4 and 25 DF, p-value: 0.008752
```

The latter, summary.lm, is much the more informative. It shows the effect sizes and their standard errors. The effect sizes are shown in the form of contrasts (as explained in detail on p. 370). The only interesting things in summary.aov are the error variance ($s^2 = 4961$) and the F ratio (4.3015) showing that there are significant differences to be explained. The first row of the summary.lm table contains a mean and all the other rows contain differences between means. Thus, the standard error in the first row (labelled (Intercept)) is the standard error of a mean $se_{\text{mean}} = \sqrt{s^2/(k \times n)}$, while the standard errors on the other rows are standard errors of the difference between two means $se_{\text{diff}} = \sqrt{2 \times s^2/n}$, where there are k factor levels each with replication $= n$.

So where do the effect sizes come from? What is 465.17 and what is 88.17? To understand the answers to these questions, we need to know how the equation for the explanatory variables is structured in a linear model when the explanatory variable, as here, is categorical. To recap, the linear regression model is written as

lm(y ~ x)

which R interprets as the two-parameter linear equation. R knows this because x is a continuous variable,

$$y = a + bx, \tag{9.1}$$

in which the values of the parameters a and b are to be estimated from the data. But what about our analysis of variance? We have one explanatory variable, $x = $ clipping, but it is a categorical variable with five levels, control, n25, n50, r10 and r5. The aov model is exactly analogous to the regression model

aov(y ~ x)

but what is the associated equation? Let's look at the equation first, then try to understand it:

$$y = a + bx_1 + cx_2 + dx_3 + ex_4 + fx_5.$$

This looks just like a multiple regression, with five explanatory variables, $x_1, \ldots, x_5$. The key point to understand is that $x_1, \ldots, x_5$ are the levels of the factor called x. The intercept, a, is the overall (or grand) mean for the whole experiment. The parameters $b, \ldots, f$ are differences between the grand mean and the mean for a given factor level. You will need to concentrate to understand this.

With a categorical explanatory variable, all the variables are coded as $x = 0$ except for the factor level that is associated with the y value in question, when x is coded as $x = 1$. You will find this hard to understand without a good deal of practice. Let's look at the first row of data in our dataframe:

comp[1,]

```
     biomass   clipping
1       551        n25
```

So the first biomass value (551) in the dataframe comes from clipping treatment n25 which, out of all the factor levels (above), comes second in the alphabet. This means that for this row of the dataframe $x_1 = 0$, $x_2 = 1$, $x_3 = 0$, $x_4 = 0$, $x_5 = 0$. The equation for the first row therefore looks like this:

$$y = a + b \times 0 + c \times 1 + d \times 0 + e \times 0 + f \times 0,$$

so the model for the fitted value at n25 is

$$\hat{y} = a + c;$$

and similarly for the other factor levels. The fitted value $\hat{y}$ is the sum of two parameters, a and c. The equation apparently does not contain an explanatory variable (there is no x in the equation as there would be in a regression equation, above). Note, too, how many parameters the full model contains: they are represented by the letters a to f and there are six of them. But we can only estimate five parameters in this experiment (one mean for each of the five factor levels). Our model contains one redundant parameter, and we need to deal with this. There are several sensible ways of doing this, and people differ about what is the best way. The writers of R agree that *treatment contrasts* represent the best solution. This method does away with parameter a, the overall mean. The mean of the factor level that comes first in the alphabet (control, in our example) is promoted to pole position, and the other effects are shown as differences (contrasts) between this mean and the other four factor level means.

An example might help make this clearer. Here are our five means:

means<-tapply(biomass,clipping,mean)
means

```
  control         n25         n50        r10          r5
465.1667    553.3333    569.3333   610.6667    610.5000
```

The idea is that the control mean (465.1667) becomes the first parameter of the model (known as the intercept). The second parameter is the difference between the second mean (n25 = 553.333) and the intercept:

means[2]-means[1]

```
      n25
88.16667
```

The third parameter is the difference between the third mean (n50 = 569.333) and the intercept:

means[3]-means[1]

```
     n50
104.1667
```

The fourth parameter is the difference between the fourth mean (r10 = 610.6667) and the intercept:

means[4]-means[1]

```
   r10
145.5
```

The fifth parameter is the difference between the fifth mean (r5 = 610.5) and the intercept:

means[5]-means[1]

```
      r5
145.3333
```

So much for the effect sizes. What about their standard errors? The first row is a mean, so we need the standard error of one factor-level mean. This mean is based on six numbers in this example, so the standard error of the mean is $\sqrt{s^2/n}$ where the error variance $s^2 = 4961$ is obtained from the summary.aov(model) above:

sqrt(4961/6)

```
[1]  28.75471
```

All the other rows have the same standard error, but it is bigger than this. That is because the effects on the 2nd and subsequent rows are *not* means, but *differences between means*. That means that the appropriate standard error is not the standard error of a mean, but rather the standard error of the difference between two means. When two samples are independent, the variance of their difference is the sum of their two variances. Thus, the formula for the standard error of a difference between two means is

$$se_{\text{diff}} = \sqrt{\frac{s_1^2}{n_1} + \frac{s_2^2}{n_2}}.$$

When the two variances and the two sample sizes are the same (as here, because our design is balanced and we are using the pooled error variance (4961) from the summary.aov table) the formula simplifies to $\sqrt{2 \times s^2/n}$:

sqrt(2*4961/6)

```
[1]  40.6653
```

With some practice, that should demystify the origin of the numbers in the summary.lm
table. But it does take lots of practice, and people do find this very difficult at first, so don't
feel bad about it.

Contrasts

Contrasts are the essence of hypothesis testing and model simplification in analysis of
variance and analysis of covariance. They are used to compare means or groups of means
with other means or groups of means, in what are known as **single-degree-of-freedom
comparisons**. There are two sorts of contrasts we might be interested in:

- contrasts we had planned to examine at the experimental design stage (these are referred
  to as *a priori* contrasts);

- contrasts that look interesting after we have seen the results (these are referred to as *a
  posteriori* contrasts).

Some people are very snooty about *a posteriori* contrasts, on the grounds that they were
unplanned. You are not supposed to decide what comparisons to make *after* you have seen
the analysis, but scientists do this all the time. The key point is that you should only do
contrasts *after* the ANOVA has established that there really are significant differences to
be investigated. It is not good practice to carry out tests to compare the largest mean with
the smallest mean, if the ANOVA has failed to reject the null hypothesis (tempting though
this may be).

There are two important points to understand about contrasts:

- there is a huge number of *possible* contrasts, and

- there are only $k - 1$ *orthogonal* contrasts,

where k is the number of factor levels. Two contrasts are said to be **orthogonal** to one
another if the comparisons are statistically independent. Technically, two contrasts are
orthogonal if the products of their contrast coefficients sum to zero (we shall see what this
means in a moment).

Let's take a simple example. Suppose we have one factor with five levels and the factor
levels are called *a, b, c, d*, and *e*. Let's start writing down the possible contrasts. Obviously
we could compare each mean singly with every other:

$$a \text{ vs. } b, \ a \text{ vs. } c, \ a \text{ vs. } d, \ a \text{ vs. } e, \ b \text{ vs. } c, \ b \text{ vs. } d, \ b \text{ vs. } e, \ c \text{ vs. } d, \ c \text{ vs. } e, \ d \text{ vs. } e.$$

But we could also compare pairs of means:

$$\{a, b\} \text{ vs. } \{c, d\}, \ \{a, b\} \text{ vs. } \{c, e\}, \{a, b\} \text{ vs. } \{d, e\}, \ \{a, c\} \text{ vs.} \{b, d\}, \ \{a, c\} \text{ vs. } \{b, e\}, \ldots$$

or triplets of means:

$$\{a, b, c\} \text{ vs. } d, \ \{a, b, c\} \text{ vs. } e, \ \{a, b, d\} \text{ vs. } c, \ \{a, b, d\} \text{ vs. } e, \{a, c, d\} \text{ vs. } b, \ldots$$

or groups of four means:

$\{a, b, c, d\}$ vs. e, $\{a, b, c, e\}$ vs. d, $\{a, b, d, e\}$ vs. c, $\{a, c, d, e\}$ vs. b, $\{b, c, d, e\}$ vs. a.

You doubtless get the idea. There are absolutely masses of possible contrasts. In practice, however, we should only compare things once, either directly or implicitly. So the two contrasts a vs. b and a vs. c implicitly contrast b vs. c. This means that if we have carried out the two contrasts a vs. b and a vs. c then the third contrast b vs. c is *not* an orthogonal contrast because you have already carried it out, implicitly. Which particular contrasts are orthogonal depends very much on your choice of the first contrast to make. Suppose there were good reasons for comparing $\{a,b,c,e\}$ vs. d. For example, d might be the placebo and the other four might be different kinds of drug treatment, so we make this our first contrast. Because $k - 1 = 4$ we only have three possible contrasts that are orthogonal to this. There may be *a priori* reasons to group $\{a,b\}$ and $\{c,e\}$ so we make this our second orthogonal contrast. This means that we have no degrees of freedom in choosing the last two orthogonal contrasts: they have to be a vs. b and c vs. e. Just remember that *with orthogonal contrasts you only compare things once*.

Contrast coefficients

Contrast coefficients are a numerical way of embodying the hypothesis we want to test. The rules for constructing contrast coefficients are straightforward:

- Treatments to be lumped together get the same sign (plus or minus).
- Groups of means to be to be contrasted get opposite sign.
- Factor levels to be excluded get a contrast coefficient of 0.
- The contrast coefficients, c, must add up to 0.

Suppose that with our five-level factor $\{a,b,c,d,e\}$ we want to begin by comparing the four levels $\{a,b,c,e\}$ with the single level d. All levels enter the contrast, so none of the coefficients is 0. The four terms $\{a,b,c,e\}$ are grouped together so they all get the same sign (minus, for example, although it makes no difference which sign is chosen). They are to be compared to d, so it gets the opposite sign (plus, in this case). The choice of what numeric values to give the contrast coefficients is entirely up to you. Most people use whole numbers rather than fractions, but it really doesn't matter. All that matters is that the cs sum to 0. The positive and negative coefficients have to add up to the same value. In our example, comparing four means with one mean, a natural choice of coefficients would be -1 for each of $\{a,b,c,e\}$ and $+4$ for d. Alternatively, with could select $+0.25$ for each of $\{a,b,c,e\}$ and -1 for d.

Factor level:	a	b	c	d	e
contrast 1 coefficients:	-1	-1	-1	4	-1

Suppose the second contrast is to compare $\{a,b\}$ with $\{c,e\}$. Because this contrast excludes d, we set its contrast coefficient to 0. $\{a,b\}$ get the same sign (say, plus) and $\{c,e\}$ get the opposite sign. Because the number of levels on each side of the contrast is equal (2 in both cases) we can use the name numeric value for all the coefficients. The value 1 is the most obvious choice (but you could use 13.7 if you wanted to be perverse):

Factor level:	a	b	c	d	e
Contrast 2 coefficients:	1	1	−1	0	−1

There are only two possibilities for the remaining orthogonal contrasts: *a* vs. *b* and *c* vs. *e*:

Factor level:	a	b	c	d	e
Contrast 3 coefficients:	1	−1	0	0	0
Contrast 4 coefficients:	0	0	1	0	−1

The variation in *y* attributable to a particular contrast is called the **contrast sum of squares**, *SSC*. The sums of squares of the $k - 1$ orthogonal contrasts add up to the total treatment sum of squares, *SSA* ($\sum_{i=1}^{k-1} SSC_i = SSA$; see p. 451). The contrast sum of squares is computed like this:

$$SSC_i = \frac{\left(\sum(c_i T_i/n_i)\right)^2}{\sum(c_i^2/n_i)},$$

where the c_i are the contrast coefficients (above), n_i are the sample sizes within each factor level and T_i are the totals of the *y* values within each factor level (often called the treatment totals). The significance of a contrast is judged by an *F* test, dividing the contrast sum of squares by the error variance. The *F* test has 1 degree of freedom in the numerator (because a contrast is a comparison of two means, and $2 - 1 = 1$) and $k(n - 1)$ degrees of freedom in the denominator (the error variance degrees of freedom).

An example of contrasts in R

The following example comes from the competition experiment we analysed on p. 155, in which the biomass of control plants is compared to the biomass of plants grown in conditions where competition was reduced in one of four different ways. There are two treatments in which the roots of neighbouring plants were cut (to 5 cm or 10 cm depth) and two treatments in which the shoots of neighbouring plants were clipped (25% or 50% of the neighbours were cut back to ground level).

```
comp<-read.table("c:\\temp\\competition.txt",header=T)
attach(comp)
names(comp)
```

```
[1] "biomass" "clipping"
```

We start with the one-way analysis of variance:

```
model1<-aov(biomass ~ clipping)
summary(model1)
```

```
            Df    Sum Sq   Mean Sq   F value    Pr(>F)
clipping     4    85356     21339    4.3015    0.008752   **
Residuals   25   124020      4961
```

Clipping treatment has a highly significant effect on biomass. But have we fully understood the result of this experiment? Probably not. For example, which factor levels had the biggest effect on biomass, and were all of the competition treatments significantly different from the controls? To answer these questions, we need to use summary.lm:

summary.lm(model1)

```
Coefficients:
              Estimate  Std. Error  t value  Pr(>|t|)
(Intercept)    465.17      28.75     16.177  9.33e-15  ***
clippingn25     88.17      40.66      2.168  0.03987    *
clippingn50    104.17      40.66      2.562  0.01683    *
clippingr10    145.50      40.66      3.578  0.00145   **
clippingr5     145.33      40.66      3.574  0.00147   **

Residual standard error: 70.43 on 25 degrees of freedom
Multiple R-Squared: 0.4077, Adjusted R-squared: 0.3129
F-statistic: 4.302 on 4 and 25 DF, p-value: 0.008752
```

This looks as if we need to keep all five parameters, because all five rows of the summary table have one or more significance stars. If fact, this is not the case. This example highlights the major shortcoming of treatment contrasts: they do not show how many significant factor levels we need to retain in the minimal adequate model because all of the rows are being compared with the intercept (with the controls in this case, simply because the factor level name for 'controls' comes first in the alphabet):

levels(clipping)

```
[1] "control" "n25" "n50" "r10" "r5"
```

A priori contrasts

In this experiment, there are several planned comparisons we should like to make. The obvious place to start is by comparing the control plants, exposed to the full rigours of competition, with all of the other treatments. That is to say, we want to contrast the first level of clipping with the other four levels. The contrast coefficients, therefore, would be $4, -1, -1, -1, -1$. The next planned comparison might contrast the shoot-pruned treatments (n25 and n50) with the root-pruned treatments (r10 and r5). Suitable contrast coefficients for this would be $0, 1, 1, -1, -1$ (because we are ignoring the control in this contrast). A third contrast might compare the two depths of root pruning; $0, 0, 0, 1, -1$. The last orthogonal contrast would therefore have to compare the two intensities of shoot pruning: $0, 1, -1, 0, 0$. Because the factor called clipping has five levels there are only $5 - 1 = 4$ orthogonal contrasts.

R is outstandingly good at dealing with contrasts, and we can associate these five user-specified *a priori* contrasts with the categorical variable called clipping like this:

contrasts(clipping)<-cbind(c(4,-1,-1,-1,-1),c(0,1,1,-1,-1),c(0,0,0,1,-1),c(0,1,-1,0,0))

We can check that this has done what we wanted by typing

contrasts(clipping)

```
          [,1]   [,2]   [,3]   [,4]
control     4      0      0      0
n25        -1      1      0      1
n50        -1      1      0     -1
r10        -1     -1      1      0
r5         -1     -1     -1      0
```

which produces the matrix of contrast coefficients that we specified. Note that all the columns add to zero (i.e. each set of contrast coefficients is correctly specified). Note also that the products of any two of the columns sum to zero (this shows that all the contrasts are orthogonal, as intended): for example, comparing contrasts 1 and 2 gives products $0 + (-1) + (-1) + 1 + 1 = 0$.

Now we can refit the model and inspect the results of our specified contrasts, rather than the default treatment contrasts:

```
model2<-aov(biomass ~ clipping)
summary.lm(model2)
```

```
Coefficients:
              Estimate   Std. Error   t value   Pr(>|t|)
(Intercept)  561.80000     12.85926    43.688     <2e-16   ***
clipping1    -24.15833      6.42963    -3.757   0.000921   ***
clipping2    -24.62500     14.37708    -1.713   0.099128     .
clipping3      0.08333     20.33227     0.004   0.996762
clipping4     -8.00000     20.33227    -0.393   0.697313
```

```
Residual standard error: 70.43 on 25 degrees of freedom
Multiple R-Squared: 0.4077, Adjusted R-squared: 0.3129
F-statistic: 4.302 on 4 and 25 DF, p-value: 0.008752
```

Instead of requiring five parameters (as suggested by out initial treatment contrasts), this analysis shows that we need only two parameters: the overall mean (561.8) and the contrast between the controls and the four competition treatments ($p = 0.000921$). All the other contrasts are non-significant.

When we specify the contrasts, the intercept is the overall (grand) mean:

```
mean(biomass)
```

```
[1]  561.8
```

The second row, labelled clipping1, estimates, like all contrasts, the difference between two means. But which two means, exactly? The means for the different factor levels are:

```
tapply(biomass,clipping,mean)
```

```
  control        n25        n50        r10         r5
465.1667   553.3333   569.3333   610.6667   610.5000
```

Our first contrast compared the controls (mean $= 465.1667$, above) with the mean of the other four treatments. The simplest way to get this other mean is to create a new factor, c_1 that has value 1 for the controls and 2 for the rest:

```
c1<-factor(1+(clipping!="control"))
tapply(biomass,c1,mean)
```

```
        1          2
465.1667   585.9583
```

The estimate in the first row, reflecting contrast 1, is the difference between the overall mean (561.8) and the mean of the four non-control treatments (585.9583):

mean(biomass)-tapply(biomass,c1,mean)

```
       1              2
96.63333    -24.15833
```

and you see the estimate on line 2 as -24.158 33. What about the second contrast on line 3? This compares the root- and shoot-pruned treatments, and c_2 is a factor that lumps together the two root and two shoot treatments

c2<-factor(2*(clipping=="n25")+2*(clipping=="n50")+(clipping=="r10")+(clipping=="r5"))

We can compute the mean biomass for the two treatments using tapply, then subtract the means from one another using the diff function, then half the differences:

diff(tapply(biomass,c2,mean))/2

```
       1              2
72.70833    -24.62500
```

So the second contrast on row $3(-24.625)$ is *half the difference* between the root- and shoot-pruned treatments. What about the third row? Contrast number 3 is between the two root-pruned treatments. We know their values already from tapply, above:

```
      r10              r5
610.6667    610.5000
```

The two means differ by 0.166 666 so the third contrast is half the difference between the two means:

(610.666666-610.5)/2

```
[1]  0.083333
```

The final contrast compares the two shoot-pruning treatments, and the contrast is half the difference between these two means:

(553.3333-569.3333)/2

```
[1]  -8
```

To recap: the first contrast compares the overall mean with the mean of the four non-control treatments, the second contrast is half the difference between the root and shoot-pruned treatment means, the third contrast is half the difference between the two root-pruned treatments, and the fourth contrast is half the difference between the two shoot-pruned treatments.

It is important to note that the first four standard errors in the summary.lm table are all different from one another. As we have just seen, the estimate in the first row of the table is a mean, while all the other rows contain estimates that are *differences between means*. The overall mean on the top row is based on 30 numbers so the standard error of the mean is $se = \sqrt{s^2/30}$, where s^2 comes from the ANOVA table:

sqrt(4961/30)

```
[1]  12.85950
```

The small difference in the fourth decimal place is due to rounding errors in calling the variance 4961.0. The next row compares two means so we need the standard error of the difference between two means. The complexity comes from the fact that the two means are each based on *different numbers of numbers*. The overall mean is based on all five factor levels (30 numbers) while the non-control mean with which it is compared is based on four means (24 numbers). Each factor level has $n = 6$ replicates, so the denominator in the standard error formula is $5 \times 4 \times 6 = 120$. Thus, the standard error of the difference between the these two means is $se = \sqrt{s^2/(5 \times 4 \times 6)}$:

```
sqrt(4961/(5*4*6))
```

```
[1]  6.429749
```

For the second contrast on row 3, each of the means is based on 12 numbers so the standard error is $se = \sqrt{2 \times (s^2/12)}$ so the standard error of half the difference is:

```
sqrt(2*(4961/12))/2
```

```
[1]  14.37735
```

The last two contrasts are both between means based on six numbers, so the standard error of the difference is $se = \sqrt{2 \times (s^2/6)}$ and the standard error of half the difference is:

```
sqrt(2*(4961/6))/2
```

```
[1]  20.33265
```

The complexity of these calculations is another reason for preferring treatment contrasts rather than user-specified contrasts as the default. The advantage of orthogonal contrasts, however, is that the summary.lm table gives us a much better idea of the number of parameters required in the minimal adequate model (2 in this case). Treatment contrasts had significance stars on all five rows (see below) because all the non-control treatments were compared to the controls (the Intercept).

Model simplification by stepwise deletion

An alternative to specifying the contrasts ourselves (as above) is to aggregate non-significant factor levels in a stepwise *a posteriori* procedure. To demonstrate this, we revert to treatment contrasts. First, we switch off our user-defined contrasts:

```
contrasts(clipping)<-NULL
options(contrasts=c("contr.treatment","contr.poly"))
```

Now we fit the model with all five factor levels as a starting point:

```
model3<-aov(biomass ~ clipping)
summary.lm(model3)
```

```
Coefficients:
              Estimate   Std. Error   t value   Pr(>|t|)
(Intercept)    465.17        28.75     16.177   9.33e-15   ***
clippingn25     88.17        40.66      2.168   0.03987     *
clippingn50    104.17        40.66      2.562   0.01683     *
clippingr10    145.50        40.66      3.578   0.00145    **
clippingr5     145.33        40.66      3.574   0.00147    **
```

Looking down the list of parameter estimates, we see that the most similar are the effects of root pruning to 10 and 5 cm (145.5 vs. 145.33). We shall begin by simplifying these to a single root-pruning treatment called root. The trick is to use the gets arrow to change the names of the appropriate factor levels. Start by copying the original factor name:

```
clip2<-clipping
```

Now inspect the level numbers of the various factor level names:

```
levels(clip2)
```

```
[1] "control"  "n25"  "n50"  "r10"  "r5"
```

The plan is to lump together r10 and r5 under the same name, 'root'. These are the fourth and fifth levels of clip2, so we write:

```
levels(clip2)[4:5]<-"root"
```

If we type

```
levels(clip2)
```

```
[1] "control"  "n25"  "n50"  "root"
```

we see that r10 and r5 have indeed been replaced by root.

The next step is to fit a new model with clip2 in place of clipping, and to test whether the new simpler model is significantly worse as a description of the data using anova:

```
model4<-aov(biomass ~ clip2)
anova(model3,model4)
```

```
Analysis of Variance Table
Model 1: biomass~clipping
Model 2: biomass~clip2
   Res.Df     RSS   Df     Sum of Sq             F Pr(>F)
1      25  124020
2      26  124020   -1    -0.0833333  0.0000168  0.9968
```

As we expected, this model simplification was completely justified.

The next step is to investigate the effects using summary.lm:

```
summary.lm(model4)
```

```
Coefficients:
              Estimate  Std. Error  t value  Pr(>|t|)
(Intercept)    465.17       28.20    16.498  2.66e-15  ***
clip2n25        88.17       39.87     2.211  0.036029    *
clip2n50       104.17       39.87     2.612  0.014744    *
clip2root      145.42       34.53     4.211  0.000269  ***
```

It looks as if the two shoot-clipping treatments (n25 and n50) are not significantly different from one another (they differ by just 16.0 with a standard error of 39.87). We can lump these together into a single shoot-pruning treatment as follows:

```
clip3<-clip2
levels(clip3)[2:3]<-"shoot"
levels(clip3)
```

```
[1] "control"  "shoot"  "root"
```

Then fit a new model with clip3 in place of clip2:

```
model5<-aov(biomass ~ clip3)
anova(model4,model5)
```

```
Analysis of Variance Table
Model 1: biomass~clip2
Model 2: biomass~clip3
   Res.Df      RSS  Df  Sum of Sq         F Pr(>F)
1      26   124020
2      27   124788  -1      -768  0.161  0.6915
```

Again, this simplification was fully justified. Do the root and shoot competition treatments differ?

```
clip4<-clip3
levels(clip4)[2:3]<-"pruned"
levels(clip4)
```

```
[1]  "control"  "pruned"
```

Now fit a new model with clip4 in place of clip3:

```
model6<-aov(biomass ~ clip4)
anova(model5,model6)
```

```
Analysis of Variance Table
```

```
Model 1: biomass~clip3
Model 2: biomass~clip4
   Res.Df      RSS  Df  Sum of Sq           F Pr(>F)
1      27  124788
2      28  139342  -1    -14553  3.1489 0.08726.
```

This simplification was close to significant, but we are ruthless ($p > 0.05$), so we accept the simplification. Now we have the minimal adequate model:

```
summary.lm(model6)
```

```
Coefficients:
                Estimate  Std. Error  t value  Pr(>|t|)
(Intercept)        465.2        28.8   16.152  1.11e-15  ***
clip4pruned        120.8        32.2    3.751  0.000815  ***
```

It has just two parameters: the mean for the controls (465.2) and the difference between the control mean and the 4 treatment means ($465.2 + 120.8 = 586.0$):

```
tapply(biomass,clip4,mean)
```

```
  control     pruned
465.1667   585.9583
```

We know that these two means are significantly different from the p value of 0.000 815, but just to show how it is done, we can make a final model7 that has no explanatory variable at all (it fits only the overall mean). This is achieved by writing y~1 in the model formula:

```
model7<-aov(biomass~1)
anova(model6,model7)
```

Analysis of Variance Table

```
Model 1: biomass~clip4
Model 2: biomass~1
  Res.Df    RSS  Df  Sum of Sq          F  Pr(>F)
1     28  139342
2     29  209377  -1    -70035  14.073  0.000815  ***
```

Note that the p value is exactly the same as in model6. The p values in R are calculated such that they avoid the need for this final step in model simplification: they are 'p on deletion' values.

Comparison of the three kinds of contrasts

In order to show the differences between treatment, Helmert and sum contrasts, we shall reanalyse the competition experiment using each in turn.

Treatment contrasts

This is the default in R. These are the contrasts you get, unless you explicitly choose otherwise.

options(contrasts=c("contr.treatment","contr.poly"))

Here are the contrast coefficients as set under treatment contrasts:

contrasts(clipping)

```
          n25   n50   r10   r5
control     0     0     0    0
n25         1     0     0    0
n50         0     1     0    0
r10         0     0     1    0
r5          0     0     0    1
```

Notice that the contrasts are *not* orthogonal (the products of the coefficients do not sum to zero).

output.treatment<-lm(biomass ~ clipping)
summary(output.treatment)

```
Coefficients:
               Estimate  Std. Error  t value  Pr(>|t|)
(Intercept)     465.17       28.75   16.177   9.33e-15  ***
clippingn25      88.17       40.66    2.168   0.03987    *
clippingn50     104.17       40.66    2.562   0.01683    *
clippingr10     145.50       40.66    3.578   0.00145   **
clippingr5      145.33       40.66    3.574   0.00147   **
```

With treatment contrasts, the factor levels are arranged in alphabetical sequence, and the level that comes first in the alphabet is made into the intercept. In our example this is control, so we can read off the control mean as 465.17, and the standard error of a mean as 28.75. The remaining four rows are differences between means, and the standard errors are standard errors of differences. Thus, clipping neighbours back to 25 cm increases biomass by 88.17 over the controls and this difference is significant at $p = 0.03987$. And so on. The downside of treatment contrasts is that all the rows appear to be significant despite the fact that rows 2–5 are actually not significantly different from one another, as we saw earlier.

Helmert contrasts

This is the default in S-PLUS, so beware if you are switching back and forth between the two languages.

```
options(contrasts=c("contr.helmert","contr.poly"))
contrasts(clipping)
```

```
           [,1]    [,2]    [,3]    [,4]
control     -1      -1      -1      -1
n25          1      -1      -1      -1
n50          0       2      -1      -1
r10          0       0       3      -1
r5           0       0       0       4
```

Notice that the contrasts are orthogonal (the products sum to zero) and their coefficients sum to zero, unlike treatment contrasts, above.

```
output.helmert<-lm(biomass ~ clipping)
summary(output.helmert)
```

```
Coefficients:
              Estimate   Std. Error   t value   Pr(>|t|)
(Intercept)    561.800       12.859    43.688     <2e-16   ***
clipping1       44.083       20.332     2.168     0.0399     *
clipping2       20.028       11.739     1.706     0.1004
clipping3       20.347        8.301     2.451     0.0216     *
clipping4       12.175        6.430     1.894     0.0699     .
```

With Helmert contrasts, the intercept is the overall mean (561.8). The first contrast (contrast 1 on row 2, labelled clipping1) compares the first mean in alphabetical sequence with the average of the first and second factor levels in alphabetical sequence (control plus n25; see above): its parameter value is the mean of the first two factor levels, minus the mean of the first factor level:

(465.16667+553.33333)/2-465.166667

[1] 44.08332

The third row contains the contrast between the third factor level (n50) and the two levels already compared (control and n25): its value is the difference between the average of the first 3 factor levels and the average of the first two factor levels:

(465.16667+553.33333+569.333333)/3-(465.166667+553.3333)/2

[1] 20.02779

The fourth row contains the contrast between the fourth factor level (r10) and the three levels already compared (control, n25 and n50): its value is the difference between the average of the first four factor levels and the average of the first three factor levels

(465.16667+553.33333+569.333333+610.66667)/
4-(553.3333+465.166667+569.3333)/3

[1] 20.34725

The fifth and final row contains the contrast between the fifth factor level (r5) and the four levels already compared (control, n25, n50 and r10): its value is the difference between the average of the first five factor levels (the grand mean), and the average of the first four factor levels:

mean(biomass)-(465.16667+553.33333+569.333333+610.66667)/4

[1] 12.175

So much for the parameter estimates. Now look at the standard errors. We have seen rather few of these values in any of the analyses we have done to date. The standard error in row 1 is the standard error of the overall mean, with s^2 taken from the overall ANOVA table: $\sqrt{s^2/kn}$.

sqrt(4961/30)

[1] 12.85950

The standard error in row 2 is a comparison of a group of two means with a single mean ($2 \times 1 = 2$). Thus 2 is multiplied by the sample size n in the denominator: $\sqrt{s^2/2n}$.

sqrt(4961/(2*6))

[1] 20.33265

The standard error in row 3 is a comparison of a group of three means with a group of two means (so $3 \times 2 = 6$ in the denominator): $\sqrt{s^2/6n}$.

sqrt(4961/(3*2*6))

[1] 11.73906

The standard error in row 4 is a comparison of a group of four means with a group of three means (so $4 \times 3 = 12$ in the denominator): $\sqrt{s^2/12n}$.

sqrt(4961/(4*3*6))

[1] 8.30077

The standard error in row 5 is a comparison of a group of five means with a group of four means (so $5 \times 4 = 20$ in the denominator): $\sqrt{s^2/20n}$.

sqrt(4961/(5*4*6))

[1] 6.429749

It is true that the parameter estimates and their standard errors are much more difficult to understand in Helmert than in treatment contrasts. But the advantage of Helmert contrasts is that they give you proper orthogonal contrasts, and hence give a much clearer picture of which factor levels need to be retained in the minimal adequate model. They do not eliminate the need for careful model simplification, however. As we saw earlier, this example requires only two parameters in the minimal adequate model, but Helmert contrasts (above) suggest the need for three (albeit only marginally significant) parameters.

Sum contrasts

Sum contrasts are the third alternative:

```
options(contrasts=c("contr.sum","contr.poly"))
```

```
output.sum<-lm(biomass ~ clipping)
summary(output.sum)
```

```
Coefficients:
              Estimate   Std. Error   t value   Pr(>|t|)
(Intercept)    561.800       12.859    43.688     <2e-16   ***
clipping1      -96.633       25.719    -3.757   0.000921   ***
clipping2       -8.467       25.719    -0.329   0.744743
clipping3        7.533       25.719     0.293   0.772005
clipping4       48.867       25.719     1.900   0.069019      .
```

As with Helmert contrasts, the first row contains the overall mean and the standard error of the overall mean. The remaining four rows are different: they are the differences between the grand mean and the first four factor means (control, n25, n50, r10 and r5):

```
tapply(biomass,clipping,mean) - 561.8
```

```
     control            n25          n50           r10            r5
-96.633333     -8.466667     7.533333     48.866667     48.700000
```

The standard errors are all the same (25.719) for all four contrasts. The contrasts compare the grand mean (based on 30 numbers) with a single treatment mean

```
sqrt(4961/30+4961/10)
```

```
[1]   25.71899
```

Aliasing

Aliasing occurs when there is no information available on which to base an estimate of a parameter value. Parameters can be aliased for one of two reasons:

- there are no data in the dataframe from which to estimate the parameter (e.g. missing values, partial designs or correlation amongst the explanatory variables), or

- the model is structured in such a way that the parameter value cannot be estimated (e.g. overspecified models with more parameters than necessary)

Intrinsic aliasing occurs when it is due to the *structure of the model*. **Extrinsic aliasing** occurs when it is due to the *nature of the data*.

Suppose that in a factorial experiment all of the animals receiving level 2 of diet (factor A) and level 3 of temperature (factor B) have died accidentally as a result of attack by a fungal pathogen. This particular combination of diet and temperature contributes no data to the response variable, so the interaction term $A(2):B(3)$ cannot be estimated. It is **extrinsically aliased**, and its parameter estimate is set to zero.

If one continuous variable is perfectly correlated with another variable that has already been fitted to the data (perhaps because it is a constant multiple of the first variable), then the second term is aliased and adds nothing to the model. Suppose that $x_2 = 0.5x_1$ then fitting a model with $x_1 + x_2$ will lead to x_2 being **intrinsically aliased** and given a zero parameter estimate.

If all the values of a particular explanatory variable are set to zero for a given level of a particular factor, then that level is **intentionally aliased**. This sort of aliasing is a useful programming trick in ANCOVA when we wish a covariate to be fitted to some levels of a factor but not to others.

Orthogonal polynomial contrasts: contr.poly

Here are the data from a randomized experiment with four levels of dietary supplement:

```
data<-read.table("c:\\temp\\poly.txt",header=T)
attach(data)
names(data)
```

```
[1] "treatment" "response"
```

We begin by noting that the factor levels are in alphabetical order (not in ranked sequence – none, low, medium, high – as we might prefer):

```
tapply(response,treatment,mean)
```

```
high    low   medium   none
4.50   5.25    7.00    2.50
```

The summary.lm table from the one-way analysis of variance looks like this

```
model<-lm(response ~ treatment)
summary(model)
```

```
Call:
lm(formula = response~treatment)

Residuals :
       Min            1Q        Median           3Q           Max
-1.250e+00    -5.000e-01    -1.388e-16    5.000e-01    1.000e+00

Coefficients:
                   Estimate   Std. Error   t value    Pr(>|t|)
(Intercept)         4.5000       0.3750     12.000    4.84e-08   ***
treatmentlow        0.7500       0.5303      1.414    0.182717
treatmentmedium     2.5000       0.5303      4.714    0.000502   ***
treatmentnone      -2.0000       0.5303     -3.771    0.002666   **

Residual standard error: 0.75 on 12 degrees of freedom
Multiple R-Squared: 0.8606, Adjusted R-squared: 0.8258
F-statistic: 24.7 on 3 and 12 DF, p-value: 2.015e-05
```

The summary.aov table looks like this:

```
summary.aov(model)
```

```
            Df   Sum Sq   Mean Sq   F value      Pr(>F)
treatment    3   41.687    13.896    24.704    2.015e-05   ***
Residuals   12    6.750     0.563
```

We can see that treatment is a factor but it is not ordered:

```
is.factor(treatment)
```

```
[1] TRUE
```

is.ordered(treatment)

```
[1] FALSE
```

To convert it into an ordered factor, we use the ordered function like this:

treatment<-ordered(treatment,levels=c("none","low","medium","high"))
levels(treatment)

```
[1] "none" "low" "medium" "high"
```

Now the factor levels appear in their ordered sequence, rather than in alphabetical order. Fitting the ordered factor makes no difference to the summary.aov table:

model2<-lm(response ~ treatment)
summary.aov(model2)

```
            Df   Sum Sq   Mean Sq   F value     Pr(>F)
treatment    3   41.687    13.896    24.704   2.015e-05   ***
Residuals   12    6.750     0.562
```

but the summary.lm table is fundamentally different when the factors are ordered. Now the contrasts are not contr.treatment but contr.poly (which stands for orthogonal polynomial contrasts):

summary(model2)

```
Call:
lm(formula = response~treatment)
Residuals:
        Min          1Q       Median          3Q         Max
-1.250e+00   -5.000e-01   -1.596e-16   5.000e-01   1.000e+00

Coefficients:
               Estimate   Std. Error   t value   Pr(>|t|)
(Intercept)      4.8125       0.1875    25.667   7.45e-12   ***
treatment.L      1.7330       0.3750     4.621   0.000589   ***
treatment.Q     -2.6250       0.3750    -7.000   1.43e-05   ***
treatment.C     -0.7267       0.3750    -1.938   0.076520    .

Residual standard error: 0.75 on 12 degrees of freedom
Multiple R-Squared: 0.8606, Adjusted R-squared: 0.8258
F-statistic: 24.7 on 3 and 12 DF, p-value: 2.015e-05
```

The levels of the factor called treatment are no longer labelled low, medium, none as with treatment contrasts (above). Instead they are labelled L, Q and C, which stand for linear, quadratic and cubic polynomial terms, respectively. But what are the coefficients, and why are they so difficult to interpret? The first thing you notice is that the intercept 4.8125 is no longer one of the treatment means:

tapply(response,treatment, mean)

```
none     low   medium   high
2.50    5.25     7.00   4.50
```

You could fit a polynomial regression model to the mean values of the response with the four levels of treatment represented by a continuous (dummy) explanatory variable (say, x<-c(1, 2, 3, 4)), then fitting terms for x x^2 and x^3 independently. This is what it would look like:

```
yv<-as.vector(tapply(response,treatment,mean))
x<-1:4
model<-lm(yv~x+1(x^2)+l(x^3))
summary(model)
```

```
Call:
lm(formula = yv~x + I(x^2) + I(x^3))

Residuals:
ALL 4 residuals are 0: no residual degrees of freedom!
Coefficients:
             Estimate  Std. Error  t value  Pr(>|t|)
(Intercept)    2.0000          NA       NA        NA
x             -1.7083          NA       NA        NA
I(x^2)         2.7500          NA       NA        NA
I(x^3)        -0.5417          NA       NA        NA

Residual standard error: NaN on 0 degrees of freedom
Multiple R-Squared: 1, Adjusted R-squared: NaN
F-statistic: NaN on 3 and 0 DF, p-value: NA Call:
lm(formula = yv~xv + I(xv^2) + I(xv^3))
```

Thus the equation for y as a function of treatment (x) could be written

$$y = 2 - 1.7083x + 2.75x^2 - 0.5417x^3.$$

Notice that the intercept is *not* one of the factor-level means (the mean of factor level 1 (none) is the equation evaluated for $x = 1$ (namely $2 - 1.7083 + 2.75 - 0.5417 = 2.5$). So why does R not do it this way? There are two main reasons: orthogonality and computational accuracy. If the linear, quadratic and cubic contrasts are orthogonal and fitted stepwise, then we can see whether adding an extra term produces significantly improved explanatory power in the model. In this case, for instance, there is no justification for retaining the cubic term $(p = 0.07652)$. Computational accuracy can become a major problem when fitting many polynomial terms, because these terms are necessarily so highly correlated:

```
x<-1:4
x2<-x^2
x3<-x^3
cor(cbind(x,x2,x3))
```

```
            x          x2          x3
x   1.0000000   0.9843740   0.9513699
x2  0.9843740   1.0000000   0.9905329
x3  0.9513699   0.9905329   1.0000000
```

Orthogonal polynomial contrasts fix both these problems simultaneously. Here is one way to obtain orthogonal polynomial contrasts for a factor with four levels. The contrasts will go up to polynomials of degree $= k - 1 = 4 - 1 = 3$.

term	x_1	x_2	x_3	x_4
linear	-3	-1	1	3
quadratic	1	-1	-1	1
cubic	-1	3	-3	1

Note that the linear x terms are equally spaced, and have a mean of zero (i.e. each point on the x axis is separated by 2). Also, note that all the rows sum to zero. The key point is that the pointwise products of the terms in any two rows also sum to zero: thus for the linear and quadratic terms we have products of $(-3, 1, -1, 3)$, for the linear and cubic terms $(3, -3, -3, 3)$ and for the quadratic and cubic terms $(1, -3, 3, 1)$. In R, the orthogonal polynomial contrasts have different numerical values, but the same properties:

t(contrasts(treatment))

```
           none          low       medium         high
.L  -0.6708204   -0.2236068    0.2236068    0.6708204
.Q   0.5000000   -0.5000000   -0.5000000    0.5000000
.C  -0.2236068    0.6708204   -0.6708204    0.2236068
```

If you wanted to be especially perverse, you could reconstruct the four estimated mean values from these polynomial contrasts and the treatment effects shown in summary.lm

```
Coefficients:
               Estimate   Std. Error   t value    Pr(>|t|)
(Intercept)      4.8125       0.1875     25.667    7.45e-12    ***
treatment.L      1.7330       0.3750      4.621    0.000589    ***
treatment.Q     -2.6250       0.3750     -7.000    1.43e-05    ***
treatment.C     -0.7267       0.3750     -1.938    0.076520    .
```

taking care with the signs in both contrasts and coefficients. The means for none, low, medium and high are respectively

4.8125 - 0.6708204*1.733 - 0.5*2.6250 + 0.2236068*0.7267

```
[1]  2.499963
```

4.8125 - 0.2236068*1.733+0.5*2.6250 - 0.6708204*0.7267

```
[1]  5.250004
```

4.8125 + 0.2236068*1.733 + 0.5*2.6250 + 0.6708204*0.7267

```
[1]  6.999996
```

4.8125 + 0.6708204*1.733 - 0.5*2.6250 - 0.2236068*0.7267

```
[1]  4.500037
```

in agreement (to 3 decimal places) with the four mean values

tapply(response,treatment,mean)

```
none    low   medium   high
2.50   5.25     7.00   4.50
```

Thus, the parameters can be interpreted as the coefficients in a polynomial model of degree 3 ($=k-1$ because there are $k=4$ levels of the factor called treatment), but only so long as the factor levels are equally spaced (and we don't know whether that is true from the information in the current dataframe, because we know only the ranking) and the class sizes are equal (that is true in the present case where $n=4$).

Because we have four data points (the treatment means) and four parameters, the fit of the model to the data is perfect (there are no residual degrees of freedom and no unexplained variation). We can see what the polynomial function looks like by drawing the smooth curve on top of a barplot for the means:

```
y<-as.vector(tapply(response,treatment,mean))
model<-lm(y~poly(x,3))
model
```

```
Call:
lm(formula = y~poly(x, 3))
```

```
Coefficients:
(Intercept)   poly(x, 3)1   poly(x, 3)2   poly(x, 3)3
     4.8125        1.7330       -2.6250       -0.7267
```

Now we can generate a smooth series of x values between 1 and 4 from which to predict the smooth polynomial function:

```
xv<-seq(1,4,0.1)
yv<-predict(model,list(x=xv))
```

The only slight difficulty is that the x axis values on the barplot do not scale exactly one-to-one with our x values, and we need to adjust the x-location of our smooth lime from

xv to $xs = -0.5 + 1.2xv$. The parameters -0.5 and 1.2 come from noting that the centres of the four bars are at

```
(bar.x<-barplot(y))
```

```
         [,1]
[1,]  0.7
[2,]  1.9
[3,]  3.1
[4,]  4.3
```

```
barplot(y,names=levels(treatment))
xs<--0.5+1.2*xv
lines(xs,yv)
```

Regression

Regression analysis is the statistical method you use when both the response variable and the explanatory variable are continuous variables (i.e. real numbers with decimal places – things like heights, weights, volumes, or temperatures). Perhaps the easiest way of knowing when regression is the appropriate analysis is to see that a scatterplot is the appropriate graphic (in contrast to analysis of variance, say, when the plot would have been a box and whisker or a bar chart). We cover seven important kinds of regression analysis in this book:

- linear regression (the simplest, and much the most frequently used);
- polynomial regression (often used to test for non-linearity in a relationship);
- piecewise regression (two or more adjacent straight lines);
- robust regression (models that are less sensitive to outliers);
- multiple regression (where there are numerous explanatory variables);
- non-linear regression (to fit a specified non-linear model to data);
- non-parametric regression (used when there is no obvious functional form).

The first five cases are covered here, nonlinear regression in Chapter 20 and non-parametric regression in Chapter 18 (where we deal with generalized additive models and non-parametric smoothing).

The essence of regression analysis is using sample data to estimate parameter values and their standard errors. First, however, we need to select a model which describes the relationship between the response variable and the explanatory variable(s). The simplest of all is the linear model

$$y = a + bx.$$

There are two variables and two parameters. The response variable is y, and x is a single continuous explanatory variable. The parameters are a and b: the intercept is a (the value of y when $x = 0$); and the slope is b (the change in y divided by the change in x which brought it about).

The R Book   Michael J. Crawley
© 2007 John Wiley & Sons, Ltd

Linear Regression

Let's start with an example which shows the growth of caterpillars fed on experimental diets differing in their tannin content:

```
reg.data<-read.table("c:\\temp\\regression.txt",header=T)
attach(reg.data)
names(reg.data)
```

```
[1]  "growth"  "tannin"
```

```
plot(tannin,growth,pch=16)
```

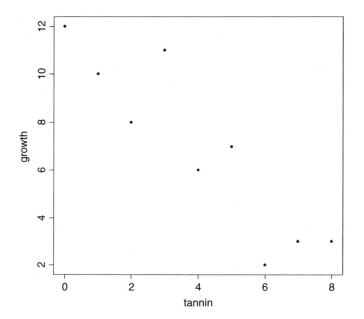

The higher the percentage of tannin in the diet, the more slowly the caterpillars grew. You can get a crude estimate of the parameter values by eye. Tannin content increased by 8 units, in response to which growth declined from about 12 units to about 2 units, a change of -10 units of growth. The slope, b, is the change in y divided by the change in x, so

$$b \approx \frac{-10}{8} = -1.25.$$

The intercept, a, is the value of y when $x = 0$, and we see by inspection of the scatterplot that growth was close to 12 units when tannin was zero. Thus, our rough parameter estimates allow us to write the regression equation as

$$y \approx 12.0 - 1.25x.$$

Of course, different people would get different parameter estimates by eye. What we want is an objective method of computing parameter estimates from the data that are in some sense the 'best' estimates of the parameters for these data and this particular model. The convention in modern statistics is to use the **maximum likelihood estimates** of the parameters as providing the 'best' estimates. That is to say that, given the data, and having selected a linear model, we want to find the values of the slope and intercept that make the data most likely. Keep re-reading this sentence until you understand what it is saying.

For the simple kinds of regression models with which we begin, we make several important assumptions:

- The variance in y is constant (i.e. the variance does not change as y gets bigger).

- The explanatory variable, x, is measured without error.

- The difference between a measured value of y and the value predicted by the model for the same value of x is called a residual.

- Residuals are measured on the scale of y (i.e. parallel to the y axis).

- The residuals are normally distributed.

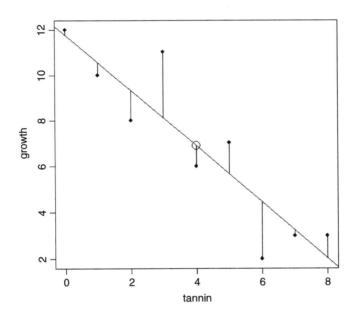

Under these assumptions, the maximum likelihood is given by the **method of least squares**. The phrase 'least squares' refers to the residuals, as shown in the figure. The residuals are the vertical differences between the data (solid circles) and the fitted model (the straight line). Each of the residuals is a distance, d, between a data point, y, and the value predicted by the fitted model, $\hat{y}$, evaluated at the appropriate value of the explanatory variable, x:

$$d = y - \hat{y}.$$

Now we replace the predicted value $\hat{y}$ by its formula $\hat{y} = a + bx$, noting the change in sign:

$$d = y - a - bx.$$

Finally, our measure of lack of fit is the sum of the squares of these distances

$$\sum d^2 = \sum (y - a - bx)^2.$$

The sum of the residuals will always be zero, because the positive and negative residuals cancel out, so $\sum d$ is no good as a measure of lack of fit (although $\sum |d|$ is useful in computationally intensive statistics; see p. 685). The best fit line is defined as passing through the point defined by the mean value of x ($\bar{x}$) and the mean value of y ($\bar{y}$). The large open circle marks the point ($\bar{x}, \bar{y}$). You can think of maximum likelihood as working as follows. Imagine that the straight line is pivoted, so that it can rotate around the point ($\bar{x}, \bar{y}$). When the line is too steep, some of the residuals are going to be very large. Likewise, if the line is too shallow, some of the residuals will again be very large. Now ask yourself what happens to the sum of the squares of the residuals as the slope is rotated from too shallow, through just right, to too steep. The sum of squares will be big at first, then decline to a minimum value, then increase again. A graph of the sum of squares against the value of the slope used in estimating it, would look like this:

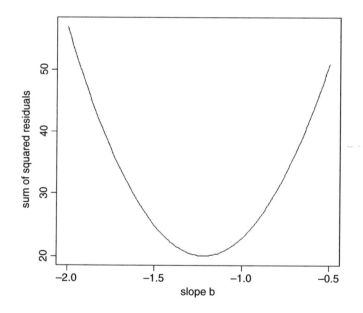

The maximum likelihood estimate of the slope is the value of b associated with the minimum value of the sum of the squares of the residuals (i.e. close to -1.25). Ideally we want an analytic solution that gives the maximum likelihood of the slope directly (this is done using calculus in Box 10.1). It turns out, however, that the least-squares estimate of b can be calculated very simply from the covariance of x and y (which we met on p. 239).

The famous five in R

We want to find the minimum value of $\sum d^2$. To work this out we need the 'famous five': these are $\sum y^2$ and $\sum y$, $\sum x^2$ and $\sum x$ and the sum of products, $\sum xy$ (introduced on p. 391). The sum of products is worked out pointwise. You can calculate the numbers from the data the long way:

sum(tannin);sum(tannin^2);sum(growth);sum(growth^2);sum(tannin*growth)

```
[1] 36
[1] 204
[1] 62
[1] 536
[1] 175
```

Alternatively, as we saw on p. 271, you can create a matrix and use matrix multiplication:

XY<-cbind(1,growth,tannin)
t(XY) %*% XY

```
         growth  tannin
       9     62      36
growth 62    536     175
tannin 36    175     204
```

Corrected sums of squares and sums of products

The next thing is to use the famous five to work out three essential 'corrected sums'. We are already familiar with corrected sums of squares, because these are used in calculating variance: s^2 is calculated as the corrected sum of squares divided by the degrees of freedom (p. 52). We shall need the corrected sum of squares of both the explanatory variable, SSX, and the response variable, SSY:

$$SSX = \sum x^2 - \frac{(\sum x)^2}{n},$$

$$SSY = \sum y^2 - \frac{(\sum y)^2}{n}.$$

The third term is the corrected sum of products, $SSXY$. The covariance of x and y is the expectation of the vector product $E[(x-\bar{x})(y-\bar{y})]$, and this depends on the value of the corrected sum of products (p. 271), which is given by

$$SSXY = \sum xy - \frac{(\sum x)(\sum y)}{n}.$$

If you look carefully you will see that the corrected sum of products has exactly the same kind of structure as SSY and SSX. For SSY, the first term is the sum of y times y and the second term contains the sum of y times the sum of y (and likewise for SSX). For $SSXY$, the first term contains the sum of x times y and the second term contains the sum of x times the sum of y.

Note that for accuracy within a computer program it is best *not* to use these shortcut formulae, because they involve differences (minus) between potentially very large numbers (sums of squares) and hence are potentially subject to rounding errors. Instead, when programming, use the following equivalent formulae:

$$SSY = \sum (y-\bar{y})^2,$$

$$SSX = \sum (x-\bar{x})^2,$$

$$SSXY = \sum (y-\bar{y})(x-\bar{x}).$$

The three key quantities *SSY, SSX* and *SSXY* can be computed the long way, substituting the values of the famous five:

$$SSY = 536 - \frac{62^2}{9} = 108.8889,$$

$$SSX = 204 - \frac{36^2}{9} = 60,$$

$$SSXY = 175 - \frac{36 \times 62}{9} = -73.$$

Alternatively, the matrix can be used (see p. 271).

The next question is how we use *SSX, SSY* and *SSXY* to find the maximum likelihood estimates of the parameters and their associated standard errors. It turns out that this step is much simpler than what has gone before. The maximum likelihood estimate of the slope, *b*, is just:

$$b = \frac{SSXY}{SSX}$$

(the detailed derivation of this is in Box 10.1). So, for our example,

$$b = \frac{-73}{60} = -1.216\ 667.$$

Compare this with our by-eye estimate of -1.25. Now that we know the value of the slope, we can use any point that we know to lie on the fitted straight line to work out the maximum likelihood estimate of the intercept, *a*. One part of the definition of the best-fit straight line is that it passes through the point $(\bar{x}, \bar{y})$ determined by the mean values of *x* and *y*. Since we know that $y = a + bx$, it must be the case that $\bar{y} = a + b\bar{x}$, and so

$$a = \bar{y} - b\bar{x} = \frac{\sum y}{n} - b\frac{\sum x}{n}$$

and, using R as a calculator, we get the value of the intercept as

mean(growth)+1.216667*mean(tannin)

[1] 11.75556

noting the change of sign. This is reasonably close to our original estimate by eye ($a \approx 12$). The function for carrying out linear regression in R is lm (which stands for 'linear model'). The response variable comes first (growth in our example), then the tilde ~, then the name of the continuous explanatory variable (tannin). R prints the values of the intercept and slope like this:

lm(growth~tannin)

Coefficients:
(Intercept) tannin
     11.756 -1.217

We can now write the maximum likelihood equation like this:

$$\text{growth} = 11.75556 - 1.216667 \times \text{tannin}.$$

Box 10.1   The least-squares estimate of the regression slope, *b*

The **best fit** slope is found by rotating the line until the *error sum of squares, SSE,* is minimized, so we want to find the minimum of $\sum(y - a - bx)^2$. We start by finding the derivative of *SSE* with respect to *b*:

$$\frac{\mathrm{d}SSE}{\mathrm{d}b} = -2\sum x(y - a - bx).$$

Now, multiplying through the bracketed term by *x* gives

$$\frac{\mathrm{d}SSE}{\mathrm{d}b} = -2\sum xy - ax - bx^2.$$

Apply summation to each term separately, set the derivative to zero, and divide both sides by -2 to remove the unnecessary constant:

$$\sum xy - \sum ax - \sum bx^2 = 0.$$

We cannot solve the equation as it stands because there are two unknowns, *a* and *b*. However, we know that the value of *a* is $\bar{y} - b\bar{x}$. Also, note that $\sum ax$ can be written as $a\sum x$, so replacing *a* and taking both *a* and *b* outside their summations gives:

$$\sum xy - \left[\frac{\sum y}{n} - b\frac{\sum x}{n}\right]\sum x - b\sum x^2 = 0.$$

Now multiply out the bracketed term by $\sum x$ to get

$$\sum xy - \frac{\sum x \sum y}{n} + b\frac{(\sum x)^2}{n} - b\sum x^2 = 0.$$

Next, take the two terms containing *b* to the right-hand side, and note their change of sign:

$$\sum xy - \frac{\sum x \sum y}{n} = b\sum x^2 - b\frac{(\sum x)^2}{n}.$$

Finally, divide both sides by $\sum x^2 - (\sum x)^2/n$ to obtain the required estimate *b*:

$$b = \frac{\sum xy - \sum x \sum y/n}{\sum x^2 - (\sum x)^2/n}.$$

Thus, the value of *b* that minimizes the sum of squares of the departures is given simply by

$$b = \frac{SSXY}{SSX}.$$

This is the **maximum likelihood estimate of the slope** of the linear regression.

Degree of scatter

There is another very important issue that needs to be considered, because two data sets with exactly the same slope and intercept could look quite different:

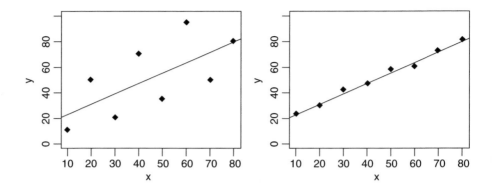

We need a way to quantify the degree of scatter, so that the graph on the left has a high value and the graph on the right has a low value. It turns out that we already have the appropriate quantity: it is the sum of squares of the residuals (p. 389). This is referred to as the *error sum of squares, SSE*. Here, **error** does not mean 'mistake', but refers to residual variation or *unexplained variation*:

$$SSE = \sum (y - a - bx)^2.$$

Graphically, you can think of *SSE* as the sum of the squares of the lengths of the vertical residuals in the plot on p. 389. By tradition, however, when talking about the degree of scatter we actually quantify the *lack* of scatter, so the graph on the left, with a perfect fit (zero scatter) gets a value of 1, and the graph on the right, which shows no relationship at all between y and x (100% scatter), gets a value of 0. This quantity used to measure the lack of scatter is officially called the coefficient of determination, but everybody refers to it as '*r* squared'. This is an important definition that you should try to memorize: r^2 is the fraction of the total variation in y that is explained by variation in x. We have already defined

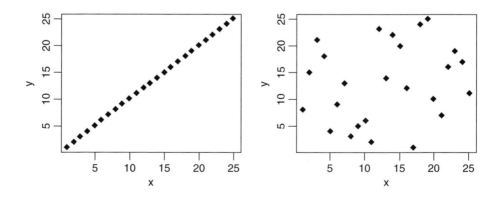

the total variation in the response variable as *SSY* (p. 273). The unexplained variation in the response variable is defined above as *SSE* (the error sum of squares) so the explained variation is simply *SSY* − *SSE*. Thus,

$$r^2 = \frac{SSY - SSE}{SSY}.$$

A value of $r^2 = 1$ means that all of the variation in the response variable is explained by variation in the explanatory variable (the left-hand graph on p. 394) while a value of $r^2 = 0$ means none of the variation in the response variable is explained by variation in the explanatory variable (the right-hand graph on p. 394).

You can get the value of *SSY* the long way as on p. 392 ($SSY = 108.8889$), or using R to fit the null model in which growth is described by a single parameter, the intercept *a*. In R, the intercept is called parameter 1, so the null model is expressed as lm(growth~1). There is a function called deviance that can be applied to a linear model which returns the sum of the squares of the residuals (in this null case, it returns $\sum(y - \bar{y})^2$, which is *SSY* as we require):

deviance(lm(growth~1))

[1] 108.8889

The value of *SSE* is worked out longhand from $\sum(y - a - bx)^2$ but this is a pain, and the value can be extracted very simply from the regression model using deviance like this:

deviance(lm(growth~tannin))

[1] 20.07222

Now we can calculate the value of r^2:

$$r^2 = \frac{SSY - SSE}{SSY} = \frac{108.8889 - 20.072\,22}{108.8889} = 0.815\,663\,3.$$

You will not be surprised that the value of r^2 can be extracted from the model:

summary(lm(growth~tannin))[[8]]

[1] 0.8156633

The **correlation coefficient**, *r*, introduced on p. 311, is given by

$$r = \frac{SSXY}{\sqrt{SSX \times SSY}}$$

Of course *r* is the square root of r^2, but we use the formula above so that we retain the sign of the correlation: *SSXY* is positive for positive correlations between *y* and *x* and *SSXY* is negative when there is a negative correlation between *y* and *x*. For our example, the correlation coefficient is

$$r = \frac{-73}{\sqrt{60 \times 108.8889}} = -0.903\,140\,7.$$

Analysis of variance in regression: $SSY = SSR + SSE$

The idea is simple: we take the total variation in y, SSY, and partition it into components that tell us about the explanatory power of our model. The variation that is explained by the model is called the regression sum of squares (denoted by SSR), and the unexplained variation is called the error sum of squares (denoted by SSE). Then $SSY = SSR + SSE$. Now, in principle, we could compute SSE because we know that it is the sum of the squares of the deviations of the data points from the fitted model, $\sum d^2 = \sum(y - a - bx)^2$. Since we know the values of a and b, we are in a position to work this out. The formula is fiddly, however, because of all those subtractions, squarings and addings-up. Fortunately, there is a very simple shortcut that involves computing SSR, the explained variation, rather than SSE. This is because

$$SSR = b.SSXY = \frac{SSXY^2}{SSX},$$

so we can immediately work out $SSR = -1.21667 \times -73 = 88.81667$. And since $SSY = SSR + SSE$ we can get SSE by subtraction:

$$SSE = SSY - SSR = 108.8889 - 88.81667 = 20.07222.$$

Using R to do the calculations, we get

```
(sse<-deviance(lm(growth~tannin)))
```

```
[1] 20.07222
```

```
(ssy<-deviance(lm(growth~1)))
```

```
[1] 108.8889
```

```
(ssr<-ssy-sse)
```

```
[1] 88.81667
```

We now have all of the sums of squares, and all that remains is to think about the degrees of freedom. We had to estimate one parameter, the overall mean, $\bar{y}$, before we could calculate $SSY = \sum(y - \bar{y})^2$, so the total degrees of freedom are $n - 1$. The error sum of squares was calculated only after two parameters had been estimated from the data (the intercept and the slope) since $SSE = \sum(y - a - bx)^2$, so the error degrees of freedom are $n - 2$. Finally, the regression model added just one parameter, the slope b, compared with the null model, so there is *one* regression degree of freedom. Thus, the ANOVA table looks like this:

Source	Sum of squares	Degrees of freedom	Mean squares	F ratio
Regression	88.817	1	88.817	30.974 ← $\frac{Reg\,Var}{Err\,Var\,(s^2)}$
Error	20.072	7	$s^2 = 2.86746$	
Total	108.889	8	Error variance	

Notice that the component degrees of freedom add up to the total degrees of freedom (this is always true, in any ANOVA table, and is a good check on your understanding of the design of the experiment). The third column, headed 'Mean Squares', contains the

H_0: Slope of line $= 0$

variances obtained by dividing the sums of squares by the degrees of freedom in the same row. In the row labelled 'Error' we obtain the very important quantity called the error variance, denoted by s^2, by dividing the error sum of squares by the error degrees of freedom. Obtaining the value of the error variance is the main reason for drawing up the ANOVA table. Traditionally, one does not fill in the bottom box (it would be the overall variance in y, $SSY/(n-1)$, although this is the basis of the adjusted r^2; value; see p. 399). Finally, the ANOVA table is completed by working out the F ratio, which is a ratio between two variances. In most simple ANOVA tables, you divide the treatment variance in the numerator (the regression variance in this case) by the error variance s^2 in the denominator. The null hypothesis under test in a linear regression is that the slope of the regression line is zero (i.e. no dependence of y on x). The two-tailed alternative hypothesis is that the slope is significantly different from zero (either positive or negative). In many applications it is not particularly interesting to reject the null hypothesis, because we are interested in the estimates of the slope and its standard error (we often know from the outset that the null hypothesis is false). To test whether the F ratio is sufficiently large to reject the null hypothesis, we compare the calculated value of F in the final column of the ANOVA table with the critical value of F, expected by chance alone (this is found from quantiles of the F distribution qf with 1 d.f. in the numerator and $n-2$ d.f. in the denominator, as described below). The table can be produced directly from the fitted model in R by using the anova function:

anova(lm(growth~tannin))

	Df	Sum Sq	Mean Sq	F value	Pr(>F)	
tannin	1	88.817	88.817	30.974	0.000846	***
Residuals	7	20.072	2.867			

The same output can be obtained using summary.aov(lm(growth~tannin)). The extra *MSR* *MSE* column given by R is the p value associated with the computed value of F.

There are two ways to assess our F ratio of 30.974. One way is to compare it with the critical value of F, with 1 d.f. in the numerator and 7 d.f. in the denominator. We have to decide on the level of uncertainty that we are willing to put up with; the traditional value for work like this is 5%, so our certainty is 0.95. Now we can use quantiles of the F distribution, qf, to find the critical value of F:

qf(0.95,1,7)

[1] 5.591448

Because our calculated value of F is much larger than this critical value, we can be confident in rejecting the null hypothesis. The other way, which is perhaps better than working rigidly at the 5% uncertainty level, is to ask what is the probability of getting a value for F as big as 30.974 or larger if the null hypothesis is true. For this we use 1-pf rather than qf:

1-pf(30.974,1,7)

[1] 0.0008460725 ← probability of getting $F=30.974$ if H_0 is true

It is very unlikely indeed ($p < 0.001$). This value is in the last column of the R output.

Unreliability estimates for the parameters

Finding the least-squares values of slope and intercept is only half of the story, however. In addition to the parameter estimates, $a = 11.756$ and $b = -1.2167$, we need to measure the unreliability associated with each of the estimated parameters. In other words, we need to calculate the standard error of the intercept and the standard error of the slope. We have already met the standard error of a mean, and we used it in calculating confidence intervals (p. 54) and in doing Student's t test (p. 294). Standard errors of regression parameters are similar in so far as they are enclosed inside a big square root term (so that the units of the standard error are the same as the units of the parameter), and they have the error variance, s^2, from the ANOVA table (above) in the numerator. There are extra components, however, which are specific to the unreliability of a slope or an intercept (see Boxes 10.2 and 10.3 for details).

Box 10.2   Standard error of the slope

The uncertainty of the estimated slope increases with increasing variance and declines with increasing number of points on the graph. In addition, however, the uncertainty is greater when the range of x values (as measured by SSX) is small:

SE↑ : ↑ variance
       ↓ x-range
SE↓ : ↑ # of points

$$se_b = \sqrt{\frac{s^2}{SSX}}$$

Box 10.3   Standard error of the intercept

The uncertainty of the estimated intercept increases with increasing variance and declines with increasing number of points on the graph. As with the slope, uncertainty is greater when the range of x values (as measured by SSX) is small. Uncertainty in the estimate of the intercept increases with the square of the distance between the origin and the mean value of x (as measured by $\sum x^2$):

$$se_a = \sqrt{\frac{s^2 \sum x^2}{n \times SSX}}$$

Longhand calculation shows that the standard error of the slope is

$$se_b = \sqrt{\frac{s^2}{SSX}} = \sqrt{\frac{2.867}{60}} = 0.2186,$$

and the standard error of the intercept is

$$se_a = \sqrt{\frac{s^2 \sum x^2}{n \times SSX}} = \sqrt{\frac{2.867 \times 204}{9 \times 60}} = 1.0408.$$

However, in practice you would always use the **summary** function applied to the fitted linear model like this:

summary(lm(growth~tannin))

```
Coefficients:
               Estimate  Std. Error  t value   Pr(>|t|)
(Intercept)     11.7556      1.0408   11.295   9.54e-06   ***
tannin          -1.2167      0.2186   -5.565   0.000846   ***
```

$\sqrt{s^2} - \binom{error}{variance}$

```
Residual standard error: 1.693 on 7 degrees of freedom
Multiple R-Squared: 0.8157,   Adjusted R-squared:   0.7893
F-statistic: 30.97 on 1 and 7 DF,   p-value:          0.000846
```

I have stripped out the details about the residuals and the explanation of the signifi-cance stars in order to highlight the parameter estimates and their standard errors (as calculated above). The residual standard error is the square root of the error variance from the ANOVA table ($1.693 = \sqrt{2.867}$). Multiple R-squared is the fraction of the total variance explained by the model ($SSR/SSY = 0.8157$). The Adjusted R-squared is close to, but different from, the value of r^2 we have just calculated. Instead of being based on the explained sum of squares SSR and the total sum of squares SSY, it is based on the overall variance (a quantity we do not typically calculate), $s_T^2 = SSY/(n-1) = 13.611$, and the error variance s^2 (from the ANOVA table, $s^2 = 2.867$) and is worked out like this:

$$\text{adjusted } R\text{-squared} = \frac{s_T^s - s^2}{s_T^2}.$$

So in this example, adjusted R-squared $= (13.611 - 2.867)/13.611 = 0.7893$. We discussed the F statistic and p value in the previous section.

The summary.lm table shows everything you need to know about the parameters and their standard errors, but there is a built-in function, confint, which produces 95% confidence intervals for the estimated parameters from the model directly like this:

confint(model)

```
                 2.5 %        97.5 %
(Intercept)   9.294457    14.2166544
tannin       -1.733601    -0.6997325
```

These values are obtained by subtracting from, and adding to, each parameter estimate an interval which is the standard error times Student's t with 7 degrees of freedom (the appropriate value of t is given by qt(.975,7) $= 2.364624$). The fact that neither interval includes 0 indicates that both parameter values are significantly different from zero, as established by the earlier F tests.

Of the two sorts of summary table, summary.lm is by far the more informative, because it shows the effect sizes (in this case the slope of the graph) and their unreliability estimates (the standard error of the slope). Generally, you should resist the temptation to put ANOVA tables in your written work. The important information such the p value and the error variance can be put in the text, or in figure legends, much more efficiently. ANOVA tables put far too much emphasis on hypothesis testing, and show nothing directly about effect sizes.

Box 10.4 Standard error for a predicted value

The standard error of a predicted value $\hat{y}$ is given by:

SE ⑃ pred
↓ w/ ↑ x-range
↓ w/ ↑ n

$$se_{\hat{y}} = \sqrt{s^2 \left[\frac{1}{n} + \frac{(x - \bar{x})^2}{SSX} \right]}.$$

It increases with the *square* of the difference between mean x and the value of x at which the prediction is made. As with the standard error of the slope, the wider the range of x values, SSX, the lower the uncertainty. The bigger the sample size, n, the lower the uncertainty. Note that the formula for the standard error of the intercept is just the special case of this for $x = 0$ (you should check the algebra of this result as an exercise).

For predictions made on the basis of the regression equation we need to know the standard error for a predicted single sample of y,

$$se_y = \sqrt{s^2 \left[1 + \frac{1}{n} + \frac{(x - \bar{x})^2}{SSX} \right]},$$

while the standard error for a predicted mean for k items at a given level of x_i is

$$se_{\bar{y}_i} = \sqrt{s^2 \left[\frac{1}{k} + \frac{1}{n} + \frac{(x - \bar{x})^2}{SSX} \right]}.$$

Prediction using the fitted model

It is good practice to save the results of fitting the model in a named object. Naming models is very much a matter of personal taste: some people like the name of the model to describe its structure, other people like the name of the model to be simple and to rely on the formula (which is part of the structure of the model) to describe what the model does. I like the second approach, so I might write

model<-lm(growth~tannin)

The object called **model** can now be used for all sorts of things. For instance, we can use the **predict** function to work out values for the response at values of the explanatory variable that we did not measure. Thus, we can ask for the predicted growth if tannin concentration was 5.5%. The value or values of the explanatory variable to be used for prediction are specified in a list like this:

predict(model,list(tannin=5.5))

[1] 5.063889

indicating a predicted growth rate of 5.06 if a tannin concentration of 5.5% had been applied. To predict growth at more than one level of tannin, the list of values for the explanatory variable is specified as a vector. Here are the predicted growth rates at 3.3, 4.4, 5.5 and 6.6% tannin:

predict(model,list(tannin=c(3.3,4.4,5.5,6.6)))

```
       1          2          3          4
7.740556   6.402222   5.063889   3.725556
```

For drawing smooth curves through a scatterplot we use predict with a vector of 100 or so closely-spaced x values, as illustrated on p. 577.

Model checking

The final thing you will want to do is to expose the model to critical appraisal. The assumptions we really want to be sure about are constancy of variance and normality of errors. The simplest way to do this is with model-checking plots. Six plots (selectable by which) are currently available: a plot of residuals against fitted values; a scale–location plot of $\sqrt{|\text{residuals}|}$ against fitted values; a normal QQ plot; a plot of Cook's distances versus row labels; a plot of residuals against leverages; and a plot of Cook's distances against leverage/(1−leverage). By default four plots are provided (the first three plus the fifth):

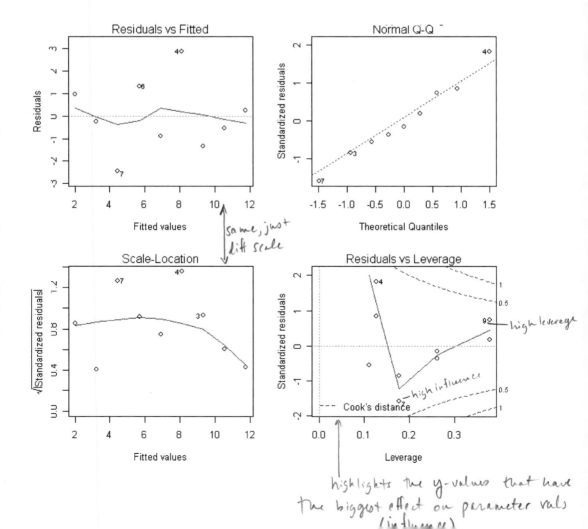

```
par(mfrow=c(2,2))
plot(model)
```

The first graph (top left) shows residuals on the *y* axis against fitted values on the *x* axis. It takes experience to interpret these plots, but what you *do not* want to see is lots of structure or pattern in the plot. Ideally, as here, the points should look like the sky at night. It is a major problem if the scatter increases as the fitted values get bigger; this would look like a wedge of cheese on its side (see p. 340). But in our present case, everything is OK on the constancy of variance front. Likewise, you do not want to see any trend on the residuals with the fitted values; we are OK here (but see p. 344 for a counter example).

The next plot (top right) shows the normal **qqnorm** plot (p. 341) which should be a straight line if the errors are normally distributed. Again, the present example looks fine. If the pattern were S-shaped or banana-shaped, we would need to fit a different model to the data.

The third plot (bottom left) is a repeat of the first, but on a different scale; it shows the square root of the standardized residuals (where all the values are positive) against the fitted values. If there was a problem, such as the variance increasing with the mean, then the points would be distributed inside a triangular shape, with the scatter of the residuals increasing as the fitted values increase. But there is no such pattern here, which is good.

The fourth and final plot (bottom right) shows standardized residuals as a function of leverage, along with Cook's distance (p. 359) for each of the observed values of the response variable. The point of this plot is to highlight those *y* values that have the biggest effect on the parameter estimates (high *influence*; p. 344). You can see that point number 9 has the highest leverage, but point number 7 is quite influential (it is closest to the Cook's distance contour). You might like to investigate how much this influential point $(6, 2)$ affected the parameter estimates and their standard errors. To do this, we repeat the statistical modelling but leave out the point in question, using **subset** like this ($!=$ means 'not equal to'):

```
model2<-update(model,subset=(tannin != 6))
summary(model2)
```

```
Coefficients:
             Estimate  Std. Error  t value   Pr(>|t|)
(Intercept)   11.6892      0.8963   13.042   1.25e-05  ***
tannin        -1.1171      0.1956   -5.712   0.00125   **

Residual standard error: 1.457 on 6 degrees of freedom
Multiple R-Squared: 0.8446,   Adjusted R-squared: 0.8188
F-statistic: 32.62 on 1 and 6 DF,  p-value: 0.001247
```

First of all, notice that we have lost one degree of freedom, because there are now eight values of *y* rather than nine. The estimate of the slope has changed from -1.2167 to -1.1171 (a difference of about 9%) and the standard error of the slope has changed from 0.2186 to 0.1956 (a difference of about 12%). What you do in response to this information depends on the circumstances. Here, we would simply note that point $(6, 2)$ was influential and stick with our first model, using all the data. In other circumstances, a data point might be so influential that the structure of the model is changed completely by leaving it out. In that case, we might gather more data or, if the study was already finished, we might publish both results (with and without the influential point) so that the reader could make up their own mind about the interpretation. The important point is that we always do model checking; the summary.lm(model) table is not the end of the process of regression analysis.

You might also want to check for lack of serial correlation in the residuals (e.g. time series effects) using the durbin.watson function from the car package (see p. 424), but there are too few data to use it with this example.

Polynomial Approximations to Elementary Functions

Elementary functions such as $\sin(x)$, $\log(x)$ and $\exp(x)$ can be expressed as Maclaurin series:

$$\sin(x) = x - \frac{x^3}{3!} + \frac{x^5}{5!} - \frac{x^7}{7!} + \cdots,$$

$$\cos(x) = 1 - \frac{x^2}{2!} + \frac{x^4}{4!} - \frac{x^6}{6!} + \cdots,$$

$$\exp(x) = \frac{x^0}{0!} + \frac{x^1}{1!} + \frac{x^2}{2!} + \frac{x^3}{3!} \cdots,$$

$$\log(x+1) = x - \frac{x^2}{2} + \frac{x^3}{3} - \frac{x^4}{4} + \frac{x^5}{5} + \cdots,$$

In fact, we can approximate any smooth continuous single-valued function by a polynomial of sufficiently high degree. To see this in action, consider the graph of $\sin(x)$ against x in the range $0 < x < \pi$ (where x is an angle measured in radians):

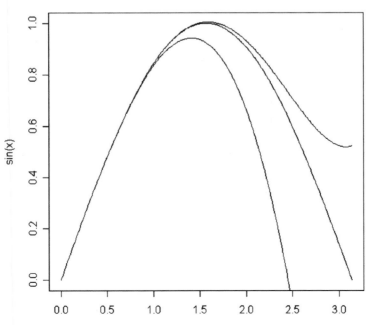

```
x<-seq(0,pi,0.01)
y<-sin(x)
plot(x,y,type="l",ylab="sin(x)")
```

Up to about $x = 0.3$ the very crude approximation $\sin(x) = x$ works reasonably well. The first approximation, including a single extra term for $-x^3/3!$, extends the reasonable fit up to about $x = 0.8$:

```
a1<-x-x^3/factorial(3)
lines(x,a1,col="green")
```

Adding the term in $x^5/5!$ captures the first peak in $\sin(x)$ quite well. And so on.

```
a2<-x-x^3/factorial(3)+x^5/factorial(5)
lines(x,a2,col="red")
```

Polynomial Regression

The relationship between y and x often turns out not to be a straight line. However, Occam's razor requires that we fit a straight-line model unless a non-linear relationship is significantly better at describing the data. So this begs the question: how do we assess the significance of departures from linearity? One of the simplest ways is to use polynomial regression.

The idea of polynomial regression is straightforward. As before, we have just one continuous explanatory variable, x, but we can fit higher powers of x, such as x^2 and x^3, to the model in addition to x to explain curvature in the relationship between y and x. It is useful to experiment with the kinds of curves that can be generated with very simple models. Even if we restrict ourselves to the inclusion of a quadratic term, x^2, there are many curves we can describe, depending upon the signs of the linear and quadratic terms:

In the top left-hand panel, there is a curve with positive but declining slope, with no hint of a hump ($y = 4 + 2x - 0.1x^2$). The top right-hand graph shows a curve with a clear maximum ($y = 4 + 2x - 0.2x^2$), and at bottom left we have a curve with a clear minimum ($y = 12 - 4x + 0.35x^2$). The bottom right-hand curve shows a positive association between y and x with the slope increasing as x increases ($y = 4 + 0.5x + 0.1x^2$). So you can see that a simple quadratic model with three parameters (an intercept, a slope for x, and a slope for x^2) is capable of describing a wide range of functional relationships between y and x. It is very important to understand that the quadratic model *describes* the relationship between y and x; it does not pretend to *explain* the mechanistic (or causal) relationship between y and x.

We can see how polynomial regression works by analysing an example where diminishing returns in output (yv) are suspected as inputs (xv) are increased:

```
poly<-read.table("c:\\temp\\diminish.txt",header=T)
attach(poly)
names(poly)
```

```
[1]  "xv"  "yv"
```

We begin by fitting a straight line model to the data:

```
plot(xv,yv,pch=16)
model1<-lm(yv~xv)
abline(model1)
```

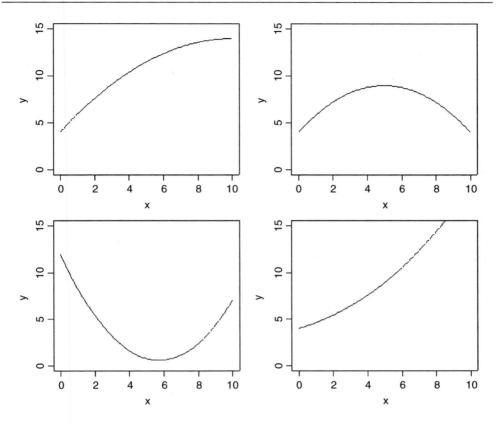

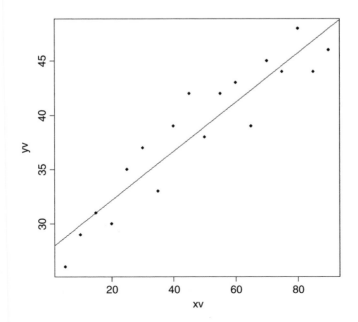

This is not a bad fit to the data ($r^2 = 0.8725$), but there is a distinct hint of curvature (diminishing returns in this case).

Next, we fit a second explanatory variable which is the square of the x value (the so-called 'quadratic term'). Note the use of I (for 'as is') in the model formula; see p. 332.

```
model2<-lm(yv~xv+I(xv^2))
```

Now we use model2 to predict the fitted values for a smooth range of x values between 0 and 90:

```
x<-0:90
y<-predict(model2,list(xv=x))
plot(xv,yv,pch=16)
lines(x,y)
```

This looks like a slightly better fit than the straight line ($r^2 = 0.9046$), but we shall choose between the two models on the basis of an F test using anova:

```
anova(model1,model2)
```

```
Analysis of Variance Table

Model 1: yv ~ xv
Model 2: yv ~ xv + I(xv^2)

     Res.Df     RSS  Df  Sum of Sq        F Pr(>F)
1        16  91.057
2        15  68.143   1     22.915   5.0441 0.0402  *
```

The more complicated curved model is a significant improvement over the linear model ($p = 0.04$) so we accept that there is evidence of curvature in these data.

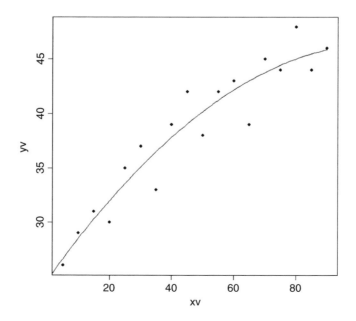

Fitting a Mechanistic Model to Data

Rather than fitting some arbitrary model for curvature (as above, with a quadratic term for inputs), we sometimes have a mechanistic model relating the value of the response variable to the explanatory variable (e.g. a mathematical model of a physical process). In the following example we are interested in the decay of organic material in soil, and our mechanistic model is based on the assumption that the fraction of dry matter lost per year is a constant. This leads to a two-parameter model of exponential decay in which the amount of material remaining (y) is a function of time (t)

$$y = y_0 e^{-bt}.$$

Here y_0 is the initial dry mass (at time $t = 0$) and b is the decay rate (the parameter we want to estimate by linear regression). Taking logs of both sides, we get

$$\log(y) = \log(y_0) - bt.$$

Now you can see that we can estimate the parameter of interest, b, as the slope of a linear regression of $\log(y)$ on t (i.e. we log-transform the y axis but not the x axis) and the value of y_0 as the antilog of the intercept. We begin by plotting our data:

```
data<-read.table("c:\\temp \\Decay.txt",header=T)
names(data)

[1]  "time"  "amount"

attach(data)
plot(time,amount,pch=16)
```

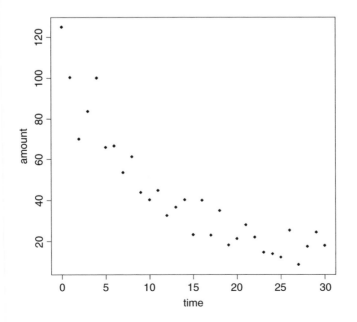

The curvature in the relationship is clearly evident from the scatterplot. Now we fit the linear model of log(amount) as a function of time:

```
model<-lm(log(amount)~time)
summary(model)
```

```
Call:
lm(formula = log(amount) ~ time)

Residuals:
      Min          1Q     Median          3Q         Max
-0.593515   -0.204324   0.006701   0.219835   0.629730

Coefficients:
              Estimate  Std. Error  t value    Pr(>|t|)
(Intercept)   4.547386    0.100295    45.34    < 2e-16   ***
time         -0.068528    0.005743   -11.93   1.04e-12   ***

Residual standard error: 0.286 on 29 degrees of freedom
Multiple R-Squared:0.8308,  Adjusted R-squared: 0.825
F-statistic: 142.4 on 1 and 29 DF,p-value: 1.038e-12
```

Thus, the slope is $-0.068\,528$ and y_0 is the antilog of the intercept: $y_0 = \exp(4.547\,386) = 94.385\,36$. The equation can now be parameterized (with standard errors in brackets) as:

$$y = e^{4.5474(\pm 0.1003) - 0.0685(\pm 0.00574)t}$$

or written in its original form, without the uncertainty estimates, as

$$y = 94.385e^{-0.0685t}$$

and we can draw the fitted line through the data, remembering to take the antilogs of the predicted values (the model predicts log(amount) and we want amount), like this:

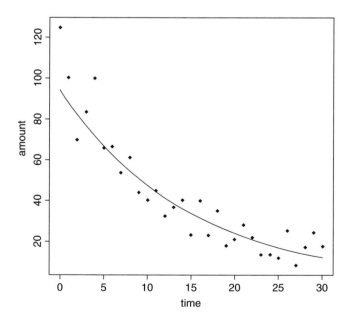

```
xv<-seq(0,30,0.2)
yv<-exp(predict(model,list(time=xv)))
lines(xv,yv)
```

Linear Regression after Transformation

Many mathematical functions that are non-linear in their parameters can be linearized by transformation (see p. 205). The most frequent transformations (in order of frequency of use), are logarithms, antilogs and reciprocals. Here is an example of linear regression associated with a power law (p. 198): this is a two-parameter function

$$y = ax^b,$$

where the parameter a describes the slope of the function for low values of x and b is the shape parameter. For $b = 0$ we have a horizontal relationship $y = a$, for $b = 1$ we have a straight line through the origin $y = ax$ with slope $= a$, for $b > 1$ the slope is positive but the slope increases with increasing x, for $0 < b < 1$ the slope is positive but the slope decreases with increasing x, while for $b < 0$ (negative powers) the curve is a negative hyperbola that is asymptotic to infinity as x approaches 0 and asymptotic to zero as x approaches infinity.

Let's load a new dataframe and plot the data:

```
power<-read.table("c:\\temp \\power.txt",header=T)
attach(power)
names(power)
```

```
[1]  "area"  "response"
```

```
par(mfrow=c(1,2))
plot(area,response,pch=16)
abline(lm(response~area))
plot(log(area),log(response),pch=16)
abline(lm(log(response)~log(area)))
```

The two plots look very similar in this case (they don't always), but we need to compare the two models.

```
model1<-lm(response~area)
model2<-lm(log(response)~log(area))
```

```
summary(model2)
```

```
Coefficients:
             Estimate  Std. Error  t value   Pr(>|t|)
(Intercept)   0.75378     0.02613   28.843    < 2e-16  ***
  log(area)   0.24818     0.04083    6.079   1.48e-06  ***

Residual standard error: 0.06171 on 28 degrees of freedom
Multiple R-Squared: 0.5689,  Adjusted R-squared: 0.5535
F-statistic: 36.96 on 1 and 28 DF, p-value: 1.480e-06
```

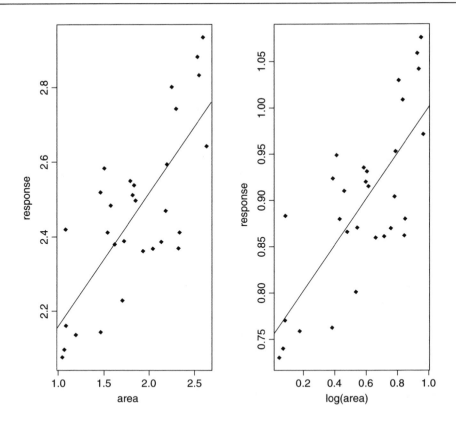

We need to do a t test to see whether the estimated shape parameter $b = 0.24818$ is significantly less than $b = 1$ (a straight line):

$$t = \frac{|0.248\,18 - 1.0|}{0.040\,83} = 18.413\,42.$$

This is highly significant ($p < 0.0001$), so we conclude that there is a non-linear relationship between response and area. Let's get a visual comparison of the two models:

```
par(mfrow=c(1,1))
plot(area,response)
abline(lm(response~area))
xv<-seq(1,2.7,0.01)
yv<-exp(0.75378)*xv^0.24818
lines(xv,yv)
```

This is a nice example of the distinction between statistical significance and scientific importance. The power law transformation shows that the curvature is highly significant ($b < 1$ with $p < 0.0001$) but over the range of the data, and given the high variance in y, the effect of the curvature is very small; the straight line and the power function are very close to one another. However, the choice of model makes an enormous difference if the function is to be used for prediction. Here are the two functions over an extended range of values for x:

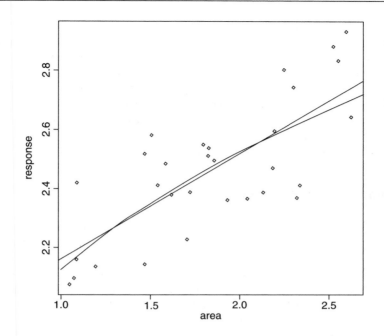

```
xv<-seq(0,5,0.01)
yv<-exp(0.75378)*xv^0.24818
plot(area,response,xlim=c(0,5),ylim=c(0,4),pch=16)
abline(model1)
lines(xv,yv)
```

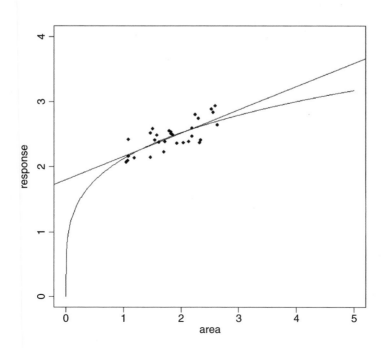

The moral is clear: you need to extremely careful when using regression models for prediction. If you know that the response *must* be zero when area $= 0$ (the graph has to pass through the origin) then obviously the power function is likely to be better for extrapolation to the left of the data. But if we have no information on non-linearity other than that contained within the data, then parsimony suggests that errors will be smaller using the simpler, linear model for prediction. Both models are equally good at describing the data (the linear model has $r^2 = 0.574$ and the power law model has $r^2 = 0.569$) but extrapolation beyond the range of the data is always fraught with difficulties. Targeted collection of new data for response at areas close to 0 and close to 5 might resolve the issue.

Prediction following Regression

The popular notion is that predicting the future is impossible, and that attempts at prediction are nothing more that crystal-gazing. However, all branches of applied science rely upon prediction. These predictions may be based on extensive experimentation (as in engineering or agriculture) or they may be based on detailed, long-term observations (as in astronomy or meteorology). In all cases, however, the main issue to be confronted in prediction is how to deal with uncertainty: uncertainty about the suitability of the fitted model, uncertainty about the representativeness of the data used to parameterize the model, and uncertainty about future conditions (in particular, uncertainty about the future values of the explanatory variables).

There are two kinds of prediction, and these are subject to very different levels of uncertainty. **Interpolation**, which is prediction *within* the measured range of the data, can often be very accurate and is not greatly affected by model choice. **Extrapolation**, which is prediction *beyond* the measured range of the data, is far more problematical, and model choice is a major issue. Choice of the wrong model can lead to wildly different predictions (see p. 411).

Here are two kinds of plots involved in prediction following regression: the first illustrates uncertainty in the parameter estimates; the second indicates uncertainty about predicted values of the response. We continue with the tannin example:

```
data<-read.table("c:\\temp\\regression.txt",header=T)
attach(data)
names(data)
```

```
[1]  "growth"  "tannin"
```

```
plot(tannin,growth,pch=16,ylim=c(0,15))
model<-lm(growth ~ tannin)
abline(model)
```

The first plot is intended to show the uncertainty associated with the estimate of the slope. It is easy to extract the slope from the vector of coefficients:

```
coef(model)[2]
```

```
     tannin
-1.216667
```

The standard error of the slope is a little trickier to find. After some experimentation, you will discover that it is in the fourth element of the list that is summary(model):

```
summary(model)[[4]][4]
```

[1] 0.2186115

Here is a function that will add dotted lines showing two extra regression lines to our existing plot – the estimated slope plus and minus one standard error of the slope:

```
se.lines<-function(model){
b1<-coef(model)[2]+ summary(model)[[4]][4]
b2<-coef(model)[2]- summary(model)[[4]][4]
xm<-mean(model[[12]][2])
ym<-mean(model[[12]][1])
a1<-ym-b1*xm
a2<-ym-b2*xm
abline(a1,b1,lty=2)
abline(a2,b2,lty=2)
}

se.lines(model)
```

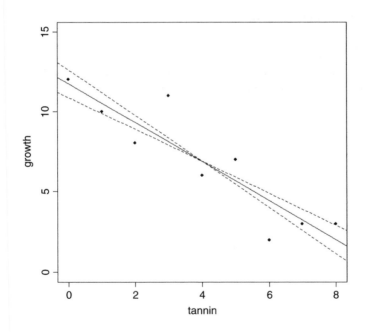

More often, however, we are interested in the uncertainty about predicted values (rather than uncertainty of parameter estimates, as above). We might want to draw the 95% confidence intervals associated with predictions of y at different values of x. As we saw on p. 400, uncertainty increases with the *square* of the difference between the mean value of x and the value of x at which the value of y is to be predicted. Before we can draw these lines we need to calculate a vector of x values; you need 100 or so values to make an attractively smooth curve. Then we need the value of Student's t (p. 222). Finally, we multiply Student's t by the standard error of the predicted value of y (p. 223) to get the confidence interval. This is added to the fitted values of y to get the upper limit and subtracted from the fitted values of y to get the lower limit. Here is the function:

```
ci.lines<-function(model){
xm<-mean(model[[12]][2])
n<-length(model[[12]][[2]])
ssx<- sum(model[[12]][2]^2)- sum(model[[12]][2])^2/n
s.t<- qt(0.975,(n-2))
xv<-seq(min(model[[12]][2]),max(model[[12]][2]),(max(model[[12]][2])-
min(model[[12]][2]))/100)
yv<- coef(model)[1]+coef(model)[2]*xv
se<-sqrt(summary(model)[[6]]^2*(1/n+(xv-xm)^2/ssx))
ci<-s.t*se
uyv<-yv+ci
lyv<-yv-ci
lines(xv,uyv,lty=2)
lines(xv,lyv,lty=2)
}
```

We replot the linear regression, then overlay the confidence intervals (Box 10.4):

```
plot(tannin,growth,pch=16,ylim=c(0,15))
abline(model)
ci.lines(model)
```

This draws attention to the points at tannin = 3 and tannin = 6 that fall outside the 95% confidence limits of our fitted values.

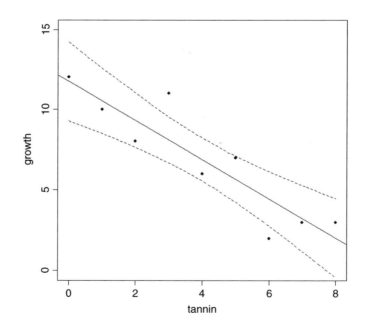

You can speed up this procedure by using the built-in ability to generate confidence intervals coupled with **matlines**. The familiar 95% confidence intervals are int="c", while prediction intervals (fitted values plus or minus 2 standard deviations) are int="p".

```
plot(tannin,growth,pch=16,ylim=c(0,15))
model<-lm(growth~tannin)
```

As usual, start by generating a series of x values for generating the curves, then create the scatterplot. The y values are predicted from the model, specifying int="c", then matlines is used to draw the regression line (solid) and the two confidence intervals (dotted), producing exactly the same graph as our last plot (above):

```
xv<-seq(0,8,0.1)
yv<-predict(model,list(tannin=xv),int="c")
matlines(xv,yv,lty=c(1,2,2),col="black")
```

A similar plot can be obtained using the effects library (see p. 178).

Testing for Lack of Fit in a Regression with Replicated Data at Each Level of x

The unreliability estimates of the parameters explained in Boxes 10.2 and 10.3 draw attention to the important issues in optimizing the efficiency of regression designs. We want to make the error variance as small as possible (as always), but in addition, we want to make SSX as large as possible, by placing as many points as possible at the extreme ends of the x axis. Efficient regression designs allow for:

- replication of least some of the levels of x;
- a preponderance of replicates at the extremes (to maximize SSX);
- sufficient different values of x to allow accurate location of thresholds.

Here is an example where replication allows estimation of pure sampling error, and this in turn allows a test of the significance of the data's departure from linearity. As the concentration of an inhibitor is increased, the reaction rate declines:

```
data<-read.delim("c:\\temp\\lackoffit.txt")
attach(data)
names(data)
```

```
[1]  "conc"  "rate"
```

```
plot(conc,rate,pch=16,ylim=c(0,8))
abline(lm(rate~conc))
```

The linear regression does not look too bad, and the slope is highly significantly different from zero:

```
model.reg<-lm(rate~conc)
summary.aov(model.reg)
```

```
           Df   Sum Sq  Mean Sq  F value     Pr(>F)
conc        1   74.298   74.298   55.333  4.853e-07  ***
Residuals  19   25.512    1.343
```

Because there is replication at each level of x we can do something extra, compared with a typical regression analysis. We can estimate what is called the **pure error variance**. This

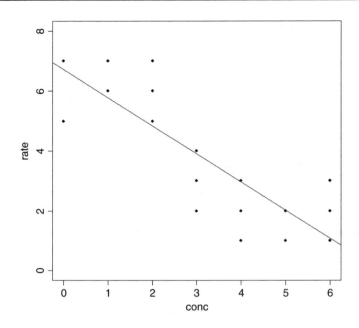

is the sum of the squares of the differences between the *y* values and the *mean* values of *y* for the relevant level of *x*. This should sound a bit familiar. In fact, it is the definition of *SSE* from a one-way analysis of variance (see p. 451). By creating a factor to represent the seven levels of *x*, we can estimate this *SSE* simply by fitting a one-way analysis of variance:

```
fac.conc<-factor(conc)
model.aov<-aov(rate~fac.conc)
summary(model.aov)
          Df   Sum Sq   Mean Sq   F value      Pr(>F)
fac.conc   6   87.810    14.635    17.074    1.047e-05   ***
Residuals 14   12.000     0.857
```

This shows that the pure error sum of squares is 12.0 on 14 degrees of freedom (three replicates, and hence 2 d.f., at each of seven levels of *x*). See if you can figure out why this sum of squares is less than the observed in the **model.reg** regression (25.512). If the means from the seven different concentrations all fell exactly on the same straight line then the two sums of squares would be identical. It is the fact that the means do *not* fall on the regression line that causes the difference. The difference between these two sums of squares $(25.512 - 12.9 = 13.512)$ is a measure of lack of fit of the rate data to the straight-line model. We can compare the two models to see if they differ in their explanatory powers:

```
anova(model.reg,model.aov)

Analysis of Variance Table

Model 1: rate ~ conc
Model 2: rate ~ fac.conc
  Res.Df      RSS  Df  Sum of Sq        F   Pr(>F)
1     19   25.512
2     14   12.000   5     13.512   3.1528  0.04106   *
```

A single anova table showing the lack-of-fit sum of squares on a separate line is obtained by fitting both the regression line (1 d.f.) and the lack of fit (5 d.f.) in the same model:

```
anova(lm(rate~conc+fac.conc))
```

```
Analysis of Variance Table

Response:  rate

           Df   Sum Sq  Mean Sq  F value     Pr(>F)
conc        1   74.298   74.298  86.6806  2.247e-07  ***
fac.conc    5   13.512    2.702   3.1528    0.04106  *
Residuals  14   12.000    0.857
```

To get a visual impression of this lack of fit we can draw vertical lines from the mean values to the fitted values of the linear regression for each level of x:

```
my<-as.vector(tapply(rate,fac.conc,mean))
for (i in 0:6)
lines(c(i,i),c(my[i+1],predict(model.reg,list(conc=0:6))[i+1]),col="red")
points(0:6,my,pch=16,col="red")
```

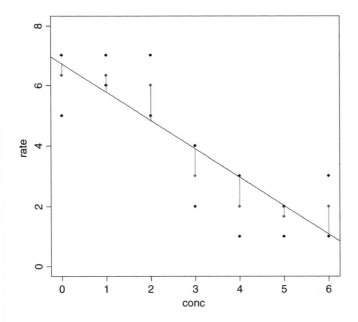

This significant lack of fit indicates that the straight line model is *not* an adequate description of these data ($p < 0.05$). A negative S-shaped function is likely to fit the data better (see p. 204).

There is an R package called lmtest on CRAN, which is full of tests for linear models.

Bootstrap with Regression

An alternative to estimating confidence intervals on the regression parameters from the pooled error variance in the ANOVA table (p. 396) is to use bootstrapping. There are two ways of doing this:

- sample cases with replacement, so that some points are left off the graph while others appear more than once in the dataframe;

- calculate the residuals from the fitted regression model, and randomize which fitted y values get which residuals.

In both cases, the randomization is carried out many times, the model fitted and the parameters estimated. The confidence interval is obtained from the quantiles of the distribution of parameter values (see p. 284).

The following dataframe contains a response variable (profit from the cultivation of a crop of carrots for a supermarket) and a single explanatory variable (the cost of inputs, including fertilizers, pesticides, energy and labour):

```
regdat<-read.table("c:\\temp\\regdat.txt",header=T)
attach(regdat)
names(regdat)
```

```
[1]  "explanatory"  "response"
```

```
plot(explanatory,response)
```

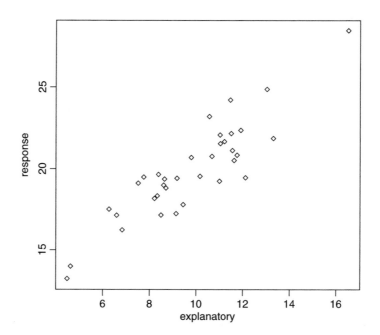

The response is a reasonably linear function of the explanatory variable, but the variance in response is quite large. For instance, when the explanatory variable is about 12, the response variable ranges between less than 20 and more than 24.

```
model<-lm(response~explanatory)
model
```

```
Coefficients:
(Intercept)      explanatory
      9.630            1.051
```

Theory suggests that the slope should be 1.0 and our estimated slope is very close to this (1.051). We want to establish a 95% confidence interval on the estimate. Here is a home-made bootstrap which resamples the data points 10 000 times and gives a bootstrapped estimate of the slope:

```
b.boot<-numeric(10000)
for (i in 1:10000){
indices<-sample(1:35,replace=T)
xv<-explanatory[indices]
yv<-response[indices]
model<-lm(yv~xv)
b.boot[i]<-coef(model)[2] }
hist(b.boot,main="")
```

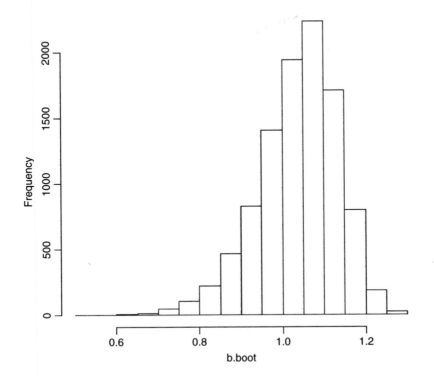

Here is the 95% interval for the bootstrapped estimate of the slope:

```
quantile(b.boot,c(0.025,0.975))
```

```
      2.5%        97.5%
0.8212288   1.1961992
```

Evidently, the bootstrapped data provide no support for the hypothesis that the slope is significantly greater than 1.0.

We now repeat the exercise, using the boot function from the boot library:

```
library(boot)
```

The first step is to write what is known as the 'statistic' function. This shows boot how to calculate the statistic we want from the resampled data. The resampling of the data is achieved by a subscript provided by boot (here called index). The point is that every time the model is fitted within the bootstrap it uses a different data set (yv and xv): we need to describe how these data are constructed and how they are used in the model fitting:

```
reg.boot<-function(regdat, index){
xv<-explanatory[index]
yv<-response[index]
model<-lm(yv~xv)
coef(model) }
```

Now we can run the boot function, then extract the intervals with the boot.ci function:

```
reg.model<-boot(regdat,reg.boot,R=10000)
boot.ci(reg.model,index=2)
```

```
BOOTSTRAP CONFIDENCE INTERVAL CALCULATIONS
Based on 10000 bootstrap replicates
```

```
CALL :
boot.ci(boot.out = reg.model, index = 2)
```

```
Intervals :
```

```
Level       Normal          Basic
95%      (0.872, 1.253)  (0.903, 1.283)
```

```
Level       Percentile        BCa
95%      (0.818, 1.198)  (0.826, 1.203)
Calculations and Intervals on Original Scale
```

All the intervals are reasonably similar: statisticians typically prefer the bias-corrected, accelerated (BCa) intervals. These indicate that if we were to repeat the data-collection exercise we can be 95% confident that the regression slope for those new data would be between 0.826 and 1.203.

The other way of bootstrapping with a model is to randomize the allocation of the residuals to fitted y values estimated from the original regression model. We start by calculating the residuals and the fitted values:

```
model<-lm(response~explanatory)
fit<-fitted(model)
res<-resid(model)
```

What we intend to do is to randomize which of the res values is added to the fit values to get a reconstructed response variable, y, which we regress as a function of the original explanatory variable. Here is the statistic function to do this:

```
residual.boot<-function(res, index){
y<-fit+res[index]
model<-lm(y~explanatory)
coef(model) }
```

Note that the data passed to the statistic function are res in this case (rather than the original dataframe regdat as in the first example, above). Now use the boot function and the boot.ci function to obtain the 95% confidence intervals on the slope (this is index = 2; the intercept is index = 1):

```
res.model<-boot(res,residual.boot,R=10000)
boot.ci(res.model,index=2)
```

```
BOOTSTRAP CONFIDENCE INTERVAL CALCULATIONS
Based on 10000 bootstrap replicates

CALL :
boot.ci(boot.out = res.model, index = 2)

Intervals :
Level       Normal              Basic
95%      (0.879, 1.223)    (0.876, 1.224)
Level      Percentile            BCa
95%      (0.877, 1.225)    (0.869, 1.214)
Calculations and Intervals on Original Scale
```

The BCa from randomizing the residuals is from 0.869 to 1.214, while from selecting random x and y points with replacement it was from 0.826 to 1.203 (above). The two rather different approaches to bootstrapping produce reassuringly similar estimates of the same parameter.

Jackknife with regression

A second alternative to alternating confidence intervals on regression parameters is to **jackknife** the data. Each point in the data set is left out, one at a time, and the parameter of interest is re-estimated. The regdat dataframe has length(response) data points:

```
names(regdat)
```

```
[1]  "explanatory"  "response"
```

```
length(response)
```

```
[1]  35
```

We create a vector to contain the 35 different estimates of the slope:

```
jack.reg<-numeric(35)
```

Now carry out the regression 35 times, leaving out a different x, y pair in each case:

```
for (i in 1:35) {
model<-lm(response[-i]~explanatory[-i])
jack.reg[i]<-coef(model)[2] }
```

Here is a histogram of the different estimates of the slope of the regression:

```
hist(jack.reg)
```

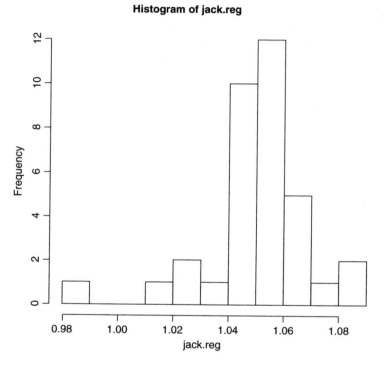

As you can see, the distribution is strongly skew to the left. The jackknife draws attention to one particularly influential point (the extreme left-hand bar) which, when omitted from the dataframe, causes the estimated slope to fall below 1.0. We say the point is **influential** because it is the only one of the 35 points whose omission causes the estimated slope to fall below 1.0. To see what is going on we should plot the data:

```
plot(explanatory,response)
```

We need to find out the identity of this influential point:

```
which(explanatory>15)
```

```
[1] 22
```

Now we can draw regression lines for the full data set (solid line) and for the model with the influential point number 22 omitted (dotted line):

```
abline(lm(response~explanatory))
abline(lm(response[-22]~explanatory[-22]),lty=2)
```

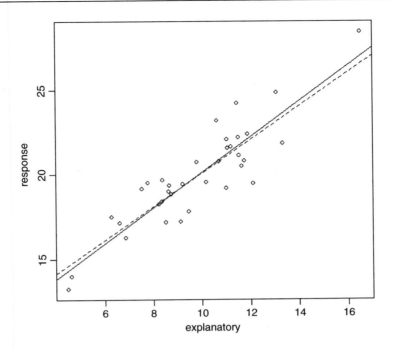

Jackknife after Bootstrap

The jack.after.boot function calculates the jackknife influence values from a bootstrap output object, and plots the corresponding jackknife-after-bootstrap plot. We illustrate its

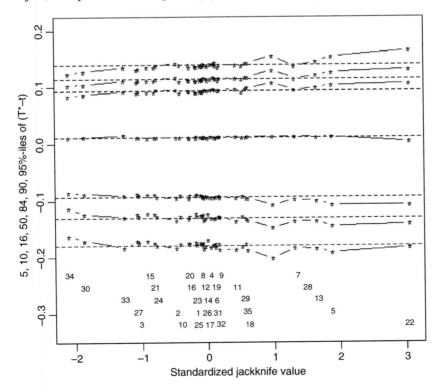

use with the boot object calculated earlier called **reg.model**. We are interested in the slope, which is **index=2**:

```
jack.after.boot(reg.model,index=2)
```

The centred jackknife quantiles for each observation are estimated from those bootstrap samples in which *the particular observation did not appear*. These are then plotted against the influence values. From the top downwards, the horizontal dotted lines show the 95th, 90th, 84th, 50th, 16th, 10th and 5th percentiles. The numbers at the bottom identify the 35 points by their index values within **regdat**. Again, the influence of point no. 22 shows up clearly (this time on the right-hand side), indicating that it has a strong positive influence on the slope.

Serial correlation in the residuals

The Durbin-Watson function is used for testing whether there is autocorrelation in the residuals from a linear model or a GLM, and is implemented as part of the **car** package (see Fox, 2002):

```
install.packages("car")
model<-lm(response~explanatory)
durbin.watson(model)
```

```
lag  Autocorrelation  D-W Statistic  p-value
  1      -0.07946739       2.049899    0.874
Alternative hypothesis: rho != 0
```

There is no evidence of serial correlation in these residuals ($p = 0.874$).

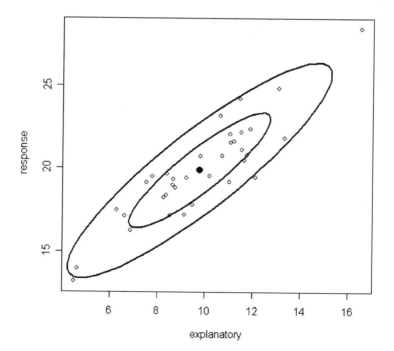

The car package also contains functions for drawing ellipses, including data ellipses, and confidence ellipses for linear and generalized linear models. Here is the data.ellipse function for the present example: by default, the ellipses are drawn at 50 and 90%:

```
data.ellipse(explanatory,response)
```

Piecewise Regression

This kind of regression fits different functions over different ranges of the explanatory variable. For example, it might fit different linear regressions to the left- and right-hand halves of a scatterplot. Two important questions arise in piecewise regression:

- how many segments to divide the line into;
- where to position the break points on the x axis.

Suppose we want to do the simplest piecewise regression, using just two linear segments. Where do we break up the x values? A simple, pragmatic view is to divide the x values at the point where the piecewise regression best fits the response variable. Let's take an example using a linear model where the response is the log of a count (the number of species recorded) and the explanatory variable is the log of the size of the area searched for the species:

```
data<-read.table("c:\\temp\\sasilwood.txt",header=T)
attach(data)
names(data)
```

```
[1]  "Species"  "Area"
```

A quick scatterplot suggests that the relationship between log(Species) and log (Area) is not linear:

```
plot(log(Species)~log(Area),pch=16)
```

The slope appears to be shallower at small scales than at large. The overall regression highlights this at the model-checking stage:

```
model1<-lm(log(Species)~log(Area))
plot(log(Area),resid(model1))
```

The residuals are very strongly U-shaped (this plot should look like the sky at night).

If we are to use piecewise regression, then we need to work out how many straight-line segments to use and where to put the breaks. Visual inspection of the scatterplot suggests that two segments would be an improvement over a single straight line and that the break point should be about $\log(\text{Area}) = 5$. The choice of break point is made more objective by choosing a range of values for the break point and selecting the break that produces the minimum deviance. We should have at least two x values for each of the pieces of the regression, so the areas associated with the first and last breaks can obtained by examination of the table of x values:

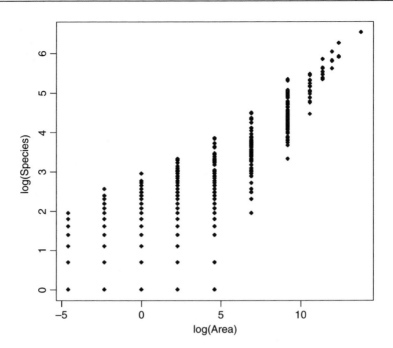

table(Area)

0.01	0.1	1	10	100	1000	10000	40000	90000	160000	250000	1000000
346	345	259	239	88	67	110	18	7	4	3	1

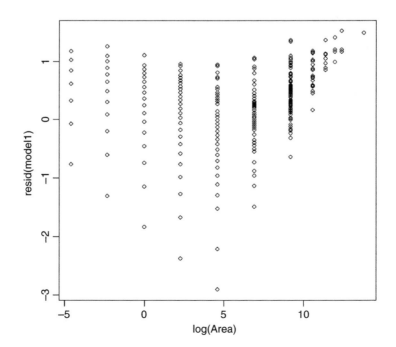

Piecewise regression is extremely simple in R: we just include a logical statement as part of the model formula, with as many logical statements as we want straight-line segments in the fit. In the present example with two linear segments, the two logical statements are (Area < Break) to define the left-hand regression and (Area ≥ Break) to define the right-hand regression. The smallest break would be at Area = 1 and the largest break at Area = 250 000. The piecewise model looks like this:

```
model2<-lm(log(Species)~log(Area)*(Area<Break)+log(Area)*(Area>=Break))
```

Note the use of the multiply operator (*) so that when the logical expression is false, the model entry evaluates to zero, and hence it is not fitted to the response (see the discussion on aliasing, p. 380). We want to fit the model for all values of Break between 1 and 250 000, so we create a vector of breaks like this:

```
sort(unique(Area))[3:11]
```

```
[1]    1   10   100   1000   10000   40000   90000   160000   250000
```

```
Break<-sort(unique(Area))[3:11]
```

Now we use a loop to fit the two-segment piecewise model nine times and to store the value of the residual standard error in a vector called d. This quantity is the sixth element of the list that is the model summary object, d[i]<-summary(model)[[6]]:

```
d<-numeric(9)
for (i in 1:9) {
model<-
lm(log(Species)~(Area<Break[i])*log(Area)+(Area>=Break[i])*log(Area))
d[i]<-summary(model)[[6]] }
```

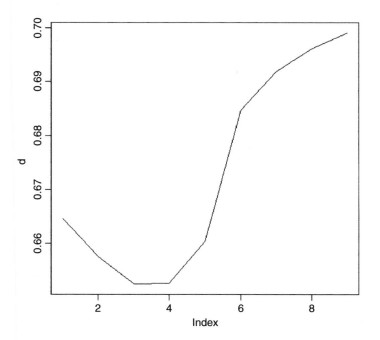

We have defined the best two-segment piecewise regression as the model with the minimal residual standard error. Here is an index plot of our vector *d*:

```
plot(d,type="l")
```

The best model is clearly at index 3 or 4, with a slight preference for index 3. This is associated with a break of

```
Break[3]
```

```
[1]  100
```

so Area = 100. We can refit the model for this value of Area and inspect the parameter values:

```
model2<-lm(log(Species)~(Area<100)*log(Area)+(Area>=100)*log(Area))
summary(model2)
```

```
Call:
lm(formula = log(Species) ~ (Area < 100) * log(Area) + (Area >=
    100) * log(Area))

Residuals:
    Min      1Q   Median      3Q     Max
-2.5058  -0.3091  0.1128  0.4822  1.3443

Coefficients: (2 not defined because of singularities)
                            Estimate  Std. Error  t value   Pr(>|t|)
(Intercept)                  0.61682     0.13059    4.723   2.54e-06  ***
Area < 100TRUE               1.07854     0.13246    8.143   8.12e-16  ***
log(Area)                    0.41019     0.01655   24.787    < 2e-16  ***
Area >= 100TRUE                   NA          NA       NA         NA
Area < 100TRUE:log(Area)    -0.25611     0.01816  -14.100    < 2e-16  ***
log(Area):Area >= 100TRUE         NA          NA       NA         NA

Residual standard error: 0.6525 on 1483 degrees of freedom
Multiple R-Squared: 0.724,  Adjusted R-squared: 0.7235
F-statistic: 1297 on 3 and 1483 DF, p-value: < 2.2e-16
```

The output needs a little getting used to. First things first: the residual standard error is down to 0.6525, a considerable improvement over the linear regression with which we began (where the deviance was 0.7021):

```
anova(model1,model2)
```

```
Analysis of Variance Table

Model 1: log(Species) ~ log(Area)
Model 2: log(Species) ~ (Area < 100) * log(Area) + (Area >= 100) *
log(Area)
    Res.Df     RSS  Df  Sum of Sq       F    Pr(>F)
1     1485  731.98
2     1483  631.36   2     100.62  118.17  < 2.2e-16  ***
```

Although there are six rows to the summary table, there are only four parameters (not six) to be estimated (two slopes and two intercepts). We have intentionally created a *singularity* in the piecewise regression between Area = 100 and Area = 1000 (the aliased parameters show up as NAs). The intercept of 0.616 82 is for the right-hand (steeper) segment of the graph,

whose slope (labelled `log(Area)`) is 0.410 19. The other two parameters (labelled `Area <
100TRUE` and `Area < 100TRUE:log(Area)`, respectively) are the difference between
the two intercepts and the difference between the two slopes. So the left-hand segment of
the graph has intercept $= 0.616\,82 + 1.078\,54 = 1.695\,36$ and slope $= 0.410\,19 - 0.256\,11 =
0.154\,08$. You might like to confirm this by fitting the two regressions separately, using the
`subset` option). Here is the piecewise regression fitted through the scatterplot:

```
area=sort(unique(Area))
plot(log(Species)~log(Area))
lines(log(area),predict(model2,list(Area=area)))
```

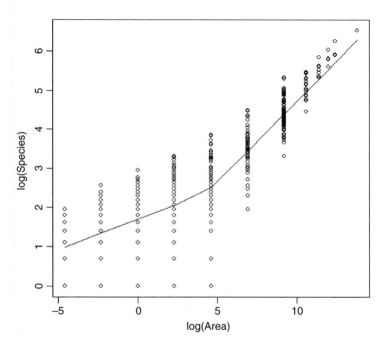

The fit is a great improvement over the single linear regression, but perhaps we could
improve the fit for the largest areas by fitting three piecewise lines (the values of log(Species)
all lie above the regression line for the four largest Areas)? We choose a break point of
Area $= 40000$ for the higher threshold, and note that now we need to use logical AND (&)
for the middle section of the regression:

```
model3<-lm(log(Species)~(Area<100)*log(Area)+
        (Area>=100 & Area < 40000)*log(Area)+(Area>=40000)*log(Area))
```

Here is the fitted model on the scatterplot:

```
plot(log(Species)~log(Area))
lines(log(area),predict(model3,list(Area=area)))
```

Visually, this is a much better fit for the largest values of Area, but because of the low
replication at these larger Areas (see above), the difference between models 2 and 3 is not
significant (`anova(model2,model3)` gives $p = 0.0963$).

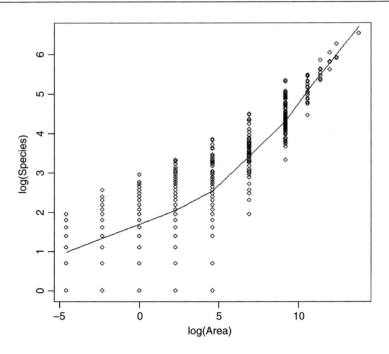

Robust Fitting of Linear Models

This uses the rlm function from the MASS library. The function allows one to fit a linear model by robust regression using an M estimator, allowing robust inference for parameters and robust model selection. The robust fit is minimally influenced by outliers in the response variable, in the explanatory variable(s) or in both.

```
robreg<-read.table("c:\\temp\\robreg.txt",header=T)
attach(robreg)
names(robreg)
```

```
[1]  "amount"  "rate"
```

```
plot(amount,rate,pch=16)
abline(lm(rate~amount))
summary(lm(rate~amount))
```

```
Coefficients:
            Estimate  Std. Error  t value   Pr(>|t|)
(Intercept)  12.1064     1.4439     8.385   1.52e-05  ***
     amount  -1.0634     0.2552    -4.166   0.00243   **

Residual standard error: 2.851 on 9 degrees of freedom
Multiple R-Squared: 0.6585, Adjusted R-squared: 0.6206
F-statistic: 17.36 on 1 and 9 DF, p-value: 0.002425
```

We can leave out the maximum and minimum values of the response or explanatory variables separately or together using subset to test for influence:

```
summary(lm(rate~amount,subset=(rate<max(rate))))
```

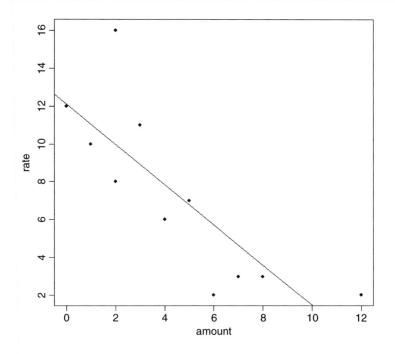

```
Coefficients:
              Estimate  Std. Error  t value  Pr(>|t|)
(Intercept)    10.8163      1.0682   10.126  7.73e-06   ***
amount         -0.9201      0.1811   -5.081  0.000952   ***
Residual standard error: 1.964 on 8 degrees of freedom
Multiple R-Squared: 0.7634, Adjusted R-squared: 0.7339
F-statistic: 25.82 on 1 and 8 DF, p-value: 0.000952
```

The intercept is lower by more than 1.0 and the slope is shallower by almost 0.1, while r^2 is more than 10% higher with the maximum value of rate removed. What about removing the maximum value of the explanatory variable?

summary(lm(rate~amount,subset=(amount<max(amount))))

```
Coefficients:
              Estimate Std.   Error t    value   Pr(>|t|)
(Intercept)     13.1415       1.5390   8.539  2.72e-05   ***
amount          -1.4057       0.3374  -4.166   0.00314    **
Residual standard error: 2.691 on 8 degrees of freedom
Multiple R-Squared: 0.6845, Adjusted R-squared: 0.645
F-statistic: 17.35 on 1 and 8 DF, p-value: 0.003141
```

The intercept is now greater by more than 1.0 and the slope is steeper by almost 0.4.

You can see that leaving out the maximum value of amount (on the x axis) has a bigger impact than leaving out the largest value of rate (which is at amount $= 2$). But all of the models are different from the original model based on all the data. Here is what rlm makes of these data:

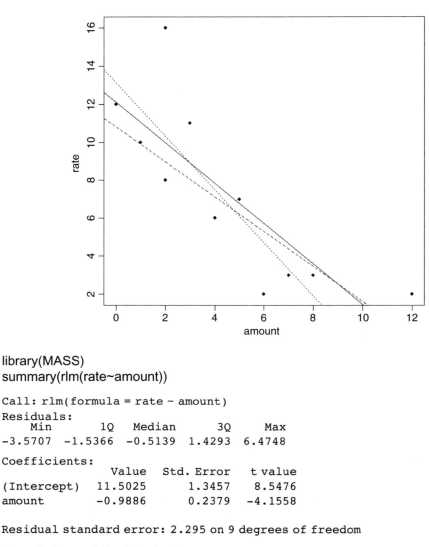

```
library(MASS)
summary(rlm(rate~amount))
```

```
Call: rlm(formula = rate ~ amount)
Residuals:
      Min      1Q   Median      3Q     Max
  -3.5707  -1.5366  -0.5139  1.4293  6.4748
```

```
Coefficients:
              Value  Std. Error  t value
(Intercept)  11.5025    1.3457    8.5476
amount       -0.9886    0.2379   -4.1558
```

Residual standard error: 2.295 on 9 degrees of freedom

```
Correlation of Coefficients:
            (Intercept)
amount       -0.8035
```

Note that rlm retains all its degrees of freedom, while our outlier deletion lost one degree of freedom per point omitted. The parameter estimates and standard errors from rlm are much closer to our model with all data points included (intercept $= 12.1$, $se = 1.44$, slope $= -1.06$, $se = 0.25$) than to either of our reduced models. Using plot(model) draws attention to data points 10 and 11 as being highly influential in both the lm and rlm, so we try leaving them both out:

```
summary(lm(rate~amount,subset=(amount<max(amount)& rate<max(rate))))
```

```
Coefficients:
              Estimate  Std. Error  t value  Pr(>|t|)
(Intercept)   11.7556     1.0408     11.295   9.54e-06  ***
amount        -1.2167     0.2186     -5.565   0.000846  ***
```

```
Residual standard error: 1.693 on 7 degrees of freedom
Multiple R-Squared: 0.8157, Adjusted R-squared: 0.7893
F-statistic: 30.97 on 1 and 7 DF, p-value: 0.000846
```

The slope is steeper than in rlm but the intercept is quite close. Our removal of one or two influential points has a much bigger effect on the estimate of the slope than does using rlm compared with lm on the full data set. You pays your money and you takes your choice.

Model Simplification

A multiple regression is a statistical model with two or more continuous explanatory variables. We contrast this kind of model with analysis of variance, where all the explanatory variables are categorical (Chapter 11) and analysis of covariance, where the explanatory variables are a mixture of continuous and categorical (Chapter 12).

The *principle of parsimony* (Occam's razor), discussed in the previous chapter on p. 325, is again relevant here. It requires that the model should be as simple as possible. This means that the model should not contain any redundant parameters or factor levels. We achieve this by fitting a maximal model and then simplifying it by following one or more of these steps:

- Remove non-significant interaction terms.

- Remove non-significant quadratic or other non-linear terms.

- Remove non-significant explanatory variables.

- Group together factor levels that do not differ from one another.

- Amalgamate explanatory variables that have similar parameter values.

- Set non-significant slopes to zero within ANCOVA.

Of course, such simplifications must make good scientific sense, and must not lead to significant reductions in explanatory power. It is likely that many of the explanatory variables are correlated with each other, and so *the order in which variables are deleted from the model* will influence the explanatory power attributed to them. The thing to remember about multiple regression is that, in principle, there is no end to it. The number of combinations of interaction terms and curvature terms is endless. There are some simple rules (like parsimony) and some automated functions (like step) to help. But, in principle, you could spend a very great deal of time in modelling a single dataframe. There are no hard-and-fast rules about the best way to proceed, but we shall typically carry out simplification of a complex model by stepwise deletion: non-significant terms are left out, and significant terms are added back (see Chapter 9).

At the data inspection stage, there are many more kinds of plots we could do:

- Plot the response against each of the explanatory variables separately.

- Plot the explanatory variables against one another (e.g. pairs).

- Plot the response against pairs of explanatory variables in three-dimensional plots.

- Plot the response against explanatory variables for different combinations of other explanatory variables (e.g. conditioning plots, coplot; see p. 16).

- Fit non-parametric smoothing functions (e.g. using generalized additive models, to look for evidence of curvature).

- Fit tree models to investigate whether interaction effects are simple or complex.

The Multiple Regression Model

There are several important issues involved in carrying out a multiple regression:

- which explanatory variables to include;

- curvature in the response to the explanatory variables;

- interactions between explanatory variables;

- correlation between explanatory variables;

- the risk of overparameterization.

The assumptions about the response variable are the same as with simple linear regression: the errors are normally distributed, the errors are confined to the response variable, and the variance is constant. The explanatory variables are assumed to be measured without error. The model for a multiple regression with two explanatory variables (x_1 and x_2) looks like this:

$$y_i = \beta_0 + \beta_1 x_{1i} + \beta_2 x_{2i} + \varepsilon_i.$$

The ith data point, y_i, is determined by the levels of the two continuous explanatory variables x_{1i} and x_{2i} by the model's three parameters (the intercept β_0 and the two slopes β_1 and β_2), and by the residual ε_i of point i from the fitted surface. More generally, the model is presented like this:

$$y_i = \sum \beta_i x_i + \varepsilon_i,$$

where the summation term is called the **linear predictor** and can involve many explanatory variables, non-linear terms and interactions.

Example

Let's begin with an example from air pollution studies. How is ozone concentration related to wind speed, air temperature and the intensity of solar radiation?

```
ozone.pollution<-read.table("c:\\temp\\ozone.data.txt",header=T)
attach(ozone.pollution)
names(ozone.pollution)
```

```
[1]  "rad"  "temp"  "wind"  "ozone"
```

In multiple regression, it is always a good idea to use pairs to look at all the correlations:

```
pairs(ozone.pollution,panel=panel.smooth)
```

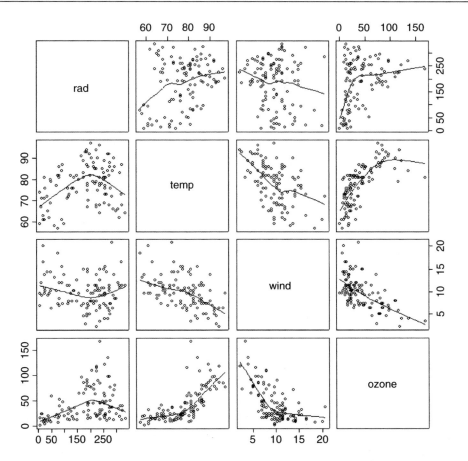

The response variable, ozone concentration, is shown on the *y* axis of the bottom row of panels: there is a strong negative relationship with wind speed, a positive correlation with temperature and a rather unclear, humped relationship with radiation.

A good way to tackle a multiple regression problem is using non-parametric smoothers in a generalized additive model like this:

```
library(mgcv)
par(mfrow=c(2,2))
model<-gam(ozone~s(rad)+s(temp)+s(wind))
plot(model)
par(mfrow=c(1,1))
```

The confidence intervals are sufficiently narrow to suggest that the curvature in the relationship between ozone and temperature is real, but the curvature of the relationship with wind is questionable, and a linear model may well be all that is required for solar radiation.

The next step might be to fit a tree model to see whether complex interactions between the explanatory variables are indicated:

```
library(tree)
model<-tree(ozone~.,data=ozone.pollution)
plot(model)
text(model)
```

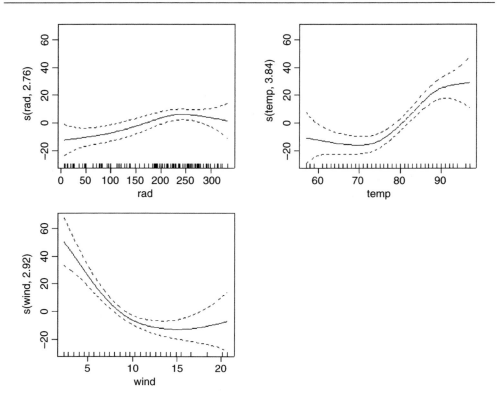

This shows that temperature is by far the most important factor affecting ozone concentration (the longer the branches in the tree, the greater the deviance explained). Wind speed is important at both high and low temperatures, with still air being associated with higher mean ozone levels (the figures at the ends of the branches). Radiation shows an interesting but subtle effect. At low temperatures, radiation matters at relatively high wind speeds (> 7.15), whereas at high temperatures, radiation matters at relatively low wind speeds (< 10.6); in both cases, however, higher radiation is associated with higher mean ozone concentration. The tree model therefore indicates that the interaction structure of the data is not complex (a reassuring finding).

Armed with this background information (likely curvature of the temperature response and an uncomplicated interaction structure), we can begin the linear modelling. We start with the most complicated model: this includes interactions between all three explanatory variables plus quadratic terms to test for curvature in response to each of the three explanatory variables. If you want to do calculations inside the model formula (e.g. produce a vector of squares for fitting quadratic terms), then you need to use the 'as is' function, I ():

```
model1<-lm(ozone~temp*wind*rad+I(rad^2)+I(temp^2)+I(wind^2))
summary(model1)
```

```
Coefficients:
                Estimate   Std.Error  t value  Pr(>|t|)
(Intercept)     5.683e+02  2.073e+02    2.741  0.00725  **
temp           -1.076e+01  4.303e+00   -2.501  0.01401   *
wind           -3.237e+01  1.173e+01   -2.760  0.00687  **
rad            -3.117e-01  5.585e-01   -0.558  0.57799
```

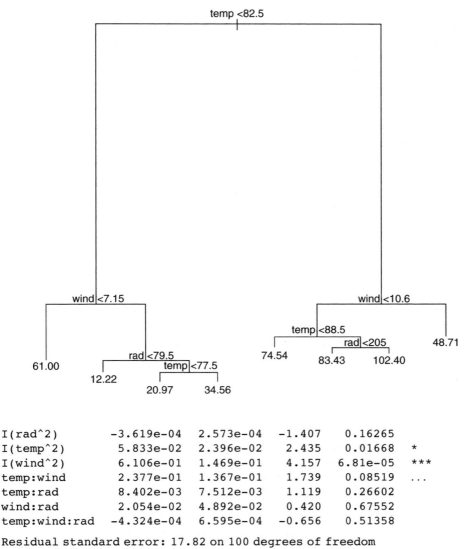

```
I(rad^2)          -3.619e-04  2.573e-04  -1.407   0.16265
I(temp^2)          5.833e-02  2.396e-02   2.435   0.01668   *
I(wind^2)          6.106e-01  1.469e-01   4.157   6.81e-05  ***
temp:wind          2.377e-01  1.367e-01   1.739   0.08519   ...
temp:rad           8.402e-03  7.512e-03   1.119   0.26602
wind:rad           2.054e-02  4.892e-02   0.420   0.67552
temp:wind:rad     -4.324e-04  6.595e-04  -0.656   0.51358
```

Residual standard error: 17.82 on 100 degrees of freedom
Multiple R-Squared: 0.7394, Adjusted R-squared: 0.7133
F-statistic: 28.37 on 10 and 100 DF, p-value: 0

The three-way interaction is clearly not significant, so we remove it to begin the process of model simplification:

```
model2<-update(model1,~. - temp:wind:rad)
summary(model2)
```

Coefficients:

	Estimate	Std.Error	t value	Pr(>|t|)	
(Intercept)	5.245e+02	1.957e+02	2.680	0.0086	**
temp	-1.021e+01	4.209e+00	-2.427	0.0170	*
wind	-2.802e+01	9.645e+00	-2.906	0.0045	**
rad	2.628e-02	2.142e-01	0.123	0.9026	
I(rad^2)	-3.388e-04	2.541e-04	-1.333	0.1855	

```
I(temp^2)    5.953e-02   2.382e-02    2.499    0.0141   *
I(wind^2)    6.173e-01   1.461e-01    4.225   5.25e-05  ***
temp:wind    1.734e-01   9.497e-02    1.825    0.0709   .
temp:rad     3.750e-03   2.459e-03    1.525    0.1303
wind:rad    -1.127e-02   6.277e-03   -1.795    0.0756   .
```

We now remove the least significant interaction term. From model1 this looks like wind:rad ($p = 0.675\,52$), but the sequence has been altered by removing the three-way term (it is now temp:rad with $p = 0.1303$):

model3<-update(model2,~. - temp:rad)
summary(model3)

Coefficients:
```
               Estimate   Std.Error  t value  Pr(>|t|)
(Intercept)   5.488e+02   1.963e+02    2.796   0.00619  **
temp         -1.144e+01   4.158e+00   -2.752   0.00702  **
wind         -2.876e+01   9.695e+00   -2.967   0.00375  **
rad           3.061e-01   1.113e-01    2.751   0.00704  **
I(rad^2)     -2.690e-04   2.516e-04   -1.069   0.28755
I(temp^2)     7.145e-02   2.265e-02    3.154   0.00211  **
I(wind^2)     6.363e-01   1.465e-01    4.343   3.33e-05 ***
temp:wind     1.840e-01   9.533e-02    1.930   0.05644  .
wind:rad     -1.381e-02   6.090e-03   -2.268   0.02541  *
```

The temp:wind interaction is close to significance, but we are ruthless in our pruning, so we take it out:

model4<-update(model3,~. - temp:wind)
summary(model4)

Coefficients:
```
               Estimate   Std.Error  t value  Pr(>|t|)
(Intercept)   2.310e+02   1.082e+02    2.135   0.035143  *
temp         -5.442e+00   2.797e+00   -1.946   0.054404  .
wind         -1.080e+01   2.742e+00   -3.938   0.000150 ***
rad           2.405e-01   1.073e-01    2.241   0.027195  *
I(rad^2)     -2.010e-04   2.524e-04   -0.796   0.427698
I(temp^2)     4.484e-02   1.821e-02    2.463   0.015432  *
I(wind^2)     4.308e-01   1.020e-01    4.225   5.16e-05 ***
wind:rad     -9.774e-03   5.794e-03   -1.687   0.094631  .
```

Now the wind:rad interaction, which looked so significant in model3 ($p = 0.02541$), is evidently not significant, so we take it out as well:

model5<-update(model4,~. - wind:rad)
summary(model5)

Coefficients:
```
               Estimate Std.    Error t    value  Pr(>|t|)
(Intercept)   2.985e+02   1.014e+02    2.942   0.00402  **
temp         -6.584e+00   2.738e+00   -2.405   0.01794  *
wind         -1.337e+01   2.300e+00   -5.810   6.89e-08 ***
rad           1.349e-01   8.795e-02    1.533   0.12820
I(rad^2)     -2.052e-04   2.546e-04   -0.806   0.42213
I(temp^2)     5.221e-02   1.783e-02    2.928   0.00419  **
I(wind^2)     4.652e-01   1.008e-01    4.617   1.12e-05 ***
```

There is no evidence to support retaining any of the two-way interactions. What about the quadratic terms: the term in `rad^2` looks insignificant, so we take it out:

```
model6<-update(model5,~. - I(rad^2))
summary(model6)
```

```
Coefficients:
              Estimate Std.    Error t   value  Pr(>|t|)
(Intercept)   291.16758  100.87723   2.886   0.00473   **
temp           -6.33955    2.71627  -2.334   0.02150   *
wind          -13.39674    2.29623  -5.834   6.05e-08  ***
rad             0.06586    0.02005   3.285   0.00139   **
I(temp^2)       0.05102    0.01774   2.876   0.00488   **
I(wind^2)       0.46464    0.10060   4.619   1.10e-05  ***
```

```
Residual standard error: 18.25 on 105 degrees of freedom
Multiple R-Squared: 0.713, Adjusted R-squared: 0.6994
F-statistic: 52.18 on 5 and 105 DF, p-value: < 2.2e-16
```

Now we are making progress. All the terms in model6 are significant. We should check the assumptions:

```
plot(model6)
```

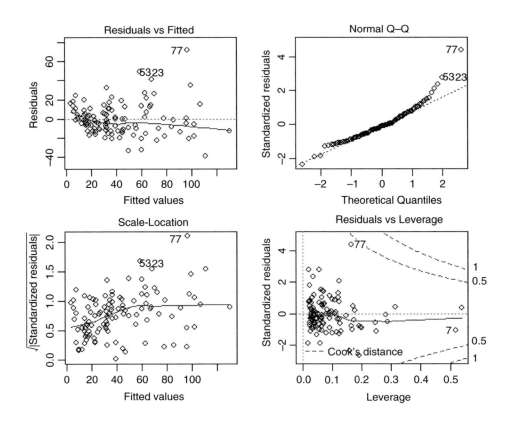

There is a clear pattern of variance increasing with the mean of the fitted values. This heteroscedasticity is bad news. Also, the normality plot is distinctly curved; again, this is bad news. Let's try transformation of the response variable. There are no zeros in the response, so a log transformation is worth trying:

```
model7<-lm(log(ozone) ~ temp + wind + rad + I(temp^2) + I(wind^2))
summary(model7)
```

```
Coefficients:
              Estimate Std.    Error t    value  Pr(>|t|)
(Intercept)   2.5538486  2.7359735  0.933   0.35274
temp         -0.0041416  0.0736703 -0.056   0.95528
wind         -0.2087025  0.0622778 -3.351   0.00112   **
rad           0.0025617  0.0005437  4.711   7.58e-06  ***
I(temp^2)     0.0003313  0.0004811  0.689   0.49255
I(wind^2)     0.0067378  0.0027284  2.469   0.01514   *

Residual standard error: 0.4949 on 105 degrees of freedom
Multiple R-Squared: 0.6882, Adjusted R-squared: 0.6734
F-statistic: 46.36 on 5 and 105 DF, p-value: 0
```

On the log(ozone) scale, there is no evidence for a quadratic term in temperature, so let's remove that:

```
model8<-update(model7,~. - I(temp^2))
summary(model8)
```

```
Coefficients:
              Estimate Std.    Error t    value  Pr(>|t|)
(Intercept)   0.7231644  0.6457316  1.120   0.26528
temp          0.0464240  0.0059918  7.748   5.94e-12  ***
wind         -0.2203843  0.0597744 -3.687   0.00036   ***
rad           0.0025295  0.0005404  4.681   8.49e-06  ***
I(wind^2)     0.0072233  0.0026292  2.747   0.00706   **

Residual standard error: 0.4936 on 106 degrees of freedom
Multiple R-Squared: 0.6868, Adjusted R-squared: 0.675
F-statistic: 58.11 on 4 and 106 DF, p-value: 0
```

```
plot(model8)
```

The heteroscedasticity and the non-normality have been cured, but there is now a highly influential data point (no. 17 on the Cook's plot). We should refit the model with this point left out, to see if the parameter estimates or their standard errors are greatly affected:

```
model9<-lm(log(ozone) ~ temp + wind + rad + I(wind^2),subset=(1:length(ozone)!=17))
summary(model9)
```

```
Coefficients:
              Estimate Std.    Error t    value  Pr(>|t|)
(Intercept)   1.1932358  0.5990022  1.992   0.048963  *
temp          0.0419157  0.0055635  7.534   1.81e-11  ***
wind         -0.2208189  0.0546589 -4.040   0.000102  ***
rad           0.0022097  0.0004989  4.429   2.33e-05  ***
I(wind^2)     0.0068982  0.0024052  2.868   0.004993  **

Residual standard error: 0.4514 on 105 degrees of freedom
Multiple R-Squared: 0.6974, Adjusted R-squared: 0.6859
F-statistic: 60.5 on 4 and 105 DF, p-value: 0
```

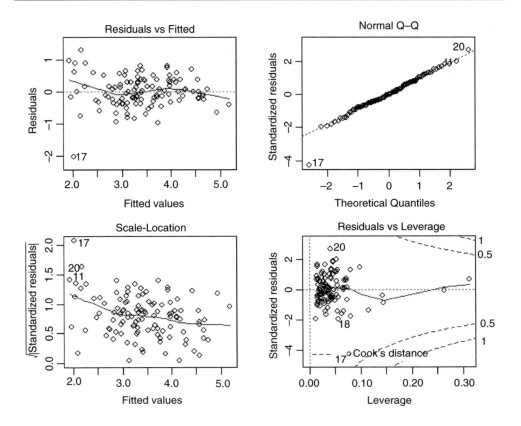

Finally, plot(model9) shows that the variance and normality are well behaved, so we can stop at this point. We have found the minimal adequate model. It is on a scale of log(ozone concentration), all the main effects are significant, but there are no interactions, and there is a single quadratic term for wind speed (five parameters in all, with 105 d.f. for error).

Let us repeat the example using the automatic model simplification function step, to see how well it does:

```
model10<-step(model1)
```

```
Start: AIC= 649.8
ozone ~ temp * wind * rad + I(rad^2) + I(temp^2) + I(wind^2)
                  Df  Sum of Sq    RSS  AIC
- temp:wind:rad    1        136  31879  648
<none>                             31742  650
- I(rad^2)         1        628  32370  650
- I(temp^2)        1       1882  33624  654
- I(wind^2)        1       5486  37228  665

Step: AIC= 648.28
ozone ~ temp + wind + rad + I(rad^2) + I(temp^2) + I(wind^2) +
temp:wind + temp:rad + wind:rad

              Df  Sum of Sq    RSS  AIC
- I(rad^2)     1        561  32440  648
<none>                         31879  648
```

```
- temp:rad    1    734  32613  649
- wind:rad    1   1017  32896  650
- temp:wind   1   1052  32930  650
- I(temp^2)   1   1971  33850  653
- I(wind^2)   1   5634  37513  664
```

```
Step: AIC= 648.21
ozone ~ temp + wind + rad + I(temp^2) + I(wind^2) + temp:wind +
temp:rad + wind:rad
```

```
              Df  Sum of Sq    RSS  AIC
- temp:rad     1        539  32978  648
<none>                        32440  648
- temp:wind    1        883  33323  649
- wind:rad     1       1034  33473  650
- I(temp^2)    1       1817  34256  652
- I(wind^2)    1       5361  37800  663
```

```
Step: AIC= 648.04
ozone ~ temp + wind + rad + I(temp^2) + I(wind^2) + temp:wind +
wind:rad
```

```
              Df  Sum of Sq    RSS  AIC
<none>                        32978  648
- temp:wind    1       1033  34011  649
- wind:rad     1       1588  34566  651
- I(temp^2)    1       2932  35910  655
- I(wind^2)    1       5781  38759  664
```

summary(model10)

Coefficients:

	Estimate Std.	Error	t value	Pr(>|t|)	
(Intercept)	514.401470	193.783580	2.655	0.00920	**
temp	-10.654041	4.094889	-2.602	0.01064	*
wind	-27.391965	9.616998	-2.848	0.00531	**
rad	0.212945	0.069283	3.074	0.00271	**
I(temp^2)	0.067805	0.022408	3.026	0.00313	**
I(wind^2)	0.619396	0.145773	4.249	4.72e-05	***
temp:wind	0.169674	0.094458	1.796	0.07538	.
wind:rad	-0.013561	0.006089	-2.227	0.02813	*

```
Residual standard error: 17.89 on 103 degrees of freedom
Multiple R-Squared: 0.7292, Adjusted R-squared: 0.7108
F-statistic: 39.63 on 7 and 103 DF, p-value: < 2.2e-16
```

This is quite typical of the **step** function: it has erred on the side of generosity (which is what you would want of an automated procedure, of course) and it has left the interaction between temperature and wind speed in the model ($p = 0.075\,38$). Manual model simplification of model10 leads to the same minimal adequate model as we obtained earlier (model9 after log transformation of the response variable).

A more complex example

In the next example we introduce two new difficulties: more explanatory variables, and fewer data points. It is another air pollution dataframe, but the response variable in this case is sulphur dioxide concentration. There are six continuous explanatory variables:

```
pollute<-read.table("c:\\temp\\sulphur.dioxide.txt",header=T)
attach(pollute)
names(pollute)
```

```
[1]  "Pollution"      "Temp"  "Industry"  "Population"  "Wind"
[6]       "Rain"  "Wet.days"
```

Here are the 36 scatterplots:

```
pairs(pollute,panel=panel.smooth)
```

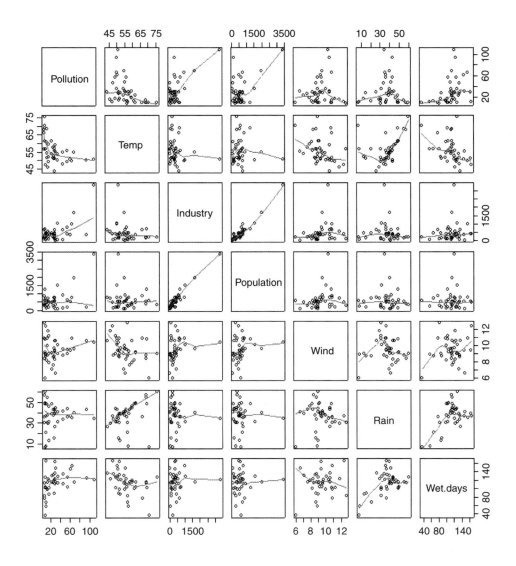

This time, let's begin with the tree model rather than the generalized additive model. A look at the **pairs** plots suggests that interactions may be more important than non-linearity in this case.

```
library(tree)
model<-tree(Pollution~.,data=pollute)
plot(model)
text(model)
```

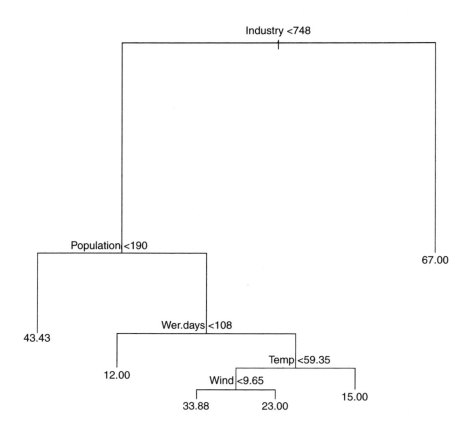

This is interpreted as follows. The most important explanatory variable is Industry, and the threshold value separating low and high values of Industry is 748. The right-hand branch of the tree indicates the mean value of air pollution for high levels of industry (67.00). The fact that this limb is unbranched means that no other variables explain a significant amount of the variation in pollution levels for high values of Industry. The left-hand limb does not show the mean values of pollution for low values of Industry, because there are other significant explanatory variables. Mean values of pollution are only shown at the extreme ends of branches. For low values of Industry, the tree shows us that Population has a significant impact on air pollution. At low values of Population (< 190) the mean level of air pollution was 43.43. For high values of Population, the number of wet days is significant. Low numbers of wet days (< 108) have mean pollution levels of 12.00 while Temperature has a significant impact on pollution for places where the number of wet days is large. At high temperatures ($> 59.35\,°F$) the mean pollution level was 15.00 while at lower temperatures the run of Wind is important. For still air (Wind < 9.65) pollution was higher (33.88) than for higher wind speeds (23.00).

We conclude that the interaction structure is highly complex. We shall need to carry out the linear modelling with considerable care. Start with some elementary calculations. With six explanatory variables, how many interactions might we fit? Well, there are $5 + 4 + 3 + 2 + 1 = 15$ two-way interactions for a start. Plus 20 three-way, 15 four-way and 6 five-way interactions, and a six-way interaction for good luck. Then there are quadratic terms for each of the six explanatory variables. So we are looking at about 70 parameters that might be estimated from the data. But how many data points have we got?

length(Pollution)

[1] 41

Oops! We are planning to estimate almost twice as many parameters as there are data points. That is taking overparameterization to new heights. We already know that you cannot estimate more parameter values than there are data points (i.e. a maximum of 41 parameters). But we also know that when we fit a saturated model to the data, it has no explanatory power (there are no degrees of freedom, so the model, by explaining everything, ends up explaining nothing at all). There is a useful rule of thumb: don't try to estimate more than $n/3$ parameters during a multiple regression. In the present case $n = 41$ so the rule of thumb is suggesting that we restrict ourselves to estimating about $41/3 \approx 13$ parameters at any one time. We know from the tree model that the interaction structure is going to be complicated so we shall concentrate on that. We begin, therefore, by looking for curvature, to see if we can eliminate this:

model1<-
lm(Pollution~Temp+I(Temp^2)+Industry+I(Industry ^2)+Population+I
(Population ^2)+ Wind+I(Wind ^2)+Rain+I(Rain ^2)+Wet.days+I(Wet.days ^2))
summary(model1)

```
Coefficients:
                    Estimate  Std. Error  t value  Pr(>|t|)
(Intercept)        -6.641e+01  2.234e+02   -0.297   0.76844
Temp                5.814e-01  6.295e+00    0.092   0.92708
I(Temp ^2)         -1.297e-02  5.188e-02   -0.250   0.80445
Industry            8.123e-02  2.868e-02    2.832   0.00847  **
I(Industry ^2)     -1.969e-05  1.899e-05   -1.037   0.30862
Population         -7.844e-02  3.573e-02   -2.195   0.03662  *
I(Population ^2)    2.551e-05  2.158e-05    1.182   0.24714
Wind                3.172e+01  2.067e+01    1.535   0.13606
I(Wind ^2)         -1.784e+00  1.078e+00   -1.655   0.10912
Rain                1.155e+00  1.636e+00    0.706   0.48575
I(Rain ^2)         -9.714e-03  2.538e-02   -0.383   0.70476
Wet.days           -1.048e+00  1.049e+00   -0.999   0.32615
I(Wet.days ^2)      4.555e-03  3.996e-03    1.140   0.26398
```

Residual standard error: 14.98 on 28 degrees of freedom
Multiple R-Squared: 0.7148, Adjusted R-squared: 0.5925
F-statistic: 5.848 on 12 and 28 DF, p-value: 5.868e-005

So that's our first bit of good news. There is no evidence of curvature for any of the six explanatory variables on their own (there may be curved interaction effects, of course).

Only the main effects of Industry and Population are significant in this (overparameterized) model. Now we need to consider the interaction terms. We do not fit interaction terms without both their component main effects, so we cannot fit all the two-way interaction terms at the same time (that would be $15 + 6 = 21$ parameters; well above the rule-of-thumb value of 13). One approach is to fit the interaction terms in randomly selected pairs. With all six main effects, we can afford to try $13 - 6 = 7$ interaction terms at a time. We'll try this. Make a vector containing the names of the 15 two-way interactions:

```
interactions<-c("ti","tp","tw","tr","td","ip","iw","ir","id","pw","pr","pd","wr","wd","rd")
```

Now shuffle the interactions into random order using sample without replacement:

```
sample(interactions)
```

```
[1]  "wr"  "wd"  "id"  "ir"  "rd"  "pr"  "tp"  "pw"  "ti"  "iw"  "tw"
     "pd"  "tr"  "td"  "ip"
```

It would be sensible and pragmatic to test the two-way interactions in three models, each containing five different two-way interaction terms:

```
model2<-
lm(Pollution~Temp+Industry+Population+Wind+Rain+Wet.days+Wind:Rain+
Wind:Wet.days+Industry:Wet.days+Industry:Rain+Rain:Wet.days)
model3<-
lm(Pollution~Temp+Industry+Population+Wind+Rain+Wet.days+Population:Rain+
Temp:Population+Population:Wind+Temp:Industry+Industry:Wind)
model4<-
lm(Pollution~Temp+Industry+Population+Wind+Rain+Wet.days+Temp:Wind+
Population:Wet.days+Temp:Rain+Temp:Wet.days+Industry:Population)
```

Extracting only the interaction terms from the three models, we see:

```
Industry:Rain         -1.616e-04   9.207e-04   -0.176   0.861891
Industry:Wet.days      2.311e-04   3.680e-04    0.628   0.534949
Wind:Rain              9.049e-01   2.383e-01    3.798   0.000690   ***
Wind:Wet.days         -1.662e-01   5.991e-02   -2.774   0.009593   **
Rain:Wet.days          1.814e-02   1.293e-02    1.403   0.171318

Temp:Industry         -1.643e-04   3.208e-03   -0.051   0.9595
Temp:Population        1.125e-03   2.382e-03    0.472   0.6402
Industry:Wind          2.668e-02   1.697e-02    1.572   0.1267
Population:Wind       -2.753e-02   1.333e-02   -2.066   0.0479   *
Population:Rain        6.898e-04   1.063e-03    0.649   0.5214

Temp:Wind              1.261e-01   2.848e-01    0.443   0.66117
Temp:Rain             -7.819e-02   4.126e-02   -1.895   0.06811
Temp:Wet.days          1.934e-02   2.522e-02    0.767   0.44949
Industry:Population    1.441e-06   4.178e-06    0.345   0.73277
Population:Wet.days    1.979e-05   4.674e-04    0.042   0.96652
```

The next step might be to put all of the significant or close-to-significant interactions into the same model, and see which survive:

```
model5<-
lm(Pollution~Temp+Industry+Population+Wind+Rain+Wet.days+Wind:Rain+
Wind:Wet.days+Population:Wind+Temp:Rain)
```

summary(model5)

```
Coefficients:
                  Estimate   Std.Error  t value  Pr(>|t|)
(Intercept)     323.054546  151.458618    2.133  0.041226  *
Temp             -2.792238    1.481312   -1.885  0.069153  .
Industry          0.073744    0.013646    5.404  7.44e-06  ***
Population        0.008314    0.056406    0.147  0.883810
Wind            -19.447031    8.670820   -2.243  0.032450  *
Rain             -9.162020    3.381100   -2.710  0.011022  *
Wet.days          1.290201    0.561599    2.297  0.028750  *
Temp:Rain         0.017644    0.027311    0.646  0.523171
Population:Wind  -0.005684    0.005845   -0.972  0.338660
Wind:Rain         0.997374    0.258447    3.859  0.000562  ***
Wind:Wet.days    -0.140606    0.053582   -2.624  0.013530  *
```

We certainly don't need `Temp:Rain`, so

model6<-update(model5,~. − Temp:Rain)

or `Population:Wind`:

model7<-update(model6,~. − Population:Wind)

All the terms in model7 are significant. Time for a check on the behaviour of the model:

plot(model7)

That's not bad at all. But what about the higher-order interactions? One way to proceed is to specify the interaction level using ^3 in the model formula, but if you do this, you will find that we run out of degrees of freedom straight away. A pragmatic option is to fit three way terms for the variables that already appear in two-way interactions: in our case, that is just one term: `Wind:Rain:Wet.days`

model8<-update(model7,~. + Wind:Rain:Wet.days)
summary(model8)

```
Coefficients:
                    Estimate  Std. Error  t value  Pr(>|t|)
(Intercept)       278.464474   68.041497    4.093  0.000282  ***
Temp               -2.710981    0.618472   -4.383  0.000125  ***
Industry            0.064988    0.012264    5.299  9.1e-06   ***
Population         -0.039430    0.011976   -3.293  0.002485  **
Wind               -7.519344    8.151943   -0.922  0.363444
Rain               -6.760530    1.792173   -3.772  0.000685  ***
Wet.days            1.266742    0.517850    2.446  0.020311  *
Wind:Rain           0.631457    0.243866    2.589  0.014516  *
Wind:Wet.days      -0.230452    0.069843   -3.300  0.002440  **
Wind:Rain:Wet.days  0.002497    0.001214    2.056  0.048247  *
```

```
Residual standard error: 11.2 on 31 degrees of freedom
Multiple R-Squared: 0.8236, Adjusted R-squared: 0.7724
F-statistic: 16.09 on 9 and 31 DF, p-value: 2.231e-009
```

That's enough for now. I'm sure you get the idea. Multiple regression is difficult, time-consuming, and always vulnerable to subjective decisions about what to include and what to leave out. The linear modelling confirms the early impression from the tree model:

for low levels of industry, the SO_2 level depends in a simple way on population (people tend to want to live where the air is clean) and in a complicated way on daily weather (the three-way interaction between wind, total rainfall and the number of wet days (i.e. on rainfall intensity)). Note that the relationship between pollution and population in the initial scatterplot suggested a positive correlation between these two variables, not the negative relationship we discovered by statistical modelling. This is one of the great advantages of multiple regression.

Note that the automatic model-simplification function step was no use in this example because we had too few data points to fit the full model to begin with.

Common problems arising in multiple regression

The following are some of the problems and difficulties that crop up when we do multiple regression:

- differences in the measurement scales of the explanatory variables, leading to large variation in the sums of squares and hence to an ill-conditioned matrix;
- multicollinearity, in which there is a near-linear relation between two of the explanatory variables, leading to unstable parameter estimates;
- rounding errors during the fitting procedure;
- non-independence of groups of measurements;
- temporal or spatial correlation amongst the explanatory variables;
- pseudoreplication.

Wetherill *et al.* (1986) give a detailed discussion of these problems. We shall encounter other examples of multiple regressions in the context of generalized linear models (Chapter 13), generalized additive models (Chapter 18), survival models (Chapter 25) and mixed-effects models (Chapter 19).

11

Analysis of Variance

Analysis of variance is the technique we use when all the explanatory variables are categorical. The explanatory variables are called **factors**, and each factor has two or more **levels**. When there is a single factor with three or more levels we use one-way ANOVA. If we had a single factor with just two levels, we would use Student's t test (see p. 294), and this would give us exactly the same answer that we would have obtained by ANOVA (remember the rule that $F = t^2$). Where there are two or more factors, then we use two-way or three-way ANOVA, depending on the number of explanatory variables. When there is replication at each combination of levels in a multi-way ANOVA, the experiment is called a **factorial** design, and this allows us to study **interactions** between variables, in which we test whether the response to one factor depends on the level of another factor.

One-Way ANOVA

There is a real paradox about analysis of variance, which often stands in the way of a clear understanding of exactly what is going on. The idea of analysis of variance is to compare two or more means, but it does this by comparing variances. How can that work?

The best way to see what is happening is to work through a simple example. We have an experiment in which crop yields per unit area were measured from 10 randomly selected fields on each of three soil types. All fields were sown with the same variety of seed and provided with the same fertilizer and pest control inputs. The question is whether soil type significantly affects crop yield, and if so, to what extent.

```
results<-read.table("c:\\temp\\yields.txt",header=T)
attach(results)
names(results)
```

```
[1]  "sand"  "clay"  "loam"
```

To see the data just type results followed by the Return key:

```
   sand clay loam
1    6   17   13
2   10   15   16
3    8    3    9
4    6   11   12
5   14   14   15
```

6	17	12	16
7	9	12	17
8	11	8	13
9	7	10	18
10	11	13	14

The function **sapply** is used to calculate the mean yields for the three soils:

```
sapply(list(sand,clay,loam),mean)
```

```
[1] 9.9 11.5 14.3
```

Mean yield was highest on loam (14.3) and lowest on sand (9.9).
It will be useful to have all of the yield data in a single vector called y:

```
y<-c(sand,clay,loam)
```

and to have a single vector called soil to contain the factor levels for soil type:

```
soil<-factor(rep(1:3,c(10,10,10)))
```

Before carrying out analysis of variance, we should check for constancy of variance (see Chapter 8) across the three soil types:

```
sapply(list(sand,clay,loam),var)
```

```
[1] 12.544444 15.388889 7.122222
```

The variances differ by more than a factor of 2. But is this significant? We test for heteroscedasticity using the Fligner–Killeen test of homogeneity of variances:

```
fligner.test(y~soil)
```

```
        Fligner-Killeen test of homogeneity of variances

data: y by soil
Fligner-Killeen:med chi-squared = 0.3651, df = 2, p-value = 0.8332
```

We could have used **bartlett.test(y~soil)**, which gives $p = 0.5283$ (but this is more a test of non-normality than of equality of variances). Either way, there is no evidence of any significant difference in variance across the three samples, so it is legitimate to continue with our one-way analysis of variance.

Because the explanatory variable is categorical (three levels of soil type), initial data inspection involves a box-and-whisker plot of y against soil like this:

```
plot(soil,y,names=c("sand","clay","loam"),ylab="yield")
```

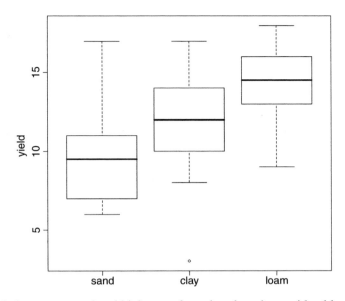

Median yield is lowest on sand and highest on loam but there is considerable variation from replicate to replicate within each soil type (there is even an outlier on clay). It looks as if yield on loam will turn out to be significantly higher than on sand (their boxes do not overlap) but it is not clear whether yield on clay is significantly greater than on sand or significantly lower than on loam. The analysis of variance will answer these questions.

The analysis of variance involves calculating the total variation in the response variable (yield in this case) and partitioning it ('analysing it') into informative components. In the simplest case, we partition the total variation into just two components: explained variation and unexplained variation:

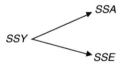

Explained variation is called the treatment sum of squares (SSA) and unexplained variation is called the error sum of squares (SSE, also known as the residual sum of squares). The unexplained variation is defined as the sum of the squares of the differences between the individual y values and the relevant treatment mean:

$$SSE = \sum_{i=1}^{k} \sum (y - \bar{y}_i)^2.$$

We compute the mean for the ith level of the factor in advance, and then add up the squares of the differences. Given that we worked it out this way, can you see how many degrees of freedom should be associated with SSE? Suppose that there were n replicates in each treatment ($n = 10$ in our example). And suppose that there are k levels of the factor ($k = 3$ in our example). If you estimate k parameters from the data before you can work out SSE, then you must have lost k degrees of freedom in the process. Since each of the k levels of the factor has n replicates, there must be $k \times n$ numbers in the whole experiment ($3 \times 10 = 30$ in our example). So the degrees of freedom associated with SSE are $kn - k = k(n - 1)$. Another way of seeing this is to say that there are n replicates in each treatment, and hence

$n − 1$ degrees of freedom for error in each treatment (because 1 d.f. is lost in estimating each treatment mean). There are k treatments (i.e. k levels of the factor) and hence there are $k \times (n − 1)$ d.f. for error in the experiment as a whole.

The component of the variation that is explained by differences between the treatment means, the treatment sum of squares, is traditionally denoted by SSA. This is because in two-way analysis of variance, with two different categorical explanatory variables, SSB is used to denote the sum of squares attributable to differences between the means of the second factor, SSC to denote the sum of squares attributable to differences between the means of the third factor, and so on.

Typically, we compute all but one of the components of the total variance, then find the value of the last component by subtraction of the others from SSY. We already have a formula for SSE, so we could obtain SSA by difference: $SSA = SSY − SSE$. Box 11.1 looks at the formula for SSA in more detail.

Box 11.1   Corrected sums of squares in one-way ANOVA

The definition of the total sum of squares, SSY, is the sum of the squares of the differences between the data points, y_{ij}, and the overall mean, $\bar{\bar{y}}$:

$$SSY = \sum_{i=1}^{k} \sum_{j=1}^{n} (y_{ij} - \bar{\bar{y}})^2,$$

where $\sum_{j=1}^{n} y_{ij}$ means the sum over the n replicates within each of the k factor levels. The error sum of squares, SSE, is the sum of the squares of the differences between the data points, y_{ij}, and their individual treatment means, $\bar{y}_i$:

$$SSE = \sum_{i=1}^{k} \sum_{j=1}^{n} (y_{ij} - \bar{y}_i)^2.$$

The treatment sum of squares, SSA, is the sum of the squares of the differences between the individual treatment means, $\bar{y}_i$, and the overall mean, $\bar{\bar{y}}$:

$$SSA = \sum_{i=1}^{k} \sum_{j=1}^{n} (\bar{y}_i - \bar{\bar{y}})^2 = n \sum_{i=1}^{k} (\bar{y}_i - \bar{\bar{y}})^2.$$

Squaring the bracketed term and applying summation gives

$$\sum \bar{y}_i^2 - 2\bar{\bar{y}} \sum \bar{y}_i + k\bar{\bar{y}}^2.$$

Let the grand total of all the values of the response variable $\sum_{i=1}^{k} \sum_{j=1}^{n} y_{ij}$ be shown as $\sum y$. Now replace $\bar{y}_i$ by T_i/n (where T is our conventional name for the k individual treatment totals) and replace $\bar{\bar{y}}$ by $\sum y/kn$ to get

$$\frac{\sum_{i=1}^{k} T_i^2}{n^2} - 2\frac{\sum y \sum_{i=1}^{k} T_i}{nkn} + k\frac{\sum y \sum y}{knkn}.$$

Note that $\sum_{i=1}^{k} T_i = \sum_{i=1}^{k} \sum_{j=1}^{n} y_{ij}$, so the right-hand positive and negative terms both have the form $(\sum y)^2/kn^2$. Finally, multiplying through by n gives

$$SSA = \frac{\sum T^2}{n} - \frac{(\sum y)^2}{kn}.$$

As an exercise, you should prove that $SSY = SSA + SSE$.

Let's work through the numbers in R. From the formula for SSY, we can obtain the total sum of squares by finding the differences between the data and the overall mean:

sum((y-mean(y))^2)

[1]   414.7

The unexplained variation, SSE, is calculated from the differences between the yields and the mean yields *for that soil type*:

sand-mean(sand)

[1] −3.9   0.1 −1.9 −3.9   4.1   7.1 −0.9   1.1 −2.9   1.1

clay-mean(clay)

[1]   5.5   3.5 −8.5 −0.5   2.5   0.5   0.5 −3.5 −1.5   1.5

loam-mean(loam)

[1] −1.3   1.7 −5.3 −2.3   0.7   1.7   2.7 −1.3   3.7 −0.3

We need the sums of the squares of these differences:

sum((sand-mean(sand))^2)

[1]   112.9
sum((clay-mean(clay))^2)

[1]   138.5
sum((loam-mean(loam))^2)

[1]   64.1

To get the sum of these totals across all soil types, we can use sapply like this:

sum(sapply(list(sand,clay,loam),function (x) sum((x-mean(x))^2)))

[1]   315.5

So SSE, the unexplained (or residual, or error) sum of squares, is 315.5.

The extent to which SSE is less than SSY is a reflection of the magnitude of the differences between the means. The greater the difference between the mean yields on the different soil types, the greater will be the difference between SSE and SSY. This is the basis of analysis of variance. We can make inferences about differences between means by looking at differences between variances (or between sums or squares, to be more precise at this stage).

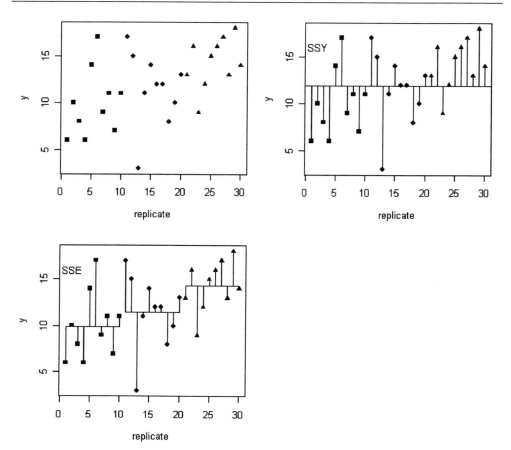

At top left we have an 'index plot' of the yields with different symbols for the different soil types: square = sand, diamond = clay, triangle = loam. At top right is a picture of the total sum of squares: *SSY* is the sum of the squares of the lengths of the lines joining each data point to the overall mean, $\bar{\bar{y}}$. At bottom left is a picture of the error sum of squares: *SSE* is the sum of the squares of the lengths of the lines joining each data point to its particular treatment mean, $\bar{y}_i$. The extent to which the lines are shorter in *SSE* than in *SSY* is a measure of the significance of the difference between the mean yields on the different soils. In the extreme case in which there was *no* variation between the replicates, then *SSY* is large, but *SSE* is zero:

This picture was created with the following code, where the *x* values, xvc, are

```
xvc<-1:15
```

and the *y* values, yvs, are

```
yvs<-rep(c(10,12,14),each=5)
```

To produce the two plots side by side, we write:

```
par(mfrow=c(1,2))
plot(xvc,yvs,ylim=c(5,16),pch=(15+(xvc>5)+(xvc>10)))
for (i in 1:15) lines(c(i,i),c(yvs[i],mean(yvs)))
```

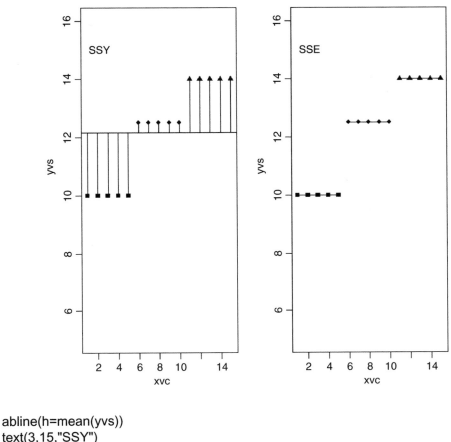

```
abline(h=mean(yvs))
text(3,15,"SSY")
plot(xvc,yvs, ylim=c(5,16),pch=(15+(xvc(>)5)+(xvc(>)10)))
lines(c(1,5),c(10,10))
lines(c(6,10),c(12,12))
lines(c(11,15),c(14,14))
text(3,15,"SSE")
```

The difference between SSY and SSE is called the treatment sum of squares, SSA: this is the amount of the variation in yield that is explained by differences between the treatment means. In our example,

$$SSA = SSY - SSE = 414.7 - 315.5 = 99.2.$$

Now we can draw up the ANOVA table. There are six columns indicating, from left to right, the source of variation, the sum of squares attributable to that source, the degrees of freedom for that source, the variance for that source (traditionally called the mean square rather than the variance), the F ratio (testing the null hypothesis that this source of variation is not significantly different from zero) and the p value associated with that F value (if $p < 0.05$ then we reject the null hypothesis). We can fill in the sums of squares just calculated, then think about the degrees of freedom:

There are 30 data points in all, so the total degrees of freedom are $30 - 1 = 29$. We lose 1 d.f. because in calculating SSY we had to estimate one parameter from the data in advance,

Source	Sum of squares	Degrees of freedom	Mean square	F ratio	p value
Soil type	99.2	2	49.6	4.24	0.025
Error	315.5	27	$s^2 = 11.685$		
Total	414.7	29			

namely the overall mean, $\bar{\bar{y}}$. Each soil type has $n = 10$ replications, so each soil type has $10 - 1 = 9$ d.f. for error, because we estimated one parameter from the data *for each soil type*, namely the treatment means $\bar{y}_i$ in calculating *SSE*. Overall, therefore, the error has $3 \times 9 = 27$ d.f. There were 3 soil types, so there are $3 - 1 = 2$ d.f. for soil type.

The mean squares are obtained simply by dividing each sum of squares by its respective degrees of freedom (in the same row). The error variance, s^2, is the residual mean square (the mean square for the unexplained variation); this is sometimes called the 'pooled error variance' because it is calculated across all the treatments; the alternative would be to have three separate variances, one for each treatment:

```
sapply(list(sand,clay,loam),var)
```

```
[1]   12.544444   15.388889   7.122222
```

You will see that the pooled error variance $s^2 = 11.685$ is simply the mean of the three separate variances, because there is equal replication in each soil type ($n = 10$):

```
mean(sapply(list(sand,clay,loam),var))
```

```
[1]   11.68519
```

By tradition, we do not calculate the total mean square, so the bottom cell of the fourth column of the ANOVA table is empty. The F ratio is the treatment variance divided by the error variance, testing the null hypothesis that the treatment means are all the same. If we reject this null hypothesis, we accept the alternative hypothesis that *at least one of the means is significantly different from the others*. The question naturally arises at this point as to whether 4.24 is a big number or not. If it is a big number then we reject the null hypothesis. If it is not a big number, then we accept the null hypothesis. As ever, we decide whether the test statistic $F = 4.24$ is big or small by comparing it with the *critical value* of F, given that there are 2 d.f. in the numerator and 27 d.f. in the denominator. Critical values in R are found from the function qf which gives us quantiles of the F distribution:

```
qf(.95,2,27)
```

```
[1]   3.354131
```

Our calculated test statistic of 4.24 is larger than the critical value of 3.35, so we reject the null hypothesis. At least one of the soils has a mean yield that is significantly different from the others. The modern approach is not to work slavishly at the 5% level but rather to calculate the *p*-value associated with our test statistic of 4.24. Instead of using the function for quantiles of the F distribution, we use the function pf for cumulative probabilities of the F distribution like this:

1-pf(4.24,2,27)

[1] 0.02503987

The *p*-value is 0.025, which means that a value of $F = 4.24$ or bigger would arise by chance alone when the null hypothesis was true about 25 times in 1000. This is a sufficiently small probability (i.e. it is less than 5%) for us to conclude that there is a significant difference between the mean yields (i.e. we reject the null hypothesis).

That was a lot of work. R can do the whole thing in a single line:

summary(aov(y~soil))

```
             Df   Sum Sq  Mean Sq  F value   Pr(>F)
soil          2   99.200   49.600   4.2447  0.02495  *
Residuals    27  315.500   11.685
```

Here you see all the values that we calculated long-hand. The error row is labelled Residuals. In the second and subsequent columns you see the degrees of freedom for treatment and error (2 and 27), the treatment and error sums of squares (99.2 and 315.5), the treatment mean square of 49.6, the error variance $s^2 = 11.685$, the *F* ratio and the *p*-value (labelled Pr($>$ F)). The single asterisk next to the *p* value indicates that the difference

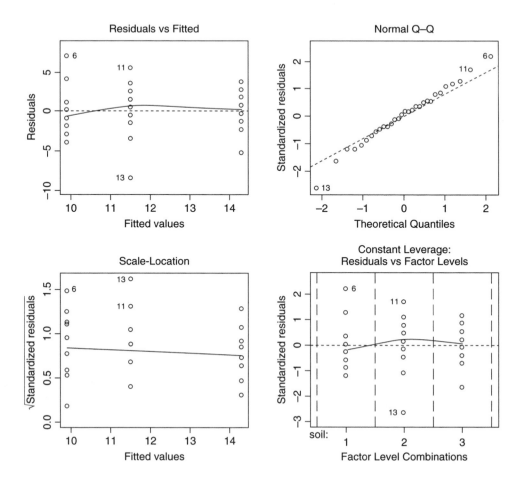

between the soil means is significant at 5% (but not at 1%, which would have merited two asterisks). Notice that R does not print the bottom row of the ANOVA table showing the total sum of squares and total degrees of freedom.

The next thing we would do is to check the assumptions of the aov model. This is done using plot like this (see Chapter 10):

plot(aov(y~soil))

The first plot (top left) checks the most important assumption of constancy of variance; there should be no pattern in the residuals against the fitted values (the three treatment means) – and, indeed, there is none. The second plot (top right) tests the assumption of normality of errors: there should be a straight-line relationship between our standardized residuals and theoretical quantiles derived from a normal distribution. Points 6, 11 and 13 lie a little off the straight line, but this is nothing to worry about (see p. 339). The residuals are well behaved (bottom left) and there are no highly influential values that might be distorting the parameter estimates (bottom right).

Effect sizes

The best way to view the effect sizes graphically is to use plot.design (which takes a formula rather than a model object):

plot.design(y~soil)

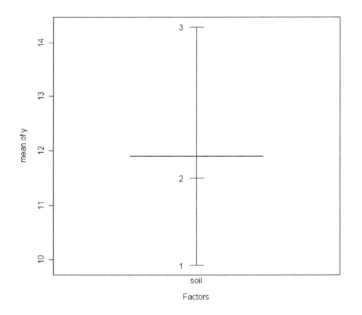

For more complicated models, you might want to use the effects library to get more attractive plots (p. 178). To see the effect sizes in tabular form use model.tables (which takes a model object as its argument) like this:

model<-aov(y~soil);model.tables(model,se=T)

Tables of effects

```
  soil
soil
   1    2    3
-2.0 -0.4  2.4
```

Standard errors of effects
```
            soil
          1.081
replic.     10
```

The effects are shown as departures from the overall mean: soil 1 (sand) has a mean yield that is 2.0 below the overall mean, and soil 3 (loam) has a mean that is 2.4 above the overall mean. The standard error of effects is 1.081 on a replication of $n = 10$ (this is the standard error of a mean). You should note that this is *not* the appropriate standard error for comparing two means (see below). If you specify **"means"** you get:

model.tables(model,"means",se=T)

Tables of means
Grand mean

11.9
```
  soil
soil
   1    2    3
 9.9 11.5 14.3
```

Standard errors for differences of means
```
            soil
          1.529
replic.     10
```

Now the three means are printed (rather than the effects) and the standard error of the difference of means is given (this *is* what you need for doing a t test to compare any two means).

Another way of looking at effect sizes is to use the **summary.lm** option for viewing the model, rather than **summary.aov** (as we used above):

summary.lm(aov(y~soil))

Coefficients:

	Estimate	Std. Error	t value	Pr(>|t|)	
(Intercept)	9.900	1.081	9.158	9.04e-10	***
soil2	1.600	1.529	1.047	0.30456	
soil3	4.400	1.529	2.878	0.00773	**

Residual standard error: 3.418 on 27 degrees of freedom
Multiple R-Squared: 0.2392, Adjusted R-squared: 0.1829
F-statistic: 4.245 on 2 and 27 DF, p-value: 0.02495

In regression analysis (p. 399) the **summary.lm** output was easy to understand because it gave us the intercept and the slope (the two parameters estimated by the model) and their standard errors. But this table has three rows. Why is that? What is an intercept in the context of analysis of variance? And why are the standard errors different for the intercept and for soil 2 and soil 3? Come to that, what are soil 2 and soil 3?

It will take a while before you feel at ease with summary.lm tables for analysis of variance. The details are explained on p. 365, but the central point is that all summary.lm tables have as many rows as there are parameters estimated from the data. There are three rows in this case because our aov model estimates three parameters; a mean yield for each of the three soil types. In the context of aov, an intercept is a mean value; in this case it is the mean yield for sand because we gave that factor level 1 when we computed the vales for the factor called soil, earlier. In general, the intercept would be the factor level whose name came lowest in alphabetical order (see p. 366). So if Intercept is the mean yield for sand, what are the other two rows labelled soil2 and soil3. This is the hardest thing to understand. All other rows in the summary.lm table for aov are differences between means. Thus row 2, labelled soil2, is the difference between the mean yields on sand and clay, and row 3, labelled soil3, is the difference between the mean yields of sand and loam:

tapply(y,soil,mean)-mean(sand)

```
   1      2      3
 0.0    1.6    4.4
```

The first row (Intercept) is a mean, so the standard error column in row 1 contains the standard error of a mean. Rows 2 and 3 are differences between means, so their standard error columns contain the standard error of the difference between two means (and this is a bigger number; see p. 367). The standard error of a mean is

$$se_{mean} = \sqrt{\frac{s^2}{n}} = \sqrt{\frac{11.685}{10}} = 1.081,$$

whereas the standard error of the difference between two means is

$$se_{diff} = \sqrt{2\frac{s^2}{n}} = \sqrt{2 \times \frac{11.685}{10}} = 1.529.$$

The summary.lm table shows that soil 3 produces significantly bigger yields than soil 1 (the intercept) with a p-value of 0.007 73. The difference between the two means was 4.400 and the standard error of the difference was 1.529. This difference is of two-star significance, meaning $0.001 < p < 0.01$. In contrast, soil 2 does not produce a significantly greater yield than soil 1; the difference is 1.600 and the standard error of the difference was 1.529 ($p = 0.304\,56$). The only remaining issue is whether soil 2 yielded significantly less than soil 3. We need to do some mental arithmetic to work this out: the difference between these two means was $4.4 - 1.6 = 2.8$ and so the t value is $2.8/1.529 = 1.83$. This is less than 2 (the rule of thumb for t) so the mean yields of soils 2 and 3 are not significantly different. To find the precise value with 10 replicates, the critical value of t is given by the function qt with 18 d.f.:

qt(0.975,18)

```
[1]  2.100922
```

Alternatively we can work out the p value associated with our calculated $t = 1.83$:

2*(1 - pt(1.83, df = 18))

```
[1]  0.0838617
```

giving $p = 0.084$. We multiply by 2 because this is a two-tailed test (see p. 208); we did not know in advance that loam would out-yield clay under the particular circumstances of this experiment.

The residual standard error in the summary.lm output is the square root of the error variance from the ANOVA table: $\sqrt{11.685} = 3.418$. R-Squared and Adjusted R-Squared are explained on p. 399. The F-statistic and the p-value come from the last two columns of the ANOVA table.

So there it is. That is how analysis of variance works. When the means are significantly different, then the sum of squares computed from the individual treatment means will be significantly smaller than the sum of squares computed from the overall mean. We judge the significance of the difference between the two sums of squares using analysis of variance.

Plots for interpreting one-way ANOVA

There are two traditional ways of plotting the results of ANOVA:

- box-and-whisker plots;
- barplots with error bars.

Here is an example to compare the two approaches. We have an experiment on plant competition with one factor and five levels. The factor is called clipping and the five levels consist of control (i.e. unclipped), two intensities of shoot pruning and two intensities of root pruning:

```
comp<-read.table("c:\\temp\\competition.txt",header=T);attach(comp);names(comp)
```

```
[1]  "biomass"  "clipping"
```

```
plot(clipping,biomass,xlab="Competition treatment",ylab="Biomass")
```

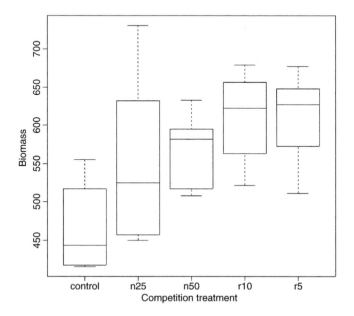

The box-and-whisker plot is good at showing the nature of the variation within each treatment, and also whether there is skew within each treatment (e.g. for the control plots, there is a wider range of values between the median and third quartile than between the median and first quartile). No outliers are shown above the whiskers, so the tops and bottoms of the bars are the maxima and minima within each treatment. The medians for the competition treatments are all higher than the third quartile of the controls, suggesting that they may be significantly different from the controls, but there is little to suggest that any of the competition treatments are significantly different from one another (see below for the analysis). We could use the notch=T option to get a visual impression of the significance of differences between the means; all the treatment medians fall outside the notch of the controls, but no other comparisons appear to be significant.

Barplots with error bars are preferred by many journal editors, and some people think that they make hypothesis testing easier. We shall see. Unlike S-PLUS, R does not have a built-in function called error.bar so we shall have to write our own. Here is a very simple version without any bells or whistles. We shall call it error.bars to distinguish it from the much more general S-PLUS function.

```
error.bars<-function(yv,z,nn){
xv<-
barplot(yv,ylim=c(0,(max(yv)+max(z))),names=nn,ylab=deparse(substitute(yv)
))
g=(max(xv)-min(xv))/50
for (i in 1:length(xv)) {
lines(c(xv[i],xv[i]),c(yv[i]+z[i],yv[i]-z[i]))
lines(c(xv[i]-g,xv[i]+g),c(yv[i]+z[i], yv[i]+z[i]))
lines(c(xv[i]-g,xv[i]+g),c(yv[i]-z[i], yv[i]-z[i]))
}}
```

To use this function we need to decide what kind of values (z) to use for the lengths of the bars. Let's use the standard error of a mean based on the pooled error variance from the ANOVA, then return to a discussion of the pros and cons of different kinds of error bars later. Here is the one-way analysis of variance:

```
model<-aov(biomass~clipping)
summary(model)
```

	Df	Sum Sq	Mean Sq	F value	Pr(>F)	
clipping	4	85356	21339	4.3015	0.008752	**
Residuals	25	124020	4961			

From the ANOVA table we learn that the pooled error variance $s^2 = 4961.0$. Now we need to know how many numbers were used in the calculation of each of the five means:

```
table(clipping)
```

```
clipping
control   n25   n50   r10   r5
      6     6     6     6    6
```

There was equal replication (which makes life easier), and each mean was based on six replicates, so the standard error of a mean is $\sqrt{s^2/n} = \sqrt{4961/6} = 28.75$. We shall draw an error bar up 28.75 from each mean and down by the same distance, so we need 5 values for z, one for each bar, each of 28.75:

```
se<-rep(28.75,5)
```

We need to provide labels for the five different bars: the factor levels should be good for this:

```
labels<-as.character(levels(clipping))
```

Now we work out the five mean values which will be the heights of the bars, and save them as a vector called ybar:

```
ybar<-as.vector(tapply(biomass,clipping,mean))
```

Finally, we can create the barplot with error bars (the function is defined on p. 462):

```
error.bars(ybar,se,labels)
```

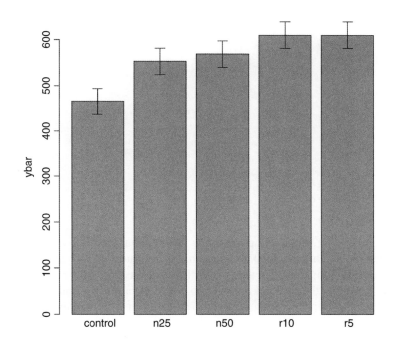

We do not get the same feel for the distribution of the values *within* each treatment as was obtained by the box-and-whisker plot, but we can certainly see clearly which means are not significantly different. If, as here, we use ±1 standard error as the length of the error bars, then *when the bars overlap this implies that the two means are not significantly different.* Remember the rule of thumb for t: significance requires 2 or more standard errors, and if the bars overlap it means that the difference between the means is less than 2 standard errors. There is another issue, too. For comparing means, we should use the standard error of the difference between two means (not the standard error of one mean) in our tests (see p. 294); these bars would be about 1.4 times as long as the bars we have drawn here. So while we can be sure that the two root-pruning treatments are not significantly different from one another, and that the two shoot-pruning treatments are not significantly different from one another (because their bars overlap), we cannot conclude from this plot that the controls have significantly lower biomass than the rest (because the error bars are not the correct length for testing differences between means).

An alternative graphical method is to use 95% confidence intervals for the lengths of the bars, rather than standard errors of means. This is easy to do: we multiply our standard errors by Student's t, qt(.975,5) = 2.570 582, to get the lengths of the confidence intervals:

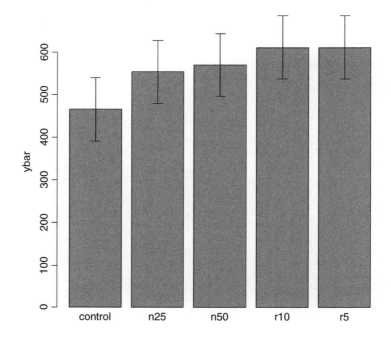

Now, all of the error bars overlap, implying visually that there are no significant differences between the means. But we know that this is not true from our analysis of variance, in which we rejected the null hypothesis that all the means were the same at $p = 0.008\,75$. If it were the case that the bars did not overlap when we are using confidence intervals (as here), then that would imply that the means differed by more than 4 standard errors, and this is a much greater difference than is required to conclude that the means are significantly different. So this is not perfect either. With standard errors we could be sure that the means were *not* significantly different when the bars *did* overlap. And with confidence intervals we can be sure that the means *are* significantly different when the bars *do not* overlap. But the alternative cases are not clear-cut for either type of bar. Can we somehow get the best of both worlds, so that the means *are* significantly different when the bars *do not* overlap, and the means are *not* significantly different when the bars *do* overlap?

The answer is yes, we can, if we use least significant difference (LSD) bars. Let's revisit the formula for Student's t test:

$$t = \frac{\text{a difference}}{\text{standard error of the difference}}$$

We say that the difference is significant when $t > 2$ (by the rule of thumb, or $t >$ qt(0.975,df) if we want to be more precise). We can rearrange this formula to find the smallest difference that we would regard as being significant. We can call this the least significant difference:

$$LSD = \text{qt(0.975,df)} \times \text{standard error of a difference} \approx 2 \times se_{\text{diff}}.$$

In our present example this is

qt(0.975,10)*sqrt(2*4961/6)

[1] 90.60794

because a difference is based on $12 - 2 = 10$ degrees of freedom. What we are saying is the two means would be significantly different if they differed by 90.61 or more. How can we show this graphically? We want overlapping bars to indicate a difference less than 90.61, and non-overlapping bars to represent a difference greater than 90.61. With a bit of thought you will realize that we need to draw bars that are *LSD*/2 in length, up and down from each mean. Let's try it with our current example:

lsd<-qt(0.975,10)*sqrt(2*4961/6)
lsdbars<-rep(lsd,5)/2
error.bars(ybar,lsdbars,labels)

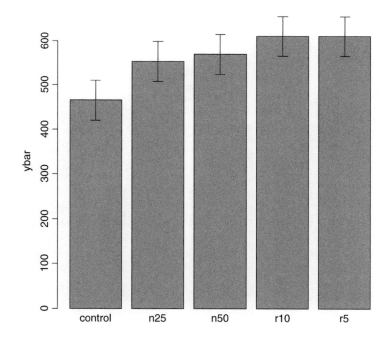

Now we can interpret the significant differences visually. The control biomass is signifi-cantly lower than any of the four treatments, but none of the four treatments is significantly different from any other. The statistical analysis of this contrast is explained in detail in Chapter 9. Sadly, most journal editors insist on error bars of 1 standard error. It is true that there are complicating issues to do with LSD bars (not least the vexed question of multiple comparisons; see p. 483), but at least they do what was intended by the error plot (i.e. overlapping bars means non-significance and non-overlapping bars means sig-nificance); neither standard errors nor confidence intervals can say that. A better option might be to use box-and-whisker plots with the notch=T option to indicate significance (see p. 159).

Factorial Experiments

A factorial experiment has two or more factors, each with two or more levels, plus replication for each combination of factors levels. This means that we can investigate statistical interactions, in which the response to one factor depends on the level of another factor. Our example comes from a farm-scale trial of animal diets. There are two factors: diet and supplement. Diet is a factor with three levels: barley, oats and wheat. Supplement is a factor with four levels: agrimore, control, supergain and supersupp. The response variable is weight gain after 6 weeks.

```
weights<-read.table("c:\\temp\\growth.txt",header=T)
attach(weights)
```

Data inspection is carried out using barplot (note the use of beside=T to get the bars in adjacent clusters rather than vertical stacks):

```
barplot(tapply(gain,list(diet,supplement),mean),
            beside=T,ylim=c(0,30),col=rainbow(3))
```

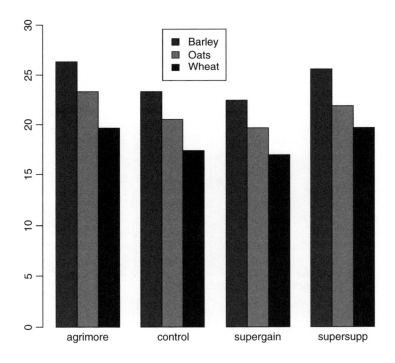

Note that the second factor in the list (supplement) appears as groups of bars from left to right in alphabetical order by factor level, from agrimore to supersupp. The second factor (diet) appears as three levels within each group of bars: red = barley, green = oats, blue = wheat, again in alphabetical order by factor level. We should really add a key to explain the levels of diet. Use locator(1) to find the coordinates for the *top left* corner of the box around the legend. You need to increase the default scale on the y axis to make enough room for the legend box.

```
labs<-c("Barley","Oats","Wheat")
legend(locator(1),labs,fill=rainbow(3))
```

We inspect the mean values using tapply as usual:

```
tapply(gain,list(diet,supplement),mean)
```

	agrimore	control	supergain	supersupp
barley	26.34848	23.29665	22.46612	25.57530
oats	23.29838	20.49366	19.66300	21.86023
wheat	19.63907	17.40552	17.01243	19.66834

Now we use aov or lm to fit a factorial analysis of variance (the choice affects whether we get an ANOVA table or a list of parameters estimates as the default output from summary). We estimate parameters for the main effects of each level of diet and each level of supplement, plus terms for the interaction between diet and supplement. Interaction degrees of freedom are the product of the degrees of freedom of the component terms (i.e. $(3 - 1) \times (4 - 1) = 6$). The model is gain~diet + supplement + diet:supplement, but this can be simplified using the asterisk notation like this:

```
model<-aov(gain~diet*supplement)
summary(model)
```

	Df	Sum Sq	Mean Sq	F value	Pr(>F)	
diet	2	287.171	143.586	83.5201	2.998e-14	***
supplement	3	91.881	30.627	17.8150	2.952e-07	**
diet:supplement	6	3.406	0.568	0.3302	0.9166	
Residuals	36	61.890	1.719			

The ANOVA table shows that there is no hint of any interaction between the two explanatory variables ($p=0.9166$); evidently the effects of diet and supplement are additive. The disadvantage of the ANOVA table is that it does not show us the effect sizes, and does not allow us to work out how many levels of each of the two factors are significantly different. As a preliminary to model simplification, summary.lm is often more useful than summary.aov:

```
summary.lm(model)
```

Coefficients:

	Estimate	Std. Error	t value	Pr(>|t|)	
(Intercept)	26.3485	0.6556	40.191	< 2e-16	***
dietoats	-3.0501	0.9271	-3.290	0.002248	**
dietwheat	-6.7094	0.9271	-7.237	1.61e-08	***
supplementcontrol	-3.0518	0.9271	-3.292	0.002237	**
supplementsupergain	-3.8824	0.9271	-4.187	0.000174	***
supplementsupersupp	-0.7732	0.9271	-0.834	0.409816	
dietoats:supplementcontrol	0.2471	1.3112	0.188	0.851571	
dietwheat:supplementcontrol	0.8183	1.3112	0.624	0.536512	
dietoats:supplementsupergain	0.2470	1.3112	0.188	0.851652	
dietwheat:supplementsupergain	1.2557	1.3112	0.958	0.344601	
dietoats:supplementsupersupp	-0.6650	1.3112	-0.507	0.615135	
dietwheat:supplementsupersupp	0.8024	1.3112	0.612	0.544381	

```
Residual standard error: 1.311 on 36 degrees of freedom
Multiple R-Squared: 0.8607, Adjusted R-squared: 0.8182
F-statistic: 20.22 on 11 and 36 DF, p-value: 3.295e-012
```

This is a rather complex model, because there are 12 estimated parameters (the number of rows in the table): six main effects and six interactions. The output re-emphasizes that none of the interaction terms is significant, but it suggests that the minimal adequate model will require five parameters: an intercept, a difference due to oats, a difference due to wheat, a difference due to control and difference due to supergain (these are the five rows with significance stars). This draws attention to the main shortcoming of using treatment contrasts as the default. If you look carefully at the table, you will see that the effect sizes of two of the supplements, control and supergain, are not significantly different from one another. You need lots of practice at doing t tests in your head, to be able to do this quickly. Ignoring the signs (because the signs are negative for both of them), we have 3.05 vs. 3.88, a difference of 0.83. But look at the associated standard errors (both 0.927); the difference is less than 1 standard error of a difference between two means. For significance, we would need roughly 2 standard errors (remember the rule of thumb, in which $t \geq 2$ is significant; see p. 228). The rows get starred in the significance column because treatments contrasts compare all the main effects in the rows with the intercept (where each factor is set to its first level in the alphabet, namely agrimore and barley in this case). When, as here, several factor levels are different from the intercept, but not different from one another, they all get significance stars. This means that you cannot count up the number of rows with stars in order to determine the number of significantly different factor levels.

We first simplify the model by leaving out the interaction terms:

```
model<-aov(gain~diet+supplement)
summary.lm(model)
```

```
Coefficients:
                      Estimate  Std. Error  t value  Pr(>|t|)
(Intercept)           26.1230    0.4408      59.258   < 2e-16   ***
dietoats              -3.0928    0.4408      -7.016   1.38e-08  ***
dietwheat             -5.9903    0.4408     -13.589   < 2e-16   ***
supplementcontrol     -2.6967    0.5090      -5.298   4.03e-06  ***
supplementsupergain   -3.3815    0.5090      -6.643   4.72e-08  ***
supplementsupersupp   -0.7274    0.5090      -1.429   0.160
```

It is clear that we need to retain all three levels of diet (oats differ from wheat by 5.99 − 3.09 = 2.90 with a standard error of 0.44). It is not clear that we need four levels of supplement, however. Supersupp is not obviously different from agrimore (0.727 with standard error 0.509). Nor is supergain obviously different from the unsupplemented control animals (3.38 − 2.70 = 0.68). We shall try a new two-level factor to replace the four-level supplement, and see if this significantly reduces the model's explanatory power. Agrimore and supersupp are recoded as best and control and supergain as worst:

```
supp2<-factor(supplement)
levels(supp2)
```

```
[1] "agrimore" "control" "supergain" "supersupp"
```

```
levels(supp2)[c(1,4)]<-"best"
levels(supp2)[c(2,3)]<-"worst"
levels(supp2)
```

```
[1] "best"  "worst"
```

Now we can compare the two models:

```
model2<-aov(gain~diet+supp2)
anova(model,model2)
```

```
Analysis of Variance Table
```

```
Model 1: gain ~ diet + supplement
Model 2: gain ~ diet + supp2
```

```
Res.Df          RSS  Df  Sum of Sq        F  Pr((>)F)
1          42  65.296
2          44  71.284  -2     -5.988  1.9257    0.1584
```

The simpler model2 has saved two degrees of freedom and is not significantly worse than the more complex model ($p = 0.158$). This is the minimal adequate model: all of the parameters are significantly different from zero and from one another:

```
summary.lm(model2)
```

```
Coefficients:
                Estimate  Std. Error  t value  Pr(>|t|)
(Intercept)      25.7593      0.3674   70.106   < 2e-16  ***
dietoats         -3.0928      0.4500   -6.873  1.76e-08  ***
dietwheat        -5.9903      0.4500  -13.311   < 2e-16  ***
supp2worst       -2.6754      0.3674   -7.281  4.43e-09  ***
```

```
Residual standard error: 1.273 on 44 degrees of freedom
Multiple R-Squared: 0.8396, Adjusted R-squared: 0.8286
F-statistic: 76.76 on 3 and 44 DF, p-value: 0
```

Model simplification has reduced our initial 12-parameter model to a four-parameter model.

Pseudoreplication: Nested Designs and Split Plots

The model-fitting functions aov and lmer have the facility to deal with complicated error structures, and it is important that you can recognize them, and hence avoid the pitfalls of pseudoreplication. There are two general cases:

- nested sampling, as when repeated measurements are taken from the same individual, or observational studies are conduced at several different spatial scales (mostly random effects);

- split-plot analysis, as when designed experiments have different treatments applied to plots of different sizes (mostly fixed effects).

Split-plot experiments

In a split-plot experiment, different treatments are applied to plots of different sizes. Each different plot size is associated with its own error variance, so instead of having one error variance (as in all the ANOVA tables up to this point), we have as many error terms as there are different plot sizes. The analysis is presented as a series of component ANOVA tables, one for each plot size, in a hierarchy from the largest plot size with the lowest replication at the top, down to the smallest plot size with the greatest replication at the bottom.

The following example refers to a designed field experiment on crop yield with three treatments: irrigation (with two levels, irrigated or not), sowing density (with three levels, low, medium and high), and fertilizer application (with three levels, low, medium and high).

```
yields<-read.table("c:\\temp\\splityield.txt",header=T)
attach(yields)
names(yields)
```

```
[1] "yield"  "block"  "irrigation"  "density"  "fertilizer"
```

The largest plots were the four whole fields (block), each of which was split in half, and irrigation was allocated at random to one half of the field. Each irrigation plot was split into three, and one of three different seed-sowing densities (low, medium or high) was allocated at random (independently for each level of irrigation and each block). Finally, each density plot was divided into three, and one of three fertilizer nutrient treatments (N, P, or N and P together) was allocated at random. The model formula is specified as a factorial, using the asterisk notation. The error structure is defined in the Error term, with the plot sizes listed from left to right, from largest to smallest, with each variable separated by the slash operator /. Note that the smallest plot size, fertilizer, does not need to appear in the Error term:

```
model<-aov(yield~irrigation*density*fertilizer+Error(block/irrigation/density))
summary(model)
```

```
Error: block
            Df   Sum Sq  Mean Sq  F value  Pr(>F)
Residuals    3  194.444   64.815
```

```
Error: block:irrigation
             Df  Sum Sq  Mean Sq  F value  Pr(>F)
irrigation    1  8277.6   8277.6   17.590  0.02473  *
Residuals     3  1411.8    470.6
```

```
Error: block:irrigation:density
                     Df   Sum Sq  Mean Sq  F value  Pr(>F)
density               2  1758.36   879.18   3.7842  0.05318  .
irrigation:density    2  2747.03  1373.51   5.9119  0.01633  *
Residuals            12  2787.94   232.33
```

```
Error: Within
                              Df   Sum Sq  Mean Sq  F value    Pr(>F)
fertilizer                     2  1977.44   988.72  11.4493  0.0001418  ***
irrigation:fertilizer          2   953.44   476.72   5.5204  0.0081078  **
density:fertilizer             4   304.89    76.22   0.8826  0.4840526
irrigation:density:fertilizer  4   234.72    58.68   0.6795  0.6106672
Residuals                     36  3108.83    86.36
```

Here you see the four ANOVA tables, one for each plot size: blocks are the biggest plots, half blocks get the irrigation treatment, one third of each half block gets a sowing density treatment, and one third of a sowing density treatment gets each fertilizer treatment. Note that the non-significant main effect for density ($p = 0.053$) does *not* mean that density is unimportant, because density appears in a significant interaction with irrigation (the density terms cancel out, when averaged over the two irrigation treatments; see below). The best way to understand the two significant interaction terms is to plot them using interaction.plot like this:

```
interaction.plot(fertilizer,irrigation,yield)
```

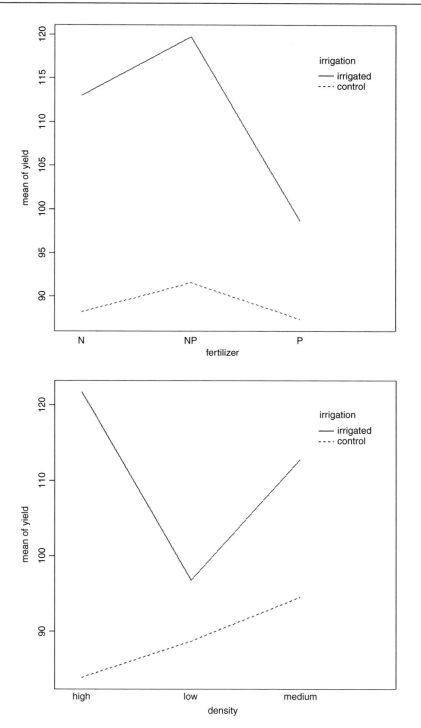

Irrigation increases yield proportionately more on the N-fertilized plots than on the P-fertilized plots. The irrigation–density interaction is more complicated:

```
interaction.plot(density,irrigation,yield)
```

On the irrigated plots, yield is minimal on the low-density plots, but on control plots yield is minimal on the high-density plots. Alternatively, you could use the **effects** package which takes a model object (a linear model or a generalized linear model) and provides attractive trellis plots of specified interaction effects (p. 178).

Missing values in a split-plot design

When there are missing values, then factors have effects in more than one stratum and the same main effect turns up in more than one ANOVA table. Suppose that the 69th yield value was missing:

yield[69]<-NA

Now the summary table looks very different:

model<-aov(yield~irrigation*density*fertilizer+Error(block/irrigation/density))
summary(model)

```
Error: block
             Df    Sum Sq  Mean Sq  F value  Pr(>F)
irrigation    1     0.075    0.075  9e-04    0.9788
Residuals     2   167.704   83.852
```

```
Error: block:irrigation
             Df   Sum Sq  Mean Sq  F value   Pr(>F)
irrigation    1   7829.9   7829.9  21.9075   0.04274  *
density       1    564.4    564.4   1.5792   0.33576
Residuals     2    714.8    357.4
```

```
Error: block:irrigation:density
                    Df    Sum Sq  Mean Sq    F value   Pr(>F)
density              2   1696.47   848.24     3.4044   0.07066  .
fertilizer           1      0.01     0.01  2.774e-05   0.99589
irrigation:density   2   2786.75  1393.37     5.5924   0.02110  *
Residuals           11   2740.72   249.16
```

```
Error: Within
                              Df   Sum Sq  Mean Sq  F value    Pr(>F)
fertilizer                     2  1959.36   979.68  11.1171   0.0001829  ***
irrigation:fertilizer          2   993.59   496.79   5.6375   0.0075447  **
density:fertilizer             4   273.56    68.39   0.7761   0.5482571
irrigation:density:fertilizer  4   244.49    61.12   0.6936   0.6014280
Residuals                     35  3084.33    88.12
```

Notice that with just one missing value, each main effect appears in two tables (not one, as above). It is recommended that when there are missing values in a split-plot experiment you use **lmer** or **lme** instead of **aov** to fit the model.

Random effects and nested designs

Mixed-effects models are so called because the explanatory variables are a mixture of fixed effects and random effects:

- fixed effects influence only the *mean* of y;
- random effects influence only the *variance* of y.

A random effect should be thought of as coming from a population of effects: the existence of this population is an extra assumption. We speak of **prediction** of random effects, rather than estimation: we **estimate** fixed effects from data, but we intend to make predictions about the population from which our random effects were sampled. Fixed effects are unknown constants to be estimated from the data. Random effects govern the variance – covariance structure of the response variable. The fixed effects are often experimental treatments that were applied under our direction, and the random effects are either categorical or continuous variables that are distinguished by the fact that we are typically not interested in the parameter values, but only in the variance they explain.

One of more of the explanatory variables represents **grouping** in time or in space. Random effects that come from the same group will be correlated, and this contravenes one of the fundamental assumptions of standard statistical models: **independence of errors**. Mixed-effects models take care of this non-independence of errors by modelling the covariance structure introduced by the grouping of the data. A major benefit of random-effects models is that they economize on the number of degrees of freedom used up by the factor levels. Instead of estimating a mean for every single factor level, the random-effects model estimates the distribution of the means (usually as the standard deviation of the differences of the factor-level means around an overall mean). Mixed-effects models are particularly useful in cases where there is temporal pseudoreplication (repeated measurements) and/or spatial pseudoreplication (e.g. nested designs or split-plot experiments). These models can allow for

- spatial autocorrelation between neighbours;

- temporal autocorrelation across repeated measures on the same individuals;

- differences in the mean response between blocks in a field experiment;

- differences between subjects in a medical trial involving repeated measures.

The point is that we really do not want to waste precious degrees of freedom in estimating parameters for each of the separate levels of the categorical random variables. On the other hand, we do want to make use of the all measurements we have taken, but because of the pseudoreplication we want to take account of both the

- correlation structure, used to model within-group correlation associated with temporal and spatial dependencies, using **correlation**, and

- variance function, used to model non-constant variance in the within-group errors using **weights**.

Fixed or random effects?

It is difficult without lots of experience to know when to use categorical explanatory variables as fixed effects and when as random effects. Some guidelines are given below.

- Am I interested in the effect sizes? Yes means fixed effects.

- Is it reasonable to suppose that the factor levels come from a population of levels? Yes means random effects.

- Are there enough levels of the factor in the dataframe on which to base an estimate of the variance of the population of effects? No means fixed effects.

- Are the factor levels informative? Yes means fixed effects.

- Are the factor levels just numeric labels? Yes means random effects.

- Am I mostly interested in making inferences about the distribution of effects, based on the random sample of effects represented in the dataframe? Yes means random effects.

- Is there hierarchical structure? Yes means you need to ask whether the data are experimental or observations.

- Is it a hierarchical experiment, where the factor levels are experimental manipulations? Yes means fixed effects in a split-plot design (see p. 469)

- Is it a hierarchical observational study? Yes means random effects, perhaps in a variance components analysis (see p. 475).

- When your model contains both fixed and random effects, use mixed-effects models.

- If your model structure is linear, use linear mixed effects, lmer.

- Otherwise, specify the model equation and use non-linear mixed effects, nlme.

Removing the pseudoreplication

The extreme response to pseudoreplication in a data set is simply to eliminate it. Spatial pseudoreplication can be averaged away and temporal pseudoreplication can be dealt with by carrying out carrying out separate ANOVAs, one at each time. This approach has two major weaknesses:

- It cannot address questions about treatment effects that relate to the longitudinal development of the mean response profiles (e.g. differences in growth rates between successive times).

- Inferences made with each of the separate analyses are not independent, and it is not always clear how they should be combined.

Analysis of longitudinal data

The key feature of longitudinal data is that the same individuals are measured repeatedly through time. This would represent temporal pseudoreplication if the data were used uncritically in regression or ANOVA. The set of observations on one individual subject will tend to be positively correlated, and this correlation needs to be taken into account in carrying out the analysis. The alternative is a cross-sectional study, with all the data gathered at a single point in time, in which each individual contributes a single data point. The advantage of longitudinal studies is that they are capable of separating *age effects* from *cohort effects*; these are inextricably confounded in cross-sectional studies. This is particularly important when differences between years mean that cohorts originating at different times experience different conditions, so that individuals of the same age in different cohorts would be expected to differ.

There are two extreme cases in longitudinal studies:

- a few measurements on a large number of individuals;

- a large number of measurements on a few individuals.

In the first case it is difficult to fit an accurate model for change within individuals, but treatment effects are likely to be tested effectively. In the second case, it is possible to get an accurate model of the way that individuals change though time, but there is less power for testing the significance of treatment effects, especially if variation from individual to individual is large. In the first case, less attention will be paid to estimating the correlation structure, while in the second case the covariance model will be the principal focus of attention. The aims are:

- to estimate the average time course of a process;

- to characterize the degree of heterogeneity from individual to individual in the rate of the process;

- to identify the factors associated with both of these, including possible cohort effects.

The response is not the individual measurement, but the *sequence of measurements* on an individual subject. This enables us to distinguish between age effects and year effects; see Diggle *et al.* (1994) for details.

Derived variable analysis

The idea here is to get rid of the pseudoreplication by reducing the repeated measures into a set of summary statistics (slopes, intercepts or means), then *analyse these summary statistics* using standard parametric techniques such as ANOVA or regression. The technique is weak when the values of the explanatory variables change through time. Derived variable analysis makes most sense when it is based on the parameters of scientifically interpretable non-linear models from each time sequence. However, the best model from a theoretical perspective may not be the best model from the statistical point of view.

There are three qualitatively different sources of random variation:

- **random effects**, where experimental units differ (e.g. genotype, history, size, physiological condition) so that there are intrinsically high responders and other low responders;

- **serial correlation**, where there may be time-varying stochastic variation within a unit (e.g. market forces, physiology, ecological succession, immunity) so that correlation depends on the time separation of pairs of measurements on the same individual, with correlation weakening with the passage of time;

- **measurement error**, where the assay technique may introduce an element of correlation (e.g. shared bioassay of closely spaced samples; different assay of later specimens).

Variance components analysis

For random effects we are often more interested in the question of how much of the variation in the response variable can be attributed to a given factor, than we are in estimating means or assessing the significance of differences between means. This procedure is called variance components analysis.

The following classic example of pseudoreplication comes from Snedecor Cochran (1980):

```
rats<-read.table("c:\\temp\\rats.txt",header=T)
attach(rats)
names(rats)
```

```
[1] "Glycogen" "Treatment" "Rat" "Liver"
```

Three experimental treatments were administered to rats, and the glycogen content of the rats' livers was analysed as the response variable. There were two rats per treatment, so the total sample was $n = 3 \times 2 = 6$. The tricky bit was that after each rat was killed, its liver was cut up into three pieces: a left-hand bit, a central bit and a right-hand bit. So now there are six rats each producing three bits of liver, for a total of $6 \times 3 = 18$ numbers. Finally, two separate preparations were made from each macerated bit of liver, to assess the measurement error associated with the analytical machinery. At this point there are $2 \times 18 = 36$ numbers in the data frame as a whole. The factor levels are numbers, so we need to declare the explanatory variables to be categorical before we begin:

```
Treatment<-factor(Treatment)
Rat<-factor(Rat)
Liver<-factor(Liver)
```

Here is the analysis done the *wrong* way:

```
model<-aov(Glycogen~Treatment)
summary(model)
```

	Df	Sum Sq	Mean Sq	F value	Pr(>F)	
Treatment	2	1557.56	778.78	14.498	3.031e−05	***
Residuals	33	1772.67	53.72			

Treatment has a highly significant effect on liver glycogen content ($p = 0.00003$). This is wrong! We have committed a classic error of pseudoreplication. Look at the error line in the ANOVA table: it says the residuals have 33 degrees of freedom. But there were only 6 rats in the whole experiment, so the error d.f. has to be $6 - 1 - 2 = 3$ (not 33)!

Here is the analysis of variance done properly, averaging away the pseudoreplication:

```
tt<-as.numeric(Treatment)
yv<-tapply(Glycogen,list(Treatment,Rat),mean)
```

```
tv<-tapply(tt,list(Treatment,Rat),mean)
model<-aov(as.vector(yv)~factor(as.vector(tv)))
summary(model)
```

	Df	Sum Sq	Mean Sq	F value	Pr(>F)
factor(as.vector(tv))	2	259.593	129.796	2.929	0.1971
Residuals	3	132.944	44.315		

Now the error degrees of freedom are correct (d.f. $= 3$, not 33), and the interpretation is completely different: there is no significant differences in liver glycogen under the three experimental treatments ($p = 0.1971$).

There are two different ways of doing the analysis properly in R: ANOVA with multiple error terms (aov) or linear mixed-effects models (lmer). The problem is that the bits of the same liver are pseudoreplicates because they are spatially correlated (they come from the same rat); they are not independent, as required if they are to be true replicates. Likewise, the two preparations from each liver bit are very highly correlated (the livers were macerated before the preparations were taken, so they are essentially the same sample (certainly not independent replicates of the experimental treatments).

Here is the correct analysis using aov with multiple error terms. In the Error term we start with the largest scale (treatment), then rats within treatments, then liver bits within rats within treatments. Finally, there were replicated measurements (two preparations) made for each bit of liver.

```
model2<-aov(Glycogen~Treatment+Error(Treatment/Rat/Liver))
summary(model2)
```

```
Error: Treatment
          Df   Sum Sq  Mean Sq
Treatment  2  1557.56   778.78

Error: Treatment:Rat
          Df  Sum Sq  Mean Sq  F value  Pr(>F)
Residuals  3  797.67   265.89

Error: Treatment:Rat:Liver
          Df  Sum Sq  Mean Sq  F value  Pr(>F)
Residuals 12   594.0     49.5

Error: Within
          Df  Sum Sq  Mean Sq  F value  Pr(>F)
Residuals 18  381.00    21.17
```

You can do the correct, non-pseudoreplicated analysis of variance from this output (Box 11.2).

Box 11.2   Sums of squares in hierarchical designs

The trick to understanding these sums of squares is to appreciate that with nested categorical explanatory variables (random effects) the correction factor, which is subtracted from the sum of squared subtotals, is *not* the conventional $(\sum y)^2 / kn$. Instead, the correction factor is the uncorrected sum of squared subtotals from the level in the hierarchy immediately above the level in question. This is very hard to see without lots of practice. The total sum of squares, SSY, and the treatment sum of squares, SSA, are computed in the usual way (see Box 11.1):

$$SSY = \sum y^2 - \frac{(\sum y)^2}{n},$$

$$SSA = \frac{\sum_{i=1}^{k} C_i^2}{n} - \frac{(\sum y)^2}{kn}.$$

The analysis is easiest to understand in the context of an example. For the rats data, the treatment totals were based on 12 numbers (two rats, three liver bits per rat and two preparations per liver bit). In this case, in the formula for SSA, above, $n = 12$ and $kn = 36$. We need to calculate sums of squares for rats within treatments, SS_{Rats}, liver bits within rats within treatments, $SS_{Liverbits}$, and preparations within liver bits within rats within treatments, $SS_{Preparations}$:

$$SS_{Rats} = \frac{\sum R^2}{6} - \frac{\sum C^2}{12},$$

$$SS_{Liverbits} = \frac{\sum L^2}{2} - \frac{\sum R^2}{6},$$

$$SS_{Preparations} = \frac{\sum y^2}{1} - \frac{\sum L^2}{2}.$$

The correction factor at any level is the *uncorrected sum of squares from the level above*. The last sum of squares could have been computed by difference:

$$SS_{Preparations} = SSY - SSA - SS_{Rats} - SS_{Liverbits}.$$

The F test for equality of the treatment means is the treatment variance divided by the 'rats within treatment variance' from the row immediately beneath: $F = 778.78/265.89 = 2.928\,956$, with 2 d.f. in the numerator and 3 d.f. in the denominator (as we obtained in the correct ANOVA, above).

To turn this into a variance components analysis we need to do a little work. The mean squares are converted into variance components like this:

Residuals = preparations within liver bits: unchanged = 21.17,

Liver bits within rats within treatments: $(49.5 - 21.17)/2 = 14.165$,

Rats within treatments: $(265.89 - 49.5)/6 = 36.065$.

You divide the difference in variance by the number of numbers in the level below (i.e. two preparations per liver bit, and six preparations per rat, in this case).

Analysis of the rats data using lmer is explained on p. 648.

What is the difference between split-plot and hierarchical samples?

Split-plot experiments have informative factor levels. Hierarchical samples have uninformative factor levels. That's the distinction. In the irrigation experiment, the factor levels were as follows:

levels(density)

[1] "high" "low" "medium"

levels(fertilizer)

[1] "N" "NP" "P"

They show the density of seed sown, and the kind of fertilizer applied: they are informative. Here are the factor levels from the rats experiment:

levels(Rat)

[1] "1" "2"

levels(Liver)

```
[1]  "1"  "2"  "3"
```

These factor levels are uninformative, because rat number 2 in treatment 1 has nothing in common with rat number 2 in treatment 2, or with rat number 2 in treatment 3. Liver bit number 3 from rat 1 has nothing in common with liver bit number 3 from rat 2. Note, however, that numbered factor levels are *not* always uninformative: treatment levels 1, 2, and 3 are informative: 1 is the control, 2 is a diet supplement, and 3 is a combination of two supplements.

When the factor levels are informative, the variable is known as a *fixed effect*. When the factor levels are uninformative, the variable is known as a *random effect*. Generally, we are interested in fixed effects as they influence the mean, and in random effects as they influence the variance. We tend not to speak of effect sizes attributable to random effects, but effect sizes and their standard errors are often the principal focus when we have fixed effects. Thus, irrigation, density and fertilizer are fixed effects, and rat and liver bit are random effects.

ANOVA with aov or lm

The difference between lm and aov is mainly in the form of the output: the summary table with aov is in the traditional form for analysis of variance, with one row for each categorical variable and each interaction term. On the other hand, the summary table for lm produces one row per estimated parameter (i.e. one row for each factor level and one row for each interaction level). If you have multiple error terms then you must use aov because lm does not support the Error term. Here is the same two-way analysis of variance fitted using aov first then using lm:

```
daphnia<-read.table("c:\\temp\\Daphnia.txt",header=T)
attach(daphnia)
names(daphnia)
```

```
[1]  "Growth.rate"  "Water"  "Detergent"  "Daphnia"
```

```
model1<-aov(Growth.rate~Water*Detergent*Daphnia)
summary(model1)
```

	Df	Sum Sq	Mean Sq	F value	Pr(>F)	
Water	1	1.985	1.985	2.8504	0.0978380	.
Detergent	3	2.212	0.737	1.0586	0.3754783	
Daphnia	2	39.178	19.589	28.1283	8.228e-09	***
Water:Detergent	3	0.175	0.058	0.0837	0.9686075	
Water:Daphnia	2	13.732	6.866	9.8591	0.0002587	***
Detergent:Daphnia	6	20.601	3.433	4.9302	0.0005323	***
Water:Detergent:Daphnia	6	5.848	0.975	1.3995	0.2343235	
Residuals	48	33.428	0.696			

```
model2<-lm(Growth.rate~Water*Detergent*Daphnia)
summary(model2)
```

```
Coefficients:
```

	Estimate	Std. Error	t value	Pr (>|t|)
(Intercept)	2.81126	0.48181	5.835	4.48e-07
WaterWear	-0.15808	0.68138	-0.232	0.81753
DetergentBrandB	-0.03536	0.68138	-0.052	0.95883
DetergentBrandC	0.47626	0.68138	0.699	0.48794
DetergentBrandD	-0.21407	0.68138	-0.314	0.75475
DaphniaClone2	0.49637	0.68138	0.728	0.46986
DaphniaClone3	2.05526	0.68138	3.016	0.00408
WaterWear:DetergentBrandB	0.46455	0.96361	0.482	0.63193
WaterWear:DetergentBrandC	-0.27431	0.96361	-0.285	0.77712
WaterWear:DetergentBrandD	0.21729	0.96361	0.225	0.82255
WaterWear:DaphniaClone2	1.38081	0.96361	1.433	0.15835
WaterWear:DaphniaClone3	0.43156	0.96361	0.448	0.65627
DetergentBrandB:DaphniaClone2	0.91892	0.96361	0.954	0.34506
DetergentBrandC:DaphniaClone2	-0.16337	0.96361	-0.170	0.86609
DetergentBrandD:DaphniaClone2	1.01209	0.96361	1.050	0.29884
DetergentBrandB:DaphniaClone3	-0.06490	0.96361	-0.067	0.94658
DetergentBrandC:DaphniaClone3	-0.80789	0.96361	-0.838	0.40597
DetergentBrandD:DaphniaClone3	-1.28669	0.96361	-1.335	0.18809
WaterWear:DetergentBrandB:DaphniaClone2	-1.26380	1.36275	-0.927	0.35837
WaterWear:DetergentBrandC:DaphniaClone2	1.35612	1.36275	0.995	0.32466
WaterWear:DetergentBrandD:DaphniaClone2	0.77616	1.36275	0.570	0.57164
WaterWear:DetergentBrandB:DaphniaClone3	-0.87443	1.36275	-0.642	0.52414
WaterWear:DetergentBrandC:DaphniaClone3	-1.03019	1.36275	-0.756	0.45337
WaterWear:DetergentBrandD:DaphniaClone3	-1.55400	1.36275	-1.140	0.25980

```
Residual standard error: 0.8345 on 48 degrees of freedom
Multiple R-Squared: 0.7147, Adjusted R-squared: 0.578
F-statistic: 5.227 on 23 and 48 DF, p-value: 7.019e-07
```

Note that two significant interactions, Water–Daphnia and Detergent–Daphnia, show up in the aov table but not in the lm summary (this is often due to the fact that the lm summary shows treatment contrasts rather than Helmert contrasts). This draws attention to the importance of model simplification rather than per-row t tests (i.e. removing the non-significant three-way interaction term in this case). In the aov table, the p values are 'on deletion' p values, which is a big advantage.

The main difference is that there are eight rows in the aov summary table (three main effects, three two-way interactions, one three-way interaction and an error term) but there are 24 rows in the lm summary table (four levels of detergent by three levels of daphnia clone by two levels of water). You can easily view the output of model1 in linear model layout, or model2 as an ANOVA table using the summary options .lm or .aov:

summary.lm(model1)
summary.aov(model2)

Effect Sizes

In complicated designed experiments, it is easiest to summarize the effect sizes with the model.tables function. This takes the name of the fitted model object as its first argument, and you can specify whether you want the standard errors (as you typically would):

model.tables(model1, "means", se = TRUE)

Tables of means
Grand mean

3.851905

  Water

Water
 Tyne   Wear
3.686   4.018

  Detergent

Detergent
BrandA   BrandB   BrandC   BrandD
 3.885   4.010   3.955   3.558

  Daphnia

Daphnia
Clone1   Clone2   Clone3
 2.840   4.577   4.139

Water:Detergent

        Detergent
Water   BrandA   BrandB   BrandC   BrandD
 Tyne   3.662   3.911   3.814   3.356
 Wear   4.108   4.109   4.095   3.760

Water:Daphnia

        Daphnia
Water   Clone1   Clone2   Clone3
 Tyne   2.868   3.806   4.383
 Wear   2.812   5.348   3.894

Detergent:Daphnia

Daphnia
Detergent   Clone1   Clone2   Clone3
   BrandA   2.732   3.919   5.003
   BrandB   2.929   4.403   4.698
   BrandC   3.071   4.773   4.019
   BrandD   2.627   5.214   2.834

Water:Detergent:Daphnia
, , Daphnia = Clone1

Detergent
Water   BrandA   BrandB   BrandC   BrandD
 Tyne   2.811   2.776   3.288   2.597
 Wear   2.653   3.082   2.855   2.656

, , Daphnia = Clone2

Detergent
Water   BrandA   BrandB   BrandC   BrandD
 Tyne   3.308   4.191   3.621   4.106
 Wear   4.530   4.615   5.925   6.322

, , Daphnia = Clone3

```
Detergent
Water     BrandA     BrandB     BrandC     BrandD
  Tyne     4.867      4.766      4.535      3.366
  Wear     5.140      4.630      3.504      2.303
```

```
Standard errors for differences of means
              Water    Detergent    Daphnia    Water:Detergent    Water:Daphnia
            0.1967       0.2782     0.2409               0.3934           0.3407
replic.        36           18         24                    9               12
            Detergent:Daphnia    Water:Detergent:Daphnia
                       0.4818                     0.6814
replic.                     6                          3
```

Note that the standard errors are standard errors of differences, and they are different in each of the different strata because the replication differs. All standard errors use the same pooled error variance $s^2 = 0.696$ (see above). For instance, the three-way interactions have $se = \sqrt{2 \times 0.696/3} = 0.681$ and the daphnia main effects have $se = \sqrt{2 \times 0.696/24} = 0.2409$.

Attractive plots of effect sizes can be obtained using the **effects** library (p. 178).

Replications

The **replications** function allows you to check the number of replicates at each level in an experimental design:

replications(Growth.rate~Daphnia*Water*Detergent,daphnia)

```
                              Daphnia                    Water
Detergent
                                   24                       36
18
                        Daphnia:Water    Daphnia:Detergent
Water:Detergent
                                   12                        6
9
Daphnia:Water:Detergent
                                    3
```

There are three replicates for the three-way interaction and for all of the two-way interactions (you need to remember the number of levels for each factor to see this: there are two water types, three daphnia clones and four detergents (see above)).

Multiple Comparisons

When comparing the means for the levels of a factor in an analysis of variance, a simple comparison using multiple t tests will inflate the probability of declaring a significant difference when there is none. This because the intervals are calculated with a given coverage probability for each *interval* but the interpretation of the coverage is usually with respect to *the entire family of intervals* (i.e. for the factor as a whole).

If you follow the protocol of model simplification recommended in this book, then issues of multiple comparisons will not arise very often. An occasional significant t test amongst a bunch of non-significant interaction terms is not likely to survive a deletion test (see p. 325). Again, if you have factors with large numbers of levels you might consider using mixed-effects models rather than ANOVA (i.e. treating the factors as random effects rather than fixed effects; see p. 627).

John Tukey introduced intervals based on the range of the sample means rather than the individual differences; nowadays, these are called Tukey's honest significant differences. The intervals returned by the TukeyHSD function are based on Studentized range statistics. Technically the intervals constructed in this way would only apply to balanced designs where the same number of observations is made at each level of the factor. This function incorporates an adjustment for sample size that produces sensible intervals for mildly unbalanced designs.

The following example concerns the yield of fungi gathered from 16 different habitats:

```
data<-read.table("c:\\temp\\Fungi.txt",header=T)
attach(data)
names(data)
```

First we establish whether there is any variation in fungus yield to explain:

```
model<-aov(Fungus.yield~Habitat)
summary(model)
```

```
            Df   Sum Sq   Mean Sq   F value      Pr(>F)
Habitat     15   7527.4    501.8    72.141   < 2.2e-16  ***
Residuals  144   1001.7      7.0
```

Yes, there is ($p < 0.000001$). But this is not of much real interest, because it just shows that some habitats produce more fungi than others. We are likely to be interested in *which* habitats produce significantly more fungi than others. Multiple comparisons are an issue because there are 16 habitats and so there are $(16 \times 15)/2 = 120$ possible pairwise comparisons. There are two options:

- apply the function TukeyHSD to the model to get Tukey's honest significant differences;

- use the function pairwise.t.test to get adjusted p values for all comparisons.

Here is Tukey's test in action: it produces a table of p values by default:

```
TukeyHSD(model)
```

```
Tukey multiple comparisons of means
  95% family-wise confidence level

Fit: aov(formula = Fungus.yield ~ Habitat)
$Habitat
                     diff         lwr         upr       p adj
Ash-Alder       3.53292777  -0.5808096   7.6466651   0.1844088
Aspen-Alder    12.78574402   8.6720067  16.8994814   0.0000000
Beech-Alder    12.32365349   8.2099161  16.4373908   0.0000000
Birch-Alder    14.11348150   9.9997441  18.2272189   0.0000000
Cherry-Alder   10.29508769   6.1813503  14.4088250   0.0000000
Chestnut-Alder 12.24107899   8.1273416  16.3548163   0.0000000
Holmoak-Alder  -1.44360558  -5.5573429   2.6701318   0.9975654
Hornbeam-Alder 10.60271044   6.4889731  14.7164478   0.0000000
Lime-Alder     19.19458205  15.0808447  23.3083194   0.0000000
Oak-Alder      20.29457340  16.1808360  24.4083108   0.0000000
Pine-Alder     14.34084715  10.2271098  18.4545845   0.0000000
```

Rowan-Alder	6.29495226	2.1812149	10.4086896	0.0000410
Spruce-Alder	−2.15119456	−6.2649319	1.9625428	0.9036592
Sycamore-Alder	2.80900108	−1.3047363	6.9227384	0.5644643
...				
Spruce-Rowan	−8.44614681	−12.5598842	−4.3324095	0.0000000
Sycamore-Rowan	−3.48595118	−7.5996885	0.6277862	0.2019434
Willow-Rowan	−3.51860059	−7.6323379	0.5951368	0.1896363
Sycamore-Spruce	4.96019563	0.8464583	9.0739330	0.0044944
Willow-Spruce	4.92754623	0.8138089	9.0412836	0.0049788
Willow-Sycamore	−0.03264941	−4.1463868	4.0810879	1.0000000

You can plot the confidence intervals if you prefer (or do both, of course):

plot(TukeyHSD(model))

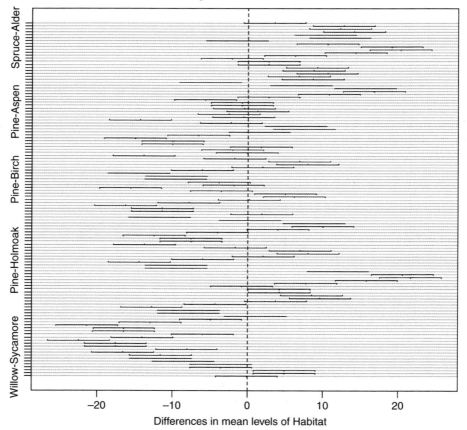

95% family-wise confidence level

Differences in mean levels of Habitat

Habitats on opposite sides of the dotted line and not overlapping it are significantly different from one another.

Alternatively, you can use the **pairwise.t.test** function in which you specify the response variable, and then the categorical explanatory variable containing the factor levels you want to be compared, separated by a comma (not a tilde):

pairwise.t.test(Fungus.yield,Habitat)

```
        Pairwise comparisons using t tests with pooled SD
data: Fungus.yield and Habitat
          Alder     Ash        Aspen     Beech     Birch     Cherry    Chestnut  Holmoak
Ash       0.10011   -          -         -         -         -         -         -
Aspen     < 2e-16   6.3e-11    -         -         -         -         -         -
Beech     < 2e-16   5.4e-10    1.00000   -         -         -         -         -
Birch     < 2e-16   1.2e-13    1.00000   1.00000   -         -         -         -
Cherry    4.7e-13   2.9e-06    0.87474   1.00000   0.04943   -         -         -
Chestnut  < 2e-16   7.8e-10    1.00000   1.00000   1.00000   1.00000   -         -
Holmoak   1.00000   0.00181    < 2e-16   < 2e-16   < 2e-16   3.9e-16   < 2e-16   -
Hornbeam  1.1e-13   8.6e-07    1.00000   1.00000   0.10057   1.00000   1.00000   < 2e-16
Lime      < 2e-16   < 2e-16    1.1e-05   1.9e-06   0.00131   3.3e-10   1.4e-06   < 2e-16
Oak       < 2e-16   < 2e-16    1.4e-07   2.0e-08   2.7e-05   1.9e-12   1.5e-08   < 2e-16
Pine      < 2e-16   3.9e-14    1.00000   1.00000   1.00000   0.02757   1.00000   < 2e-16
Rowan     1.8e-05   0.51826    8.5e-06   4.7e-05   3.9e-08   0.03053   6.2e-05   5.3e-08
Spruce    1.00000   0.00016    < 2e-16   < 2e-16   < 2e-16   < 2e-16   < 2e-16   1.00000
Sycamore  0.50084   1.00000    2.1e-12   1.9e-11   3.3e-15   1.5e-07   2.7e-11   0.01586
Willow    0.51826   1.00000    1.9e-12   1.6e-11   2.8e-15   1.4e-07   2.4e-11   0.01702
          Hornbeam  Lime       Oak       Pine      Rowan     Spruce    Sycamore
Ash       -         -          -         -         -         -         -
Aspen     -         -          -         -         -         -         -
Beech     -         -          -         -         -         -         -
Birch     -         -          -         -         -         -         -
Cherry    -         -          -         -         -         -         -
Chestnut  -         -          -         -         -         -         -
Holmoak   -         -          -         -         -         -         -
Hornbeam  -         -          -         -         -         -         -
Lime      1.3e-09   -          -         -         -         -         -
Oak       8.4e-12   1.00000    -         -         -         -         -
Pine      0.05975   0.00253    6.1e-05   -         -         -         -
Rowan     0.01380   < 2e-16    < 2e-16   1.5e-08   -         -         -
Spruce    < 2e-16   < 2e-16    < 2e-16   < 2e-16   2.5e-09   -         -
Sycamore  4.2e-08   < 2e-16    < 2e-16   1.1e-15   0.10218   0.00187   -
Willow    3.8e-08   < 2e-16    < 2e-16   9.3e-16   0.10057   0.00203   1.00000

P value adjustment method: holm
```

As you see, the default method of adjustment of the p-values is holm, but other adjustment methods include hochberg, hommel, bonferroni, BH, BY, fdr and none. Without adjustment of the p values, the rowan–willow comparison looks highly significant ($p = 0.003\,35$), as you can see if you try

pairwise.t.test(Fungus.yield,Habitat,p.adjust.method="none")

I like TukeyHSD because it is conservative without being ridiculously so (in contrast to Bonferroni). For instance, Tukey gives the birch–cherry comparison as non-significant ($p = 0.101\,102\,7$) while Holm makes this difference significant ($p = 0.049\,43$). Tukey had Willow-Holm Oak as significant ($p = 0.038\,091\,0$), whereas Bonferroni throws this baby out with the bathwater ($p = 0.056\,72$). You need to decide how circumspect you want to be in the context of your particular question.

There is a useful package for multiple comparisons called multcomp:

install.packages("multcomp")

You can see at once how contentious the issue of multiple comparisons is, just by looking at the length of the list of different multiple comparisons methods supported in this package

- the many-to-one comparisons of Dunnett
- the all-pairwise comparisons of Tukey
- Sequen
- AVE
- changepoint
- Williams
- Marcus
- McDermott
- Tetrade
- Bonferroni correction
- Holm
- Hochberg
- Hommel
- Benjamini–Hochberg
- Benjamini–Yekutieli

The old-fashioned Bonferroni correction is highly conservative, because the p values are multiplied by the number of comparisons. Instead of using the usual Bonferroni and Holm procedures, the adjustment methods include less conservative corrections that take the exact correlations between the test statistics into account by use of the multivariate t-distribution. The resulting procedures are therefore substantially more powerful (the Bonferroni and Holm adjusted p values are reported for reference). There seems to be no reason to use the unmodified Bonferroni correction because it is dominated by Holm's method, which is valid under arbitrary assumptions.

The tests are designed to suit multiple comparisons within the general linear model, so they allow for covariates, nested effects, correlated means and missing values. The first four methods are designed to give strong control of the familywise error rate. The methods of Benjamini, Hochberg, and Yekutieli control the false discovery rate, which is the expected proportion of false discoveries amongst the rejected hypotheses. The false discovery rate is a less stringent condition than the familywise error rate, so these methods are more powerful than the others.

Projections of Models

If you want to see how the different factor levels contribute their additive effects to each of the observed values of the response, then use the proj function like this:

library(help="multcomp")

```
                (Intercept)      Water   Detergent     Daphnia Water:Detergent
Water:Daphnia
1                  3.851905  −0.1660431  0.03292724  −1.0120302      −0.05698158
0.1941404
2                  3.851905  −0.1660431  0.03292724  −1.0120302      −0.05698158
0.1941404
3                  3.851905  −0.1660431  0.03292724  −1.0120302      −0.05698158
0.1941404
...
```

The name proj comes from the fact that the function returns a matrix or list of matrices giving the 'projections of the data onto the terms of a linear model'.

Multivariate Analysis of Variance

Two or more response variables are sometimes measured in the same experiment. Of course you can analyse each response variable separately, and that is the typical way to proceed. But there are occasions where you want to treat the group of response variables as one multivariate response. The function for this is manova, the multivariate analysis of variance. Note that manova does not support multi-stratum analysis of variance, so the formula must not include an Error term.

```
data<-read.table("c:\\temp\\manova.txt",header=T)
attach(data)
names(data)
```

```
[1] "tear" "gloss" "opacity" "rate" "additive"
```

First, create a multivariate response variable, Y, by binding together the three separate response variables (tear, gloss and opacity), like this:

```
Y <- cbind(tear, gloss, opacity)
```

Then fit the multivariate analysis of variance using the manova function:

```
model<-manova(Y~rate*additive)
```

There are two ways to inspect the output. First, as a multivariate analysis of variance:

```
summary(model)
```

	Df	Pillai	approx F	num Df	den Df	Pr(>F)	
rate	1	0.6181	7.5543	3	14	0.003034	**
additive	1	0.4770	4.2556	3	14	0.024745	*
rate:additive	1	0.2229	1.3385	3	14	0.301782	
Residuals	16						

This shows significant main effects for both rate and additive, but no interaction. Note that the F tests are based on 3 and 14 degrees of freedom (not 1 and 16). The default method in summary.manova is the Pillai–Bartlett statistic. Other options include Wilks, Hotelling–Lawley and Roy. Second, you will want to look at each of the three response variables separately:

summary.aov(model)

Response tear :

	Df	Sum Sq	Mean Sq	F value	Pr(>F)	
rate	1	1.74050	1.74050	15.7868	0.001092	**
additive	1	0.76050	0.76050	6.8980	0.018330	*
rate:additive	1	0.00050	0.00050	0.0045	0.947143	
Residuals	16	1.76400	0.11025			

Response gloss :

	Df	Sum Sq	Mean Sq	F value	Pr(>F)	
rate	1	1.30050	1.30050	7.9178	0.01248	*
additive	1	0.61250	0.61250	3.7291	0.07139	.
rate:additive	1	0.54450	0.54450	3.3151	0.08740	.
Residuals	16	2.62800	0.16425			

Response opacity :

	Df	Sum Sq	Mean Sq	F value	Pr(>F)
rate	1	0.421	0.421	0.1036	0.7517
additive	1	4.901	4.901	1.2077	0.2881
rate:additive	1	3.961	3.961	0.9760	0.3379
Residuals	16	64.924	4.058		

Notice that one of the three response variables, opacity, is not significantly associated with either of the explanatory variables.

12

Analysis of Covariance

Analysis of covariance (ANCOVA) combines elements from regression and analysis of variance. The response variable is continuous, and there is at least one continuous explanatory variable and at least one categorical explanatory variable. The procedure works like this:

- Fit two or more linear regressions of y against x (one for each level of the factor).
- Estimate different slopes and intercepts for each level.
- Use model simplification (deletion tests) to eliminate unnecessary parameters.

For example, we could use ANCOVA in a medical experiment where the response variable was 'days to recovery' and the explanatory variables were 'smoker or not' (categorical) and 'blood cell count' (continuous). In economics, local unemployment rate might be modelled as a function of country (categorical) and local population size (continuous). Suppose we are modelling weight (the response variable) as a function of sex and age. Sex is a factor with two levels (male and female) and age is a continuous variable. The maximal model therefore has four parameters: two slopes (a slope for males and a slope for females) and two intercepts (one for males and one for females) like this:

$$weight_{male} = a_{male} + b_{male} \times age,$$
$$weight_{female} = a_{female} + b_{female} \times age.$$

This maximal model is shown in the top left-hand panel. Model simplification is an essential part of analysis of covariance, because the principle of parsimony requires that we keep as few parameters in the model as possible.

There are six possible models in this case, and the process of model simplification begins by asking whether we need all four parameters (top left). Perhaps we could make do with 2 intercepts and a common slope (top right), or a common intercept and two different slopes (centre left). There again, age may have no significant effect on the response, so we only need two parameters to describe the main effects of sex on weight; this would show up as two separated, horizontal lines in the plot (one mean weight for each sex; centre right). Alternatively, there may be no effect of sex at all, in which case we only need two parameters (one slope and one intercept) to describe the effect of age on weight (bottom

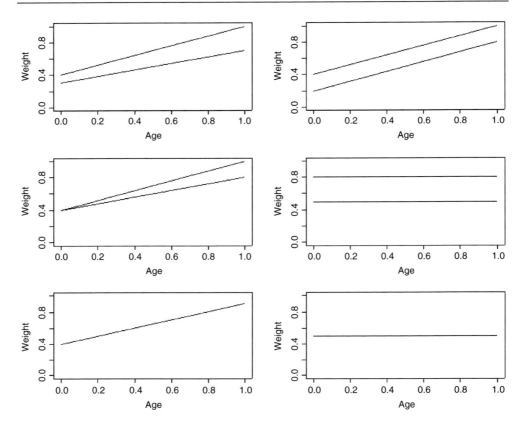

left). In the limit, neither the continuous nor the categorical explanatory variables might have any significant effect on the response, in which case model simplification will lead to the one-parameter null model $\hat{y} = \bar{y}$ (a single horizontal line; bottom right).

Analysis of Covariance in R

We could use either lm or aov; the choice affects only the format of the summary table. We shall use both and compare their output. Our worked example concerns an experiment on the impact of grazing on the seed production of a biennial plant. Forty plants were allocated to two treatments, grazed and ungrazed, and the grazed plants were exposed to rabbits during the first two weeks of stem elongation. They were then protected from subsequent grazing by the erection of a fence and allowed to regrow. Because initial plant size was thought likely to influence fruit production, the diameter of the top of the rootstock was measured before each plant was potted up. At the end of the growing season, the fruit production (dry weight in milligrams) was recorded on each of the 40 plants, and this forms the response variable in the following analysis.

```
regrowth<-read.table("c:\\temp\\ipomopsis.txt",header=T)
attach(regrowth)
names(regrowth)
```

```
[1] "Root" "Fruit" "Grazing"
```

The object of the exercise is to estimate the parameters of the minimal adequate model for these data. We begin by inspecting the data with a plot of fruit production against root size for each of the two treatments separately: the diamonds are ungrazed plants and the triangles are grazed plants:

```
plot(Root,Fruit,
     pch=16+as.numeric(Grazing),col=c("blue","red")[as.numeric(Grazing)])
```

where red diamonds represent the ungrazed plants and blue triangles represent the grazed plants. Note the use of **as.numeric** to select the plotting symbols and colours. How are the grazing treatments reflected in the factor levels?

```
levels(Grazing)
```

```
[1] "Grazed"  "Ungrazed"
```

Now we can use logical subscripts (p. 21) to draw linear regression lines for the two grazing treatments separately, using **abline** (we could have used **subset** instead):

```
abline(lm(Fruit[Grazing=="Grazed"]~Root[Grazing=="Grazed"]),lty=2,col="blue")
abline(lm(Fruit[Grazing=="Ungrazed"]~Root[Grazing=="Ungrazed"]),lty=2,col="red")
```

Note the use of **as.numeric** to select the plotting symbols and colours, and the use of subscripts within the **abline** function to fit linear regression models separately for each level of the grazing treatment (we could have used **subset** instead).

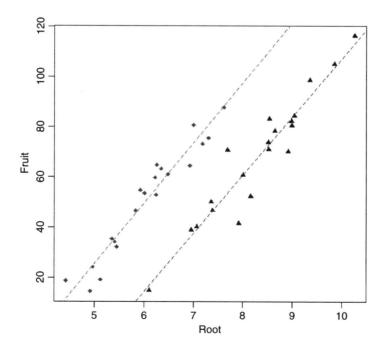

The odd thing about these data is that grazing seems to *increase* fruit production, a highly counter-intuitive result:

tapply(Fruit,Grazing, mean)

```
 Grazed   Ungrazed
67.9405    50.8805
```

This difference is statistically significant ($p = 0.027$) if you do a t test (although this is the *wrong* thing to do in this case, as explained below):

t.test(Fruit~Grazing)

```
       Welch Two Sample t-test
```

```
data: Fruit by Grazing
t = 2.304, df = 37.306, p-value = 0.02689
alternative hypothesis: true difference in means is not equal to 0
95 percent confidence interval:
2.061464  32.058536
sample estimates:
mean in group Grazed mean in group Ungrazed
           67.9405                 50.8805
```

Several important points are immediately apparent from this initial analysis:

- Different sized plants were allocated to the two treatments.
- The grazed plants were bigger at the outset.
- The regression line for the ungrazed plants is above the line for the grazed plants.
- The regression lines are roughly parallel.
- The intercepts (not shown to the left) are likely to be significantly different.

Each of these points will be addressed in detail.

To understand the output from analysis of covariance it is useful to work through the calculations by hand. We start by working out the sums, sums of squares and sums of products for the whole data set combined (40 pairs of numbers), and then for each treatment separately (20 pairs of numbers). We shall fill in a table of totals, because it helps to be really well organized for these calculations. Check to see where (and why) the sums and the sums of squares of the root diameters (the x values) and the fruit yields (the y values) have gone in the table: First, we shall work out the overall totals based on all 40 data points.

sum(Root);sum(Root^2)

```
[1] 287.246
[1] 2148.172
```

sum(Fruit);sum(Fruit^2)

```
[1] 2376.42
[1] 164928.1
```

sum(Root*Fruit)

```
[1] 18263.16
```

These are the famous five, which we shall make use of shortly, and complete the overall data summary. Now we select the root diameters for the grazed and ungrazed plants and then the fruit yields for the grazed and ungrazed plants:

sum(Root[Grazing=="Grazed"]);sum(Root[Grazing=="Grazed"]^2)

```
[1] 166.188
[1] 1400.834
```

sum(Root[Grazing=="Ungrazed"]);sum(Root[Grazing=="Ungrazed"]^2)

```
[1] 121.058
[1] 747.3387
```

sum(Fruit[Grazing=="Grazed"]);sum(Fruit[Grazing=="Grazed"]^2)

```
[1] 1358.81
[1] 104156.0
```

sum(Fruit[Grazing=="Ungrazed"]);sum(Fruit[Grazing=="Ungrazed"]^2)

```
[1] 1017.61
[1] 60772.11
```

Finally, we want the sums of products: first for the grazed plants and then for the ungrazed plants:

sum(Root[Grazing=="Grazed"]*Fruit[Grazing=="Grazed"])

```
[1] 11753.64
```

sum(Root[Grazing=="Ungrazed"]*Fruit[Grazing=="Ungrazed"])

```
[1] 6509.522
```

Here is our table:

	Sums	Squares and products
x ungrazed	121.058	747.3387
y ungrazed	1017.61	60772.11
xy ungrazed		6509.522
x grazed	166.188	1400.834
y grazed	1358.81	104156.0
xy grazed		11753.64
x overall	287.246	2148.172
y overall	2376.42	164928.1
xy overall		18263.16

Now we have all of the information necessary to carry out the calculations of the corrected sums of squares and products, *SSY*, *SSX* and *SSXY*, for the whole data set ($n = 40$) and for the two separate treatments (with 20 replicates in each). To get the right answer you will need to be extremely methodical, but there is nothing mysterious or difficult about the process. First, calculate the regression statistics for the whole experiment, ignoring the grazing treatment, using the famous five which we have just calculated:

$$SSY = 164\,928.1 - \frac{2376.42^2}{40} = 23\,743.84,$$

$$SSX = 2148.172 - \frac{287.246^2}{40} = 85.4158,$$

$$SSXY = 18\,263.16 - \frac{287.246 \times 2376.42}{40} = 1197.731,$$

$$SSR = \frac{1197.731^2}{85.4158} = 16\,795,$$

$$SSE = 23\,743.84 - 16\,795 = 6948.835.$$

The effect of differences between the two grazing treatments, SSA, is

$$SSA = \frac{1358.81^2 + 1017.61^2}{20} - \frac{2376.42^2}{40} = 2910.436.$$

Next calculate the regression statistics for each of the grazing treatments separately. First, for the grazed plants:

$$SSY_g = 104\,156 - \frac{1358.81^2}{20} = 11\,837.79,$$

$$SSX_g = 1400.834 - \frac{166.188^2}{20} = 19.9111,$$

$$SSXY_g = 11\,753.64 - \frac{1358.81 \times 166.188}{20} = 462.7415,$$

$$SSR_g = \frac{462.7415^2}{19.9111} = 10\,754.29,$$

$$SSE_g = 11837.79 - 10754.29 = 1083.509,$$

so the slope of the graph of Fruit against Root for the grazed plants is given by

$$b_g = \frac{SSXY_g}{SSX_g} = \frac{462.7415}{19.9111} = 23.240.$$

Now for the ungrazed plants:

$$SSY_u = 60\,772.11 - \frac{1017.61^2}{20} = 8995.606,$$

$$SSX_u = 747.3387 - \frac{121.058^2}{20} = 14.58677,$$

$$SSXY_u = 6509.522 - \frac{121.058 \times 1017.61}{20} = 350.0302,$$

$$SSR_u = \frac{350.0302^2}{14.58677} = 8399.466,$$

$$SSE_u = 8995.606 - 8399.466 = 596.1403,$$

so the slope of the graph of Fruit against Root for the ungrazed plants is given by

$$b_u = \frac{SSXY_u}{SSX_u} = \frac{350.0302}{14.58677} = 23.996$$

Now add up the regression statistics across the factor levels (grazed and ungrazed):

$$SSY_{g+u} = 11\,837.79 + 8995.606 = 20\,833.4,$$

$$SSX_{g+u} = 19.9111 + 14.58677 = 34.49788,$$

$$SSXY_{g+u} = 462.7415 + 350.0302 = 812.7717,$$

$$SSR_{g+u} = 10\,754.29 + 8399.436 = 19\,153.75,$$

$$SSE_{g+u} = 1083.509 + 596.1403 = 1684.461.$$

The SSR for a model with a single common slope is given by

$$SSR_c = \frac{(SSXY_{g+u})^2}{SSX_{g+u}} = \frac{812.7717^2}{34.49788} = 19\,148.94,$$

and the value of the single common slope is

$$b = \frac{SSXY_{g+u}}{SSX_{g+u}} = \frac{812.7717}{34.49788} = 23.560$$

The difference between the two estimates of SSR ($SSR_{diff} = SSR_{g+u} - SSR_c = 19153.75 - 19148.94 = 4.81$) is a measure of the significance of the difference between the two slopes estimated separately for each factor level. Finally, SSE is calculated by difference:

$$SSE = SSY - SSA - SSR_c - SSR_{diff}$$

$$= 23743.84 - 2910.44 - 19148.94 - 4.81 = 1679.65.$$

Now we can complete the ANOVA table for the full model:

Source	SS	d.f.	MS	F
Grazing	2910.44	1		
Root	19148.94	1		
Different slopes	4.81	1	4.81	n.s.
Error	1679.65	36	46.66	
Total	23743.84	39		

Degrees of freedom for error are $40 - 4 = 36$ because we have estimated four parameters from the data: two slopes and two intercepts. So the error variance is $46.66(= SSE/36)$. The difference between the slopes is clearly not significant ($F = 4.81/46.66 = 0.10$) so we can fit a simpler model with a common slope of 23.56. The sum of squares for differences between the slopes (4.81) now becomes part of the error sum of squares:

Source	SS	d.f.	MS	F
Grazing	2910.44	1	2910.44	63.9291
Root	19148.94	1	19148.94	420.6156
Error	1684.46	37	45.526	
Total	23743.84	39		

This is the minimal adequate model. Both of the terms are highly significant and there are no redundant factor levels.

The next step is to calculate the intercepts for the two parallel regression lines. This is done exactly as before, by rearranging the equation of the straight line to obtain $a = y - bx$. For each line we can use the mean values of x and y, with the common slope in each case. Thus:

$$a_1 = \overline{Y}_1 - b\overline{X}_1 = 50.88 - 23.56 \times 6.0529 = -91.7261,$$

$$a_2 = \overline{Y}_2 - b\overline{X}_2 = 67.94 - 23.56 \times 8.309 = -127.8294.$$

This demonstrates that the grazed plants produce, on average, 36.1 mg of fruit *less* than the ungrazed plants ($127.83 - 91.73$).

Finally, we need to calculate the standard errors for the common regression slope and for the difference in mean fecundity between the treatments, based on the error variance in the minimal adequate model, given in the table above:

$$s^2 = \frac{1684.46}{37} = 45.526$$

The standard errors are obtained as follows. The standard error of the common slope is found in the usual way:

$$se_b = \sqrt{\frac{s^2}{SSX_{g+u}}} = \sqrt{\frac{45.526}{19.9111 + 14.45667}} = 1.149.$$

The standard error of the intercept of the regression for the grazed treatment is also found in the usual way:

$$se_a = \sqrt{s^2 \left[\frac{1}{n} + \frac{(0 - \overline{x})^2}{SSX_{g+u}}\right]} = \sqrt{45.526 \left[\frac{1}{20} + \frac{8.3094^2}{34.498}\right]} = 9.664.$$

It is clear that the intercept of -127.829 is very significantly less than zero ($t = 127.829/9.664 = 13.2$), suggesting that there is a threshold rootstock size before reproduction can begin. Finally, the standard error of the difference between the elevations of the two lines (the grazing effect) is given by

$$se_{\hat{y}_u - \hat{y}_g} = \sqrt{s^2 \left[\frac{2}{n} + \frac{(\overline{x}_1 - \overline{x}_2)^2}{SSX_{g+u}}\right]}$$

which, substituting the values for the error variance and the mean rootstock sizes of the plants in the two treatments, becomes:

$$se_{\hat{y}_u - \hat{y}_g} = \sqrt{45.526 \left[\frac{2}{20} + \frac{(6.0529 - 8.3094)^2}{34.498} \right]} = 3.357.$$

This suggests that any lines differing in elevation by more than about $2 \times 3.357 = 6.66\,\text{mg}$ dry weight would be regarded as significantly different. Thus, the present difference of 36.09 clearly represents a highly significant reduction in fecundity caused by grazing ($t = 10.83$).

The hand calculations were convoluted, but ANCOVA is exceptionally straightforward in R using lm. The response variable is fecundity, and there is one experimental factor (Grazing) with two levels (Ungrazed and Grazed) and one covariate (initial rootstock diameter). There are 40 values for each of these variables. As we saw earlier, the largest plants were allocated to the grazed treatments, but for a *given* rootstock diameter (say, 7 mm) the scatterplot shows that the grazed plants produced *fewer* fruits than the ungrazed plants (not more, as a simple comparison of the means suggested). This is an excellent example of the value of analysis of covariance. Here, the correct analysis using ANCOVA completely reverses our interpretation of the data.

The analysis proceeds in the following way. We fit the most complicated model first, then simplify it by removing non-significant terms until we are left with a minimal adequate model, in which all the parameters are significantly different from zero. For ANCOVA, the most complicated model has different slopes and intercepts for each level of the factor. Here we have a two-level factor (Grazed and Ungrazed) and we are fitting a linear model with two parameters ($y = a + bx$) so the most complicated mode has four parameters (two slopes and two intercepts). To fit different slopes and intercepts we use the asterisk * notation:

ancova <- lm(Fruit~Grazing*Root)

You should realize that *order matters*: we would get a different output if the model had been written Fruit ~ Root * Grazing (more of this on p. 507).

summary(ancova)

```
Coefficients:
                     Estimate  Std. Error  t value  Pr (>|t|)
(Intercept)          -125.173      12.811   -9.771   1.15e-11  ***
GrazingUngrazed        30.806      16.842    1.829   0.0757    .
Root                   23.240       1.531   15.182   < 2e-16   ***
GrazingUngrazed:Root    0.756       2.354    0.321   0.7500
```

This shows that initial root size has a massive effect on fruit production ($t = 15.182$), but there is no indication of any difference in the slope of this relationship between the two grazing treatments (this is the Grazing by Root interaction with $t = 0.321$, $p \gg 0.05$). The ANOVA table for the maximal model looks like this:

anova(ancova)

```
Analysis of Variance Table
```

Response: Fruit

	Df	Sum Sq	Mean Sq	F value	Pr(>F)
Grazing	1	2910.4	2910.4	62.3795	2.262e-09 ***
Root	1	19148.9	19148.9	410.4201	< 2.2e-16 ***
Grazing:Root	1	4.8	4.8	0.1031	0.75
Residuals	36	1679.6	46.7		

The next step is to delete the non-significant interaction term from the model. We can do this manually or automatically: here we shall do both for the purposes of demonstration. The function for manual model simplification is update. We update the current model (here called ancova) by deleting terms from it. The syntax is important: the punctuation reads 'comma tilde dot minus'. We define a new name for the simplified model:

ancova2<-update(ancova, ~ . - Grazing:Root)

Now we compare the simplified model with just three parameters (one slope and two intercepts) with the maximal model using anova like this:

anova(ancova,ancova2)

```
Analysis of Variance Table

Model 1:  Fruit ~ Grazing * Root
Model 2:  Fruit ~ Grazing + Root
   Res.Df       RSS  Df   Sum of Sq        F    Pr (>F)
1       36  1679.65
2       37  1684.46  -1       -4.81   0.1031     0.75
```

This says that model simplification was justified because it caused a negligible reduction in the explanatory power of the model ($p = 0.75$; to retain the interaction term in the model we would need $p < 0.05$).

The next step in model simplification involves testing whether or not grazing had a significant effect on fruit production once we control for initial root size. The procedure is similar: we define a new model and use update to remove Grazing from ancova2 like this:

ancova3<-update(ancova2, ~ . - Grazing)

Now we compare the two models using anova:

anova(ancova2,ancova3)

```
Analysis of Variance Table

Model 1:  Fruit ~ Grazing + Root
Model 2:  Fruit ~ Root
   Res.Df     RSS  Df   Sum of Sq        F       Pr(>F)
1       37  1684.5
2       38  6948.8  -1     -5264.4   115.63   6.107e-13   ***
```

This model simplification is a step too far. Removing the Grazing term causes a massive reduction in the explanatory power of the model, with an F value of 115.63 and a vanishingly small p value. The effect of grazing in reducing fruit production is highly significant and needs to be retained in the model. Thus ancova2 is our minimal adequate model, and we should look at its summary table to compare with our earlier calculations carried out by hand:

summary(ancova2)

```
Coefficients:
                  Estimate   Std. Error   t value   Pr (>|t|)
(Intercept)       -127.829        9.664    -13.23   1.35e-15 ***
GrazingUngrazed     36.103        3.357     10.75   6.11e-13 ***
Root                23.560        1.149     20.51   < 2e-16 ***
```

```
Residual standard error: 6.747 on 37 degrees of freedom
Multiple R-Squared: 0.9291, Adjusted R-squared: 0.9252
F-statistic: 242.3 on 2 and 37 DF, p-value: < 2.2e-16
```

You know when you have got the minimal adequate model, because every row of the coefficients table has one or more significance stars (three in this case, because the effects are all so strong). In contrast to our initial interpretation based on mean fruit production, grazing is associated with a 36.103 mg *reduction* in fruit production.

anova(ancova2)

```
Analysis of Variance Table

Response: Fruit
           Df   Sum Sq  Mean Sq   F value     Pr(>F)
Grazing     1   2910.4   2910.4    63.929  1.397e-09  ***
Root        1  19148.9  19148.9   420.616  < 2.2e-16  ***
Residuals  37   1684.5     45.5
```

These are the values we obtained the long way on p. 495.

Now we repeat the model simplification using the automatic model-simplification function called step. It couldn't be easier to use. The full model is called ancova:

step(ancova)

This function causes all the terms to be tested to see whether they are needed in the minimal adequate model. The criterion used is AIC, Akaike's information criterion (p. 353). In the jargon, this is a 'penalized log-likelihood'. What this means in simple terms is that it weighs up the inevitable trade-off between degrees of freedom and fit of the model. You can have a perfect fit if you have a parameter for every data point, but this model has zero explanatory power. Thus **deviance goes down as degrees of freedom in the model go up**. The AIC adds 2 times the number of parameters in the model to the deviance (to penalize it). Deviance, you will recall, is twice the log-likelihood of the current model. Anyway, AIC is a measure of lack of fit; big AIC is bad, small AIC is good. The full model (four parameters: two slopes and two intercepts) is fitted first, and AIC calculated as 157.5:

```
Start: AIC = 157.5
Fruit ~ Grazing * Root

                 Df  Sum of Sq      RSS     AIC
- Grazing: Root   1       4.81  1684.46  155.61
<none>                          1679.65  157.50

Step: AIC = 155.61
Fruit ~ Grazing + Root

           Df  Sum of Sq      RSS    AIC
<none>                       1684.5  155.6
- Grazing   1     5264.4   6948.8  210.3
- Root      1    19148.9  20833.4  254.2
```

```
Call:
lm(formula = Fruit ~ Grazing + Root)

Coefficients:
(Intercept)    GrazingUngrazed         Root
    -127.83              36.10        23.56
```

Then step tries removing the most complicated term (the Grazing by Root interaction). This reduces AIC to 155.61 (an improvement, so the simplification is justified). No further simplification is possible (as we saw when we used update to remove the Grazing term from the model) because AIC goes up to 210.3 when Grazing is removed and up to 254.2 if Root size is removed. Thus, step has found the minimal adequate model (it doesn't always, as we shall see later; it is good, but not perfect).

ANCOVA and Experimental Design

There is an extremely important general message in this example for experimental design. No matter how carefully we randomize at the outset, our experimental groups are likely to be heterogeneous. Sometimes, as in this case, we may have made initial measurements that we can use as covariates later on, but this will not always be the case. There are bound to be important factors that we did not measure. If we had not measured initial root size in this example, we would have come to entirely the wrong conclusion about the impact of grazing on plant performance.

A far better design for this experiment would have been to measure the rootstock diameters of all the plants at the beginning of the experiment (as was done here), but then to place the plants in matched pairs with rootstocks of similar size. Then, one of the plants would be picked at random and allocated to one of the two grazing treatments (e.g. by tossing a coin); the other plant of the pair then receives the unallocated gazing treatment. Under this scheme, the size ranges of the two treatments would overlap, and the analysis of covariance would be unnecessary.

A More Complex ANCOVA: Two Factors and One Continuous Covariate

The following experiment, with Weight as the response variable, involved Genotype and Sex as two categorical explanatory variables and Age as a continuous covariate. There are six levels of Genotype and two levels of Sex.

```
Gain <-read.table("c:\\temp\\Gain.txt",header=T)
attach(Gain)
names(Gain)
```

```
[1] "Weight"  "Sex"  "Age"  "Genotype"  "Score"
```

We begin by fitting the maximal model with its 24 parameters: different slopes and intercepts for every combination of Sex and Genotype.

```
m1<-lm(Weight~Sex*Age*Genotype)
summary(m1)
```

Coefficients:

	Estimate	Std. Error	t value	Pr (>|t|)	
(Intercept)	7.80053	0.24941	31.276	< 2e-16	***
Sexmale	-0.51966	0.35272	-1.473	0.14936	
Age	0.34950	0.07520	4.648	4.39e-05	***
GenotypeCloneB	1.19870	0.35272	3.398	0.00167	**
GenotypeCloneC	-0.41751	0.35272	-1.184	0.24429	
GenotypeCloneD	0.95600	0.35272	2.710	0.01023	*
GenotypeCloneE	-0.81604	0.35272	-2.314	0.02651	*
GenotypeCloneF	1.66851	0.35272	4.730	3.41e-05	***
Sexmale:Age	-0.11283	0.10635	-1.061	0.29579	
Sexmale:GenotypeCloneB	-0.31716	0.49882	-0.636	0.52891	
Sexmale:GenotypeCloneC	-1.06234	0.49882	-2.130	0.04010	*
Sexmale:GenotypeCloneD	-0.73547	0.49882	-1.474	0.14906	
Sexmale:GenotypeCloneE	-0.28533	0.49882	-0.572	0.57087	
Sexmale:GenotypeCloneF	-0.19839	0.49882	-0.398	0.69319	
Age:GenotypeCloneB	-0.10146	0.10635	-0.954	0.34643	
Age:GenotypeCloneC	-0.20825	0.10635	-1.958	0.05799	.
Age:GenotypeCloneD	-0.01757	0.10635	-0.165	0.86970	
Age:GenotypeCloneE	-0.03825	0.10635	-0.360	0.72123	
Age:GenotypeCloneF	-0.05512	0.10635	-0.518	0.60743	
Sexmale:Age:GenotypeCloneB	0.15469	0.15040	1.029	0.31055	
Sexmale:Age:GenotypeCloneC	0.35322	0.15040	2.349	0.02446	*
Sexmale:Age:GenotypeCloneD	0.19227	0.15040	1.278	0.20929	
Sexmale:Age:GenotypeCloneE	0.13203	0.15040	0.878	0.38585	
Sexmale:Age:GenotypeCloneF	0.08709	0.15040	0.579	0.56616	

Residual standard error: 0.2378 on 36 degrees of freedom
Multiple R-Squared: 0.9742, Adjusted R-squared: 0.9577
F-statistic: 59.06 on 23 and 36 DF, p-value: < 2.2e-16

There are one or two significant parameters, but it is not at all clear that the three-way or two-way interactions need to be retained in the model. As a first pass, let's use step to see how far it gets with model simplification:

m2<-step(m1)

Start: AIC= -155.01

Weight ~ Sex * Age * Genotype

	Df	Sum of Sq	RSS	AIC
- Sex:Age:Genotype	5	0.349	2.385	-155.511
<none>			2.036	-155.007

Step: AIC= -155.51
Weight ~ Sex + Age + Genotype + Sex:Age + Sex:Genotype +
Age:Genotype

	Df	Sum of Sq	RSS	AIC
- Sex:Genotype	5	0.147	2.532	-161.924
- Age:Genotype	5	0.168	2.553	-161.423
- Sex:Age	1	0.049	2.434	-156.292
<none>			2.385	-155.511

```
Step: AIC= -161.92
Weight ~ Sex + Age + Genotype + Sex:Age + Age:Genotype
                 Df    Sum of Sq    RSS         AIC
- Age:Genotype    5       0.168    2.700    -168.066
- Sex:Age         1       0.049    2.581    -162.776
<none>                             2.532    -161.924
Step: AIC= -168.07
Weight ~ Sex + Age + Genotype + Sex:Age
                 Df    Sum of Sq    RSS         AIC
- Sex:Age         1       0.049    2.749    -168.989
<none>                             2.700    -168.066
- Genotype        5      54.958   57.658       5.612
Step: AIC= -168.99
Weight ~ Sex + Age + Genotype
                 Df    Sum of Sq    RSS         AIC
<none>                             2.749    -168.989
- Sex             1      10.374   13.122     -77.201
- Age             1      10.770   13.519     -75.415
- Genotype        5      54.958   57.707       3.662
Call:
lm(formula = Weight ~ Sex + Age + Genotype)

Coefficients:
   (Intercept)    Sexmale      Age    GenotypeCloneB
GenotypeCloneC
        7.9370   -0.8316    0.2996           0.9678
    -1.0436
GenotypeCloneD   GenotypeCloneE   GenotypeCloneF
        0.8240          -0.8754           1.5346
```

We definitely do not need the three-way interaction, despite the effect of Sexmale:Age:GenotypeCloneC which gave a significant t test on its own. How about the three 2-way interactions? The step function leaves out Sex by Genotype and then assesses the other two. No need for Age by Genotype. Try removing Sex by Age. Nothing. What about the main effects? They are all highly significant. This is R's idea of the minimal adequate model: three main effects but no interactions. That is to say, the slope of the graph of weight gain against age does not vary with sex or genotype, but the intercepts *do* vary. It would be a good idea to look at the summary.lm table for this model:

summary(m2)

```
Coefficients:
                  Estimate    Std. Error    t value    Pr(>|t|)
(Intercept)        7.93701       0.10066     78.851    < 2e-16  ***
Sexmale           -0.83161       0.05937    -14.008    < 2e-16  ***
Age                0.29958       0.02099     14.273    < 2e-16  ***
GenotypeCloneB     0.96778       0.10282      9.412   8.07e-13  ***
GenotypeCloneC    -1.04361       0.10282    -10.149   6.21e-14  ***
GenotypeCloneD     0.82396       0.10282      8.013   1.21e-10  ***
GenotypeCloneE    -0.87540       0.10282     -8.514   1.98e-11  ***
GenotypeCloneF     1.53460       0.10282     14.925    < 2e-16  ***
```

```
Residual standard error: 0.2299 on 52 degrees of freedom
Multiple R-Squared: 0.9651, Adjusted R-squared: 0.9604
F-statistic: 205.7 on 7 and 52 DF, p-value: < 2.2e-16
```

This is where Helmert contrasts would actually come in handy (see p. 378). Everything is three-star significantly different from Genotype[1] Sex[1], but it is not obvious that the intercepts for genotypes B and D need different values (+0.96 and +0.82 above genotype A with $se_{diff} = 0.1028$), nor is it obvious that C and E have different intercepts (-1.043 and -0.875). Perhaps we could reduce the number of factor levels of Genotype from the present six to four without any loss of explanatory power ?

We create a new categorical variable called newGenotype with separate levels for clones A and F, and for B and D combined and C and E combined.

newGenotype<-Genotype
levels(newGenotype)

```
[1] "CloneA"  "CloneB"  "CloneC"  "CloneD"  "CloneE"  "CloneF"
```

levels(newGenotype)[c(3,5)]<-"ClonesCandE"
levels(newGenotype)[c(2,4)]<-"ClonesBandD"
levels(newGenotype)

```
[1] "CloneA"  "ClonesBandD"  "ClonesCandE"  "CloneF"
```

Then we redo the modelling with newGenotype (4 levels) instead of Genotype (6 levels):

m3<-lm(Weight~Sex+Age+newGenotype)

and check that the simplification was justified

anova(m2,m3)

```
Analysis of Variance Table

Model 1:  Weight ~ Sex + Age + Genotype
Model 2:  Weight ~ Sex + Age + newGenotype
   Res.Df       RSS   Df   Sum of Sq       F    Pr(>F)
1      52   2.74890
2      54   2.99379   -2    -0.24489   2.3163   0.1087
```

Yes, it was. The p value was 0.1087 so we accept the simpler model m3:

```
Coefficients:
                            Estimate  Std. Error   t value  Pr(>|t|)
(Intercept)                  7.93701     0.10308    76.996   < 2e-16 ***
Sexmale                     -0.83161     0.06080   -13.679   < 2e-16 ***
Age                          0.29958     0.02149    13.938   < 2e-16 ***
newGenotypeClonesBandD       0.89587     0.09119     9.824  1.28e-13 ***
newGenotypeClonesCandE      -0.95950     0.09119   -10.522  1.10e-14 ***
newGenotypeCloneF            1.53460     0.10530    14.574   < 2e-16 ***

Residual standard error: 0.2355 on 54 degrees of freedom
Multiple R-Squared: 0.962, Adjusted R-squared: 0.9585
F-statistic: 273.7 on 5 and 54 DF, p-value: < 2.2e-16
```

After an analysis of covariance, it is useful to draw the fitted lines through a scatterplot, with each factor level represented by different plotting symbols and line types (see p. 167):

```
plot(Age,Weight,col=as.numeric(newGenotype),pch=(15+as.numeric(Sex)))

xv<-c(1,5)

for (i in 1:2) {
for (j in 1:4){

a<-coef(m3)[1]+(i>1)* coef(m3)[2]+(j>1)*coef(m3)[j+2];b<-coef(m3)[3]
yv<-a+b*xv
lines(xv,yv,lty=2)
}}
```

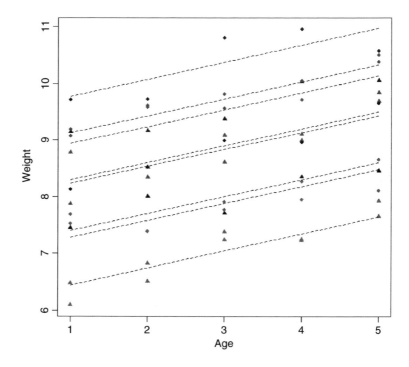

Note the use of colour to represent the four genotypes col = as.numeric(newGenotype) and plotting symbols to represent the two sexes pch=(15+as.numeric(Sex)). You can see that the males (circles) are heavier than the females (triangles) in all of the genotypes. Other functions to be considered in plotting the results of ANCOVA are split and augPred in lattice graphics.

Contrasts and the Parameters of ANCOVA Models

In analysis of covariance, we estimate a slope and an intercept for each level of one or more factors. Suppose we are modelling weight (the response variable) as a function of sex and age, as illustrated on p. 490. The difficulty arises because there are several different ways of expressing the values of the four parameters in the summary.lm table:

- two slopes, and two intercepts (as in the equations on p. 490);

- one slope and one difference between slopes, and one intercept and one difference between intercepts;

- the overall mean slope and the overall mean intercept, and one difference between slopes and one difference between intercepts.

In the second case (two estimates and two differences) a decision needs to be made about which factor level to associate with the estimate, and which level with the difference (e.g. should males be expressed as the intercept and females as the difference between intercepts, or vice versa)? When the factor levels are unordered (the typical case), then R takes the factor level that comes first in the alphabet as the estimate and the others are expressed as differences. In our example, the parameter estimates would be female, and male parameters would be expressed as differences from the female values, because 'f' comes before 'm' in the alphabet. This should become clear from an example:

Ancovacontrasts <-read.table("c:\\temp\\Ancovacontrasts.txt",header=T)
attach(Ancovacontrasts)
names(Ancovacontrasts)

```
[1] "weight"  "sex"  "age"
```

First we work out the two regressions separately so that we know the values of the two slopes and the two intercepts:

lm(weight[sex=="male"]~age[sex=="male"])

```
Coefficients:
(Intercept)    age[sex == "male"]
      3.115                 1.561
```

lm(weight[sex=="female"]~age[sex=="female"])

```
Coefficients:
(Intercept)    age[sex == "female"]
     1.9663                   0.9962
```

So the intercept for males is 3.115 and the intercept for females is 1.966. The difference between the first (female) and second intercepts (male) is therefore

$$3.115 - 1.9266 = +1.1884.$$

Now we can do an overall regression, ignoring gender:

lm(weight~age)

```
Coefficients:
(Intercept)       age
      2.541     1.279
```

This tells us that the average intercept is 2.541 and the average slope is 1.279.

Next we can carry out an analysis of covariance and compare the output produced by each of the three different contrast options allowed by R: treatment (the default in R and in Glim), Helmert (the default in S-PLUS), and sum. First, the analysis using **treatment contrasts** as used by R and by Glim:

```
options(contrasts=c("contr.treatment", "contr.poly"))
model1<-lm(weight~age*sex)
summary(model1)
```

```
Coefficients:
              Estimate   Std. Error   t value   Pr(>|t|)
(Intercept)     1.9663       0.6268     3.137    0.00636  ***
age             0.9962       0.1010     9.862    3.33e-08 ***
sexmale         1.1489       0.8864     1.296    0.21331
age:sexmale     0.5646       0.1429     3.952    0.00114  ***
```

The intercept (1.9663) is the intercept for females (because f comes before m in the alphabet). The age parameter (0.9962) is the slope of the graph of weight against age for females. The sex parameter (1.1489) is the difference between the (female) intercept and the male intercept ($1.9663 + 1.1489 = 3.1152$). The age – sex interaction term is the difference between slopes of the female and male graphs ($0.9962 + 0.5646 = 1.5608$). So with treatment contrasts, the parameters (in order 1 to 4) are an intercept, a slope, a difference between two intercepts, and a difference between two slopes. In the standard error column we see, from row 1 downwards, the standard error of an intercept for a regression with females only (0.6268 with $n = 10$, $\sum x^2 = 385$ and $SSX = 82.5$), the standard error of a slope for females only (0.1010, with $SSX = 82.5$), the standard error of the difference between two intercepts each based on $n = 10$ data points ($\sqrt{2 \times 0.6268^2} = 0.8864$) and the standard error of the difference between two slopes each based on $n = 10$ data points ($\sqrt{2 \times 0.1010^2} = 0.1429$). The formulas for these standard errors are on p. 496. Many people are more comfortable with this method of presentation than they are with Helmert or sum contrasts.

We now turn to the analysis using Helmert contrasts:

```
options(contrasts=c("contr.helmert", "contr.poly"))
model2<-lm(weight~age*sex)
summary(model2)
```

```
Coefficients:
              Estimate   Std. Error   t value   Pr(>|t|)
(Intercept)    2.54073      0.44319     5.733    3.08e-05 ***
age            1.27851      0.07143    17.899    5.26e-12 ***
sex1           0.57445      0.44319     1.296    0.21331
age:sex1       0.28230      0.07143     3.952    0.00114  ***
```

Let's see if we can work out what the four parameter values represent. The first parameter, 2.540 73 (labelled Intercept), is the intercept of the overall regression, ignoring sex (see above). The parameter labelled age (1.278 51) is a *slope* because age is our continuous explanatory variable. Again, you will see that it is the slope for the regression of weight against age, ignoring sex. The third parameter, labelled sex (0.574 45), must have something to do with intercepts because sex is our categorical variable. If we want to reconstruct the second intercept (for males) we need to add 0.5744 to the overall intercept: $2.540\,73 + 0.574\,45 = 3.115\,18$. To get the intercept for females we need to subtract it: $2.540\,73 - 0.574\,45 = 1.966\,28$. The fourth parameter (0.282 30), labelled age:sex, is the difference between the overall mean slope (1.279) and the male slope: $1.278\,51 + 0.282\,30 = 1.560\,81$. To get the slope of weight against age for females we need to subtract the interaction term from the age term: $1.278\,51 - 0.282\,30 = 0.996\,21$.

In the standard errors column, from the top row downwards, you see the standard error of an intercept based on a regression with all 20 points (the overall regression, ignoring sex, $0.443\,19$) and the standard error of a slope based on a regression with all 20 points ($0.071\,43$). The standard errors of differences (both intercept and slope) involve *half* the difference between the male and female values, because with Helmert contrasts the difference is between the male value and the overall value, rather than between the male and female values. Thus the third row has the standard error of a difference between the overall intercept and the intercept for males based on a regression with 10 points ($0.443\,19 = 0.8864/2$), and the bottom row has the standard error of a difference between the overall slope and the slope for males, based on a regression with 10 points ($0.1429/2 = 0.071\,43$). Thus the values in the bottom two rows of the Helmert table are simply half the values in the same rows of the treatment table.

The advantage of Helmert contrasts is in hypothesis testing in more complicated models than this, because it is easy to see which terms we need to retain in a simplified model by inspecting their significance levels in the summary.lm table. The disadvantage is that it is much harder to reconstruct the slopes and the intercepts from the estimated parameters values (see also p. 378).

Finally, we look at the third option which is sum contrasts:

```
options(contrasts=c("contr.sum", "contr.poly"))
model3<-lm(weight~age*sex)
summary(model3)
```

```
Coefficients:
              Estimate  Std. Error  t value  Pr(>|t|)
(Intercept)    2.54073    0.44319    5.733   3.08e-05   ***
age            1.27851    0.07143   17.899   5.26e-12   ***
sex1          -0.57445    0.44319   -1.296   0.21331
age:sex1      -0.28230    0.07143   -3.952   0.00114    ***
```

The first two estimates are the same as those produced by Helmert contrasts: the overall intercept and slope of the graph relating weight to age, ignoring sex. The sex parameter ($-0.574\,45$) is *sign reversed* compared with the Helmert option: it shows how to calculate the female (the *first*) intercept from the overall intercept $2.540\,73 - 0.574\,45 = 1.966\,28$. The interaction term also has reversed sign: to get the slope for females, add the interaction term to the slope for age: $1.278\,51 - 0.282\,30 = 0.996\,21$.

The four standard errors for the sum contrasts are exactly the same as those for Helmert contrasts (explained above).

Order matters in summary.aov

People are often disconcerted by the ANOVA table produced by summary.aov in analysis of covariance. Compare the tables produced for these two models:

```
summary.aov(lm(weight~sex*age))
```

	Df	Sum Sq	Mean Sq	F value	Pr(>F)	
sex	1	90.492	90.492	107.498	1.657e-08	***
age	1	269.705	269.705	320.389	5.257e-12	***
sex:age	1	13.150	13.150	15.621	0.001141	***
Residuals	16	13.469	0.842			

summary.aov(lm(weight~age*sex))

	Df	Sum Sq	Mean Sq	F value	Pr(>F)	
age	1	269.705	269.705	320.389	5.257e-12	***
sex	1	90.492	90.492	107.498	1.657e-08	***
age:sex	1	13.150	13.150	15.621	0.001141	***
Residuals	16	13.469	0.842			

Exactly the same sums of squares and p values. No problem. But look at these two models from the plant compensation example analysed in detail earlier (p. 490):

summary.aov(lm(Fruit~Grazing*Root))

	Df	Sum Sq	Mean Sq	F value	Pr(>F)	
Grazing	1	2910.4	2910.4	62.3795	2.262e-09	***
Root	1	19148.9	19148.9	410.4201	<2.2e-16	***
Grazing:Root	1	4.8	4.8	0.1031	0.75	
Residuals	36	1679.6	46.7			

summary.aov(lm(Fruit~Root*Grazing))

	Df	Sum Sq	Mean Sq	F value	Pr(>F)	
Root	1	16795.0	16795.0	359.9681	< 2.2e-16	***
Grazing	1	5264.4	5264.4	112.8316	1.209e-12	***
Root:Grazing	1	4.8	4.8	0.1031	0.75	
Residuals	36	1679.6	46.7			

In this case the order of variables within the model formula has a huge effect: it changes the sum of squares associated with the two main effects (root size is continuous and grazing is categorical, grazed or ungrazed) and alters their p values. The interaction term, the residual sum of squares and the error variance are unchanged. So what is the difference between the two cases?

In the first example, where order was irrelevant, the x values for the continuous variable (age) were identical for both sexes (there is one male and one female value at each of the ten experimentally controlled ages):

table(sex,age)

		age								
sex	1	2	3	4	5	6	7	8	9	10
female	1	1	1	1	1	1	1	1	1	1
male	1	1	1	1	1	1	1	1	1	1

In the second example, the x values (root size) were different in the two treatments, and mean root size was greater for the grazed plants than for the ungrazed ones:

tapply(Root,Grazing, mean)

```
Grazed   Ungrazed
8.3094    6.0529
```

Whenever the x values are different in different factor levels, and/or there is different replication in different factor levels, then SSX and $SSXY$ will vary from level to level and this will affect the way the sum of squares is distributed across the main effects. It is of no consequence in terms of your interpretation of the model, however, because the effect sizes and standard errors in the summary.lm table are unaffected:

summary(lm(Fruit~Root*Grazing))

```
Coefficients:
                    Estimate  Std. Error  t value  Pr(>|t|)
(Intercept)         -125.173      12.811   -9.771  1.15e-11  ***
Root                  23.240       1.531   15.182   < 2e-16  ***
GrazingUngrazed       30.806      16.842    1.829    0.0757  .
Root:GrazingUngrazed   0.756       2.354    0.321    0.7500
```

summary(lm(Fruit~Grazing*Root))

```
Coefficients:
                    Estimate  Std. Error  t value  Pr(>|t|)
(Intercept)         -125.173      12.811   -9.771  1.15e-11  ***
GrazingUngrazed       30.806      16.842    1.829    0.0757  .
Root                  23.240       1.531   15.182   < 2e-16  ***
GrazingUngrazed:Root   0.756       2.354    0.321    0.7500
```

13

Generalized Linear Models

We can use generalized linear models (GLMs) pronounced 'glims' – when the variance is not constant, and/or when the errors are not normally distributed. Certain kinds of response variables invariably suffer from these two important contraventions of the standard assumptions, and GLMs are excellent at dealing with them. Specifically, we might consider using GLMs when the response variable is:

- count data expressed as proportions (e.g. logistic regressions);
- count data that are not proportions (e.g. log-linear models of counts);

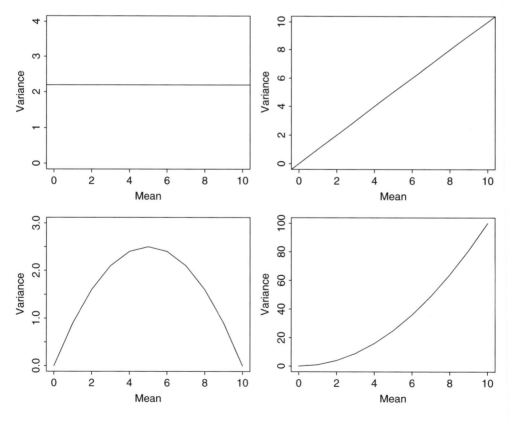

- binary response variables (e.g. dead or alive);
- data on time to death where the variance increases faster than linearly with the mean (e.g. time data with gamma errors).

The central assumption that we have made up to this point is that variance was constant (top left-hand graph). In count data, however, where the response variable is an integer and there are often lots of zeros in the dataframe, the variance may increase linearly with the mean (top tight). With proportion data, where we have a count of the number of failures of an event as well as the number of successes, the variance will be an inverted U-shaped function of the mean (bottom left). Where the response variable follows a gamma distribution (as in time-to-death data) the variance increases faster than linearly with the mean (bottom right). Many of the basic statistical methods such as regression and Student's t test assume that variance is constant, but in many applications this assumption is untenable. Hence the great utility of GLMs.

A generalized linear model has three important properties:

- the error structure;
- the linear predictor;
- the link function.

These are all likely to be unfamiliar concepts. The ideas behind them are straightforward, however, and it is worth learning what each of the concepts involves.

Error Structure

Up to this point, we have dealt with the statistical analysis of data with normal errors. In practice, however, many kinds of data have non-normal errors: for example:

- errors that are strongly skewed;
- errors that are kurtotic;
- errors that are strictly bounded (as in proportions);
- errors that cannot lead to negative fitted values (as in counts).

In the past, the only tools available to deal with these problems were transformation of the response variable or the adoption of non-parametric methods. A GLM allows the specification of a variety of different error distributions:

- Poisson errors, useful with count data;
- binomial errors, useful with data on proportions;
- gamma errors, useful with data showing a constant coefficient of variation;
- exponential errors, useful with data on time to death (survival analysis).

The **error structure** is defined by means of the family directive, used as part of the model formula. Examples are

glm(y ~ z, family = poisson)

which means that the response variable y has Poisson errors, and

glm(y ~ z, family = binomial)

which means that the response is binary, and the model has binomial errors. As with previous models, the explanatory variable z can be continuous (leading to a regression analysis) or categorical (leading to an ANOVA-like procedure called analysis of deviance, as described below).

Linear Predictor

The structure of the model relates each observed y value to a predicted value. The predicted value is obtained by transformation of the value emerging from the **linear predictor**. The linear predictor, η (eta), is a linear sum of the effects of one or more explanatory variables, x_j:

$$\eta_i = \sum_{j=1}^{p} x_{ij} \beta_j$$

where the xs are the values of the p different explanatory variables, and the βs are the (usually) unknown parameters to be estimated from the data. The right-hand side of the equation is called the **linear structure**.

There are as many terms in the linear predictor as there are parameters, p, to be estimated from the data. Thus, with a simple regression, the linear predictor is the sum of two terms whose parameters are the intercept and the slope. With a one-way ANOVA with four treatments, the linear predictor is the sum of four terms leading to the estimation of the mean for each level of the factor. If there are covariates in the model, they add one term each to the linear predictor (the slope of each relationship). Interaction terms in a factorial ANOVA add one or more parameters to the linear predictor, depending upon the degrees of freedom of each factor (e.g. there would be three extra parameters for the interaction between a two-level factor and a four-level factor, because $(2-1) \times (4-1) = 3$).

To determine the fit of a given model, a GLM evaluates the linear predictor for each value of the response variable, then compares the predicted value with a *transformed* value of y. The transformation to be employed is specified in the link function, as explained below. The fitted value is computed by applying the reciprocal of the link function, in order to get back to the original scale of measurement of the response variable.

Link Function

One of the difficult things to grasp about GLMs is the relationship between the values of the response variable (as measured in the data and predicted by the model in fitted values) and the linear predictor. The thing to remember is that the **link function** relates the mean value of y to its linear predictor. In symbols, this means that

$$\eta = g(\mu),$$

which is simple, but needs thinking about. The linear predictor, η, emerges from the linear model as a sum of the terms for each of the p parameters. This is not a value of y (except in the special case of the **identity link** that we have been using (implicitly) up to now). The value of η is obtained by transforming the value of y by the link function, and the predicted value of y is obtained by applying the inverse link function to η.

The most frequently used link functions are shown below. An important criterion in the choice of link function is to ensure that the fitted values stay within reasonable bounds. We would want to ensure, for example, that counts were all greater than or equal to 0 (negative count data would be nonsense). Similarly, if the response variable was the proportion of individuals that died, then the fitted values would have to lie between 0 and 1 (fitted values greater than 1 or less than 0 would be meaningless). In the first case, a log link is appropriate because the fitted values are antilogs of the linear predictor, and all antilogs are greater than or equal to 0. In the second case, the logit link is appropriate because the fitted values are calculated as the antilogs of the log odds, $\log(p/q)$.

By using different link functions, the performance of a variety of models can be compared directly. The total deviance is the same in each case, and we can investigate the consequences of altering our assumptions about precisely how a given change in the linear predictor brings about a response in the fitted value of y. The most appropriate link function is the one which produces the minimum residual deviance.

Canonical Link Functions

The canonical link functions are the default options employed when a particular error structure is specified in the family directive in the model formula. Omission of a link directive means that the following settings are used:

Error	Canonical link
normal	*identity*
poisson	*log*
binomial	*logit*
Gamma	*reciprocal*

You should try to memorize these canonical links and to understand why each is appropriate to its associated error distribution. Note that only gamma errors have a capital initial letter in R.

Choosing between using a link function (e.g. log link) and transforming the response variable (i.e. having $\log(y)$ as the response variable rather than y) takes a certain amount of experience. The decision is usually based on *whether the variance is constant* on the original scale of measurement. If the variance was constant, you would use a link function. If the variance increased with the mean, you would be more likely to log transform the response.

Proportion Data and Binomial Errors

Proportion data have three important properties that affect the way the data should be analysed:

- the data are strictly bounded;

- the variance is non-constant;

- errors are non-normal.

You cannot have a proportion greater than 1 or less than 0. This has obvious implications for the kinds of functions fitted and for the distributions of residuals around these fitted functions. For example, it makes no sense to have a linear model with a negative slope for proportion data because there would come a point, with high levels of the x variable, that negative proportions would be predicted. Likewise, it makes no sense to have a linear model with a positive slope for proportion data because there would come a point, with high levels of the x variable, that proportions greater than 1 would be predicted.

With proportion data, if the probability of success is 0, then there will be no successes in repeated trials, all the data will be zeros and hence the variance will be zero. Likewise, if the probability of success is 1, then there will be as many successes as there are trials, and again the variance will be 0. For proportion data, therefore, the variance increases with the mean up to a maximum (when the probability of success is one half) then declines again towards zero as the mean approaches 1. The variance–mean relationship is humped, rather than constant as assumed in the classical tests.

The final assumption is that the errors (the differences between the data and the fitted values estimated by the model) are normally distributed. This cannot be so in proportional data because the data are bounded above and below: no matter how big a negative residual might be at high predicted values, $\hat{y}$, a positive residual cannot be bigger than $1 - \hat{y}$. Similarly, no matter how big a positive residual might be for low predicted values $\hat{y}$, a negative residual cannot be greater than $\hat{y}$ (because you cannot have negative proportions). This means that confidence intervals must be asymmetric whenever $\hat{y}$ takes large values (close to 1) or small values (close to 0).

All these issues (boundedness, non-constant variance, non-normal errors) are dealt with by using a generalized linear model with a binomial error structure. It could not be simpler to deal with this. Instead of using a linear model and writing

lm(y~x)

we use a generalized linear model and specify that the error family is binomial like this:

glm(y~x,family=binomial)

That's all there is to it. In fact, it is even easier than that, because we don't need to write family=:

glm(y~x,binomial)

Count Data and Poisson Errors

Count data have a number of properties that need to be considered during modelling:

- count data are bounded below (you cannot have counts less than zero);

- variance is not constant (variance increases with the mean);

- errors are not normally distributed;

- the fact that the data are whole numbers (integers) affects the error distribution.

It is very simple to deal with all these issues by using a GLM. All we need to write is

glm(y~x,poisson)

and the model is fitted with a log link (to ensure that the fitted values are bounded below) and Poisson errors (to account for the non-normality).

Deviance: Measuring the Goodness of Fit of a GLM

The fitted values produced by the model are most unlikely to match the values of the data perfectly. The size of the discrepancy between the model and the data is a measure of the inadequacy of the model; a small discrepancy may be tolerable, but a large one will not be. The measure of discrepancy in a GLM to assess the goodness of fit of the model to the data is called the **deviance**. Deviance is defined as −2 times the difference in log-likelihood between the current model and a saturated model (i.e. a model that fits the data perfectly). Because the latter does not depend on the parameters of the model, minimizing the deviance is the same as maximizing the likelihood.

Deviance is estimated in different ways for different families within glm (Table 13.1). Numerical examples of the calculation of deviance for different glm families are given in Chapter 14 (Poisson errors), Chapter 15 (binomial errors), and Chapter 24 (gamma errors). Where there is grouping structure in the data, leading to spatial or temporal pseudoreplication, you will want to use generalized mixed models (lmer) with one of these error families (p. 590).

Table 13.1. Deviance formulae for different GLM families where y is observed data, $\bar{y}$ the mean value of y, μ are the fitted values of y from the maximum likelihood model, and n is the binomial denominator in a binomial GLM.

Family (error structure)	Deviance	Variance function
normal	$\sum (y - \bar{y})^2$	1
poisson	$2 \sum y \ln(y/\mu) - (y - \mu)$	μ
binomial	$2 \sum y \ln(y/\mu) + (n - y) \ln(n - y)/(n - \mu)$	$\dfrac{\mu(n - \mu)}{n}$
Gamma	$2 \sum (y - \mu)/y - \ln(y/\mu)$	μ^2
inverse. gaussian	$\sum (y - \mu)^2/(\mu^2 y)$	μ^3

Quasi-likelihood

The precise relationship between the variance and the mean is well established for all the GLM error families (Table 13.1). In some cases, however, we may be uneasy about specifying the precise form of the error distribution. We may know, for example, that it is not normal (e.g. because the variance increases with the mean), but we don't know with any confidence that the underlying distribution is, say, negative binomial.

There is a very simple and robust alternative known as quasi-likelihood, introduced by Wedderburn (1974), which uses only the most elementary information about the response variable, namely the variance–mean relationship (see Taylor's power law, p. 198). It is extraordinary that this information alone is often sufficient to retain close to the full efficiency of maximum likelihood estimators.

Suppose that we know that the response is always positive, the data are invariably skew to the right, and the variance increases with the mean. This does not enable us to specify a particular distribution (e.g. it does not discriminate between Poisson or negative binomial errors), and hence we cannot use techniques like maximum likelihood or likelihood ratio tests. Quasi-likelihood frees us from the need to specify a particular distribution, and requires us only to specify the mean-to-variance relationship up to a proportionality constant, which can be estimated from the data:

$$\text{var}(y_i) \propto \nu(\mu_i).$$

An example of the principle at work compares quasi-likelihood with maximum likelihood in the case of Poisson errors (full details are in McCulloch and Searle, 2001). This means that the maximum quasi-likelihood (MQL) equations for β are

$$\frac{\partial}{\partial \beta} \sum (y_i \log \mu_i - \mu_i) = 0,$$

which is exactly the same as the maximum likelihood equation for the Poisson (see p. 250). In this case, MQL and maximum likelihood give precisely the same estimates, and MQL would therefore be fully efficient. Other cases do not work out quite as elegantly as this, but MQL estimates are generally robust and efficient. Their great virtue is the simplicity of the central premise that $\text{var}(y_i) \propto \nu(\mu_i)$, and the lack of the need to assume a specific distributional form.

If we take the original GLM density function and find the derivative of the log-likelihood with respect to the mean,

$$\frac{\partial l_i}{\partial \mu_i} = \frac{\partial l_i}{\partial \theta_i}\frac{\partial \theta_i}{\partial \mu_i} = \frac{y_i b'(\theta_i)}{a_i(\phi)}\frac{1}{b''(\theta_i)} = \frac{y_i - \mu_i}{\text{var}(y_i)}$$

(where the primes denote differentiation), the quasi-likelihood Q is defined as

$$Q(y, \mu) = \int_y^\mu \frac{y - \mu}{\phi V(\mu)}\,d\mu.$$

Here, the denominator is the variance of y, $\text{var}(y) = \phi V(\mu)$, where ϕ is called the scale parameter (or the dispersion parameter) and $V(\mu)$ is the variance function. We need only specify the two moments (mean μ and variance $\phi V(\mu)$) and maximize Q to find the MQL estimates of the parameters.

The scale parameter is estimated from the generalized Pearson statistic rather than from the residual deviance (as when correcting for overdispersion with Poisson or binomial errors):

$$\hat{\phi} = \frac{\sum_i \{(y_i - \hat{\mu}_i)/V_i(\hat{\mu}_i)\}}{n - p} = \frac{\chi^2}{n - p}.$$

For normally distributed data, the residual sum of squares SSE is chi-squared distributed.

Generalized Additive Models

Generalized additive models (GAMs) are like GLMs in that they can have different error structures and different link functions to deal with count data or proportion data. What makes them different is that the shape of the relationship between y and a continuous variable x is not specified by some explicit functional form. Instead, non-parametric smoothers are used to describe the relationship. This is especially useful for relationships that exhibit complicated shapes, such as hump-shaped curves (see p. 666). The model looks just like a GLM, except that the relationships we want to be smoothed are prefixed by s: thus, if we had a three-variable multiple regression (with three continuous explanatory variables w, x and z) on count data and we wanted to smooth all three explanatory variables, we would write:

```
model<-gam(y~s(w)+s(x)+s(z),poisson)
```

These are hierarchical models, so the inclusion of a high-order interaction (such as $A{:}B{:}C$) necessarily implies the inclusion of all the lower-order terms marginal to it (i.e. $A{:}B$, $A{:}C$ and $B{:}C$, along with main effects for A, B and C).

Because the models are nested, the more complicated model will necessarily explain at least as much of the variation as the simpler model (and usually more). What we want to know is whether the extra parameters in the more complex model are justified in the sense that they add significantly to the models explanatory power. If they do not, then parsimony requires that we accept the simpler model.

Offsets

An **offset** is a component of the linear predictor that is known in advance (typically from theory, or from a mechanistic model of the process) and, because it is known, requires no parameter to be estimated from the data. For linear models with normal errors an offset is redundant, since you can simply subtract the offset from the values of the response variable, and work with the residuals instead of the y values. For GLMs, however, it is necessary to specify the offset; this is held constant while other explanatory variables are evaluated. Here is an example from the timber data.

The background theory is simple. We assume the logs are roughly cylindrical (i.e. that taper is negligible between the bottom and the top of the log). Then volume, v, in relation to girth, g, and height, h, is given by

$$v = \frac{g^2}{4\pi} h.$$

Taking logarithms gives

$$\log(v) = \log\left(\frac{1}{4\pi}\right) + 2\log(g) + \log(h).$$

We would expect, therefore, that if we did a multiple linear regression of $\log(v)$ on $\log(h)$ and $\log(g)$ we would get estimated slopes of 1.0 for $\log(h)$ and 2.0 for $\log(g)$. Let's see what happens:

```
data <-read.delim("c:\\temp\\timber.txt")
attach(data)
names(data)
```

[1] "volume" "girth" "height"

The girths are in centimetres but all the other data are in metres, so we convert the girths to metres at the outset:

girth<-girth/100

Now fit the model:

model1<-glm(log(volume)~log(girth)+log(height))
summary(model1)

```
Coefficients:
             Estimate  Std. Error  t value  Pr (>|t|)
(Intercept)  -2.89938    0.63767    -4.547   9.56e-05  ***
log(girth)    1.98267    0.07503    26.426    < 2e-16  ***
log(height)   1.11714    0.20448     5.463   7.83e-06  ***
```

Residual deviance: 0.18555 on 28 degrees of freedom
AIC: -62.697

The estimates are reasonably close to expectation (1.11714 rather than 1.0 for $\log(h)$ and 1.98267 rather than 2.0 for $\log(g)$).

Now we shall use offset to specify the theoretical response of $\log(v)$ to $\log(h)$; i.e. a slope of 1.0 rather than the estimated 1.11714:

model2<-glm(log(volume)~log(girth)+offset(log(height)))
summary(model2)

```
Coefficients:
             Estimate  Std. Error  t value  Pr (>|t|)
(Intercept)  -2.53419    0.01457    -174.0    <2e-16  ***
log(girth)    2.00545    0.06287      31.9    <2e-16  ***
```

Residual deviance: 0.18772 on 29 degrees of freedom
AIC: -64.336

Naturally the residual deviance is greater, but only by a very small amount. The AIC has gone down from -62.697 to -64.336, so the model simplification was justified.

Let us try including the theoretical slope (2.0) for $\log(g)$ in the offset as well:

model3<-glm(log(volume)~1+offset(log(height)+2*log(girth)))
summary(model3)

```
Coefficients:
             Estimate  Std. Error  t value  Pr(>|t|)
(Intercept)  -2.53403    0.01421    -178.3    <2e-16  ***
```

Residual deviance: 0.18777 on 30 degrees of freedom
AIC: -66.328

Again, the residual deviance is only marginally greater, and AIC is smaller, so the simplification is justified.

What about the intercept? If our theoretical model of cylindrical logs is correct then the intercept should be

log(1/(4*pi))

[1] -2.531024

This is almost exactly the same as the intercept estimated by GLM in model3, so we are justified in putting the entire model in the offset and informing GLM not to estimate an intercept from the data (y~-1):

```
model4<-glm(log(volume) ~ offset(log(1/(4*pi))+log(height)+2*log(girth))-1)
summary(model4)
```

```
No Coefficients
```

```
Residual deviance:  0.18805  on  31  degrees  of  freedom
AIC: -68.282
```

This is a rather curious model with no estimated parameters, but it has a residual deviance of just 0.18805 (compared with model1, where all three parameters were estimated from the data, which had a deviance of 0.18555). Because we were saving one degree of freedom with each step in the procedure, AIC became smaller with each step, justifying all of the model simplifications.

Residuals

After fitting a model to data, we should investigate how well the model describes the data. In particular, we should look to see if there are any systematic trends in the goodness of fit. For example, does the goodness of fit increase with the observation number, or is it a function of one or more of the explanatory variables? We can work with the raw residuals:

$$\text{residuals} = \text{response variable} - \text{fitted values}.$$

With normal errors, the identity link, equal weights and the default scale factor, the raw and standardized residuals are identical. The standardized residuals are required to correct for the fact that with non-normal errors (like count or proportion data) we violate the fundamental assumption that the variance is constant (p. 389) because the residuals tend to change in size as the mean value the response variable changes.

For **Poisson** errors, the standardized residuals are

$$\frac{y - \text{fitted values}}{\sqrt{\text{fitted values}}}.$$

For **binomial** errors they are

$$\frac{y - \text{fitted values}}{\sqrt{\text{fitted values} \times \left[1 - \dfrac{\text{fitted values}}{\text{binomial denominator}}\right]}}$$

where the binomial denominator is the size of the sample from which the y successes were drawn. For **Gamma** errors they are

$$\frac{y - \text{fitted values}}{\text{fitted values}}$$

In general, we can use several kinds of standardized residuals

$$\text{standardized residuals} = (y - \text{fitted values}) \sqrt{\frac{\text{prior weight}}{\text{scale parameter} \times \text{variance funcion}}}$$

where the prior weights are optionally specified by you to give individual data points more or less influence (see p. 345), the scale parameter measures the degree of overdispersion (see p. 573), and the variance function describes the relationship between the variance and the mean (e.g. equality for a Poisson process; see Table 13.1).

Misspecified Error Structure

A common problem with real data is that the variance increases with the mean. The assumption in previous chapters has been of normal errors with constant variance at all values of the response variable. For continuous measurement data with non-constant errors we can specify a generalized linear model with gamma *errors*. These are discussed in Chapter 25 along with worked examples, and we need only note at this stage that they assume a **constant coefficient of variation** (see Taylor's power law, p. 198).

With count data, we often assume Poisson errors, but the data may exhibit overdispersion (see below and p. 540), so that the variance is actually greater than the mean (rather than equal to it, as assumed by the Poisson distribution). An important distribution for describing aggregated data is the *negative binomial*. While R has no direct facility for specifying negative binomial errors, we can use quasi-likelihood to specify the variance function in a GLM with family = quasi (see p. 517).

Misspecified Link Function

Although each error structure has a canonical link function associated with it (see p. 514), it is quite possible that a different link function would give a better fit for a particular model specification. For example, in a GLM with normal errors we might try a log link or a reciprocal link using quasi to improve the fit (for examples, see p. 513). Similarly, with binomial errors we might try a complementary log-log link instead of the default logit link function (see p. 572).

An alternative to changing the link function is to transform the values of the response variable. The important point to remember here is that changing the scale of y will alter the error structure (see p. 327). Thus, if you take logs of y and carry out regression with normal errors, then you will be assuming that the errors in y were log normally distributed. This may well be a sound assumption, but a bias will have been introduced if the errors really were additive on the original scale of measurement. If, for example, theory suggests that there is an exponential relationship between y and x,

$$y = ae^{bx},$$

then it would be reasonable to suppose that the log of y would be linearly related to x:

$$\ln y = \ln a + bx.$$

Now suppose that the errors ε in y are multiplicative with a mean of 0 and constant variance, like this:

$$y = ae^{bx}(1 + \varepsilon).$$

Then they will also have a mean of 0 in the transformed model. But if the errors are additive,

$$y = ae^{bx} + \varepsilon,$$

then the error variance in the transformed model will depend upon the expected value of y. In a case like this, it is much better to analyse the untransformed response variable and to employ the log link function, because this retains the assumption of additive errors.

When both the error distribution and functional form of the relationship are unknown, there is no single specific rationale for choosing any given transformation in preference to another. The aim is pragmatic, namely to find a transformation that gives:

- constant error variance;

- approximately normal errors;

- additivity;

- a linear relationship between the response variables and the explanatory variables;

- straightforward scientific interpretation.

The choice is bound to be a compromise and, as such, is best resolved by quantitative comparison of the deviance produced under different model forms (see Chapter 9).

Overdispersion

Overdispersion is the polite statistician's version of Murphy's law: if something can go wrong, it will. Overdispersion can be a problem when working with Poisson or binomial errors, and tends to occur because you have not measured one or more of the factors that turn out to be important. It may also result from the underlying distribution being non-Poisson or non-binomial. This means that the probability you are attempting to model is not constant within each cell, but behaves like a random variable. This, in turn, means that the residual deviance is inflated. In the worst case, all the predictor variables you have measured may turn out to be unimportant so that you have no information at all on any of the genuinely important predictors. In this case, the minimal adequate model is just the overall mean, and all your 'explanatory' variables provide no extra information.

The techniques of dealing with overdispersion are discussed in detail when we consider Poisson errors (p. 527) and binomial errors (p. 569). Here it is sufficient to point out that there are two general techniques available to us:

- use F tests with an empirical scale parameter instead of chi-squared;

- use quasi-likelihood to specify a more appropriate variance function.

It is important, however, to stress that these techniques introduce another level of uncertainty into the analysis. Overdispersion happens for real, scientifically important reasons, and these reasons may throw doubt upon our ability to interpret the experiment in an unbiased way. It means that something we did not measure turned out to have an important impact on the results. If we did not measure this factor, then we have no confidence that our randomization process took care of it properly and we may have introduced an important bias into the results.

Bootstrapping a GLM

There are two contrasting ways of using bootstrapping with statistical models:

- Fit the model lots of times by selecting cases for inclusion at random with replacement, so that some data points are excluded and others appear more than once in any particular model fit.

- Fit the model once and calculate the residuals and the fitted values, then shuffle the residuals lots of times and add them to the fitted values in different permutations, fitting the model to the many different data sets.

In both cases, you will obtain a distribution of parameter values for the model from which you can derive confidence intervals. Here we use the timber data (a multiple regression with two continuous explanatory variables, introduced on p. 336) to illustrate the two approaches (see p. 284 for an introduction to the bootstrap).

library(boot)

The GLM model with its parameter estimates and standard errors is on p. 519. The hard part of using boot is writing the sampling function correctly. It has at least two arguments: the first *must* be the data on which the resampling is to be carried out (in this case, the whole dataframe called trees), and the second *must* be the index (the randomized subscripts showing which data values are to be used in a given realization; some cases will be repeated, others will be omitted). Inside the function we create a new dataframe based on the randomly selected indices, then fit the model to this new data set. Finally, the function should return the coefficients of the model. Here is the 'statistic' function in full:

```
model.boot<-function(data,indices){
sub.data<-data[indices,]
model<-glm(log(volume)~log(girth)+log(height),data=sub.data)
coef(model) }
```

Now run the bootstrap for 2000 resamplings using the boot function:

```
glim.boot<-boot(trees,model.boot,R=2000)
glim.boot
```

```
ORDINARY NONPARAMETRIC BOOTSTRAP

Call:
boot(data = trees, statistic = model.boot, R = 2000)
```

```
Bootstrap Statistics :
        original                  bias  std. error
t1*   −2.899379   −0.046089511    0.6452832
t2*    1.982665   −0.001071986    0.0603073
t3*    1.117138    0.014858487    0.2082793
```

There is very little bias in any of the three parameters, and the bootstrapped standard errors are close to their parametric estimates.

The other way of bootstrapping with a model, mentioned above, is to include all the original cases (rather than a subset of them with repeats, as we did above) but to randomize the residuals that are associated with each case. The raw residuals are y − fitted(model) and it is these values that are shuffled and allocated to cases at random. The model is then refitted and the coefficients extracted. The new y values, therefore, are fitted(model)+ sample(y − fitted(model)). Here is a home-made version:

```
model<-glm(log(volume)~log(girth)+log(height))
yhat<-fitted(model)
residuals<- log(volume)- yhat
coefs<-numeric(6000)
coefs<-matrix(coefs,nrow=2000)
```

We shuffle the residuals 2000 times to get different vectors of y values:

```
for (i in 1:2000){
y<-yhat+sample(residuals)
boot.model<-glm(y~log(girth)+log(height))
coefs[i,]<-coef(boot.model) }
```

Extracting the means and standard deviations of the coefficients gives

```
apply(coefs,2,mean)
```

```
[1]  -2.898088  1.982693  1.116724
```

```
apply(coefs,2,sd)
```

```
[1]  0.60223281  0.07231379  0.19317107
```

These values are close to the estimates obtained by other means earlier. Next, we use the boot function to carry out this method. The preliminaries involve fitting the GLM and extracting the fitted values (yhat which will be the same each time) and the residuals (resids) which will be independently shuffled each time:

```
model<-glm(log(volume)~log(girth)+log(height))
yhat<-fitted(model)
resids<-resid(model)
```

Now make a dataframe that will be fed into the bootstrap, containing the residuals to be shuffled, along with the two explanatory variables:

```
res.data<-data.frame(resids,girth,height)
```

Now for the only hard part: writing the 'statistic' function to do the work within boot. The first argument is always the dataframe and the second is always the index *i*, which controls the shuffling:

```
bf<-function(res.data,i) {
y<-yhat+res.data[i,1]
ndv-data.frame(y,girth,height)
model<-glm(y~log(girth)+log(height),data=nd)
coef(model) }
```

Inside the function we create a particular vector of y values by adding the shuffled residuals res.data[i,1] to the fitted values, then put this vector, y, along with the explanatory variables into a new dataframe nd that will be different each time GLM the is fitted. The function returns the three coefficients from the particular fitted model coef(model); the coefficients are the 'statistics' of the bootstrap, hence the name of the function.

Finally, because we want to shuffle the residuals rather than sample them with replacement, we specify sim="permutation" in the call to the boot function:

boot(res.data, bf, R=2000, sim="permutation")

```
DATA PERMUTATION
Call:
boot(data = res.data, statistic = bf, R = 2000, sim = "permutation")
Bootstrap Statistics :
        original          bias    std. error
t1*   -2.899379     0.014278399   0.62166875
t2*    1.982665     0.001601178   0.07064475
t3*    1.117138    -0.004586529   0.19938992
```

Again, the parameter values and their standard errors are very close to those obtained by our other bootstrapping methods. Here are the confidence intervals for the three parameters, specified by index = 1 fo the intercept, index = 2 for the slope of the regression on $\log(g)$ and index = 3 for the slope of the regression on $\log(h)$:

perms<- boot(res.data, bf, R=2000, sim="permutation")
boot.ci(perms,index=1)

```
BOOTSTRAP CONFIDENCE INTERVAL CALCULATIONS
Based on 2000 bootstrap replicates

CALL :
boot.ci(boot.out = perms, index = 1)

Intervals :
Level       Normal              Basic
95%    (-4.117, -1.692)  (-4.118, -1.680)

Level      Percentile            BCa
95%    (-4.119, -1.681)  (-4.302, -1.784)
Calculations and Intervals on Original Scale
There were 32 warnings (use warnings() to see them)
```

boot.ci(perms,index=2)

```
BOOTSTRAP CONFIDENCE INTERVAL CALCULATIONS
Based on 2000 bootstrap replicates

CALL :
boot.ci(boot.out = perms, index = 2)
```

```
Intervals :
Level        Normal                   Basic
95%      ( 1.837, 2.125 )    ( 1.836, 2.124 )

Level        Percentile                 BCa
95%      ( 1.841, 2.129 )    ( 1.827, 2.115 )
Calculations and Intervals on Original Scale
There were 32 warnings (use warnings( ) to see them)
```

boot.ci(perms,index=3)

```
BOOTSTRAP CONFIDENCE INTERVAL CALCULATIONS
Based on 2000 bootstrap replicates

CALL :
boot.ci(boot.out = perms, index = 3)

Intervals :
Level        Normal                   Basic
95%      ( 0.730, 1.508 )    ( 0.726, 1.509 )

Level        Percentile                 BCa
95%      ( 0.725, 1.508 )    ( 0.758, 1.566 )
Calculations and Intervals on Original Scale
There were 32 warnings (use warnings( ) to see them)
```

You can see that all the intervals for the slope on $\log(g)$ include the value 2.0 and all the intervals for the slope on $\log(h)$ include 1.0, consistent with the theoretical expectation that the logs are cylindrical, and that the volume of usable timber can be estimated from the length of the log and the square of its girth.

14

Count Data

Up to this point, the response variables have all been continuous measurements such as weights, heights, lengths, temperatures, and growth rates. A great deal of the data collected by scientists, medical statisticians and economists, however, is in the form of **counts** (whole numbers or integers). The number of individuals that died, the number of firms going bankrupt, the number of days of frost, the number of red blood cells on a microscope slide, and the number of craters in a sector of lunar landscape are all potentially interesting variables for study. With count data, the number 0 often appears as a value of the response variable (consider, for example, what a 0 would mean in the context of the examples just listed). In this chapter we deal with data on **frequencies**, where we count how many times something happened, but we have no way of knowing how often it did *not* happen (e.g. lightning strikes, bankruptcies, deaths, births). This is in contrast to count data on **proportions**, where we know the number doing a particular thing, but also the number not doing that thing (e.g. the proportion dying, sex ratios at birth, proportions of different groups responding to a questionnaire).

Straightforward linear regression methods (assuming constant variance, normal errors) are not appropriate for count data for four main reasons:

- The linear model might lead to the prediction of negative counts.

- The variance of the response variable is likely to increase with the mean.

- The errors will not be normally distributed.

- Zeros are difficult to handle in transformations.

In R, count data are handled very elegantly in a generalized linear model by specifying family=poisson which sets errors = Poisson and link = log (see p. 515). The log link ensures that all the fitted values are positive, while the Poisson errors take account of the fact that the data are integer and have variances that are equal to their means.

A Regression with Poisson Errors

The following example has a count (the number of reported cancer cases per year per clinic) as the response variable, and a single continuous explanatory variable (the distance from a nuclear plant to the clinic in km). The question is whether or not proximity to the reactor affects the number of cancer cases.

```
clusters<-read.table("c:\\temp\\clusters.txt",header=T)
attach(clusters)
names(clusters)
```

```
[1] "Cancers"  "Distance"
```

```
plot(Distance,Cancers)
```

There seems to be a downward trend in cancer cases with distance (see the plot below). But is the trend significant? We do a regression of cases against distance, using a GLM with Poisson errors:

```
model1<-glm(Cancers~Distance,poisson)
summary(model1)
```

```
Coefficients:
              Estimate  Std. Error  z value  Pr(>|z|)
(Intercept)   0.186865    0.188728    0.990    0.3221
Distance     -0.006138    0.003667   -1.674    0.0941  .
```

```
(Dispersion parameter for poisson family taken to be 1)
```

```
    Null deviance: 149.48 on 93 degrees of freedom
Residual deviance: 146.64 on 92 degrees of freedom
AIC: 262.41
```

The trend does not look to be significant, but look at the residual deviance. It is assumed that this is the same as the residual degrees of freedom. The fact that residual deviance is larger than residual degrees of freedom indicates that we have overdispersion (extra, unexplained variation in the response). We compensate for the overdispersion by refitting the model using quasi-Poisson rather than Poisson errors:

```
model2<-glm(Cancers~Distance,quasipoisson)
summary(model2)
```

```
Coefficients:

              Estimate  Std. Error  t value  Pr(>|t|)
(Intercept)   0.186865    0.235341    0.794    0.429
Distance     -0.006138    0.004573   -1.342    0.183
```

```
(Dispersion parameter for quasipoisson family taken to be 1.555271)
```

```
    Null deviance: 149.48 on 93 degrees of freedom
Residual deviance: 146.64 on 92 degrees of freedom
AIC: NA
```

Compensating for the overdispersion has increased the p value to 0.183, so there is no compelling evidence to support the existence of a trend in cancer incidence with distance from the nuclear plant. To draw the fitted model through the data, you need to understand that the GLM with Poisson errors uses the log link, so the parameter estimates and the predictions from the model (the 'linear predictor') are in logs, and need to be antilogged before the (non-significant) fitted line is drawn.

```
xv<-seq(0,100,.1
yv<-predict(model2,list(Distance=xv))
lines(xv,exp(yv))
```

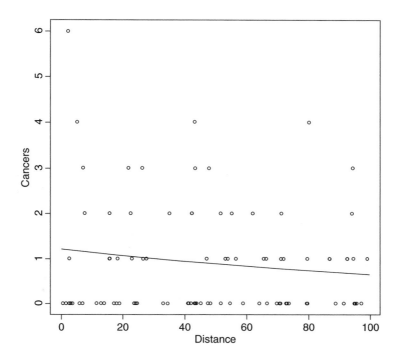

Analysis of Deviance with Count Data

In our next example the response variable is a count of infected blood cells per mm² on microscope slides prepared from randomly selected individuals. The explanatory variables are smoker (logical, yes or no), age (three levels, under 20, 21 to 59, 60 and over), sex (male or female) and body mass score (three levels, normal, overweight, obese).

```
count<-read.table("c:\\temp\\cells.txt",header=T)
attach(count)
names(count)
```

```
[1] "cells"  "smoker"  "age"  "sex"  "weight"
```

It is always a good idea with count data to get a feel for the overall frequency distribution of counts using table:

```
table(cells)
```

```
  0    1    2    3    4    5   6   7
314   75   50   32   18   13   7   2
```

Most subjects (314 of them) showed no damaged cells, and the maximum of 7 was observed in just two patients.

We begin data inspection by tabulating the main effect means:

tapply(cells,smoker,mean)

```
    FALSE        TRUE
0.5478723   1.9111111
```

tapply(cells,weight,mean)

```
   normal       obese        over
0.5833333   1.2814371   0.9357143
```

tapply(cells,sex,mean)

```
   female        male
0.6584507   1.2202643
```

tapply(cells,age,mean)

```
      mid         old       young
0.8676471   0.7835821   1.2710280
```

It looks as if smokers have a substantially higher mean count than non-smokers, that overweight and obese subjects had higher counts than normal weight, males had a higher count that females, and young subjects had a higher mean count than middle-aged or older people. We need to test whether any of these differences are significant and to assess whether there are interactions between the explanatory variables.

model1<-glm(cells~smoker*sex*age*weight,poisson)
summary(model1)

```
    Null deviance: 1052.95 on 510 degrees of freedom
Residual deviance: 736.33 on 477 degrees of freedom
AIC: 1318
```

```
Number of Fisher Scoring iterations: 6
```

The residual deviance (736.33) is much greater than the residual degrees of freedom (477), indicating overdispersion, so before interpreting any of the effects, we should refit the model using quasi-Poisson errors:

model2<-glm(cells~smoker*sex*age*weight,quasipoisson)
summary(model2)

```
Call:
glm(formula = cells ~ smoker * sex * age * weight, family = quasipoisson)

Deviance Residuals:
   Min       1Q  Median       3Q      Max
-2.236   -1.022  -0.851   0.520    3.760

Coefficients: (2 not defined because of singularities)
Estimate                 Std. Error   t value  Pr(>|t|)
(Intercept)                 -0.8329   0.4307   -1.934   0.0537  .
smokerTRUE                  -0.1787   0.8057   -0.222   0.8246
sexmale                     0.1823   0.5831    0.313   0.7547
ageold                     -0.1830   0.5233   -0.350   0.7267
ageyoung                    0.1398   0.6712    0.208   0.8351
weightobese                 1.2384   0.8965    1.381   0.1678
weightover                 -0.5534   1.4284   -0.387   0.6986
smokerTRUE:sexmale          0.8293   0.9630    0.861   0.3896
smokerTRUE:ageold          -1.7227   2.4243   -0.711   0.4777
```

smokerTRUE:ageyoung	1.1232	1.0584	1.061	0.2892
sexmale:ageold	−0.2650	0.9445	−0.281	0.7791
sexmale:ageyoung	−0.2776	0.9879	−0.281	0.7788
smokerTRUE:weightobese	3.5689	1.9053	1.873	0.0617 .
smokerTRUE:weightover	2.2581	1.8524	1.219	0.2234
sexmale:weightobese	−1.1583	1.0493	−1.104	0.2702
sexmale:weightover	0.7985	1.5256	0.523	0.6009
ageold:weightobese	−0.9280	0.9687	−0.958	0.3386
ageyoung:weightobese	−1.2384	1.7098	−0.724	0.4693
ageold:weightover	1.0013	1.4776	0.678	0.4983
ageyoung:weightover	0.5534	1.7980	0.308	0.7584
smokerTRUE:sexmale:ageold	1.8342	2.1827	0.840	0.4011
smokerTRUE:sexmale:ageyoung	−0.8249	1.3558	−0.608	0.5432
smokerTRUE:sexmale: weightobese	−2.2379	1.7788	−1.258	0.2090
smokerTRUE:sexmale:weightover	−2.5033	2.1120	−1.185	0.2365
smokerTRUE:ageold: weightobese	0.8298	3.3269	0.249	0.8031
smokerTRUE:ageyoung: weightobese	−2.2108	1.0865	−2.035	0.0424 *
smokerTRUE:ageold: weightover	1.1275	1.6897	0.667	0.5049
smokerTRUE:ageyoung:weightover	−1.6156	2.2168	−0.729	0.4665
sexmale:ageold:weightobese	2.2210	1.3318	1.668	0.0960 .
sexmale:ageyoung:weightobese	2.5346	1.9488	1.301	0.1940
sexmale:ageold:weightover	−1.0641	1.9650	−0.542	0.5884
sexmale:ageyoung:weightover	−1.1087	2.1234	−0.522	0.6018
smokerTRUE:sexmale:ageold: weightobese	−1.6169	3.0561	−0.529	0.5970
smokerTRUE:sexmale:ageyoung weightobese	NA	NA	NA	NA
smokerTRUE:sexmale:ageold: weightover	NA	NA	NA	NA
smokerTRUE:sexmale:ageyoung: weightover	2.4160	2.6846	0.900	0.3686

(Dispersion parameter for quasipoisson family taken to be 1.854815)

```
    Null deviance: 1052.95 on 510 degrees of freedom
Residual deviance: 736.33 on 477 degrees of freedom
AIC: NA

Number of Fisher Scoring iterations: 6
```

There is an apparently significant three-way interaction between smoking, age and obesity ($p = 0.0424$). There were too few subjects to assess the four-way interaction (see the NAs in the table), so we begin model simplification by removing the highest-order interaction:

```
model3<-update(model2, ~. -smoker:sex:age:weight)
summary(model3)

Call:
glm(formula = cells ~ smoker + sex + age + weight + smoker:sex +
    smoker:age + sex:age + smoker:weight + sex:weight + age:weight +
```

```
    smoker:sex:age + smoker:sex:weight + smoker:age:weight +
    sex:age:weight, family = quasipoisson)
```

Deviance Residuals:
```
    Min       1Q    Median      3Q      Max
-2.2442   -1.0477   -0.8921   0.5195   3.7613
```

Coefficients:

	Estimate	Std. Error	t value	Pr(>|t|)	
(Intercept)	-0.897195	0.436988	-2.053	0.04060	*
smokerTRUE	0.030263	0.735386	0.041	0.96719	
sexmale	0.297192	0.570009	0.521	0.60234	
ageold	-0.118726	0.528165	-0.225	0.82224	
ageyoung	0.289259	0.639618	0.452	0.65130	
weightobese	1.302660	0.898307	1.450	0.14768	
weightover	-0.005052	1.027198	-0.005	0.99608	
smokerTRUE:sexmale	0.527345	0.867294	0.608	0.54345	
smokerTRUE:ageold	-0.566584	1.700590	-0.333	0.73915	
smokerTRUE:ageyoung	0.757297	0.939746	0.806	0.42073	
sexmale:ageold	-0.379884	0.935365	-0.406	0.68483	
sexmale:ageyoung	-0.610703	0.920969	-0.663	0.50758	
smokerTRUE:weightobese	3.924591	1.475476	2.660	0.00800	**
smokerTRUE:weightover	1.192159	1.259888	0.946	0.34450	
sexmale:weightobese	-1.273202	1.040701	-1.223	0.22178	
sexmale:weightover	0.154097	1.098781	0.140	0.88853	
ageold:weightobese	-0.993355	0.970484	-1.024	0.30656	
ageyoung:weightobese	-1.346913	1.459454	-0.923	0.35653	
ageold:weightover	0.454217	1.090260	0.417	0.67715	
ageyoung:weightover	-0.483955	1.300866	-0.372	0.71004	
smokerTRUE:sexmale:ageold	0.771116	1.451512	0.531	0.59549	
smokerTRUE:sexmale:ageyoung	-0.210317	1.140384	-0.184	0.85376	
smokerTRUE:sexmale:weightobese	-2.500668	1.369941	-1.825	0.06857	.
smokerTRUE:sexmale:weightover	-1.110222	1.217531	-0.912	0.36230	
smokerTRUE:ageold:weightobese	-0.882951	1.187871	-0.743	0.45766	
smokerTRUE:ageyoung:weightobese	-2.453315	1.047067	-2.343	0.01954	*
smokerTRUE:ageold:weightover	0.823018	1.528233	0.539	0.59045	
smokerTRUE:ageyoung:weightover	0.040795	1.223664	0.033	0.97342	
sexmale:ageold:weightobese	2.338617	1.324805	1.765	0.07816	.
sexmale:ageyoung:weightobese	2.822032	1.623849	1.738	0.08288	.
sexmale:ageold:weightover	-0.442066	1.545451	-0.286	0.77497	
sexmale:ageyoung:weightover	0.357807	1.291194	0.277	0.78181	

(Dispersion parameter for quasipoisson family taken to be 1.847991)

```
    Null deviance: 1052.95 on 510 degrees of freedom
Residual deviance: 737.87 on 479 degrees of freedom
AIC: NA
```
Number of Fisher Scoring iterations: 6

The remaining model simplification is left to you as an exercise. Your minimal adequate model might look something like this:

summary(model18)

Call:
glm(formula = cells ~ smoker + weight + smoker:weight, family =
quasipoisson)

Deviance Residuals:
```
    Min       1Q    Median      3Q      Max
-2.6511   -1.1742   -0.9148   0.5533   3.6436
```

```
Coefficients:
                         Estimate  Std. Error  t value  Pr(>|t|)
(Intercept)               -0.8712      0.1760   -4.950  1.01e-06  ***
smokerTRUE                 0.8224      0.2479    3.318  0.000973  ***
weightobese                0.4993      0.2260    2.209  0.027598  *
weightover                 0.2618      0.2522    1.038  0.299723
smokerTRUE:weightobese     0.8063      0.3105    2.597  0.009675  **
smokerTRUE:weightover      0.4935      0.3442    1.434  0.152226
```

(Dispersion parameter for quasipoisson family taken to be 1.827927)

```
    Null deviance: 1052.95 on 510 degrees of freedom
Residual deviance: 737.87 on 479 degrees of freedom
AIC: NA
Number of Fisher Scoring iterations: 6
```

This model shows a highly significant interaction between smoking and weight in determining the number of damaged cells, but there are no convincing effects of age or sex. In a case like this, it is useful to produce a summary table to highlight the effects:

tapply (cells,list(smoker,weight),mean)

```
           normal       obese         over
FALSE   0.4184397   0.689394   0.5436893
TRUE    0.9523810   3.514286   2.0270270
```

The interaction arises because the response to smoking depends on body weight: smoking adds a mean of about 0.5 damaged cells for individuals with normal body weight, but adds 2.8 damaged cells for obese people.

It is straightforward to turn the summary table into a barplot:

barplot(tapply(cells,list(smoker,weight),mean),col=c(2,7),beside=T)
legend(1.2,3.4,c("non","smoker"),fill=c(2,7))

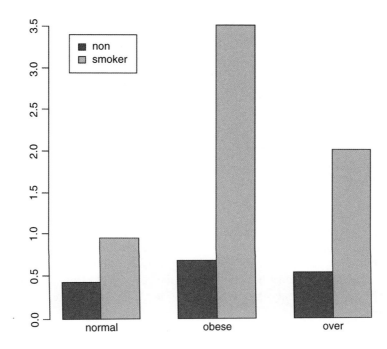

Analysis of Covariance with Count Data

In this next example the response is a count of the number of plant species on plots that have different biomass (a continuous explanatory variable) and different soil pH (a categorical variable with three levels: high, mid and low).

```
species<-read.table("c:\\temp\\species.txt",header=T)
attach(species)
names(species)
```

```
[1]  "pH"  "Biomass"  "Species"
```

```
plot(Biomass,Species,type="n")
spp<-split(Species,pH)
bio<-split(Biomass,pH)
points(bio[[1]],spp[[1]],pch=16)
points(bio[[2]],spp[[2]],pch=17)
points(bio[[3]],spp[[3]])
```

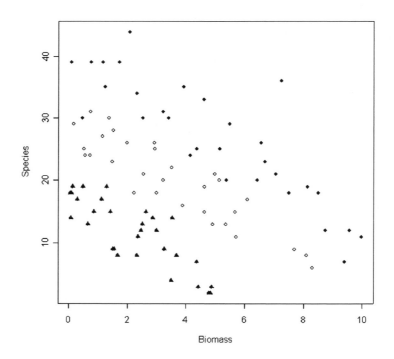

Note the use of split to create separate lists of plotting coordinates for the three levels of pH. It is clear that Species declines with Biomass, and that soil pH has a big effect on Species, but does the slope of the relationship between Species and Biomass depend on pH? The lines look reasonably parallel from the scatterplot. This is a question about interaction effects, and in analysis of covariance, interaction effects are about differences between slopes:

```
model1<-glm(Species~ Biomass*pH,poisson)
summary(model1)
```

```
Coefficients:
                Estimate  Std. Error  z value  Pr(>|z|)
(Intercept)      3.76812    0.06153    61.240   < 2e-16   ***
Biomass         -0.10713    0.01249    -8.577   < 2e-16   ***
pHlow           -0.81557    0.10284    -7.931   2.18e-15  ***
pHmid           -0.33146    0.09217    -3.596   0.000323  ***
Biomass:pHlow   -0.15503    0.04003    -3.873   0.000108  ***
Biomass:pHmid   -0.03189    0.02308    -1.382   0.166954
(Dispersion parameter for poisson family taken to be 1)
```

```
    Null deviance: 452.346 on 89 degrees of freedom
Residual deviance: 83.201 on 84 degrees of freedom
AIC: 514.39
```

```
Number of Fisher Scoring iterations: 4
```

We can test for the need for different slopes by comparing this maximal model (with six parameters) with a simpler model with different intercepts but the same slope (four parameters):

```
model2<-glm(Species~Biomass+pH,poisson)
anova(model1,model2,test="Chi")
```

```
Analysis of Deviance Table
```

```
Model 1: Species ~ Biomass * pH
Model 2: Species ~ Biomass + pH
```

```
  Resid. Df  Resid. Dev  Df  Deviance  P(>|Chi|)
1       84      83.201
2       86      99.242   -2  -16.040   0.0003288
```

The slopes are very significantly different ($p = 0.000\,33$), so we are justified in retaining the more complicated model1.

Finally, we draw the fitted lines through the scatterplot, using predict:

```
xv<-seq(0,10,0.1)
levels(pH)
```

```
[1]  "high"  "low"  "mid"
```

```
length(xv)
```

```
[1]  101
```

```
phv<-rep("high",101)
yv<-predict(model1,list(pH=factor(phv),Biomass=xv),type="response")
lines(xv,yv)
phv<-rep("mid",101)
yv<-predict(model1,list(pH=factor(phv),Biomass=xv),type="response")
lines(xv,yv)
phv<-rep("low",101)
yv<-predict(model1,list(pH=factor(phv),Biomass=xv),type="response")
lines(xv,yv)
```

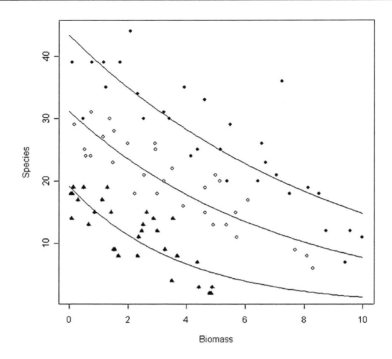

Note the use of type="response" in the predict function. This ensures that *yv* is calculated as Species rather than log(Species), and means we do not need to back-transform using antilogs before drawing the lines (compare with the example on p. 579). You could make the R code more elegant by writing a function to plot any number of lines, depending on the number of levels of the factor (three levels of pH in this case).

Frequency Distributions

Here are data on the numbers of bankruptcies in 80 districts. The question is whether there is any evidence that some districts show greater than expected numbers of cases. What would we expect? Of course we should expect some variation, but how much, exactly? Well that depends on our model of the process. Perhaps the simplest model is that absolutely nothing is going on, and that every singly bankruptcy case is absolutely independent of every other. That leads to the prediction that the numbers of cases per district will follow a Poisson process, a distribution in which the variance is equal to the mean (see p. 250). Let's see what the data show.

```
case.book<-read.table("c:\\temp\\cases.txt",header=T)
attach(case.book)
names(case.book)
```

```
[1]  "cases"
```

First we need to count the numbers of districts with no cases, one case, two cases, and so on. The R function that does this is called table:

```
frequencies<-table(cases)
frequencies
```

```
cases
  0   1   2  3  4  5  6  7  8  9  10
 34  14  10  7  4  5  2  1  1  1   1
```

There were no cases at all in 34 districts, but one district had 10 cases. A good way to proceed is to compare our distribution (called frequencies) with the distribution that would be observed if the data really did come from a Poisson distribution as postulated by our model. We can use the R function **dpois** to compute the probability density of each of the 11 frequencies from 0 to 10 (we multiply the probability produced by **dpois** by the total sample of 80 to obtain the predicted frequencies). We need to calculate the mean number of cases per district: this is the Poisson distribution's only parameter:

mean(cases)

```
[1]  1.775
```

The plan is to draw two distributions side by side, so we set up the plotting region:

par(mfrow=c(1,2))

Now we plot the observed frequencies in the left-hand panel and the predicted, Poisson frequencies in the right-hand panel:

barplot(frequencies,ylab="Frequency",xlab="Cases",col="red")

barplot(dpois(0:10,1.775)*80,names=as.character(0:10),
                        ylab="Frequency",xlab="Cases",col="red")

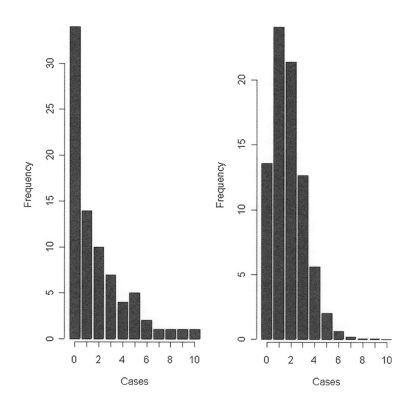

The distributions are very different: the mode of the observed data is 0, but the mode of the Poisson with the same mean is 1; the observed data contained examples of 8, 9 and 10 cases, but these would be highly unlikely under a Poisson process. We would say that the observed data are highly **aggregated**; they have a variance–mean ratio much greater than 1 (the Poisson distribution, of course, has a variance–mean ratio of 1):

```
var(cases)/mean(cases)
```

```
[1]  2.99483
```

So, if the data are not Poisson distributed, how are they distributed? A good candidate distribution where the variance–mean ratio is this big (*c*. 3.0) is the negative binomial distribution (see p. 252). This is a two-paramter distribution: the first parameter is the mean number of cases (1.775), and the second is called the clumping parameter, k (measuring the degree of aggregation in the data: small values of $k(k < 1)$ show high aggregation, while large values of $k(k > 5)$ show randomness). We can get an approximate estimate of the magnitude of k from

$$\hat{k} = \frac{\bar{x}^2}{s^2 - \bar{x}}.$$

We can work this out:

```
mean(cases)^2/(var(cases)-mean(cases))
```

```
[1]  0.8898003
```

so we shall work with $k = 0.89$. How do we compute the expected frequencies? The density function for the negative binomial distribution is dnbinom and it has three arguments: the frequency for which we want the probability (in our case 0 to 10), the number of successes (in our case 1), and the mean number of cases (1.775); we multiply by the total number of cases (80) to obtain the expected frequencies

```
exp<-dnbinom(0:10,1,mu=1.775)*80
```

We will draw a single figure in which the observed and expected frequencies are drawn side by side. The trick is to produce a new vector (called both) which is twice as long as the observed and expected frequency vectors ($2 \times 11 = 22$). Then, we put the observed frequencies in the odd-numbered elements (using modulo 2 to calculate the values of the subscripts), and the expected frequencies in the even-numbered elements:

```
both<-numeric(22)
both[1:22 %% 2 != 0]<-frequencies
both[1:22 %% 2 == 0]<-exp
```

On the x axis, we intend to label only every other bar:

```
labels<-character(22)
labels[1:22 %% 2 == 0]<-as.character(0:10)
```

Now we can produce the barplot, using white for the observed frequencies and grey for the negative binomial frequencies:

```
par(mfrow=c(1,1))
barplot(both,col=rep(c("white","grey"),11),names=labels,ylab="Frequency",
     xlab="Cases")
```

Now we need to add a legend to show what the two colours of the bars mean. You can locate the legend by trial and error, or by left-clicking mouse when the cursor is in the correct position, using the locator(1) function (see p. 257):

legend(16,30,c("Observed","Expected"), fill=c("white","grey"))

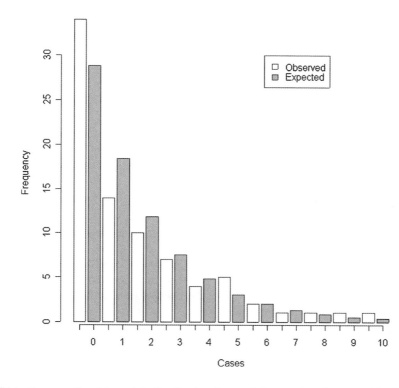

The fit to the negative binomial distribution is much better than it was with the Poisson distribution, especially in the right-hand tail. But the observed data have too many 0s and too few 1s to be represented perfectly by a negative binomial distribution. If you want to quantify the lack of fit between the observed and expected frequency distributions, you can calculate Pearson's chi-squared $\sum (O - E)^2 / E$ based on the number of comparisons that have expected frequency greater than 4:

exp

```
[1]   28.8288288   18.4400617   11.7949944   7.5445460   4.8257907   3.0867670
[7]    1.9744185    1.2629164    0.8078114   0.5167082   0.3305070
```

If we accumulate the rightmost six frequencies, then all the values of exp will be bigger than 4. The degrees of freedom are then given by the number of comparisons (6) - the number of parameters estimated from the data (2 in our case) -1 (for contingency, because the total frequency must add up to 80) $= 3$. We use a gets arrow to reduce the lengths of the observed and expected vectors, creating an upper interval called 5+ for '5 or more':

```
cs<-factor(0:10)
levels(cs)[6:11]<-"5+"
levels(cs)
```

```
[1]   "0"   "1"   "2"   "3"   "4"   "5+"
```

Now make the two shorter vectors 'of' and 'ef' (for observed and expected frequencies):

```
ef<-as.vector(tapply(exp,cs,sum))
of<-as.vector(tapply(frequencies,cs,sum))
```

Finally we can compute the chi-squared value measuring the difference between the observed and expected frequency distributions, and use 1-pchisq to work out the *p* value:

```
sum((of-ef)^2/ef)
```

```
[1]   3.594145
```

```
1-pchisq(3.594145,3)
```

```
[1]   0.3087555
```

We conclude that a negative binomial description of these data is reasonable (the observed and expected distributions are not significantly different; $p = 0.31$).

Overdispersion in Log-linear Models

The data analysed in this section refer to children from Walgett, New South Wales, Australia, who were classified by sex (with two levels: male (M) and female (F)), culture (also with two levels: Aboriginal (A) and not (N)), age group (with four levels: F0 (primary), F1, F2 and F3) and learner status (with two levels: average (AL) and slow (SL)). The response variable is a count of the number of days absent from school in a particular school year.

```
library(MASS)
data(quine)
attach(quine)
names(quine)
```

```
[1]   "Eth"   "Sex"   "Age"   "Lrn"   "Days"
```

We begin with a log-linear model for the counts, and fit a maximal model containing all the factors and all their interactions:

```
model1<-glm(Days~Eth*Sex*Age*Lrn,poisson)
summary(model1)
```

```
(Dispersion parameter for poisson family taken to be 1)

    Null deviance: 2073.5 on 145 degrees of freedom
Residual deviance: 1173.9 on 118 degrees of freedom
AIC: 1818.4
```

Next, we check the residual deviance to see if there is overdispersion. Recall that the residual deviance should be equal to the residual degrees of freedom if the Poisson errors assumption is appropriate. Here it is 1173.9 on 118 d.f., indicating overdispersion by a factor of roughly 10. This is much too big to ignore, so before embarking on model simplification we try a different approach, using quasi-Poisson errors to account for the overdispersion:

model2<-glm(Days~Eth*Sex*Age*Lrn,quasipoisson)
summary(model2)

Deviance Residuals:

Min	1Q	Median	3Q	Max
-7.3872	-2.5129	-0.4205	1.7424	6.6783

Coefficients: (4 not defined because of singularities)

	Estimate	Std. Error	t value	Pr(>|t|)	
(Intercept)	3.0564	0.3346	9.135	2.22e-15	***
EthN	-0.1386	0.4904	-0.283	0.7780	
SexM	-0.4914	0.5082	-0.967	0.3356	
AgeF1	-0.6227	0.5281	-1.179	0.2407	
AgeF2	-2.3632	2.2066	-1.071	0.2864	
AgeF3	-0.3784	0.4296	-0.881	0.3802	
LrnSL	-1.9577	1.8120	-1.080	0.2822	
EthN:SexM	-0.7524	0.8272	-0.910	0.3649	
EthN:AgeF1	0.1029	0.7427	0.139	0.8901	
EthN:AgeF2	-0.5546	3.8094	-0.146	0.8845	
EthN:AgeF3	0.0633	0.6194	0.102	0.9188	
SexM:AgeF1	0.4092	0.9372	0.437	0.6632	
SexM:AgeF2	3.1098	2.2506	1.382	0.1696	
SexM:AgeF3	1.1145	0.6173	1.806	0.0735	.
EthN:LrnSL	2.2588	1.9474	1.160	0.2484	
SexM:LrnSL	1.5900	1.9448	0.818	0.4152	
AgeF1:LrnSL	2.6421	1.8688	1.414	0.1601	
AgeF2:LrnSL	4.8585	2.8413	1.710	0.0899	.
AgeF3:LrnSL	NA	NA	NA	NA	
EthN:SexM:AgeF1	-0.3105	1.6756	-0.185	0.8533	
EthN:SexM:AgeF2	0.3469	3.8928	0.089	0.9291	
EthN:SexM:AgeF3	0.8329	0.9629	0.865	0.3888	
EthN:SexM:LrnSL	-0.1639	2.1666	-0.076	0.9398	
EthN:AgeF1:LrnSL	-3.5493	2.0712	-1.714	0.0892	.
EthN:AgeF2:LrnSL	-3.3315	4.2739	-0.779	0.4373	
EthN:AgeF3:LrnSL	NA	NA	NA	NA	
SexM:AgeF1:LrnSL	-2.4285	2.1901	-1.109	0.2697	
SexM:AgeF2:LrnSL	-4.1914	2.9472	-1.422	0.1576	
SexM:AgeF3:LrnSL	NA	NA	NA	NA	
EthN:SexM:AgeF1:LrnSL	2.1711	2.7527	0.789	0.4319	
EthN:SexM:AgeF2:LrnSL	2.1029	4.4203	0.476	0.6351	
EthN:SexM:AgeF3:LrnSL	NA	NA	NA	NA	

- - -

Signif. codes: 0 '***'0.001 '**'0.01 '*'0.05 '.'0.1 ' '1
(Dispersion parameter for quasipoisson family taken to be 9.514226)

    Null deviance: 2073.5 on 145 degrees of freedom
Residual deviance: 1173.9 on 118 degrees of freedom
AIC: NA

Number of Fisher Scoring iterations: 5

Notice that certain interactions have not been estimated because of missing factor-level combinations, as indicated by the zeros in the following table:

ftable(table(Eth,Sex,Age,Lrn))

```
Eth  Sex  Age  Lrn  AL  SL
A    F    F0        4   1
          F1        5  10
          F2        1   8
          F3        9   0
     M    F0        5   3
          F1        2   3
          F2        7   4
          F3        7   0
N    F    F0        4   1
          F1        6  11
          F2        1   9
          F3       10   0
     M    F0        6   3
          F1        2   7
          F2        7   3
          F3        7   0
```

This occurs because slow learners never get into Form 3.

Unfortunately, AIC is not defined for this model, so we cannot automate the simplification using stepAIC. We need to do the model simplification long-hand, therefore, remembering to do F tests (not chi-squared) because of the overdispersion. Here is the last step of the simplification before obtaining the minimal adequate model. Do we need the age by learning interaction?

```
model4<-update(model3,~. - Age:Lrn)
anova(model3,model4,test="F")
```

Analysis of Deviance Table

```
  Resid. Df  Res.Dev  Df  Deviance       F  Pr(>-F)
1        127  1280.52
2        129  1301.08  -2    -20.56  1.0306   0.3598
```

No we don't. So here is the minimal adequate model with quasi-Poisson errors:

```
summary(model4)
```

Coefficients:

	Estimate	Std. Error	t value	Pr(>|t|)	
(Intercept)	2.83161	0.30489	9.287	4.98e-16	***
EthN	0.09821	0.38631	0.254	0.79973	
SexM	-0.56268	0.38877	-1.447	0.15023	
AgeF1	-0.20878	0.35933	-0.581	0.56223	
AgeF2	0.16223	0.37481	0.433	0.66586	
AgeF3	-0.25584	0.37855	-0.676	0.50036	
LrnSL	0.50311	0.30798	1.634	0.10479	
EthN:SexM	-0.24554	0.37347	-0.657	0.51206	
EthN:AgeF1	-0.68742	0.46823	-1.468	0.14450	
EthN:AgeF2	-1.07361	0.42449	-2.529	0.01264	*
EthN:AgeF3	0.01879	0.42914	0.044	0.96513	
EthN:LrnSL	-0.65154	0.45857	-1.421	0.15778	
SexM:AgeF1	-0.26358	0.50673	-0.520	0.60385	

```
SexM:AgeF2          0.94531   0.43530    2.172   0.03171   *
SexM:AgeF3          1.35285   0.42933    3.151   0.00202   *
SexM:LrnSL         -0.29570   0.41144   -0.719   0.47363
EthN:SexM:LrnSL     1.60463   0.57112    2.810   0.00573   *
```

(Dispersion parameter for quasipoisson family taken to be 9.833426)

```
    Null deviance: 2073.5 on 145 degrees of freedom
Residual deviance: 1301.1 on 129 degrees of freedom
```

There is a very significant three-way interaction between ethnic origin, sex and learning difficulty; non-Aboriginal slow-learning boys were more likely to be absent than non-aboriginal boys without learning difficulties.

ftable(tapply(Days,list(Eth,Sex,Lrn),mean))

```
            AL          SL
A   F    14.47368    27.36842
    M    22.28571    20.20000
N   F    13.14286     7.00000
    M    13.36364    17.00000
```

Note, however, that amongst the pupils without learning difficulties it is the Aboriginal boys who miss the most days, and it is Aboriginal girls with learning difficulties who have the highest rate of absenteeism overall.

Negative binomial errors

Instead of using quasi-Poisson errors (as above) we could use a negative binomial model. This is in the **MASS** library and involves the function **glm.nb**. The modelling proceeds in exactly the same way as with a typical GLM:

model.nb1<-glm.nb(Days~Eth*Sex*Age*Lrn)
summary(model.nb1,cor=F)

```
Call:
glm.nb(formula = Days ~ Eth * Sex * Age * Lrn, init.theta =
1.92836014510701, link = log)

(Dispersion parameter for Negative Binomial(1.9284) family taken to be 1)

    Null deviance: 272.29 on 145 degrees of freedom
Residual deviance: 167.45 on 118 degrees of freedom
AIC: 1097.3
            Theta: 1.928
        Std. Err.: 0.269
2 x log-likelihood: -1039.324
```

The output is slightly different than a conventional GLM: you see the estimated negative binomial parameter (here called theta, but known to us as k and equal to 1.928) and its approximate standard error (0.269) and 2 times the log-likelihood (contrast this with the residual deviance from our quasi-Poisson model, which was 1301.1; see above). Note that the residual deviance in the negative binomial model (167.45) is not 2 times the log-likelihood.

An advantage of the negative binomial model over the quasi-Poisson is that we can automate the model simplification with stepAIC:

model.nb2<-stepAIC(model.nb1)
summary(model.nb2,cor=F)

```
Coefficients: (3 not defined because of singularities)
                  Estimate  Std. Error  z value  Pr(>|z|)
(Intercept)        3.1693      0.3411     9.292    < 2e-16  ***
EthN              -0.3560      0.4210    -0.845    0.397848
SexM              -0.6920      0.4138    -1.672    0.094459  .
AgeF1             -0.6405      0.4638    -1.381    0.167329
AgeF2             -2.4576      0.8675    -2.833    0.004612  **
AgeF3             -0.5880      0.3973    -1.480    0.138885
LrnSL             -1.0264      0.7378    -1.391    0.164179
EthN:SexM         -0.3562      0.3854    -0.924    0.355364
EthN:AgeF1         0.1500      0.5644     0.266    0.790400
EthN:AgeF2        -0.3833      0.5640    -0.680    0.496746
EthN:AgeF3         0.4719      0.4542     1.039    0.298824
SexM:AgeF1         0.2985      0.6047     0.494    0.621597
SexM:AgeF2         3.2904      0.8941     3.680    0.000233  ***
SexM:AgeF3         1.5412      0.4548     3.389    0.000702  ***
EthN:LrnSL         0.9651      0.7753     1.245    0.213255
SexM:LrnSL         0.5457      0.8013     0.681    0.495873
AgeF1:LrnSL        1.6231      0.8222     1.974    0.048373  *
AgeF2:LrnSL        3.8321      1.1054     3.467    0.000527  ***
AgeF3:LrnSL          NA          NA        NA        NA
EthN:SexM:LrnSL    1.3578      0.5914     2.296    0.021684  *
EthN:AgeF1:LrnSL  -2.1013      0.8728    -2.408    0.016058  *
EthN:AgeF2:LrnSL  -1.8260      0.8774    -2.081    0.037426  *
EthN:AgeF3:LrnSL     NA          NA        NA        NA
SexM:AgeF1:LrnSL  -1.1086      0.9409    -1.178    0.238671
SexM:AgeF2:LrnSL  -2.8800      1.1550    -2.493    0.012651  *
SexM:AgeF3:LrnSL     NA          NA        NA        NA
```

(Dispersion parameter for Negative Binomial(1.8653) family taken to be 1)

```
    Null deviance: 265.27 on 145 degrees of freedom
Residual deviance: 167.44 on 123 degrees of freedom
AIC: 1091.4
              Theta: 1.865
          Std. Err.: 0.258
2 x log-likelihood: -1043.409
```

model.nb3<-update(model.nb2,~. - Sex:Age:Lrn)
anova(model.nb3,model.nb2)

```
Likelihood ratio tests of Negative Binomial Models

    theta Resid.    df   2 x log-lik.   Test   df   LR stat.     Pr(Chi)
1  1.789507     125     -1049.111
2  1.865343     123     -1043.409    1 vs 2    2   5.701942   0.05778817
```

The sex-by-age-by-learning interaction does not survive a deletion test ($p = 0.058$), nor does ethnic-origin-by-age-by-learning ($p = 0.115$) nor age-by-learning ($p = 0.150$):

```
model.nb4<-update(model.nb3,~. - Eth:Age:Lrn)
anova(model.nb3,model.nb4)
```

Likelihood ratio tests of Negative Binomial Models

	theta	Resid. df	2 x log-lik.	Test	df	LR stat.	Pr(Chi)
1	1.724987	127	-1053.431				
2	1.789507	125	-1049.111	1 vs 2	2	4.320086	0.1153202

```
model.nb5<-update(model.nb4,~. - Age:Lrn)
anova(model.nb4,model.nb5)
```

Likelihood ratio tests of Negative Binomial Models

	theta	Resid. df	2 x log-lik.	Test	df	LR stat.	Pr(Chi)
1	1.678620	129	-1057.219				
2	1.724987	127	-1053.431	1 vs 2	2	3.787823	0.150482

```
summary(model.nb5,cor=F)
```

Coefficients:

	Estimate	Std. Error	z value	Pr(>|z|)	
(Intercept)	2.91755	0.32626	8.942	< 2e-16	***
EthN	0.05666	0.39515	0.143	0.88598	
SexM	-0.55047	0.39014	-1.411	0.15825	
AgeF1	-0.32379	0.38373	-0.844	0.39878	
AgeF2	-0.06383	0.42046	-0.152	0.87933	
AgeF3	-0.34854	0.39128	-0.891	0.37305	
LrnSL	0.57697	0.33382	1.728	0.08392	.
EthN:SexM	-0.41608	0.37491	-1.110	0.26708	
EthN:AgeF1	-0.56613	0.43162	-1.312	0.18965	
EthN:AgeF2	-0.89577	0.42950	-2.086	0.03702	*
EthN:AgeF3	0.08467	0.44010	0.192	0.84744	
SexM:AgeF1	-0.08459	0.45324	-0.187	0.85195	
SexM:AgeF2	1.13752	0.45192	2.517	0.01183	*
SexM:AgeF3	1.43124	0.44365	3.226	0.00126	**
EthN:LrnSL	-0.78724	0.43058	-1.828	0.06750	.
SexM:LrnSL	-0.47437	0.45908	-1.033	0.30147	
EthN:SexM:LrnSL	1.75289	0.58341	3.005	0.00266	**

(Dispersion parameter for Negative Binomial(1.6786) family taken to be 1)

```
    Null deviance: 243.98 on 145 degrees of freedom
Residual deviance: 168.03 on 129 degrees of freedom
AIC: 1093.2
            Theta: 1.679
          Std. Err.: 0.22
2 x log-likelihood: -1057.219
```

The minimal adequate model, therefore, contains exactly the same terms as we obtained with quasi-Poisson, but the significance levels are higher (e.g. the three-way interaction has $p = 0.00266$ compared with $p = 0.00573$). We need to plot the model to check assumptions:

```
par(mfrow=c(1,2))
plot(model.nb5)
par(mfrow=c(1,1))
```

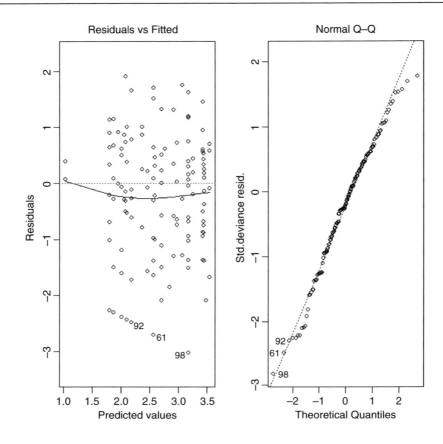

The variance is well behaved and the residuals are close to normally distributed. The combination of low *p* values plus the ability to use **stepAIC** makes **glm.nb** a very useful modelling function for count data such as these.

Use of **lmer** with Complex Nesting

In this section we have count data (snails) so we want to use **family = poisson**. But we have complicated spatial pseudoreplication arising from a split-plot design, so we cannot use a GLM. The answer is to use generalized mixed models, **lmer**. The default method for a generalized linear model fit with **lmer** has been switched from PQL to the Laplace method. The Laplace method is more reliable than PQL, and is not so much slower to as to preclude its routine use (Doug Bates, personal communication).

The syntax is extended in the usual way to accommodate the random effects (Chapter 19), with slashes showing the nesting of the random effects, and with the factor associated with the largest plot size on the left and the smallest on the right. We revisit the split-plot experiment on biomass (p. 469) and analyse the count data on snails captured from each plot. The model we want to fit is a generalized mixed model with Poisson errors (because the data are counts) with complex nesting to take account of the four-level split-plot design (Rabbit exclusion within Blocks, Lime treatment within Rabbit plots,

3 Competition treatments within each Lime plot and 4 nutrient regimes within each Competition plot):

```
counts<-read.table("c:\\temp\\splitcounts.txt",header=T)
attach(counts)
names(counts)
```

```
[1]  "vals"          "Block"      "Rabbit"   "Lime"
[5]  "Competition"   "Nutrient"
```

The syntax within lmer is very straightforward: fixed effects after the tilde ~, then random effects inside brackets, then the GLM family:

```
library(lme4)
model<-
lmer(vals~Nutrient+(1|Block/Rabbit/Lime/Competition),family=poisson)
summary(model)
```

```
Generalized linear mixed model fit using Laplace
Formula: vals ~ Nutrient + (1 | Block/Rabbit/Lime/Competition)
 Family: poisson(log link)
  AIC    BIC  logLik  deviance
420.2  451.8  -202.1     404.2
```

Random effects:

```
Groups                              Name          Variance    Std.Dev.
Competition:(Lime:(Rabbit:Block))  (Intercept)   2.2660e-03  4.7603e-02
Lime:(Rabbit:Block)                (Intercept)   5.0000e-10  2.2361e-05
Rabbit:Block                       (Intercept)   5.0000e-10  2.2361e-05
Block                              (Intercept)   5.0000e-10  2.2361e-05
number of obs: 384, groups: Competition:(Lime:(Rabbit:Block)),96;
Lime:(Rabbit:Block), 32; Rabbit:Block, 16; Block, 8
Estimated scale (compare to 1) 0.974339
```

```
Fixed effects:
             Estimate  Std. Error  z value  Pr(>|z|)
(Intercept)   1.10794     0.05885   18.826    <2e-16  ***
NutrientNP    0.11654     0.08063    1.445     0.148
NutrientO    -0.02094     0.08338   -0.251     0.802
NutrientP    -0.01047     0.08316   -0.126     0.900
```

Correlation of Fixed Effects:

```
            (Intr)   NtrnNP   NtrntO
NutrientNP  -0.725
NutrientO   -0.701   0.512
NutrientP   -0.703   0.513    0.496
```

There are no significant differences in snail density under any of the four nutrient treatments (Fixed effects, minimum $p = 0.148$) and only Competition within Lime within Rabbit within Block has an appreciable variance component (standard deviation 0.047 603). Note that because we are using Poisson errors, the fixed effects are on the log scale (the scale of the linear predictor; see p. 513). You might want to compare these best linear unbiased predictors with the logs of the arithmetic mean snail counts:

log(tapply(vals,Nutrient,mean))

```
        N          NP           O           P
 1.108975    1.225612    1.088141    1.098612
```

The values are so close because in this case the random effects are so slight (see p. 627). Note, too, that there is no evidence of overdispersion once the random effects have been incorporated, and the estimated scale parameter is 0.974 339 (it would be 1 in a perfect Poisson world).

15

Count Data in Tables

The analysis of count data with categorical explanatory variables comes under the heading of contingency tables. The general method of analysis for contingency tables involves log-linear modelling, but the simplest contingency tables are often analysed by Pearson's chi-squared, Fisher's exact test or tests of binomial proportions (see p. 308)

A Two-Class Table of Counts

You count 47 animals and find that 29 of them are males and 18 are females. Are these data sufficiently male-biased to reject the null hypothesis of an even sex ratio? With an even sex ratio the expected number of males and females is $47/2 = 23.5$. The simplest test is Pearson's chi-squared in which we calculate

$$\chi^2 = \sum \frac{(\text{observed} - \text{expected})^2}{\text{expected}}.$$

Substituting our observed and expected values, we get

$$\chi^2 = \frac{(29 - 23.5)^2 + (18 - 23.5)^2}{23.5} = 2.574\,468.$$

This is less than the critical value for chi-squared with 1 degree of freedom (3.841), so we conclude that the sex ratio is not significantly different from 50:50. There is a built-in function for this

```
observed<-c(29,18)
chisq.test(observed)
```

        Chi-squared test for given probabilities

```
data:  observed
X-squared = 2.5745, df = 1, p-value = 0.1086
```

which indicates that a sex ratio of this size or more extreme than this would arise by chance alone about 10% of the time ($p = 0.1086$). Alternatively, you could carry out a binomial test:

```
binom.test(observed)
```

```
        Exact binomial test
data: observed
number of successes = 29, number of trials = 47, p-value = 0.1439
alternative hypothesis: true probability of success is not equal to 0.5
95 percent confidence interval:
0.4637994  0.7549318
sample estimates:
probability of success
              0.6170213
```

You can see that the 95% confidence interval for the proportion of males (0.46, 0.75) contains 0.5, so there is no evidence against a 50:50 sex ratio in these data. The p value is slightly different than it was in the chi-squared test, but the interpretation is exactly the same.

Sample Size for Count Data

How many samples do you need before you have any chance of detecting a significant departure from equality? Suppose you are studying sex ratios in families. How many female children would you need to discover in a family with no males before you could conclude that a father's sex-determining chromosomes were behaving oddly? What about five females and no males? This is not significant because it can occur by chance when $p = 0.5$ with probability $2 \times 0.5^5 = 0.0625$ (note that this is a two-tailed test). The smallest sample that gives significance is a family of six children, all of one sex: $2 \times 0.5^6 = 0.03125$. How big would the sample need to be to reject the null hypothesis if one of the children was of the opposite sex? One out of seven is no good, as is one out of eight. You need a sample of at least nine children before you can reject the hypothesis that $p = 0.5$ when one of the children is of the opposite sex. Here is that calculation using the binom.test function:

binom.test(1,9)

```
        Exact binomial test
data: 1 and 9
number of successes = 1, number of trials = 9, p-value = 0.03906
alternative hypothesis: true probability of success is not equal to 0.5
95 percent confidence interval:
0.002809137  0.482496515
sample estimates:
probability of success
              0.1111111
```

A Four-Class Table of Counts

Mendel's famous peas produced 315 yellow round phenotypes, 101 yellow wrinkled, 108 green round and 32 green wrinkled offspring (a total of 556):

observed<-c(315,101,108,32)

The question is whether these data depart significantly from the 9:3:3:1 expectation that would arise if there were two independent 3:1 segregations (with round seeds dominating wrinkled, and yellow seeds dominating to green)?

Because the null hypothesis is not a 25:25:25:25 distribution across the four categories, we need to calculate the expected frequencies explicitly:

(expected<-556*c(9,3,3,1)/16)

```
312.75 104.25 104.25 34.75
```

The expected frequencies are very close to the observed frequencies in Mendel's experiment, but we need to quantify the difference between them and ask how likely such a difference is to arise by chance alone:

chisq.test(observed,p=c(9,3,3,1),rescale.p=TRUE)

```
        Chi-squared test for given probabilities

data:   observed
X-squared = 0.47, df = 3, p-value = 0.9254
```

Note the use of different probabilities for the four phenotypes p=c(9,3,3,1). Because these values do not sum to 1.0, we require the extra argument rescale.p=TRUE. A difference as big as or bigger than the one observed will arise by chance alone in more than 92% of cases and is clearly not statistically significant. The chi-squared value is

sum((observed-expected)^2/expected)

```
[1]   0.470024
```

and the p-value comes from the right-hand tail of the cumulative probability function of the chi-squared distribution 1-pchisq with 3 degrees of freedom (4 comparisons -1 for contingency; the total count must be 556)

1-pchisq(0.470024,3)

```
[1]   0.9254259
```

exactly as we obtained using the built-in chisq.test function, above.

Two-by-Two Contingency Tables

Count data are often classified by more than one categorical explanatory variable. When there are two explanatory variables and both have just two levels, we have the famous two-by-two contingency table (see p. 309). We can return to the example of Mendel's peas. We need to convert the vector of observed counts into a matrix with two rows:

observed<-matrix(observed,nrow=2)
observed

```
       [,1]  [,2]
[1,]    315   108
[2,]    101    32
```

Fisher's exact test (p. 308) can take such a matrix as its sole argument:

fisher.test(observed)

```
                Fisher's Exact Test for Count Data
data: observed
p-value = 0.819
alternative hypothesis: true odds ratio is not equal to 1
95 percent confidence interval:
0.5667874  1.4806148
sample estimates:
odds ratio
 0.9242126
```

Alternatively we can use Pearson's chi-squared test with Yates' continuity correction:

chisq.test(observed)

```
            Pearson's Chi-squared test with Yates' continuity correction
data: observed
X-squared = 0.0513, df = 1, p-value = 0.8208
```

Again, the p-values are different with different tests, but the interpretation is the same: these pea plants behave in accordance with Mendel's predictions of two independent traits, coat colour and seed shape, each segregating 3:1.

Using Log-linear Models for Simple Contingency Tables

It is worth repeating these simple examples with a log-linear model so that when we analyse more complex cases you have a feel for what the GLM is doing. Recall that the deviance for a log-linear model of count data (p. 516) is

$$\text{deviance} = 2 \sum O \ln \left(\frac{O}{E} \right),$$

where O is a vector of observed counts and E is a vector of expected counts. Our first example had 29 males and 18 females and we wanted to know if the sex ratio was significantly male-biased:

observed<-c(29,18)
summary(glm(observed~1,poisson))

```
     Null deviance: 2.5985 on 1 degrees of freedom
Residual deviance: 2.5985 on 1 degrees of freedom
AIC: 14.547
Number of Fisher Scoring iterations: 4
```

Only the bottom part of the summary table is informative in this case. The residual deviance is compared to the critical value of chi-squared in tables with 1 d.f.:

1-pchisq(2.5985,1)

```
[1] 0.1069649
```

We accept the null hypothesis that the sex ratio is 50:50 ($p=0.10696$).

In the case of Mendel's peas we had a four-level categorical variable (i.e. four phenotypes) and the null hypothesis was a 9:3:3:1 distribution of traits:

```
observed<-c(315,101,108,32)
```

We need vectors of length 4 for the two seed traits, shape and colour:

```
shape<-factor(c("round","round","wrinkled","wrinkled"))
colour<-factor(c("yellow","green","yellow","green"))
```

Now we fit a saturated model (model1) and a model without the interaction term (model2) and compare the two models using anova with a chi-squared test:

```
model1<-glm(observed~shape*colour,poisson)
model2<-glm(observed~shape+colour,poisson)
anova(model1,model2,test="Chi")
```

```
Analysis of Deviance Table

Model 1: observed ~ shape * colour
Model 2: observed ~ shape + colour
  Resid. Df  Resid. Dev  Df  Deviance  P(>|Chi|)
1         0   1.021e-14
2         1     0.11715  -1  -0.11715    0.73215
```

There is no interaction between seed colour and seed shape ($p = 0.732\,15$) so we conclude that the two traits are independent and the phenotypes are distributed 9:3:3:1 as predicted. The p-value is slightly different because the ratios of the two dominant traits are not exactly 3:1 in the data: round to wrinkled is $\exp(1.089\,04) = 2.971\,42$ and yellow to green is $\exp(1.157\,02) = 3.180\,441$:

```
summary(model2)
```

```
Coefficients:

               Estimate  Std. Error  z value  Pr(>|z|)
(Intercept)     4.60027     0.09013    51.04    <2e-16  ***
shapewrinkled  -1.08904     0.09771   -11.15    <2e-16  ***
colouryellow    1.15702     0.09941    11.64    <2e-16  ***
```

To summarize, the log-linear model involves fitting a saturated model with zero residual deviance (a parameter is estimated for every row of the dataframe) and then simplifying the model by removing the highest-order interaction term. The increase in deviance gives the chi-squared value for testing the hypothesis of independence. The minimal model must contain all the nuisance variables necessary to constrain the marginal totals (i.e. main effects for shape and colour in this example), as explained on p. 302.

The Danger of Contingency Tables

We have already dealt with simple contingency tables and their analysis using Fisher's exact test or Pearson's chi-squared (see p. 308). But there is an important further issue to be dealt with. In observational studies we quantify only a limited number of explanatory variables. It is inevitable that we shall fail to measure a number of factors that have an important influence on the behaviour of the system in question. That's life, and given that we make every effort to note the important factors, there is little we can do about it. The problem comes when we ignore factors that have an important influence on the response variable.

This difficulty can be particularly acute if we aggregate data over important explanatory variables. An example should make this clear.

Suppose we are carrying out a study of induced defences in trees. A preliminary trial has suggested that early feeding on a leaf by aphids may cause chemical changes in the leaf which reduce the probability of that leaf being attacked later in the season by hole-making insects. To this end we mark a large cohort of leaves, then score whether they were infested by aphids early in the season and whether they were holed by insects later in the year. The work was carried out on two different trees and the results were as follows:

Tree	Aphids	Holed	Intact	Total leaves	Proportion holed
Tree 1	Absent	35	1750	1785	0.0196
	Present	23	1146	1169	0.0197
Tree 2	Absent	146	1642	1788	0.0817
	Present	30	333	363	0.0826

There are four variables: the response variable, count, with eight values (highlighted above), a two-level factor for late season feeding by caterpillars (holed or intact), a two-level factor for early season aphid feeding (aphids present or absent) and a two-level factor for tree (the observations come from two separate trees, imaginatively named Tree1 and Tree2).

```
induced<-read.table("C:\\temp\\induced.txt",header=T)
attach(induced)
names(induced)
```

```
[1]  "Tree"  "Aphid"  "Caterpillar"  "Count"
```

We begin by fitting what is known as a **saturated model**. This is a curious thing, which has as many parameters as there are values of the response variable. The fit of the model is perfect, so there are no residual degrees of freedom and no residual deviance. The reason why we fit a saturated model is that it is always the best place to start modelling complex contingency tables. If we fit the saturated model, then there is no risk that we inadvertently leave out important interactions between the so-called 'nuisance variables'. These are the parameters that need to be in the model to ensure that the marginal totals are properly constrained.

```
model<-glm(Count~Tree*Aphid*Caterpillar,family=poisson)
```

The asterisk notation ensures that the saturated model is fitted, because all of the main effects and two-way interactions are fitted, along with the 3-way interaction Tree by Aphid by Caterpillar. The model fit involves the estimation of $2 \times 2 \times 2 = 8$ parameters, and exactly matches the eight values of the response variable, Count. Looking at the saturated model in any detail serves no purpose, because the reams of information it contains are all superfluous.

The first real step in the modelling is to use update to remove the three-way interaction from the saturated model, and then to use anova to test whether the three-way interaction is significant or not:

```
model2<-update(model , ~ . - Tree:Aphid:Caterpillar)
```

The punctuation here is very important (it is comma, tilde, dot, minus), and note the use of colons rather than asterisks to denote interaction terms rather than main effects plus interaction terms. Now we can see whether the three-way interaction was significant by specifying test="Chi" like this:

anova(model,model2,test="Chi")

```
Analysis of Deviance Table
```

```
Model 1: Count~Tree * Aphid * Caterpillar
Model 2: Count~ Tree + Aphid + Caterpillar + Tree:Aphid +
Tree:Caterpillar + Aphid:Caterpillar
```

	Resid. Df	Resid. Dev	Df	Deviance	P(>|Chi|)
1	0	-9.97e-14			
2	1	0.00079	-1	-0.00079	0.97756

This shows clearly that the interaction between caterpillar attack and leaf holing does not differ from tree to tree ($p = 0.977\,56$). Note that if this interaction had been significant, then we would have stopped the modelling at this stage. But it wasn't, so we leave it out and continue.

What about the main question? Is there an interaction between aphid attack and leaf holing? To test this we delete the Caterpillar–Aphid interaction from the model, and assess the results using anova:

model3<-update(model2 , ~ . - Aphid:Caterpillar)
anova(model3,model2,test="Chi")

```
Analysis of Deviance Table
```

```
Model 1: Count~ Tree + Aphid + Caterpillar + Tree:Aphid +
Tree:Caterpillar
Model 2: Count~ Tree + Aphid + Caterpillar + Tree:Aphid +
Tree:Caterpillar + Aphid:Caterpillar
```

	Resid. Df	Resid. Dev	Df	Deviance	P(>|Chi|)
1	2	0.00409			
2	1	0.00079	1	0.00329	0.95423

There is absolutely no hint of an interaction ($p = 0.954$). The interpretation is clear: this work provides no evidence for induced defences caused by early season caterpillar feeding.

But look what happens when we do the modelling the wrong way. Suppose we went straight for the interaction of interest, Aphid–Caterpillar. We might proceed like this:

wrong<-glm(Count~Aphid*Caterpillar,family=poisson)
wrong1<-update (wrong,~. - Aphid:Caterpillar)
anova(wrong,wrong1,test="Chi")

```
Analysis of Deviance Table
Model 1: Count  ~ Aphid * Caterpillar
Model 2: Count  ~ Aphid + Caterpillar
```

	Resid. Df	Resid. Dev	Df	Deviance	P(>|Chi|)
1	4	550.19			
2	5	556.85	-1	-6.66	0.01

The Aphid by Caterpillar interaction is highly significant ($p = 0.01$), providing strong evidence for induced defences. This is wrong! By failing to include Tree in the model we have omitted an important explanatory variable. As it turns out, and as we should really have determined by more thorough preliminary analysis, the trees differ enormously in their average levels of leaf holing:

as.vector(tapply(Count,list(Caterpillar,Tree),sum))[1]/tapply(Count,Tree,sum) [1]

```
      Tree1
0.01963439
```

as.vector(tapply(Count,list(Caterpillar,Tree),sum))[3]/tapply(Count,Tree,sum) [2]

```
      Tree2
0.08182241
```

Tree2 has more than four times the proportion of its leaves holed by caterpillars. If we had been paying more attention when we did the modelling the wrong way, we should have noticed that the model containing only Aphid and Caterpillar had massive overdispersion, and this should have alerted us that all was not well.

The moral is simple and clear. Always fit a saturated model first, containing all the variables of interest and all the interactions involving the nuisance variables (Tree in this case). Only delete from the model those interactions that involve the variables of interest (Aphid and Caterpillar in this case). Main effects are meaningless in contingency tables, as are the model summaries. Always test for overdispersion. It will never be a problem if you follow the advice of simplifying down from a saturated model, because you only ever leave out non-significant terms, and you never delete terms involving any of the nuisance variables.

Quasi-Poisson and Negative Binomial Models Compared

The data on red blood cell counts were introduced on p. 187. Here we read similar count data from a file:

```
data<-read.table("c:\\temp\\bloodcells.txt",header=T)
attach(data)
names(data)
```

```
[1]  "count"
```

Now we need to create a vector for gender containing 5000 repeats of 'female' and then 5000 repeats of 'male':

```
gender<-factor(rep(c("female","male"),c(5000,5000)))
```

The idea is to test the significance of the difference in mean cell counts for the two genders, which is slightly higher in males than in females:

```
tapply(count,gender,mean)
```

```
female    male
1.1986   1.2408
```

We begin with the simplest log-linear model – a GLM with Poisson errors:

```
model<-glm(count~gender,poisson)
summary(model)
```

You should check for overdispersion before drawing any conclusions about the significance of the gender effect. It turns out that there is substantial overdispersion (scale parameter = 23 154/9998 = 2.315 863), so we repeat the modelling using quasi-Poisson errors instead:

```
model<-glm(count~gender,quasipoisson)
summary(model)
```

```
Call:
glm(formula = count ~ gender, family = quasipoisson)

Deviance Residuals:
     Min       1Q    Median       3Q      Max
  -1.5753  -1.5483  -1.5483   0.6254   7.3023

Coefficients:
              Estimate  Std. Error  t value  Pr(>|t|)
(Intercept)    0.18115     0.02167    8.360   <2e-16  ***
gendermale     0.03460     0.03038    1.139    0.255

(Dispersion parameter for quasipoisson family taken to be 2.813817)

    Null deviance: 23158  on 9999  degrees of freedom
Residual deviance: 23154  on 9998  degrees of freedom
AIC: NA

Number of Fisher Scoring iterations: 6
```

As you see, the gender effect falls well short of significance ($p = 0.255$).

Alternatively, you could use a GLM with negative binomial errors. The function is in the MASS library:

```
library(MASS)
model<-glm.nb(count~gender)
summary(model)
```

```
Call:
glm.nb(formula = count ~ gender, init.theta = 0.667624600666417,
       link = log)

Deviance Residuals:

     Min       1Q    Median       3Q      Max
  -1.1842  -1.1716  -1.1716   0.3503   3.1523

Coefficients:
              Estimate  Std. Error  z value  Pr(>|z|)
(Intercept)    0.18115     0.02160    8.388   <2e-16  ***
gendermale     0.03460     0.03045    1.136    0.256

Dispersion parameter for Negative Binomial(0.6676) family taken to be 1

    Null deviance: 9610.8  on 9999  degrees of freedom
Residual deviance: 9609.5  on 9998  degrees of freedom
AIC: 30362

Number of Fisher Scoring iterations: 1
Correlation of Coefficients:
           (Intercept)
```

```
gendermale -0.71
```

$$\text{Theta: } 0.6676$$
$$\text{Std. Err.: } 0.0185$$

```
2 x log-likelihood: -30355.6010
```

You would come to the same conclusion, although the p value is slightly different ($p = 0.256$).

A Contingency Table of Intermediate Complexity

We start with a three-dimenstional table of count data from college records. It is a contingency table with two levels of year (freshman and sophomore), two levels of discipline (arts and science), and two levels of gender (male and female):

```
numbers<-c(24,30,29,41,14,31,36,35)
```

The statistical question is whether the relationship between gender and discipline varies between freshmen and sophomores (i.e. we want to know the significance of the three-way interaction between year, discipline and gender).

The first task is to define the dimensions of numbers using the dim function.

```
dim(numbers)<-c(2,2,2)
numbers
```

```
, , 1

     [,1]  [,2]
[1,]   24    29
[2,]   30    41

, , 2

     [,1]  [,2]
[1,]   14    36
[2,]   31    35
```

The top table refers to the males [„1] and the bottom table to the females [„2]. Within each table, the rows are the year groups and the columns are the disciplines. It would make the table much easier to understand if we provided these dimensions with names using the dimnames function:

```
dimnames(numbers)[[3]] <- list("male", "female")
dimnames(numbers)[[2]] <- list("arts", "science")
dimnames(numbers)[[1]] <- list("freshman", "sophomore")
```

To see this as a flat table, use the ftable function like this

```
ftable(numbers)
```

```
                    male  female

freshman   arts       24      14
           science    29      36
sophomore  arts       30      31
           science    41      35
```

The thing to understand is that the dimnames are the factor levels (e.g. male or female), not the names of the factors (e.g. gender).

We convert this table into a dataframe using the **as.data.frame.table** function. This saves us from having to create separate vectors to describe the levels of gender, year and discipline associated with each count:

as.data.frame.table(numbers)

```
       Var1       Var2     Var3  Freq
1   freshman      arts     male    24
2  sophomore      arts     male    30
3   freshman   science     male    29
4  sophomore   science     male    41
5   freshman      arts   female    14
6  sophomore      arts   female    31
7   freshman   science   female    36
8  sophomore   science   female    35
```

You can see that R has generated reasonably sensible variable names for the four columns, but we want to use our own names:

frame<-as.data.frame.table(numbers)
names(frame)<-c("year","discipline","gender","count")
frame

```
       year  discipline  gender  count
1   freshman       arts    male     24
2  sophomore       arts    male     30
3   freshman    science    male     29
4  sophomore    science    male     41
5   freshman       arts  female     14
6  sophomore       arts  female     31
7   freshman    science  female     36
8  sophomore    science  female     35
```

Now we can do the statistical modelling. The response variable is count, and we begin by fitting a saturated model with eight estimated parameters (i.e. the model generates the observed counts exactly, so the deviance is zero and there are no degrees of freedom):

attach(frame)
model1<-glm(count~year*discipline*gender,poisson)

We test for the significance of the year-by-discipline-by-gender interaction by deleting the year by discipline by gender interaction from model1 to make model2 using **update**

model2<-update(model1,~. - year:discipline:gender)

then comparing model1 and model2 using **anova** with a chi-squared test:

anova(model1,model2,test="Chi")

```
Analysis of Deviance Table
Model 1: count ~ year * discipline * gender
Model 2: count ~ year + discipline + gender + year:discipline +
year:gender +
     discipline:gender
   Resid. Df   Resid. Dev   Df  Deviance  P(>|Chi|)
1          0  -5.329e-15
2          1     3.08230   -1  -3.08230    0.07915
```

The interaction is not significant ($p = 0.079$), indicating similar gender-by-discipline relationships in the two year groups. We finish the analysis at this point because we have answered the question that we were asked to address.

Schoener's Lizards: A Complex Contingency Table

In this section we are interested in whether lizards show any niche separation across various ecological factors and, in particular, whether there are any interactions – for example, whether they show different habitat separation at different times of day.

```
lizards<-read.table("c:\\temp\\lizards.txt",header=T)
attach(lizards)
names(lizards)
```

```
[1] "n" "sun" "height" "perch" "time" "species"
```

The response variable is n, the count for each contingency. The explanatory variables are all categorical: sun is a two-level factor (Sun and Shade), height is a two-level factor (High and Low), perch is a two-level factor (Broad and Narrow), time is a three-level factor (Afternoon, Mid.day and Morning), and there are two lizard species both belonging to the genus *Anolis* (*A. grahamii* and *A. opalinus*). As usual, we begin by fitting a saturated model, fitting all the interactions and main effects:

```
model1<-glm(n~sun*height*perch*time*species,poisson)
```

Model simplification begins with removal of the highest-order interaction effect: the sun-by-height-by-perch-by-time-by-species interaction (!):

```
model2<-update(model1, ~.- sun:height:perch:time:species)
anova(model1,model2,test="Chi")
```

```
Analysis of Deviance Table
```

	Resid. Df	Resid. Dev	Df	Deviance	P(>|Chi|)
1	0	3.348e-10			
2	2	2.181e-10	-2	1.167e-10	1

It is a considerable relief that this interaction is not significant (imagine trying to explain what it meant in the Discussion section of your paper). The key point to understand in this kind of analysis is that the only interesting terms are interactions involving species. All of the other interactions and main effects are nuisance variables that have to be retained in the model to constrain the marginal totals (see p. 302 for an explanation of what this means).

There are four 4-way interactions of interest – species by sun by height by perch, species by sun by height by time, species by sun by perch by time, and species by height by perch by time – and we should test their significance by deleting them from model2 which contains all of the four-way interactions. Here goes:

```
model3<-update(model2, ~.-sun:height:perch:species)
anova(model2,model3,test="Chi")
```

```
Analysis of Deviance Table
```

	Resid. Df	Resid. Dev	Df	Deviance	P(>|Chi|)
1	2	2.181e-10			
2	3	2.7088	-1	-2.7088	0.0998

Close, but not significant ($p = 0.0998$).

```
model4<-update(model2, ~.-sun:height:time:species)
anova(model2,model4,test="Chi")
```

```
Analysis of Deviance Table
   Resid. Df  Resid. Dev   Df  Deviance  P(>|Chi|)
1          2   2.181e-10
2          4     0.44164   -2  -0.44164    0.80186
```

Nothing at all ($p = 0.802$).

```
model5<-update(model2, ~.-sun:perch:time:species)
anova(model2,model5,test="Chi")
```

```
Analysis of Deviance Table

   Resid. Df  Resid. Dev   Df  Deviance  P(>|Chi|)
1          2   2.181e-10
2          4     0.81008   -2  -0.81008    0.66695
```

Again, nothing there ($p = 0.667$). Finally,

```
model6<-update(model2, ~.-height:perch:time:species)
anova(model2,model6,test="Chi")
```

```
Analysis of Deviance Table

   Resid. Df  Resid. Dev   Df  Deviance  P(>|Chi|)
1          2   2.181e-10
2          4     3.2217    -2  -3.2217     0.1997
```

This means that none of the four-way interactions involving species need be retained.

Now we have to assess all six of the three-way interactions involving species: species by height by perch, species by height by time, species by perch by time, species by sun by height, species by sun by time, and species by sun by perch. Can we speed this up by using automatic deletion? Yes and no. Yes, we can use step, to remove terms assessed by AIC to be non-significant (p. 501). No, unless we are very careful. We must not allow step to remove any interactions that do not involve species (because these are the nuisance variables). We do this with the lower argument:

```
model7<-step(model1,lower=~sun*height*perch*time)
```

```
                              Df  Deviance    AIC
<none>                            3.340   246.589
- sun:height:perch:time       2   7.529   246.778
- sun:height:perch:species    1   5.827   247.076
- height:perch:time:species   2   8.542   247.791
```

You can see that step has been very forgiving, and has left two of the four-way interactions involving species in the model. What we can do next is to take out all of the four-way interactions and start step off again with this simpler starting point. We want to start at the lower model plus all the three-way interactions involving species with sun, height, perch and time:

```
model8<-
glm(n ~sun*height*perch*time+(species+sun+height+perch+time)^3,poisson)
model9<-step(model8,lower= ~sun*height*perch*time)
```

```
                                   Df   Deviance        AIC
<none>                                  11.984   237.233
- sun:height:species           1        14.205   237.453
- sun:height:perch:time        2        17.188   238.436
- time:species                 2        23.714   244.962
- perch:species                1        24.921   248.170
```

Again, we need to be harsh and to test whether these terms really to deserve to stay in model9. The most complex term is the interaction sun by height by perch by time, but we don't want to remove this because it is a nuisance variable (the interaction does not involve species). We should start by removing sun by height by species:

```
model10<-update(model9, ~.-sun:height:species)
anova(model9,model10,test="Chi")
```

```
Analysis of Deviance Table
    Resid. Df   Resid. Dev   Df   Deviance   P(>|Chi|)
1        17      11.9843
2        18      14.2046    -1    -2.2203      0.1362
```

This interaction is not significant, so we leave it out. Model10 contains no three- or four-way interactions involving species.

Let's try deleting the two-way interactions in turn from model10:

```
model11<-update(model10, ~.-sun:species)
model12<-update(model10, ~.-height:species)
model13<-update(model10, ~.-perch:species)
model14<-update(model10, ~.-time:species)
anova(model10,model11,test="Chi")
```

```
Analysis of Deviance Table
    Resid. Df   Resid. Dev   Df   Deviance   P(>|Chi|)
1        18      14.2046
2        19      21.8917    -1    -7.6871      0.0056
```

We need to retain a main effect for sun ($p = 0.0056$).

```
anova(model10,model12,test="Chi")
```

```
Analysis of Deviance Table
    Resid. Df   Resid. Dev   Df   Deviance   P(>|Chi|)
1        18      14.205
2        19      36.271    -1   -22.066    2.634e-06
```

We need to retain a main effect for height ($p < 0.0001$)

```
anova(model10,model13,test="Chi")
```

```
Analysis of Deviance Table
    Resid. Df   Resid. Dev   Df   Deviance   P(>|Chi|)
1        18      14.2046
2        19      27.3346    -1   -13.1300     0.0003
```

We need to retain a main effect for perch ($p = 0.0003$).

anova(model10,model14,test="Chi")

```
Analysis of Deviance Table
    Resid. Df  Resid. Dev  Df  Deviance  P(>|Chi|)
1       18        14.205
2       20        25.802  -2  -11.597      0.003
```

We need to retain a main effect for time of day ($p = 0.003$). To see where we are, we should produce a summary table of the counts:

ftable(tapply(n,list(species,sun,height,perch,time),sum))

				Afternoon	Mid.day	Morning
grahamii	Shade	High	Broad	4	1	2
			Narrow	3	1	3
		Low	Broad	0	0	0
			Narrow	1	0	0
	Sun	High	Broad	10	20	11
			Narrow	8	32	15
		Low	Broad	3	4	5
			Narrow	4	5	1
opalinus	Shade	High	Broad	4	8	20
			Narrow	5	4	8
		Low	Broad	12	8	13
			Narrow	1	0	6
	Sun	High	Broad	18	69	34
			Narrow	8	60	17
		Low	Broad	13	55	31
			Narrow	4	21	12

The modelling has indicated that species differ in their responses to all four explanatory variables, but that there are no interactions between the factors. The only remaining question for model simplification is whether we need to keep all three levels for time of day, or whether two levels would do just as well (we lump together Mid.Day and Morning):

tod<-factor(1+(time=="Afternoon"))
model15<-update(model10, ~.-species:time+species:tod)
anova(model10,model15,test="Chi")

```
Analysis of Deviance Table
    Resid. Df  Resid. Dev  Df  Deviance  P(>|Chi|)
1       18        14.2046
2       19        15.0232  -1  -0.8186      0.3656
```

That simplification was justified, so we keep time in the model but as a two-level factor.

That was hard, I think you will agree. You need to be extremely well organized to do this sort of analysis without making any mistakes. A high degree of serenity is required throughout. What makes it difficult is keeping track of the interactions that are in the model and those that have been excluded, and making absolutely sure that no nuisance variables have been omitted unintentionally. It turns out that life can be made much more straightforward if the analysis can be reformulated as an exercise in proportions rather than counts, because if it can, then all of the problems with nuisance variables disappear. On p. 584 the example is reanalysed with the proportion of all lizards that belong to species *A. opalinus* as the response variable in a GLM with binomial errors.

Plot Methods for Contingency Tables

The departures from expectations of the observed frequencies in a contingency table can be regarded as $(O-E)/\sqrt{E}$. The R function called assocplot produces a Cohen–Friendly association plot indicating deviations from independence of rows and columns in a two-dimensional contingency table.

Here are data on hair colour and eye colour:

```
data(HairEyeColor)
(x <- margin.table(HairEyeColor, c(1, 2)) )
```

```
       Eye
Hair  Brown  Blue  Hazel  Green
Black    68    20     15      5
Brown   119    84     54     29
Red      26    17     14     14
Blond     7    94     10     16
```

```
assocplot(x, main = "Relation between hair and eye color")
```

The plot shows the excess (black bars) of people with black hair that have brown eyes, the excess of people with blond hair that have blue eyes, and the excess of redheads that have green eyes. The red bars show categories where fewer people were observed than expected under the null hypothesis of independence of hair colour and eye colour.

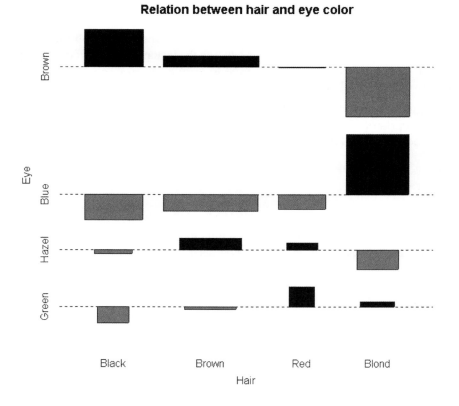

Here are the same data plotted as a mosaic plot:

```
mosaicplot(HairEyeColor, shade = TRUE)
```

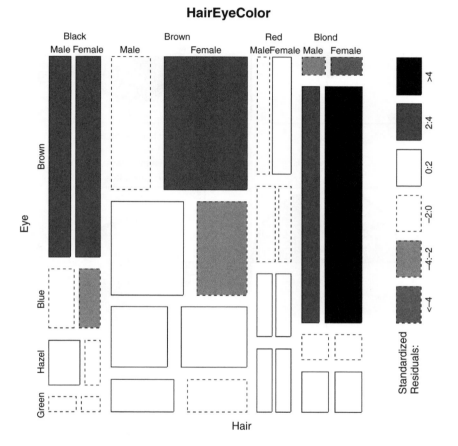

The plot indicates that there are significantly more blue-eyed blond females than expected in the case of independence, and too few brown-eyed blond females. Extended mosaic displays show the standardized residuals of a log-linear model of the counts from by the color and outline of the mosaic's tiles. Negative residuals are drawn in shades of red and with broken outlines, while positive residuals are drawn in shades of blue with solid outlines.

Where there are multiple 2×2 tables (dataframes with three or more categorical explanatory variables), then **fourfoldplot** might be useful. It allows a visual inspection of the association between two dichotomous variables in one or several populations (known as strata). Here are the college admissions data plotted for each department separately:

```
data(UCBAdmissions)
x <- aperm(UCBAdmissions, c(2, 1, 3))
names(dimnames(x)) <- c("Sex", "Admit?", "Department")
ftable(x)
```

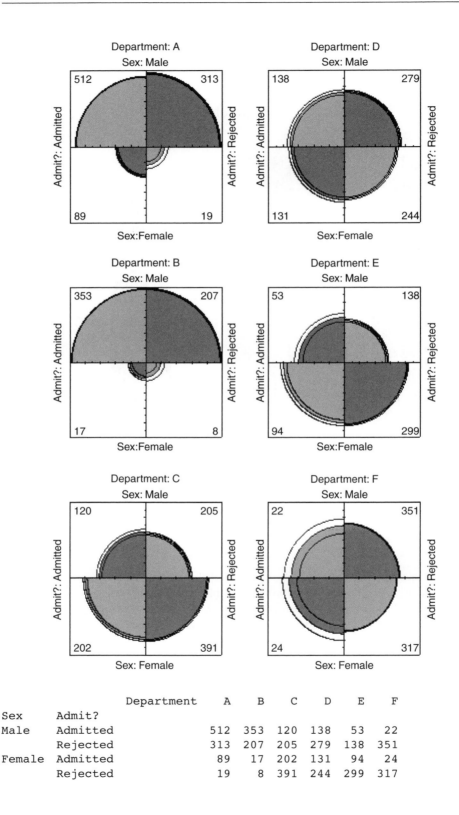

		Department	A	B	C	D	E	F
Sex	Admit?							
Male	Admitted		512	353	120	138	53	22
	Rejected		313	207	205	279	138	351
Female	Admitted		89	17	202	131	94	24
	Rejected		19	8	391	244	299	317

fourfoldplot(x, margin = 2)

You will need to compare the graphs with the frequency table (above) to see what is going on. The central questions are whether the rejection rate for females is different from the rejection rate for males, and whether any such difference varies from department to department. The log-linear model suggests that the difference does vary with department ($p=0.0011$; see below). That Department B attracted a smaller number of female applicants is very obvious. What is less clear (but in many ways more interesting) is that they rejected proportionally fewer of the female applicants (32%) than the male applicants (37%). You may find these plots helpful, but I must admit that I do not.

Here were use gl to generate factor levels for department, sex and admission, then fit a saturated contingency table model for the counts, x. We then use anova with test="Chi" to assess the significance of the three-way interaction

```
dept<-gl(6,4)
sex<-gl(2,1,24)
admit<-gl(2,2,24)
model1<-glm(as.vector(x) ~dept*sex*admit,poisson)
model2<-update(model1, ~. -dept:sex:admit)
anova(model1,model2,test="Chi")
```

```
Analysis of Deviance Table
```

```
Model 1: as.vector(x) ~ dept * sex * admit
Model 2: as.vector(x) ~ dept + sex + admit + dept:sex + dept:admit +
sex:admit
```

	Resid. Df	Resid. Dev	Df	Deviance	P(>|Chi|)
1	0	-5.551e-14			
2	5	20.2043	-5	-20.2043	0.0011

Sex	Admit?	Department	A	B	C	D	E	F
Male	Admitted		512	353	120	138	53	22
	Rejected		313	207	205	279	138	351
Female	Admitted		89	17	202	131	94	24
	Rejected		19	8	391	244	299	317

The interaction is highly significant, indicating that the admission rates of the two sexes differ from department to department.

16

Proportion Data

An important class of problems involves data on proportions such as:

- studies on percentage mortality,

- infection rates of diseases,

- proportion responding to clinical treatment,

- proportion admitting to particular voting intentions,

- sex ratios, or

- data on proportional response to an experimental treatment.

What all these have in common is that we know how many of the experimental objects are in one category (dead, insolvent, male or infected) and we also know how many are in another (alive, solvent, female or uninfected). This contrasts with Poisson count data, where we knew how many times an event occurred, but *not* how many times it did not occur (p. 527).

We model processes involving proportional response variables in R by specifying a generalized linear model with family=binomial. The only complication is that whereas with Poisson errors we could simply specify family=poisson, with binomial errors we must give the number of failures as well as the numbers of successes in a two-vector response variable. To do this we bind together two vectors using cbind into a single object, *y*, comprising the numbers of successes and the number of failures. The **binomial denominator**, *n*, is the total sample, and

number.of.failures = binomial.denominator − number.of.successes
y <- cbind(number.of.successes, number.of.failures)

The old fashioned way of modelling this sort of data was to use the percentage mortality as the response variable. There are four problems with this:

- The errors are not normally distributed.

- The variance is not constant.

- The response is bounded (by 1 above and by 0 below).

The R Book   Michael J. Crawley
© 2007 John Wiley & Sons, Ltd

- By calculating the percentage, we lose information of the size of the sample, n, from which the proportion was estimated.

R carries out weighted regression, using the individual sample sizes as weights, and the logit link function to ensure linearity. There are some kinds of proportion data, such as **percentage cover**, which are best analysed using conventional models (normal errors and constant variance) following **arcsine transformation**. The response variable, y, measured in radians, is $\sin^{-1} \sqrt{0.01 \times p}$, where p is percentage cover. If, however, the response variable takes the form of a **percentage change** in some continuous measurement (such as the percentage change in weight on receiving a particular diet), then rather than arcsine-transforming the data, it is usually better treated by either

- analysis of covariance (see p. 489), using final weight as the response variable and initial weight as a covariate, or

- by specifying the response variable as a relative growth rate, measured as log(final weight/initial weight),

both of which can be analysed with normal errors without further transformation.

Analyses of Data on One and Two Proportions

For comparisons of one binomial proportion with a constant, use binom.test (see p. 300). For comparison of two samples of proportion data, use prop.test (see p. 301). The methods of this chapter are required only for more complex models of proportion data, including regression and contingency tables, where GLMs are used.

Count Data on Proportions

The traditional transformations of proportion data were arcsine and probit. The arcsine transformation took care of the error distribution, while the probit transformation was used to linearize the relationship between percentage mortality and log dose in a bioassay. There is nothing wrong with these transformations, and they are available within R, but a simpler approach is often preferable, and is likely to produce a model that is easier to interpret.

The major difficulty with modelling proportion data is that the responses are *strictly bounded*. There is no way that the percentage dying can be greater than 100% or less than 0%. But if we use simple techniques such as regression or analysis of covariance, then the fitted model could quite easily predict negative values or values greater than 100%, especially if the variance was high and many of the data were close to 0 or close to 100%.

The **logistic** curve is commonly used to describe data on proportions, because, unlike the straight-line model, it asymptotes at 0 and 1 so that negative proportions and responses of more than 100% cannot be predicted. Throughout this discussion we shall use p to describe the proportion of individuals observed to respond in a given way. Because much of their jargon was derived from the theory of gambling, statisticians call these **successes**, although to a demographer measuring death rates this may seem somewhat macabre. The proportion of individuals that respond in other ways (the statistician's **failures**) is therefore $1 - p$, and we shall call this proportion q. The third variable is the size of the sample, n, from which p was estimated (this is the binomial denominator, and the statistician's **number of attempts**).

An important point about the binomial distribution is that the variance is not constant. In fact, the variance of a binomial distribution with mean np is

$$s^2 = npq,$$

so that the variance changes with the mean like this:

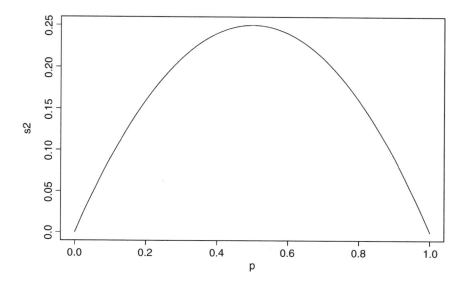

The variance is low when p is very high or very low, and the variance is greatest when $p = q = 0.5$. As p gets smaller, so the binomial distribution gets closer and closer to the Poisson distribution. You can see why this is so by considering the formula for the variance of the binomial (above). Remember that for the Poisson, the variance is equal to the mean: $s^2 = np$. Now, as p gets smaller, so q gets closer and closer to 1, so the variance of the binomial converges to the mean:

$$s^2 = npq \approx np \qquad (q \approx 1).$$

Odds

The logistic model for p as a function of x is given by

$$p = \frac{e^{a+bx}}{1 + e^{a+bx}},$$

and there are no prizes for realizing that the model is not linear. But if $x = -\infty$, then $p = 0$, and if $x = +\infty$ then $p = 1$, so the model is strictly bounded. If $x = 0$, then $p = \exp(a)/[1 + \exp(a)]$. The trick of linearizing the logistic model actually involves a very simple transformation. You may have come across the way in which bookmakers specify probabilities by quoting the **odds** against a particular horse winning a race (they might give odds of 2 to 1 on a reasonably good horse or 25 to 1 on an outsider). This is a rather different way of presenting information on probabilities than scientists are used to dealing with. Thus, where the scientist might state a proportion as 0.667 (2 out of 3), the bookmaker would give odds of 2 to 1 (2 successes to 1 failure). In symbols, this is the difference between the scientist stating the probability p, and the bookmaker stating the odds p/q. Now if we take the odds p/q and substitute this into the formula for the logistic, we get

$$\frac{p}{q} = \frac{e^{a+bx}}{1+e^{a+bx}} \left[1 - \frac{e^{a+bx}}{1+e^{a+bx}} \right]^{-1}$$

which looks awful. But a little algebra shows that

$$\frac{p}{q} = \frac{e^{a+bx}}{1+e^{a+bx}} \left[\frac{1}{1+e^{a+bx}} \right]^{-1} = e^{a+bx}.$$

Now, taking natural logs and recalling that $\ln(e^x) = x$ will simplify matters even further, so that

$$\ln\left(\frac{p}{q}\right) = a + bx.$$

This gives a **linear predictor**, $a + bx$, not for p but for the *logit* transformation of p, namely $\ln(p/q)$. In the jargon of R, the logit is the *link function* relating the linear predictor to the value of p.

Here are p as a function of x (left panel) and logit(p) as a function of x (right panel) for the logistic model with $a = 0.2$ and $b = 0.1$:

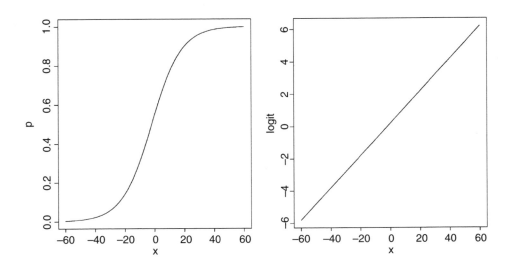

You might ask at this stage: 'why not simply do a linear regression of $\ln(p/q)$ against the explanatory x-variable?' R has three great advantages here:

- It allows for the non-constant binomial variance.

- It deals with the fact that logits for ps near 0 or 1 are infinite.

- It allows for differences between the sample sizes by weighted regression.

Overdispersion and Hypothesis Testing

All the different statistical procedures that we have met in earlier chapters can also be used with data on proportions. Factorial analysis of variance, multiple regression, and a variety of models in which different regression lines are fitted in each of several levels of one or more factors, can be carried out. The only difference is that we assess the significance of terms on the basis of chi-squared – the increase in scaled deviance that results from removal of the term from the current model.

The important point to bear in mind is that hypothesis testing with binomial errors is less clear-cut than with normal errors. While the chi-squared approximation for changes in scaled deviance is reasonable for large samples (i.e. larger than about 30), it is poorer with small samples. Most worrisome is the fact that the degree to which the approximation is satisfactory is itself unknown. This means that considerable care must be exercised in the interpretation of tests of hypotheses on parameters, especially when the parameters are marginally significant or when they explain a very small fraction of the total deviance. With binomial or Poisson errors we cannot hope to provide exact p-values for our tests of hypotheses.

As with Poisson errors, we need to address the question of overdispersion (see p. 522). When we have obtained the minimal adequate model, *the residual scaled deviance should be roughly equal to the residual degrees of freedom*. When the residual deviance is larger than the residual degrees of freedom there are two possibilities: either the model is misspecified, or the probability of success, p, is not constant within a given treatment level. The effect of randomly varying p is to increase the binomial variance from npq to

$$s^2 = npq + n(n-1)\sigma^2,$$

leading to a large residual deviance. This occurs even for models that would fit well if the random variation were correctly specified.

One simple solution is to assume that the variance is not npq but $npq\phi$, where ϕ is an unknown *scale parameter* ($\phi > 1$). We obtain an estimate of the scale parameter by dividing the Pearson chi-squared by the degrees of freedom, and use this estimate of ϕ to compare the resulting scaled deviances. To accomplish this, we use family = quasibinomial rather than family = binomial when there is overdispersion.

The most important points to emphasize in modelling with binomial errors are as follows:

- Create a two-column object for the response, using cbind to join together the two vectors containing the counts of success and failure.

- Check for overdispersion (residual deviance greater than the residual degrees of freedom), and correct for it by using family=quasibinomial rather than binomial if necessary.

- Remember that you do not obtain exact p-values with binomial errors; the chi-squared approximations are sound for large samples, but small samples may present a problem.

- The fitted values are counts, like the response variable.

- The linear predictor is in logits (the log of the odds $= \ln(p/q)$).
- You can back-transform from logits (z) to proportions (p) by $p = 1/[1 + 1/\exp(z)]$.

Applications

You can do as many kinds of modelling in a GLM as in a linear model. Here we show examples of:

- regression with binomial errors (continuous explanatory variables);
- analysis of deviance with binomial errors (categorical explanatory variables);
- analysis of covariance with binomial errors (both kinds of explanatory variables).

Logistic Regression with Binomial Errors

This example concerns sex ratios in insects (the proportion of all individuals that are males). In the species in question, it has been observed that the sex ratio is highly variable, and an experiment was set up to see whether population density was involved in determining the fraction of males.

```
numbers <-read.table("c:\\temp\\sexratio.txt",header=T)
numbers
```

```
  density  females  males
1       1        1      0
2       4        3      1
3      10        7      3
4      22       18      4
5      55       22     33
6     121       41     80
7     210       52    158
8     444       79    365
```

It certainly looks as if there are proportionally more males at high density, but we should plot the data as proportions to see this more clearly:

```
attach(numbers)
par(mfrow=c(1,2))
p<-males/(males+females)
plot(density,p,ylab="Proportion male")
plot(log(density),p,ylab="Proportion male")
```

Evidently, a logarithmic transformation of the explanatory variable is likely to improve the model fit. We shall see in a moment.

The question is whether increasing population density leads to a significant increase in the proportion of males in the population – or, more briefly, whether the sex ratio is density-dependent. It certainly looks from the plot as if it is.

The response variable is a matched pair of counts that we wish to analyse as proportion data using a GLM with binomial errors. First, we bind together the vectors of male and female counts into a single object that will be the response in our analysis:

```
y<-cbind(males,females)
```

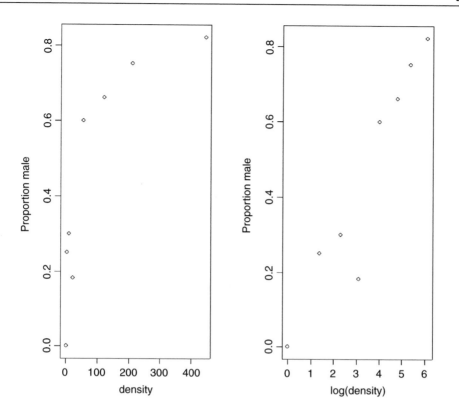

This means that y will be interpreted in the model as the proportion of all individuals that were male. The model is specified like this:

model<-glm(y~density,binomial)

This says that the object called model gets a generalized linear model in which y (the sex ratio) is modelled as a function of a single continuous explanatory variable (called density), using an error distribution from the binomial family. The output looks like this:

summary(model)

```
Coefficients:
              Estimate  Std. Error  z value  Pr(>| z |)
(Intercept)   0.0807368  0.1550376   0.521      0.603
density       0.0035101  0.0005116   6.862    6.81e-12  ***

    Null deviance: 71.159 on 7 degrees of freedom
Residual deviance: 22.091 on 6 degrees of freedom
AIC: 54.618
```

The model table looks just as it would for a straightforward regression. The first parameter is the intercept and the second is the slope of the graph of sex ratio against population density. The slope is highly significantly steeper than zero (proportionately more males at higher population density; $p = 6.81 \times 10^{-12}$). We can see if log transformation of the explanatory variable reduces the residual deviance below 22.091:

```
model<-glm(y~log(density),binomial)
summary(model)
```

```
Coefficients:
              Estimate  Std. Error  z value  Pr(>|z|)
(Intercept)   -2.65927     0.48758   -5.454  4.92e-08  ***
log(density)   0.69410     0.09056    7.665  1.80e-14  ***
```

(Dispersion parameter for binomial family taken to be 1)

```
    Null deviance: 71.1593 on 7 degrees of freedom
Residual deviance: 5.6739 on 6 degrees of freedom
AIC: 38.201
```

This is a big improvement, so we shall adopt it. There is a technical point here, too. In a GLM like this, it is assumed that the residual deviance is the same as the residual degrees of freedom. If the residual deviance is larger than the residual degrees of freedom, this is called overdispersion. It means that there is extra, unexplained variation, over and above the binomial variance assumed by the model specification. In the model with log(density) there is no evidence of overdispersion (residual deviance $= 5.67$ on 6 d.f.), whereas the lack of fit introduced by the curvature in our first model caused substantial overdispersion (residual deviance $= 22.09$ on 6 d.f.).

Model checking involves the use of plot(model). As you will see, there is no pattern in the residuals against the fitted values, and the normal plot is reasonably linear. Point no. 4 is highly influential (it has a large Cook's distance), but the model is still significant with this point omitted.

We conclude that the proportion of animals that are males increases significantly with increasing density, and that the logistic model is linearized by logarithmic transformation of the explanatory variable (population density). We finish by drawing the fitted line though the scatterplot:

```
xv<-seq(0,6,0.1)
plot(log(density),p,ylab="Proportion male")
lines(xv,predict(model,list(density=exp(xv)),type="response"))
```

Note the use of type="response" to back-transform from the logit scale to the S-shaped proportion scale.

Estimating LD50 and LD90 from bioassay data

The data consist of numbers dead and initial batch size for five doses of pesticide application, and we wish to know what dose kills 50% of the individuals (or 90% or 95%, as required). The tricky statistical issue is that one is using a value of y (50% dead) to predict a value of x (the relevant dose) and to work out a standard error on the x axis.

```
data<-read.table("c:\\temp\\bioassay.txt",header=T)
attach(data)
names(data)
```

```
[1]  "dose"  "dead"  "batch"
```

The logistic regression is carried out in the usual way:

```
y<-cbind(dead,batch-dead)
model<-glm(y~log(dose),binomial)
```

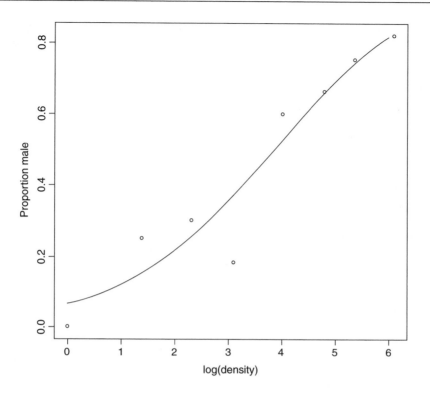

Then the function **dose.p** from the **MASS** library is run with the model object, specifying the proportion killed ($p = 0.5$ is the default for LD50):

```
library(MASS)
dose.p(model,p=c(0.5,0.9,0.95))
```

```
                Dose           SE
p = 0.50: 2.306981   0.07772065
p = 0.90: 3.425506   0.12362080
p = 0.95: 3.805885   0.15150043
```

Here the logs of LD50, LD90 and LD95 are printed, along with their standard errors.

Proportion data with categorical explanatory variables

This next example concerns the germination of seeds of two genotypes of the parasitic plant *Orobanche* and two extracts from host plants (bean and cucumber) that were used to stimulate germination. It is a two-way factorial analysis of deviance.

```
germination<-read.table("c:\\temp\\germination.txt",header=T)
attach(germination)
names(germination)
```

```
[1]  "count"  "sample"  "Orobanche"  "extract"
```

Count is the number of seeds that germinated out of a batch of size = sample. So the number that did not germinate is sample – count, and we construct the response vector like this:

y<-cbind(count, sample-count)

Each of the categorical explanatory variables has two levels:

levels(Orobanche)

[1] "a73" "a75"

levels(extract)

[1] "bean" "cucumber"

We want to test the hypothesis that there is no interaction between *Orobanche* genotype (a73 or a75) and plant extract (bean or cucumber) on the germination rate of the seeds. This requires a factorial analysis using the asterisk * operator like this:

model<-glm(y ~ Orobanche * extract, binomial)

summary(model)

```
Coefficients:
                            Estimate Std. Error z value Pr(>|z|)
(Intercept)                  -0.4122     0.1842  -2.238   0.0252 *
Orobanchea75                 -0.1459     0.2232  -0.654   0.5132
extractcucumber               0.5401     0.2498   2.162   0.0306 *
Orobanchea75:extractcucumber  0.7781     0.3064   2.539   0.0111 *

(Dispersion parameter for binomial family taken to be 1)

    Null deviance: 98.719 on 20 degrees of freedom
 Residual deviance: 33.278 on 17 degrees of freedom
AIC: 117.87
```

At first glance, it looks as if there is a highly significant interaction ($p = 0.0111$). But we need to check that the model is sound. The first thing is to check for is overdispersion. The residual deviance is 33.278 on 17 d.f. so the model is quite badly overdispersed:

33.279 / 17

[1] 1.957588

The overdispersion factor is almost 2. The simplest way to take this into account is to use what is called an 'empirical scale parameter' to reflect the fact that the errors are not binomial as we assumed, but were larger than this (overdispersed) by a factor of 1.9576. We refit the model using **quasibinomial** to account for the overdispersion:

model<-glm(y ~ Orobanche * extract, quasibinomial)

Then we use **update** to remove the interaction term in the normal way:

model2<-update(model, ~ . - Orobanche:extract)

The only difference is that we use an *F* test instead of a chi-squared test to compare the original and simplified models:

anova(model,model2,test="F")

```
Analysis of Deviance Table

Model 1: y ~ Orobanche * extract
Model 2: y ~ Orobanche + extract

     Resid. Df   Resid. Dev   Df   Deviance        F    Pr(>F)
1           17       33.278
2           18       39.686   -1    -6.408   3.4418   0.08099   .
```

Now you see that the interaction is not significant ($p = 0.081$). There is no compelling evidence that different genotypes of *Orobanche* respond differently to the two plant extracts. The next step is to see if any further model simplification is possible.

anova(model2,test="F")

```
Analysis of Deviance Table

Model: quasibinomial, link: logit
Response: y

            Df  Deviance  Resid. Df  Resid. Dev        F      Pr(>F)
NULL                             20      98.719
Orobanche    1     2.544         19      96.175   1.1954      0.2887
extract      1    56.489         18      39.686  26.5412   6.692e-05   ***
```

There is a highly significant difference between the two plant extracts on germination rate, but it is not obvious that we need to keep *Orobanche* genotype in the model. We try removing it:

model3<-update(model2, ~ . - Orobanche)
anova(model2,model3,test="F")

```
Analysis of Deviance Table

Model 1: y ~ Orobanche + extract
Model 2: y ~ extract
     Resid. Df   Resid. Dev   Df   Deviance        F    Pr(>F)
1           18       39.686
2           19       42.751   -1    -3.065   1.4401   0.2457
```

There is no justification for retaining *Orobanche* in the model. So the minimal adequate model contains just two parameters:

coef(model3)

```
(Intercept)     extract
 -0.5121761   1.0574031
```

What, exactly, do these two numbers mean? Remember that the coefficients are from the linear predictor. They are on the transformed scale, so because we are using binomial errors, they are in logits ($\ln(p/(1 - p))$). To turn them into the germination rates for the two plant extracts requires a little calculation. To go from a logit x to a proportion p, you need to do the following sum

$$p = \frac{1}{1 + 1/e^x}.$$

So our first x value is -0.5122 and we calculate

```
1/(1+1/(exp(−0.5122)))
```

```
[1]  0.3746779
```

This says that the mean germination rate of the seeds with the first plant extract was 37%. What about the parameter for extract (1.0574). Remember that with categorical explanatory variables the parameter values are differences between means. So to get the second germination rate we add 1.057 to the intercept before back-transforming:

```
1/(1+1/(exp(−0.5122+1.0574)))
```

```
[1]  0.6330212
```

This says that the germination rate was nearly twice as great (63%) with the second plant extract (cucumber). Obviously we want to generalize this process, and also to speed up the calculations of the estimated mean proportions. We can use predict to help here, because type="response" makes predictions on the back-transformed scale automatically:

```
tapply(predict(model3,type="response"),extract,mean)
```

```
      bean     cucumber
0.3746835   0.6330275
```

It is interesting to compare these figures with the averages of the raw proportions. First we need to calculate the proportion germinating, p, in each sample:

```
p<-count/sample
```

Then we can find the average germination rates for each extract:

```
tapply(p,extract,mean)
```

```
      bean     cucumber
0.3487189   0.6031824
```

You see that this gives different answers. Not too different in this case, it's true, but different none the less. The correct way to average proportion data is to add up the total counts for the different levels of abstract, and only then to turn them into proportions:

```
tapply(count,extract,sum)
```

```
bean    cucumber
 148       276
```

This means that 148 seeds germinated with bean extract and 276 with cucumber. But how many seeds were involved in each case?

```
tapply(sample,extract,sum)
```

```
bean    cucumber
 395       436
```

This means that 395 seeds were treated with bean extract and 436 seeds were treated with cucumber. So the answers we want are 148/395 and 276/436 (i.e. the correct mean proportions). We automate the calculation like this:

```
as.vector(tapply(count,extract,sum))/as.vector(tapply(sample,extract,sum))
```

```
[1]  0.3746835  0.6330275
```

These are the correct mean proportions that were produced by the GLM. The moral here is that you calculate the average of proportions by using total counts and total samples and not by averaging the raw proportions.

To summarize this analysis:

- Make a two-column response vector containing the successes and failures.

- Use glm with family=binomial (you can omit family=).

- Fit the maximal model (in this case it had four parameters).

- Test for overdispersion.

- If, as here, you find overdispersion then use quasibinomial rather than binomial errors.

- Begin model simplification by removing the interaction term.

- This was non-significant once we had adjusted for overdispersion.

- Try removing main effects (we didn't need *Orobanche* genotype in the model).

- Use plot to obtain your model-checking diagnostics.

- Back-transform using predict with the option type="response" to obtain means.

Analysis of covariance with binomial data

We now turn to an example concerning flowering in five varieties of perennial plants. Replicated individuals in a fully randomized design were sprayed with one of six doses of a controlled mixture of growth promoters. After 6 weeks, plants were scored as flowering or not flowering. The count of flowering individuals forms the response variable. This is an ANCOVA because we have both continuous (dose) and categorical (variety) explanatory variables. We use logistic regression because the response variable is a count (flowered) that can be expressed as a proportion (flowered/number).

```
props<-read.table("c:\\temp\\flowering.txt",header=T)
attach(props)
names(props)
```

```
[1]  "flowered"  "number"  "dose"  "variety"
```

```
y<-cbind(flowered,number-flowered)
pf<-flowered/number
pfc<-split(pf,variety)
dc<-split(dose,variety)
```

```
plot(dose,pf,type="n",ylab="Proportion flowered")
points(dc[[1]],pfc[[1]],pch=16)
points(dc[[2]],pfc[[2]],pch=1)
points(dc[[3]],pfc[[3]],pch=17)
points(dc[[4]],pfc[[4]],pch=2)
points(dc[[5]],pfc[[5]],pch=3)
```

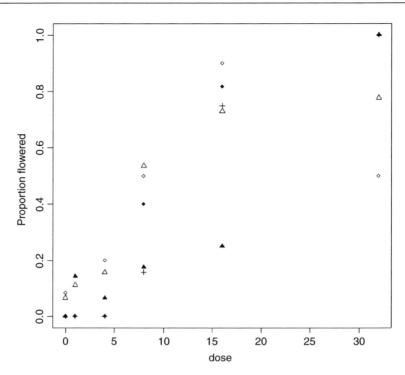

There is clearly a substantial difference between the plant varieties in their response to the flowering stimulant. The modelling proceeds in the normal way. We begin by fitting the maximal model with different slopes and intercepts for each variety (estimating ten parameters in all):

```
model1<-glm(y~dose*variety,binomial)
summary(model1)
```

```
Coefficients:
                 Estimate   Std. Error   z value   Pr(>|z|)
(Intercept)     -4.591189    1.021236    -4.496    6.93e-06   ***
dose             0.412564    0.099107     4.163    3.14e-05   ***
varietyB         3.061504    1.082866     2.827    0.004695   **
varietyC         1.232022    1.178527     1.045    0.295842
varietyD         3.174594    1.064689     2.982    0.002866   **
varietyE        -0.715041    1.537320    -0.465    0.641844
dose:varietyB   -0.342767    0.101188    -3.387    0.000706   ***
dose:varietyC   -0.230334    0.105826    -2.177    0.029515   *
dose:varietyD   -0.304762    0.101374    -3.006    0.002644   **
dose:varietyE   -0.006443    0.131786    -0.049    0.961006
```

```
(Dispersion parameter for binomial family taken to be 1)

    Null deviance: 303.350 on 29 degrees of freedom
Residual deviance: 51.083 on 20 degrees of freedom
AIC: 123.55
```

The model exhibits substantial overdispersion, but this is probably due to poor model selection rather than extra, unmeasured variability. Here is the mean proportion flowered at each dose for each variety:

```
p<-flowered/number
tapply(p,list(dose,variety),mean)
```

	A	B	C	D	E
0	0.0000000	0.08333333	0.00000000	0.06666667	0.0000000
1	0.0000000	0.00000000	0.14285714	0.11111111	0.0000000
4	0.0000000	0.20000000	0.06666667	0.15789474	0.0000000
8	0.4000000	0.50000000	0.17647059	0.53571429	0.1578947
16	0.8181818	0.90000000	0.25000000	0.73076923	0.7500000
32	1.0000000	0.50000000	1.00000000	0.77777778	1.0000000

There are several ways to plot the five different curves on the scatterplot, but perhaps the simplest is to fit the regression model separately for each variety (see the book's website):

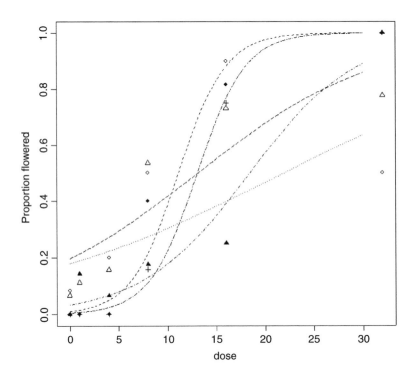

As you can see, the model is reasonable for two of the genotypes (A and E, represented by open and solid diamonds, respectively), moderate for one genotype (C, solid triangles) but poor for two of them, B (open circles) and D (the open triangles). For both of the latter, the model overestimates the proportion flowering at zero dose, and for genotype B there seems to be some inhibition of flowering at the highest dose because the graph falls from 90% flowering at dose 16 to just 50% at dose 32. Variety D appears to be asymptoting at less than 100% flowering. These failures of the model focus attention for future work.

The moral is that the fact that we have proportion data does not mean that the data will necessarily be well described by the logistic model. For instance, in order to describe the response of genotype B, the model would need to have a hump, rather than to asymptote at $p = 1$ for large doses.

Converting Complex Contingency Tables to Proportions

In this section we show how to remove the need for all of the nuisance variables that are involved in complex contingency table modelling (see p. 302) by converting the response variable from a count to a proportion. We work thorough the analysis of Schoener's lizards, which we first encountered in Chapter 15. Instead of analysing the *counts* of the numbers of *Anolis grahamii* and *A. opalinus*, we restructure the data to represent the *proportions* of all lizards that were *A. grahamii* under the various contingencies.

```
lizards<-read.table("c:\\temp\\lizards.txt",header=T)
attach(lizards)
names(lizards)
```

```
[1]  "n"  "sun"  "height"  "perch"  "time"  "species"
```

First, we need to make absolutely sure that all the explanatory variables are in exactly the same order for both species of lizards. The reason for this is that we are going to cbind the counts for one of the lizard species onto the half dataframe containing the other species counts and all of the explanatory variables. Any mistakes here would be disastrous because the count would be lined up with the wrong combination of explanatory variables, and the analysis would be wrong and utterly meaningless.

```
sorted<-lizards[order(species,sun,height,perch,time),]
sorted
```

```
      n    sun   height   perch        time    species
41    4  Shade     High   Broad   Afternoon   grahamii
33    1  Shade     High   Broad     Mid.day   grahamii
25    2  Shade     High   Broad     Morning   grahamii
43    3  Shade     High  Narrow   Afternoon   grahamii
35    1  Shade     High  Narrow     Mid.day   grahamii
27    3  Shade     High  Narrow     Morning   grahamii
42    0  Shade      Low   Broad   Afternoon   grahamii
34    0  Shade      Low   Broad     Mid.day   grahamii
26    0  Shade      Low   Broad     Morning   grahamii
44    1  Shade      Low  Narrow   Afternoon   grahamii
36    0  Shade      Low  Narrow     Mid.day   grahamii
28    0  Shade      Low  Narrow     Morning   grahamii
45   10    Sun     High   Broad   Afternoon   grahamii
37   20    Sun     High   Broad     Mid.day   grahamii
...
24    4    Sun      Low  Narrow   Afternoon   opalinus
16   21    Sun      Low  Narrow     Mid.day   opalinus
 8   12    Sun      Low  Narrow     Morning   opalinus
```

Next we need to take the top half of this dataframe (i.e. rows 1–24):

```
short<-sorted[1:24,]
short
```

```
      n    sun   height   perch      time   species
41    4   Shade    High   Broad  Afternoon  grahamii
33    1   Shade    High   Broad    Mid.day  grahamii
25    2   Shade    High   Broad    Morning  grahamii
43    3   Shade    High  Narrow  Afternoon  grahamii
35    1   Shade    High  Narrow    Mid.day  grahamii
27    3   Shade    High  Narrow    Morning  grahamii
42    0   Shade     Low   Broad  Afternoon  grahamii
34    0   Shade     Low   Broad    Mid.day  grahamii
26    0   Shade     Low   Broad    Morning  grahamii
44    1   Shade     Low  Narrow  Afternoon  grahamii
36    0   Shade     Low  Narrow    Mid.day  grahamii
28    0   Shade     Low  Narrow    Morning  grahamii
45   10     Sun    High   Broad  Afternoon  grahamii
37   20     Sun    High   Broad    Mid.day  grahamii
29   11     Sun    High   Broad    Morning  grahamii
47    8     Sun    High  Narrow  Afternoon  grahamii
39   32     Sun    High  Narrow    Mid.day  grahamii
31   15     Sun    High  Narrow    Morning  grahamii
46    3     Sun     Low   Broad  Afternoon  grahamii
38    4     Sun     Low   Broad    Mid.day  grahamii
30    5     Sun     Low   Broad    Morning  grahamii
48    4     Sun     Low  Narrow  Afternoon  grahamii
40    5     Sun     Low  Narrow    Mid.day  grahamii
32    1     Sun     Low  Narrow    Morning  grahamii
```

Note that this loses all of the data for *A. opalinus*. Also, the name for the left-hand variable, *n*, is no longer appropriate. It is the count for *A. grahamii*, so we should rename it Ag, say (with the intention of adding another column called Ao in a minute to contain the counts of *A. opalinus*):

```
names(short)[1]<-"Ag"
names(short)
```

```
[1] "Ag"  "sun"  "height"  "perch"  "time"  "species"
```

The right-hand variable, species, is redundant now (all the entries are grahamii), so we should drop it:

```
short<-short[,-6]
short
```

```
      Ag    sun   height   perch      time
41     4  Shade    High   Broad  Afternoon
33     1  Shade    High   Broad    Mid.day
25     2  Shade    High   Broad    Morning
43     3  Shade    High  Narrow  Afternoon
...
```

The counts for each row of *A. opalinus* are in the variable called *n* in the bottom half of the dataframe called sorted. We extract them like this:

```
sorted$n[25:48]
```

```
[1]    4  8  20   5  4  8  12  8  13  1  0  6  18  69  34  8  60  17  13  55
      31  4  21  12
```

The idea is to create a new dataframe with these counts for *A. opalinus* lined up alongside the equivalent counts for *A. grahamii*:

```
new.lizards<-data.frame(sorted$n[25:48], short)
```

The first variable needs an informative name, like Ao:

```
names(new.lizards)[1]<-"Ao"
new.lizards
```

```
      Ao  Ag    sun  height   perch       time
41     4   4  Shade    High   Broad  Afternoon
33     8   1  Shade    High   Broad    Mid.day
25    20   2  Shade    High   Broad    Morning
43     5   3  Shade    High  Narrow  Afternoon
35     4   1  Shade    High  Narrow    Mid.day
27     8   3  Shade    High  Narrow    Morning
...
```

That completes the editing of the dataframe. Notice, however, that we have got three dataframes, all of different configurations, but each containing the same variable names (sun, height, perch and time) – look at objects() and search(). We need to do some housekeeping:

```
detach(lizards)
rm(short,sorted)
attach(new.lizards)
```

Analysing Schoener's Lizards as Proportion Data

```
names(new.lizards)
```

```
[1]  "Ao"  "Ag"  "sun"  "height"  "perch"  "time"
```

The response variable is a two-column object containing the counts of the two species:

```
y<-cbind(Ao,Ag)
```

We begin by fitting the saturated model with all possible interactions:

```
model1<-glm(y~sun*height*perch*time,binomial)
```

Since there are no nuisance variables, we can use step directly to begin the model simplification (compare this with p. 560 with a log-linear model of the same data):

```
model2<-step(model1)
```

```
Start: AIC= 102.82
y ~ sun * height * perch * time
```

```
                           Df    Deviance        AIC
- sun:height:perch:time    1    2.180e-10    100.82
<none>                          3.582e-10    102.82
```

Out goes the four-way interaction (with a sigh of relief):

```
Step: AIC= 100.82
                         Df   Deviance        AIC
- sun:height:time        2      0.442     97.266
- sun:perch:time         2      0.810     97.634
- height:perch:time      2      3.222    100.046
<none>                         2.18e-10   100.824
- sun:height:perch       1      2.709    101.533
```

Next, we wave goodbye to three of the three-way interactions

```
Step: AIC= 97.27
                       Df   Deviance        AIC
- sun:perch:time       2      1.071     93.896
<none>                        0.442     97.266
- height:perch:time    2      4.648     97.472
- sun:height:perch     1      3.111     97.936
```

```
Step: AIC= 93.9
                     Df   Deviance        AIC
- sun:time           2      3.340     92.165
<none>                      1.071     93.896
- sun:height:perch   1      3.302     94.126
- height:perch:time  2      5.791     94.615
```

and the two-way interaction of sun by time

```
Step: AIC= 92.16
                     Df   Deviance        AIC
<none>                      3.340     92.165
- sun:height:perch   1      5.827     92.651
- height:perch:time  2      8.542     93.366
```

summary(model2)

```
Call:
glm(formula = y ~ sun + height + perch + time + sun:height + sun:perch
        + height:perch + height:time + perch:time + sun:height:perch +
        height:perch:time, family = binomial)
```

We need to test whether we would have kept the two 3-way interactions and the five 2-way interactions by hand:

```
model3<-update(model2,~. - height:perch:time)
model4<-update(model2,~. - sun:height:perch)
anova(model2,model3,test="Chi")
```

```
Analysis of Deviance Table
     Resid. Df  Resid. Dev  Df  Deviance  P(>|Chi|)
1          7       3.3403
2          9       8.5418  -2   -5.2014    0.0742
```

```
anova(model2,model4,test="Chi")
```

```
Analysis of Deviance Table
     Resid. Df  Resid. Dev  Df  Deviance  P(>|Chi|)
1          7       3.3403
2          8       5.8273  -1   -2.4869    0.1148
```

No. We would not keep either of those three-way interactions. What about the two-way interactions? We need to start with a simpler base model than model2:

```
model5<-glm(y~(sun+height+perch+time)^2-sun:time,binomial)
model6<-update(model5,~. - sun:height)
anova(model5,model6,test="Chi")
```

```
Analysis of Deviance Table
     Resid. Df  Resid. Dev  Df  Deviance  P(>|Chi|)
1         10      10.9032
2         11      13.2543  -1   -2.3511    0.1252
```

```
model7<-update(model5,~. - sun:perch)
anova(model5,model7,test="Chi")
```

```
Analysis of Deviance Table
     Resid. Df  Resid. Dev  Df  Deviance  P(>|Chi|)
1         10      10.9032
2         11      10.9268  -1   -0.0236    0.8779
```

```
model8<-update(model5,~. - height:perch)
anova(model5,model8,test="Chi")
```

```
Analysis of Deviance Table
     Resid. Df  Resid. Dev  Df  Deviance  P(>|Chi|)
1         10      10.9032
2         11      11.1432  -1   -0.2401    0.6242
```

```
model9<-update(model5,~. - time:perch)
anova(model5,model9,test="Chi")
```

```
Analysis of Deviance Table
     Resid. Df  Resid. Dev  Df  Deviance  P(>|Chi|)
1         10      10.9032
2         12      10.9090  -2   -0.0058    0.9971
```

```
model10<-update(model5,~. - time:height)
anova(model5,model10,test="Chi")
```

Analysis of Deviance Table

```
     Resid. Df   Resid. Dev   Df   Deviance   P(>|Chi|)
1          10       10.9032
2          12       11.7600   -2    -0.8568      0.6516
```

So we do not need any of the two-way interactions. What about the main effects?

model11<-glm(y~sun+height+perch+time,binomial)
summary(model11)

Coefficients:

```
                Estimate   Std. Error   z value   Pr(>|z|)
(Intercept)       1.2079       0.3536     3.416   0.000634   ***
sunSun           -0.8473       0.3224    -2.628   0.008585   **
heightLow         1.1300       0.2571     4.395   1.11e-05   ***
perchNarrow      -0.7626       0.2113    -3.610   0.000306   ***
timeMid.day       0.9639       0.2816     3.423   0.000619   ***
timeMorning       0.7368       0.2990     2.464   0.013730   *
```

All the main effects are significant and so must be retained.

Just one last point. We might not need all three levels for time, since the summary suggests that Mid.day and Morning are not significantly different (parameter difference of $0.9639 - 0.7368 = 0.2271$ with a standard error of the difference of 0.29). We lump them together in a new factor called t2:

t2<-time
levels(t2)[c(2,3)]<-"other"
levels(t2)

[1] "Afternoon" "other"

model12<-glm(y~sun+height+perch+t2,binomial)
anova(model11,model12,test="Chi")

Analysis of Deviance Table

Model 1: y ~ sun + height + perch + time
Model 2: y ~ sun + height + perch + t2
```
     Resid. Df   Resid. Dev   Df   Deviance   P(>|Chi|)
1          17       14.2046
2          18       15.0232   -1    -0.8186      0.3656
```

summary(model12)

Coefficients:

```
                Estimate   Std. Error   z value   Pr(>| z | )
(Intercept)       1.1595       0.3484     3.328    0.000874   ***
sunSun           -0.7872       0.3159    -2.491    0.012722   *
heightLow         1.1188       0.2566     4.360    1.30e-05   ***
perchNarrow      -0.7485       0.2104    -3.557    0.000375   ***
t2other           0.8717       0.2611     3.338    0.000844   ***
```

All the parameters are significant, so this is the minimal adequate model. There are just five parameters, and the model contains no nuisance variables (compare this with the massive contingency table model on p. 560). The ecological interpretation is straightforward: the two lizard species differ significantly in their niches on all the niche axes that were measured. However, there were no significant interactions (nothing subtle was happening such as swapping perch sizes at different times of day).

Generalized mixed models lmer with proportion data

Generalized mixed models using lmer are introduced on p. 546. The data concern the proportion of insects killed by pesticide application in four pseudoreplicated plots within each randomly selected half-field in six different farms (blocks A to F):

```
data<-read.table("c:\\temp\\insects.txt",header=T)
attach(data)
names(data)
```

```
[1]  "block"  "treatment"  "replicate"  "dead"  "alive"
```

The response variable for the binomial analysis is created like this:

```
y<-cbind(dead,alive)
```

We intend to use the lmer function, which is part of the lme4 library:

```
library(lme4)
```

The model is fitted by specifying the fixed effects (block and treatment), the random effects (replicates within treatments within blocks), the error family (binomial) and the dataframe name (data) like this:

```
model<-lmer(y~block+treatment+(1|block/treatment),binomial,data=data)
summary(model)
```

This summary indicates a very high level of overdispersion (estimated scale (compare to 1) $= 2.19$) so we refit the model with quasi-binomial rather than binomial errors:

```
model<-lmer(y~block+treatment+(1|block/treatment),quasibinomial,data=data)
summary(model)
```

```
Generalized linear mixed model fit using Laplace
Formula: y ~ block + treatment + (1 | block/treatment)
   Data: data
 Family: quasibinomial(logit link)
  AIC   BIC   logLik   deviance
255.2   272   -118.6      237.2

Random effects:

Groups            Name          Variance     Std.Dev.
treatment:block   (Intercept)   2.4134e-09   4.9127e-05
block             (Intercept)   2.4134e-09   4.9127e-05
Residual                        4.8268e+00   2.1970e+00

number of obs: 48, groups: treatment:block, 12; block, 6
```

```
Fixed effects:
                   Estimate Std.    Error   t value
(Intercept)         -0.5076   0.1624   -3.126
blockB              -0.8249   0.4562   -1.808
blockC              -0.2981   0.2854   -1.044
blockD              -1.3308   0.3270   -4.069
blockE              -1.2758   0.7244   -1.761
blockF              -0.4250   0.3025   -1.405
treatmentsrayed      3.2676   0.2597   12.582
```

The treatment (sprayed) effect is highly significant, but there are also significant differences between the blocks (largely due to blocks D and E). Here is how to calculate the mean proportions dead in the sprayed and unsprayed halves of each block:

tapply(dead,list(treatment,block),sum)/tapply(dead+alive,list(treatment,block),sum)

```
              A           B           C           D           E           F
control  0.3838798   0.2035398   0.3176101   0.1209150   0.2105263   0.2611940
srayed   0.9169960   0.8815789   0.9123377   0.8309179   0.7428571   0.9312715
```

Does the spray differ significantly in effectiveness across different blocks? A simple measure of insecticide effectiveness is the ratio of the proportion killed by the pesticide to the proportion dying without the insecticide. Note the use of the two rows of the tapply object to calculate this ratio:

effectiveness-tapply(dead,list(treatment,block),sum)/
    tapply(dead+alive,list(treatment,block),sum)

effectiveness[2,]/effectiveness[1,]

```
       A           B           C           D           E           F
2.388758    4.331236    2.872509    6.871915    3.528571    3.565439
```

The effectiveness was highest in block D and lowest in block A. We have replication within each block, so we can test this by fitting an interaction term treatment:block and comparing the two models using ANOVA. Because we are changing the fixed-effects structure of the model we need to switch to maximum likelihood:

model2<-
lmer(y~block*treatment+(1|block/treatment),quasibinomial,data=data,method="ML")
model1<-
lmer(y~block+treatment+(1|block/treatment),quasibinomial,data=data,method="ML")
anova(model1,model2,test="F")

```
Data: data
models:
model1:   y ~ block + treatment + (1 | block/treatment)
model2:   y ~ block * treatment + (1 | block/treatment)
          Df    AIC     BIC    logLik   Chisq Chi Df  Pr(>Chisq)
model1    9   255.16  272.00  -118.58
model2   14   256.26  282.46  -114.13  8.8981      5      0.1132
```

The model containing interaction effects uses up five more degrees of freedom, and its AIC is greater than that of the simpler model, so we accept that there is no significant interaction between blocks and spray effectiveness ($p = 0.1132$).

17

Binary Response Variables

Many statistical problems involve binary response variables. For example, we often classify individuals as

- dead or alive,

- occupied or empty,

- healthy or diseased,

- wilted or turgid,

- male or female,

- literate or illiterate,

- mature or immature,

- solvent or insolvent, or

- employed or unemployed.

It is interesting to understand the factors that are associated with an individual being in one class or the other. Binary analysis will be a useful option when at least one of your explanatory variables is continuous (rather than categorical). In a study of company insolvency, for instance, the data would consist of a list of measurements made on the insolvent companies (their age, size, turnover, location, management experience, workforce training, and so on) and a similar list for the solvent companies. The question then becomes which, if any, of the explanatory variables increase the probability of an individual company being insolvent.

The response variable contains only 0s or 1s; for example, 0 to represent dead individuals and 1 to represent live ones. Thus, there is only a single column of numbers for the response, in contrast to proportion data where two vectors (successes and failures) were bound together to form the response (see Chapter 16). The way that R treats binary data is to assume that the 0s and 1s come from *a binomial trial with sample size 1*. If the probability that an individual is dead is p, then the probability of obtaining y (where y is either dead or alive,

0 or 1) is given by an abbreviated form of the binomial distribution with $n = 1$, known as the Bernoulli distribution:

$$P(y) = p^y (1 - p)^{(1-y)}.$$

The random variable y has a mean of p and a variance of $p(1 - p)$, and the objective is to determine how the explanatory variables influence the value of p. The trick to using binary response variables effectively is to know when it is worth using them, and when it is better to lump the successes and failures together and analyse the *total counts* of dead individuals, occupied patches, insolvent firms or whatever. The question you need to ask yourself is: do I have unique values of one or more explanatory variables for each and every individual case?

If the answer is 'yes', then analysis with a binary response variable is likely to be fruitful. If the answer is 'no', then there is nothing to be gained, and you should reduce your data by aggregating the counts to the resolution at which each count *does* have a unique set of explanatory variables. For example, suppose that all your explanatory variables were categorical – sex (male or female), employment (employed or unemployed) and region (urban or rural). In this case there is nothing to be gained from analysis using a binary response variable because none of the individuals in the study have *unique* values of any of the explanatory variables. It might be worthwhile if you had each individual's body weight, for example, then you could ask whether, when you control for sex and region, heavy people are more likely to be unemployed than light people. In the absence of *unique* values for any explanatory variables, there are two useful options:

- Analyse the data as a contingency table using Poisson errors, with the count of the total number of individuals in each of the eight contingencies ($2 \times 2 \times 2$) as the response variable (see Chapter 15) in a dataframe with just eight rows.

- Decide which of your explanatory variables is the key (perhaps you are interested in gender differences), then express the data as proportions (the number of males and the number of females) and recode the binary response as a count of a two-level factor. The analysis is now of proportion data (the proportion of all individuals that are female, for instance) using binomial errors (see Chapter 16).

If you *do* have unique measurements of one or more explanatory variables for each individual, these are likely to be continuous variables such as body weight, income, medical history, distance to the nuclear reprocessing plant, geographic isolation, and so on. This being the case, successful analyses of binary response data tend to be multiple regression analyses or complex analyses of covariance, and you should consult Chapters 9 and 10 for details on model simplification and model criticism.

In order to carry out modelling on a binary response variable we take the following steps:

- Create a single vector containing 0s and 1s as the response variable.

- Use glm with family=binomial.

- Consider changing the link function from default logit to complementary log-log.

- Fit the model in the usual way.

- Test significance by deletion of terms from the maximal model, and compare the change in deviance with chi-squared.

Note that there is no such thing as overdispersion with a binary response variable, and hence no need to change to using **quasibinomial** when the residual deviance is large. The choice of link function is generally made by trying both links and selecting the link that gives the lowest deviance. The logit link that we used earlier is symmetric in p and q, but the complementary log-log link is asymmetric. You may also improve the fit by transforming one or more of the explanatory variables. Bear in mind that you can fit non-parametric smoothers to binary response variables using generalized additive models (as described in Chapter 18) instead of carrying out parametric logistic regression.

Incidence functions

In this example, the response variable is called incidence; a value of 1 means that an island was occupied by a particular species of bird, and 0 means that the bird did not breed there. The explanatory variables are the area of the island (km^2) and the isolation of the island (distance from the mainland, km).

```
island<-read.table("c:\\temp\\isolation.txt",header=T)
attach(island)
names(island)
```

```
[1]  "incidence"  "area"  "isolation"
```

There are two continuous explanatory variables, so the appropriate analysis is multiple regression. The response is binary, so we shall do logistic regression with binomial errors.

We begin by fitting a complex model involving an interaction between isolation and area:

```
model1<-glm(incidence~area*isolation,binomial)
```

Then we fit a simpler model with only main effects for isolation and area:

```
model2<-glm(incidence~area+isolation,binomial)
```

We now compare the two models using ANOVA:

```
anova(model1,model2,test="Chi")
```

```
Analysis of Deviance Table
```

```
Model 1: incidence ~ area * isolation
Model 2: incidence ~ area + isolation
     Resid. Df   Resid. Dev   Df   Deviance   P(>|Chi|)
1          46      28.2517
2          47      28.4022   -1    -0.1504      0.6981
```

The simpler model is not significantly worse, so we accept this for the time being, and inspect the parameter estimates and standard errors:

```
summary(model2)
```

```
Call:
glm(formula = incidence ~ area + isolation, family = binomial)
```

```
Deviance Residuals:
    Min        1Q    Median        3Q       Max
-1.8189   -0.3089    0.0490    0.3635    2.1192
```

```
Coefficients:
               Estimate   Std. Error   z value   Pr(>|z|)
(Intercept)     6.6417       2.9218     2.273    0.02302   *
area            0.5807       0.2478     2.344    0.01909   *
isolation      -1.3719       0.4769    -2.877    0.00401   **
```

(Dispersion parameter for binomial family taken to be 1)

```
        Null deviance: 68.029 on 49 degrees of freedom
    Residual deviance: 28.402 on 47 degrees of freedom
```

The estimates and their standard errors are in logits. Area has a significant positive effect (larger islands are more likely to be occupied), but isolation has a very strong negative effect (isolated islands are much less likely to be occupied). This is the minimal adequate model. We should plot the fitted model through the scatterplot of the data. It is much easier to do this for each variable separately, like this:

```
modela<-glm(incidence~area,binomial)
modeli<-glm(incidence~isolation,binomial)
par(mfrow=c(2,2))
xv<-seq(0,9,0.01)
yv<-predict(modela,list(area=xv),type="response")
plot(area,incidence)
lines(xv,yv)
xv2<-seq(0,10,0.1)
yv2<-predict(modeli,list(isolation=xv2),type="response")
plot(isolation,incidence)
lines(xv2,yv2)
```

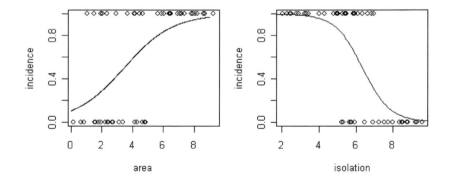

Graphical Tests of the Fit of the Logistic to Data

The logistic plots above are all well and good, but it is very difficult to know how good the fit of the model is when the data are shown only as 0s or 1s. Some people have argued for putting histograms instead of rugs on the top and bottom axes, but there are issues here about the arbitrary location of the bins (see p. 162). **Rugs** are a one-dimensional addition to the bottom (or top) of the plot showing the locations of the data points along the x axis. The idea is to indicate the extent to which the values are clustered at certain values of the explanatory variable, rather than evenly spaced out along it. If there are many values at the same value of x, it will be useful to use the jitter function to spread them out (by randomly selected small distances from x).

A different tack is to cut the data into a number of sectors and plot empirical probabilities (ideally with their standard errors) as a guide to the fit of the logistic curve, but this, too, can be criticized on the arbitrariness of the boundaries to do the cutting, coupled with the fact that there are often too few data points to give acceptable precision to the empirical probabilities and standard errors in any given group.

For what it is worth, here is an example of this approach: the response is occupation of territories and the explanatory variable is resource availability in each territory:

```
occupy<-read.table("c:\\temp\\occupation.txt",header=T)
attach(occupy)
names(occupy)
```

```
[1]  "resources"  "occupied"
```

```
plot(resources,occupied,type="n")
rug(jitter(resources[occupied==0]))
rug(jitter(resources[occupied==1]),side=3)
```

Now fit the logistic regression and draw the line:

```
model<-glm(occupied~resources,binomial)
xv<-0:1000
yv<-predict(model,list(resources=xv),type="response")
lines(xv,yv)
```

The idea is to cut up the ranked values on the x axis (resources) into five categories and then work out the mean and the standard error of the proportions in each group:

```
cutr<-cut(resources,5)
tapply(occupied,cutr,sum)
```

(13.2,209]	(209,405]	(405,600]	(600,796]	(796,992]
0	10	25	26	31

```
table(cutr)
cutr
```

(13.2,209]	(209,405]	(405,600]	(600,796]	(796,992]
31	29	30	29	31

So the empirical probabilities are given by

```
probs<-tapply(occupied,cutr,sum)/table(cutr)
probs
```

(13.2,209]	(209,405]	(405,600]	(600,796]	(796,992]
0.0000000	0.3448276	0.8333333	0.8965517	1.0000000

```
probs<-as.vector(probs)
resmeans<-tapply(resources,cutr,mean)
resmeans<-as.vector(resmeans)
```

We can plot these as big points on the graph – the closer they fall to the line, the better the fit of the logistic model to the data:

```
points(resmeans,probs,pch=16,cex=2)
```

We need to add a measure of unreliability to the points. The standard error of a binomial proportion will do: $se = \sqrt{p(1-p)/n}$.

```
se<-sqrt(probs*(1-probs)/table(cutr))
```

Finally, draw lines up and down from each point indicating 1 standard error:

```
up<-probs+as.vector(se)
down<-probs-as.vector(se)
for (i in 1:5){
    lines(c(resmeans[i],resmeans[i]),c(up[i],down[i]))}
```

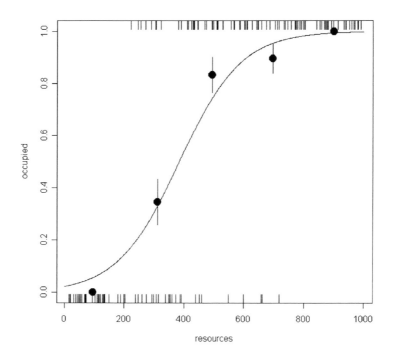

Evidently, the logistic is a good fit to the data above resources of 800 (not surprising, though, given that there were no unoccupied patches in this region), but it is rather a poor fit for resources between 400 and 800, as well as below 200, despite the fact that there were no occupied patches in the latter region (empirical $p = 0$).

ANCOVA with a Binary Response Variable

In our next example the binary response variable is parasite infection (infected or not) and the explanatory variables are weight and age (continuous) and sex (categorical). We begin with data inspection:

```
infection<-read.table("c:\\temp\\infection.txt",header=T)
attach(infection)
names(infection)
```

```
[1] "infected" "age" "weight" "sex"
```

```
par(mfrow=c(1,2))
plot(infected,weight,xlab="Infection",ylab="Weight")
plot(infected,age,xlab="Infection",ylab="Age")
```

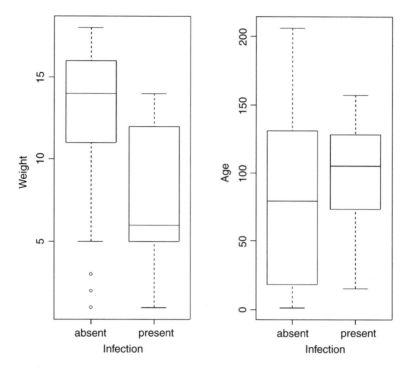

Infected individuals are substantially lighter than uninfected individuals, and occur in a much narrower range of ages. To see the relationship between infection and gender (both categori cal variables) we can use table:

table(infected,sex)

```
table(infected,sex)
             sex
infected   female   male
  absent       17     47
  present      11      6
```

This indicates that the infection is much more prevalent in females (11/28) than in males (6/53).

We now proceed, as usual, to fit a maximal model with different slopes for each level of the categorical variable:

```
model<-glm(infected~age*weight*sex,family=binomial)
summary(model)
```

```
Coefficients:
                      Estimate  Std. Error  z value  Pr(>|z|)
(Intercept)          -0.109124   1.375388   -0.079    0.937
age                   0.024128   0.020874    1.156    0.248
weight               -0.074156   0.147678   -0.502    0.616
sexmale              -5.969109   4.278066   -1.395    0.163
age:weight           -0.001977   0.002006   -0.985    0.325
age:sexmale           0.038086   0.041325    0.922    0.357
weight:sexmale        0.213830   0.343265    0.623    0.533
age:weight:sexmale   -0.001651   0.003419   -0.483    0.629
```

```
(Dispersion parameter for binomial family taken to be 1)
```

```
    Null deviance: 83.234 on 80 degrees of freedom
Residual deviance: 55.706 on 73 degrees of freedom
AIC: 71.706
```

```
Number of Fisher Scoring iterations: 6
```

It certainly does not look as if any of the high-order interactions are significant. Instead of using update and anova for model simplification, we can use step to compute the AIC for each term in turn:

```
model2<-step(model)
```

```
Start: AIC= 71.71
```

First, it tests whether the three-way interaction is required:

```
infected ~ age * weight * sex
```

```
                   Df  Deviance    AIC
- age:weight:sex    1    55.943   69.943
<none>                   55.706   71.706
```

This causes a reduction in AIC of just $71.7 - 69.9 = 1.8$ and hence is not significant. Next, it looks at the three 2-way interactions and decides which to delete first:

```
Step: AIC= 69.94
   infected ~ age + weight + sex + age:weight + age:sex + weight:sex
```

```
               Df  Deviance    AIC
- weight:sex    1    56.122   68.122
- age:sex       1    57.828   69.828
<none>               55.943   69.943
- age:weight    1    58.674   70.674
```

```
Step: AIC= 68.12
   infected ~ age + weight + sex + age:weight + age:sex
```

```
               Df  Deviance    AIC
<none>               56.122   68.122
- age:sex       1    58.142   68.142
- age:weight    1    58.899   68.899
```

Only the removal of the weight–sex interaction causes a reduction in AIC, so this interaction is deleted and the other two interactions are retained. Let's see if we would have been this lenient:

summary(model2)

```
Call:
glm(formula = infected ~ age + weight + sex + age:weight + age:sex,
    family = binomial)

Coefficients:
              Estimate  Std. Error  z value  Pr(>|z|)
(Intercept)  -0.391566    1.265230   -0.309    0.7570
age           0.025764    0.014921    1.727    0.0842   .
weight       -0.036494    0.128993   -0.283    0.7772
sexmale      -3.743771    1.791962   -2.089    0.0367   *
age:weight   -0.002221    0.001365   -1.627    0.1038
age:sexmale   0.020464    0.015232    1.343    0.1791

(Dispersion parameter for binomial family taken to be 1)

    Null deviance: 83.234 on 80 degrees of freedom
Residual deviance: 56.122 on 75 degrees of freedom
AIC: 68.122

Number of Fisher Scoring iterations: 6
```

Neither of the two interactions retained by step would figure in our model ($p > 0.10$). We shall use update to simplify model2:

model3<-update(model2,~.-age:weight)
anova(model2,model3,test="Chi")

```
Analysis of Deviance Table

Model 1: infected ~ age + weight + sex + age:weight + age:sex
Model 2: infected ~ age + weight + sex + age:sex
  Resid. Df   Resid. Dev   Df   Deviance   P(>|Chi|)
1        75       56.122
2        76       58.899   -1    -2.777      0.096
```

So there is no really persuasive evidence of an age–weight term ($p = 0.096$).

model4<-update(model2,~.-age:sex)
anova(model2,model4,test="Chi")

Note that we are testing all the two-way interactions by deletion from the model that contains all two-way interactions (model2): $p = 0.155$, so nothing there, then.

```
Analysis of Deviance Table

Model 1: infected ~ age + weight + sex + age:weight + age:sex
Model 2: infected ~ age + weight + sex + age:weight

   Resid. Df   Resid. Dev   Df   Deviance   P(>|Chi|)
1         75       56.122
2         76       58.142   -1    -2.020      0.155
```

What about the three main effects?

```
model5<-glm(infected~age+weight+sex,family=binomial)
summary(model5)
```

```
Coefficients:
               Estimate  Std. Error   z value  Pr(>|z|)
(Intercept)    0.609369    0.803288     0.759  0.448096
age            0.012653    0.006772     1.868  0.061701  .
weight        -0.227912    0.068599    -3.322  0.000893  ***
sexmale       -1.543444    0.685681    -2.251  0.024388  *
```

```
(Dispersion parameter for binomial family taken to be 1)

    Null deviance:  83.234  on 80  degrees of freedom
Residual deviance:  59.859  on 77  degrees of freedom
AIC: 67.859
```

```
Number of Fisher Scoring iterations: 5
```

Weight is highly significant, as we expected from the initial boxplot, sex is quite significant, and age is marginally significant. It is worth establishing whether there is any evidence of non-linearity in the response of infection to weight or age. We might begin by fitting quadratic terms for the two continuous explanatory variables:

```
model6<-
glm(infected~age+weight+sex+I(weight^2)+I(age^2),family=binomial)
summary(model6)
```

```
Coefficients:
               Estimate   Std. Error   z value   Pr(>|z|)
(Intercept)  -3.4475839    1.7978359    -1.918     0.0552  .
age           0.0829364    0.0360205     2.302     0.0213  *
weight        0.4466284    0.3372352     1.324     0.1854
sexmale      -1.2203683    0.7683288    -1.588     0.1122
I(weight^2)  -0.0415128    0.0209677    -1.980     0.0477  *
I(age^2)     -0.0004009    0.0002004    -2.000     0.0455  *
```

```
(Dispersion parameter for binomial family taken to be 1)

    Null deviance: 83.234 on 80 degrees of freedom
Residual deviance: 48.620 on 75 degrees of freedom
AIC: 60.62
```

Evidently, both relationships are significantly non-linear. It is worth looking at these non-linearities in more detail, to see if we can do better with other kinds of models (e.g. non-parametric smoothers, piecewise linear models or step functions). A generalized additive model is often a good way to start when we have continuous covariates:

```
library(mgcv)
model7<-gam(infected~sex+s(age)+s(weight),family=binomial)
par(mfrow=c(1,2))
plot.gam(model7)
```

These non-parametric smoothers are excellent at showing the humped relationship between infection and age, and at highlighting the possibility of a threshold at weight ≈ 8 in the relationship between weight and infection. We can now return to a GLM to incorporate

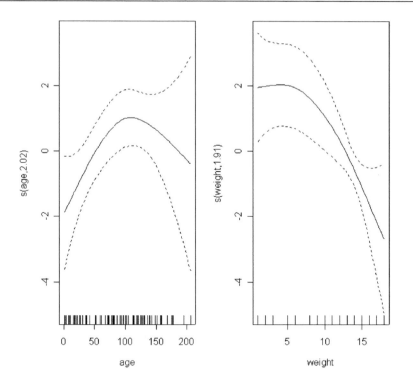

these ideas. We shall fit age and age^2 as before, but try a piecewise linear fit for weight, estimating the threshold weight at a range of values (say 8–14) and selecting the threshold that gives the lowest residual deviance; this turns out to be a threshold of 12. The piecewise regression is specified by the term:

I((weight - 12) * (weight > 12))

The I ('as is') is necessary to stop the * as being evaluated as an interaction term in the model formula. What this expression says is 'regress infection on the value of weight -12, but only do this when (weight > 12) is true' (see p. 429). Otherwise, assume that infection is independent of weight.

model8<-glm(infected~sex+age+I(age^2)+I((weight-12)*(weight>12)),family=binomial)
summary(model8)

```
Coefficients:
                                  Estimate Std. Error z value Pr(>|z|)
(Intercept)                     -2.7511382  1.3678824  -2.011   0.0443 *
sexmale                         -1.2864683  0.7349201  -1.750   0.0800 .
age                              0.0798629  0.0348184   2.294   0.0218 *
I(age^2)                        -0.0003892  0.0001955  -1.991   0.0465 *
I((weight - 12) * (weight > 12)) -1.3547520  0.5350853  -2.532   0.0113 *

(Dispersion parameter for binomial family taken to be 1)

    Null deviance: 83.234 on 80 degrees of freedom
Residual deviance: 48.687 on 76 degrees of freedom
```

```
AIC: 58.687
```

```
Number of Fisher Scoring iterations: 7
```

model9<-update(model8,~.-sex)
anova(model8,model9,test="Chi")
model10<-update(model8,~.-I(age^2))
anova(model8,model10,test="Chi")

The effect of sex on infection is not quite significant ($p = 0.071$ for a chi-squared test on deletion), so we leave it out. The quadratic term for age does not look highly significant here, but a deletion test gives $p = 0.011$, so we retain it. The minimal adequate model is therefore model9:

summary(model9)

```
Coefficients:
                                  Estimate  Std. Error  z value  Pr(>|z|)
(Intercept)                     -3.1207552   1.2665593   -2.464    0.0137 *
age                              0.0765784   0.0323376    2.368    0.0179 *
I(age^2)                        -0.0003843   0.0001846   -2.081    0.0374 *
I((weight - 12) * (weight > 12)) -1.3511706   0.5134681   -2.631    0.0085 **
```

```
(Dispersion parameter for binomial family taken to be 1)

    Null deviance: 83.234 on 80 degrees of freedom
Residual deviance: 51.953 on 77 degrees of freedom
AIC: 59.953
```

```
Number of Fisher Scoring iterations: 7
```

We conclude there is a humped relationship between infection and age, and a threshold effect of weight on infection. The effect of sex is marginal, but might repay further investigation ($p = 0.071$).

Binary Response with Pseudoreplication

In the bacteria dataframe, which is part of the MASS library, we have repeated assessment of bacterial infection (yes or no, coded as y or n) in a series of patients allocated at random to one of three treatments: placebo, drug and drug plus supplement. The trial lasted for 11 weeks and different patients were assessed different numbers of times. The question is whether the two treatments significantly reduced bacterial infection.

library(MASS)
attach(bacteria)
names(bacteria)

```
[1] "y"  "ap"  "hilo"  "week"  "ID"  "trt"
```

table(y)

```
y
  n     y
 43   177
```

The data are binary, so we need to use family=binomial. There is temporal pseudoreplication (repeated measures on the same patients) so we cannot use glm. The ideal solution is the generalized mixed models function lmer. Unlike glm, the lmer function cannot take text (e.g. a two-level factor like y) as the response variable, so we need to convert the y and n into a vector of 1s and 0s:

```
y<-1*(y=="y")
table(y,trt)
```

```
      trt
  y   placebo   drug   drug+
  0        12     18      13
  1        84     44      49
```

Preliminary data inspection suggests that the drug might be effective because only 12 out of 96 patient visits were bacteria-free in the placebos, compared with 31 out of 124 for the treated individuals. We shall see. The modelling goes like this: the lmer function is in the lme4 package and the random effects appear in the same formula as the fixed effects, but defined by the brackets and the 'given' operator |

```
library(lme4)
model1<-lmer(y~trt+(week|ID),family=binomial,method="PQL")
summary(model1)
```

```
Generalized linear mixed model fit using PQL
Formula: y ~ trt + (week | ID)
  Family: binomial(logit link)
    AIC     BIC   logLik   deviance
  217.4   237.7   -102.7      205.4

Random effects:
Groups            Name    Variance   Std.Dev.    Corr
ID         (Intercept)     2.78887    1.66999
                  week     0.17529    0.41868   -0.221

number of obs: 220, groups: ID, 50

Estimated scale (compare to 1 ) 0.7029861


Fixed effects:
                Estimate   Std. Error   z value   Pr(>|z|)
(Intercept)       2.4802       0.5877     4.220   2.44e-05   ***
trtdrug          -1.0422       0.8479    -1.229      0.219
trtdrug+         -0.4054       0.8736    -0.464      0.643

Correlation of Fixed Effects:
            (Intr)   trtdrg
trtdrug     -0.693
trtdrug+    -0.673    0.466
```

There is no indication of a significant drug effect ($p = 0.219$) and the random effect for week had a standard deviation of just 0.41868. We can simplify the model by removing the dependence of infection on week, retaining only the intercept as a random effect +(1|ID)

```
model2<-lmer(y~trt+(1|ID),family=binomial,method="PQL")
anova(model1,model2)
```

```
Data:
Models:
```

```
model2: y ~ trt + (1 | ID)
model1: y ~ trt + (week | ID)
         Df      AIC      BIC    logLik    Chisq  Chi Df  Pr(>Chisq)
model2    4   216.76   230.33   -104.38
model1    6   217.39   237.75   -102.69   3.3684       2      0.1856
```

The simpler model2 is not significantly worse ($p = 0.1856$) so we accept it (it has a lower AIC than model1).

There is a question about the factor levels: perhaps the drug effect would be more significant if we combine the drug and drug plus treatments?

```
drugs<-factor(1+(trt!="placebo"))
table(y,drugs)
```

```
    drugs
y      1    2
   0  12   31
   1  84   93
```

```
model3<-lmer(y~drugs+(1|ID),family=binomial,method="PQL")
summary(model3)
```

```
Generalized linear mixed model fit using PQL
Formula: y ~ drugs + (1 | ID)
 Family: binomial(logit link)
   AIC     BIC   logLik   deviance
 215.5   225.7   -104.7      209.5
```

```
Random effects:
Groups  Name             Variance  Std.Dev.
ID      (Intercept)      2.1316    1.46
number of obs: 220, groups: ID, 50
```

```
Estimated scale (compare to 1 ) 0.8093035
```

```
Fixed effects:
              Estimate   Std. Error   z value   Pr(>|z|)
(Intercept)     2.2085       0.4798     4.603   4.16e-06   ***
drugs2         -0.9212       0.6002    -1.535      0.125
```

```
Correlation of Fixed Effects:
        (Intr)
drugs2 -0.799
```

The interpretation is straightforward: there is no evidence in this experiment that either treatment significantly reduces bacterial infection. Note that this is not the same as saying that the drug does not work. It is simply that this trial is too small to demonstrate the significance of its efficacy.

It is also important to appreciate the importance of the pseudoreplication. If we had ignored the fact that there were multiple measures per patient we should have concluded wrongly that the drug effect was significant. Here are the raw data on the counts:

```
table(y,trt)
```

```
    trt

y      placebo   drug   drug+
   0        12     18      13
   1        84     44      49
```

and here is the *wrong* way of testing for the significance of the treatment effect:

prop.test(c(12,18,13),c(96,62,62))

```
3-sample test for equality of proportions without continuity

correction data: c(12, 18, 13) out of c(96, 62, 62)
X-squared = 6.6585, df = 2, p-value = 0.03582
alternative hypothesis: two.sided
sample estimates:
     prop 1      prop 2      prop 3
0.1250000   0.2903226   0.2096774
```

It appears that the drug has increased the rate of non-infection from 0.125 in the placebos to 0.29 in the treated patients, and that this effect is significant ($p = 0.035\,82$). As we have seen, however, when we remove the pseudoreplication by using the appropriate mixed model with **lmer** the response is non-significant.

Another way to get rid of the pseudoreplication is to restrict the analysis to the patients that were there at the end of the experiment. We just use subset(week==11) and this removes all the pseudoreplication because no subjects were measured twice within any week – we can check this with the **any** function:

any(table(ID,week) >1)

```
[1] FALSE
```

The model is a straightforward GLM with a binary response variable and a single explanatory variable (the three-level factor called trt):

model<-glm(y~trt,binomial,subset=(week==11))
summary(model)

```
Coefficients:
              Estimate   Std. Error   z value   Pr(>|z|)
(Intercept)    1.3863       0.5590      2.480     0.0131    *
trtdrug       -0.6931       0.8292     -0.836     0.4032
trtdrug+      -0.6931       0.8292     -0.836     0.4032

(Dispersion parameter for binomial family taken to be 1)

    Null deviance: 51.564 on 43 degrees of freedom
Residual deviance: 50.569 on 41 degrees of freedom
AIC: 56.569
```

Neither drug treatment has anything approaching a significant effect in lowering bacterial infection rates compared with the placebos ($p = 0.4032$). The supplement was expected to *increase* bacterial control over the drug treatment, so perhaps the interpretation will be modified lumping together the two drug treatments:

drugs<-factor(1+(trt=="placebo"))

so there are placebos plus patients getting one drug treatment or the other:

table(drugs[week==11])

```
 1    2
24   20
```

Thus there were 24 patients receiving one drug or the other, and 20 placebos (at 11 weeks).

```
model <-glm(y~drugs,binomial,subset=(week==11))
summary(model)
```

```
Call:
glm(formula = y ~ drugs, family = binomial, subset = (week == 11))
```

```
Deviance Residuals:
    Min        1Q    Median        3Q       Max
-1.7941   -1.4823   0.6680    0.9005    0.9005
```

```
Coefficients:
              Estimate   Std. Error   z value   Pr(>|z|)
(Intercept)     0.6931       0.4330     1.601      0.109
drugs2          0.6931       0.7071     0.980      0.327
```

(Dispersion parameter for binomial family taken to be 1)

```
    Null deviance: 51.564 on 43 degrees of freedom
Residual deviance: 50.569 on 42 degrees of freedom
AIC: 54.569
```

Clearly, there is no convincing effect for the drug treatments on bacterial infection when we use this subset of the data ($p=0.327$).

An alternative way of analysing all the data (including the pseudoreplication) is to ask what proportion of tests on each patient scored positive for the bacteria. The response variable now becomes a proportion, and the pseudoreplication disappears because we only have one number for each patient (i.e. a count of the number of occasions on which each patient scored positive for the bacteria, with the binomial denominator as the total number of tests on that patient).

There are some preliminary data-shortening tasks. We need to create a vector of length 50 containing the drug treatments of each patient (tss) and a table (ys, with elements of length 50) scoring how many times each patient was infected and uninfected by bacteria. Finally, we use cbind to create a two-column response variable, yv:

```
dss<-data.frame(table(trt,ID))
tss<-dss[dss[,3]>0,]$trt
ys<- table(y,ID)
yv<-cbind(ys[2,],ys[1,])
```

Now we can fit a very simple model for the binomial response (glm with binomial errors):

```
model<-glm(yv~tss,binomial)
summary(model)
```

```
Coefficients:
              Estimate   Std. Error   z value   Pr(>|z|)
(Intercept)     1.9459       0.3086     6.306    2.87e-10   ***
tssdrug        -1.0521       0.4165    -2.526      0.0115   *
tssdrug+       -0.6190       0.4388    -1.411      0.1583
```

(Dispersion parameter for binomial family taken to be 1)

```
    Null deviance: 86.100 on 49 degrees of freedom
Residual deviance: 79.444 on 47 degrees of freedom
AIC: 130.9
```

Drug looks to be significant here, but note that the residual deviance is much bigger than the residual degrees of freedom so we should correct for overdispersion by using `quasibinomial` instead of binomial errors (p. 522):

```
model<-glm(yv~tss,quasibinomial)
summary(model)
```

```
Coefficients:
             Estimate   Std. Error   t value   Pr(>|t|)
(Intercept)    1.9459       0.3837     5.071   6.61e-06   ***
tssdrug       -1.0521       0.5180    -2.031   0.0479     *
tssdrug+      -0.6190       0.5457    -1.134   0.2624
```

```
(Dispersion parameter for quasibinomial family taken to be 1.546252)

    Null deviance: 86.100 on 49 degrees of freedom
Residual deviance: 79.444 on 47 degrees of freedom
AIC: NA
```

There is a marginally significant effect of drug, but no significant difference between the two drug treatments, so we aggregate them into a single drug treatment:

```
tss2<-factor(1+(tss=="placebo"))
model<-glm(yv~tss2,quasibinomial)
summary(model)
```

```
Coefficients:
             Estimate   Std. Error   t value   Pr(>|t|)
(Intercept)    1.0986       0.2582     4.255   9.63e-05   ***
tss22          0.8473       0.4629     1.830   0.0734     .
```

Again, the treatment effect is not significant, as we concluded in the generalized mixed-effects model (p. 606).

18

Generalized Additive Models

Up to this point, continuous explanatory variables have been added to models as linear functions, linearized parametric transformations, or through various link functions. In all cases, an explicit or implicit assumption was made about the parametric form of the function to be fitted to the data (whether quadratic, logarithmic, exponential, logistic, reciprocal or whatever). In many cases, however, you have one or more continuous explanatory variables, but you have no *a priori* reason to choose one particular parametric form over another for describing the shape of the relationship between the response variable and the explanatory variable(s). Generalized additive models (GAMs) are useful in such cases because they allow you to capture the shape of a relationship between y and x without prejudging the issue by choosing a particular parametric form.

Generalized additive models (implemented in R by the function gam) extend the range of application of generalized linear models (glm) by allowing non-parametric smoothers in addition to parametric forms, and these can be associated with a range of link functions. All of the error families allowed with glm are available with gam (binomial, Poisson, gamma, etc.). Indeed, gam has many of the attributes of both glm and lm, and the output can be modified using update. You can use all of the familiar methods such as print, plot, summary, anova, predict, and fitted after a GAM has been fitted to data. The gam function used in this book is in the mgcv library contributed by Simon Wood:

library(mgcv)

There are many ways of specifying the model in a GAM: all of the continuous explanatory variables x, w and z can enter the model as non-parametrically smoothed functions $s(x)$, $s(w)$, and $s(z)$:

y~s(x) + s(w) + s(z)

Alternatively, the model can contain a mix of parametrically estimated parameters (x and z) and smoothed variables $s(w)$:

y~x + s(w) + z

Formulae can involve nested (two-dimensional) terms in which the smoothing s() terms have more than one argument, implying an isotropic smooth:

y~s(x) + s(z) + s(x,z)

Alternatively the smoothers can have overlapping terms such as

y~s(x,z) + s(z,w)

The R Book   Michael J. Crawley
© 2007 John Wiley & Sons, Ltd

The user has a high degree of control over the way that interactions terms can be fitted, and te() smooths are provided as an effective means for modelling smooth interactions of any number of variables via scale-invariant tensor product smooths. Here is an example of a model formula with a fully nested tensor product te(x,z,k=6):

y ~ s(x,bs="cr",k=6) + s(z,bs="cr",k=6) + te(x,z,k=6)

The optional arguments to the smoothers are bs="cr",k=6, where bs indicates the basis to use for the smooth ("cr" is a cubic regression spline; the default is thin plate bs="tp"), and k is the dimension of the basis used to represent the smooth term (it defaults to k = 10*3^ (d-1) where d is the number of covariates for this term).

Non-parametric Smoothers

You can see non-parametric smoothers in action for fitting a curve through a scatterplot in Chapter 5 (p. 151). Here we are concerned with using non-parametric smoothers in statistical modelling where the object is to assess the relative merits of a range of different models in explaining variation in the response variable. One of the simplest model-fitting functions is loess (which replaces its predecessor called lowess).

The following example shows population change, Delta $= \log(N(t+1)/N(t))$ as a function of population density $(N(t))$ in an investigation of density dependence in a sheep population. This is what the data look like:

```
soay<-read.table("c:\\temp\\soaysheep.txt",header=T)
attach(soay)
names(soay)
```

```
[1] "Year"  "Population"  "Delta"
```

```
plot(Population,Delta)
```

Broadly speaking, population change is positive at low densities (Delta > 0) and negative at high densities (Delta < 0) but there is a great deal of scatter, and it is not at all obvious what shape of smooth function would best describe the data. Here is the default loess:

```
model<-loess(Delta~Population)
summary(model)
```

```
Call:
loess(formula = Delta~Population)
```

```
Number of Observations: 44
Equivalent Number of Parameters: 4.66
Residual Standard Error: 0.2616
Trace of smoother matrix: 5.11
Control settings:
 normalize:TRUE
 span :     0.75
 degree :   2
 family :   gaussian
 surface :  interpolate   cell = 0.2
```

Now draw the smoothed line using predict to extract the predicted values from model:

```
xv<-seq(600,2000,1)
yv<-predict(model,data.frame(Population=xv))
lines(xv,yv)
```

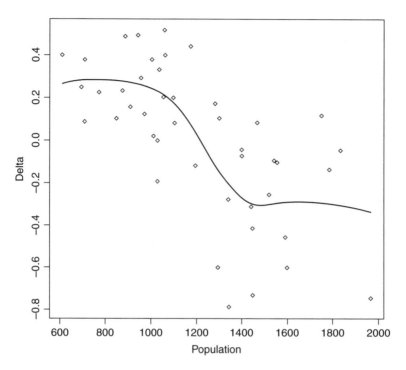

The smooth curve looks rather like a step function. We can compare this smooth function with a step function, using a tree model (p. 686) as an objective way of determining the threshold for splitting the data into low- and high-density parts:

```
library(tree)
thresh<-tree(Delta~Population)
print(thresh)
```

The threshold for the first split of the tree model is at Population = 1289.5 so we define this as the threshold density:

```
th<-1289.5
```

Then we can use this threshold to create a two-level factor for fitting two constant rates of population change using aov

```
model2<-aov(Delta~(Population>th))
summary(model2)
```

	Df	Sum Sq	Mean Sq	F value	Pr(> F)	
Population > th	1	2.80977	2.80977	47.636	2.008e-08	***
Residuals	42	2.47736	0.05898			

showing a residual sum of squares of 0.059. This compares with the residual of $0.2616^2 = 0.068$ from the loess (above). To draw the step function we need the average low-density population increase and the average high-density population decline:

```
tapply(Delta[-45],(Population[-45]>th),mean)
```

```
     FALSE                TRUE
0.2265084    -0.2836616
```

Note the use of negative subscripts to drop the NA from the last value of Delta. Then use these figures to draw the step function:

```
lines(c(600,th),c(0.2265,0.2265),lty=2)
lines(c(th,2000),c(-0.2837,-0.2837),lty=2)
lines(c(th,th),c(-0.2837,0.2265),lty=2)
```

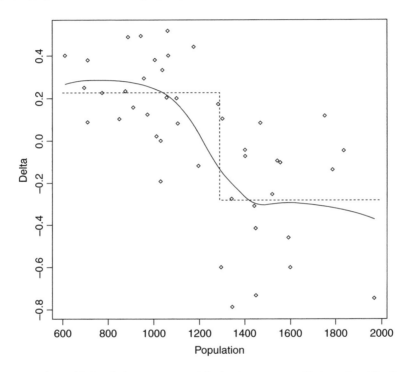

It is a moot point which of these two models is the most realistic scientifically, but the step function involved three estimated parameters (two averages and a threshold), while the loess is based on 4.66 degrees of freedom, so parsimony favours the step function (it also has a slightly lower residual sum of squares).

Generalized Additive Models

This dataframe contains measurements of radiation, temperature, wind speed and ozone concentration. We want to model ozone concentration as a function of the three continuous explanatory variables using non-parametric smoothers rather than specified nonlinear functions (the parametric multiple regression analysis is on p. 434):

```
ozone.data<-read.table("c:\\temp\\ozone.data.txt",header=T)
attach(ozone.data)
names(ozone.data)
```

```
[1] "rad"  "temp"  "wind"  "ozone"
```

For data inspection we use pairs with a non-parametric smoother, lowess:

```
pairs(ozone.data, panel=function(x,y) { points(x,y); lines(lowess(x,y))} )
```

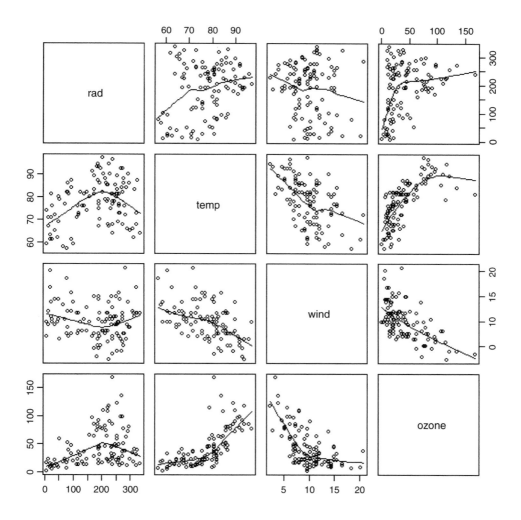

Now fit all three explanatory variables using the non-parametric smoother s():

```
model<-gam(ozone~s(rad)+s(temp)+s(wind))
summary(model)
```

```
Family: gaussian
Link function: identity
```

```
Formula:
ozone ~ s(rad) + s(temp) + s(wind)
Parametric coefficients:
```

	Estimate	Std. Error	t value	Pr(> |t|)	
(Intercept)	42.10	1.66	25.36	< 2e-16	***

```
Approximate significance of smooth terms:
           edf   Est.rank        F    p-value
s(rad)   2.763      6.000    2.830     0.0138   *
s(temp)  3.841      8.000    8.080   2.27e-08   ***
s(wind)  2.918      6.000    8.973   7.62e-08   ***

R-sq.(adj) = 0.724 Deviance explained = 74.8%
GCV score = 338  Scale est. = 305.96  n = 111
```

Note that the intercept is estimated as a parametric coefficient (upper table) and the three explanatory variables are fitted as smooth terms. All three are significant, but radiation is the least significant at $p = 0.0138$. We can compare a GAM with and without a term for radiation using ANOVA in the normal way:

```
model2<-gam(ozone~s(temp)+s(wind))
anova(model,model2,test="F")
```

```
Analysis of Deviance Table

Model 1: ozone~s(rad) + s(temp) + s(wind)
Model 2: ozone~s(temp) + s(wind)

   Resid. Df  Resid. Dev       Df  Deviance        F    Pr(>F)
1   100.4779       30742
2   102.8450       34885  -2.3672     -4142   5.7192  0.002696   **
```

Clearly, radiation should remain in the model, since deletion of radiation caused a highly significant increase in deviance ($p = 0.0027$), emphasizing the fact that deletion is a better test than inspection of parameters (the p values in the full model table were *not* deletion p values).

We should investigate the possibility that there is an interaction between wind and temperature:

```
model3<-gam(ozone~s(temp)+s(wind)+s(rad)+s(wind,temp))
summary(model3)
```

```
Family: gaussian
Link function: identity

Formula:
ozone~s(temp) + s(wind) + s(rad) + s(wind, temp)

Parametric coefficients:
             Estimate  Std. Error  t value  Pr(>|t|)
(Intercept)    42.099       1.361    30.92    <2e-16   ***

Approximate significance of smooth terms:
                edf   Est.rank          F   p-value
s(temp)       1.000          1   0.000292   0.98640
s(wind)       5.613          9      5.349  9.19e-06   ***
s(rad)        1.389          3      4.551   0.00528   **
s(wind,temp) 18.246         27      3.791  1.53e-06   ***

R-sq.(adj) = 0.814  Deviance explained = 85.9%
GCV score = 272.66  Scale est. = 205.72  n = 111
```

The interaction appears to be highly significant, but the main effect of temperature is cancelled out. We can inspect the fit of model3 like this:

```
par(mfrow=c(2,2))
plot(model3,residuals=T,pch=16)
```

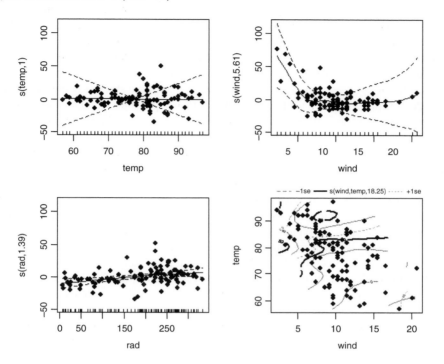

You need to press the Return key to see each of the four graphs in turn. The etchings on the x axis are called rugs (see p. 596) and indicate the locations of measurements of x values on each axis. The default option is rug=T. The bottom right-hand plot shows the complexity of the interaction between temperature and wind speed.

Technical aspects

The degree of smoothness of model terms is estimated as part of fitting; isotropic or scale-invariant smooths of any number of variables are available as model terms. Confidence or credible intervals are readily available for any quantity predicted using a fitted model. In mgcv, gam solves the smoothing parameter estimation problem by using the generalized cross validation (GCV) criterion

$$CGCV = \frac{nD}{(n - DoF)^2}$$

or an unbiased risk estimator (UBRE) criterion

$$UBRE = \frac{D}{n} + 2\phi\frac{DoF}{n} - \phi$$

where D is the deviance, n the number of data, ϕ the scale parameter and DoF the effective degrees of freedom of the model. Notice that UBRE is effectively just AIC rescaled, but is only used when ϕ is known. It is also possible to replace D by the Pearson statistic (see

?gam.method), but this can lead to oversmoothing. Smoothing parameters are chosen to minimize the GCV or UBRE score for the model, and the main computational challenge solved by the mgcv package is to do this efficiently and reliably. Various alternative numerical methods are provided: see ?gam.method. Smooth terms are represented using penalized regression splines (or similar smoothers) with smoothing parameters selected by GCV/UBRE or by regression splines with fixed degrees of freedom (mixtures of the two are permitted). Multi-dimensional smooths are available using penalized thin plate regression splines (isotropic) or tensor product splines (when an isotropic smooth is inappropriate).

This gam function is *not* a clone of what S-PLUS provides – there are three major differences: First, by default, estimation of the degree of smoothness of model terms is part of model fitting. Second, a Bayesian approach to variance estimation is employed that makes for easier confidence interval calculation (with good coverage probabilities). Third, the facilities for incorporating smooths of more than one variable are different.

If absolutely any smooth functions were allowed in model fitting then maximum likelihood estimation of such models would invariably result in complex overfitting estimates of the smoothed functions $s(x)$ and $s(z)$. For this reason the models are usually fitted by penalized likelihood maximization, in which the model (negative log-)likelihood is modified by the addition of a penalty for each smooth function, penalizing what gam's author, Simon Wood, calls its 'wiggliness'. To control the trade-off between penalizing wiggliness and penalizing badness of fit, each penalty is multiplied by an associated smoothing parameter: how to estimate these parameters and how to practically represent the smooth functions are the main statistical questions introduced by moving from GLMs to GAMs.

The built-in alternatives for univariate smooths terms are: a conventional penalized cubic regression spline basis, parameterized in terms of the function values at the knots; a cyclic cubic spline with a similar parameterization; and thin plate regression splines. The cubic spline bases are computationally very efficient, but require knot locations to be chosen (automatically by default). The thin plate regression splines are optimal low-rank smooths which do not have knots, but are computationally more costly to set up. Smooths of several variables can be represented using thin plate regression splines, or tensor products of any available basis, including user-defined bases (tensor product penalties are obtained automatically form the marginal basis penalties).

Thin plate regression splines are constructed by starting with the basis for a full thin plate spline and then truncating this basis in an optimal manner, to obtain a low-rank smoother. Details are given in Wood (2003). One key advantage of the approach is that it avoids the knot placement problems of conventional regression spline modelling, but it also has the advantage that smooths of lower rank are nested within smooths of higher rank, so that it is legitimate to use conventional hypothesis testing methods to compare models based on pure regression splines. The thin plate regression spline basis can become expensive to calculate for large data sets. In this case the user can supply a reduced set of knots to use in basis construction (see knots in the argument list), or use tensor products of cheaper bases. In the case of the cubic regression spline basis, knots of the spline are placed evenly throughout the covariate values to which the term refers. For example, if fitting 101 data points with an 11-knot spline of x then there would be a knot at every 10th (ordered) x value. The parameterization used represents the spline in terms of its values at the knots. The values at neighbouring knots are connected by sections of cubic polynomial constrained to be continuous up to and including second derivatives at the knots. The resulting curve is a natural cubic spline through the values at the knots (given two extra conditions specifying that the second derivative of the curve should be zero at the two end knots). This parameterization gives the parameters a nice interpretability. Details of the underlying fitting methods are given in Wood (2000, 2004).

You must have more unique combinations of covariates than the model has total parameters. (Total parameters is the sum of basis dimensions plus the sum of non-spline terms less the number of spline terms.) Automatic smoothing parameter selection is not likely to work well when fitting models to very few response data. With large data sets (more than a few thousand data) the tp basis gets very slow to use: use the knots argument as discussed above and shown in the examples. Alternatively, for low-density smooths you can use the cr basis and for multi-dimensional smooths use te smooths.

For data with many zeros clustered together in the covariate space it is quite easy to set up GAMs which suffer from identifiability problems, particularly when using Poisson or binomial families. The problem is that with log or logit links, for example, mean value zero corresponds to an infinite range on the linear predictor scale.

Another situation that occurs quite often is the one in which we would like to find out if the model

$$E(y) = f(x, z)$$

is really necessary, or whether

$$E(y) = f_1(x) + f_2(z)$$

would not do just as well. One way to do this is to look at the results of fitting

y~s(x)+s(z)+s(x,z)

gam automatically generates side conditions to make this model identifiable. You can also estimate overlapping models such as

y~s(x,z)+s(z,v)

Sometimes models of the form

$$E(y) = b_0 + f(x)z$$

need to be estimated (where f is a smooth function, as usual). The appropriate formula is

y~z+s(x,by=z)

where the by argument ensures that the smooth function gets multiplied by covariate z, but GAM smooths are centred (average value zero), so the parametric term for z is needed as well (f is being represented by a constant plus a centred smooth). If we had wanted

$$E(y) = f(x)z$$

then the appropriate formula would be

y~z+s(x,by=z)-1

The by mechanism also allows models to be estimated in which the form of a smooth depends on the level of a factor, but to do this the user must generate the dummy variables for each level of the factor. Suppose, for example, that fac is a factor with three levels 1,

2, 3, and at each level of this factor the response depends smoothly on a variable x in a manner that is level-dependent. Three dummy variables fac.1, fac.2, fac.3, can be generated for the factor (e.g. fac.1<-as.numeric(fac==1)). Then the model formula would be:

y~fac+s(x,by=fac.1)+s(x,by=fac.2)+s(x,by=fac.3)

In the above examples the smooths of more than one covariate have all employed single-penalty thin plate regression splines. These isotropic smooths are not always appropriate: if variables are not naturally well scaled relative to each other then it is often preferable to use tensor product smooths, with a wiggliness penalty for each covariate of the term. See ?te for examples.

The most logically consistent method to use for deciding which terms to include in the model is to compare GCV/UBRE scores for models with and without the term. More generally, the score for the model with a smooth term can be compared to the score for the model with the smooth term replaced by appropriate parametric terms. Candidates for removal can be identified by reference to the approximate p values provided by summary.gam. Candidates for replacement by parametric terms are smooth terms with estimated degrees of freedom close to their minimum possible.

An example with strongly humped data

The ethanol dataframe contains 88 sets of measurements for variables from an experiment in which ethanol was burned in a single cylinder automobile test engine. The response variable, NOx, is the concentration of nitric oxide (NO) and nitrogen dioxide (NO_2) in engine exhaust, normalized by the work done by the engine, and the two continuous explanatory variables are C (the compression ratio of the engine), and E (the equivalence ratio at which the engine was run, which is a measure of the richness of the air–ethanol mix).

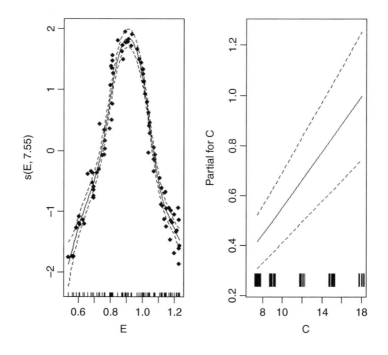

```
install.packages("SemiPar")
library(SemiPar)
data(ethanol)
attach(ethanol)
names(ethanol)
```

```
[1] "NOx" "C" "E"
```

Because NOx is such a strongly humped function of the equivalence ratio, E, we start with a model, NOx ~ s(E) + C, that fits this as a smoothed term and estimates a parametric term for the compression ratio:

```
model<-gam(NOx~s(E)+C)
par(mfrow=c(1,2))
plot.gam(model,residuals=T,pch=16,all.terms=T)
```

The coplot function is helpful in showing where the effect of C on NOx was most marked:

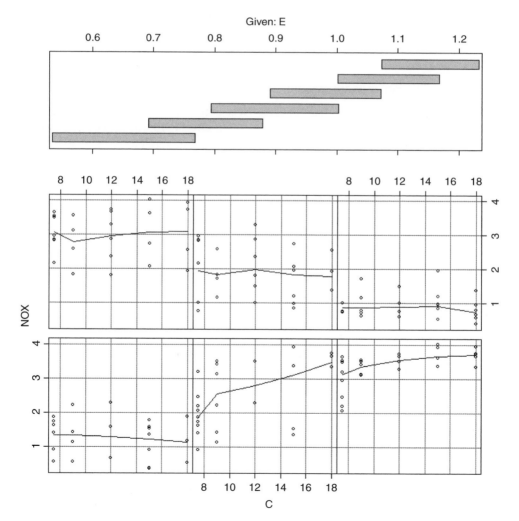

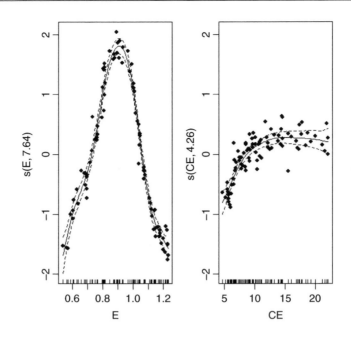

```
coplot(NOx~C|E,panel=panel.smooth)
```

There is a pronounced positive effect of C on NOx only in panel 2 (ethanol $0.7 < E < 0.9$ from the shingles in the upper panel), but only slight effects elsewhere. You can estimate the interaction between E and C from the product of the two variables:

```
CE<-E*C
model2<-gam(NOx~s(E)+s(CE))
plot.gam(model2,residuals=T,pch=16,all.terms=T)

summary(model2)

Family: gaussian

Link function: identity

Formula:
NOx ~ s(E) + s(CE)

Parametric coefficients:
              Estimate   Std. Error   t value   Pr( > |t|)
(Intercept)    1.95738      0.02126     92.07     < 2e-16   ***

Approximate significance of smooth terms:
          edf   Est.rank        F    p-value
s(E)    7.636          9   267.26    < 2e-16   ***
s(CE)   4.261          9    15.75   4.12e-14   ***

R-sq.(adj) = 0.969   Deviance explained = 97.3%
GCV score = 0.0466   Scale est. = 0.039771   n = 88
```

The summary of this GAM shows highly significant terms for both smoothed terms: the effect of ethanol, $s(E)$, on 7.6 estimated degrees of freedom, and the interaction between

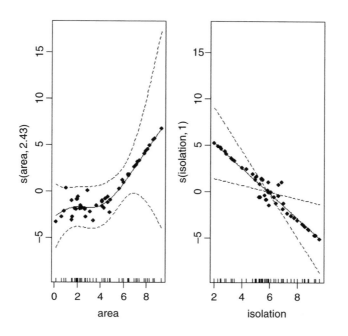

E and C, $s(CE)$, on 4.3 estimated degrees of freedom. The model explains 97.3% of the deviance in NOx concentration.

Generalized Additive Models with Binary Data

GAMs are particularly valuable with binary response variables (for background, see p. 593). To illustrate the use of **gam** for modelling binary response data, we return to the example analysed by logistic regression on p. 595. We want to understand how the isolation of an island and its area influence the probability that the island is occupied by our study species.

```
island<-read.table("c:\\temp\\isolation.txt",header=T)
attach(island)
names(island)
```

```
[1] "incidence" "area" "isolation"
```

In the logistic regression, isolation had a highly significant negative effect on the probability that an island will be occupied by our species ($p = 0.004$), and area (island size) had a significant positive effect on the likelihood of occupancy ($p = 0.019$). But we have no *a priori* reason to believe that the logit of the probability should be linearly related to either of the explanatory variables. We can try using a GAM to fit smoothed functions to the incidence data:

```
model3<-gam(incidence~s(area)+s(isolation),binomial)
summary(model3)
```

```
Family: binomial
Link function: logit
```

```
Formula:
incidence ~ s(area) + s(isolation)
```

```
Parametric coefficients:
               Estimate    Std. Error   z value   Pr(>|z|)
(Intercept)    1.6371        0.8545       1.916     0.0554    .
```

```
Approximate significance of smooth terms:
                 edf    Est.rank    Chi.sq   p-value
s(area)         2.429        5        6.335   0.27494
s(isolation)    1.000        1        7.532   0.00606    **
```

```
R-sq.(adj) = 0.63  Deviance explained = 63.1%
UBRE score = -0.32096   Scale est. = 1   n = 50
```

This indicates a highly significant effect of isolation on occupancy ($p = 0.006\,06$) but no effect of area ($p = 0.274\,94$). We plot the model to look at the residuals:

```
par(mfrow=c(1,2))
plot.gam(model3,residuals=T,pch=16)
```

This suggests a strong effect of area, with very little scatter, above a threshold of about area $= 5$. We assess the significance of area by deletion and compare a model containing $s(\text{area}) + s(\text{isolation})$ with a model containing $s(\text{isolation})$ alone:

```
model4<-gam(incidence~s(isolation),binomial)
anova(model3,model4,test="Chi")
```

```
Analysis of Deviance Table
Model 1: incidence ~ s(area) + s(isolation)
Model 2: incidence ~ s(isolation)
```

	Resid. Df	Resid. Dev	Df	Deviance	P(>|Chi|)
1	45.5710	25.094			
2	48.0000	36.640	-2.4290	-11.546	0.005

This shows the effect of area to be highly significant ($p = 0.005$), despite the non-significant p value in the summary table of model3. An alternative is to fit area as a parametric term and isolation as a smoothed term:

```
model5<-gam(incidence~area+s(isolation),binomial)
summary(model5)
```

```
Family: binomial
Link function: logit
```

```
Formula:
incidence~area + s(isolation)
```

```
Parametric coefficients:
               Estimate    Std. Error   z value   Pr(>|z|)
(Intercept)    -1.3928       0.9002      -1.547     0.1218
area            0.5807       0.2478       2.344     0.0191    *
```

```
Approximate significance of smooth terms:
                 edf    Est.rank    Chi.sq   p-value
s(isolation)      1         1       8.275   0.00402    **
```

```
R-sq. (adj) = 0.597 Deviance explained = 58.3%
UBRE score = -0.31196  Scale est. = 1  n = 50
```

Again, this shows a significant effect of area on occupancy. The lesson here is that a term can appear to be significant when entered into the model as a parametric term (area has $p=0.019$ in model5) but not come close to significance when entered as a smoothed term (s(area) has $p=0.275$ in model3). Also, the comparison of model3 and model4 draws attention to the benefits of using deletion with anova in assessing the significance of model terms.

Three-Dimensional Graphic Output from gam

Here is an example by Simon Wood which shows the kind of three-dimensional graphics that can be obtained from gam using vis.gam when there are two continuous explanatory variables. Note that in this example the smother works on both variables together, y~s(x,z):

```
par(mfrow=c(1,1))
test1<-function(x,z,sx=0.3,sz=0.4)
  {(pi**sx*sz)*(1.2*exp(-(x-0.2)^2/sx^2-(z-0.3)^2/sz^2)+
  0.8*exp(-(x-0.7)^2/sx^2-(z-0.8)^2/sz^2))
  }
  n<-500
x<-runif(n);z-runif(n);
y<-test1(x,z)+rnorm(n)*0.1
b4<-gam(y~s(x,z))
vis.gam(b4)
```

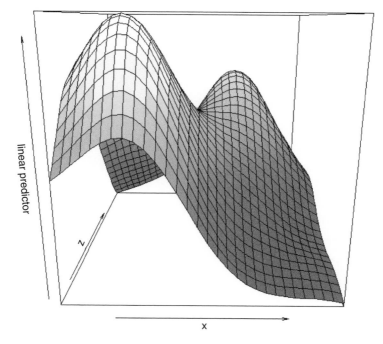

Note also that the vertical scale of the graph is the linear predictor, not the response.

19

Mixed-Effects Models

Up to this point, we have treated all categorical explanatory variables as if they were the same. This is certainly what R. A. Fisher had in mind when he invented the analysis of variance in the 1920s and 1930s. It was Eisenhart (1947) who realized that there were actually two fundamentally different sorts of categorical explanatory variables: he called these **fixed effects** and **random effects**. It will take a good deal of practice before you are confident in deciding whether a particular categorical explanatory variable should be treated as a fixed effect or a random effect, but in essence:

- fixed effects influence only the **mean** of y;

- random effects influence only the **variance** of y.

Fixed effects are unknown constants to be estimated from the data. Random effects govern the variance–covariance structure of the response variable (see p. 473). Nesting (or hierarchical structure) of random effects is a classic source of pseudoreplication, so it important that you are able to recognize it and hence not fall into its trap. Random effects that come from the same group will be correlated, and this contravenes one of the fundamental assumptions of standard statistical models: *independence of errors*. Random effects occur in two contrasting kinds of circumstances:

- observational studies with hierarchical structure;

- designed experiments with different spatial or temporal scales.

Fixed effects have informative factor levels, while random effects often have uninformative factor levels. The distinction is best seen by an example. In most mammal species the categorical variable sex has two levels: male and female. For any individual that you find, the knowledge that it is, say, female conveys a great deal of information about the individual, and this information draws on experience gleaned from many other individuals that were female. A female will have a whole set of attributes (associated with her being female) no matter what population that individual was drawn from. Take a different categorical variable like genotype. If we have two genotypes in a population we might label them A and B. If we take two more genotypes from a *different* population we might label them A and B as well. In a case like this, the label A does not convey any information at all about the genotype, other than that it is probably different from genotype B. In the case of sex, the factor level (male or female) is informative: sex is a fixed effect. In the case of genotype, the factor level (A or B) is uninformative: genotype is a random effect.

Random effects have factor levels that are drawn from a large (potentially very large) population in which the individuals differ in many ways, but we do not know exactly how or why they differ. To get a feel for the difference between fixed effects and random effects here are some more examples:

Fixed effects	Random effects
Drug administered or not	Genotype
Insecticide sprayed or not	Brood
Nutrient added or not	Block within a field
One country versus another	Split plot within a plot
Male or female	History of development
Upland or lowland	Household
Wet versus dry	Individuals with repeated measures
Light versus shade	Family
One age versus another	Parent

The important point is that because the random effects come from a large population, there is not much point in concentrating on estimating means of our small subset of factor levels, and no point at all in comparing individual pairs of means for different factor levels. Much better to recognize them for what they are, random samples from a much larger population, and to concentrate on their variance. This is the *added* variation caused by differences between the levels of the random effects. Variance components analysis is all about estimating the size of this variance, and working out its percentage contribution to the overall variation. There are five fundamental assumptions of linear mixed-effects models:

- Within-group errors are independent with mean zero and variance σ^2.

- Within-group errors are independent of the random effects.

- The random effects are normally distributed with mean zero and covariance matrix Ψ.

- The random effects are independent in different groups.

- The covariance matrix does not depend on the group.

The validity of these assumptions needs to be tested by employing a series of plotting methods involving the residuals, the fitted values and the predicted random effects. The tricks with mixed-effects models are

- learning which variables are random effects;

- specifying the fixed and random effects in two model formulae;

- getting the nesting structure of the random effects right;

- remembering to get library(lme4) or library(nlme) at the outset.

The issues fall into two broad categories: questions about experimental design and the management of experimental error (e.g. where does most of the variation occur, and where would increased replication be most profitable?); and questions about hierarchical structure, and the relative magnitude of variation at different levels within the hierarchy (e.g. studies on the genetics of individuals within families, families within parishes, and parishes with counties, to discover the relative importance of genetic and phenotypic variation).

Most ANOVA models are based on the assumption that there is a single error term. But in hierarchical studies and nested experiments, where the data are gathered at two or more different spatial scales, there is *a different error variance for each different spatial scale*. There are two reasonably clear-cut sets of circumstances where your first choice would be to use a linear mixed-effects model: you want to do variance components analysis because all your explanatory variables are categorical random effects and you don't have any fixed effects; or you do have fixed effects, but you also have pseudoreplication of one sort or another (e.g. temporal pseudoreplication resulting from repeated measurements on the same individuals; see p. 645). To test whether one should use a model with mixed effects or just a plain old linear model, Douglas Bates wrote in the R help archive: 'I would recommend the likelihood ratio test against a linear model fit by lm. The *p*-value returned from this test will be conservative because you are testing on the boundary of the parameter space.'

Replication and Pseudoreplication

To qualify as replicates, measurements must have the following properties:

- They must be independent.

- They must not form part of a time series (data collected from the same place on successive occasions are not independent).

- They must not be grouped together in one place (aggregating the replicates means that they are not spatially independent).

- They must be of an appropriate spatial scale;

- Ideally, one replicate from each treatment ought to be grouped together into a block, and each treatment repeated in many different blocks.

- Repeated measures (e.g. from the same individual or the same spatial location) are not replicates (this is probably the commonest cause of pseudoreplication in statistical work).

Pseudoreplication occurs when you analyse the data as if you had more degrees of freedom than you really have. There are two kinds of pseudoreplication:

- temporal pseudoreplication, involving repeated measurements from the same individual;

- spatial pseudoreplication, involving several measurements taken from the same vicinity.

Pseudoreplication is a problem because one of the most important assumptions of standard statistical analysis is *independence of errors*. Repeated measures through time on the same individual will have non-independent errors because peculiarities of the individual will be reflected in all of the measurements made on it (the repeated measures will be temporally correlated with one another). Samples taken from the same vicinity will have non-independent errors because peculiarities of the location will be common to all the samples (e.g. yields will all be high in a good patch and all be low in a bad patch).

Pseudoreplication is generally quite easy to spot. The question to ask is this. How many degrees of freedom for error does the experiment really have? If a field experiment appears to have lots of degrees of freedom, it is probably pseudoreplicated. Take an example from pest control of insects on plants. There are 20 plots, 10 sprayed and 10 unsprayed. Within each plot there are 50 plants. Each plant is measured five times during the growing season. Now this experiment generates $20 \times 50 \times 5 = 5000$ numbers. There are two spraying treatments, so there must be 1 degree of freedom for spraying and 4998 degrees of freedom for error. Or must there? Count up the replicates in this experiment. Repeated measurements on the same plants (the five sampling occasions) are certainly not replicates. The 50 individual plants within each quadrat are not replicates either. The reason for this is that conditions within each quadrat are quite likely to be unique, and so all 50 plants will experience more or less the same unique set of conditions, irrespective of the spraying treatment they receive. In fact, there are 10 replicates in this experiment. There are 10 sprayed plots and 10 unsprayed plots, and each plot will yield only one independent datum for the response variable (the proportion of leaf area consumed by insects, for example). Thus, there are 9 degrees of freedom within each treatment, and $2 \times 9 = 18$ degrees of freedom for error in the experiment as a whole. It is not difficult to find examples of pseudoreplication on this scale in the literature (Hurlbert 1984). The problem is that it leads to the reporting of masses of spuriously significant results (with 4998 degrees of freedom for error, it is almost impossible *not* to have significant differences). The first skill to be acquired by the budding experimenter is the ability to plan an experiment that is properly replicated. There are various things that you can do when your data are pseudoreplicated:

- Average away the pseudoreplication and carry out your statistical analysis on the means.
- Carry out separate analyses for each time period.
- Use proper time series analysis or mixed-effects models.

The lme and lmer Functions

Most of the examples in this chapter use the linear mixed model formula lme. This is to provide compatibility with the excellent book by Pinheiro and Bates (2000) on *Mixed-Effects Models in S and S-PLUS*. More recently, however, Douglas Bates has released the generalized mixed model function lmer as part of the lme4 package, and you may prefer to use this in your own work (see the Index for worked examples of lmer in this book; all of the analyses in this chapter using lme are repeated using lmer on the book's website). Here, I provide a simple comparison of the basic syntax of the two functions.

lme

Specifying the fixed and random effects in the model formula is done with two formulae. Suppose that there are no fixed effects, so that all of the categorical variables are random effects. Then the fixed effect simply estimates the intercept (parameter 1):

fixed = y~1

The fixed effect (a compulsory part of the lme structure) is just the overall mean value of the response variable y ~ 1. The fixed = part of the formula is optional. The random effects show the identities of the random variables and their relative locations in the hierarchy. The random effects are specified like this:

random = ~ 1 | a/b/c

and in this case the phrase random = is *not* optional. An important detail to notice is that the name of the response variable (y) is not repeated in the random-effects formula: there is a blank space to the left of the tilde ~. In most mixed-effects models we assume that the random effects have a mean of zero and that we are interested in quantifying variation in the intercept (this is parameter 1) caused by differences between the factor levels of the random effects. After the intercept comes the vertical bar | which is read as 'given the following spatial arrangement of the random variables'. In this example there are three random effects with 'c nested within b which in turn is nested within a'. The factors are separated by forward slash characters, and the variables are listed from left to right in declining order of spatial (or temporal) scale. This will only become clear with practice, but it is a simple idea. The formulae are put together like this:

lme(fixed = y~1, random = ~ 1 | a/b/c)

lmer

There is just one formula in lmer, not separate formulae for the fixed and random effects. The fixed effects are specified first, to the right of the tilde, in the normal way. Next comes a plus sign, then one or more random terms enclosed in parentheses (in this example there is just one random term, but we might want separate random terms for the intercept and for the slopes, for instance). R can identify the random terms because they must contain a 'given' symbol |, to the right of which are listed the random effects in the usual way, from largest to smallest scale, left to right. So the lmer formula for this example is

lmer(y~1+(1 | a/b/c))

Best Linear Unbiased Predictors

In aov, the effect size for treatment i is defined as $\bar{y}_i - \mu$, where μ is the overall mean. In mixed-effects models, however, correlation between the pseudoreplicates within a group causes what is called **shrinkage**. The best linear unbiased predictors (BLUPs, denoted by a_i) are smaller than the effect sizes ($\bar{y}_i - \mu$), and are given by

$$a_i = (\bar{y}_i - \mu) \left(\frac{\sigma_a^2}{\sigma_a^2 + \sigma^2/n} \right),$$

where σ^2 is the residual variance and σ_a^2 is the between-group variance which introduces the correlation between the pseudoreplicates within each group. Thus, the parameter estimate a_i is 'shrunk' compared to the fixed effect size ($\bar{y}_i - \mu$). When σ_a^2 is estimated to be large compared with the estimate of σ^2/n, then the fixed effects and the BLUP are similar (i.e. when most of the variation is between classes and there is little variation within classes). On the other hand, when σ_a^2 is estimated to be small compared with the estimate of σ^2/n, then the fixed effects and the BLUP can be very different (p. 547).

A Designed Experiment with Different Spatial Scales: Split Plots

The important distinction in models with categorical explanatory variables is between cases where the data come from a designed experiment, in which treatments were allocated to locations or subjects at random, and cases where the data come from an observational study in which the categorical variables are associated with an observation before the study. Here, we call the first case split-plot experiments and the second case hierarchical designs. The point is that their dataframes look identical, so it is easy to analyse one case wrongly as if it were the other. You need to be able to distinguish between fixed effects and random effects in both cases. Here is the linear model for a split-plot experiment analysed in Chapter 11 by aov (see p. 470).

```
yields<-read.table("c:\\temp\\splityield.txt",header=T)
attach(yields)
names(yields)
```

```
[1] "yield" "block" "irrigation" "density" "fertilizer"
```

```
library(nlme)
```

The fixed-effects part of the model is specified in just the same way as in a straightforward factorial experiment: yield~irrigation*density*fertilizer. The random-effects part of the model says that we want the random variation to enter via effects on the intercept (which is parameter 1) as random=~1. Finally, we define the spatial structure of the random effects after the 'given' symbol | as: block/irrigation/density. There is no need to specify the smallest spatial scale (fertilizer plots in this example).

```
model<-lme(yield~irrigation*density*fertilizer,random=~1|block/irrigation/density)
summary(model)
```

```
Linear mixed-effects model fit by REML
  Data: NULL
       AIC       BIC      logLik
   481.6212  525.3789  -218.8106

Random effects:
Formula: ~ 1 | block
        (Intercept)
StdDev: 0.0006601056

Formula: ~ 1 | irrigation %in% block
    (Intercept)
StdDev: 1.982461

Formula: ~ 1 | density %in% irrigation %in% block
      (Intercept)  Residual
StdDev: 6.975554   9.292805
```

```
Fixed effects: yield ~ irrigation * density * fertilizer
                        Value     Std.Error    DF      t-value
(Intercept)             80.50     5.893741     36      13.658558
irrigationirrigated     31.75     8.335008      3       3.809234
densitylow               5.50     8.216282     12       0.669403
```

densitymedium	14.75	8.216282	12	1.795216
fertilizerNP	5.50	6.571005	36	0.837010
fertilizerP	4.50	6.571005	36	0.684827
irrigationirrigated:densitylow	-39.00	11.619577	12	-3.356404
irrigationirrigated:densitymedium	-22.25	11.619577	12	-1.914872
irrigationirrigated:fertilizerNP	13.00	9.292805	36	1.398932
irrigationirrigated:fertilizerP	5.50	9.292805	36	0.591856
densitylow:fertilizerNP	3.25	9.292805	36	0.349733
densitymedium:fertilizerNP	-6.75	9.292805	36	-0.726368
densitylow:fertilizerP	-5.25	9.292805	36	-0.564953
densitymedium:fertilizerP	-5.50	9.292805	36	-0.591856
irrigationirrigated:densitylow:fertilizerNP	7.75	13.142011	36	0.589712
irrigationirrigated:densitymedium:fertilizerNP	3.75	13.142011	36	0.285344
irrigationirrigated:densitylow:fertilizerP	20.00	13.142011	36	1.521837
irrigationirrigated:densitymedium:fertilizerP	4.00	13.142011	36	0.304367

	p-value
(Intercept)	0.0000
irrigationirrigated	0.0318
densitylow	0.5159
densitymedium	0.0978
fertilizerNP	0.4081
fertilizerP	0.4978
irrigationirrigated:densitylow	0.0057
irrigationirrigated:densitymedium	0.0796
irrigationirrigated:fertilizerNP	0.1704
irrigationirrigated:fertilizerP	0.5576
densitylow:fertilizerNP	0.7286
densitymedium:fertilizerNP	0.4723
densitylow:fertilizerP	0.5756
densitymedium:fertilizerP	0.5576
irrigationirrigated:densitylow:fertilizerNP	0.5591
irrigationirrigated:densitymedium:fertilizerNP	0.7770
irrigationirrigated:densitylow:fertilizerP	0.1368
irrigationirrigated:densitymedium:fertilizerP	0.7626

This output suggests that the only significant effects are the main effect of irrigation ($p = 0.0318$) and the irrigation by density interaction ($p = 0.0057$). The three-way interaction is not significant so we remove it, fitting all terms up to two-way interactions:

```
model<-
lme(yield~(irrigation+density+fertilizer)^2,random=~1|block/irrigation/density)
summary(model)
```

```
Linear mixed-effects model fit by REML
  Data: NULL
      AIC        BIC      logLik
  503.1256   540.2136   -233.5628

Random effects:
Formula: ~ 1 | block
       (Intercept)
StdDev: 0.000563668
```

```
Formula: ~ 1 | irrigation %in% block
       (Intercept)
StdDev: 1.982562
```

```
Formula: ~ 1 | density %in% irrigation %in% block
       (Intercept)  Residual
StdDev: 7.041303 9.142696
```

```
Fixed effects: yield ~ (irrigation + density + fertilizer)^2
                                        Value  Std.Error  DF    t-value  p-value
(Intercept)                          82.47222   5.443438  40  15.150760   0.0000
irrigationirrigated                  27.80556   7.069256   3   3.933307   0.0293
densitylow                            0.87500   7.256234  12   0.120586   0.9060
densitymedium                        13.45833   7.256234  12   1.854727   0.0884
fertilizerNP                          3.58333   5.278538  40   0.678850   0.5011
fertilizerP                           0.50000   5.278538  40   0.094723   0.9250
irrigationirrigated:densitylow      -29.75000   8.800165  12  -3.380618   0.0055
irrigationirrigated:densitymedium   -19.66667   8.800165  12  -2.234807   0.0452
irrigationirrigated:fertilizerNP     16.83333   5.278538  40   3.189014   0.0028
irrigationirrigated:fertilizerP      13.50000   5.278538  40   2.557526   0.0144
densitylow:fertilizerNP               7.12500   6.464862  40   1.102112   0.2770
densitymedium:fertilizerNP           -4.87500   6.464862  40  -0.754076   0.4552
densitylow:fertilizerP                4.75000   6.464862  40   0.734741   0.4668
densitymedium:fertilizerP            -3.50000   6.464862  40  -0.541388   0.5912
```

The fertilizer by density interaction is not significant, so we remove it:

```
model<-
lme(yield~irrigation*density+irrigation*fertilizer,random=~1|block/irrigation/density)
summary(model)
```

```
Linear mixed-effects model fit by REML
Data: NULL
     AIC        BIC      logLik
519.9035   549.6834   -245.9517
```

```
Random effects:
Formula: ~ 1 | block
         (Intercept)
StdDev: 0.0005569251
```

```
Formula: ~ 1 | irrigation %in% block
       (Intercept)
StdDev:  1.982615
```

```
Formula: ~ 1 | density %in% irrigation %in% block
       (Intercept)  Residual
StdDev:  7.057132  9.105995
```

```
Fixed effects: yield ~ irrigation * density + irrigation * fertilizer
```

	Value	Std.Error	DF	t-value	p-value
(Intercept)	82.08333	4.994999	44	16.433103	0.0000
irrigationirrigated	27.80556	7.063995	3	3.936236	0.0292
densitylow	4.83333	6.222653	12	0.776732	0.4524
densitymedium	10.66667	6.222653	12	1.714167	0.1122
fertilizerNP	4.33333	3.717507	44	1.165656	0.2500
fertilizerP	0.91667	3.717507	44	0.246581	0.8064
irrigationirrigated:densitylow	-29.75000	8.800160	12	-3.380620	0.0055
irrigationirrigated:densitymedium	-19.66667	8.800160	12	-2.234808	0.0452
irrigationirrigated:fertilizerNP	16.83333	5.257349	44	3.201867	0.0025
irrigationirrigated:fertilizerP	13.50000	5.257349	44	2.567834	0.0137

The moral is that you must do the model simplification to get the appropriate p values.

Remember, too, that if you want to use anova to compare mixed models with different fixed-effects structures, then you must use maximum likelihood (method = "ML") rather than the default restricted maximum likelihood (REML). Here is the analysis again, but this time using anova to compare models with progressively simplified fixed effects.

```
model.lme<-lme(yield~irrigation*density*fertilizer,
                    random=~ 1| block/irrigation/density,method="ML")

model.lme.2<-update(model.lme,~. - irrigation:density:fertilizer)
anova(model.lme,model.lme.2)
```

	Model	df	AIC	BIC	logLik	Test	L.Ratio	p-value
model.lme	1	22	573.5108	623.5974	-264.7554			
model.lme.2	2	18	569.0046	609.9845	-266.5023	1 vs 2	3.493788	0.4788

```
model.lme.3<-update(model.lme.2,~. - density:fertilizer)
anova(model.lme.3,model.lme.2)
```

	Model	df	AIC	BIC	logLik	Test	L.Ratio	p-value
model.lme.3	1	14	565.1933	597.0667	-268.5967			
model.lme.2	2	18	569.0046	609.9845	-266.5023	1 vs 2	4.188774	0.3811

```
model.lme.4<-update(model.lme.3,~. - irrigation:fertilizer)
anova(model.lme.3,model.lme.4)
```

	Model	df	AIC	BIC	logLik	Test	L.Ratio	p-value
model.lme.3	1	14	565.1933	597.0667	-268.5967			
model.lme.4	2	12	572.3373	599.6573	-274.1687	1 vs 2	11.14397	0.0038

```
model.lme.5<-update(model.lme.2,~. - irrigation:density)
anova(model.lme.5,model.lme.2)
```

	Model	df	AIC	BIC	logLik	Test	L.Ratio	p-value
model.lme.5	1	16	576.7134	613.1400	-272.3567			
model.lme.2	2	18	569.0046	609.9845	-266.5023	1 vs 2	11.70883	0.0029

The irrigation–fertilizer interaction is more significant ($p = 0.0038$ compared to $p = 0.0081$) under this mixed-effects model than it was in the linear model earlier, as is the irrigation–density interaction ($p = 0.0029$ compared to $p = 0.01633$). You need to do the model simplification in lme to uncover the significance of the main effect and interaction terms, but it is worth it, because the lme analysis can be more powerful. The minimal adequate model under the lme is:

summary(model.lme.3)

```
Linear mixed-effects model fit by maximum likelihood
Data: NULL
      AIC         BIC      logLik
  565.1933    597.0667   -268.5967

Random effects:
Formula: ~ 1 | block
        (Intercept)
StdDev:  0.0005261129

Formula: ~ 1 | irrigation %in% block
      (Intercept)
StdDev:  1.716889

Formula: ~ 1 | density %in% irrigation %in% block
      (Intercept)  Residual
StdDev:  5.722413  8.718327

Fixed effects: yield ~ irrigation + density + fertilizer +
irrigation:density + irrigation:fertilizer
```

	Value	Std.Error	DF	t-value	p-value
(Intercept)	82.08333	4.756285	44	17.257867	0.0000
irrigationirrigated	27.80556	6.726403	3	4.133793	0.0257
densitylow	4.83333	5.807347	12	0.832279	0.4215
densitymedium	10.66667	5.807347	12	1.836754	0.0911
fertilizerNP	4.33333	3.835552	44	1.129781	0.2647
fertilizerP	0.91667	3.835552	44	0.238992	0.8122
irrigationirrigated:densitylow	-29.75000	8.212829	12	-3.622382	0.0035
irrigationirrigated:densitymedium	-19.66667	8.212829	12	-2.394628	0.0338
irrigationirrigated:fertilizerNP	16.83333	5.424290	44	3.103325	0.0033
irrigationirrigated:fertilizerP	13.50000	5.424290	44	2.488805	0.0167

Note that the degrees of freedom are not pseudoreplicated: d.f. $= 12$ for testing the irrigation by density interaction and d.f. $= 44$ for irrigation by fertilizer (this is $36 + 4 + 4 = 44$ after model simplification). Also, remember that you must do your model simplification using maximum likelihood (method = "ML") because you cannot compare models with different fixed-effect structures using REML.

Model-checking plots show that the residuals are well behaved:

plot(model.lme.3)

The response variable is a reasonably linear function of the fitted values:

plot(model.lme.3,yield~fitted(.))

and the errors are reasonably close to normally distributed in all four blocks:

qqnorm(model.lme.3,~ resid(.)| block)

When, as here, the experiment is balanced and there are no missing values, then it is much simpler to interpret the aov using an Error term to describe the structure of the spatial pseudoreplication (p. 470). Without balance, however, you will need to use lme and to use model simplification to estimate the p values of the significant interaction terms.

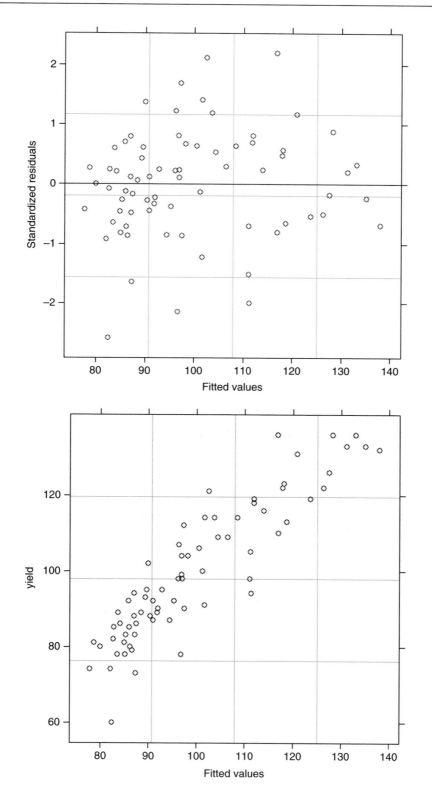

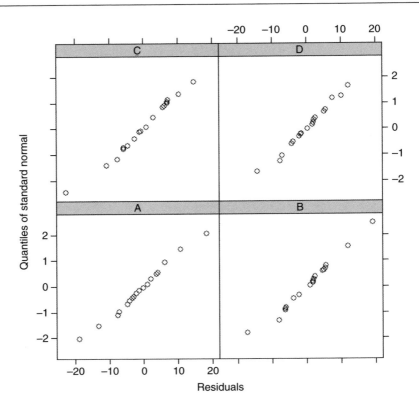

Hierarchical Sampling and Variance Components Analysis

Hierarchical data are often encountered in observational studies where information is collected at a range of different spatial scales. Consider an epidemiological study of childhood diseases in which blood samples were taken for individual children, households, streets, districts, towns, regions, and countries. All these categorical variables are random effects. The spatial scale increases with each step in the hierarchy. The interest lies in discovering where most of the variation originates: is it between children within households or between districts within the same town? When it comes to testing hypotheses at larger spatial scales (such as town or regions), such data sets contain huge amounts of pseudoreplication.

The following example has a slightly simpler spatial structure than this: infection is measured for two replicate males and females within each of three families within four streets within three districts within five towns (720 measurements in all). We want to carry out a variance components analysis. Here are the data:

```
hierarchy<-read.table("c:\\temp\\hre.txt",header=T)
attach(hierarchy)
names(hierarchy)
```

```
[1]  "subject"  "town"       "district"  "street"  "family"
[6]  "gender"   "replicate"
```

```
library(nlme)
library(lattice)
```

model1<-lme(subject~1,random=~1|town/district/street/family/gender)
summary(model1)

```
Linear mixed-effects model fit by REML
Data: NULL
       AIC        BIC      logLik
  3351.294   3383.339   -1668.647

Random effects:
Formula: ~1 | town
        (Intercept)
StdDev:   1.150604

Formula: ~1 | district %in% town
        (Intercept)
StdDev:   1.131932

Formula: ~1 | street %in% district %in% town
        (Intercept)
StdDev:   1.489864

Formula: ~1 | family %in% street %in% district %in% town
        (Intercept)
StdDev:   1.923191

Formula: ~1 | gender %in% family %in% street %in% district %in% town
        (Intercept)   Residual
StdDev:   3.917264   0.9245321

Fixed effects: subject ~ 1
                 Value   Std.Error   DF   t-value   p-value
(Intercept)   8.010941   0.6719753  360  11.92148         0

Standardized Within-Group Residuals:
       Min           Q1          Med          Q3          Max
-2.64600654  -0.47626815  -0.06009422  0.47531635  2.35647504

Number of Observations: 720
Number of Groups:
                                                            town
                                                               5
                                           district %in% town
                                                              15
                              street %in% district %in% town
                                                              60
                 family %in% street %in% district %in% town
                                                             180
gender %in% family %in% street %in% district %in% town
                                                             360
```

Notice that the model was fitted by REML rather than by the more familiar maximum likelihood. REML methods differ because they allow for the degrees of freedom used up in estimating the fixed effects. Thus, the variance components are estimated without being affected by the fixed effects (they are invariant to the values of the fixed effects). Also, REML estimators are less sensitive to outliers than are ML estimators.

To calculate the variance components we need to extract the standard deviations of the random effects from the model summary, square them to get the variances, then express each as a percentage of the total:

```
sds<-c(1.150604,1.131932,1.489864,1.923191,3.917264,0.9245321)
vars<-sds^2
100*vars/sum(vars)

[1]   5.354840   5.182453   8.978173   14.960274   62.066948   3.457313
```

This indicates that the gender effect (62%) is much the most important component of overall variance. Next most important is variation from family to family (15%).

For comparison, here is the layout of the output for the same analysis using lmer:

```
library(lme4)
model1<-lmer(subject~1+(1|town/district/street/family/gender))
summary(model1)

Linear mixed-effects model fit by REML
Formula: subject ~ 1 + (1 | town/district/street/family/gender)
  AIC    BIC    logLik   MLdeviance   REMLdeviance
  3349   3377   -1669      3338          3337
```

Random effects:

Groups	Name	Variance	Std.Dev.
gender:(family:(street:(district:town)))	(Intercept)	15.3387	3.91647
family:(street:(district:town))	(Intercept)	3.7008	1.92375
street:(district:town)	(Intercept)	2.2283	1.49274
district:town	(Intercept)	1.2796	1.13121
town	(Intercept)	1.3238	1.15056
Residual		0.8548	0.92456

```
number of obs: 720, groups: gender:(family:(street:(district:town))),
360; family:(street:(district:town)), 180; street:(district:town), 60;
district:town, 15; town, 5
```

Fixed effects:

	Estimate	Std. Error	t value
(Intercept)	8.011	0.672	11.92

You will see that the variance components are given in the penultimate column. Fixed effects in this model are discussed on p. 656.

Model Simplification in Hierarchical Sampling

We need to know whether all of the random effects are required in the model. The key point to grasp here is that you will need to recode the factor levels if you want to leave out a random effect from a larger spatial scale. Suppose we want to test the effect of leaving out the identity of the towns. Because the districts were originally coded with the same names within each town,

```
levels(district)

[1]   "d1"   "d2"   "d3"
```

we shall need to create 15 new, unique district names. Much the simplest way to do this is to use paste to combine the town names and the district names:

```
newdistrict<-factor(paste(town,district,sep=""))
levels(newdistrict)
```

```
 [1]  "Ad1"  "Ad2"  "Ad3"  "Bd1"  "Bd2"  "Bd3"  "Cd1"  "Cd2"  "Cd3"  "Dd1"
[11]  "Dd2"  "Dd3"  "Ed1"  "Ed2"  "Ed3"
```

In model2 we leave out the random effect for towns and include the new factor for districts:

```
model2<-lme(subject~1,random=~1|newdistrict/street/family/gender)
anova(model1,model2)
```

	Model	df	AIC	BIC	logLik	Test	L.Ratio	p-value
model1	1	7	3351.294	3383.339	-1668.647			
model2	2	6	3350.524	3377.991	-1669.262	1 vs 2	1.229803	0.2674

Evidently there is no significant effect attributable to differences between towns ($p = 0.2674$).

The next question concerns differences between the districts. Because the streets within districts were all coded in the same way in the original dataframe, we need to create 60 unique codes for the different streets:

```
newstreet<-factor(paste(newdistrict,street,sep=""))
levels(newstreet)
```

```
 [1]  "Ad1s1"  "Ad1s2"  "Ad1s3"  "Ad1s4"  "Ad2s1"  "Ad2s2"  "Ad2s3"  "Ad2s4"  "Ad3s1"
[10]  "Ad3s2"  "Ad3s3"  "Ad3s4"  "Bd1s1"  "Bd1s2"  "Bd1s3"  "Bd1s4"  "Bd2s1"  "Bd2s2"
[19]  "Bd2s3"  "Bd2s4"  "Bd3s1"  "Bd3s2"  "Bd3s3"  "Bd3s4"  "Cd1s1"  "Cd1s2"  "Cd1s3"
[28]  "Cd1s4"  "Cd2s1"  "Cd2s2"  "Cd2s3"  "Cd2s4"  "Cd3s1"  "Cd3s2"  "Cd3s3"  "Cd3s4"
[37]  "Dd1s1"  "Dd1s2"  "Dd1s3"  "Dd1s4"  "Dd2s1"  "Dd2s2"  "Dd2s3"  "Dd2s4"  "Dd3s1"
[46]  "Dd3s2"  "Dd3s3"  "Dd3s4"  "Ed1s1"  "Ed1s2"  "Ed1s3"  "Ed1s4"  "Ed2s1"  "Ed2s2"
[55]  "Ed2s3"  "Ed2s4"  "Ed3s1"  "Ed3s2"  "Ed3s3"  "Ed3s4"
```

Now fit the new model leaving out both towns and districts,

```
model3<-lme(subject~1,random=~1|newstreet/family/gender)
```

and compare this with model2 from which towns had been removed:

```
anova(model2,model3)
```

	Model	df	AIC	BIC	logLik	Test	L.Ratio	p-value
model2	1	6	3350.524	3377.991	-1669.262			
model3	2	5	3354.084	3376.973	-1672.042	1 vs 2	5.559587	0.0184

This simplification was not justified ($p = 0.0184$) so we conclude that there is significant variation from district to district. Model-checking plots are illustrated on p. 657.

Mixed-Effects Models with Temporal Pseudoreplication

A common cause of temporal pseudoreplication in growth experiments with fixed effects is when each individual is measured several times as it grows during the course of an experiment. The next example is as simple as possible: we have a single fixed effect (a two-level categorical variable: with fertilizer added or not) and six replicate plants in each treatment, with each plant measured on five occasions (after 2, 4, 6, 8 or 10 weeks of growth). The response variable is root length. The fixed-effect formula looks like this:

fixed = root~fertilizer

The random-effects formula needs to indicate that the week of measurement (a continuous random effect) represents pseudoreplication within each individual plant:

random = ~week|plant

Because we have a continuous random effect (weeks) we write ~week in the random-effects formula rather than the ~1 that we used with categorical random effects (above). Here are the data:

```
results<-read.table("c:\\temp\\fertilizer.txt",header=T)
attach(results)
names(results)
```

```
[1]   "root"   "week"   "plant"   "fertilizer"
```

We begin with data inspection. For the kind of data involved in mixed-effects models there are some excellent built-in plotting functions (variously called panel plots, trellis plots, or lattice plots).

```
library(nlme)
library(lattice)
```

To use trellis plotting, we begin by turning our dataframe called results (created by read.table) into a groupedData object (p. 668). To do this we specify the nesting structure of the random effects, and indicate the fixed effect by defining fertilizer as outer to this nesting:

results<-groupedData(root~week|plant,outer = ~ fertilizer,results)

Because results is now a groupedData object, the plotting is fantastically simple:

plot(results)

Here you get separate time series plots for each of the individual plants, ranked from bottom left to top right on the basis of mean root length. In terms of understanding the fixed effects, it is often more informative to group together the six replicates within each treatment, and to have two panels, one for the fertilized plants and one for the controls. This is easy:

plot(results,outer=T)   2 R×S   6 replicate plants

You can see that by week 10 there is virtually no overlap between the two treatment groups. The largest control plant has about the same root length as the smallest fertilized plant (c.9 cm).

   Now for the statistical modelling. Ignoring the pseudoreplication, we should have 1 d.f. for fertilizer and $2 \times (6 - 1) = 10$ d.f. for error.

```
model<-lme(root~fertilizer,random=~week|plant)
summary(model)
```

```
Linear mixed-effects model fit by REML
Data: NULL
       AIC          BIC        logLik
171.0236   183.3863   -79.51181
```

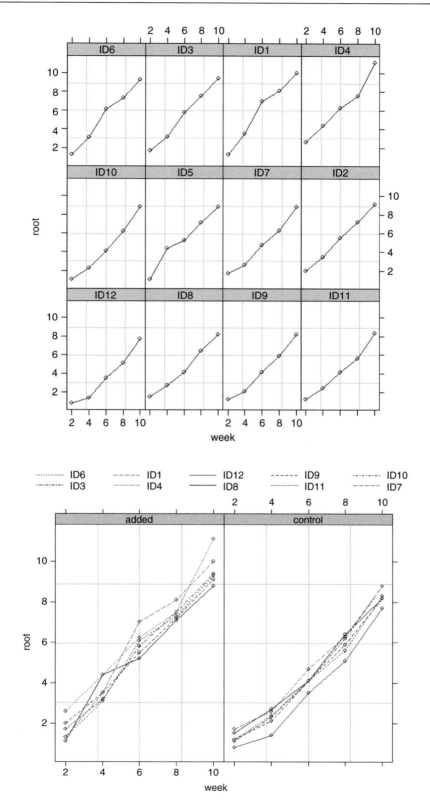

Random effects:
  Formula: ~week | plant
  Structure: General positive-definite, Log-Cholesky parametrization
             StdDev   Corr
(Intercept) 2.8639832   (Intr)
week   0.9369412   -0.999
Residual   0.4966308

Fixed effects: root ~ fertilizer
                    Value   Std.Error   DF   t-value   p-value
(Intercept)   2.799710   0.1438367   48   19.464499   0e+00
fertilizercontrol -1.039383 0.2034158   10   -5.109645   5e-04

Correlation:
                    (Intr)
fertilizercontrol -0.707

Standardized Within-Group Residuals:

       Min   Q1   Med   Q3   Max
  -1.9928118   -0.6586834   -0.1004301   0.6949714   2.0225381

Number of Observations: 60
Number of Groups: 12

The output looks dauntingly complex, but once you learn your way around it, the essential information is relatively easy to extract. The mean reduction in root size associated with the unfertilized controls is $-1.039\,383$ and this has a standard error of $0.203\,415\,8$ based on the correct 10 residual d.f. (six replicates per factor level). Can you see why the intercept has 48 d.f.?

Here is a simple one-way ANOVA for the non-pseudoreplicated data from week 10:

model2<-aov(root~fertilizer,subset=(week==10))
summary(model2)

             Df   Sum Sq   Mean Sq   F value   Pr(>F)
fertilizer   1   4.9408   4.9408   11.486   0.006897   **
Residuals   10   4.3017   0.4302

summary.lm(model2)

Coefficients:
                    Estimate Std.   Error   t value   Pr(>|t|)
(Intercept)   9.6167   0.2678   35.915   6.65e-12   ***
fertilizercontrol   -1.2833   0.3787   -3.389   0.0069   **

We can compare this with the output from the lme. The effect size in the lme is slightly smaller (-1.039393 compared to -1.2833) but the standard error is appreciably lower ($0.203\,415\,8$ compared to 0.3787), so the significance of the result is higher in the lme than in the aov. You get increased statistical power as a result of going to the trouble of fitting the mixed-effects model. And, crucially, you do not need to make potentially arbitrary judgements about which time period to select for the non-pseudoreplicated analysis. You use all of the data in the model, and you specify its structure appropriately so that the hypotheses are tested with the correct degrees of freedom (10 in this case, not 48).

The reason why the effect sizes are different in the lm and lme models is that linear models use maximum likelihood estimates of the parameters based on arithmetic means. The linear mixed models, however, use the wonderfully named BLUPs.

Time Series Analysis in Mixed-Effects Models

It is common to have repeated measures on subjects in observational studies, where we would expect that the observation on an individual at time $t + 1$ would be quite strongly correlated with the observation on the same individual at time t. This contravenes one of the central assumptions of mixed-effects models (p. 627), that the within-group errors are independent. However, we often observe significant serial correlation in data such as these.

The following example comes from Pinheiro and Bates (2000) and forms part of the nlme library. The data refer to the numbers of ovaries observed in repeated measures on 11 mares (their oestrus cycles have been scaled such that ovulation occurred at time 0 and at time 1). The issue is how best to model the correlation structure of the data. We know from previous work that the fixed effect can be modelled as a three-parameter sine–cosine function

$$y = a + b \sin(2\pi x) + d \cos(2\pi x) + \varepsilon_{ij},$$

and we want to assess different structures for modelling the within-class correlation structure. The dataframe is of class groupedData which makes the plotting and error checking much simpler.

```
library(nlme)
library(lattice)
data(Ovary)
attach(Ovary)
names(Ovary)
```

```
[1]  "Mare"  "Time"  "follicles"
```

```
plot(Ovary)
```

The panel plot has ranked the horses from bottom left to top right on the basis of their mean number of ovules (mare 4 with the lowest number, mare 8 with the highest). Some animals show stronger cyclic behaviour than others.

We begin by fitting a mixed-effects model making no allowance for the correlation structure, and investigate the degree of autocorrelation that is exhibited by the residuals (recall that the assumption of the model is that there is no correlation).

```
model<-lme(follicles~sin(2*pi*Time)+cos(2*pi*Time),
                              data=Ovary,random=~ 1| Mare)
summary(model)
```

```
Linear mixed-effects model fit by REML
Data: Ovary
      AIC        BIC        logLik
1669.360   1687.962   -829.6802

Random effects:
Formula: ~ 1 | Mare
        (Intercept)   Residual
StdDev:  3.041344    3.400466

Fixed effects: follicles ~ sin(2 * pi * Time) + cos(2 * pi * Time)
```

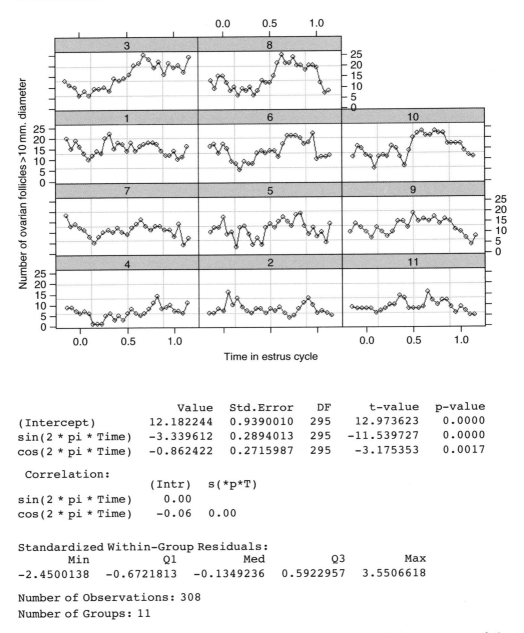

	Value	Std.Error	DF	t-value	p-value
(Intercept)	12.182244	0.9390010	295	12.973623	0.0000
sin(2 * pi * Time)	-3.339612	0.2894013	295	-11.539727	0.0000
cos(2 * pi * Time)	-0.862422	0.2715987	295	-3.175353	0.0017

```
 Correlation:
                     (Intr)   s(*p*T)
sin(2 * pi * Time)    0.00
cos(2 * pi * Time)   -0.06    0.00
```

```
Standardized Within-Group Residuals:
      Min            Q1           Med            Q3           Max
-2.4500138    -0.6721813    -0.1349236     0.5922957     3.5506618
```

```
Number of Observations: 308
Number of Groups: 11
```

The function ACF allows us to calculate the empirical autocorrelation structure of the residuals from this model:

```
plot(ACF(model),alpha=0.05)
```

You can see that there is highly significant autocorrelation at lags 1 and 2 and marginally significant autocorrelation at lags 3 and 4. We model the autocorrelation structure using one of the standard corStruct classes (p. 701). For time series data like this, we typically choose between 'moving average', 'autoregressive' or 'autoregressive moving average' classes. Again, experience with horse biology suggests that a simple moving average model might be appropriate, so we start with this. The class is called corARMA and we need to specify

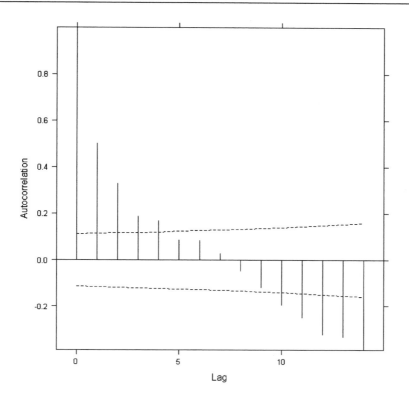

the order of the model (the lag of the moving average part): the simplest assumption is that only the first two lags exhibit non-zero correlations (q = 2):

```
model2<-update(model,correlation=corARMA(q=2))
anova(model,model2)
```

	Model	df	AIC	BIC	logLik	Test	L.Ratio	p-value
model	1	5	1669.360	1687.962	-829.6802			
model2	2	7	1574.895	1600.937	-780.4476	1 vs 2	98.4652	<.0001

This is a great improvement over the original model, which assumed no correlation in the residuals. But what about a different time series assumption? Let us compare the moving average assumption with a simple first-order autoregressive model corAR1():

```
model3<-update(model2,correlation=corAR1())
anova(model2,model3)
```

	Model	df	AIC	BIC	logLik	Test	L.Ratio	p-value
model2	1	7	1574.895	1600.937	-780.4476			
model3	2	6	1573.453	1595.775	-780.7264	1 vs 2	0.5577031	0.4552

There is nothing to chose between the models on the basis of ANOVA, $p = 0.455$, so we choose the corAR1() because it has the lowest AIC (it also uses fewer degrees of freedom, d.f. $= 6$). Error checking on model3 might proceed like this:

```
plot(model3,resid(.,type="p")~fitted(.)|Mare)
```

The residuals appear to be reasonably well behaved. And the normality assumption?

```
qqnorm(model3,~resid(.)|Mare)
```

The model is well behaved, so we accept a first-order autocorrelation structure corAR1().

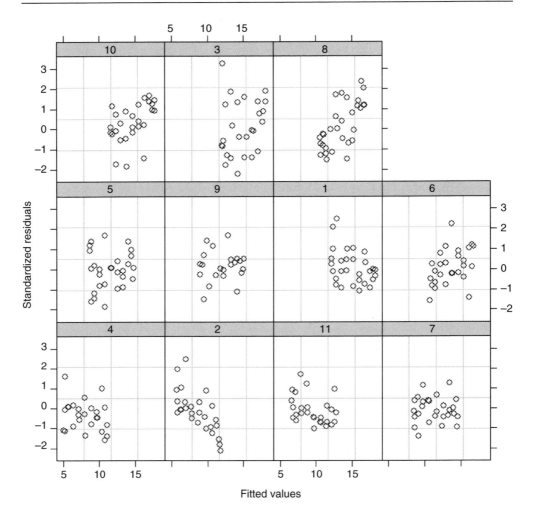

Random Effects in Designed Experiments

The rats example, studied by **aov** with an **Error** term on p. 476, can be repeated as a linear mixed-effects model. This example works much better with **lmer** than with **lme**.

```
dd<-read.table("c:\\temp\\rats.txt",h=T)
attach(dd)
names(dd)
```

```
[1] "Glycogen"  "Treatment"  "Rat"  "Liver"
```

```
Treatment<-factor(Treatment)
Liver<-factor(Liver)
Rat<-factor(Rat)
```

There is a single fixed effect (Treatment), and pseudoreplication enters the dataframe because each rat's liver is cut into three pieces and each separate liver bit produces two readings.

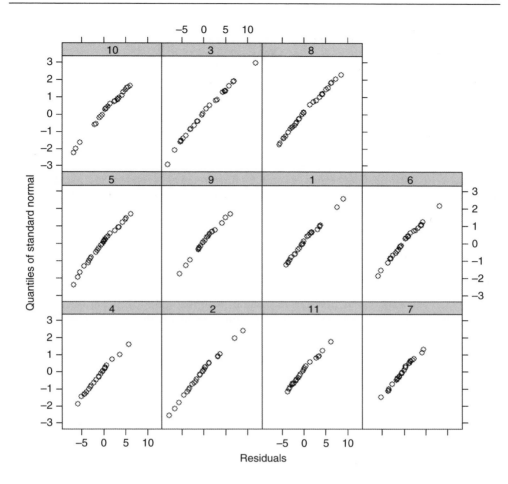

The rats are numbered 1 and 2 within each treatment, so we need Treatment as the largest scale of the random effects.

```
model<-lmer(Glycogen~Treatment+(1|Treatment/Rat/Liver))
summary(model)
```

```
Linear mixed-effects model fit by REML
Formula: Glycogen ~ Treatment + (1 | Treatment/Rat/Liver)
  AIC    BIC   logLik  MLdeviance  REMLdeviance
231.6  241.1  -109.8      234.9        219.6
```

```
Random effects:
Groups                    Name           Variance   Std.Dev.
Liver:(Rat:Treatment)     (Intercept)    14.1617    3.7632
Rat:Treatment             (Intercept)    36.0843    6.0070
Treatment                 (Intercept)     4.7039    2.1689
Residual                                 21.1678    4.6008
```

```
number of obs: 36, groups: Liver:(Rat:Treatment), 18; Rat:Treatment, 6;
Treatment, 3
```

```
Fixed effects:
              Estimate   Std. Error   t value
(Intercept)    140.500        5.184    27.104
Treatment2      10.500        7.331     1.432
Treatment3      -5.333        7.331    -0.728

Correlation of Fixed Effects:
             (Intr)    Trtmn2
Treatment2   -0.707
Treatment3   -0.707     0.500
```

You can see that the treatment effect is correctly interpreted as being non-significant ($t < 2$). The variance components (p. 478) can be extracted by squaring the standard deviations, then expressing them as percentages:

```
vars<- c(14.1617,36.0843,21.1678)
100*vars/sum(vars)
```

```
[1]  19.83048  50.52847  29.64105
```

so 50.5% of the variation is between rats within treatments, 19.8% is between liver bits within rats and 29.6% is between readings within liver bits within rats (see p. 333). You can extract the variance components with the VarCorr(model) function.

Regression in Mixed-Effects Models

The next example involves a regression of plant size against local point measurements of soil nitrogen (N) at five places within each of 24 farms. It is expected that plant size and soil nitrogen will be positively correlated. There is only one measurement of plant size and soil nitrogen at any given point (i.e. there is no temporal pseudoreplication; cf. p. 629):

```
yields<-read.table("c:\\temp\\farms.txt",header=T)
attach(yields)
names(yields)
```

```
[1]  "N"  "size"  "farm"
```

Here are the data in aggregate, with different plotting colours for each farm:

```
plot(N,size,pch=16,col=farm)
```

The most obvious pattern is that there is substantial variation in mean values of both soil nitrogen and plant size across the farms: the minimum (yellow) fields have a mean y value of less than 80, while the maximum (red) fields have a mean y value above 110.

The key distinction to understand is between fitting lots of linear regression models (one for each farm) and fitting one mixed-effects model, taking account of the differences between farms in terms of their contribution to the variance in response as measured by a standard deviation in intercept and a standard deviation in slope. We investigate these differences by contrasting the two fitting functions, lmList and lme. We begin by fitting 24 separate linear models, one for each farm:

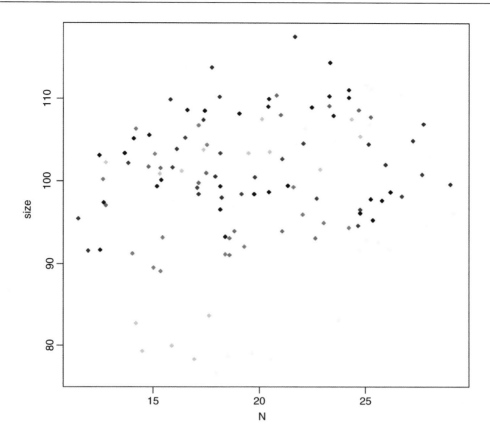

```
linear.models<-lmList(size~N|farm,yields)
coef(linear.models)
```

	(Intercept)	N
1	67.46260	1.5153805
2	118.52443	-0.5550273
3	91.58055	0.5551292
4	87.92259	0.9212662
5	92.12023	0.5380276
6	97.01996	0.3845431
7	68.52117	0.9339957
8	91.54383	0.8220482
9	92.04667	0.8842662
10	85.08964	1.4676459
11	114.93449	-0.2689370
12	82.56263	1.0138488
13	78.60940	0.1324811
14	80.97221	0.6551149
15	84.85382	0.9809902
16	87.12280	0.3699154
17	52.31711	1.7555136
18	83.40400	0.8715070
19	88.91675	0.2043755

```
20   93.08216   0.8567066
21   90.24868   0.7830692
22   78.30970   1.1441291
23   59.88093   0.9536750
24   89.07963   0.1091016
```

You see very substantial variations in the value of the intercept from 118.52 on farm 2 to 52.32 on farm 17. Slopes are also dramatically different, from negative -0.555 on farm 2 to steep and positive 1.7555 on farm 17. This is a classic problem in regression analysis when (as here) the intercept is a long way from the average value of x (see p. 398); large values of the intercept are almost bound to be correlated with low values of the slope.

Here are the slopes and intercepts from the model specified entirely in terms of random effects: a population of regression slopes predicted within each farm with nitrogen as the continuous explanatory variable:

random.model<-lme(size~1,random=~N|farm)
coef(random.model)

```
      (Intercept)              N
1        85.98139    0.574205307
2       104.67366   -0.045401473
3        95.03442    0.331080922
4        98.62679    0.463579823
5        95.00270    0.407906211
6        99.82294    0.207203698
7        85.57345    0.285520353
8        96.09461    0.520896471
9        95.22186    0.672262924
10       93.14157    1.017995727
11      108.27200    0.015213748
12       87.36387    0.689406409
13       80.83933    0.003617002
14       89.84309    0.306402249
15       93.37050    0.636778709
16       92.10914    0.145772153
17       94.93395    0.084935465
18       85.90160    0.709943262
19       92.00628    0.052485986
20       95.26296    0.738029400
21       93.35069    0.591151955
22       87.66161    0.673119269
23       70.57827    0.432993915
24       90.29151    0.036747120
```

Variation in the intercepts explains 97.26% of the variance, differences in slope a mere 0.245%, with a residual variance of 2.49% (see the summary table). The thing you notice is that the random effects are less extreme (i.e. closer to the mean) than the fixed effects. This is an example of shrinkage (p. 631), and is most clear from a graphical comparison of the coefficients of the linear and mixed models:

mm<-coef(random.model)
ll<-coef(linear.models)

```
par(mfrow=c(2,2))
plot(ll[,1],mm[,1],pch=16,xlab="linear",ylab="random effects")
abline(0,1)
plot(ll[,2],mm[,2],pch=16,xlab="linear",ylab="random effects")
abline(0,1)
```

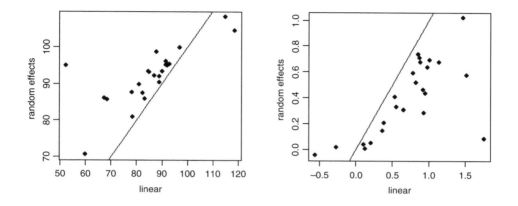

Most of the random-effects intercepts (left) are greater than their linear model equivalents (they are above the 45 degree line) while most of the random-effects slopes (right) are shallower than their linear model equivalents (i.e. below the line). For farm 17 the linear model had an intercept of 52.317 11 while the random-effects model had an intercept of 94.933 95. Likewise, the linear model for farm 17 had a slope of 1.755 513 6 while the random-effects model had a slope of 0.084 935 465.

We can fit a mixed model with both fixed and random effects. Here is a model in which size is modelled as a function of nitrogen and farm as fixed effects, and farm as a random effect. Because we intend to compare models with different fixed effect structures we need to specify method="ML" in place of the default REML.

```
farm<-factor(farm)
mixed.model1<-lme(size~N*farm,random=~1|farm,method="ML")
mixed.model2<-lme(size~N+farm,random=~1|farm,method="ML")
mixed.model3<-lme(size~N,random=~1|farm,method="ML")
mixed.model4<-lme(size~1,random=~1|farm,method="ML")
anova(mixed.model1,mixed.model2,mixed.model3,mixed.model4)
```

	Model	df	AIC	BIC	logLik	Test	L.Ratio	p-value
mixed.model1	1	50	542.9035	682.2781	-221.4518			
mixed.model2	2	27	524.2971	599.5594	-235.1486	1 vs 2	27.39359	0.2396
mixed.model3	3	4	614.3769	625.5269	-303.1885	2 vs 3	136.07981	<.0001
mixed.model4	4	3	658.0058	666.3683	-326.0029	3 vs 4	45.62892	<.0001

The first model contains a full factorial, with different slopes and intercepts for each of the 25 farms (using up 50 degrees of freedom). The second model has a common slope but different intercepts for the 25 farms (using 27 degrees of freedom); model2 does not have significantly lower explanatory power than model1 ($p = 0.2396$). The main effects of farm and of nitrogen application (model3 and model4) are both highly significant ($p < 0.0001$).

Finally, we could do an old-fashioned analysis of covariance, fitting a different two-parameter model to each and every farm without any random effects:

```
model<-lm(size~N*factor(farm))
summary(model)
```

```
Call:
lm(formula = size ~ N * factor(farm))
```

```
Residuals:
     Min       1Q    Median       3Q       Max
-3.60765  -1.29473   0.04789   1.07322   4.12972
```

Coefficients:

	Estimate	Std. Error	t value	Pr(>|t|)	
(Intercept)	67.46260	14.43749	4.673	1.35e-05	***
N	1.51538	0.73395	2.065	0.0426	*
factor(farm)2	51.06183	22.86930	2.233	0.0287	*
factor(farm)3	24.11794	16.54029	1.458	0.1492	
factor(farm)4	20.45999	34.59610	0.591	0.5561	
factor(farm)5	24.65762	17.29578	1.426	0.1583	
factor(farm)6	29.55736	17.74007	1.666	0.1000	
factor(farm)7	1.05856	20.53771	0.052	0.9590	
factor(farm)8	24.08122	16.23722	1.483	0.1424	
factor(farm)9	24.58407	15.45967	1.590	0.1162	
factor(farm)10	17.62703	16.68467	1.056	0.2943	
factor(farm)11	47.47189	18.24214	2.602	0.0112	*
factor(farm)12	15.10002	15.77085	0.957	0.3415	
factor(farm)13	11.14680	17.82896	0.625	0.5338	
factor(farm)14	13.50961	19.36739	0.698	0.4877	
factor(farm)15	17.39122	20.74850	0.838	0.4047	
factor(farm)16	19.66019	18.72739	1.050	0.2973	
factor(farm)17	-15.14550	49.01250	-0.309	0.7582	
factor(farm)18	15.94140	15.15371	1.052	0.2963	
factor(farm)19	21.45414	17.99214	1.192	0.2370	
factor(farm)20	25.61956	15.50019	1.653	0.1027	
factor(farm)21	22.78608	15.65699	1.455	0.1499	
factor(farm)22	10.84710	17.69820	0.613	0.5419	
factor(farm)23	-7.58167	16.89435	-0.449	0.6549	
factor(farm)24	21.61703	17.28697	1.250	0.2152	
N:factor(farm)2	-2.07041	0.98369	-2.105	0.0388	*
N:factor(farm)3	-0.96025	0.89786	-1.069	0.2884	
N:factor(farm)4	-0.59411	1.52204	-0.390	0.6974	
N:factor(farm)5	-0.97735	0.84718	-1.154	0.2525	
N:factor(farm)6	-1.13084	0.97207	-1.163	0.2485	
N:factor(farm)7	-0.58138	0.92164	-0.631	0.5302	
N:factor(farm)8	-0.69333	0.87773	-0.790	0.4322	
N:factor(farm)9	-0.63111	0.81550	-0.774	0.4415	
N:factor(farm)10	-0.04773	0.86512	-0.055	0.9562	
N:factor(farm)11	-1.78432	0.87838	-2.031	0.0459	*
N:factor(farm)12	-0.50153	0.84820	-0.591	0.5562	
N:factor(farm)13	-1.38290	0.98604	-1.402	0.1651	
N:factor(farm)14	-0.86027	0.89294	-0.963	0.3386	
N:factor(farm)15	-0.53439	0.94640	-0.565	0.5741	
N:factor(farm)16	-1.14547	0.91070	-1.258	0.2125	
N:factor(farm)17	0.24013	1.97779	0.121	0.9037	
N:factor(farm)18	-0.64387	0.79080	-0.814	0.4182	

```
N:factor(farm)19   -1.31100      0.90886   -1.442    0.1535
N:factor(farm)20   -0.65867      0.78956   -0.834    0.4069
N:factor(farm)21   -0.73231      0.81990   -0.893    0.3747
N:factor(farm)22   -0.37125      0.89597   -0.414    0.6798
N:factor(farm)23   -0.56171      0.85286   -0.659    0.5122
N:factor(farm)24   -1.40628      0.95103   -1.479    0.1436
```

```
Residual standard error: 1.978 on 72 degrees of freedom
Multiple R-Squared: 0.9678,  Adjusted R-squared: 0.9468
F-statistic: 46.07 on 47 and 72 DF, p-value: < 2.2e-16
```

There is a marginally significant overall effect of soil nitrogen on plant size (N has $p = 0.0426$) and (compared to farm 1) farms 2 and 11 have significantly higher intercepts and shallower slopes. The problem, of course, is that this model, with its 24 slopes and 24 intercepts, is vastly overparameterized. Let's fit a greatly simplified model with a common slope but different intercepts for the different farms:

```
model2<-lm(size~N+factor(farm))
anova(model,model2)
```

```
Analysis of Variance Table
```

```
Model 1: size ~ N * factor(farm)
Model 2: size ~ N + factor(farm)
     Res.Df      RSS   Df   Sum of Sq       F   Pr(>F)
1        72   281.60
2        95   353.81  -23     -72.21  0.8028    0.717
```

This analysis provides no support for any significant differences between slopes. What about differences between farms in their intercepts?

```
model3<-lm(size~N)
anova(model2,model3)
```

```
Analysis of Variance Table
```

```
Model 1: size ~ N + factor(farm)
Model 2: size ~ N
     Res.Df      RSS   Df   Sum of Sq       F      Pr(>F)
1        95    353.8
2       118   8454.9  -23    -8101.1  94.574   < 2.2e-16   ***
```

This shows that there are highly significant differences in intercepts between farms. The interpretation of the analysis of covariance is exactly the same as the interpretation of the mixed model in this case where there is balanced structure and equal replication, but lme is vastly superior to the linear model when there is unequal replication.

Generalized Linear Mixed Models

Pseudoreplicated data with non-normal errors lead to a choice of generalized linear mixed-effects models using lmer with a specified error family. These were previously handled by the function glmmPQL which is part of the MASS library (see Venables and Ripley, 2002). That function fitted a generalized linear mixed model with multivariate normal random effects, using penalized quasi-likelihood (hence the 'PQL'). The default method for a generalized linear model fit with lmer has been switched from PQL to the more reliable

Laplace method, as explained in Chapter 14. The lmer function can deal with the same error structures as a generalized linear model, namely Poisson (for count data), binomial (for binary data or proportion data) or gamma (for continuous data where the variance increase with the square of the mean). The model call is just like a mixed-effects model but with the addition of the name of the error family, like this:

lmer(y~fixed+(time | random), family=binomial)

For a worked example, involving patients who were tested for the presence of a bacterial infection on a number of occasions (the number varying somewhat from patient to patient), see pp. 604–609. The response variable is binary: yes for infected patients or no for patients not scoring as infected, so the family is binomial. There is a single categorical explanatory variable (a fixed effect) called treatment, which has three levels: drug, drug plus supplement, and placebo. The week numbers in which the repeated assessments were made is also recorded.

Fixed Effects in Hierarchical Sampling

Given that the gender effect in our hierarchical sampling example on p. 639 was so large, and that gender makes a sensible fixed effect (it has informative factor levels: male and female), we might fit gender as a main effect. The important point to note is that when you want to compare models with different fixed effects using lme you must change the fitting method from the default REML to the optional maximum likelihood method="ML". This then allows you to use anova to compare lme models with different fixed effects:

model10<-lme(subject~gender,random=~1|town/district/street/family/gender,
method="ML")
model11<-lme(subject~1,random=~1|town/district/street/family/gender,
method="ML")
anova(model10,model11)

	Model	df	AIC	BIC	logLik	Test	L.Ratio	p-value
model10	1	8	3331.584	3368.218	-1657.792			
model11	2	7	3352.221	3384.276	-1669.111	1 vs 2	22.63755	<.0001

It is clear that the model with gender as a fixed effect (model10) is vastly superior to the model with out any fixed effects ($p < 0.0001$). It has a much lower AIC, despite its extra parameter. The variance components have been little affected by fitting gender as a fixed effect, and the effect size of gender is given by:

summary(model10)

Fixed effects: subject ~ gender

	Value	Std.Error	DF	t-value	p-value
(Intercept)	8.976328	0.6332402	360	14.175234	0
gendermale	-1.930773	0.3936610	179	-4.904659	0

You can see what the parameter values are by looking at the treatment means:

tapply(subject,gender,mean)

```
   female        male
8.976328   7.045555
```

The intercept in the lme is the mean for females and the gender.male effect is the difference between the male and female means: $7.045\,555 - 8.976\,328 = -1.930\,773$.

Error Plots from a Hierarchical Analysis

You will find the syntax of model checking in lme both complicated and difficult to remember.

```
library(nlme)
library(lattice)
trellis.par.set(col.whitebg())
```

If you use the standard plot(model) with lme you get a single panel showing the standardized residuals as a function of the fitted values. For more comprehensive model checking, it is useful to make the dataframe into a groupedData object, then refit the model. Here we investigate the REML model with gender as a fixed effect:

```
hs<-groupedData(subject~gender|town/district/street/family/gender/replicate,
                         outer=~gender,data=hierarchy)
model<-
lme(subject~gender,random=~1|town/district/street/family/gender,data=hs)
plot(model,gender~resid(.))
```

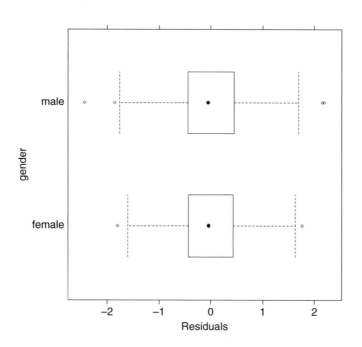

To inspect the constancy of variance across towns and check for heteroscedasticity:

```
plot(model,resid(.,type="p")~fitted(.)|town)
```

It should be clear that this kind of plot only makes sense for those variables with informative factor levels such as gender and town; it would make no sense to group together the streets labelled s1 or s3 or the families labelled f1, f2 or f3.

Tests for normality use the familiar QQ plots, but applied to panels:

```
qqnorm(model,~resid(.)|gender)
```

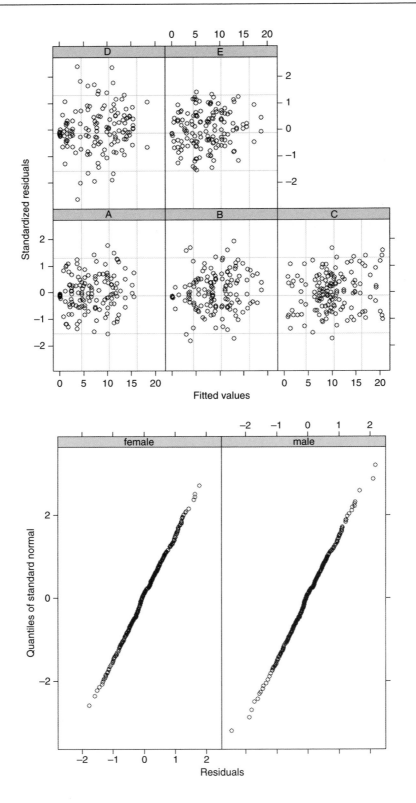

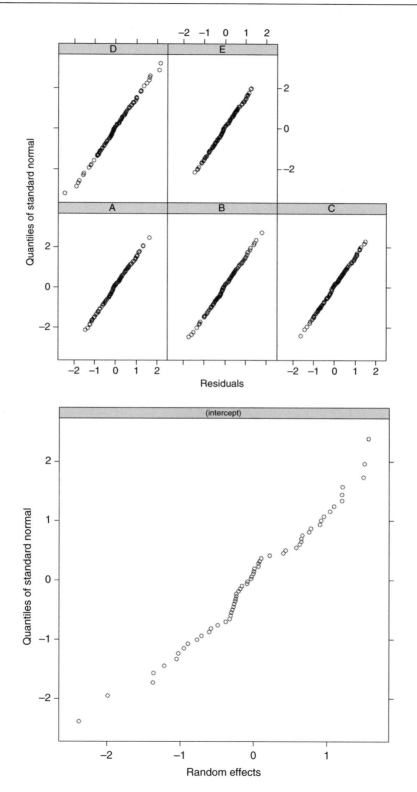

The residuals are normally distributed for both genders, and within each town:

qqnorm(model,~resid(.)|town)

To assess the normality of the random effects, you need to specify the level of interest. In this example we have five levels, and you can experiment with the others:

qqnorm(model,~ranef(.,level=3))

By level 5, the random effects are beautifully normal, but at level 1 (towns) the data are too sparse to be particularly informative. Here at level 3 there is reasonable, but not perfect, normality of the random effects.

Finally, we plot the response variable plotted against the fitted values:

plot(model,subject~fitted(.))

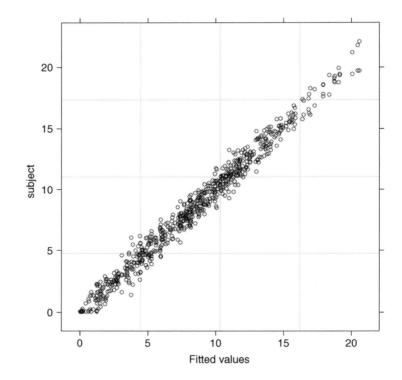

20

Non-linear Regression

Sometimes we have a mechanistic model for the relationship between y and x, and we want to estimate the parameters and standard errors of the parameters of a specific non-linear equation from data. Some frequently used non-linear models are shown in Table 20.1. What we mean in this case by 'non-linear' is not that the relationship is curved (it was curved in the case of polynomial regressions, but these were linear models), but that the relationship cannot be linearized by transformation of the response variable or the explanatory variable (or both). Here is an example: it shows jaw bone length as a function of age in deer. Theory indicates that the relationship is an asymptotic exponential with three parameters:

$$y = a - b\mathrm{e}^{-cx}.$$

In R, the main difference between linear models and non-linear models is that we have to tell R the exact nature of the equation as part of the model formula when we use non-linear modelling. In place of **lm** we write **nls** (this stands for 'non-linear least squares'). Then, instead of y~x, we write y~a-b*exp(-c*x) to spell out the precise nonlinear model we want R to fit to the data.

The slightly tedious thing is that R requires us to specify initial guesses for the values of the parameters a, b and c (note, however, that some common non-linear models have 'self-starting' versions in R which bypass this step; see p. 675). Let's plot the data to work out sensible starting values. It always helps in cases like this to work out the equation's 'behaviour at the limits' – that is to say, to find the values of y when $x = 0$ and when $x = \infty$ (p. 195). For $x = 0$, we have $\exp(-0)$ which is 1, and $1 \times b = b$, so $y = a - b$. For $x = \infty$, we have $\exp(-\infty)$ which is 0, and $0 \times b = 0$, so $y = a$. That is to say, the asymptotic value of y is a, and the intercept is $a - b$.

```
deer<-read.table("c:\\temp\\jaws.txt",header=T)
attach(deer)
names(deer)
```

```
[1] "age" "bone"
```

```
plot(age,bone,pch=16)
```

Table 20.1. Useful non-linear functions.

Name	Equation
Asymptotic functions	
Michaelis–Menten	$y = \dfrac{ax}{1 + bx}$
2-parameter asymptotic exponential	$y = a(1 - e^{-bx})$
3-parameter asymptotic exponential	$y = a - be^{-cx}$
S-shaped functions	
2-parameter logistic	$y = \dfrac{e^{a+bx}}{1 + e^{a+bx}}$
3-paramerter logistic	$y = \dfrac{a}{1 + be^{-cx}}$
4-parameter logistic	$y = a + \dfrac{b - a}{1 + e^{(c-x)/d}}$
Weibull	$y = a - be^{-(cx^d)}$
Gompertz	$y = ae^{-be^{-cx}}$
Humped curves	
Ricker curve	$y = axe^{-bx}$
First-order compartment	$y = k\exp(-\exp(a)x) - \exp(-\exp(b)x)$
Bell-shaped	$y = a\exp(-|bx|^2)$
Biexponential	$y = ae^{bx} - ce^{-dx}$

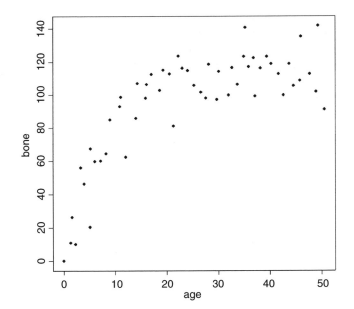

Inspection suggests that a reasonable estimate of the asymptote is $a \approx 120$ and intercept ≈ 10, so $b = 120 - 10 = 110$. Our guess at the value of c is slightly harder. Where the curve is rising most steeply, jaw length is about 40 where age is 5. Rearranging the equation gives

$$c = -\frac{\log((a-y)/b)}{x} = -\frac{\log(120-40)/110)}{5} = 0.063\,690\,75.$$

Now that we have the three parameter estimates, we can provide them to R as the starting conditions as part of the nls call like this:

```
model<-nls(bone~a-b*exp(-c*age),start=list(a=120,b=110,c=0.064))
summary(model)
```

```
Formula: bone~a - b * exp(-c * age)

Parameters:
     Estimate  Std. Error  t value  Pr(>|t|)
a    115.2528      2.9139    39.55   < 2e-16  ***
b    118.6875      7.8925    15.04   < 2e-16  ***
c      0.1235      0.0171     7.22  2.44e-09  ***
```

Residual standard error: 13.21 on 51 degrees of freedom

All the parameters appear to be significantly different from zero at $p < 0.001$. Beware, however. This does not necessarily mean that all the parameters need to be retained in the model. In this case, $a = 115.2528$ with standard error 2.9139 is clearly not significantly different from $b = 118.6875$ with standard error 7.8925 (they would need to differ by more than 2 standard errors to be significant). So we should try fitting the simpler two-parameter model

$$y = a(1 - e^{-cx}).$$

```
model2<-nls(bone~a*(1-exp(-c*age)),start=list(a=120,c=0.064))
anova(model,model2)
```

```
Analysis of Variance Table

Model 1: bone~a - b * exp(-c * age)
Model 2: bone~a * (1 - exp(-c * age))

   Res.Df  Res.Sum Sq   Df   Sum Sq  F value  Pr(>F)
1      51      8897.3
2      52      8929.1   -1    -31.8   0.1825   0.671
```

Model simplification was clearly justified ($p = 0.671$), so we accept the two-parameter version, model2, as our minimal adequate model. We finish by plotting the curve through the scatterplot. The age variable needs to go from 0 to 50 in smooth steps:

```
av<-seq(0,50,0.1)
```

and we use predict with model2 to generate the predicted bone lengths:

```
bv<-predict(model2,list(age=av))
lines(av,bv)
```

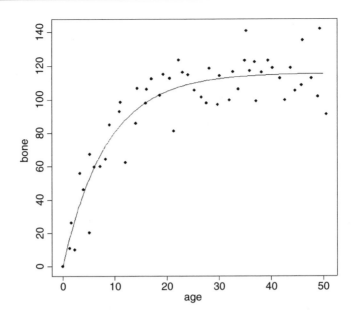

The parameters of this curve are obtained from model2:

summary(model2)

```
Parameters:
     Estimate   Std. Error   t value   Pr(>|t|)
a    115.58056     2.84365    40.645    < 2e-16   ***
c      0.11882     0.01233     9.635   3.69e-13   ***
```

Residual standard error: 13.1 on 52 degrees of freedom

which we could write as $y = 115.58(1 - e^{-0.1188x})$ or as $y = 115.58(1 - \exp(-0.1188x))$ according to taste or journal style. If you want to present the standard errors as well as the parameter estimates, you could write: 'The model $y = a(1 - \exp(-bx))$ had $a = 115.58 \pm 2.84$ (1 standard error) and $b = 0.1188 \pm 0.0123$ (1 standard error, $n = 54$) and explained 84.6% of the total variation in bone length'. Note that because there are only two parameters in the minimal adequate model, we have called them a and b (rather than a and c as in the original formulation).

Comparing Michaelis–Menten and Asymptotic Exponential

Model choice is always an important issue in curve fitting. We shall compare the fit of the asymptotic exponential (above) with a Michaelis–Menten with parameter values estimated from the same deer jaws data. As to starting values for the parameters, it is clear that a reasonable estimate for the asymptote would be 100 (this is a/b; see p. 202). The curve passes close to the point $(5, 40)$ so we can guess a value of a of $40/5 = 8$ and hence $b = 8/100 = 0.08$. Now use nls to estimate the parameters:

(model3<-nls(bone~a*age/(1+b*age),start=list(a=8,b=0.08)))

```
Nonlinear regression model
  model: bone~a * age/(1 + b * age)
```

```
    data: parent.frame()
         a              b
18.7253859   0.1359640
```

```
residual sum-of-squares: 9854.409
```

Finally, we can add the line for Michaelis–Menten to the original plot. You could draw the best-fit line by transcribing the parameter values

```
ymm<-18.725*av/(1+0.13596*av)
lines(av,ymm,lty=2)
```

Alternatively, you could use **predict** with the model name, using list to allocate *x* values to age:

```
ymm<-predict(model3, list(age=av))
lines(av,ymm,lty=2)
```

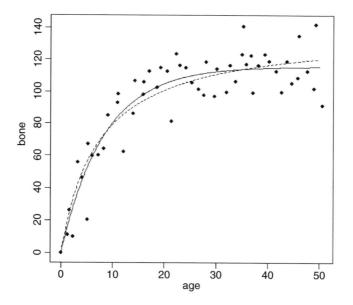

You can see that the asymptotic exponential (solid line) tends to get to its asymptote first, and that the Michaelis–Menten (dotted line) continues to increase. Model choice, therefore would be enormously important if you intended to use the model for prediction to ages much greater than 50 months.

Generalized Additive Models

Sometimes we can see that the relationship between *y* and *x* is non-linear but we don't have any theory or any mechanistic model to suggest a particular functional form (mathematical equation) to describe the relationship. In such circumstances, generalized additive models GAMs are particularly useful because they fit non-parametric smoothers to the data without requiring us to specify any particular mathematical model to describe the non-linearity (background and more examples are given in Chapter 18).

```
humped<-read.table("c:\\temp\\hump.txt",header=T)
attach(humped)
names(humped)
```

```
[1] "y"  "x"
```

```
plot(x,y,pch=16)
library(mgcv)
```

The model is specified very simply by showing which explanatory variables (in this case just *x*) are to be fitted as smoothed functions using the notation y~s(x):

```
model<-gam(y~s(x))
```

Now we can use predict in the normal way to fit the curve estimated by gam:

```
xv<-seq(0.5,1.3,0.01)
yv<-predict(model,list(x=xv))
lines(xv,yv)
```

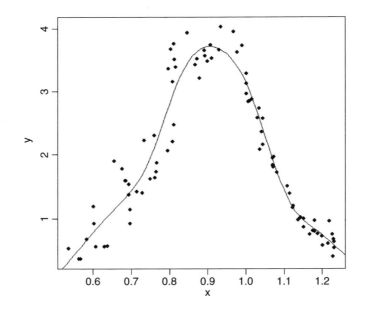

```
summary(model)
```

```
Family: gaussian
Link function: identity
```

```
Formula:
y ~ s(x)
```

```
Parametric coefficients:
            Estimate  Std. Error  t value  Pr(>|t|)
(Intercept)  1.95737     0.03446     56.8   <2e-16  ***
```

```
Approximate significance of smooth terms:
          edf   Est.rank      F    p-value
s(x)    7.452          9   110.0   <2e-16   ***

R-sq.(adj) = 0.919   Deviance explained = 92.6%
GCV score = 0.1156   Scale est. = 0.1045  n = 88
```

Fitting the curve uses up 7.452 degrees of freedom (i.e. it is quite expensive) but the resulting fit is excellent and the model explains more than 92% of the deviance in *y*.

Grouped Data for Non-linear Estimation

Here is a dataframe containing experimental results on reaction rates as a function of enzyme concentration for five different bacterial strains, with reaction rate measured just once for each strain at each of ten enzyme concentrations. The idea is to fit a family of five Michaelis–Menten functions with parameter values depending on the strain.

```
reaction<-read.table("c:\\temp\\reaction.txt",header=T)
attach(reaction)
names(reaction)
```

```
[1] "strain" "enzyme" "rate"
```

```
plot(enzyme,rate,pch=as.numeric(strain))
```

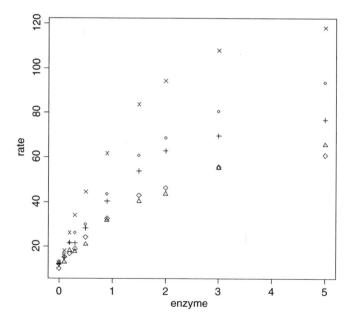

Clearly the different strains will require different parameter values, but there is a reasonable hope that the same functional form will describe the response of the reaction rate of each strain to enzyme concentration.

```
library(nlme)
```

The function we need is nlsList which fits the same functional form to a group of subjects (as indicated by the 'given' operator |):

```
model<-nlsList(rate~c+a*enzyme/(1+b*enzyme)|strain,
                          data=reaction,start=c(a=20,b=0.25,c=10))
```

Note the use of the groupedData style formula rate~enzyme | strain.

```
summary(model)
```

```
Call:
    Model: rate ~ c + a * enzyme/(1 + b * enzyme) | strain
     Data: reaction

Coefficients:
  a
    Estimate   Std. Error    t value       Pr(>|t|)
A  51.79746    4.093791   12.652686   1.943005e-06
B  26.05893    3.063474    8.506335   2.800344e-05
C  51.86774    5.086678   10.196781   7.842353e-05
D  94.46245    5.813975   16.247482   2.973297e-06
E  37.50984    4.840749    7.748767   6.462817e-06
  b
    Estimate    Std. Error    t value       Pr(>|t|)
A  0.4238572   0.04971637    8.525506   2.728565e-05
B  0.2802433   0.05761532    4.864041   9.173722e-04
C  0.5584898   0.07412453    7.534479   5.150210e-04
D  0.6560539   0.05207361   12.598587   1.634553e-05
E  0.5253479   0.09354863    5.615774   5.412405e-05

  c
    Estimate   Std. Error    t value       Pr(>|t|)
A  11.46498    1.194155    9.600916   1.244488e-05
B  11.73312    1.120452   10.471780   7.049415e-06
C  10.53219    1.254928    8.392663   2.671651e-04
D  10.40964    1.294447    8.041768   2.909373e-04
E  10.30139    1.240664    8.303123   4.059887e-06

Residual standard error: 1.81625 on 35 degrees of freedom
```

There is substantial variation from strain to strain in the values of *a* and *b*, but we should test whether a model with a common intercept of, say, 11.0 might not fit equally well.

The plotting is made much easier if we convert the dataframe to a grouped data object:

```
reaction<-groupedData(rate~enzyme|strain,data=reaction)
library(lattice)
plot(reaction)
```

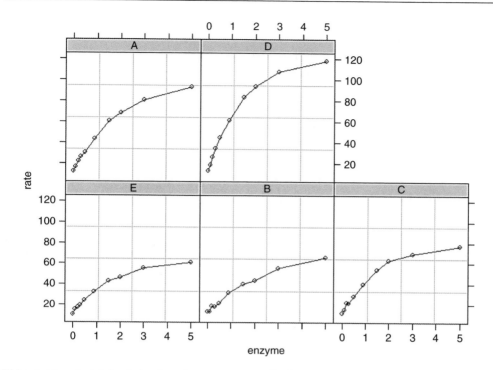

This plot has just joined the dots, but we want to fit the separate non-linear regressions. To do this we fit a non-linear mixed-effects model with nlme, rather than use nlsList:

```
model<-nlme(rate~c+a*enzyme/(1+b*enzyme),fixed=a+b+c~1,
        random=a+b+c~1|strain,data=reaction,start=c(a=20,b=0.25,c=10))
```

Now we can employ the very powerful augPred function to fit the curves to each panel:

```
plot(augPred(model))
```

Here is the summary of the non-linear mixed model:

```
summary(model)
```

```
Nonlinear mixed-effects model fit by maximum likelihood
Model: rate ~ c + a * enzyme/(1 + b * enzyme)
 Data: reaction
      AIC       BIC      logLik
 253.4805  272.6007  -116.7403

Random effects:
  Formula: list(a ~ 1, b ~ 1, c ~ 1)
  Level: strain
  Structure: General positive-definite, Log-Cholesky parametrization
           StdDev        Corr
a          22.9151522    a         b
b          0.1132367     0.876
c          0.4230049    -0.537    -0.875
Residual   1.7105945
```

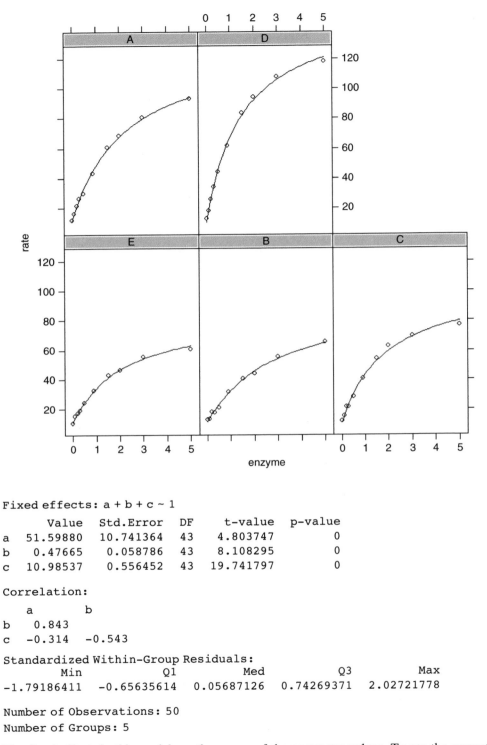

```
Fixed effects: a + b + c ~ 1
        Value     Std.Error   DF      t-value    p-value
a    51.59880    10.741364    43     4.803747         0
b     0.47665     0.058786    43     8.108295         0
c    10.98537     0.556452    43    19.741797         0

Correlation:
      a         b
b    0.843
c   -0.314    -0.543

Standardized Within-Group Residuals:
        Min              Q1            Med              Q3             Max
-1.79186411    -0.65635614    0.05687126    0.74269371    2.02721778

Number of Observations: 50
Number of Groups: 5
```

The fixed effects in this model are the means of the parameter values. To see the separate parameter estimates for each strain use coef:

```
coef(model)
           a              b              c
E    34.09051    0.4533456       10.81722
B    28.01273    0.3238688       11.54813
C    49.63892    0.5193772       10.67189
A    53.20468    0.4426243       11.23613
D    93.04715    0.6440384       10.65348
```

Note that the rows of this table are no longer in alphabetical order but sequenced in the way they appeared in the panel plot (i.e. ranked by their asymptotic values). The parameter estimates are close to, but not equal to, the values estimated by nlsList (above) as a result of 'shrinkage' in the restricted maximum likelihood estimates (see p. 631).

Non-linear Time Series Models (Temporal Pseudoreplication)

The previous example was a designed experiment in which there was no pseudoreplication. However, we often want to fit non-linear models to growth curves where there is temporal pseudoreplication across a set of subjects, each providing repeated measures on the response variable. In such a case we shall want to model the temporal autocorrelation.

```
nl.ts<-read.table("c:\\temp\\nonlinear.txt",header=T)
attach(nl.ts)
names(nl.ts)
```

```
[1] "time" "dish" "isolate" "diam"
```

```
growth<-groupedData(diam~time|dish,data=nl.ts)
```

Here, we model the temporal autocorrelation as first-order autoregessive, corAR1():

```
model<-nlme(diam~a+b*time/(1+c*time),
fixed=a+b+c~1,
random=a+b+c~1,
data=growth,
correlation=corAR1(),
start=c(a=0.5,b=5,c=0.5))
```

```
summary(model)
```

```
Nonlinear mixed-effects model fit by maximum likelihood
Model: diam ~ a + b * time/(1 + c * time)
Data: growth
      AIC        BIC       logLik
129.7694    158.3157    -53.88469

Random effects:
Formula: list(a ~ 1, b ~ 1, c ~ 1)
Level: dish
Structure: General positive-definite, Log-Cholesky parametrization
```

```
          StdDev       Corr
a         0.1014474    a          b
b         1.2060379    -0.557
c         0.1095790    -0.958    0.772
Residual  0.3150068
```

Correlation Structure: AR(1)
Formula: ~1 | dish
Parameter estimate(s):
```
          Phi
-0.03344944
```
Fixed effects: a + b + c ~ 1
```
          Value    Std.Error   DF   t-value    p-value
a         1.288262  0.1086390  88   11.85819         0
b         5.215251  0.4741954  88   10.99810         0
c         0.498222  0.0450644  88   11.05578         0
```

Correlation:
```
     a         b
b   -0.506
c   -0.542    0.823
```

Standardized Within-Group Residuals:
```
       Min              Q1            Med             Q3           Max
-1.74222882    -0.64713657    -0.03349834    0.70298805    2.24686653
```

Number of Observations: 99
Number of Groups: 9

coef(model)

```
         a            b            c
5   1.288831    3.348752    0.4393772
4   1.235632    5.075219    0.5373948
1   1.252725    5.009538    0.5212435
3   1.285847    4.843221    0.4885947
9   1.111135    7.171305    0.7061053
7   1.272570    5.361570    0.5158167
6   1.435784    4.055242    0.3397510
2   1.348523    5.440494    0.4553723
8   1.363310    6.631920    0.4803384
```

It could not be simpler to plot the family of non-linear models in a panel of scatterplots. We just use **augPred** like this:

```
plot(augPred(model))
```

To get all the curves in a single panel we could use **predict** instead of **augPred**:

```
xv<-seq(0,10,0.1)
plot(time,diam,pch=16,col=as.numeric(dish))
sapply(1:9,function(i) lines(xv,predict(model,list(dish=i,time=xv)),lty=2))
```

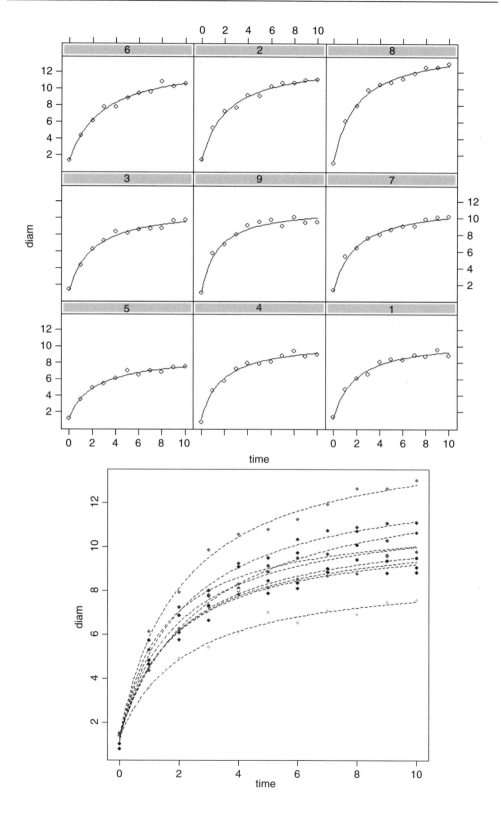

Self-starting Functions

One of the most likely things to go wrong in non-linear least squares is that the model fails because your initial guesses for the starting parameter values were too far off. The simplest solution is to use one of R's 'self-starting' models, which work out the starting values for you automatically. These are the most frequently used self-starting functions:

SSasymp   asymptotic regression model

SSasympOff   asymptotic regression model with an offset

SSasympOrig   asymptotic regression model through the origin

SSbiexp   biexponential model

SSfol   first-order compartment model

SSfpl   four-parameter logistic model

SSgompertz   Gompertz growth model

SSlogis   logistic model

SSmicmen   Michaelis–Menten model

SSweibull   Weibull growth curve model

Self-starting Michaelis–Menten model

In our next example, reaction rate is a function of enzyme concentration; reaction rate increases quickly with concentration at first but asymptotes once the reaction rate is no longer enzyme-limited. R has a self-starting version called SSmicmen parameterized as

$$y = \frac{ax}{b+x},$$

where the two parameters are a (the asymptotic value of y) and b (which is the x value at which half of the maximum response, $a/2$, is attained). In the field of enzyme kinetics a is called the Michaelis parameter (see p. 202; in R help the two parameters are called Vm and K respectively).

Here is SSmicmen in action:

```
data<-read.table("c:\\temp\\mm.txt",header=T)
attach(data)
names(data)
```

```
[1]  "conc"  "rate"
```

```
plot(rate~conc,pch=16)
```

To fit the non-linear model, just put the name of the response variable (rate) on the left of the tilde ~ then put SSmicmen(conc,a,b)) on the right of the tilde, with the name of your explanatory variable first in the list of arguments (conc in this case), then your names for the two parameters (a and b, above):

```
model<-nls(rate~SSmicmen(conc,a,b))
summary(model)
```

```
Formula: rate ~ SSmicmen(conc, a, b)
Parameters:
      Estimate   Std. Error   t value   Pr(>|t|)
a   2.127e+02   6.947e+00    30.615    3.24e-11   ***
b   6.412e-02   8.281e-03     7.743    1.57e-05   ***
```

So the equation looks like this:

$$rate = \frac{212.7 \times conc}{0.064\,12 + conc}$$

and we can plot it like this:

```
xv<-seq(0,1.2,.01)
yv<-predict(model,list(conc=xv))
lines(xv,yv)
```

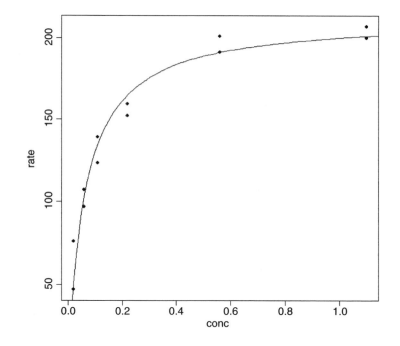

Self-starting asymptotic exponential model

In Chapter 7 we wrote the three-paramter asymptotic exponential like this:

$$y = a - be^{-cx}.$$

In R's self-starting version **SSasymp**, the parameters are:

- a is the horizontal asymptote on the right-hand side (called **Asym** in R help);
- $b = a - R0$ where $R0$ is the intercept (the response when x is zero);
- c is the rate constant (the log of **lrc** in R help).

Here is SSasymp applied to the jaws data (p. 151):

```
deer<-read.table("c:\\temp\\jaws.txt",header=T)
attach(deer)
names(deer)
```

```
[1] "age"  "bone"
```

```
model<-nls(bone~SSasymp(age,a,b,c))
plot(age,bone,pch=16)
xv<-seq(0,50,0.2)
yv<-predict(model,list(age=xv))
lines(xv,yv)
summary(model)
```

```
Formula: bone~SSasymp(age, a, b, c)
```

```
Parameters:
    Estimate   Std. Error   t value   Pr(>|t|)
a   115.2527      2.9139      39.553    <2e-16  ***
b    -3.4348      8.1961      -0.419     0.677
c    -2.0915      0.1385     -15.101    <2e-16  ***
```

```
Residual standard error: 13.21 on 51 degrees of freedom
```

The plot of this fit is on p. 664 along with the simplified model without the non-significant parameter b.

Alternatively, one can use the two-parameter form that passes through the origin, SSasympOrig, which fits the function $y = a(1 - \exp(-bx))$. The final form of the asymptotic exponential allows one to specify the function with an offset, d, on the x values, using SSasympOff, which fits the function $y = a - b\exp(-c(x - d))$.

Profile likelihood

The profile function is a generic function for profiling models, by investigating the behaviour of the objective function near the solution represented by the model's fitted values. In the case of nls, it investigates the profile log-likelihood function:

```
par(mfrow=c(2,2))
plot(profile(model))
```

The profile t-statistic (tau) is defined as the square root of change in sum-of-squares divided by residual standard error with an appropriate sign.

Self-starting logistic

This is one of the most commonly used three-parameter growth models, producing a classic S-shaped curve:

```
sslogistic<-read.table("c:\\temp\\sslogistic.txt",header=T)
attach(sslogistic)
names(sslogistic)
```

```
[1] "density"  "concentration"
```

```
plot(density~log(concentration),pch=16)
```

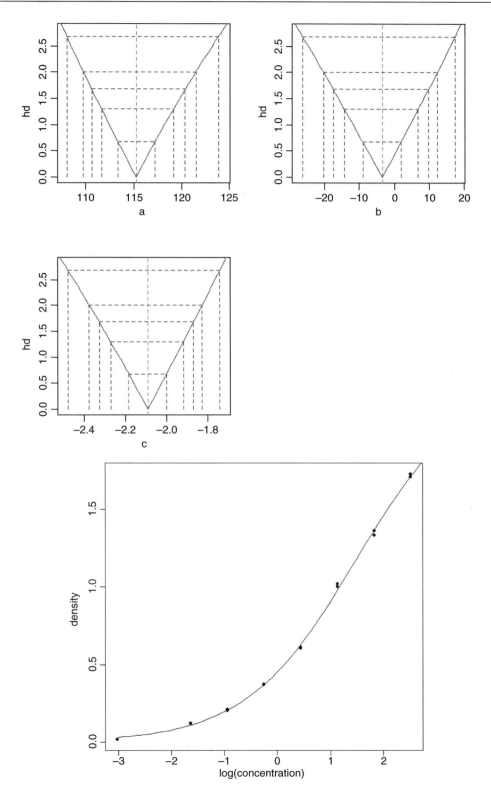

We estimate the three parameters (a, b, c) using the self-starting function SSlogis:

```
model<-nls( density ~ SSlogis(log(concentration), a, b, c ))
```

Now draw the fitted line using predict (note the antilog of xv in list):

```
xv<-seq(-3,3,0.1)
yv<-predict(model,list(concentration=exp(xv)))
lines(xv,yv)
```

The fit is excellent, and the parameter values and their standard errors are given by:

```
summary(model)
```

```
Parameters:
      Estimate   Std. Error   t value   Pr(>|t|)
a     2.34518    0.07815      30.01     2.17e-13   ***
b     1.48309    0.08135      18.23     1.22e-10   ***
c     1.04146    0.03227      32.27     8.51e-14   ***
```

Here a is the asymptotic value, b is the mid-value of x when y is $a/2$, and c is the scale.

Self-starting four-parameter logistic

This model allows a lower asymptote (the fourth parameter) as well as an upper:

```
data<-read.table("c:\\temp\\chicks.txt",header=T)
attach(data)
names(data)
```

```
[1]  "weight"  "Time"
```

```
model <- nls(weight~SSfpl(Time, a, b, c, d))
xv<-seq(0,22,.2)
yv<-predict(model,list(Time=xv))
plot(weight~Time,pch=16)
lines(xv,yv)
```

```
summary(model)
Formula: weight~SSfpl(Time, a, b, c, d)
```

```
Parameters:
      Estimate   Std. Error   t value   Pr(>|t|)
a     27.453     6.601        4.159     0.003169   **
b     348.971    57.899       6.027     0.000314   ***
c     19.391     2.194        8.836     2.12e-05   ***
d     6.673      1.002        6.662     0.000159   ***
```

```
Residual standard error: 2.351 on 8 degrees of freedom
```

The four-parameter logistic is given by

$$y = A + \frac{B - A}{1 + e^{(D-x)/C}}.$$

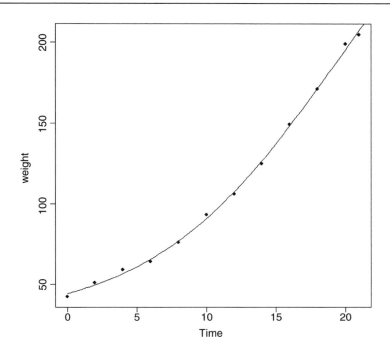

This is the same formula as we used in Chapter 7, but note that C above is $1/c$ on p. 203. A is the horizontal asymptote on the left (for low values of x), B is the horizontal asymptote on the right (for large values of x), D is the value of x at the point of inflection of the curve (represented by xmid in our model for the chicks data), and C is a numeric scale parameter on the x axis (represented by scal). The parameterized model would be written like this:

$$y = 27.453 + \frac{348.971 - 27.453}{1 + \exp((19.391 - x)/6.673)}.$$

Self-starting Weibull growth function

R's parameterization of the Weibull growth function is

$$\text{Asym-Drop*exp(-exp(lrc)*x\^pwr)}$$

where Asym is the horizontal asymptote on the right, Drop is the difference between the asymptote and the intercept (the value of y at $x = 0$), lrc is the natural logarithm of the rate constant, and pwr is the power to which x is raised.

```
weights<-read.table("c:\\temp\\weibull.growth.txt",header=T)
attach(weights)
model <- nls(weight ~ SSweibull(time, Asym, Drop, lrc, pwr))
summary(model)
```

```
Formula: weight ~ SSweibull(time, Asym, Drop, lrc, pwr)
```

```
Parameters:
        Estimate   Std. Error   t value   Pr(>|t|)
Asym    158.5012       1.1769    134.67   3.28e-13   ***
Drop    110.9971       2.6330     42.16   1.10e-09   ***
lrc      -5.9934       0.3733    -16.06   8.83e-07   ***
pwr       2.6461       0.1613     16.41   7.62e-07   ***
```

Residual standard error: 2.061 on 7 degrees of freedom

```
plot(time,weight,pch=16)
xt<-seq(2,22,0.1)
yw<-predict(model,list(time=xt))
lines(xt,yw)
```

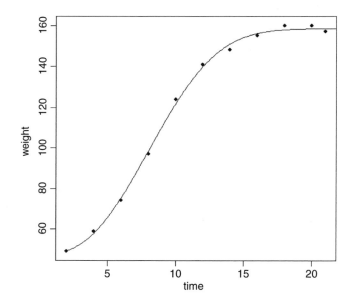

The fit is good, but the model cannot accommodate a drop in y values once the asymptote has been reached (you would need some kind of humped function).

Self-starting first-order compartment function

In the following, the response, drug concentration in the blood, is modelled as a function of time after the dose was administered. There are three parameters (a, b, c) to be estimated:

```
foldat<-read.table("c:\\temp\\fol.txt",header=T)
attach(foldat)
```

The model looks like this:

$$y = k \exp(-\exp(a)x) - \exp(-\exp(b)x),$$

where $k = Dose \times \exp(a + b - c)/(\exp(b) - \exp(a))$ and Dose is a vector of identical values provided to the fit (4.02 in this example):

```
model<-nls(conc~SSfol(Dose,Time,a,b,c))
summary(model)
```

```
Formula: conc ~ SSfol(Dose, Time, a, b, c)
```

```
Parameters:
     Estimate   Std. Error    t value   Pr(>| t| )
a     -2.9196       0.1709    -17.085     1.40e-07   ***
b      0.5752       0.1728      3.328       0.0104   *
c     -3.9159       0.1273    -30.768     1.35e-09   ***
```

```
plot(conc~Time,pch=16)
xv<-seq(0,25,0.1)
yv<-predict(model,list(Time=xv))
lines(xv,yv)
```

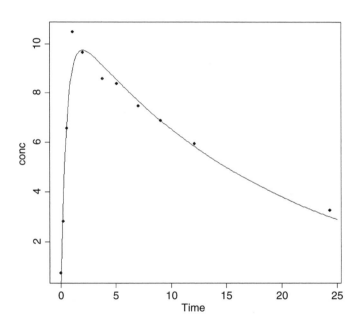

As you can see, this is a rather poor model for predicting the value of the peak concentration, but a reasonable description of the ascending and declining sections.

Bootstrapping a Family of Non-linear Regressions

There are two broad applications of bootstrapping to the estimation of parameters in non-linear models:

- Select certain of the data points at random with replacement, so that, for any given model fit, some data points are duplicated and others are left out.

- Fit the model and estimate the residuals, then allocate the residuals at random, adding them to different fitted values in different simulations

Our next example involves the viscosity data from the MASS library, where sinking time is measured for three different weights in fluids of nine different viscosities:

$$\text{Time} = \frac{b \times \text{Viscosity}}{Wt - c}.$$

We need to estimate the two parameters b and c and their standard errors.

```
library(MASS)
data(stormer)
attach(stormer)
```

Here are the results of the straightforward non-linear regression:

```
model<-nls(Time~b*Viscosity/(Wt-c),start=list(b=29,c=2))
summary(model)
```

```
Formula: Time ~ b * Viscosity/(Wt - c)

Parameters:
     Estimate   Std. Error   t value   Pr(>|t|)
b     29.4013      0.9155      32.114    < 2e-16   ***
c      2.2182      0.6655       3.333    0.00316   **

Residual standard error: 6.268 on 21 degrees of freedom
```

Here is a home-made bootstrap which leaves out cases at random. The idea is to sample the indices (subscripts) of the 23 cases at random with replacement:

```
sample(1:23,replace=T)
 [1]  4  4 10 10 12  3 23 22 21 13  9 14  8  5 15 14 21 14 12  3 20 14 19
```

In this realization cases 1 and 2 were left out, case 3 appeared twice, and so on. We call the subscripts ss as follows, and use the subscripts to select values for the response (y_1) and the two explanatory variables $(x_1$ and $x_2)$ like this:

```
ss<-sample(1:23,replace=T)
y<-Time[ss]
x1<-Viscosity[ss]
x2<-Wt[ss]
```

Now we put this in a loop and fit the model

```
model<-nls(y~b*x1/(x2-c),start=list(b=29,c=2))
```

one thousand times, storing the coefficients in vectors called bv and cv:

```
bv<-numeric(1000)
cv<-numeric(1000)
for(i in 1:1000){
ss<-sample(1:23,replace=T)
y<-Time[ss]
x1<-Viscosity[ss]
x2<-Wt[ss]
model<-nls(y~b*x1/(x2-c),start=list(b=29,c=2))
bv[i]<-coef(model)[1]
```

```
cv[i]<-coef(model)[2]
}
```

This took 7 seconds for 1000 iterations. The 95% confidence intervals for the two parameters are obtained using the **quantile** function:

```
quantile(bv,c(0.025,0.975))
```

```
     2.5%      97.5%
27.91842   30.74411
```

```
quantile(cv,c(0.025,0.975))
```

```
      2.5%       97.5%
0.9084572   3.7694501
```

Alternatively, you can randomize the locations of the residuals while keeping all the cases in the model for every simulation. We use the built-in functions in the **boot** library to illustrate this procedure.

```
library(boot)
```

First, we need to calculate the residuals and the fitted values from the **nls** model we fitted on p. 682:

```
rs<-resid(model)
fit<-fitted(model)
```

and make the fit along with the two explanatory variables Viscosity and *Wt* into a new dataframe called storm that will be used inside the 'statistic' function

```
storm<-data.frame(fit,Viscosity,Wt)
```

Next, you need to write a statistic function (p. 320) to describe the model fitting:

```
statistic<-function(rs,i){
storm$y<-storm$fit+rs[i]
coef(nls(y~b*Viscosity/(Wt-c),storm,start=coef(model)))}
```

The two arguments to **statistic** are the vector of residuals, *rs*, and the randomized indices, *i*. Now we can run the boot function over 1000 iterations:

```
boot.model<-boot(rs,statistic,R=1000)
boot.model
```

```
ORDINARY NONPARAMETRIC BOOTSTRAP
```

```
Call:
boot(data = rs, statistic = statistic, R = 1000)
```

```
Bootstrap Statistics :
```

```
         original         bias    std. error
t1*    29.401294     0.6915554    0.8573951
t2*     2.218247    -0.2552968    0.6200594
```

The parametric estimates for *b* (t1) and *c* (t2) in boot.model are reasonably unbiased, and the bootstrap standard errors are slightly smaller than when we used **nls**. We get the boot-strapped confidence intervals with the **boot.ci** function: *b* is index = 1 and *c* is index = 2:

boot.ci(boot.model,index=1)

```
BOOTSTRAP CONFIDENCE INTERVAL CALCULATIONS
Based on 1000 bootstrap replicates

CALL :
boot.ci(boot.out = storm.boot)

Intervals :
Level       Normal              Basic           Studentized
95%     (26.33, 29.65 )   (26.43, 29.78 )   (25.31, 29.63 )


Level       Percentile            BCa
95%     (27.65, 31.00 )   (26.92, 29.60 )
```

boot.ci(boot.model,index=2)

```
BOOTSTRAP CONFIDENCE INTERVAL CALCULATIONS
Based on 1000 bootstrap replicates

CALL :
boot.ci(boot.out = boot.model, index = 2 )

Intervals :
Level       Normal              Basic
95%     ( 1.258, 3.689 )   ( 1.278, 3.637 )


Level       Percentile            BCa
95%     ( 0.800, 3.159 )   ( 1.242, 3.534 )
```

For comparison, here are the parametric confidence intervals (from model): for b, from 28.4858 to 30.3168; and for c, from 0.8872 to 3.5492.

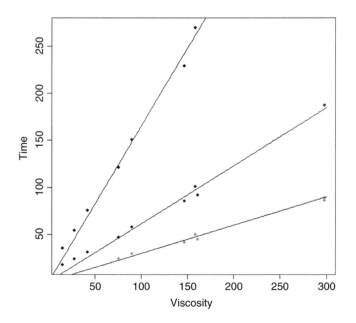

21

Tree Models

Tree models are computationally intensive methods that are used in situations where there are many explanatory variables and we would like guidance about which of them to include in the model. Often there are so many explanatory variables that we simply could not test them all, even if we wanted to invest the huge amount of time that would be necessary to complete such a complicated multiple regression exercise. Tree models are particularly good at tasks that might in the past have been regarded as the realm of multivariate statistics (e.g. classification problems). The great virtues of tree models are as follows:

- They are very simple.

- They are excellent for initial data inspection.

- They give a very clear picture of the structure of the data.

- They provide a highly intuitive insight into the kinds of interactions between variables.

It is best to begin by looking at a tree model in action, before thinking about how it works. Here is the air pollution example that we have worked on already as a multiple regression (see p. 311):

```
install.packages("tree")
library(tree)
Pollute<-read.table("c:\\temp\\Pollute.txt",header=T)
attach(Pollute)
names(Pollute)
```

```
[1]  "Pollution"  "Temp"   "Industry"  "Population"  "Wind"
[6]  "Rain"       "Wet.days"
```

```
model<-tree(Pollute)
plot(model)
text(model)
```

You follow a path from the top of the tree (called, in defiance of gravity, the **root**) and proceed to one of the terminal nodes (called a **leaf**) by following a succession of rules (called **splits**). The numbers at the tips of the leaves are the mean values in that subset of the data (mean SO_2 concentration in this case). The details are explained below.

The R Book   Michael J. Crawley
© 2007 John Wiley & Sons, Ltd

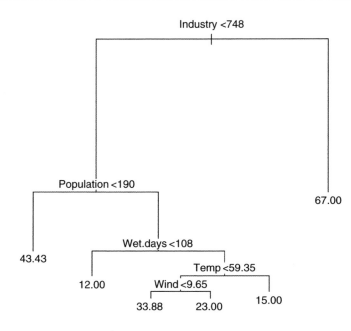

Background

The model is fitted using **binary recursive partitioning**, whereby the data are successively split along coordinate axes of the explanatory variables so that, at any node, the split which maximally distinguishes the response variable in the left and the right branches is selected. Splitting continues until nodes are pure or the data are too sparse (fewer than six cases, by default).

Each explanatory variable is assessed in turn, and the variable explaining the greatest amount of the deviance in y is selected. Deviance is calculated on the basis of a threshold in the explanatory variable; this threshold produces two mean values for the response (one mean above the threshold, the other below the threshold).

```
low<-(Industry<748)
tapply(Pollution,low,mean)
```

```
    FALSE        TRUE
67.00000   24.91667
```

```
plot(Industry,Pollution,pch=16)
abline(v=748,lty=2)
lines(c(0,748),c(24.92,24.92))
lines(c(748,max(Industry)),c(67,67))
```

The procedure works like this. For a given explanatory variable (say Industry, above):

- Select a threshold value of the explanatory variable (the vertical dotted line at Industry = 748).

- Calculate the mean value of the response variable above and below this threshold (the two horizontal solid lines).

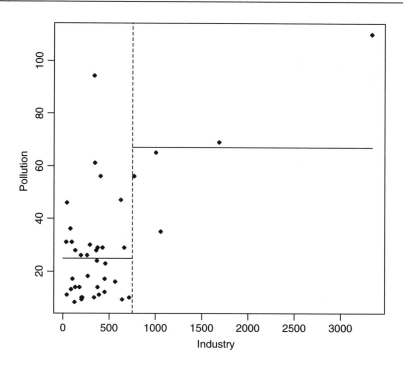

- Use the two means to calculate the deviance (as with *SSE*, see p. 451).

- Go through all possible values of the threshold (values on the *x* axis).

- Look to see which value of the threshold gives the lowest deviance.

- Split the data into high and low subsets on the basis of the threshold for this variable.

- Repeat the whole procedure on each subset of the data on either side of the threshold.

- Keep going until no further reduction in deviance is obtained, or there are too few data points to merit further subdivision (e.g. the right-hand side of the Industry split, above, is too sparse to allow further subdivision).

The deviance is defined as

$$D = \sum_{\text{cases}\,j} (y_j - \mu_{[j]})^2,$$

where $\mu_{[j]}$ is the mean of all the values of the response variable assigned to node j and this sum of squares is add up over all the nodes. The *value* of any split is defined as the reduction in this residual sum of squares. The probability model used in R is that the values of the response variable are normally distributed within each leaf of the tree with mean μ_i and variance σ^2. Note that because this assumption applies to the terminal nodes, the interior nodes represent a mixture of different normal distributions, so the deviance is only appropriate at the terminal nodes (i.e. for the leaves).

If the twigs of the tree are categorical (i.e. levels of a factor like names of particular species) then we have a **classification tree**. On the other hand, if the terminal nodes of the tree are predicted values of a continuous variable, then we have a **regression tree**.

The key questions are these:

- Which variables to use for the division.

- How best to achieve the splits for each selected variable.

It is important to understand that tree models have a tendency to over-interpret the data: for instance, the occasional 'ups' in a generally negative correlation probably don't mean anything substantial.

Regression Trees

In this case the response variable is a continuous measurement, but the explanatory variables can be any mix of continuous and categorical variables. You can think of regression trees as analogous to multiple regression models. The difference is that a regression tree works by forward selection of variables, whereas we have been used to carrying out regression analysis by deletion (backward selection).

For our air pollution example, the regression tree is fitted by stating that the continuous response variable Pollution is to be estimated as a function of *all* of the explanatory variables in the dataframe called Pollute by use of the 'tilde dot' notation like this:

model<-tree(Pollution ~ . , Pollute)

For a regression tree, the print method produces the following kind of output:

print(model)

```
node), split, n, deviance, yval
   * denotes terminal node

1) root 41 22040 30.05
   2) Industry < 748 36 11260 24.92
     4) Population < 190 7 4096 43.43 *
     5) Population > 190 29 4187 20.45
      10) Wet.days < 108 11   96 12.00 *
      11) Wet.days > 108 18 2826 25.61
        22) Temp < 59.35 13 1895 29.69
          44) Wind < 9.65 8 1213 33.88 *
          45) Wind > 9.65 5  318 23.00 *
        23) Temp > 59.35 5  152 15.00 *
3) Industry > 748 5 3002 67.00 *
```

The terminal nodes (the leaves) are denoted by * (there are six of them). The node number is on the left, labelled by the variable on which the split at that node was made. Next comes the 'split criterion' which shows the threshold value of the variable that was used to create the split. The number of cases going *into* the split (or into the terminal node) comes next. The penultimate figure is the deviance at that node. Notice how the deviance goes down as non-terminal nodes are split. In the root, based on all $n = 41$ data points, the deviance is *SSY* (see p. 311) and the y value is the overall mean for Pollution. The last figure on the right is the mean value of the response variable within that node or at that that leaf. The highest mean pollution (67.00) was in node 3 and the lowest (12.00) was in node 10.

Note how the nodes are nested: within node 2, for example, node 4 is terminal but node 5 is not; within node 5 node 10 is terminal but node 11 is not; within node 11, node 23 is terminal but node 22 is not, and so on.

Tree models lend themselves to circumspect and critical analysis of complex dataframes. In the present example, the aim is to understand the causes of variation in air pollution levels from case to case. The interpretation of the regression tree would proceed something like this:

- The five most extreme cases of Industry stand out (mean $= 67.00$) and need to be considered separately.

- For the rest, Population is the most important variable but, interestingly, it is low populations that are associated with the highest levels of pollution (mean $= 43.43$). Ask yourself which might be cause, and which might be effect.

- For high levels of population (> 190), the number of wet days is a key determinant of pollution; the places with the fewest wet days (less than 108 per year) have the lowest pollution levels of anywhere in the dataframe (mean $= 12.00$).

- For those places with more than 108 wet days, it is temperature that is most important in explaining variation in pollution levels; the warmest places have the lowest air pollution levels (mean $= 15.00$).

- For the cooler places with lots of wet days, it is wind speed that matters: the windier places are less polluted than the still places.

This kind of complex and contingent explanation is much easier to see, and to understand, in tree models than in the output of a multiple regression.

Tree models as regressions

To see how a tree model works when there is a single, continuous response variable, it is useful to compare the output with a simple linear regression. Take the relationship between mileage and weight in the car.test.frame data:

```
car.test.frame<-read.table("c:\\temp\\car.test.frame.txt",header=T)
attach(car.test.frame)
names(car.test.frame)
```

```
[1]  "Price"  "Country"  "Reliability"  "Mileage"
[5]  "Type"   "Weight"   "Disp."        "HP"
```

```
plot(Weight,Mileage)
```

The heavier cars do fewer miles per gallon, but there is a lot of scatter. The **tree** model starts by finding the weight that splits the mileage data in a way that explains the maximum deviance. This weight turns out to be 2567.5.

```
a<-mean(Mileage[Weight<2567.5])
b<-mean(Mileage[Weight>=2567.5])
lines(c(1500,2567.5,2567.5,4000),c(a,a,b,b))
```

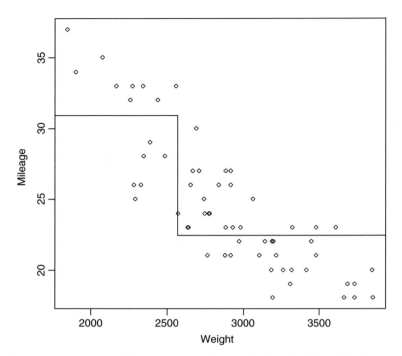

The next thing the tree model does is to work out the threshold weight that would best split the mileage data for the lighter cars: this turns out to be 2280. It then works out the threshold split for the heavier cars: this turns out to be 3087.5. And so the process goes on, until there are too few cars in each split to justify continuation (five or fewer by default). To see the full regression tree as a function plot we can use the predict function with the regression tree object car.model like this:

```
car.model<-tree(Mileage~Weight)
plot(Weight,Mileage)
wt<-seq(1500,4000,10)
y<-predict(car.model,list(Weight=wt))
lines(wt,y)
```

You would not normally do this, of course (and you *could not* do it with more than two explanatory variables) but it is a good way of showing how tree models work with a continuous response variable.

Model simplification

Model simplification in regression trees is based on a **cost–complexity measure**. This reflects the trade-off between fit and explanatory power (a model with a perfect fit would have as many parameters as there were data points, and would consequently have no explanatory power at all. We return to the air pollution example, analysed earlier where we fitted the tree model object called 'model'.

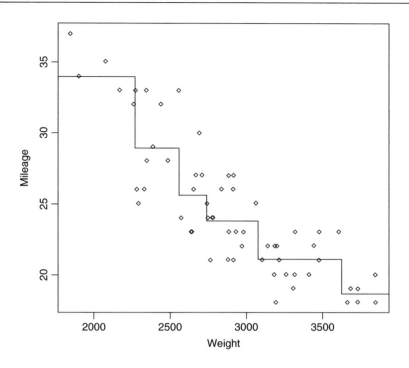

Regression trees can be over-elaborate and can respond to random features of the data (the so-called **training set**). To deal with this, R contains a set of procedures to prune trees on the basis of the cost–complexity measure. The function **prune.tree** determines a nested sequence of sub-trees of the supplied tree by recursively 'snipping' off the least important splits, based upon the cost–complexity measure. The **prune.tree** function returns an object of class **tree.sequence**, which contains the following components:

prune.tree(model)

$size:

[1] 6 5 4 3 2 1

This shows the number of terminal nodes in each tree in the cost–complexity pruning sequence: the most complex model had six terminal nodes (see above)

$dev:

[1] 8876.589 9240.484 10019.992 11284.887 14262.750 22037.902

This is the total deviance of each tree in the cost–complexity pruning sequence.

$k:

[1] -Inf 363.8942 779.5085 1264.8946 2977.8633 7775.1524

This is the value of the cost–complexity pruning parameter of each tree in the sequence. If determined algorithmically (as here, k is not specified as an input), its first value defaults to $-\infty$, its lowest possible bound.

plot(prune.tree(model))

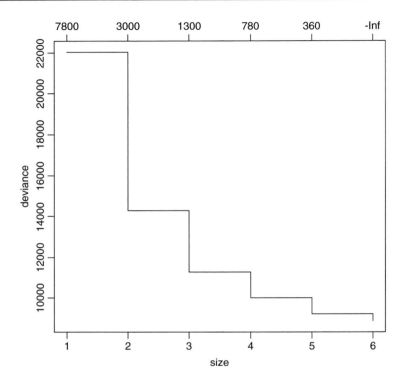

This shows the way that deviance declines as complexity is increased. The total deviance is 22 037.902 (size = 1), and this is reduced to as the complexity of the tree increases up to 6 nodes. An alternative is to specify the number of nodes to which you want the tree to be pruned; this uses the **"best ="** option. Suppose we want the best tree with four nodes.

```
model2<-prune.tree(model,best=4)
plot(model2)
text(model2)
```

In printed form, this is

```
model2
node), split, n, deviance, yval
      * denotes terminal node
1) root 41 22040 30.05
 2) Industry<748 36 11260  24.92
  4) Population<190 7 4096 43.43 *
  5) Population>190 29 4187 20.45
   10) Wet.days<108 11   96 12.00 *
   11) Wet.days>108 18 2826 25.61 *
 3) Industry>748  5  3002  67.00 *
```

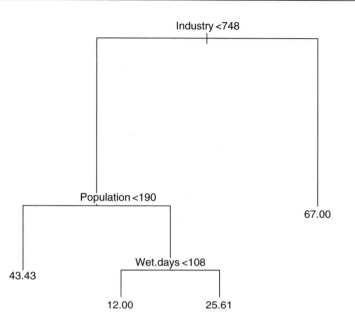

Subtrees and subscripts

It is straightforward to remove parts of trees, or to select parts of trees, using subscripts. For example, a negative subscript [−3] leaves off everything above node 3, while a positive subscript [3] selects only that part of the tree above node 3.

Classification trees with categorical explanatory variables

Tree models are a superb tool for helping to write efficient and effective taxonomic keys.

Suppose that all of our explanatory variables are categorical, and that we want to use tree models to write a dichotomous key. There is only one entry for each species, so we want the twigs of the tree to be the individual rows of the dataframe (i.e. we want to fit a tree perfectly to the data). To do this we need to specify two extra arguments: minsize = 2 and mindev = 0. In practice, it is better to specify a very small value for the minimum deviance (say, 10^{-6}) rather than zero (see below).

The following example relates to the nine lowland British species in the genus *Epilobium* (Onagraceae). We have eight categorical explanatory variables and we want to find the optimal dichotomous key. The dataframe looks like this:

```
epilobium<-read.table("c:\\temp\\epilobium.txt",header=T)
attach(epilobium)
epilobium
```

	species	stigma	stem.hairs	glandular.hairs	seeds	pappilose
1	hirsutum	lobed	spreading	absent	none	uniform
2	parviflorum	lobed	spreading	absent	none	uniform
3	montanum	lobed	spreading	present	none	uniform
4	lanceolatum	lobed	spreading	present	none	uniform
5	tetragonum	clavate	appressed	present	none	uniform

6	obscurum	clavate	appressed	present	none	uniform
7	roseum	clavate	spreading	present	none	uniform
8	palustre	clavate	spreading	present	appendage	uniform
9	ciliatum	clavate	spreading	present	appendage	ridged

	stolons	petals	base
1	absent	>9mm	rounded
2	absent	<10mm	rounded
3	absent	<10mm	rounded
4	absent	<10mm	cuneate
5	absent	<10mm	rounded
6	stolons	<10mm	rounded
7	absent	<10mm	cuneate
8	absent	<10mm	rounded
9	absent	<10mm	rounded

Producing the key could not be easier:

```
model<-tree(species ~ .,epilobium,mindev=1e-6,minsize=2)
plot(model)
text(model)
```

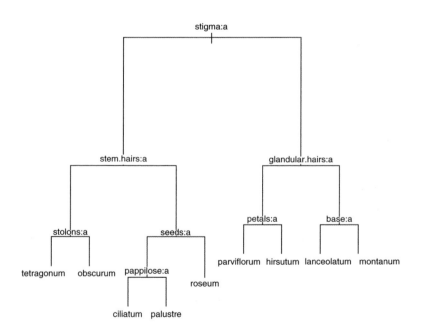

The window in which the above tree appears has been stretched to make it easier to read. Here is the tree written as a dichotomous key:

1. Stigma entire and club-shaped	2
1. Stigma four lobed	6
2. Stem hairs all appressed	3
2. At least some stem hairs spreading	4

3. Glandular hairs present on hypanthium	*E. obscurum*
3. No glandular hairs on hypanthium	*E. tetragonum*
4. Seeds with a terminal appendage	5
4. Seeds without terminal appendage	*E. roseum*
5. Surface of seed with longitudinal papillose ridges	*E. ciliatum*
5. Surface of seed uniformly papillose	*E. palustre*
6. At least some spreading hairs non-glandular	7
6. Spreading hairs all glandular	8
7. Petals large (>9 mm)	*E. hirsutum*
7. Petals small (<10 mm)	*E. parviflorum*
8. Leaf base cuneate	*E. lanceolatum*
8. Leaf base rounded	*E. montanum*

The computer has produced a working key to a difficult group of plants. The result stands as testimony to the power and usefulness of tree models. The same principle underlies good key-writing as is used in tree models: find the characters that explain most of the variation, and use these to split the cases into roughly equal-sized groups at each dichotomy.

Classification trees for replicated data

In this next example from plant taxonomy, the response variable is a four-level, categorical variable called Taxon (it is a label expressed as Roman numerals I to IV). The aim is to use the measurements from the seven morphological explanatory variables to construct the best key to separate these four taxa (the 'best' key is the one with the lowest error rate – the key that misclassifies the smallest possible number of cases).

```
taxonomy<-read.table("c:\\temp\\taxonomy.txt",header=T)
attach(taxonomy)
names(taxonomy)
```

```
[1]  "Taxon"   "Petals"  "Internode"  "Sepal"  "Bract"  "Petiole"
[7]  "Leaf"    "Fruit"
```

Using the tree model for classification could not be simpler:

```
model1<-tree(Taxon~.,taxonomy)
```

We begin by looking at the plot of the tree:

```
plot(model1)
text(model1)
```

With only a small degree of rounding on the suggested break points, the tree model suggests a simple (and for these 120 plants, completely error-free) key for distinguishing the four taxa:

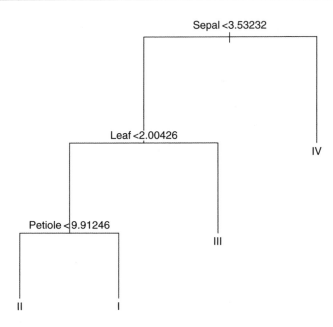

1. Sepal length > 4.0   *Taxon IV*
1. Sepal length < = 4.0   2.

2. Leaf width > 2.0   *Taxon III*
2. Leaf width < = 2.0   3.

3. Petiole length < 10   *Taxon II*
3. Petiole length > = 10   *Taxon I*

The summary option for classification trees produces the following:

summary(model1)

```
Classification tree:
tree(formula = Taxon ~ ., data = taxonomy)
Variables actually used in tree construction:
[1] "Sepal" "Leaf" "Petiole"
Number of terminal nodes: 4
Residual mean deviance: 0 = 0 / 116
Misclassification error rate: 0 = 0 / 120
```

Three of the seven variables were chosen for use (Sepal, Leaf and Petiole); four variables were assessed and rejected (Petals, Internode, Bract and Fruit). The key has four nodes and hence three dichotomies. As you see, the misclassification error rate was an impressive 0 out of 120. It is noteworthy that this classification tree does much better than the multivariate classification methods described in Chapter 23.

For classification trees, the print method produces a great deal of information

print(model1)

```
node), split, n, deviance, yval, (yprob)
      * denotes terminal node
```

```
1) root 120 332.70 I ( 0.2500 0.2500 0.2500 0.25 )
  2) Sepal<3.53232 90 197.80 I ( 0.3333 0.3333 0.3333 0.00 )
    4) Leaf<2.00426 60 83.18 I ( 0.5000 0.5000 0.0000 0.00 )
      8) Petiole<9.91246 30 0.00 II ( 0.0000 1.0000 0.0000 0.00 ) *
      9) Petiole>9.91246 30 0.00 I ( 1.0000 0.0000 0.0000 0.00 ) *
    5) Leaf>2.00426 30 0.00 III ( 0.0000 0.0000 1.0000 0.00 ) *
  3) Sepal>3.53232 30 0.00 IV ( 0.0000 0.0000 0.0000 1.00 ) *
```

The **node** number is followed by the **split criterion** (e.g. Sepal < 3.53 at node 2). Then comes the number of cases passed through that node (90 in this case, versus 30 going into node 3, which is the terminal node for Taxon IV). The remaining deviance within this node is 197.8 (compared with zero in node 3 where all the individuals are alike; they are all Taxon IV). Next is the name of the factor level(s) left in the split (I, II and III in this case, with the convention that the first in the alphabet is listed), then a list of the empirical probabilities: the fractions of all the cases at that node that are associated with each of the levels of the response variable (in this case the 90 cases are equally split between taxa I, II and III and there are no individuals of taxon IV at this node, giving 0.33, 0.33, 0.33 and 0 as the four probabilities).

There is quite a useful plotting function for classification trees called partition.tree but it is only sensible to use it when the model has two explanatory variables. Its use is illustrated here by taking the two most important explanatory variables, Sepal and Leaf:

```
model2<-tree(Taxon~Sepal+Leaf,taxonomy);
partition.tree(model2)
```

This shows how the phase space defined by sepal length and leaf width has been divided up between the four taxa, but it does not show where the data fall. We could use points(Sepal,Leaf) to overlay the points, but for illustration we shall use text. We create a vector called labels that has 'a' for taxon I, 'b' for II, and so on:

```
label<-ifelse(Taxon=="I", "a", ifelse(Taxon=="II","b",ifelse(Taxon=="III","c","d")))
```

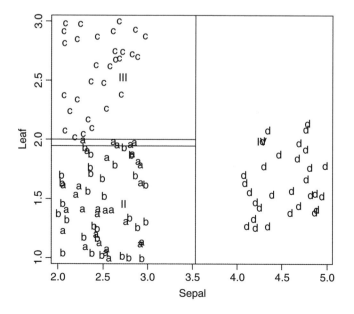

Then we use these letters as a text overlay on the partition.tree like this:

text(Sepal,Leaf,label)

You see that taxa III and IV are beautifully separated on the basis of sepal length and leaf width, but taxa I and II are all jumbled up (recall that they are separated from one another on the basis of petiole length).

Testing for the existence of humps

Tree models can be useful in assessing whether or not there is a hump in the relationship between y and x. This is difficult to do using other kinds of regression, because linear models seldom distinguish between humps and asymptotes. If a tree model puts a lower section at the right of the graph than in the centre, then this hints at the presence of a hump in the data. Likewise, if it puts an elevated section at the left-hand end of the x axis then that is indicative of a U-shaped function.

Here is a function called hump which extracts information from a tree model to draw the stepped function through a scatterplot:

```
hump<-function(x,y){
library(tree)
model<-tree(y~x)
xs<-grep("[0-9]",model[[1]][[5]])
xv<-as.numeric(substring(model[[1]][[5]][xs],2,10))
xv<-xv[1:(length(xv)/2)]
xv<-c(min(x),sort(xv),max(x))
yv<-model[[1]][[4]][model[[1]][[1]]=="<leaf>"]
plot(x,y, xlab=deparse(substitute(x)),ylab=deparse(substitute(y)))
i<-1
j<-2
k<-1
b<-2*length(yv)+1
for (a in 1:b){
lines(c(xv[i],xv[j]),c(yv[k],yv[i]))
if (a %% 2 == 0 ){
j<-j+1
k<-k+1 }
else{
i<-i+1
}}}
```

We shall test it on the ethanol data which are definitely humped (p. 840):

```
library(lattice)
attach(ethanol)
names(ethanol)
```

```
[1] "NOx"  "C"  "E"
```

hump(E,NOx)

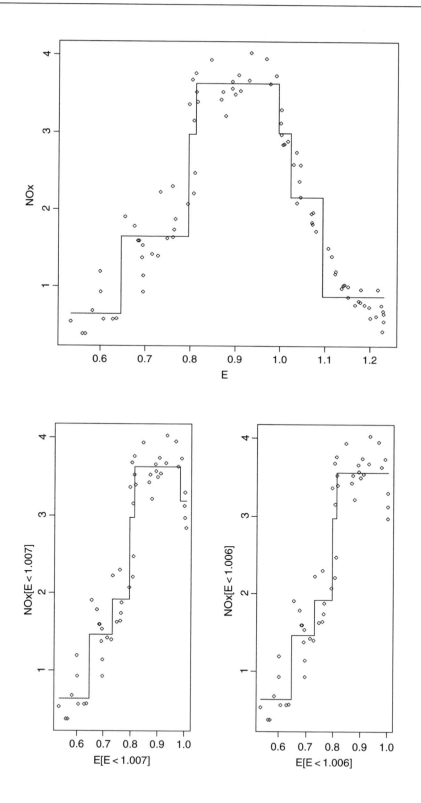

There is a minimum number of points necessary for creating a new step ($n = 5$), and a minimum difference in the mean of one group and the next. To see this, you should contrast these two fits:

```
hump(E[E<1.007],NOx[E<1.007])
hump(E[E<1.006],NOx[E<1.006])
```

The data set on the left has evidence of a hump, but the one on the right does not.

22

Time Series Analysis

Time series data are vectors of numbers, typically regularly spaced in time. Yearly counts of animals, daily prices of shares, monthly means of temperature, and minute-by-minute details of blood pressure are all examples of time series, but they are measured on different time scales. Sometimes the interest is in the time series itself (e.g. whether or not it is cyclic, or how well the data fit a particular theoretical model), and sometimes the time series is incidental to a designed experiment (e.g. repeated measures). We cover each of these cases in turn.

The three key concepts in time series analysis are

- trend,
- serial dependence, and
- stationarity

Most time series analyses assume that the data are untrended. If they do show a consistent upward or downward trend, then they can be detrended before analysis (e.g. by differencing). Serial dependence arises because the values of adjacent members of a time series may well be correlated. Stationarity is a technical concept, but it can be thought of simply as meaning that the time series has the same properties wherever you start looking at it (e.g. white noise is a sequence of mutually independent random variables each with mean zero and variance $\sigma^2 > 0$).

Nicholson's Blowflies

The Australian ecologist, A.J. Nicholson, reared blowfly larvae on pieces of liver in laboratory cultures that his technicians kept running continuously for almost 7 years (361 weeks, to be exact). The time series for numbers of adult flies looks like this:

```
blowfly<-read.table("c:\\temp\\blowfly.txt",header=T)
attach(blowfly)
names(blowfly)
```

```
[1] "flies"
```

First, make the flies variable into a time series object:

```
flies<-ts(flies)
```

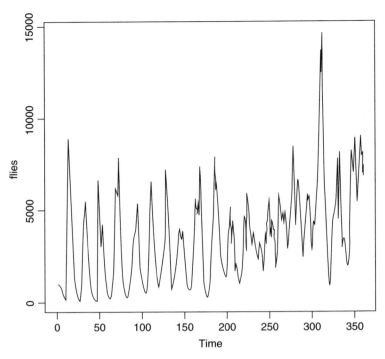

and plot it:

plot(flies)

This classic time series has two clear features:

- For the first 200 weeks the system exhibits beautifully regular cycles.

- After week 200 things change (perhaps a genetic mutation had arisen); the cycles become much less clear-cut, and the population begins a pronounced upward trend.

There are two important ideas to understand in time series analysis: **autocorrelation** and **partial autocorrelation**. The first describes how this week's population is related to last week's population. This is the autocorrelation at lag 1. The second describes the relationship between this week's population and the population at lag t once we have controlled for the correlations between all of the successive weeks between this week and week t. This should become clear if we draw the scatterplots from which the first four autocorrelation terms are calculated (lag 1 to lag 4).

There is a snag, however. The vector of flies at lag 1 is shorter (by one) than the original vector because the first element of the lagged vector is the second element of flies. The coordinates of the first data point to be drawn on the scatterplot are (flies[1],flies[2]) and the coordinates of the last plot that can be drawn are (flies[360], flies[361]) because the original vector is 361 element long:

length(flies)

```
[1]   361
```

Thus, the lengths of the vectors that can be plotted go down by one for every increase in the lag of one. We can produce the four plots for lags 1 to 4 in a function like this:

```
par(mfrow=c(2,2))
sapply(1:4, function(x) plot(flies[-c(361: (361-x+1))], flies[-c(1:x)] ) )
```

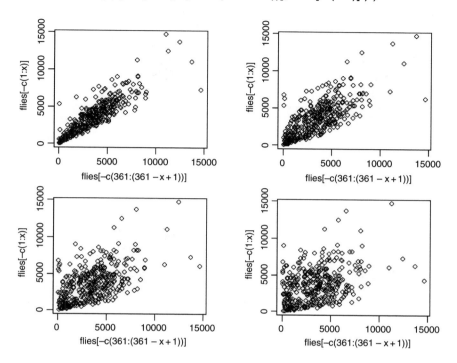

The correlation is very strong at lag 1, but notice how the variance increases with population size: small populations this week are invariably correlated with small populations next week, but large populations this week may be associated with large or small populations next week. The striking pattern here is the way that the correlation fades away as the size of the lag increases. Because the population is cyclic, the correlation goes to zero, then becomes weakly negative and then becomes strongly negative. This occurs at lags that are half the cycle length. Looking back at the time series, the cycles look to be about 20 weeks in length. So let's repeat the exercise by producing scatterplots at lags of 7, 8, 9 and 10 weeks:

```
sapply(7:10, function(x) plot(flies[-c((361-x+1):361)], flies[-c(1:x)] ) )
par(mfrow=c(1,1))
```

The negative correlation at lag 10 gradually emerges from the fog of no correlation at lag 7.

More formally, the autocorrelation function $\rho(k)$ at lag k is

$$\rho(k) = \frac{\gamma(k)}{\gamma(0)},$$

where $\gamma(k)$ is the autocovariance function at lag k of a stationary random function $\{Y(t)\}$ given by

$$\gamma(k) = \text{cov}\{Y(t), Y(t-k)\}.$$

The most important properties of the autocorrelation coefficient are these:

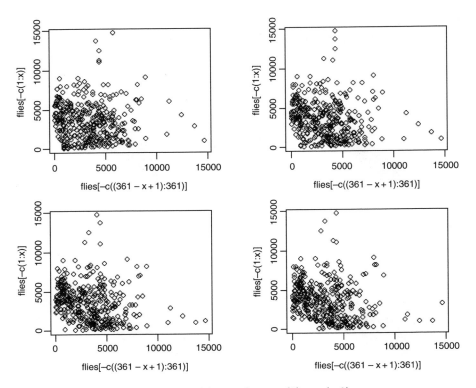

- They are symmetric backwards and forwards, so $\rho(k) = \rho(-k)$.

- The limits are $-1 \leq \rho(k) \leq 1$.

- When $Y(t)$ and $Y(t-k)$ are independent, then $\rho(k) = 0$.

- The converse of this is not true, so that $\rho(k) = 0$ does not imply that $Y(t)$ and $Y(t-k)$ are independent (look at the scatterplot for $k = 7$ in the scatterplots above).

A first-order autoregressive process is written as

$$Y_t = \alpha Y_{t-1} + Z_t.$$

This says that this week's population is α times last week's population plus a random term Z_t. (The randomness is **white noise**; the values of Z are **serially independent**, they have a **mean of zero**, and they have **finite variance** σ^2.)

In a stationary times series $-1 < \alpha < 1$. In general, then, the autocorrelation function of $\{Y(t)\}$ is

$$\rho_k = \alpha^k, \qquad k = 0, 1, 2, \ldots.$$

Partial autocorrelation is the relationship between this week's population and the population at lag t when we have controlled for the correlations between all of the successive weeks between this week and week t. That is to say, the partial autocorrelation is the correlation between $Y(t)$ and $Y(t+k)$ after regression of $Y(t)$ on $Y(t+1)$, $Y(t+2)$, $Y(t+3)$, ... $Y(t+k-1)$. It is obtained by solving the Yule–Walker equation

$$\rho_k = \sum_1^p \alpha_i \rho_{k-i}, \qquad k > 0,$$

with the ρ replaced by r (correlation coefficients estimated from the data). Suppose we want the partial autocorrelation between time 1 and time 3. To calculate this, we need the three ordinary correlation coefficients $r_{1,2}$, $r_{1,3}$ and $r_{2,3}$. The partial $r_{13,2}$ is then

$$r_{13,2} = \frac{r_{13} - r_{12}r_{23}}{\sqrt{(1 - r_{12}^2)(1 - r_{23}^2)}}.$$

For more on the partial correlation coefficients, see p. 715.

Let's look at the correlation structure of the blowfly data. The R function for calculating autocorrelations and partial autocorrelations is **acf** (the 'autocorrelation function'). First, we produce the autocorrelation plot to look for evidence of cyclic behaviour:

acf(flies,main="")

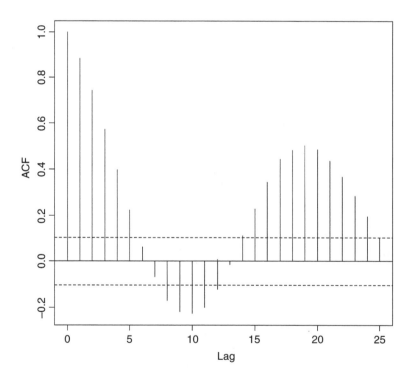

You will not see more convincing evidence of cycles than this. The blowflies exhibit highly significant, regular cycles with a period of 19 weeks. What kind of time lags are involved in the generation of these cycles? We use partial autocorrelation (**type="p"**) to find this out:

acf(flies,type="p",main="")

The significant density-dependent effects are manifest at lags of 2 and 3 weeks, with other, marginally significant negative effects at lags of 4 and 5 weeks. These lags reflect the duration of the larval and pupal period (1 and 2 periods, respectively). The cycles are clearly

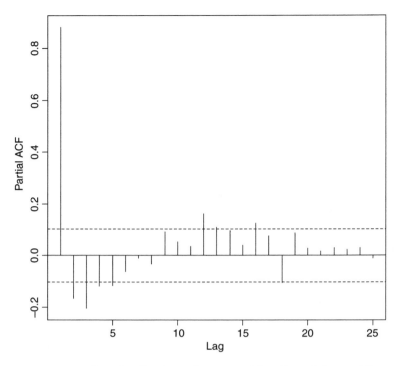

caused by overcompensating density dependence, resulting from intraspecific competition between the larvae for food (what Nicholson christened 'scramble competition'). There is a curious positive feedback at a lag of 12 weeks (12–16 weeks, in fact). Perhaps you can think of a possible cause for this?

We should investigate the behaviour of the second half of the time series separately. Let's say it is from week 201 onwards:

```
second<-flies[201:361]
```

Now test for a linear trend in mean fly numbers against day number, from 1 to length(second):

```
summary(lm(second~I(1:length(second))))
```

Note the use of I in the model formula (for 'as is') to tell R that the colon we have used is to generate a sequence of x values for the regression (and not an interaction term as it would otherwise have assumed).

```
Coefficients:
                      Estimate Std. Error  t value  Pr(>|t|)
(Intercept)          2827.531     336.661    8.399  2.37e-14  ***
I(1:length(second))    21.945       3.605    6.087  8.29e-09  ***

Residual standard error: 2126 on 159 degrees of freedom
Multiple R-Squared: 0.189,  Adjusted R-squared: 0.1839
F-statistic: 37.05 on 1 and 159 DF,  p-value: 8.289e-09
```

This shows that there is a highly significant upward trend of about 22 extra flies on average each week in the second half of time series. We can detrend the data by subtracting the fitted values from the linear regression of second on day number:

```
detrended<-second - predict(lm(second~I(1:length(second))))
ts.plot(detrended)
```

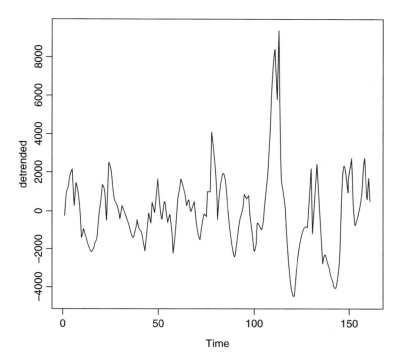

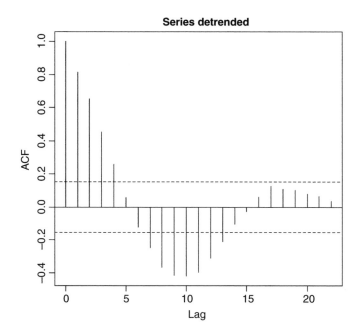

There are still cycles there, but they are weaker and less regular. We repeat the correlation analysis on the detrended data:

acf(detrended)

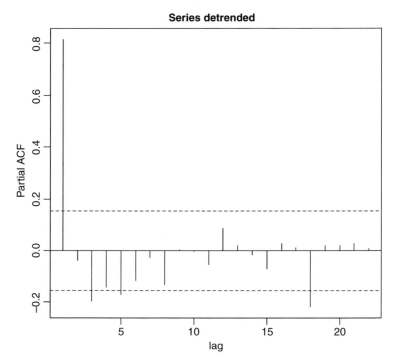

These look more like damped oscillations than repeated cycles. What about the partials?

acf(detrended, type="p")

There are still significant negative partial autocorrelations at lags 3 and 5, but now there is a curious extra negative partial at lag 18. It looks, therefore, as if the main features of the ecology are the same (a scramble contest for food between the larvae, leading to negative partials at 3 and 5 weeks after 1 and 2 generation lags), but population size is drifting upwards and the cycles are showing a tendency to dampen out.

Moving Average

The simplest way of seeing pattern in time series data is to plot the moving average. A useful summary statistic is the three-point moving average:

$$y_i' = \frac{y_{i-1} + y_i + y_{i+1}}{3}.$$

The function ma3 will compute the three-point moving average for any input vector x:

```
ma3<-function (x) {
    y<-numeric(length(x)-2)
```

```
    for (i in 2:(length(x)-1)) {
      y[i]<-(x[i-1]+x[i]+x[i+1])/3
    }
    y }
```

A time series of mean monthly temperatures will illustrate the use of the moving average:

```
temperature<-read.table("c:\\temp\\temp.txt",header=T)
attach(temperature)
tm<-ma3(temps)
plot(temps)
lines(tm[2:158])
```

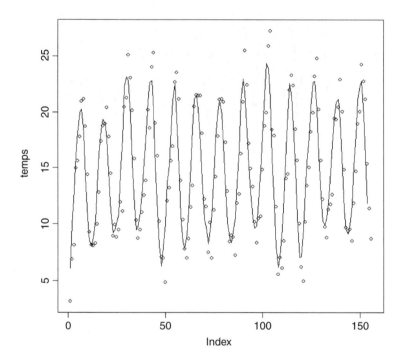

The seasonal pattern of temperature change over the 13 years of the data is clear. Note that a moving average can never capture the maxima or minima of a series (because they are averaged away). Note also that the three-point moving average is undefined for the first and last points in the series.

Seasonal Data

Many time series applications involve data that exhibit seasonal cycles. The commonest applications involve weather data: here are daily maximum and minimum temperatures from Silwood Park in south-east England over the period 1987–2005 inclusive:

```
weather<-read.table("c:\\temp\\SilwoodWeather.txt",header=T)
attach(weather)
names(weather)
```

```
[1] "upper" "lower" "rain" "month" "yr"
```

```
plot(upper,type="l")
```

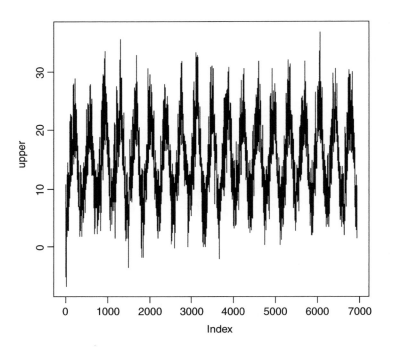

The seasonal pattern of temperature change over the 19-year period is clear, but there is no clear trend (e.g. warming, see p. 715). Note that the x axis is labelled by the day number of the time series (Index).

We start by modelling the seasonal component. The simplest models for cycles are scaled so that a complete annual cycle is of length 1.0 (rather than 365 days). Our series consists of 6940 days over a 19-year span, so we write:

```
length(upper)
```

```
[1] 6940
```

```
index<-1:6940
6940/19
```

```
[1] 365.2632
```

```
time<-index/365.2632
```

The equation for the seasonal cycle is this:

$$y = \alpha + \beta \sin(2\pi t) + \gamma \cos(2\pi t) + \varepsilon.$$

This is a linear model, so we can estimate its three parameters very simply:

```
model<-lm(upper~sin(time*2*pi)+cos(time*2*pi))
```

To investigate the fit of this model we need to plot the scattergraph using very small symbols (otherwise the fitted line will be completely obscured). The smallest useful plotting symbol is the dot ".".

```
plot(time, upper, pch=".")
lines(time, predict(model))
```

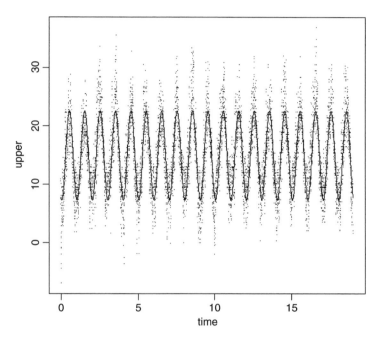

The three parameters of the model are all highly significant:

```
summary(model)
```

```
Coefficients:
                   Estimate  Std. Error  t value  Pr(>|t|)
(Intercept)        14.95647     0.04088   365.86    <2e-16  ***
sin(time * 2 * pi)  -2.53883     0.05781   -43.91    <2e-16  ***
cos(time * 2 * pi)  -7.24017     0.05781  -125.23    <2e-16  ***
Residual standard error: 3.406 on 6937 degrees of freedom
Multiple R-Squared: 0.7174,  Adjusted R-squared: 0.7173
F-statistic: 8806 on 2 and 6937 DF,  p-value: < 2.2e-16
```

We can investigate the residuals to look for patterns (e.g. trends in the mean, or autocorrelation structure). Remember that the residuals are stored as part of the model object:

```
plot(model$resid,pch=".")
```

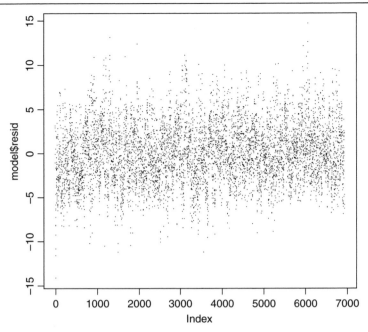

There looks to be some periodicity in the residuals, but no obvious trends. To look for serial correlation in the residuals, we use the acf function like this:

```
par(mfrow=c(1,2))
acf(model$resid)
acf(model$resid,type="p")
```

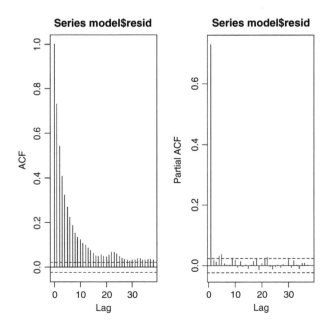

There is very strong serial correlation in the residuals, and this drops off roughly exponentially with increasing lag (left-hand graph). The partial autocorrelation at lag 1 is very large (0.7317), but the correlations at higher lags are much smaller. This suggests that an AR(1) model (autoregressive model with order 1) might be appropriate. This is the statistical justification behind the old joke about the weather forecaster who was asked what tomorrow's weather would be. 'Like today's', he said.

Pattern in the monthly means

The monthly average upper temperatures show a beautiful seasonal pattern when analysed by acf:

```
temp<-ts(as.vector(tapply(upper,list(month,yr),mean)))
par(mfrow=c(1,1))
acf(temp)
```

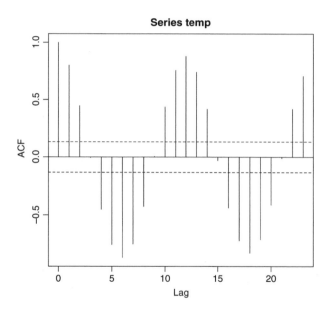

There is a perfect cycle with period 12 (as expected). What about patterns across years?

```
ytemp<-ts(as.vector(tapply(upper,yr,mean)))
acf(ytemp)
```

Nothing! The pattern you may (or may not) see depends upon the scale at which you look for it. As for spatial patterns (Chapter 24), so it is with temporal patterns. There is strong pattern between days within months (tomorrow will be like today). There is very strong pattern from month to month within years (January is cold, July is warm). But there is no pattern at all from year to year (there may be progressive global warming, but it is not apparent within this recent time series (see below), and there is absolutely no evidence for untrended serial correlation).

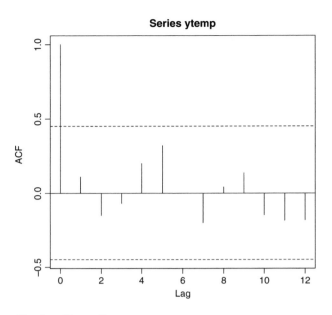

Built-in Time Series Functions

The analysis is simpler, and the graphics are better labelled, if we convert the temperature data into a regular time series object using ts. We need to specify the first date (January 1987) as start=c(1987,1), and the number of data points per year as frequency=365.

```
high<-ts(upper,start=c(1987,1),frequency=365)
```

Now use plot to see a plot of the time series, correctly labelled by years:

```
plot(high)
```

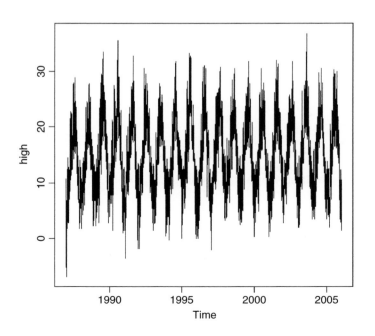

Decompositions

It is useful to be able to turn a time series into components. The function stl (letter L, not numeral one) performs seasonal decomposition of a time series into seasonal, trend and irregular components using loess. First, we make a time series object, specifying the start date and the frequency (above), then use stl to decompose the series:

```
up<-stl(high,"periodic")
```

The plot function produces the data series, the seasonal component, the trend and the residuals in a single frame:

```
plot(up)
```

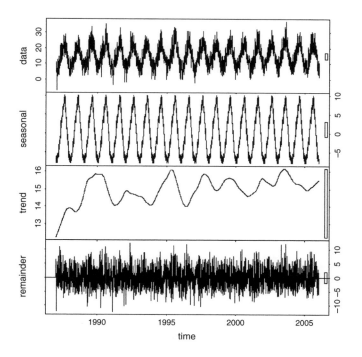

The remainder component is the residuals from the seasonal plus trend fit. The bars at the right-hand side are of equal heights (in user coordinates).

Testing for a Trend in the Time Series

It is important to know whether these data provide any evidence for global warming. The trend part of the figure indicates a fluctuating increase, but is it significant? The mean temperature in the last 9 years was 0.71° C higher degrees than in the first 10 years:

```
ys<-factor(1+(yr>1996))
tapply(upper,ys,mean)
        1          2
14.62056  15.32978
```

We cannot test for a trend with linear regression because of the massive temporal pseudoreplication. Suppose we tried this:

```
model1<-lm(upper~index+sin(time*2*pi)+cos(time*2*pi))
summary(model1)
```

```
Coefficients:
                    Estimate  Std. Error   t value  Pr(>|t|)
(Intercept)        1.433e+01   8.136e-02   176.113   <2e-16  ***
index              1.807e-04   2.031e-05     8.896   <2e-16  ***
sin(time * 2 * pi) -2.518e+00  5.754e-02   -43.758   <2e-16  ***
cos(time * 2 * pi) -7.240e+00  5.749e-02  -125.939   <2e-16  ***
```

It would suggest (wrongly, as we shall see) that the warming was highly significant (index p value $< 2 \times 10^{-16}$ for a slope of 0.000 180 7 degrees of warming per day, leading to a predicted increase in mean temperature of 1.254 degrees over the 6940 days of the time series).

Since there is so much temporal pseudoreplication we should use a mixed model (lmer, p. 640), and because we intend to compare two models with different fixed effects we use the method of maximum likelihood ("ML" rather than "REML"). The explanatory variable for any trend is index, and we fit the model with and without this variable, allowing for different intercepts for the different years as a random effect:

```
library(lme4)
model2<-
lmer(upper~index+sin(time*2*pi)+cos(time*2*pi)+(1|factor(yr)),method="ML")
model3<-lmer(upper~sin(time*2*pi)+cos(time*2*pi)+(1|factor(yr)),method="ML")
anova(model2,model3)
```

```
Data:
Models:
model3: upper ~ sin(time * 2 * pi) + cos(time * 2 * pi) + (1 |
factor(yr))
model2: upper ~ index + sin(time * 2 * pi) + cos(time * 2 * pi) + (1 |
model3: factor(yr))
        Df   AIC    BIC   logLik  Chisq  Chi Df  Pr(>Chisq)
model3   4 36450  36477  -18221
model2   5 36451  36485  -18220  1.0078       1      0.3154
```

Clearly, the trend is not significant ($p = 0.3154$). If you are prepared to ignore all the within-year variation (day-to-day and month-to-month), then you can get rid of the pseudoreplication by averaging and test for trend in the yearly mean values: these show a significant trend if the first year (1987) is included, but not if it is omitted:

```
means<-as.vector(tapply(upper,yr,mean))
model<-lm(means~I(1:19))
summary(model)
```

```
Coefficients:
             Estimate  Std. Error  t value  Pr(>|t|)
(Intercept)  14.27105    0.32220    44.293   <2e-16  ***
I(1:19)       0.06858    0.02826     2.427   0.0266  *
```

```
model<-lm(means[-1]~I(1:18))
```

summary(model)

```
Coefficients:
              Estimate  Std. Error  t value  Pr(>|t|)
(Intercept)   14.59826     0.30901   47.243    <2e-16  ***
I(1:18)        0.04761     0.02855    1.668     0.115
```

Obviously, you need to be circumspect when interpreting trends in time series.

Spectral Analysis

There is an alternative approach to time series analysis, which is based on the analysis of **frequencies** rather than fluctuations of numbers. Frequency is the reciprocal of cycle period. Ten-year cycles would have a frequency 0.1 per year. Here are the famous Canadian lynx data:

```
numbers<-read.table("c:\\temp\\lynx.txt",header=T)
attach(numbers)
names(numbers)
```

```
[1]  "Lynx"
```

plot.ts(Lynx)

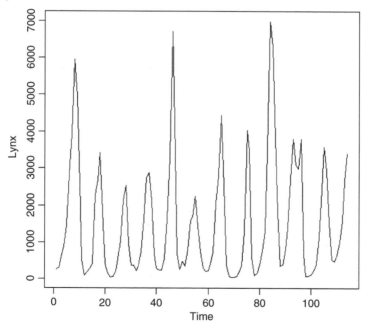

The fundamental tool of spectral analysis is the **periodogram**. This is based on the squared correlation between the time series and sine/cosine waves of frequency ω, and conveys exactly the same information as the autocovariance function. It may (or may not) make the information easier to interpret. Using the function is straightforward; we employ the **spectrum** function like this:

spectrum(Lynx)

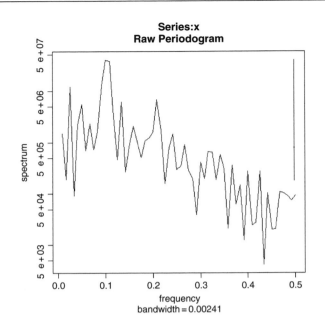

The plot is on a log scale, in units of decibels, and the sub-title on the x axis shows the bandwidth, while the 95% confidence interval in decibels is shown by the vertical bar in the top right-hand corner. The figure is interpreted as showing strong cycles with a frequency of about 0.1, where the maximum value of spectrum occurs. That is to say, it indicates cycles with a period of $1/0.1 = 10$ years. There is a hint of longer period cycles (the local peak at frequency 0.033 would produce cycles of length $1/0.033 = 30$ years) but no real suggestion of any shorter-term cycles.

Multiple Time Series

When we have two or more time series measured over the same period, the question naturally arises as to whether or not the ups and downs of the different series are correlated. It may be that we suspect that change in one of the variables causes changes in the other (e.g. changes in the number of predators may cause changes in the number of prey, because more predators means more prey eaten). We need to be careful, of course, because it will not always be obvious which way round the causal relationship might work (e.g. predator numbers may go up because prey numbers are higher; ecologists call this a numerical response). Suppose we have the following sets of counts:

```
twoseries<-read.table("c:\\temp\\twoseries.txt",header=T)
attach(twoseries)
names(twoseries)
```

```
[1]  "x"  "y"
```

We start by inspecting the two time series one above the other:

```
plot.ts(cbind(x,y),main="")
```

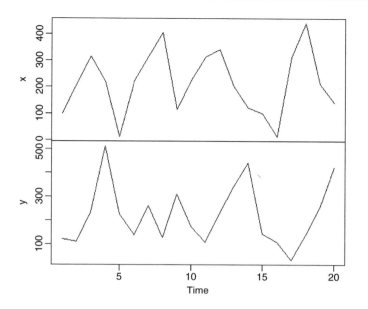

There is some evidence of periodicity (at least in *x*) and it looks as if *y* lags behind *x* by roughly 2 periods (sometimes 1). Now let's carry out straightforward analyses on each time series separately:

```
par(mfrow=c(1,2))
acf(x,type="p")
acf(y,type="p")
```

As we suspected, the evidence for periodicity is stronger in *x* than in *y*: the partial autocorrelation is significant and negative at lag 2 for *x*, but none of the partial autocorrelations are significant for *y*.

To look at the cross-correlation between the two series we use the same acf function, but we provide it with a matrix containing both *x* and *y* as its argument:

```
par(mfrow=c(1,1))
acf(cbind(x,y))
```

The four plots show the autocorrelations for *x*, *x* and *y*, *y* and *x*, and *y*. The plots for *x* and *y* are exactly the same as in the single-series case. Note the reversal of the time lag scale on the plot of *y* against *x*. In the case of partial autocorrelations, things are a little different, because we now see in the *xx* column the values of the autocorrelation of *x* controlling for *x and for y* (not just for *x* as in the earlier figure).

```
acf(cbind(x,y),type="p")
```

This indicates a hint (not quite significant) of a positive partial autocorrelation for *x* at lag 4 that we did not see in the analysis of the single time series of *x*. There are significant negative cross-correlations of *x* and *y* at lags of 1 and 2, strongly suggestive of some sort of relationship between the two time series.

To get a visual assessment of the relationship between *x* and *y* we can plot the change in *y* against the change in *x* using the diff function like this:

```
plot(diff(x),diff(y))
```

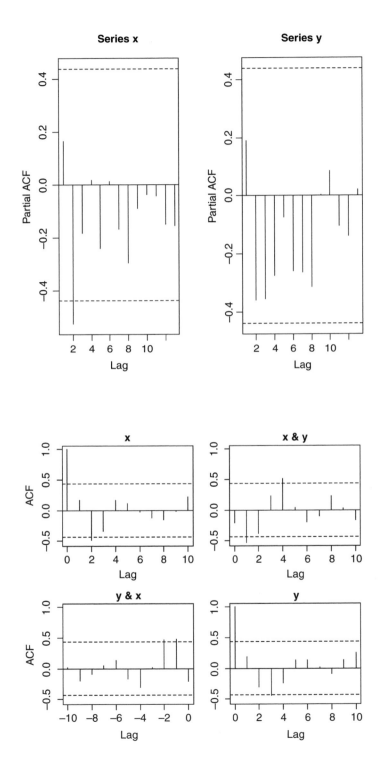

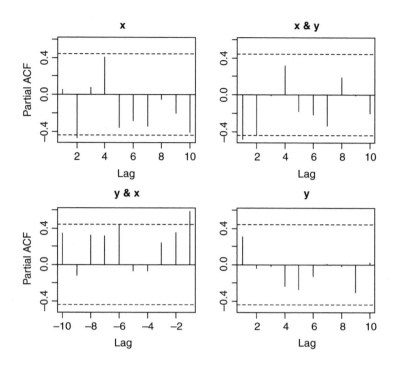

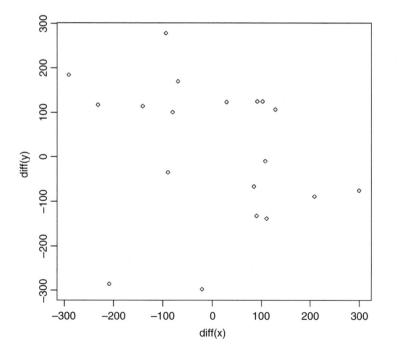

The negative correlation is clear enough, but there are two conspicuous outliers, making it clear that not all negative changes in x are associated with large positive changes in y.

Simulated Time Series

To see how the correlation structure of an AR(1) depends on the value of α, we can simulate the process over, say, 250 time periods using different values of α. We generate the white noise Z_t using the random number generator rnorm(n,0,s) which gives n random numbers with a mean of 0 and a standard deviation of s. To simulate the time series we evaluate

$$Y_t = \alpha Y_{t-1} + Z_t,$$

multiplying last year's population by α then adding the relevant random number from Z_t.

We begin with the special case of $\alpha = 0$ so $Y_t = Z_t$ and the process is pure white noise:

```
Y<-rnorm(250,0,2)
par(mfrow=c(1,2))
plot.ts(Y)
acf(Y)
```

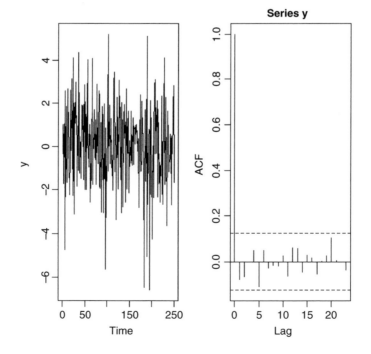

The time series is bound to be stationary because each value of Z is independent of the value before it. The correlation at lag 0 is 1 (of course), but there is absolutely no hint of any correlations at higher lags.

To generate the time series for non-zero values of α we need to use recursion: this year's population is last years population times α plus the white noise. We begin with a negative value of $\alpha = -0.5$. First we generate all the noise values (by definition, these *don't* depend on population size):

```
Z<-rnorm(250,0,2)
```

Now the initial population at time 0 is set to 0 (remember that the population is stationary, so we can think of the Y values as departures from the long-term mean population size). This means that $Y_1 = Z_1$. Thus, Y_2 will be whatever Y_1 was, times -0.5, plus Z_2. And so on.

```
Y<-numeric(250)
Y[1]<-Z[1]
for (i in 2:250) Y[i]<- - 0.5*Y[i-1]+Z[i]
plot.ts(Y)
acf(Y)
```

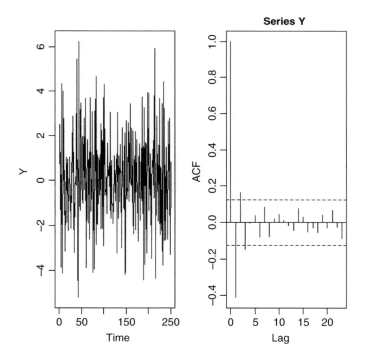

The time series shows rapid return to equilibrium following random departures from it. There is a highly significant negative autocorrelation at lag 1, significant positive autocorrelation at lag 2 and so on, with the size of the correlation gradually damping away.

Let's simulate a time series with a positive value of, say, $\alpha = 0.5$:

```
Z<-rnorm(250,0,2)
Y[1]<-Z[1]
```

```
for (i in 2:250) Y[i]<- 0.5*Y[i-1]+Z[i]
plot.ts(Y)
acf(Y)
```

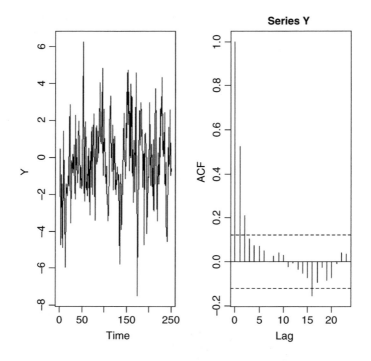

Now the time series plot looks very different, with protracted periods spent drifting away from the long-term average. The autocorrelation plot shows significant positive correlations for the first three lags.

Finally, we look at the special case of $\alpha = 1$. This means that the time series is a classic **random walk**, given by

$$Y_t = Y_{t-1} + Z_t.$$

```
Z<-rnorm(250,0,2)
Y[1]<-Z[1]
for (i in 2:250) Y[i]<- Y[i-1]+Z[i]
plot.ts(Y)
acf(Y)
```

The time series wanders about and strays far away from the long-term average. The acf plot shows positive correlations dying away very slowly, and still highly significant at lags of more than 20. Of course, if you do another realization of the process, the time series will look very different, but the autocorrelations will be similar:

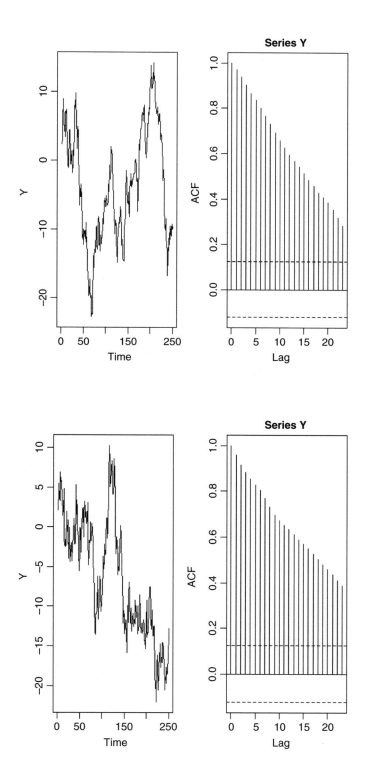

Time Series Models

Time series models come in three kinds (Box and Jenkins 1976):

- Moving average (MA) models where

$$X_t = \sum_0^q \beta_j \varepsilon_{t-j};$$

- autoregressive (AR) models where

$$X_t = \sum_1^p \alpha_i X_{t-i} + \varepsilon_t;$$

- autoregressive moving average (ARMA) models where

$$X_t = \sum_1^p \alpha_i X_{t-i} + \sum_0^q \beta_j \varepsilon_{t-j}.$$

A moving average of order q averages the random variation over the last q time periods. An autoregressive model of order p computes X_t as a function of the last p values of X, so, for a second-order process, we would use

$$X_t = \alpha_1 X_{t-1} + \alpha_2 X_{t-2} + \varepsilon_t.$$

Typically, we would use the partial autocorrelation plot (above) to determine the order. So, for the lynx data (p. 717) we would use order 2 or 4, depending on taste. Other things being equal, parsimony suggests the use of order 2. The fundamental difference is that a set of random components (ε_{t-j}) influences the current value of a MA process, whereas only the current random effect (ε_t) affects an AR process. Both kinds of effects are at work in an ARMA processes. Ecological models of population dynamics are typically AR models. For instance,

$$N_t = \lambda N_{t-1}$$

(the discrete-time version of exponential growth ($\lambda > 1$) or decay ($\lambda < 1$)) looks just like an first order AR process with the random effects missing. This is somewhat misleading, however, since time series are supposed to be stationary, which would imply a long-term average value of $\lambda = 1$. But, in the absence of density dependence (as here), this is impossible. The α of the AR model is *not* the λ of the population model.

Models are fitted using the **arima** function, and their performances are compared using AIC (see p. 728). The most important component of the model is order. This is a vector of length 3 specifying the order of the autoregressive operators, the number of differences, and the order of moving average operators. Thus **order=c(1,3,2)** is based on a first-order autoregressive process, three differences, and a second-order moving average. The Canadian lynx data are used as an example of **arima** in time series modelling.

Time series modelling on the Canadian lynx data

Records of the number of skins of predators (lynx) and prey (snowshoe hares) returned by trappers were collected over many years by the Hudson's Bay Company. The lynx numbers are shown on p. 717 and exhibit a clear 10-year cycle. We begin by plotting the autocorrelation and partial autocorrelation functions:

```
par(mfrow=c(1,2))
acf(Lynx,main="")
acf(Lynx,type="p",main="")
```

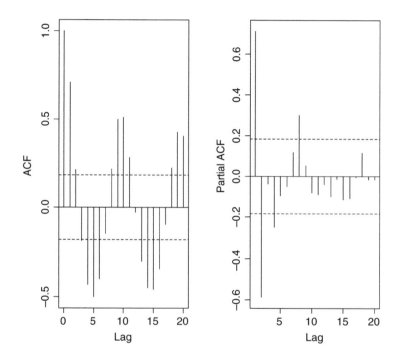

The population is very clearly cyclic, with a period of 10 years. The dynamics appear to be driven by strong, negative density dependence (a partial autocorrelation of −0.588) at lag 2. There are other significant partials at lag 1 and lag 8 (positive) and lag 4 (negative). Of course you cannot infer the mechanism by observing the dynamics, but the lags associated with significant negative and positive feedbacks are extremely interesting and highly suggestive. The main prey species of the lynx is the snowshoe hare and the negative feedback at lag 2 may reflect the timescale of this predator–prey interaction. The hares are known to cause medium-term induced reductions in the quality of their food plants as a result of heavy browsing pressure when the hares at high density, and this could map through to lynx populations with lag 4.

The order vector specifies the non-seasonal part of the ARIMA model: the three components (p, d, q) are the AR order, the degree of differencing, and the MA order. We start by investigating the effects of AR order with no differencing and no moving average terms, comparing models on the basis of AIC:

```
model10<-arima(Lynx,order=c(1,0,0))
model20<-arima(Lynx,order=c(2,0,0))
model30<-arima(Lynx,order=c(3,0,0))
model40<-arima(Lynx,order=c(4,0,0))
model50<-arima(Lynx,order=c(5,0,0))
model60<-arima(Lynx,order=c(6,0,0))
AIC(model10,model20,model30,model40,model50,model60)
```

```
        df      AIC
model10  3  1926.991
model20  4  1878.032
model30  5  1879.957
model40  6  1874.222
model50  7  1875.276
model60  8  1876.858
```

On the basis of AR alone, it appears that order 4 is best (AIC = 1874.222). What about MA?

```
model01<-arima(Lynx,order=c(0,0,1))
model02<-arima(Lynx,order=c(0,0,2))
model03<-arima(Lynx,order=c(0,0,3))
model04<-arima(Lynx,order=c(0,0,4))
model05<-arima(Lynx,order=c(0,0,5))
model06<-arima(Lynx,order=c(0,0,6))
AIC(model01,model02,model03,model04,model05,model06)
```

```
        df      AIC
model01  3  1917.947
model02  4  1890.061
model03  5  1887.770
model04  6  1888.279
model05  7  1885.698
model06  8  1885.230
```

The AIC values are generally higher than given by the AR models. Perhaps there is a combination of AR and MA terms that is better than either on their own?

```
model40<-arima(Lynx,order=c(4,0,0))
model41<-arima(Lynx,order=c(4,0,1))
model42<-arima(Lynx,order=c(4,0,2))
model43<-arima(Lynx,order=c(4,0,3))
AIC(model40,model41,model42,model43)
```

```
        df      AIC
model40  6  1874.222
model41  7  1875.351
model42  8  1862.435
model43  9  1880.432
```

Evidently there is no need for a moving average term (model40 is best). What about the degree of differencing?

```
model400<-arima(Lynx,order=c(4,0,0))
```

```
model401<-arima(Lynx,order=c(4,1,0))
model402<-arima(Lynx,order=c(4,2,0))
model403<-arima(Lynx,order=c(4,3,0))
AIC(model400,model401,model402,model403)
```

```
          df        AIC
model400   6   1874.222
model401   5   1890.961
model402   5   1917.882
model403   5   1946.143
```

The model with no differencing performs best. The lowest AIC is 1874.222, which suggests that a model with an AR lag of 4, no differencing and no moving average terms is best. This implies that a rather complex ecological model is required which takes account of both the significant partial correlations at lags of 2 and 4 years, and not just the 2-year lag (i.e. plant–herbivore effects may be necessary to explain the dynamics, in addition to predator–prey effects).

23

Multivariate Statistics

This class of statistical methods is fundamentally different from the others in this book because there is no response variable. Instead of trying to understand variation in a response variable in terms of explanatory variables, in multivariate statistics we look for **structure in the data**. The problem is that structure is rather easy to find, and all too often it is a feature of that particular data set alone. The real challenge is to find *general* structure that will apply to other data sets as well. Unfortunately, there is no guaranteed means of detecting pattern, and a great deal of ingenuity has been shown by statisticians in devising means of pattern recognition in multivariate data sets. The main division is between methods that **assume a given structure** and seek to divide the cases into groups, and methods that seek to **discover structure** from inspection of the dataframe.

The multivariate techniques implemented in R include:

- principal components analysis (prcomp);
- factor analysis (factanal);
- cluster analysis (hclust, kmeans);
- discriminant analysis (lda, qda);
- neural networks (nnet).

These techniques are *not* recommended unless you know exactly what you are doing, and exactly *why* you are doing it. Beginners are sometimes attracted to multivariate techniques because of the complexity of the output they produce, making the classic mistake of confusing the opaque for the profound.

Principal Components Analysis

The idea of principal components analysis (PCA) is to find a small number of **linear combinations** of the variables so as to capture most of the variation in the dataframe as a whole. With a large number of variables it may be easier to consider a small number of combinations of the original data rather than the entire dataframe. Suppose, for example, that you had three variables measured on each subject, and you wanted to distil the essence of each individual's performance into a single number. An obvious solution is the arithmetic mean of the three numbers $1/3v_1 + 1/3v_2 + 1/3v_3$ where v_1, v_2 and v_3 are the three variables (e.g. maths score, physics score and chemistry score for pupils' exam results). The vector of coefficients $l = (1/3, 1/3, 1/3)$ is called a linear combination. Linear combinations where $\sum l^2 = 1$ are called standardized linear combinations. Principal components analysis finds a set of orthogonal standardized linear combinations which together explain all of the variation in the original data. There are as many principal components as there are variables, but typically it is only the first few that explain important amounts of the total variation.

Calculating principal components is easy. Interpreting what the components mean in scientific terms is hard, and potentially equivocal. You need to be more than usually circumspect when evaluating multivariate statistical analyses.

The following dataframe contains mean dry weights (g) for 54 plant species on 89 plots, averaged over 10 years; see Crawley *et al.* (2005) for species names and more background. The question is what are the principal components and what environmental factors are associated with them?

```
pgdata<-read.table("c:\\temp\\pgfull.txt",header=T)
names(pgdata)
```

```
 [1]  "AC"        "AE"      "AM"    "AO"    "AP"    "AR"      "AS"
 [8]  "AU"        "BH"      "BM"    "CC"    "CF"    "CM"      "CN"
[15]  "CX"        "CY"      "DC"    "DG"    "ER"    "FM"      "FP"
[22]  "FR"        "GV"      "HI"    "HL"    "HP"    "HS"      "HR"
[29]  "KA"        "LA"      "LC"    "LH"    "LM"    "LO"      "LP"
[36]  "OR"        "PL"      "PP"    "PS"    "PT"    "QR"      "RA"
[43]  "RB"        "RC"      "SG"    "SM"    "SO"    "TF"      "TG"
[50]  "TO"        "TP"      "TR"    "VC"    "VK"    "plot"    "lime"
[57]  "species"   "hay"     "pH"
```

We need to extract the 54 variables that refer to the species' abundances and leave behind the variables containing the experimental treatments (plot and lime) and the covariates (species richness, hay biomass and soil pH). This creates a smaller dataframe containing all 89 plots (i.e. all the rows) but only columns 1 to 54 (see p. 742 for details):

```
pgd<-pgdata[,1:54]
```

There are two functions for carrying out PCA in R. The generally preferred method for numerical accuracy is **prcomp** (where the calculation is done by a singular value decomposition of the centred and scaled data matrix, not by using **eigen** on the covariance matrix, as in the alternative function **princomp**).

The aim is to find linear combinations of a set of variables that maximize the variation contained within them, thereby displaying most of the original variation in fewer dimensions. These principal components have a value for every one of the 89 rows of the dataframe. By contrast, in factor analysis (see p. 735), each factor contains a contribution (which may in some cases be zero) from each variable, so the length of each factor is the total number of variables (54 in the current example). This has practical implications, because you can plot the principal components against other explanatory variables from the dataframe, but you cannot do this for factors because the factors are of length 54 while the covariates are of length 89.

You need to use the option **scale=TRUE** because the variances are significantly different for the 54 plant species:

```
model<-prcomp(pgd,scale=TRUE)
```

summary(model)

Importance of components:

	PC1	PC2	PC3	PC4	PC5	PC6	PC7	PC8
Standard deviation	3.005	2.336	1.9317	1.786	1.7330	1.5119	1.5088	1.3759
Proportion of Variance	0.167	0.101	0.0691	0.059	0.0556	0.0423	0.0422	0.0351
Cumulative Proportion	0.167	0.268	0.3373	0.396	0.4520	0.4943	0.5365	0.5716

...

	PC53	PC54
Standard deviation	0.11255	0.01721
Proportion of Variance	0.00023	0.00001
Cumulative Proportion	0.99999	1.00000

You can see that the first principal component (PC1) explains 16.7% of the total variation, and only the next four (PC2–PC5) explain more than 5% of the total variation. Here is the plot of this model, showing the relative importance of PC1.

plot(model,main="")

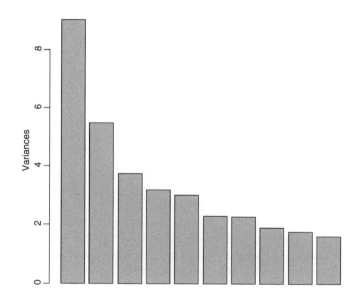

This is called a **scree plot** in PCA because it is supposed to look like a cliff face on a mountainside (on the left), with a scree slope below it (the tail on the right). The standard practice is to assume that you need sufficient principal components to account for 90 % of the total variation (but that would take 24 components in the present case). Principal component loadings show how the original variables (the 54 different species in our example) influence the principal components.

   In a biplot, the original variables are shown by arrows (54 of them in this case):

biplot(model)

The numbers represent the rows in the original dataframe, and the directions of the arrows show the relative loadings of the species on the first and second principal components.

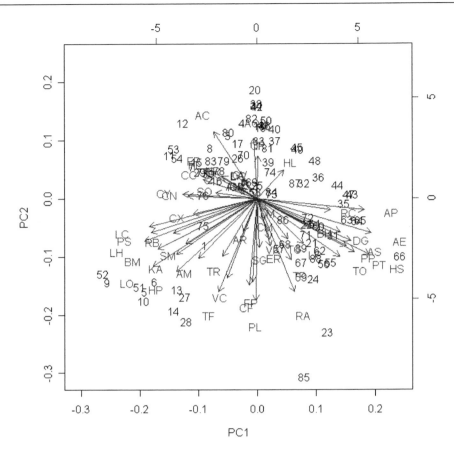

Thus, the species AP, AE and HS have strong positive loadings on PC1 and LC, PS BM and LO have strong negative loadings. PC2 has strong positive loadings from species AO and AC and negative loadings from TF, PL and RA.

If there are explanatory variables available, you can plot these against the principal components to look for patterns. In this example, it looks as if the first principal component is associated with increasing biomass (and hence increasing competition for light) and as if the second principal component is associated with declining soil pH (increasing acidity):

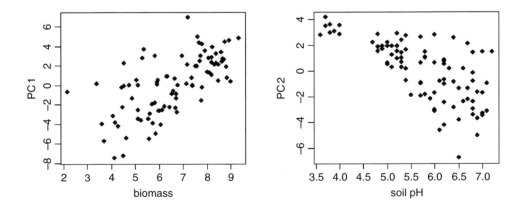

```
yv<-predict(model)[,1]
yv2<-predict(model)[,2]
par(mfrow=c(2,2))
plot(pgdata$hay,yv,pch=16,xlab="biomass",ylab="PC 1")
plot(pgdata$pH,yv2,pch=16,xlab="soil pH",ylab="PC 2")
```

Factor Analysis

With principal components analysis we were fundamentally interested in the variables and their contributions. Factor analysis aims to provide usable numerical values for quantities such as intelligence or social status that are not directly measurable. The idea is to use correlations between observable variables in terms of underlying 'factors'. Note that 'factors' in factor analysis are *not* the same as the categorical explanatory variables we have been calling factors throughout the rest of this book.

Compared with PCA, the variables themselves are of relatively little interest in factor analysis; it is an understanding of the hypothesized underlying factors that is the main aim. The idea is that the correlations amongst the variables are explained by the common factors. The function factanal performs maximum likelihood factor analysis on a covariance matrix or data matrix. The pgd dataframe is introduced on p. 732. You need to specify the number of factors you want to estimate – we begin with 8:

```
factanal(pgd,8)

Call:
factanal(x = pgd, factors = 8)
```

```
Uniquenesses:
    AC      AE      AM      AO      AP      AR      AS      AU      BH      BM
 0.638   0.086   0.641   0.796   0.197   0.938   0.374   0.005   0.852   0.266
    CC      CF      CM      CN      CX      CY      DC      DG      ER      FM
 0.056   0.574   0.786   0.579   0.549   0.733   0.837   0.408   0.072   0.956
    FP      FR      GV      HI      HL      HP      HS      HR      KA      LA
 0.371   0.815   0.971   0.827   0.921   0.218   0.332   0.915   0.319   0.305
    LC      LH      LM      LO      LP      OR      PL      PP      PS      PT
 0.349   0.333   0.927   0.121   0.403   0.005   0.286   0.606   0.336   0.401
    QR      RA      RB      RC      SG      SM      SO      TF      TG      TO
 0.913   0.491   0.005   0.754   0.341   0.212   0.825   0.428   0.476   0.469
    TP      TR      VC      VK
 0.309   0.611   0.651   0.170
```

```
Loadings:
     Factor1  Factor2  Factor3  Factor4  Factor5  Factor6  Factor7  Factor8
AC    -0.512   -0.268                               0.121
AE     0.925   -0.107            -0.146            -0.118
AM    -0.206    0.413    0.213             0.163    0.115    0.153    0.186
AO    -0.312   -0.196   -0.151   -0.105            -0.148   -0.102
AP     0.827   -0.173   -0.195   -0.167            -0.123
AR               0.150             0.111                     0.127
AS     0.778
AU                                                                    0.996
```

	Factor1	Factor2	Factor3	Factor4	Factor5	Factor6	Factor7	Factor8
BH	0.380							
BM	−0.116	0.292		0.695		0.380		
CC	−0.152			0.159		0.943		
CF		0.539		0.342				
CM			0.434	−0.110				
CN	−0.276	0.143			0.541	0.147		
CX				0.628	0.169	0.146		
CY	−0.211		−0.162	0.340			0.270	
DC		−0.125			0.372			
DG	0.738			−0.127	0.145			
ER					0.960			
FM	−0.108				0.133			
FP	0.245	0.226		0.478	0.493		−0.176	
FR	−0.386		−0.144					
GV	−0.134							
HI	−0.202	−0.129	−0.163	0.182			0.216	
HL		−0.157		−0.127		−0.139		
HP	−0.155	0.832					0.240	
HS	0.746	−0.102	0.257	−0.152				
HR	−0.155	−0.107	−0.122	0.101			0.150	
KA	−0.167	0.774	−0.169	0.139				
LA						0.829		
LC	−0.306	0.378	−0.125	0.529				0.328
LH	−0.256	0.556	−0.132	0.421	0.223		0.195	
LM				0.112	0.221			
LO	−0.129	0.432		0.781			0.251	
LP	0.115		0.745					
OR								0.996
PL		0.369	0.675	0.337				
PP	0.527		0.226	−0.167		−0.175		
PS	−0.212	0.301	−0.130	0.681	0.150		0.158	
PT	0.741			−0.100	0.150	−0.105		
QR	−0.194	−0.135						
RA	0.195	0.227	0.578		0.205	−0.166	−0.107	
RB	−0.122	0.158		0.272			0.934	
RC	0.361			−0.198		−0.176	−0.152	
SG					0.806			
SM		0.388					0.787	
SO			−0.100	0.386				
TF		0.702	0.260					
TG	0.141		0.583	−0.110		0.367	0.107	
TO	0.418		0.567	−0.158				
TP			0.818					
TR		0.141	0.306	0.238			0.458	
VC		0.403	0.246	0.309		−0.169		
VK					0.909			

	Factor1	Factor2	Factor3	Factor4	Factor5	Factor6	Factor7	Factor8
SS loadings	5.840	3.991	3.577	3.540	3.028	2.644	2.427	2.198
Proportion Var	0.108	0.074	0.066	0.066	0.056	0.049	0.045	0.041
Cumulative Var	0.108	0.182	0.248	0.314	0.370	0.419	0.464	0.505

Test of the hypothesis that 8 factors are sufficient.
The chi-squared statistic is 1675.57 on 1027 degrees of freedom.
The p-value is 5.92e-34

On factor 1 you see strong positive correlations with AE, AP and AS and negative correlations with AC, AO and FR: this has a natural interpretation as a gradient from tall neutral grassland (positive correlations) to short, acidic grasslands (negative correlations). On factor 2, low-growing species associated with moderate to high soil pH (AM, CF, HP, KA) have large positive values and low-growing acid-loving species (AC and AO) have negative values. Factor 3 picks out the key nitrogen-fixing (legume) species LP and TP with high positive values. And so on.

Note that the loadings are of length 54 (the number of variables) not 89 (the number of rows in the dataframe representing the different plots in the experiment), so we cannot plot the loadings against the covariates as we did with PCA (p. 734). However, we can plot the factor loadings against one another:

```
par(mfrow=c(2,2))
plot(loadings(model)[,1],loadings(model)[,2],pch=16,xlab="Factor 1",ylab="Factor 2")
plot(loadings(model)[,1],loadings(model)[,3],pch=16,xlab="Factor 1",ylab="Factor 3")
plot(loadings(model)[,1],loadings(model)[,4],pch=16,xlab="Factor 1",ylab="Factor 4")
plot(loadings(model)[,1],loadings(model)[,5],pch=16,xlab="Factor 1",ylab="Factor 5")
```

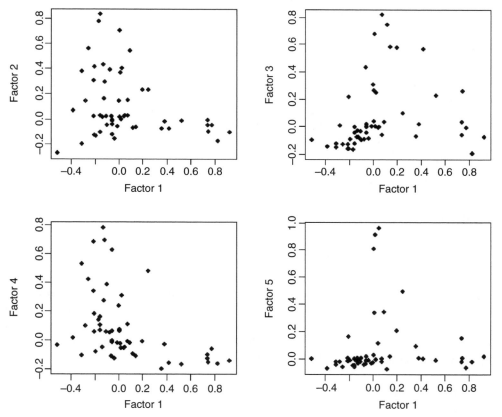

What factanal does would conventionally be described as exploratory, not confirmatory, factor analysis. For the latter, try the sem package:

```
install.packages("sem")
library(sem)
?sem
```

Cluster Analysis

Cluster analysis is a set of techniques that look for groups (clusters) in the data. Objects belonging to the same group resemble each other. Objects belonging to different groups are dissimilar. Sounds simple, doesn't it? The problem is that there is usually a huge amount of redundancy in the explanatory variables. It is not obvious which measurements (or combinations of measurements) will turn out to be the ones that are best for allocating individuals to groups. There are three ways of carrying out such allocation:

- partitioning into a number of clusters specified by the user, with functions such as kmeans

- hierarchical, starting with each individual as a separate entity and ending up with a single aggregation, using functions such as hclust

- divisive, starting with a single aggregate of all the individuals and splitting up clusters until all the individuals are in different groups.

Partitioning

The kmeans function operates on a dataframe in which the columns are variables and the rows are the individuals. Group membership is determined by calculating the centroid for each group. This is the multidimensional equivalent of the mean. Each individual is assigned to the group with the nearest centroid. The kmeans function fits a *user-specified* number of cluster centres, such that the within-cluster sum of squares from these centres is minimized, based on Euclidian distance. Here are data from six groups with two continuous explanatory variables (x and y):

```
kmd<-read.table("c:\\temp\\kmeansdata.txt",header=T)
attach(kmd)
names(kmd)
```

```
[1] "x" "y" "group"
```

The raw data show hints of clustering (top left) but the clustering becomes clear only after the groups have been coloured differently using col=group (top right). kmeans does an excellent job when it is told that there are six clusters (open symbols, bottom left) but, of course, there can be no overlap in assignation (as there was in the original data). When just four clusters are estimated, it is the centre cluster and the south-east cluster that disappear (open symbols, bottom right). The plots were produced like this:

```
par(mfrow=c(2,2))
plot(x,y,pch=16)
plot(x,y,col=group,pch=16)
model<-kmeans(data.frame(x,y),6)
plot(x,y,col=model[[1]])
model<-kmeans(data.frame(x,y),4)
plot(x,y,col=model[[1]])
par(mfrow=c(1,1))
```

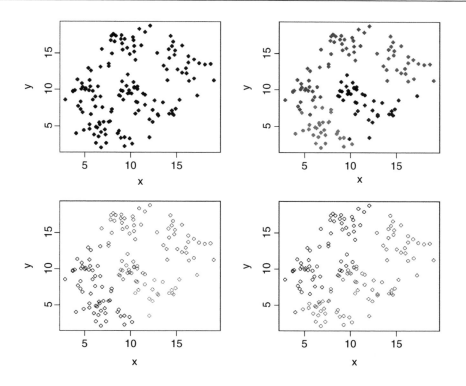

To see the rate of misclassification we can tabulate the real groups against the groups determined by kmeans:

```
model<-kmeans(data.frame(x,y),6)
table(model[[1]],group)
```

```
  group
      1   2   3   4   5   6
1    20   0   0   1   0   0
2     0  24   0   0   3   0
3     0   0   0  18   0   0
4     0   0  25   0   0   0
5     0   0   0   0   0  25
6     0   1   0   1  27   0
```

Group 1 were associated perfectly (20 out of 20) but one member of group 4 was included, group 2 had one omission (1 out of 25 was allocated to group 6) and three members of group 5 were included, group 3 was associated perfectly (out of 25), there were 2 omissions from group 4 (one to group 1 and one to group 6 out of 20), there were three omissions from group 5 (3 out of 30 were allocated to group 2) and the group was tainted by one from group 2 and one from group 4, while group 6 was matched perfectly (25 out of 25). This is impressive, given that there were several obvious overlaps in the original data.

Taxonomic use of kmeans

In the next example we have measurements of seven variables on 120 individual plants. The question is which of the variables (fruit size, bract length, internode length, petal width, sepal length, petiole length or leaf width) are the most useful taxonomic characters.

```
taxa<-read.table("c:\\temp\\taxon.txt",header=T)
attach(taxa)
names(taxa)
```

```
[1] "Petals"  "Internode"  "Sepal"  "Bract"  "Petiole"  "Leaf"
[7] "Fruit"
```

A simple and sensible way to start is by looking at the dataframe as a whole, using pairs to plot every variable against every other:

```
pairs(taxa)
```

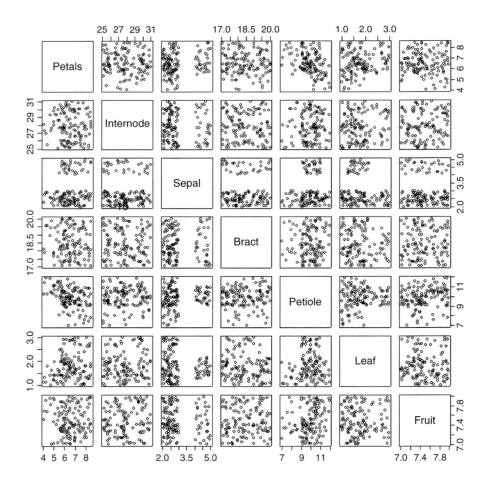

There appears to be excellent data separation on sepal length, and reasonable separation on petiole length and leaf width, but nothing obvious for the other variables.

These data actually come from four taxa (labelled I–IV), so in this contrived case we *know* that there are four groups. In reality, of course, we would not know this, and finding out the number of groups would be one of the central aims of the study. We begin, therefore, by seeing how well **kmeans** allocates individuals to four groups:

kmeans(taxa,4)

```
$cluster
  [1]  4 4 4 2 2 2 4 4 4 2 4 2 4 4 4 4 4 2 4 4 1 4 2 4 4 1 4 4 1 4
       3 3 1 1 4 3 3
 [38]  2 3 3 1 1 3 3 2 1 3 3 4 3 3 3 2 3 1 3 1 1 1 3 1 2 2 2 1 1 2
       2 2 1 1 2 1 2
 [75]  4 2 2 1 1 1 3 2 2 3 1 1 1 2 4 1 3 1 3 1 3 3 1 4 3 3 1 3 2 3
       1 3 3 3 3 3 3
[112]  1 1 3 3 3 1 3 1 1
```

```
$centers
     Petals  Internode    Sepal     Bract   Petiole     Leaf     Fruit
1  6.710503   29.95147  3.058517  18.29739   9.671325  2.022993  7.518510
2  6.544988   26.13485  2.523379  18.53230  10.297043  2.156161  7.482513
3  7.122162   26.69160  3.580120  18.37431   9.174025  1.570292  7.418814
4  5.421399   28.21695  2.548485  18.65893  10.645695  1.468021  7.540113
```

```
$withinss
[1]  132.59054  78.67365  186.29369  67.54199
```

```
$size
[1]  35  24  36  25
```

Because we *know* what the answer ought to be (the data are arranged so that the first 30 numbers are taxon I, the next 30 are taxon II, and so on) we can see that **kmeans** has made lots of mistakes. The output of the clustering vector should really look like this:

```
  [1]  1 1 1 1 1 1 1 1 1 1 1 1 1 1 1 1 1 1 1 1 1 1 1 1 1
 [26]  1 1 1 1 1 2 2 2 2 2 2 2 2 2 2 2 2 2 2 2 2 2 2 2 2
 [51]  2 2 2 2 2 2 2 2 2 2 3 3 3 3 3 3 3 3 3 3 3 3 3 3 3
 [76]  3 3 3 3 3 3 3 3 3 3 3 3 3 3 3 4 4 4 4 4 4 4 4 4 4
[101]  4 4 4 4 4 4 4 4 4 4 4 4 4 4 4 4 4 4 4 4
```

Let's try three groups:

kmeans(taxa,3)

```
$cluster
  [1]  3 3 3 3 1 1 3 3 3 3 3 3 3 3 3 3 3 3 3 3 3 2 3 3 3 3 2 3 3 3 2 3 1 1 2 2 3 1 1
 [38]  1 1 1 2 2 1 2 1 2 1 2 1 1 3 1 1 1 1 1 1 2 1 2 2 2 1 2 1 1 1 1 2 2 1 1 1 2 2 1 2 1
 [75]  3 1 1 2 2 2 1 3 1 1 2 2 2 1 3 2 1 2 1 2 1 2 2 3 1 1 2 1 1 1 1 2 1 1 1 1 1 1
[112]  2 2 1 1 1 2 1 2 2
```

```
$centers
     Petals  Internode    Sepal     Bract   Petiole     Leaf     Fruit
1  6.974714   26.42633  3.208105  18.30580   9.460762  1.804778  7.427979
2  6.784111   29.86369  3.088054  18.37981   9.642643  2.006003  7.532579
3  5.489770   27.83010  2.552457  18.74768  10.751460  1.539130  7.527354
```

```
$withinss
```

```
[1] 249.1879 152.1044 105.0409
```

```
$size
```

```
[1] 52 37 31
```

Not very impressive at all. Of course, the computer was doing its classification blind. But we know the four islands from which the plants were collected. Can we write a key that ascribes taxa to islands better than the blind scheme adopted by kmeans? The answer is definitely yes. When we used a classification tree model on the same data, it discovered a faultless key on its own (see p. 695).

Hierarchical cluster analysis

The idea behind hierarchical cluster analysis is to show which of a (potentially large) set of samples are most similar to one another, and to group these similar samples in the same limb of a tree. Groups of samples that are distinctly different are placed in other limbs. The trick is in defining what we mean by 'most similar'. Each of the samples can be thought of a sitting in an m-dimensional space, defined by the m variables (columns) in the dataframe. We define similarity on the basis of the **distance** between two samples in this m-dimensional space. Several different distance measures could be used, but the default is Euclidean distance (for the other options, see ?dist), and this is used to work out the distance from every sample to every other sample. This quantitative dissimilarity structure of the data is stored in a matrix produced by the dist function. Initially, each sample is assigned to its own cluster, and then the hclust algorithm proceeds iteratively, at each stage joining the two most similar clusters, continuing until there is just a single cluster (see ?hclust for details).

The following data (introduced on p. 732) show the distribution of 54 plant species over 89 plots receiving different experimental treatments. The aim is to see which plots are most similar in their botanical composition, and whether there are reasonably homogeneous groups of plots that might represent distinct plant communities.

```
pgdata<-read.table("c:\\temp\\pgfull.txt",header=T)
attach(pgdata)
labels<-paste(plot,letters[lime],sep="")
```

The first step is to turn the matrix of measurements on individuals into a dissimilarity matrix. In the dataframe, the columns are variables and the rows are the individuals, and we need to calculate the 'distances' between each row in the dataframe and every other using dist(pgdata[,1:54]). These distances are then used to carry out hierarchical cluster analysis using the hclust function:

```
hpg<-hclust(dist(pgdata[,1:54]))
```

We can plot the object called hpg, and we specify that the leaves of the hierarchy are labelled by their plot numbers (pasted together from the plot number and lime treatment):

```
plot(hpg,labels=labels,main="")
```

If you view this object in full-screen mode within R you will be able to read all the plot labels, and to work out the groupings. It turns out that the groupings have very natural scientific interpretations. The highest break, for instance, separates the two plots dominated by *Holcus lanatus* (11.1d and 11.2d) from the other 87 plots. The plots on the right-hand

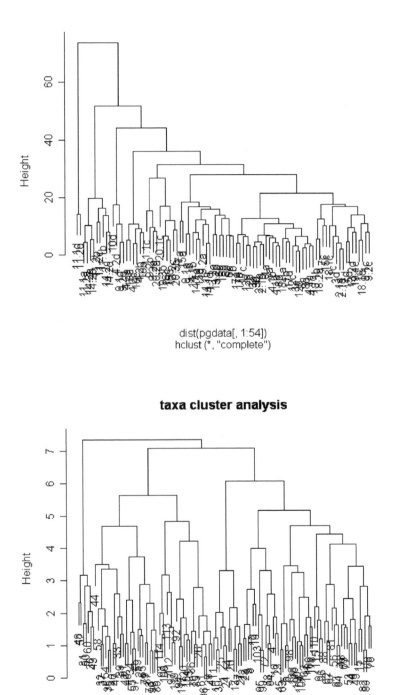

dist(pgdata[, 1:54])
hclust (*, "complete")

taxa cluster analysis

dist(taxa)
hclust (*, "complete")

side all have soils that exhibit phosphorus deficiency. The leftmost groups are all from plots receiving high rates of nitrogen and phosphorus input. More subtly, plots 12a and 3a are supposed to be the same (they are replicates of the no-fertilizer, high-lime treatment), yet they are separated at a break at height approximately 15. And so on.

Let's try hierarchical clustering on the taxonomic data (p. 740).

```
plot(hclust(dist(taxa)),main="taxa cluster analysis")
```

Because in this artificial example we know that the first 30 rows in the dataframe come from group 1, rows 31–60 from group 2, rows 61–90 from group 3 and rows 91–120 from group 4, we can see than the grouping produced by hclust is pretty woeful. Most of the rows in the leftmost major split are from group 2, but the rightmost split contains members from groups 1, 4 and 3. Neither kmeans nor hclust is up to the job in this case. When we *know* the group identities, then it is easy to use tree models to devise the optimal means of distinguishing and classifying individual cases (see p. 695).

So there we have it. When we *know* the identity of the species, then tree models are wonderfully efficient at constructing keys to distinguish between the individuals, and at allocating them to the relevant categories. When we do *not* know the identities of the individuals, then the statistical task is much more severe, and inevitably ends up being much more error-prone. The best that cluster analysis could achieve through kmeans was with five groups (one too many, as we know, having constructed the data) and in realistic cases we have no way of knowing whether five groups was too many, too few or just right. Multivariate clustering without a response variable is fundamentally difficult and equivocal.

Discriminant analysis

In this case, you know the identity of each individual (unlike cluster analysis) and you want to know how the explanatory variables contribute to the correct classification of individuals. The method works by uncovering relationships among the groups' covariance matrices to discriminate between groups. With k groups you will need $k - 1$ discriminators. The functions you will need for discriminant analysis are available in the MASS library. Returning to the taxon dataframe (see p. 740), we will illustrate the use of lda to carry out a linear discriminant analysis. For further relevant information, see ?predict.lda and ?qda in the MASS library.

```
library(MASS)
model<-lda(Taxon~.,taxa)
plot(model)
```

The linear discriminators LD1 and LD2 clearly separate taxon IV without error, but this is easy because there is no overlap in sepal length between this taxon and the others. LD2 and LD3 are quite good at finding taxon II (upper right), and LC1 and LC3 are quite good at getting taxon I (bottom right). Taxon III would be what was left over. Here is the printed model:

```
model
```

```
Call:
lda(Taxon~., data = taxa)

Prior probabilities of groups:
    I     II    III    IV
 0.25   0.25   0.25   0.25
```

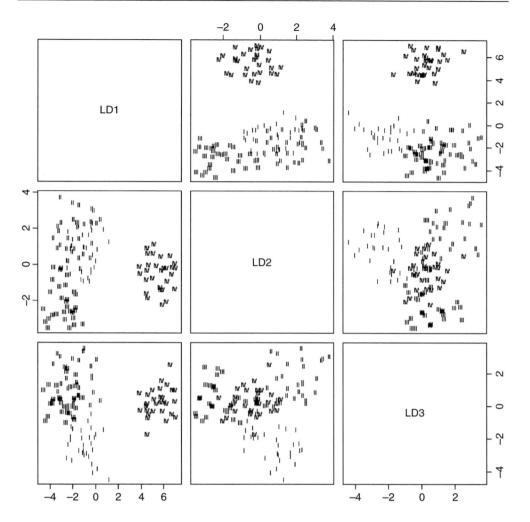

Group means:

	Petals	Internode	Sepal	Bract	Petiole	Leaf	Fruit
I	5.476128	27.91886	2.537955	18.60268	10.864184	1.508029	7.574642
II	7.035078	27.69834	2.490336	18.47557	8.541085	1.450260	7.418702
III	6.849666	27.99308	2.446003	18.26330	9.866983	2.588555	7.482349
IV	6.768464	27.78503	4.532560	18.42953	10.128838	1.645945	7.467917

Coefficients of linear discriminants:

	LD1	LD2	LD3
Petals	−0.01891137	0.034749952	0.559080267
Internode	0.03374178	0.009670875	0.008808043
Sepal	3.45605170	−0.500418135	0.401274694
Bract	0.07557480	0.068774714	−0.024930728
Petiole	0.25041949	−0.343892260	−1.249519047
Leaf	−1.13036429	−3.008335468	0.647932763
Fruit	0.18285691	−0.208370808	−0.269924935

```
Proportion of trace:
   LD1      LD2      LD3
0.7268   0.1419   0.1313
```

So you would base your key on sepal first (3.45) then leaf (−3.008) then petiole (−1.249). Compare this with the key uncovered by the tree model on p. 695. Here are the predictions from the linear discriminant analysis:

predict(model)

```
$class
  [1] I    I    I    I    I    I    I    I    I    I    I    I    I    I    I    I    I    I
 [19] I    I    III  I    I    I    I    I    I    I    I    I    II   II   II   II   II   II
 [37] II   II   II   II   II   II   II   II   II   II   II   II   II   II   II   II   II   II
 [55] II   II   II   II   II   II   III  III  III  III  III  III  III  III  III  III  III  III
 [73] III  III  III  III  III  III  III  III  III  III  III  III  III  III  III  III  III  III
 [91] IV   IV   IV   IV   IV   IV   IV   IV   IV   IV   IV   IV   IV   IV   IV   IV   IV   IV
[109] IV   IV   IV   IV   IV   IV   IV   IV   IV   IV   IV   IV

Levels:   I    II   III   IV
```

One of the members of taxon I is misallocated to taxon III (case 21), but otherwise the discrimination is perfect. You can train the model with a random subset of the data (say, half of it):

train<-sort(sample(1:120,60))
table(Taxon[train])

```
 I    II   III   IV
13   18    16   13
```

This set has only 13 members of taxon I and IV but reasonable representation of the other two taxa.

model2<-lda(Taxon~.,taxa,subset=train)
predict(model2)

```
$class
  [1] I    I    I    I    I    I    I    I    I    I    I    I    I    II   II   II   II   II   II
 [20] II   II   II   II   II   II   II   II   II   II   II   II   III  III  III  III  III  III  III
 [39] III  III  III  III  III  III  III  III  III  IV   IV   IV   IV   IV   IV   IV   IV   IV   IV
 [58] IV   IV   IV

Levels: I II III IV
```

This is still very good: the first 13 should be I, the next 18 II, and so on. The discrimination is perfect in this randomization. You can use the model based on the training data to predict the unused data:

unused<-taxa[-train,]
predict(model,unused)$class

```
  [1] I    I    I    I    I    I    I    I    I    I    I    III  I    I    I    I    I    II   II
 [20] II   II   II   II   II   II   II   II   II   II   III  III  III  III  III  III  III  III  III
 [39] III  III  III  III  III  IV   IV   IV   IV   IV   IV   IV   IV   IV   IV   IV   IV   IV   IV
 [58] IV   IV   IV

Levels: I II III IV
```

table(unused$Taxon)

```
 I   II  III   IV
17   12   14   17
```

As you can see, one of the first 17 that should have been Taxon I was misclassified as Taxon III, but all the other predictions were spot on.

Neural Networks

These are computationally intensive methods for finding pattern in data sets that are so large, and contain so many explanatory variables, that standard methods such as multiple regression are impractical (they would simply take too long to plough through). The key feature of neural network models is that they contain a *hidden layer*: each node in the hidden layer receives information from each of many inputs, sums the inputs, adds a constant (the bias) then transforms the result using a fixed function. Neural networks can operate like multiple regressions when the outputs are continuous variables, or like classifications when the outputs are categorical. They are described in detail by Ripley (1996). Facilities for analysing neural networks in the **MASS** library.

24

Spatial Statistics

There are three kinds of problems that you might tackle with spatial statistics:

- point processes (locations and spatial patterns of individuals);
- maps of a continuous response variable (kriging);
- spatially explicit responses affected by the identity, size and proximity of neighbours.

Point Processes

There are three broad classes of spatial pattern on a continuum from complete regularity (evenly spaced hexagons where every individual is the same distance from its nearest neighbour) to complete aggregation (all the individuals clustered into a single clump): we call these regular, random and aggregated patterns and they look like this:

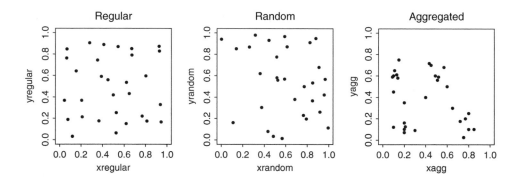

In their simplest form, the data consist of sets of x and y coordinates within some sampling frame such as a square or a circle in which the individuals have been mapped. The first question is often whether there is any evidence to allow rejection of the null hypothesis of **complete spatial randomness** (CSR). In a **random** pattern the distribution of each individual is completely independent of the distribution of every other. Individuals neither inhibit nor promote one another. In a **regular** pattern individuals are more spaced out than in a random one, presumably because of some mechanism (such as competition) that eliminates individuals that are too close together. In an **aggregated** pattern, individual are more clumped than in a random one, presumably because of some process such as reproduction with limited dispersal, or because of underlying spatial heterogeneity (e.g. good patches and bad patches).

Counts of individuals within sample areas (quadrats) can be analysed by comparing the frequency distribution of counts with a Poisson distribution with the same mean. Aggregated spatial patterns (in which the variance is greater than the mean) are often well described by a negative binomial distribution with aggregation parameter k (see p. 76). The main problem with quadrat-based counts is that they are highly **scale-dependent**. The same spatial pattern could appear to be regular when analysed with small quadrats, aggregated when analysed with medium-sized quadrats, yet random when analysed with large quadrats.

Distance measures are of two broad types: measures from individuals to their nearest neighbours, and measures from random points to the nearest individual. Recall that the nearest individual to a random point is *not* a randomly selected individual: this protocol favours selection of isolated individuals and individuals on the edges of clumps.

In other circumstances, you might be willing to take the existence of patchiness for granted, and to carry out a more sophisticated analysis of the spatial attributes of the patches themselves, their mean size and the variance in size, spatial variation in the spacing of patches of different sizes, and so on.

Nearest Neighbours

Suppose that we have been set the problem of drawing lines to join the nearest neighbour pairs of any given set of points (x, y) that are mapped in two dimensions. There are three steps to the computing: we need to

- compute the distance to every neighbour;
- identify the smallest neighbour distance for each individual;
- use these minimal distances to identify all the nearest neighbours.

We start by generating a random spatial distribution of 100 individuals by simulating their x and y coordinates from a uniform probability distribution:

```
x<-runif(100)
y<-runif(100)
```

The graphics parameter pty="s" makes the plotting area square, as we would want for a map like this:

```
par(pty="s")
plot(x,y,pch=16)
```

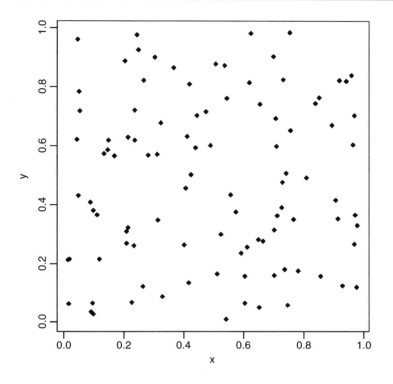

Computing the distances is straightforward: for each individual we use Pythagoras to calculate the distance to every other plant. The distance between two points with coordinates (x_1, y_1) and (x_2, y_2) is d:

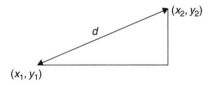

The square on the hypotenuse (d^2) is the sum of the squares on the two adjacent sides: $(x_2 - x_1)^2 + (y_2 - y_1)^2$ so the distance d is given by

$$d = \sqrt{(y_2 - y_1)^2 + (x_2 - x_1)^2}.$$

We write a function for this as follows:

```
distance<-function(x1, y1, x2, y2) sqrt((x2 − x1)^2 + (y2 − y1)^2)
```

Now we loop through each individual i and calculate a vector of distances, d, from every other individual. The nearest neighbour distance is the minimum value of d, and the identity of the nearest neighbour, nn, is found using the which function, which(d==min(d[-i])), which gives the subscript of the minimum value of d (the [-i] is necessary to exclude the distance 0 which results from the ith individual's distance from itself). Here is the complete code to compute nearest neighbour distances, *r*, and identities, *nn*, for all 100 individuals on the map:

```
r<-numeric(100)
nn<-numeric(100)
d<-numeric(100)
for (i in 1:100) {
for (k in 1:100) d[k]<-distance(x[i],y[i],x[k],y[k])
r[i]<-min(d[-i])
nn[i]<-which(d==min(d[-i]))
}
```

Now we can fulfil the brief, and draw lines to join each individual to its nearest neighbour, like this:

```
for (i in 1:100) lines(c(x[i],x[nn[i]]),c(y[i],y[nn[i]]))
```

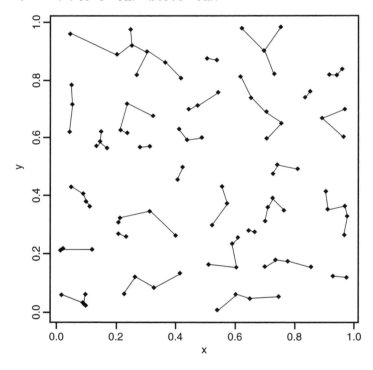

Note that when two points are very close together, and each is the nearest neighbour of the other, it can look as if a single point is not joined to any neighbours.

The next task is to work out how many of the individuals are closer to the edge of the area than they are to their nearest neighbour. Because the bottom and left margins are at $y = 0$ and $x = 0$ respectively, the y coordinate of any point gives the distance from the bottom edge of the area and the x coordinate gives the distance from the left-hand margin. We need only work out the distance of each individual from the top and right-hand margins of the area:

```
topd <- 1-y
rightd <- 1-x
```

Now we use the parallel minimum function pmin to work out the distance to the nearest margin for each of the 100 individuals:

```
edge<-pmin(x,y,topd,rightd)
```

Finally, we count the number of cases where the distance to the edge is less than the distance to the nearest neighbour:

```
sum(edge<r)
```

```
[1] 18
```

and identify these points on the map by circling them in red:

```
id<-which(edge<r)
points(x[id],y[id],col="red",cex=1.5)
```

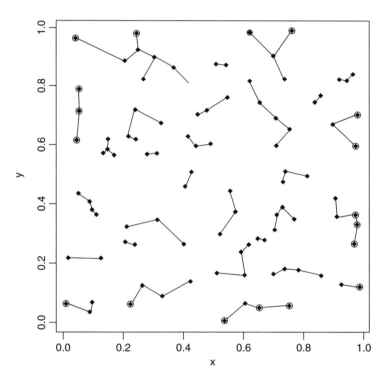

Edge effects are potentially very important in spatial point processes, especially when there are few individuals or the mapped area is long and thin (rather than square or circular). Excluding the individuals that are closer to the edge reduces the mean nearest neighbour distance:

```
mean(r)
```

```
[1]  0.05404338
```

```
mean(r[-id])
```

```
[1]  0.04884806
```

Tests for Spatial Randomness

Clark and Evans (1954) give a very simple test of spatial randomness. Making the strong assumption that you know the population density of the individuals, ρ (generally you do not know this, and would need to estimate it independently), then the expected mean distance to the nearest neighbour is

$$E(r) = \frac{\sqrt{\rho}}{2}.$$

In our example we have 100 individuals in a unit square, so $\rho = 0.01$ and $E(r) = 0.05$. The actual mean nearest neighbour distance was

mean(r)

[1] 0.05404338

which is very close to expectation: this clearly is a random distribution of individuals (as we constructed it to be). An index of randomness is given by the ratio $\bar{r}/E(r) = 2\bar{r}\sqrt{\rho}$. This takes the value 1 for random patterns, more than 1 for regular (spaced-out) patterns, and less than 1 for aggregated patterns.

One problem with such **first-order** estimates of spatial pattern (including measures such as the variance–mean ratio) is that they can give no feel for the way that spatial distribution changes *within* an area.

Ripley's K

The **second-order** properties of a spatial point process describe the way that spatial interactions *change* through space. These are computationally intensive measures that take a range of distances within the area, calculate a pattern measure, then plot a graph of the function against distance, to show how the pattern measure changes with scale. The most widely used second order measure is the K function, which is defined as

$$K(d) = \frac{1}{\lambda} E[\text{number of points} \le \text{distance } d \text{ of an arbitrary point}],$$

where λ is the mean number of points per unit area (the **intensity** of the pattern). If there is *no spatial dependence*, then the expected number of points that are within a distance d of an arbitrary point is πd^2 times the mean density. So, if the mean density is 2 points per square metre ($\lambda = 2$), then the expected number of points within a 5 m radius is $\lambda \pi d^2 = 2 \times \pi \times 5^2 = 50\pi = 157.1$. If there is clustering, then we expect an excess of points at short distances (i.e. $K(d) > \pi d^2$ for small d). Likewise, for a regularly spaced pattern, we expect an excess of long distances, and hence few individuals at short distances (i.e. $K(d) < \pi d^2$). Ripley's K (published in 1976) is calculated as follows:

$$\hat{K}(d) = \frac{1}{n^2} |A| \sum \sum_{i \ne j} \frac{I_d(d_{ij})}{w_{ij}}.$$

Here n is the number of points in region A with area $|A|$, d_{ij} are the distances between points (the distance between the ith and jth points, to be precise). To account for edge effects, the model includes the term w_{ij} which is the fraction of the area, centred on i and passing through j, that lies within the area A (all the w_{ij} are 1 for points that lie well away from the edges of the area). $I_d(d_{ij})$ is an indicator function to show which points are to be counted as neighbours at this value of d: it takes the value 1 if $d_{ij} \leq d$ and zero otherwise (i.e. points with $d_{ij} > d$ are omitted from the summation). The pattern measure is obtained by plotting $\hat{K}(d)$ against d. This is then compared with the curve that would be observed under complete spatial randomness (namely, a plot of πd^2 against d). When clustering occurs, $K(d) > \pi d^2$ and the curve lies *above* the CSR curve, while regular patterns produce a curve *below* the CSR curve.

You can see *why* you need the edge correction, from this simple simulation experiment. For individual number 1, with coordinates (x_1, y_1), calculate the distances to all the other individuals, using the function dist that we wrote earlier (p. 751):

```
distances<-numeric(100)
for(i in 1:100) distances[i]<-distance(x[1],y[1],x[i],y[i])
```

Now find out how many other individuals are within a distance d of this individual. Take as an example $d = 0.1$.

```
sum(distances<0.1) - 1
```

```
[1]  1
```

There was one other individual within a distance $d = 0.1$ of the first individual (the distance 0 from itself is included in the sum so we have to correct for this by subtracting 1). The next step is to generalize the procedure from this one individual to all the individuals. We make a two-dimensional matrix called dd to contain all the distances from every individual (rows) to every other individual (columns):

```
dd<-numeric(10000)
dd<-matrix(dd,nrow=100)
```

The matrix of distances is computed within loops for both individual (j) and neighbour (i) like this:

```
for (j in 1:100) {for(i in 1:100) dd[j,i]<-distance(x[j],y[j],x[i],y[i])}
```

Alternatively, you could use sapply with an anonymous function like this:

```
dd<-sapply(1:100,function (i,j=1:100) distance(x[j],y[j],x[i],y[i]))
```

We should check that the number of individuals within 0.1 of individual 1 is still 1 under this new notation. Note the use of blank subscripts [1,] to mean 'all the individuals in row number 1':

```
sum(dd[1,]<0.1)-1
```

```
[1]  1
```

So that's OK. We want to calculate the sum of this quantity over all individuals, not just individual number 1.

```
sum(dd<0.1)-100
```

```
[1]   252
```

This means that there are 252 cases in which other individuals are counted within $d = 0.1$ of focal individuals. Next, create a vector containing a range of different distances, d, over which we want to calculate $K(d)$ by counting the number of individuals within distance d, summed over all individuals:

```
d<-seq(0.01,1,0.01)
```

For each of these distances we need to work out the total number of neighbours of all individuals. So, in place of 0.1 (in the sum, above), we need to put each of the d values in turn. The count of individuals is going to be a vector of length 100 (one for each d):

```
count<-numeric(100)
```

Calculate the count for each distance d:

```
for (i in 1:100) count[i]<-sum(dd<d[i])-100
```

The expected count increases with d as πd^2 so we scale our count by dividing by the square of the total number of individuals $n^2 = 100^2 = 10000$.

```
K<-count/10000
```

Finally, plot a graph of K against d:

```
plot(d,K,type="l")
```

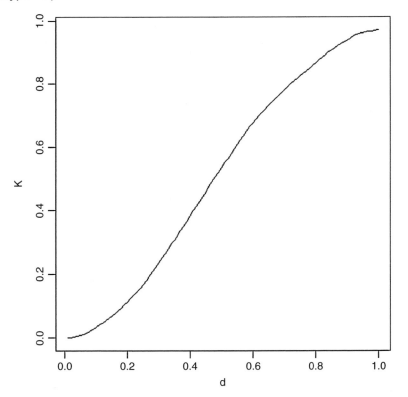

Not surprisingly, when we sample the whole area ($d = 1$), we count all of the individuals in every neighbourhood ($K = 1$). For CSR the graph should follow πd^2 so we add a line to show this

```
lines(d,pi*d^2)
```

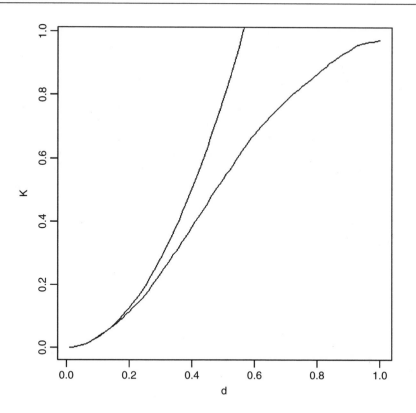

Up to about $d = 0.2$ the agreement between the two lines is reasonably good, but for longer distances our algorithm is counting far too few neighbours. This is because much of the area scanned around marginal individuals is invisible, since it lies outside the study area (there may well be individuals out there, but we shall never know). This demonstrates that the *edge correction* is a fundamental part of Ripley's K.

Fortunately, we don't have to write a function to work out a corrected value for K; it is available as **Kfn** in the built-in **spatial** library. Here we use it to analyse the pattern of trees in the dataframe called pines. The library function **ppinit** reads the data from a library file called pines.dat which is stored in directory spatial/ppdata. It then converts this into a list with names $x, $y and $area. The first row of the file contains the number of trees (71) the second row has the name of the data set (PINES), the third row has the four boundaries of the region plus the scaling factor (0, 96, 0, 100, 10 so that the coordinates of lower and upper x are computed as 0 and 9.6, and the coordinates of lower and upper y are 0 and 10). The remaining rows of the data file contain x and y coordinates for each individual tree, and these are converted into a list of x values and a separate list of y values. You need to know these details about the structure of the data files in order to use these library functions with your own data (see p. 102).

```
library(spatial)
pines<-ppinit("pines.dat")
```

First, set up the plotting area with two square frames:

```
par(mfrow=c(1,2),pty="s")
```

On the left, make a map using the x and y locations of the trees, and on the right make a plot of $L(t)$ (the pattern measure) against distance:

```
plot(pines,pch=16)
plot(Kfn(pines,5),type="s",xlab="distance",ylab="L(t)")
```

Recall that if there was CSR, then the expected value of K would be πd^2; to linearize this, we could divide by π and then take the square root. This is the measure used in the function Kfn, where it is called $L(t) = \sqrt{K(t)/\pi}$. Now for the simulated upper and lower bounds: the first argument in Kenvl (calculating envelopes for K) is the maximum distance (half the length of one side), the second is the number of simulations (100 is usually sufficient), and the third is the number of individuals within the mapped area (71 pine trees in this case).

```
lims<-Kenvl(5,100,Psim(71))
lines(lims$x,lims$lower,lty=2,col="green")
lines(lims$x,lims$upper,lty=2,col="green")
```

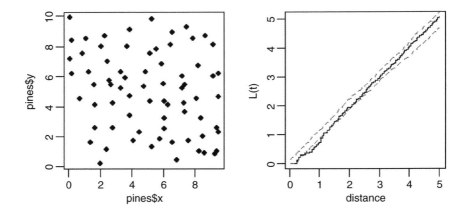

There is a suggestion that at relatively small distances (around 1 or so), the trees are rather regularly distributed (more spaced out than random), because the plot of $L(t)$ against distance falls below the lower envelope of the CSR line (it should lie between the two limits for its whole length if there was CSR). The mechanism underlying this spatial regularity (e.g. non-random recruitment or mortality, competition between growing trees, or underlying non-randomness in the substrate) would need to be investigated in detail. With an aggregated pattern, the line would fall above the upper envelope (see p. 766).

Quadrat-based methods

Another approach to testing for spatial randomness is to count the number of individuals in quadrats of different sizes. Here, the quadrats have an area of 0.01, so the expected number per quadrat is 1.

```
x<-runif(100)
y<-runif(100)
plot(x,y,pch=16)
grid(10,10,lty=1)
```

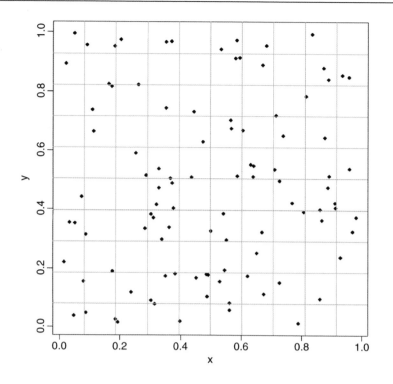

The trick here is to use cut to convert the *x* and *y* coordinates of the map into bin numbers (between 1 and 10 for the quadrat size we have drawn here). To achieve this, the break points are generated by the sequence (0,1,0.1)

```
xt<-cut(x,seq(0,1,0.1))
yt<-cut(y,seq(0,1,0.1))
```

which creates vectors of integer subscripts between 1 and 10 for *xt* and *yt*. Now all we need to do is use table to count up the number of individuals in every cell (i.e. in every combination of *xt* and *yt*):

```
count<-as.vector (table(xt,yt))
table(count)
```

```
count
 0   1   2   3   4
36  39  17   5   3
```

This shows that 36 cells are empty, 3 cells had four individuals, but no cells contained five or more individuals. Now we need to see what this distribution would look like under a particular null hypothesis. For a Poisson process (see p. 250), for example,

$$P(x) = \frac{e^{-\lambda}\lambda^x}{x!}.$$

Note that the mean depends upon the quadrat size we have chosen. With 100 individuals in the whole area, the expected number in any one of our 100 cells, λ, is 1.0. The expected frequencies of counts between 0 and 5 are therefore given by

```
(expected<-100*exp(-1)/sapply(0:5,factorial))
```

```
[1]    36.7879441  36.7879441  18.3939721  6.1313240  1.5328310
[6]    0.3065662
```

The fit between observed and expected is almost perfect (as we should expect, of course, having generated the random pattern ourselves). A test of the significance of the difference between an observed and expected frequency distribution is shown on p. 762.

Aggregated pattern and quadrat count data

Here is an example of a quadrat-based analysis of an aggregated spatial pattern:

```
trees<-read.table("c:\\temp\\trees.txt",header=T)
attach(trees)
names(trees)
```

```
[1]   "x"  "y"
```

```
plot(x,y,pch=16)
grid(10,10,lty=1)
```

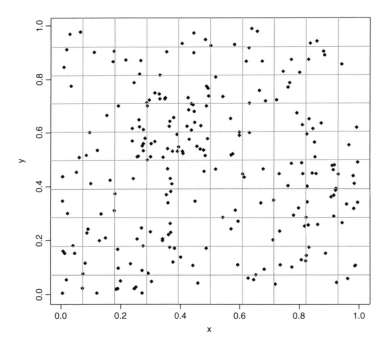

Now there are quadrats with as many as eight individuals, and despite the fact that the mean number is greater than one individual per square, there are still 14 completely empty squares. We cut up the data and tabulate the counts:

```
xt<-cut(x,seq(0,1,0.1))
yt<-cut(y,seq(0,1,0.1))
count<-as.vector(table(xt,yt))
table(count)
```

```
count
 0   1   2   3   4   5   6   7   8
14  26  19  21   6   7   1   3   3
```

The expected frequencies under the null hypothesis of a random pattern depend only on the mean number per cell

mean(count)

[1] 2.37

and as a preliminary estimate of the departure from randomness we calculate the variance–mean ratio (recall that with the Poisson distribution the variance is equal to the mean):

var(count)/mean(count)

[1] 1.599582

These data are distinctly aggregated (variance/mean $\gg$ 1), so we might compare the counts with a negative binomial distribution (p. 251). The expected frequencies are estimated by multiplying our total number of squares (100) by the probability densities from a negative binomial distribution generated by the function dnbinom. This has three arguments: the counts for which we want the probabilities (0:8), the mean (mu=2.37) and the aggregation parameter $k = $ mu^2/(var-mu) = size=3.95:

expected<-dnbinom(0:8, size=3.95, mu=2.37)*100

The observed (black) and expected frequencies (white) are to be shown as paired bars – you need to click when the cursor is in the right position on the graph, then the locator(1) function will return the x and y coordinates of the bottom left-hand corner of the legend:

```
ht<-numeric(18)
observed<-table(count)
ht[seq(1,17,2)]<-observed
ht[seq(2,18,2)]<-expected
names<-rep("",18)
names[seq(1,17,4)]<-as.character(seq(0,8,2))
barplot(ht,col=c(1,0),names=names)
legend(locator(1),c("observed","expected"),fill=c(1,0))
```

The fit is reasonably good, but we need a quantitative estimate of the lack of agreement between the observed and expected distributions. Pearson's chi-squared is perhaps the simplest (p. 303). We need to trim the observed and expected vectors so that none of the expected frequencies is less than 4:

```
expected[7]<-sum(expected[7:9])
expected<-expected[-c(8:9)]
observed[7]<-sum(observed[7:9])
observed<-observed[-c(8:9)]
```

Now calculate Pearson's chi-squared as $\sum [(O - E)^2/E]$:

sum((observed-expected)^2/expected)

[1] 4.442496

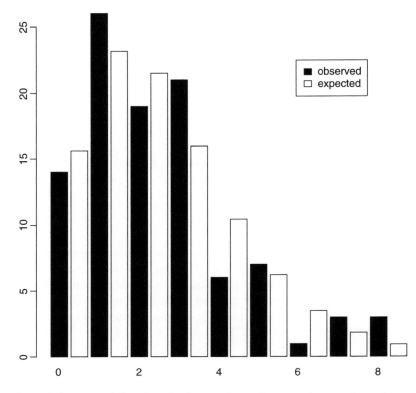

The number of degrees of freedom is the number of comparisons minus the number of parameters estimated minus 1 (i.e. $7 - 2 - 1 = 4$ d.f.). So the probability of obtaining a chi-squared value of this size (4.44) or greater is

1-pchisq(4.44,4)

```
[1]  0.3497213
```

We conclude that the negative binomial is a reasonable description of these quadrat data (because p is much greater than 0.05).

Libraries for spatial statistics

In addition to the built-in library **spatial** there are two substantial contributed packages for analysing spatial data. The **spatstat** library is what you need for the statistical analysis of spatial point patterns (below left), while the **spdep** library is good for the spatial analysis of data from mapped regions (right).

With point patterns the things you will want to do include

- creation, manipulation and plotting of point patterns,

- exploratory data analysis,

- simulation of point process models,

- parametric model fitting,

- hypothesis tests and diagnostics,

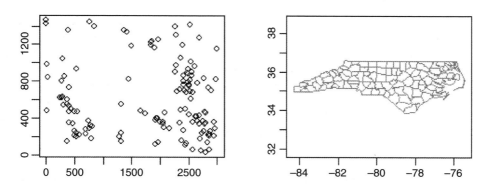

whereas with maps you might

- compute basic spatial statistics such as Moran's I and Geary's C,
- create neighbour objects of class nb,
- create weights list objects of class lw,
- work out neighbour relations from polygons (outlines of regions),
- colour mapped regions on the basis of derived statistics.

You need to take time to master the different formats of the data objects used by the two libraries. You will waste a lot of time if you try to use the functions in these libraries with your own, unreconstructed data files.

Here is the code for installing and reading about spatstat and spdep:

```
install.packages("spatstat")
library(help=spatstat)
library(spatstat)
demo(spatstat)
install.packages("spdep")
library(help=spdep)
library(spdep)
```

The spatstat library

You need to use the function ppp to convert your coordinate data into an object of class ppp representing a point pattern data set in the two-dimensional plane. Our next dataframe contains information on the locations and sizes of 3359 ragwort plants in a 30 m × 15 m map:

```
data<-read.table("c:\\temp\\ragwortmap2006.txt",header=T)
attach(data)
names(data)
```

```
 [1]  "ROW"           "COLUMN"   "X"         "Y"        "rosette"
 [6]  "regenerating"  "stems"    "diameter"  "xcoord"   "ycoord"
[11]  "type"
```

The plants are classified as belonging to one of four types: skeletons are dead stems of plants that flowered the year before, regrowth are skeletons that have live shoots at the base), seedlings are small plants (a few weeks old) and rosettes are larger plants (one or more years old) destined to flower this year. The function ppp requires separate vectors for the *x* and *y* coordinates: these are in our file under the names xcoord and ycoord. The third and fourth arguments to ppp are the boundary coordinates for *x* and *y* respectively (in this example c(0,3000) for *x* and c(0,1500) for *y*). The final argument to ppp contains a vector of what are known as 'marks': these are the factor levels associated with each of the points (in this case, type is either skeleton, regrowth, seedling or rosette). You give a name to the ppp object (ragwort) and define it like this:

```
ragwort<-ppp(xcoord,ycoord,c(0,3000),c(0,1500),marks=type)
```

You can now use the object called ragwort in a host of different functions for plotting and statistical modelling within the **spatstat** library. For instance, here are maps of the point patterns for the four plant types separately:

```
plot(split(ragwort),main="")
```

regrowth

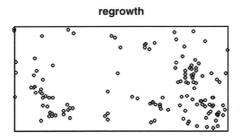

rosette

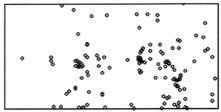

seedling

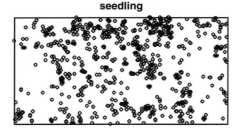

skeleton

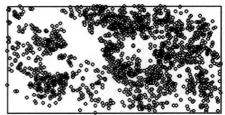

Point patterns are summarized like this:

```
summary(ragwort)
```

```
Marked planar point pattern: 3359 points
Average intensity 0.000746 points per unit area
Marks:
```

	frequency	proportion	intensity
regrowth	135	0.0402	3.00e-05
rosette	146	0.0435	3.24e-05
seedling	1100	0.3270	2.44e-04
skeleton	1980	0.5890	4.40e-04

```
Window: rectangle = [ 0 , 3000 ] × [ 0 , 1500 ]
Window area = 4500000
```

which computes the frequency and intensity for each mark ('intensity' is the mean density of points per unit area). In this case, where distances are in cem centimetres, the intensity is the mean number of plants per square centimetre (the highest intensity is skeletons, with $0.00044/cm^2$). The function quadratcount produces a useful summary of counts:

plot(quadratcount(ragwort))

quadratcount(ragwort)

60	147	261	292	76
105	101	188	155	141
64	95	151	228	147
98	129	115	177	154
62	77	88	145	103

This is the default, but you can specify either the numbers of quadrats in the x and y directions (default 5 and 5), or provide numeric vectors giving the x and y coordinates of the boundaries of the quadrats. If we want counts in 0.5 m squares, then

plot(quadratcount(ragwort,
    xbreaks=c(0,500,1000,1500,2000,2500,3000),
    ybreaks=c(0,500,1000,1500)),main="")

100	145	261	388	267	149
100	169	157	261	259	171
83	169	96	190	216	178

There are functions for producing density plots of the point pattern:

Z <- density.ppp(ragwort)
plot(Z,main="")

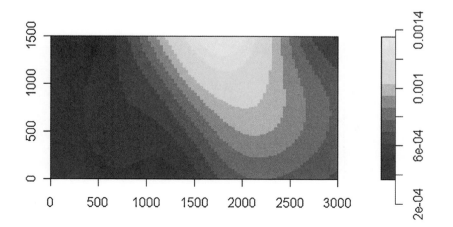

The classic graphical description of spatial point patterns is Ripley's K (see p. 757).

```
K <- Kest(ragwort)
plot(K, main = "K function")
```

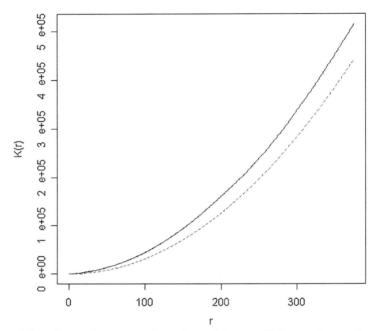

The red dotted line shows the expected number of plants within a radius r of a plant under the assumption of complete spatial randomness. The observed curve (black) lies above this line, indicating spatial aggregation at all spatial scales up to more than 300 cm.

The pair correlation function for the ragwort data looks like this:

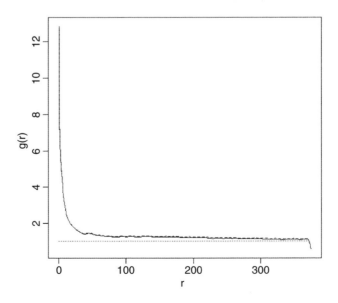

```
pc <- pcf(ragwort)
plot(pc, main = "Pair correlation function")
```

There is strong correlation between pairs of plants at small scales but much less above $r = 20$ cm. The function distmap shows the distance map around individual plants:

```
Z <- distmap(ragwort)
plot(Z,main="")
```

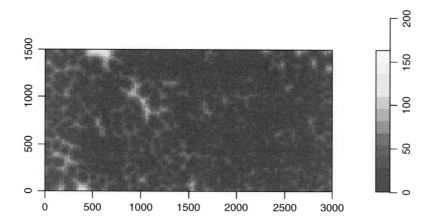

You can use **spatstat** to generate a wide range of patterns of random points, including independent uniform random points, inhomogeneous Poisson point processes, inhibition processes, and Gibbs point processes using Metropolis–Hastings (see **?spatstat** for details). Some useful functions on point-to-point distances in **spatstat** include

nndist	nearest neighbour distances
nnwhich	find nearest neighbours
pairdist	distances between all pairs of points
crossdist	distances between points in two patterns
exactdt	distance from any location to nearest data point
distmap	distance map image
density.ppp	kernel smoothed density.

There are several summary statistics for a multitype point pattern with a component **$marks** which is a factor.

Gcross,Gdot,Gmulti	multitype nearest neighbour distributions.
Kcross,Kdot, Kmulti	multitype K-functions.
Jcross,Jdot,Jmulti	multitype J-functions.
alltypes	estimates of the above for all i, j pairs

lest   multitype *I*-function

Kcross.inhom   inhomogeneous counterpart of Kcross

Kdot.inhom   inhomogeneous counterpart of Kdot.

Point process models are fitted using the ppm function like this:

```
model <- ppm(ragwort, ~marks + polynom(x, y, 2), Poisson())
plot(model)
summary(model)
```

The spdep library

The key to using this library is to understand the differences between the various formats in which the spatial data can be held:

- *x* and *y* coordinates (in a two-column matrix, with *x* in column 1 and *y* in 2);

- lists of the regions that are neighbours to each region, with (potentially) unequal numbers of neighbours in different cases (this is called a neighbour file and belongs to class nb);

- dataframes containing a region, its neighbour and the statistical weight of the association between the two regions on each row (class data.frame);

- lists containing the identities of the *k* nearest neighbours (class knn);

- a weights list object suitable for computing Moran's *I* or Geary's *C* (class lw);

- lists of polygons, defining the outlines of regions on a map (class polylist).

Unlike spatstat (above) where the *x* and *y* coordinates were in separate vectors, spdep wants the *x* and *y* coordinates in a single two-column matrix. For the ragwort data (p. 763) we need to write:

```
library(spdep)
myco<-cbind(xcoord,ycoord)
myco<-matrix(myco,ncol=2)
```

A raw list of coordinates contains no information about neighbours, but we can use the knearneigh function to convert a matrix of coordinates into an object of class knn. Here we ask for the four nearest neighbours of each plant:

```
myco. knn <- knearneigh(myco, k=4)
```

This list object has the following structure:

```
str(myco.knn)
```

```
List of 5
$ nn        : int [1:3359, 1:4] 2 1 4 3 7 4 8 7 10 9 ...
$ np        : int 3359
$ k         : num 4
$ dimension: int 2
$ x         : num [1:3359, 1:2] 27 29 20 20 78 25 89 97 253 259 ...
```

```
- attr(*, "class")= chr "knn"
- attr(*, "call")= language knearneigh(x = myco, k = 4)
```

- $nn contains 3359 lists, each a vector of length 4, containing the identities of the four points that are the nearest neighbours of each of the points from 1 to 3359.

- $np (an integer) is the number of points in the pattern.

- $k is the number of neighbours of each point.

- $dimension is 2.

- $x is the matrix of coordinates of each point (x in the first column, y in the second).

Before you can do much with a knn object you will typically want to convert it to a neighbour object (nb) using the knn2nb function like this:

myco.nb<-knn2nb(myco.knn)

You can do interesting things with nb objects. Here is a plot with each point joined to its four nearest neighbours – you specify the nb object and the matrix of coordinates:

plot(myco.nb,myco)

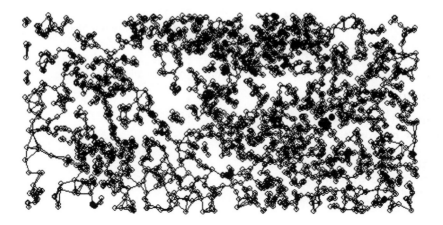

The essential concept for using the spdep package is the neighbour object (with class nb). For a given location, typically identified by the (x, y) coordinates of its centroid, the neighbour object is a list, with the elements of the list numbered from 1 to the number of locations, and each element of the list contains a vector of integers representing the identities of the locations that share a boundary with that location. The important point is that different vectors are likely to be of different lengths.

The simplest way to create an nb object is to read a text file containing one row for each neighbour relationship, using the special input function read.gwt2nb. The header row can take one of two forms. The simplest (called 'Old-style GWT') is a single integer giving the number of locations in the file. There will always be many more rows in the data file than this number, because each location will typically have several neighbours. The second form of the header row has four elements: the first is set arbitrarily to zero, the second is the integer number of locations, the third is the name of the shape object and the fourth is the vector of names identifying the locations. An example should make this clear. These are the contents of a text file called naydf.txt:

```
5
1   2   1
1   3   1
2   1   1
2   3   1
2   4   1
3   1   1
3   2   1
3   5   1
4   2   1
4   3   1
4   5   1
5   3   1
5   4   1
```

The 5 in the first row indicates that this file contains information on five locations. On subsequent lines the first number identifies the location, the second number identifies one of its neighbours, and the third number is the weight of that relationship. Thus, location 5 has just two neighbours, and they are locations 3 and 4 (the last two rows of the file). We create a neighbour object for these data with the **read.gwt2nb** function like this:

dd<-read.gwt2nb("c:\\temp\\naydf.txt")

Here is a summary of the newly-created neighbour object called *dd*:

summary(dd)

```
Neighbour list object:
Number of regions: 5
Number of nonzero links: 13
Percentage nonzero weights: 52
Average number of links: 2.6
Non-symmetric neighbours list
Link number distribution:

2 3
2 3
2 least connected regions:
1 5 with 2 links
3 most connected regions:
2 3 4 with 3 links
```

Here are the five vectors of neighbours:

dd[[1]]

```
[1]  2  3
```

dd[[2]]

```
[1]  1  3  4
```

dd[[3]]

```
[1]  1  2  5
```

dd[[4]]

```
[1]  2  3  5
```

dd[[5]]

```
[1]  3  4
```

The coordinates of the five locations need to be specified:

```
coox<-c(1,2,3,4,5)
cooy<-c(3,1,2,0,3)
```

and the vectors of coordinates need to be combined into a two-column matrix. Now we can use plot with *dd* and the coordinate matrix to indicate the neighbour relations of all five locations like this:

```
plot(dd,matrix(cbind(coox,cooy),ncol=2))
text(coox,cooy,as.character(1:5),pos=rep(3,5))
```

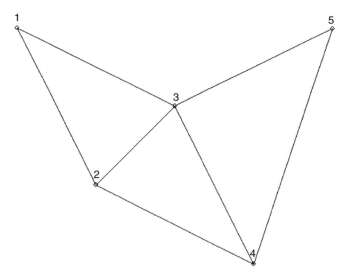

Note the use of pos = 3 to position the location numbers 1 to 5 *above* their points. You can see that locations 1 and 5 are the least connected (two neighbours) and location 3 is the most connected (four neighbours). Note that the specification in the data file was not fully reciprocal, because location 4 was defined as a neighbour of location 3 but not vice versa. There is a comment, Non-symmetric neighbours list, in the output to summary(dd) to draw attention to this. A function make.sym.nb(dd) is available to convert the object *dd* into a symmetric neighbours list.

For calculating indices much as Moran's *I* and Geary's *C* you need a 'weights list' object. This is created most simply from a neighbour object using the function nb2listw. For the ragwort data, we have already created a neighbour object called myco.nb (p. 768) and we create the weights list object myco.lw like this:

```
myco.lw<-nb2listw(myco.nb, style="W")
myco.lw
```

```
Characteristics of weights list object:
Neighbour list object:
Number of regions: 3359
Number of nonzero links: 13436
Percentage nonzero weights: 0.1190831
Average number of links: 4
Non-symmetric neighbours list

Weights style: W
Weights constants summary:

       n         nn       S0        S1          S2
W   3359   11282881    3359    1347.625    10385.25
```

There are three classic tests based on spatial cross products $C(i, j)$, where $z(i) = (x(i) - \text{mean}(x))/\text{sd}(x)$:

- Moran $(C(i,j) = z(i)z(j))$;
- Geary $(C(i,j) = (z(i) - z(j))^2)$;
- Sokal $(C(i,j) = |z(i) - z(j)|)$.

Here is the Moran I test for the ragwort data, using the weights list object myco.lw:

```
moran(1:3359,myco.lw,length(myco.nb),Szero(myco.lw))
```

```
$I
[1]  0.9931224

$K
[1]  1.800000
```

Here is Geary's C for the same data:

```
geary(1:3359,myco.lw,length(myco.nb),length(myco.nb)-1,Szero(myco.lw))
```

```
$C
[1]  0.004549794

$K

[1]  1.800000
```

Here is Mantel's permutation test:

```
sp.mantel.mc(1:3359,myco.lw,nsim=99)
```

```
    Mantel permutation test for moran measure

data: 1:3359
weights: myco.lw
number of simulations + 1: 100
```

```
statistic = 3334.905, observed rank = 100, p-value = 0.01
alternative hypothesis: greater
sample estimates:
mean of permutations    sd of permutations
            3.291485                35.557043
```

In all cases, the first argument is a vector of location numbers (1 to 3359 in the rag-wort example), the second argument is the weight list object **myco.lw**. For **moran**, the third argument is the length of the neighbours object, **length(myco.nb)** and the fourth is **Szero(myco.lw)**, the global sum of weights, both of which evaluate to 3359 in this case. The function **geary** has an extra argument, **length(myco.nb)-1** and **sp.mantel.mc** specifies the number of simulations.

Polygon lists

Perhaps the most complex spatial data handled by **spdep** comprise digitized outlines (sets of x and y coordinates) defining multiple regions, each of which can be interpreted by R as a polygon. Here is such a list from the built-in **columbus** data set:

```
data(columbus)
polys
```

The **polys** object is of class **polylist** and comprises a list of 49 polygons. Each element of the list contains a two-column matrix with x coordinates in column 1 and y coordinates in column 2, with as many rows as there are digitized points on the outline of the polygon in question. After the matrix of coordinates come the boundary box, and various plotting options:

```
attr(,"bbox")
[1]   10.42425   11.63387   11.20483   12.03754
attr(,"ringDir")
[1]   1
attr(,"after")
[1]   NA
attr(,"plotOrder")
[1]   1
attr(,"nParts")
[1]   1
attr(,"pstart")
attr(,"pstart")$from
[1]   1
attr(,"pstart")$to
[1]   27
```

There is an extremely useful function **poly2nb** that takes the list of polygons and works out which regions are neighbours of one another by looking for shared boundaries. The result is an **nb** object (here called **colnbs**), and we can get a visual check of how well **poly2nb** has worked by overlaying the neighbour relations on a map of the polygon outlines:

```
plot(polys, border="grey", forcefill=FALSE)
colnbs<- poly2nb(polys)
plot(colnbs,coords,add=T)
```

The agreement is perfect. Obviously, creating a polygon list is likely to be a huge amount of work, especially if there are many regions, each with complicated outlines. Before you

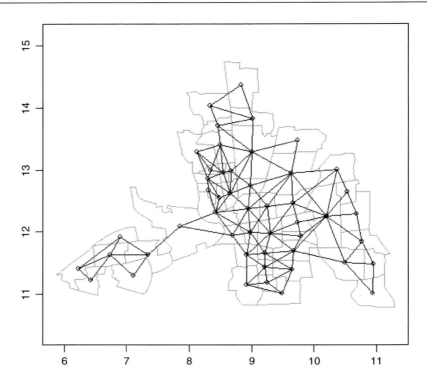

start making one, you should check that it has not been done already by someone else who might be willing to share it with you. To create polygon lists and bounding boxes from imported shape files you should use one of read.shapefile or Map2poly

?read.shapefile
?Map2poly

Subtleties include the facts that lakes are digitized anti-clockwise and islands are digitized clockwise.

Geostatistical data

Mapped data commonly show the value of a continuous response variable (e.g. the concentration of a mineral ore) at different spatial locations. The fundamental problem with this kind of data is spatial pseudoreplication. Hot spots tend to generate lots of data, and these data tend to be rather similar because they come from essentially the same place. Cold spots are poorly represented and typically widely separated. Large areas between the cold spots have no data at all.

Spatial statistics takes account of this spatial autocorrelation in various ways. The fundament tool of spatial statistics is the **variogram** (or **semivariogram**). This measures how quickly spatial autocorrelation $\gamma(h)$ falls off with increasing distance:

$$\gamma(h) = \frac{1}{2\,|N(h)|} \sum_{N(h)} (z_i - z_j)^2$$

where $N(h)$ is the set of all pairwise Euclidean distances $i - j = h$, $|N(h)|$ is the number of distinct pairs within $N(h)$, and z_i and z_j are values of the response variable at spatial locations i and j. There are two important rules of thumb: (1) the distance of reliability of the variogram is less than half the maximum distance over the entire field of data; and (2) you should only consider producing an empirical variogram when you have more than 30 data points on the map.

Plots of the empirical variogram against distance are characterized by some quaintly named features which give away its origin in geological prospecting:

- **nugget**, small-scale variation plus measurement error;

- **sill**, the asymptotic value of $\gamma(h)$ as $h \to \infty$, representing the variance of the random field;

- **range**, the threshold distance (if such exists) beyond which the data are no longer autocorrelated

Variogram plots that do not asymptote may be symptomatic of trended data or a non-stationary stochastic process. The **covariogram** $C(h)$ is the covariance of z values at separation h, for all i and $i + h$ within the maximum distance over the whole field of data:

$$\text{cov}(Z(i + h), Z(i)) = C(h)$$

The **correlogram** is a ratio of covariances:

$$\rho(h) = \frac{C(h)}{C(0)} = 1 - \frac{\gamma(h)}{C(0)}.$$

Here $C(0)$ is the variance of the random field and $\gamma(h)$ is the variogram. Where the variogram increases with distance, the correlogram and covariogram decline with distance.

The variogram assumes that the data are untrended. If there are trends, then one option is median polishing. This involves modelling row and column effects from the map like this:

y ~ overall mean + row effect + column effect + residual

This two-way model assumes additive effects and would not work if there was an interaction between the rows and columns of the map. An alternative would be to use a generalized additive model (p. 611) with non-parametric smoothers for latitude and longitude.

Anisotropy occurs when spatial autocorrelation changes with direction. If the sill changes with direction, this is called zonal anisotropy. When it is the range that changes with direction, the process is called geometric anisotropy.

Geographers have a wonderful knack of making the simplest ideas sound complicated. **Kriging** is nothing more than linear interpolation through space. Ordinary kriging uses a random function model of spatial correlation to calculate a weighted linear combination of the available samples to predict the response for an unmeasured location. Universal kriging is a modification of ordinary kriging that allows for spatial trends. We say no more about models for spatial prediction here; details can be found in Kaluzny *et al.* (1998). Our concern is with using spatial information in the interpretation of experimental or observational studies that have a single response variable. The emphasis is on using location-specific measurements to model the spatial autocorrelation structure of the data.

The idea of a variogram is to illustrate the way in which spatial variance increases with spatial scale (or alternatively, how correlation between neighbours falls off with distance).

Confusingly, R has two functions with the same name: variogram (lower-case v) is in the spatial library and Variogram (upper-case V) is in nlme. Their usage is contrasted here for the ragwort data (p. 763).

To use variogram from the spatial library, you need to create a trend surface or a kriging object with columns x, y and z. The first two columns are the spatial coordinates, while the third contains the response variable (basal stem diameter in the case of the ragwort data):

```
library(spatial)
data<-read.table("c:\\temp\\ragwortmap2006.txt",header=T)
attach(data)
names(data)
```

```
[1]  "stems"  "diameter"  "xcoord"  "ycoord"
```

```
dts<-data.frame(x=xcoord,y=ycoord,z=diameter)
```

Next, you need to create a trend surface using a function such as surf.ls:

```
surface<-surf.ls(2,dts)
```

This trend surface object is then the first argument to variogram, followed by the number of bins (here 300). The function computes the average squared difference for pairs with separation in each bin, returning results for bins that contain six or more pairs:

```
variogram(surface,300)
```

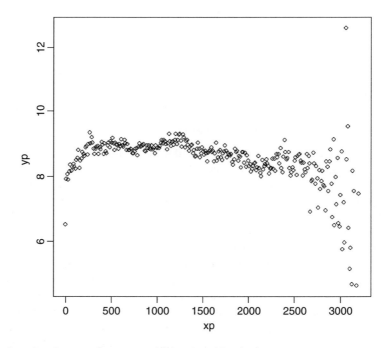

The sister function is correlogram, which takes identical arguments:

```
correlogram(surface,300)
```

The positive correlations have disappeared by about 100 cm.

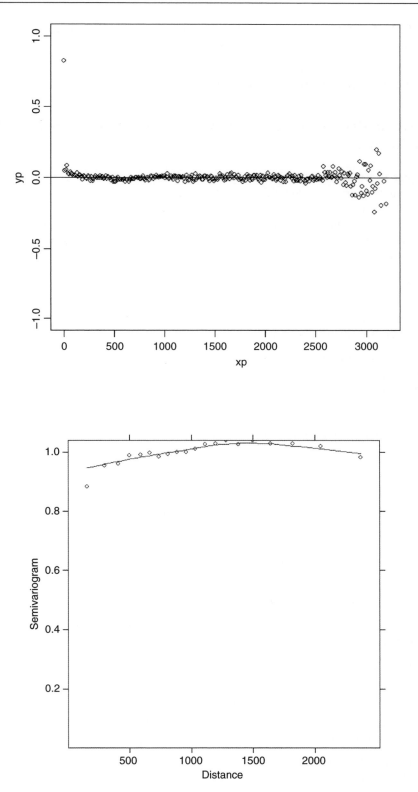

For the Variogram function in the nlme library, you need to fit a model (typically using gls or lme), then provide the model object along with a form function in the call:

```
model<-gls(diameter~xcoord+ycoord)
plot(Variogram(model,form= ~xcoord+ycoord))
```

Regression Models with Spatially Correlated Errors: Generalized Least Squares

In Chapter 19 we looked at the use of linear mixed-effects models for dealing with random effects and temporal pseudoreplication. Here we illustrate the use of generalized least squares (GLS) for regression modelling where we would expect neighbouring values of the response variable to be correlated. The great advantage of the gls function is that the errors are allowed to be correlated and/or to have unequal variances. The gls function is part of the nlme package:

```
library(nlme)
```

The following example is a geographic-scale trial to compare the yields of 56 different varieties of wheat. What makes the analysis more challenging is that the farms carrying out the trial were spread out over a wide range of latitudes and longitudes.

```
spatialdata<-read.table("c:\\temp\\spatialdata.txt",header=T)
attach(spatialdata)
names(spatialdata)
```

```
[1] "Block"  "variety"  "yield"  "latitude"  "longitude"
```

We begin with graphical data inspection to see the effect of location on yield:

```
par(mfrow=c(1,2))
plot(latitude,yield)
plot(longitude,yield)
```

There are clearly big effects of latitude and longitude on both the mean yield and the variance in yield. The latitude effect looks like a threshold effect, with little impact for latitudes less than 30. The longitude effect looks more continuous but there is a hint of non-linearity (perhaps even a hump). The varieties differ substantially in their mean yields:

```
par(mfrow=c(1,1))
barplot(sort(tapply(yield,variety,mean)))
```

The lowest-yielding varieties are producing about 20 and the highest about 30 kg of grain per unit area. There are also substantial block effects on yield:

```
tapply(yield,Block,mean)
```

```
        1           2           3           4
   27.575    28.81091    24.42589    21.42807
```

Here is the simplest possible analysis – a one-way analysis of variance using variety as the only explanatory variable:

```
model1<-aov(yield~variety)
summary(model1)
```

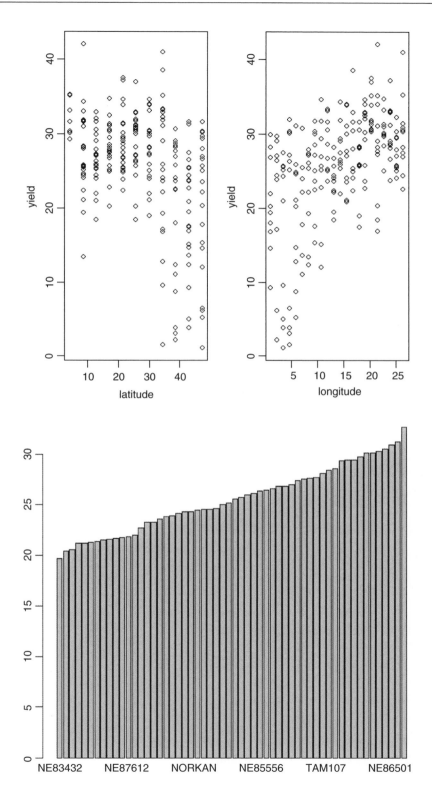

```
            Df   Sum Sq   Mean Sq   F value   Pr(>F)
variety     55   2387.5     43.4      0.73     0.912
Residuals  168   9990.2     59.5
```

This says that there are no significant differences between the yields of the 56 varieties. We can try a split plot analysis (see p. 470) using varieties nested within blocks:

```
Block<-factor(Block)
model2<-aov(yield~Block+variety+Error(Block))
summary(model2)
```

```
Error: Block
          Df    Sum Sq   Mean Sq
Block      3   1853.57    617.86

Error: Within
            Df   Sum Sq   Mean Sq   F value   Pr(>F)
variety     55   2388.8     43.4     0.8809    0.7025
Residuals  165   8135.3     49.3
```

This has made no difference to our interpretation. We could fit latitude and longitude as covariates:

```
model3<-aov(yield~Block+variety+latitude+longitude)
summary(model3)
```

```
            Df   Sum Sq   Mean Sq   F value      Pr(>F)
Block        3   1853.6    617.9   19.8576    5.072e-11   ***
variety     55   2388.8     43.4    1.3959      0.05652   .
latitude     1    686.1    686.1   22.0515    5.603e-06   ***
longitude    1   2377.6   2377.6   76.4137    2.694e-15   ***
Residuals  163   5071.6     31.1
```

This makes an enormous difference. Now the differences between varieties are close to significance ($p = 0.05652$).

Finally, we could use a GLS model to introduce spatial covariance between yields from locations that are close together. We begin by making a grouped data object:

```
space<-groupedData(yield~variety|Block)
```

Now we use this to fit a model using gls which allows the errors to be correlated and to have unequal variances. We shall add these sophistications later:

```
model4<-gls(yield~variety-1,space)
summary(model4)
```

```
Generalized least squares fit by REML
Model: yield ~ variety - 1
Data: space

     AIC        BIC      logLik
 1354.742   1532.808   -620.3709

Coefficients:
                  Value   Std.Error   t-value   p-value
varietyARAPAHOE  29.4375   3.855687   7.634827        0
varietyBRULE     26.0750   3.855687   6.762738        0
varietyBUCKSKIN  25.5625   3.855687   6.629818        0
```

and so on, for all 56 varieties. The variety means are given, rather than differences between means, because we removed the intercept from the model by using yield~variety-1 rather than yield~variety in the model formula (see p. 333).

Now we want to include the spatial covariance. The Variogram function is applied to model4 like this:

plot(Variogram(model4,form=~latitude+longitude))

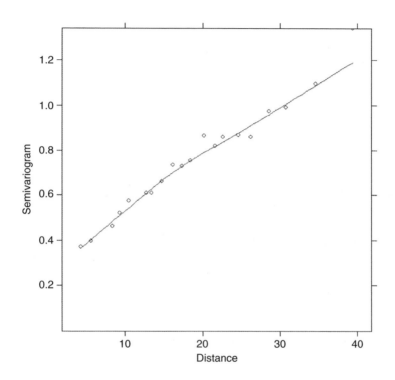

The sample variogram increases with distance, illustrating the expected spatial correlation. Extrapolating back to zero distance, there appears to be a nugget of about 0.2. There are several assumptions we could make about the spatial correlation in these data. For instance, we could try a spherical correlation structure, using the corSpher class (the range of options

Table 24.1. Spatial correlation structures. Options for specifying the form and distance-dependence of spatial correlation in generalized least squares models. For more detail, see the help on ?corClasses and on the individual correlation structures (e.g. ?corExp).

corExp	exponential spatial correlation
corGaus	Gaussian spatial correlation
corLin	linear spatial correlation
corRatio	rational quadratic spatial correlation
corSpher	spherical spatial correlation
corSymm	general correlation matrix, with no additional structure

for spatial correlation structure is shown in Table 24.1). We need to specify the distance at which the semivariogram first reaches 1. Inspection shows this distance to be about 28. We can update model4 to include this information:

```
model5<-update(model4,
corr=corSpher(c(28,0.2),form=~latitude+longitude,nugget=T))
summary(model5)
```

```
Generalized least squares fit by REML
Model: yield ~ variety - 1
Data: space

      AIC       BIC      logLik
 1185.863  1370.177  -533.9315

Correlation Structure: Spherical spatial correlation
Formula: ~latitude + longitude
Parameter estimate(s):
     range        nugget
27.4574777    0.2093144

Coefficients:
                    Value   Std.Error    t-value  p-value
varietyARAPAHOE   26.65898   3.437352    7.755672        0
varietyBRULE      25.84956   3.441792    7.510496        0
varietyBUCKSKIN   34.84837   3.478290   10.018822        0
```

This is a big improvement, and AIC has dropped from 1354.742 to 1185.863. The range (27.46) and nugget (0.209) are very close to our visual estimates. There are other kinds of spatial model, of course. We might try a rational quadratic model (corRatio); this needs an estimate of the distance at which the semivariogram is $(1 + \text{nugget})/2 = 1.2/2 = 0.6$, as well as an estimate of the nugget. Inspection gives a distance of about 12.5, so we write

```
model6<-update(model4,
corr=corRatio(c(12.5,0.2),form=~latitude+longitude,nugget=T))
```

We can use **anova** to compare the two spatial models:

```
anova(model5,model6)
```

```
        Model  df       AIC       BIC     logLik
model5      1  59  1185.863  1370.177  -533.9315
model6      2  59  1183.278  1367.592  -532.6389
```

The rational quadratic model (model6) has the lower AIC and is therefore preferred to the spherical model. To test for the significance of the spatial correlation parameters we need to compare the preferred spatial model6 with the non-spatial model4 (which assumed spatially independent errors):

```
anova(model4,model6)
```

```
        Model df       AIC       BIC     logLik   Test  L.Ratio  p-value
model4      1 57  1354.742  1532.808  -620.3709
model6      2 59  1183.278  1367.592  -532.6389  1 vs 2  175.464   <.0001
```

The two extra degrees of freedom used up in accounting for the spatial structure are clearly justified. We need to check the adequacy of the **corRatio** model. This is done by inspection of the sample variogram for the normalized residuals of model6:

plot(Variogram(model6,resType="n"))

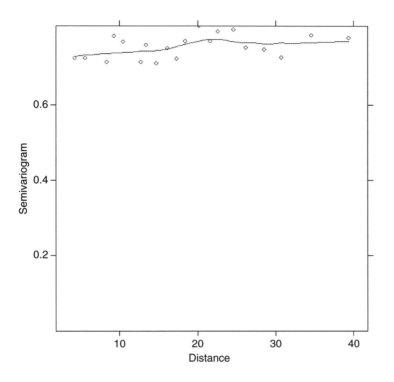

There is no pattern in the plot of the sample variogram, so we conclude that the rational quadratic is adequate. To check for constancy of variance, we can plot the normalized residuals against the fitted values like this:

plot(model6,resid(., type="n")~fitted(.),abline=0)

and the normal plot is obtained in the usual way:

qqnorm(model6,~resid(.,type="n"))

The model looks fine.

The next step is to investigate the significance of any differences between the varieties. Use update to change the structure of the model from yield~variety-1 to yield~variety:

model7<-update(model6,model=yield~variety)
anova(model7)

```
Denom. DF: 168

             numDF   F-value  p-value
(Intercept)      1  30.40490  <.0001
    variety     55   1.85092  0.0015
```

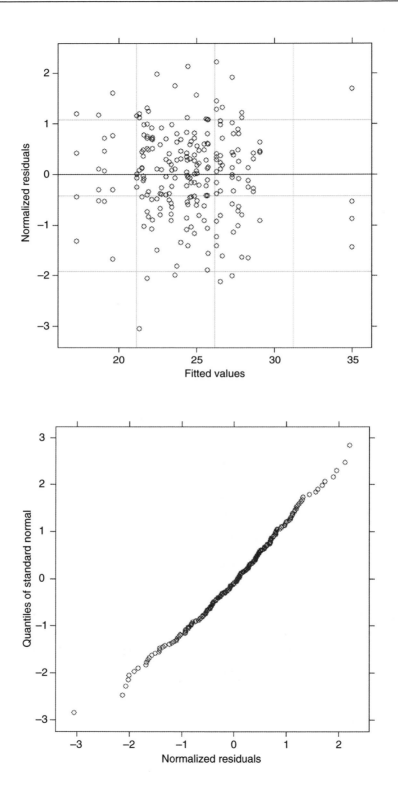

The differences between the varieties now appear to be highly significant (recall that they were only marginally significant with our linear model3 using analysis of covariance to take account of the latitude and longitude effects). Specific contrasts between varieties can be carried out using the L argument to **anova**. Suppose that we want to compare the mean yields of the first and third varieties. To do this, we set up a vector of contrast coefficients c(-1,0,1) and apply the contrast like this:

anova(model6,L=c(-1,0,1))

```
Denom. DF: 168
F-test for linear combination(s)
varietyARAPAHOE varietyBUCKSKIN
                -1             1
    numDF    F-value  p-value
1       1   7.696578   0.0062
```

Note that we use model6 (with all the variety means), not model7 (with an intercept and Helmert contrasts). The specified varieties, Arapahoe and Buckskin, exhibit highly significant differences in mean yield.

25

Survival Analysis

A great many studies in statistics deal with deaths or with failures of components: they involve the numbers of deaths, the timing of death, or the risks of death to which different classes of individuals are exposed. The analysis of survival data is a major focus of the statistics business (see Kalbfleisch and Prentice, 1980; Miller, 1981; Fleming and Harrington 1991), for which R supports a wide range of tools. The main theme of this chapter is the analysis of data that take the form of measurements of the **time to death**, or the **time to failure** of a component. Up to now, we have dealt with mortality data by considering the proportion of individuals that were dead *at a given time*. In this chapter each individual is followed until it dies, then the time of death is recorded (this will be the response variable). Individuals that survive to the end of the experiment will die at an unknown time in the future; they are said to be **censored** (as explained below).

A Monte Carlo Experiment

With data on time to death, the most important decision to be made concerns the error distribution. The key point to understand is that the variance in age at death is almost certain to increase with the mean, and hence standard models (assuming constant variance and normal errors) will be inappropriate. You can see this at once with a simple Monte Carlo experiment. Suppose that the per-week probability of failure of a component is 0.1 from one factory but 0.2 from another. We can simulate the fate of an individual component in a given week by generating a uniformly distributed random number between 0 and 1. If the value of the random number is less than or equal to 0.1 (or 0.2 for the second factory), then the component fails during that week and its lifetime can be calculated. If the random number is larger than 0.1, then the component survives to the next week. The lifetime of the component is simply the number of the week in which it finally failed. Thus, a component that failed in the first week has an age at failure of 1 (this convention means that there are no zeros in the dataframe).

The simulation is very simple. We create a vector of random numbers, rnos, that is long enough to be certain to contain a value that is less than our failure probabilities of 0.1 and 0.2. Remember that the mean life expectancy is the reciprocal of the failure rate, so our mean lifetimes will be $1/0.1 = 10$ and $1/0.2 = 5$ weeks, respectively. A length of 100 should be more than sufficient:

```
rnos<-runif(100)
```

The trick is to find the week number in which the component failed; this is the lowest subscript for which rnos ≤ 0.1 for factory 1. We can do this very efficiently using the which function: which returns a vector of subscripts for which the specified logical condition is true. So for factory 1 we would write

```
which(rnos<= 0.1)
```

```
[1]  5  8  9  19  29  33  48  51  54  63  68  74  80  83  94  95
```

This means that 16 of my first set of 100 random numbers were less than or equal to 0.1. The important point is that the *first* such number occurred in week 5. So the simulated value of the age of death of this *first* component is 5 and is obtained from the vector of failure ages using the subscript [1]:

```
which(rnos<= 0.1)[1]
```

```
[1]  5
```

All we need to do to simulate the life spans of a sample of 30 components, death1, is to repeat the above procedure 30 times:

```
death1<-numeric(30)
```

```
for (i in 1:30) {
rnos<-runif(100)
death1[i]<- which(rnos<= 0.1)[1]
}
```

```
death1
```

```
 [1]  5  8   7  23  5   4  18  2  6  4  10  12  7  3  5  17  1  3  2  1  12
[22]  8  2  12   6  3  13  16  3  4
```

The fourth component survived for a massive 23 weeks but the 17th component failed during its first week. The simulation has roughly the right average weekly failure rate:

```
1/mean(death1)
```

```
[1]  0.1351351
```

which is as close to 0.1 as we could reasonably expect from a sample of only 30 components.
   Now we do the same for the second factory with its failure rate of 0.2:

```
death2<-numeric(30)
```

```
for (i in 1:30) {
rnos<-runif(100)
death2[i]<- which(rnos<= 0.2)[1]
}
```

The sample mean is again quite reasonable:

```
1/mean(death2)
```

```
[1]  0.2205882
```

We now have the simulated raw data to carry out a comparison in age at death between factories 1 and 2. We combine the two vectors into one, and generate a vector to represent the factory identities:

```
death<-c(death1,death2)
factory<-factor(c(rep(1,30),rep(2,30)))
```

We get a visual assessment of the data as follows:

```
plot(factory,death)
```

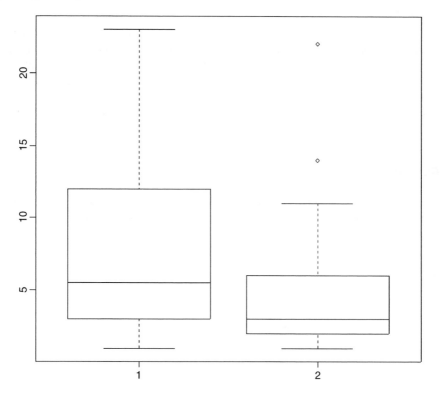

The median age at death for factory 1 is somewhat greater, but the variance in age at death is much higher than from factory 2. For data like this we expect the variance to be proportional to the square of the mean, so an appropriate error structure is the gamma (as explained below). We model the data very simply as a one-way analysis of deviance using glm with family = Gamma (note the upper-case G):

```
model1<-glm(death~factory,Gamma)
summary(model1)
```

```
Call:
glm(formula = death ~ factory, family = Gamma)

Deviance Residuals:
    Min        1Q    Median        3Q       Max
-1.5077   -0.7366   -0.3772    0.2998    2.1323

Coefficients:
               Estimate   Std. Error   t value   Pr(>|t|)
(Intercept)     0.13514      0.02218     6.092    9.6e-08   ***
factory2        0.08545      0.04246     2.013     0.0488   *
```

(Dispersion parameter for Gamma family taken to be 0.8082631)

    Null deviance: 44.067 on 59 degrees of freedom
Residual deviance: 40.501 on 58 degrees of freedom
AIC: 329.62

Number of Fisher Scoring iterations: 6

We conclude that the factories are just marginally significantly different in mean age at failure of these components ($p = 0.0488$). So, even with a twofold difference in the true failure rate, it is hard to detect a significant difference in mean age at death with samples of size $n = 30$. The moral is that for data like this on age at death you are going to need really large sample sizes in order to find significant differences.

    It is good practice to remove variable names (like death) that you intend to use later in the same session (see rm on p. 804).

Background

Since everything dies eventually, it is often not possible to analyse the results of survival experiments in terms of the proportion that were killed (as we did in Chapter 16); in due course, they *all* die. Look at the following figure:

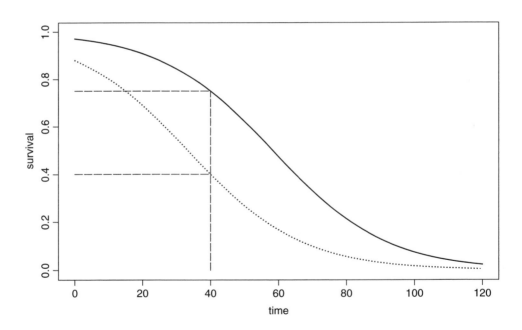

It is clear that the two treatments caused different patterns of mortality, but both start out with 100% survival and both end up with zero. We could pick some arbitrary point in the middle of the distribution at which to compare the percentage survival (say at time = 40), but this may be difficult in practice, because one or both of the treatments might have few observations at the same location. Also, the choice of when to measure the difference is entirely subjective and hence open to bias. It is much better to use R's powerful facilities for the analysis of survival data than it is to pick an arbitrary time at which to compare two proportions.

Demographers, actuaries and ecologists use three interchangeable concepts when dealing with data on the timing of death: **survivorship, age at death** and **instantaneous risk of death**. There are three broad patterns of survivorship: Type I, where most of the mortality occurs late in life (e.g. humans); Type II, where mortality occurs at a roughly constant rate throughout life; and Type III, where most of the mortality occurs early in life (e.g. salmonid fishes). On a log scale, the numbers surviving from an initial cohort of 1000, say, would look like this:

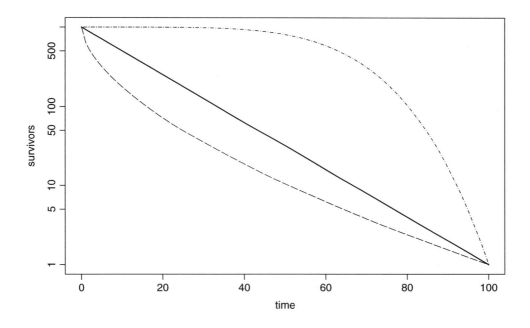

The survivor function

The survivorship curve plots the natural log of the proportion of a cohort of individuals starting out at time 0 that is still alive at time t. For the so-called Type II survivorship curve, there is a linear decline in log numbers with time (see above). This means that a constant proportion of the individuals alive at the beginning of a time interval will die during that time interval (i.e. the proportion dying is independent of density and constant for all ages). When the death rate is highest for the younger age classes we get Type III survivorship curve, which descends steeply at first, with the rate of descent easing later on. When it is the oldest animals that have the highest risk of death (as in the case of human populations in affluent societies where there is low infant mortality) we obtain the Type I curve, which has a shallow descent to start, becoming steeper later.

The density function

The density function describes the fraction of all deaths from our initial cohort that are likely to occur in a given brief instant of time. For the Type II curve this is a negative exponential. Because the fraction of individuals dying is constant with age, the number dying declines exponentially as the number of survivors (the number of individuals at risk of death) declines exponentially with the passage of time. The density function declines more steeply than exponentially for Type III survivorship curves. In the case of Type I curves, however, the density function has a maximum at the time when the product of the risk of death and the number of survivors is greatest (see below).

The hazard function

The hazard is the instantaneous risk of death – that is, the derivative of the survivorship curve. It is the instantaneous rate of change in the log of the number of survivors per unit time. Thus, for the Type II survivorship the hazard function is a horizontal line, because the risk of death is constant with age. Although this sounds highly unrealistic, it is a remarkably robust assumption in many applications. It also has the substantial advantage of parsimony. In some cases, however, it is clear that the risk of death changes substantially with the age of the individuals, and we need to be able to take this into account in carrying out our statistical analysis. In the case of Type III survivorship, the risk of death declines with age, while for Type I survivorship (as in humans) the risk of death increases with age.

The Exponential Distribution

This is a one-parameter distribution in which the hazard function is independent of age (i.e. it describes a Type II survivorship curve). The exponential is a special case of the gamma distribution in which the shape parameter α is equal to 1.

Density function

The density function is the probability of dying in the small interval of time between t and $t + dt$, and a plot of the number dying in the interval around time t as a function of t (i.e. the proportion of the original cohort dying at a given age) declines exponentially:

$$f(t) = \frac{e^{-t/\mu}}{\mu},$$

where both μ and $t > 0$. Note that the density function has an intercept of $1/\mu$ (remember that e^0 is 1). The number from the initial cohort dying per unit time declines exponentially with time, and a fraction $1/\mu$ dies during the first time interval (and, indeed, during every subsequent time interval).

Survivor function

This shows the proportion of individuals from the initial cohort that are still alive at time t:

$$S(t) = e^{-t/\mu}.$$

The survivor function has an intercept of 1 (i.e. all the cohort is alive at time 0), and shows the probability of surviving at least as long as t.

Hazard function

This is the statistician's equivalent of the ecologist's *instantaneous death rate*. It is defined as the ratio between the density function and the survivor function, and is the conditional density function at time t, given survival up to time t. In the case of Type II curves this has an extremely simple form:

$$h(t) = \frac{f(t)}{S(t)} = \frac{e^{-t/\mu}}{\mu e^{-t/\mu}} = \frac{1}{\mu},$$

because the exponential terms cancel out. Thus, with the exponential distribution the *hazard is the reciprocal of the mean time to death*, and vice versa. For example, if the mean time to death is 3.8 weeks, then the hazard is 0.2632; if the hazard were to increase to 0.32, then the mean time of death would decline to 3.125 weeks. The **survivor**, **density** and **hazard** functions of the exponential distribution are as follows (note the changes of scale on the y axes):

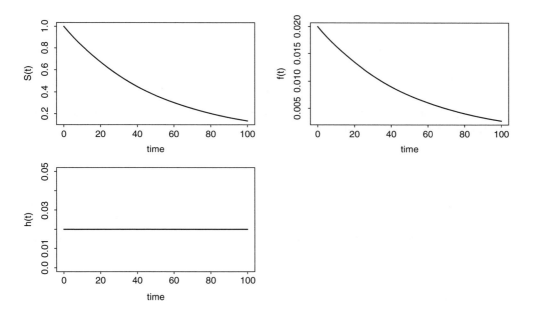

Of course, the death rate may not be a linear function of age. For example, the death rate may be high for very young as well as for very old animals, in which case the survivorship curve is like an S shape on its side.

Kaplan–Meier Survival Distributions

This is a discrete stepped survivorship curve that adds information as each death occurs. Suppose we had $n = 5$ individuals and that the times at death were 12, 17, 29, 35 and 42 weeks after the beginning of a trial. Survivorship is 1 at the outset, and stays at 1 until time 12, when it steps down to $4/5 = 0.8$. It stays level until time 17, when it drops to $0.8 \times 3/4 = 0.6$. Then there is a long level period until time 29, when survivorship drops to $0.6 \times 2/3 = 0.4$, then drops at time 35 to $0.4 \times 1/2 = 0.2$, and finally to zero at time 42.

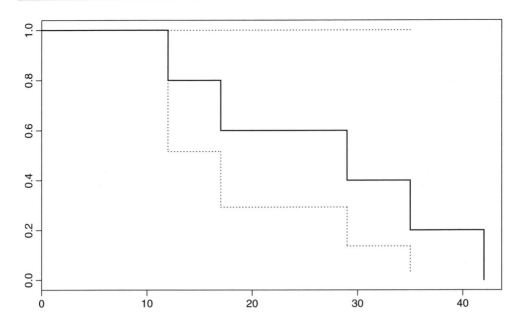

The solid line shows the survival distribution and the dotted lines show the confidence intervals (see below). In general, therefore, we have two groups at any one time: the number of deaths $d(t_i)$ and the number at risk $r(t_i)$ (i.e. those that have not yet died: the survivors). The Kaplan–Meier survivor function is

$$\hat{S}_{\mathrm{KM}} = \prod_{t_i < t} \frac{r(t_i) - d(t_i)}{r(t_i)}$$

which, as we have seen, produces a step at every time at which one or more deaths occurs. Censored individuals that survive beyond the end of the study are shown by a + on the plot or after their age in a dataframe (thus 65 means died at time 65, but 65+ means still alive when last seen at age 65).

Age-Specific Hazard Models

In many circumstances, the risk of death increases with age. There are many models to chose from:

Distribution	Hazard
Exponential	constant $= \dfrac{1}{\mu}$
Weibull	$\alpha\lambda(\lambda t)^{\alpha-1}$
Gompertz	be^{ct}
Makeham	$a + be^{ct}$
Extreme value	$\dfrac{1}{\sigma}e^{(t-\eta)/\sigma}$
Rayleigh	$a + bt$

The Rayleigh is obviously the simplest model in which hazard increases with time, but the Makeham is widely regarded as the best description of hazard for human subjects. After infancy, there is a constant hazard (a) which is due to age-independent accidents, murder, suicides, etc., with an exponentially increasing hazard in later life. The Gompertz assumption was that 'the average exhaustion of a man's power to avoid death is such that at the end of equal infinitely small intervals of time he has lost equal portions of his remaining power to oppose destruction which he had at the commencement of these intervals'. Note that the Gompertz differs from the Makeham only by the omission of the extra background hazard (a), and this becomes negligible in old age.

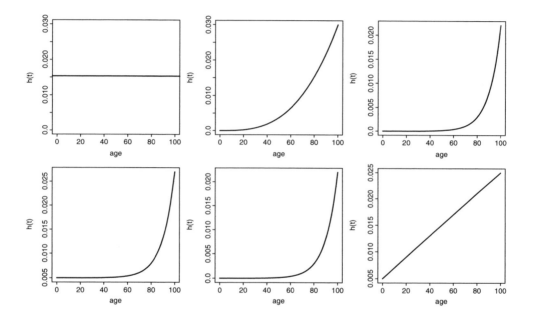

These plots show how hazard changes with age for the following distributions: from top left to bottom right: **exponential, Weibull, Gompertz, Makeham, extreme value** and **Rayleigh**.

Survival analysis in R

There are three cases that concern us here:

- constant hazard and no censoring;
- constant hazard with censoring;
- age-specific hazard, with or without censoring.

The first case is dealt with very simply in R by specifying a generalized linear model with gamma errors. The second case involves the use of exponential survival models with a *censoring indicator* (1 indicates that the response is a time at death, 0 indicates that the individual was alive when last seen; see below, and p. 801). The third case involves a choice between **parametric** models, based on the **Weibull** distribution, and **non-parametric** techniques, based on the **Cox proportional hazard** model.

Parametric models

We are unlikely to know much about the error distribution in advance of the study, except that it will certainly not be normal. In R we are offered several choices for the analysis of survival data:

- gamma;
- exponential;
- piecewise exponential;
- extreme value;
- log-logistic;
- lognormal;
- Weibull.

In practice, it is often difficult to choose between them. In general, the best solution is to try several distributions and to pick the error structure that produces the minimum error deviance.

Cox proportional hazards model

This is the most widely used regression model for survival data. It assumes that the hazard is of the form

$$\lambda(t; Z_i) = \lambda_0(t) r_i(t),$$

where $Z_i(t)$ is the set of explanatory variables for individual i at time t. The *risk score* for subject i is

$$r_i(t) = e^{\beta Z_i(t)},$$

in which β is a vector of parameters from the linear predictor and $\lambda_0(t)$ is an *unspecified baseline hazard function* that will cancel out in due course. The antilog guarantees that λ is positive for any regression model $\beta Z_i(t)$. If a death occurs at time t^*, then, conditional on this death occurring, the likelihood that it is individual i that dies, rather than any other individual at risk, is

$$L_i(\beta) = \frac{\lambda_0(t^*) r_i(t^*)}{\sum_j Y_j(t^*) \lambda_0(t^*) r_j(t^*)} = \frac{r_i(t^*)}{\sum_j Y_j(t^*) r_j(t^*)}.$$

The product of these terms over all times of death, $L(\beta) = \prod L_i(\beta)$, was christened a partial likelihood by Cox (1972). This is clever, because maximizing $\log(L(\beta))$ allows an estimate of β without knowing anything about the baseline hazard function ($\lambda_0(t)$ is a nuisance variable in this context). The proportional hazards model is nonparametric in the sense that it depends only on the **ranks** of the survival times.

Cox's proportional hazard or a parametric model?

In cases where you have censoring, or where you want to use a more complex error structure, you will need to choose between a parametric model, fitted using survreg, and a non-parametric model, fitted using coxph. If you want to use the model for prediction, then you have no choice: you must use the parametric survreg because coxph does not extrapolate beyond the last observation. Traditionally, medical studies use coxph while engineering studies use survreg (so-called accelerated failure-time models), but both disciples could fruitfully use either technique, depending on the nature of the data and the precise question being asked. Here is a typical question addressed with coxph: 'How much does the risk of dying decrease if a new drug treatment is given to a patient?' In contrast, parametric techniques are typically used for questions like this: 'What proportion of patients will die in 2 years based on data from an experiment that ran for just 4 months?'

Parametric analysis

The following example concerns survivorship of two cohorts of seedlings. All the seedlings died eventually, so there is no censoring in this case. There are two questions:

- Was survivorship different in the two cohorts?

- Was survivorship affected by the size of the canopy gap in which germination occurred?

Here are the data:

```
seedlings<-read.table("c:\\temp\\seedlings.txt",header=T)
attach(seedlings)
names(seedlings)
```

```
[1] "cohort" "death" "gapsize"
```

We need to load the survival library:

```
library(survival)
```

We begin by creating a variable called status to indicate which of the data are censored:

```
status<-1*(death>0)
```

There are no cases of censoring in this example, so all of the values of status are equal to 1.
   The fundamental object in survival analysis is Surv(death,status), the Kaplan–Meier survivorship object. We can plot it out using survfit with plot like this:

```
plot(survfit(Surv(death,status)),ylab="Survivorship",xlab="Weeks")
```

This shows the overall survivorship curve with the confidence intervals. All the seedlings were dead by week 21. Were there any differences in survival between the two cohorts?

```
model<-survfit(Surv(death,status)~cohort)
summary(model)
```

```
Call: survfit(formula = Surv(death, status) ~ cohort)

        cohort=October
```

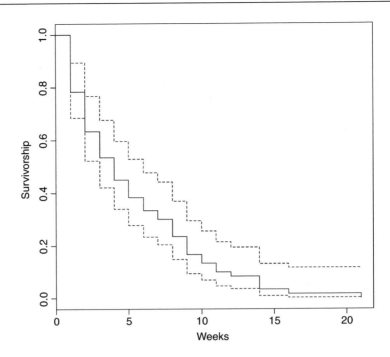

time	n.risk	n.event	survival	std.err	lower 95% CI	upper 95% CI
1	30	7	0.7667	0.0772	0.62932	0.934
2	23	3	0.6667	0.0861	0.51763	0.859
3	20	3	0.5667	0.0905	0.41441	0.775
4	17	2	0.5000	0.0913	0.34959	0.715
5	15	3	0.4000	0.0894	0.25806	0.620
6	12	1	0.3667	0.0880	0.22910	0.587
8	11	2	0.3000	0.0837	0.17367	0.518
9	9	4	0.1667	0.0680	0.07488	0.371
10	5	1	0.1333	0.0621	0.05355	0.332
11	4	1	0.1000	0.0548	0.03418	0.293
14	3	1	0.0667	0.0455	0.01748	0.254
16	2	1	0.0333	0.0328	0.00485	0.229
21	1	1	0.0000	NA	NA	NA

cohort=September

time	n.risk	n.event	survival	std.err	lower 95% CI	upper 95% CI
1	30	6	0.8000	0.0730	0.6689	0.957
2	24	6	0.6000	0.0894	0.4480	0.804
3	18	3	0.5000	0.0913	0.3496	0.715
4	15	3	0.4000	0.0894	0.2581	0.620
5	12	1	0.3667	0.0880	0.2291	0.587
6	11	2	0.3000	0.0837	0.1737	0.518
7	9	2	0.2333	0.0772	0.1220	0.446
8	7	2	0.1667	0.0680	0.0749	0.371
10	5	1	0.1333	0.0621	0.0535	0.332
11	4	1	0.1000	0.0548	0.0342	0.293
12	3	1	0.0667	0.0455	0.0175	0.254
14	2	2	0.0000	NA	NA	NA

To plot these figures use plot(model) like this:

plot(model,lty=c(1,3),ylab="Survivorship",xlab="week")

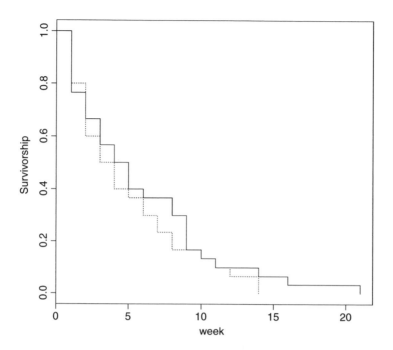

The solid line is for the October cohort and the dashed line is for September. To see the median times at death for the two cohorts, just type

model

```
Call: survfit(formula = Surv(death, status) ~ cohort)

                  n  events  median  0.95LCL  0.95UCL
cohort=October   30      30     4.5        3        9
cohort=September 30      30     3.5        2        7
```

Cox's Proportional Hazards

The median age at death was one week later in the October cohort, but look at the width of the confidence intervals: 3 to 9 versus 2 to 7. Clearly there is no significant effect of cohort on time of death. What about gap size? We start with a full analysis of covariance using coxph rather than survfit.

model1<-coxph(Surv(death,status)~strata(cohort)*gapsize)
summary(model1)

```
Call:
coxph(formula = Surv(death, status) ~ strata(cohort) * gapsize)
   n= 60
```

	coef	exp(coef)	se(coef)	z	p
gapsize	-1.19	0.305	0.621	-1.910	0.056
gapsize:strata(cohort)	0.58	1.785	0.826	0.701	0.480
cohort=September					

	exp(coef)	exp(-coef)	lower .95	upper .95
gapsize	0.305	3.27	0.0904	1.03
gapsize:strata(cohort)	1.785	0.56	0.3534	9.02
cohort=September				

```
Rsquare= 0.076 (max possible= 0.993 )
Likelihood ratio test= 4.73 on 2 df, p=0.0937
Wald test        = 4.89 on 2 df, p=0.0868
Score (logrank) test = 5.04 on 2 df, p=0.0805
```

There is no evidence of any interaction ($p = 0.480$) and the main effect of gap size is not quite significant in this model ($p = 0.056$). We fit the simpler model with no interaction:

```
model2<-coxph(Surv(death,status)~strata(cohort)+gapsize)
anova(model1,model2)
```

```
Analysis of Deviance Table
```

```
Model 1: Surv(death, status)~ strata(cohort) * gapsize
Model 2: Surv(death, status)~ strata(cohort) + gapsize
```

	Resid. Df	Resid. Dev	Df	Deviance
1	58	293.898		
2	59	294.392	-1	-0.494

There is no significant difference in explanatory power, so we accept the simpler model without an interaction term. Note that removing the interaction makes the main effect of gap size significant ($p = 0.035$):

```
summary(model2)
```

```
Call:
coxph(formula = Surv(death, status)~ strata(cohort) + gapsize)
n= 60
```

	coef	exp(coef)	se(coef)	z	p
gapsize	-0.855	0.425	0.405	-2.11	0.035

	exp(coef)	exp(-coef)	lower .95	upper .95
gapsize	0.425	2.35	0.192	0.942

```
Rsquare= 0.068 (max possible= 0.993 )
Likelihood ratio test = 4.24  on 1 df,  p=0.0395
Wald test         = 4.44  on 1 df,  p=0.0350
Score (logrank) test = 4.54  on 1 df,  p=0.0331
```

We conclude that risk of seedling death is lower in bigger gaps ($coef = -0.855$) but this effect is similar in the September and October-germinating cohorts.

You see that the modelling methodology is exactly the same as usual: fit a complicated model and simplify it to find a minimal adequate model. The only difference is the use of Surv(death,status) if that the response is a Kaplan–Meier object.

```
detach(seedlings)
rm(status)
```

Models with Censoring

Censoring occurs when we do not know the time of death for all of the individuals. This comes about principally because some individuals outlive the experiment, while others leave the experiment before they die. We know when we last saw them alive, but we have no way of knowing their age at death. These individuals contribute something to our knowledge of the survivor function, but nothing to our knowledge of the age at death. Another reason for censoring occurs when individuals are lost from the study: they may be killed in accidents, they may emigrate, or they may lose their identity tags.

In general, then, our survival data may be a mixture of times at death and times after which we have no more information on the individual. We deal with this by setting up an extra vector called the **censoring indicator** to distinguish between the two kinds of numbers. If a time really is a time to death, then the censoring indicator takes the value 1. If a time is just the last time we saw an individual alive, then the censoring indicator is set to 0. Thus, if we had the time data T and censoring indicator W on seven individuals,

```
T   4   7   8   8   12   15   22
W   1   1   0   1   1    0    1
```

this would mean that five of the times were times at death while in two cases, one at time 8 and another at time 15, individuals were seen alive but never seen again.

With repeated sampling in survivorship studies, it is usual for the degree of censoring to decline as the study progresses. Early on, many of the individuals are alive at the end of each sampling interval, whereas few if any survive to the end of the last study period. The following example comes from a study of cancer patients undergoing one of four drug treatment programmes (drugs A, B and C and a placebo):

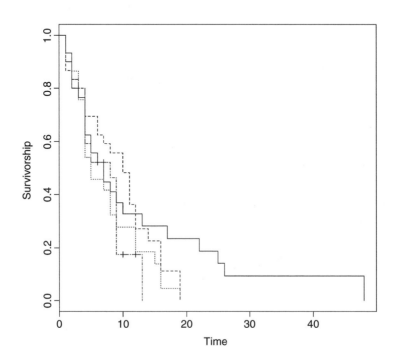

```
cancer<-read.table("c:\\temp\\cancer.txt",header=T)
attach(cancer)
names(cancer)
```

```
[1] "death"  "treatment"  "status"
```

```
plot(survfit(Surv(death,status)~treatment),lty=c(1:4),ylab="Surivorship",xlab="Time")
tapply(death[status==1],treatment[status==1],mean)
```

```
   DrugA       DrugB       DrugC      placebo
9.480000    8.360000    6.800000    5.238095
```

The long tail is for drug A. The latest deaths in the other treatments were at times 14 and 19. The variances in age at death are dramatically different under the various treatments:

```
tapply(death[status==1],treatment[status==1],var)
```

```
    DrugA        DrugB       DrugC       placebo
117.51000    32.65667    27.83333     11.39048
```

Parametric models

The simplest model assumes a constant hazard: dist="exponential".

```
model1<-survreg(Surv(death,status)~treatment,dist="exponential")
summary(model1)
```

```
Call:
survreg(formula = Surv(death, status) ~ treatment, dist = "exponential")

                  Value   Std. Error      z        p
(Intercept)       2.448      0.200     12.238   1.95e-34
treatmentDrugB   -0.125      0.283     -0.443   6.58e-01
treatmentDrugC   -0.430      0.283     -1.520   1.28e-01
treatmentplacebo -0.333      0.296     -1.125   2.61e-01

Scale fixed at 1

Exponential distribution
Loglik(model)= -310.1  Loglik(intercept only)= -311.5
        Chisq= 2.8 on 3 degrees of freedom, p= 0.42
Number of Newton-Raphson Iterations: 4
n= 120
```

Under the assumption of exponential errors there are no significant effects of drug treatment on survivorship (all $p > 0.1$). How about modelling non-constant hazard using Weibull errors instead (these are the default for survreg)?

```
model2<-survreg(Surv(death,status)~treatment)
summary(model2)
```

```
Call:
survreg(formula = Surv(death, status)~ treatment)
                  Value   Std. Error      z        p
(Intercept)       2.531      0.1572    16.102   2.47e-58
treatmentDrugB   -0.191      0.2193    -0.872   3.83e-01
treatmentDrugC   -0.475      0.2186    -2.174   2.97e-02
treatmentplacebo -0.454      0.2313    -1.963   4.96e-02
Log(scale)       -0.260      0.0797    -3.264   1.10e-03
```

```
Scale= 0.771
```

```
Weibull distribution
Loglik(model)= -305.4 Loglik(intercept only)= -308.3
        Chisq= 5.8 on 3 degrees of freedom, p= 0.12
Number of Newton-Raphson Iterations: 5
n= 120
```

The scale parameter 0.771, being less than 1, indicates that hazard *decreases* with age in this study. Drug B is not significantly different from drug A ($p = 0.38$), but drug C and the placebo are significantly poorer ($p < 0.05$). We can use anova to compare model1 and model2:

anova(model1,model2)

```
        Terms  Resid. Df     -2*LL  Test Df  Deviance      P>|Chi|)
1 treatment          116  620.1856       NA        NA            NA
2 treatment          115  610.7742      = 1    9.4114  0.002156405
```

model2 with Weibull errors is significant improvement over model1 with exponential errors ($p = 0.002$).

We can try amalgamating the factor levels – the analysis suggests that we begin by grouping A and B together:

treat2<-treatment
levels(treat2)

```
[1] "DrugA" "DrugB" "DrugC" "placebo"
```

levels(treat2)[1:2]<-"DrugsAB"
levels(treat2)

```
[1] "DrugsAB" "DrugC" "placebo"
```

model3<-survreg(Surv(death,status)~treat2)
anova(model2,model3)

```
          Terms  Resid. Df     -2*LL    Test Df    Deviance   P>|Chi|)
1   treatment          115  610.7742         NA          NA         NA
2      treat2          116  611.5190  1 vs. 2 -1  -0.744833  0.3881171
```

That model simplification was justified. What about drug C? Can we lump it together with the placebo?

levels(treat2)[2:3]<-"placeboC"
model4<-survreg(Surv(death,status)~treat2)
anova(model3,model4)

```
      Terms  Resid. Df     -2*LL  Test Df     Deviance   P>|Chi|)
1    treat2          116  611.5190       NA           NA         NA
2    treat2          117  611.5301     = -1  -0.01101309  0.9164208
```

Yes we can. That simplification was clearly justified ($p = 0.916$):

summary(model4)

```
Call:
survreg(formula = Surv(death, status)~ treat2)
```

```
                    Value   Std. Error        z          p
(Intercept)         2.439        0.112    21.76  5.37e-105
treat2placeboC     -0.374        0.160    -2.33   1.96e-02
Log(scale)         -0.249        0.078    -3.20   1.39e-03

Scale= 0.78

Weibull distribution
Loglik(model)= -305.8 Loglik(intercept only)= -308.3
        Chisq= 5.05 on 1 degrees of freedom, p= 0.025
Number of Newton-Raphson Iterations: 5
n= 120
```

We can summarize the results in terms of the mean age at death, taking account of the censoring:

```
tapply(predict(model4,type="response"),treat2,mean)
```

```
  DrugsAB    placeboC
11.459885    7.887685
```

```
detach(cancer)
rm(death, status)
```

Comparing coxph and survreg survival analysis

Finally, we shall compare the methods, parametric and non-parametric, by analysing the same data set both ways. It is an analysis of covariance with one continuous explanatory variable (initial weight) and one categorical explanatory variable (group):

```
insects<-read.table("c:\\temp\\roaches.txt",header=T)
attach(insects)
names(insects)
```

```
[1] "death"  "status"  "weight"  "group"
```

First, we plot the survivorship curves of the three groups:

```
plot(survfit(Surv(death,status)~group),lty=c(1,3,5),ylab="Survivorship",xlab="Time")
```

The crosses + at the end of the survivorship curves for groups A and B indicate that there was censoring in these groups (not all of the individuals were dead at the end of the experiment).

We begin the modelling with parametric methods (survreg). We shall compare the default error distribution (Weibull, which allows for non-constant hazard with age) with the simpler exponential (assuming constant hazard):

```
model1<-survreg(Surv(death,status)~weight*group,dist="exponential")
summary(model1)
```

```
Call:
survreg(formula = Surv(death, status) ~ weight * group, dist =
"exponential")
```

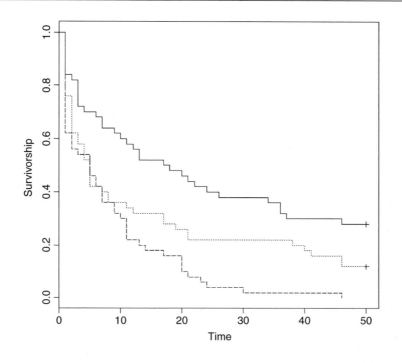

	Value	Std. Error	z	p
(Intercept)	3.8702	0.3854	10.041	1.00e-23
weight	-0.0803	0.0659	-1.219	2.23e-01
groupB	-0.8853	0.4508	-1.964	4.95e-02
groupC	-1.7804	0.4386	-4.059	4.92e-05
weight:groupB	0.0643	0.0674	0.954	3.40e-01
weight:groupC	0.0796	0.0674	1.180	2.38e-01

Scale fixed at 1

Exponential distribution
Loglik(model)= -480.6 Loglik(intercept only)= -502.1
      Chisq= 43.11 on 5 degrees of freedom, p= 3.5e-08
Number of Newton-Raphson Iterations: 5
n= 150

model2 employs the default Weibull distribution allowing non-constant hazard:

```
model2<-survreg(Surv(death,status)~weight*group)
summary(model2)
```

Call:
survreg(formula = Surv(death, status) ~ weight * group)

	Value	Std. Error	z	p
(Intercept)	3.9506	0.5308	7.443	9.84e-14
weight	-0.0973	0.0909	-1.071	2.84e-01
groupB	-1.1337	0.6207	-1.826	6.78e-02
groupC	-1.9841	0.6040	-3.285	1.02e-03

```
weight:groupB   0.0826   0.0929   0.889   3.74e-01
weight:groupC   0.0931   0.0930   1.002   3.16e-01
Log(scale)      0.3083   0.0705   4.371   1.24e-05

Scale= 1.36

Weibull distribution
Loglik(model)= -469.6 Loglik(intercept only)= -483.3
        Chisq= 27.42 on 5 degrees of freedom, p= 4.7e-05
Number of Newton-Raphson Iterations: 5
n= 150
```

The fact that the scale parameter is greater than 1 indicates that the risk of death increases with age in this case. We compare the two models in the usual way, using **anova**:

anova(model1,model2)

```
            Terms  Resid. Df     -2*LL  Test Df  Deviance      P(>|Chi|)
1 weight * group        144   961.1800       NA        NA             NA
2 weight * group        143   939.2261     = 1  21.95388   2.792823e-06
```

The Weibull model (2) is vastly superior to the exponential ($p < 0.00001$) so we continue with model2. There is no evidence in model2 (summary above) of any interaction between weight and group ($p = 0.374$) so we simplify using **step**:

model3<-step(model2)

```
Start:  AIC= 953.23
  Surv(death, status) ~ weight * group

                 Df      AIC
- weight:group    2   950.30
<none>                953.23

Step:  AIC= 950.3
  Surv(death, status) ~ weight + group

            Df      AIC
- weight     1   949.01
<none>           950.30
-group       2   967.75

Step:  AIC= 949.01
  Surv(death, status) ~ group

           Df      AIC
 <none>         949.01
 -group     2   970.64
```

After eliminating the non-significant interaction (AIC $= 950.30$), R has removed the main effect of weight (AIC $= 949.01$), but has kept the main effect of group (AIC of <none> is less than AIC of -group). The minimal model with **survreg** is this:

summary(model3)

```
Call:
survreg(formula = Surv(death, status) ~ group)
```

```
                Value    Std. Error      z          p
(Intercept)     3.459      0.2283     15.15    7.20e-52
groupB         -0.822      0.3097     -2.65    7.94e-03
groupC         -1.540      0.3016     -5.11    3.28e-07
Log(scale)      0.314      0.0705      4.46    8.15e-06

Scale= 1.37

Weibull distribution
Loglik(model)= -470.5 Loglik(intercept only)= -483.3
        Chisq= 25.63 on 2 degrees of freedom, p= 2.7e-06
Number of Newton-Raphson Iterations: 5
n= 150
```

It is clear that all three groups are required (B and C differ by 0.72, with standard error 0.31), so this is the minimal adequate model. Here are the predicted mean ages at death:

tapply(predict(model3),group,mean)

```
       A          B          C
31.796137   13.972647   6.814384
```

You can compare these with the mean ages of those insects that died

tapply(death[status==1],group[status==1],mean)

```
       A          B          C
12.611111   9.568182   8.020000
```

and the ages when insects were last seen (dead or alive)

tapply(death,group,mean)

```
   A       B       C
23.08   14.42   8.02
```

The predicted ages at death are substantially greater than the observed ages at last sighting when there is lots of censoring (e.g. 31.8 vs. 23.08 for group A).

Here are the same data analysed with the Cox proportional hazards model:

model10<-coxph(Surv(death,status)~weight*group)
summary(model10)

```
Call:
coxph(formula = Surv(death, status) ~ weight * group)

n= 150
```

	coef	exp(coef)	se(coef)	z	p
weight	0.0633	1.065	0.0674	0.940	0.3500
groupB	0.7910	2.206	0.4564	1.733	0.0830
groupC	1.2863	3.620	0.4524	2.843	0.0045
weight:groupB	-0.0557	0.946	0.0688	-0.809	0.4200
weight:groupC	-0.0587	0.943	0.0690	-0.851	0.3900

	exp(coef)	exp(-coef)	lower .95	upper .95
weight	1.065	0.939	0.934	1.22
groupB	2.206	0.453	0.902	5.40
groupC	3.620	0.276	1.491	8.79
weight:groupB	0.946	1.057	0.827	1.08
weight:groupC	0.943	1.060	0.824	1.08

```
Rsquare= 0.135 (max possible= 0.999 )
Likelihood ratio test= 21.8 on 5 df, p=0.000564
Wald test              = 20.8 on 5 df, p=0.000903
Score (logrank) test = 22.1 on 5 df, p=0.000513
```

As you see, the interaction terms are not significant ($p > 0.39$) so we simplify using **step** as before:

model11<-step(model10)

```
Start: AIC= 1113.54
Surv(death, status) ~ weight * group

                   Df      AIC
- weight:group      2    1110.3
<none>                   1113.5

Step: AIC= 1110.27
Surv(death, status) ~ weight + group

            Df      AIC
- weight     1    1108.8
<none>            1110.3
- group      2    1123.7

Step: AIC= 1108.82
Surv(death, status) ~ group

            Df      AIC
<none>            1108.8
- group      2    1125.4
```

Note that the AIC values are different than they were with the parametric model. The interaction term is dropped because this simplification reduces AIC to 1110.3. Then the covariate (weight) is dropped because this simplification also reduces AIC (to 1108.8). But removing group would *increase* AIC to 1125.4, so this is not done. The minimal model contains a main effect for group but no effect of initial weight:

summary(model11)

```
Call:
coxph(formula = Surv(death, status) ~ group)

n= 150
          coef   exp(coef)   se(coef)      z        p
groupB   0.561      1.75       0.226    2.48   1.3e-02
groupC   1.008      2.74       0.226    4.46   8.3e-06

          exp(coef)   exp(-coef)   lower .95   upper .95
groupB       1.75       0.571        1.13        2.73
groupC       2.74       0.365        1.76        4.27
```

```
Rsquare= 0.128 (max possible= 0.999 )
Likelihood ratio test= 20.6 on 2 df, p=3.45e-05
Wald test              = 19.9 on 2 df, p=4.87e-05
Score (logrank) test = 21 on 2 df,   p=2.77e-05
```

To see what these numbers mean, it is a good idea to go back to the raw data on times of death (or last sighting for the censored individuals). Here are the mean values:

tapply(death,group,mean)

```
    A       B       C
23.08   14.42    8.02
```

Evidently, individuals in group A lived a lot longer than those in group C. The ratio of their mean ages at death is 23.08/8.02 which evaluates to:

23.08/8.02

```
[1]  2.877805
```

Likewise, individuals in group A lived linger than individuals in group B by a ratio

23.08/14.42

```
[1]  1.600555
```

These figures are the approximate hazards for an individual in group C or group B relative to an individual in group A. In the coxph output of model11 they are labelled exp(coef). The model values are slightly different from the raw means because of the way that the model has dealt with censoring (14 censored individuals in group A, 6 in group B and none in group C): 1.6 vs. 1.75 and 2.8778 vs. 2.74

You should compare the outputs from the two functions coxph and survreg to make sure you understand their similarities and their differences. One fundamental difference is that the parametric Kaplan–Meier survivorship curves refer to the population, whereas Cox proportional hazards refer to an individual in a particular group.

plot(survfit(model11))
legend(35,0.8,c("Group A","Group B","Group C"),lty=c(2,1,2))

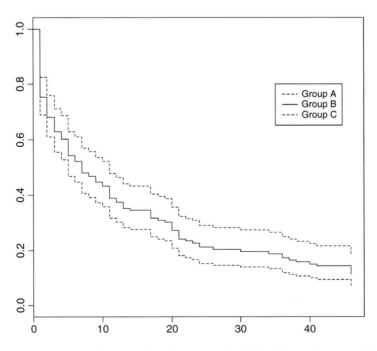

This plot shows the survivorship curve for the average individual in each group with the covariates held at their average values (but not in this example, since we have eliminated them).

26

Simulation Models

Simulation modelling is an enormous topic, and all I intend here is to demonstrate a few very simple temporal and spatial simulation techniques that give the flavour of what is possible in R.

Simulation models are used for investigating dynamics in time, in space, or in both space and time together. For temporal dynamics we might be interested in:

- the transient dynamics (the behaviour after the start but before equilibrium is attained – if indeed equilibrium is ever attained);

- equilibrium behaviour (after the transients have damped away);

- chaos (random-looking, but actually deterministic temporal dynamics that are extremely sensitive to initial conditions).

For spatial dynamics, we might use simulation models to study:

- metapopulation dynamics (where local extinction and recolonization of patches characterize the long-term behaviour, with constant turnover of occupied patches);

- neighbour relations (in spatially explicit systems where the performance of individuals is determined by the identity and attributes of their immediate neighbours);

- pattern generation (dynamical processes that lead to the generation of emergent, but more or less coherent patterns).

Temporal Dynamics: Chaotic Dynamics in Population Size

Biological populations typically increase exponentially when they are small, but individuals perform less well as population density rises, because of competition, predation or disease. In aggregate, these effects on birth and death rates are called **density-dependent processes**, and it is the nature of the density-dependent processes that determines the temporal pattern of population dynamics. The simplest density-dependent model of population dynamics is known as the **quadratic map**. It is a first-order non-linear difference equation,

$$N(t+1) = \lambda N(t) \left[1 - N(t) \right],$$

where $N(t)$ is the population size at time t, $N(t+1)$ is the population size at time $t+1$, and the single parameter, λ, is known as the per-capita multiplication rate. The population can only increase when the population is small if $\lambda > 1$, the so-called **invasion criterion**. But how does the system behave as λ increases above 1?

We begin by simulating time series of populations for different values of λ and plotting them to see what happens. First, here are the dynamics with $\lambda = 2$:

```
par(mfrow=c(2,2))
lambda<-2
x<-numeric(40)
x[1]<-0.6
for (t in 2 : 40) x[t] <- lambda * x[t-1] * (1 - x[t-1])
plot(1:40,x,type="l",ylim=c(0,1),ylab="population",xlab="time",main="2.0")
```

The population falls very quickly from its initial value (0.6) to equilibrium ($N^* = 0.5$) and stays there; this system has a stable point equilibrium. What if λ were to increase to 3.3?

```
lambda<-3.3
x<-numeric(40)
x[1]<-0.6
for (t in 2 : 40) x[t] <- lambda * x[t-1] * (1 - x[t-1])
plot(1:40,x,type="l",ylim=c(0,1),ylab="population",xlab="time",main="3.3")
```

Now the dynamics show persistent two-point cycles. What about $\lambda = 3.5$?

```
lambda<-3.5
x<-numeric(40)
x[1]<-0.6
for (t in 2 : 40) x[t] <- lambda * x[t-1] * (1 - x[t-1])
plot(1:40,x,type="l",ylim=c(0,1),ylab="population",xlab="time",main="3.5")
```

The outcome is qualitatively different. Now we have persistent four-point cycles. Suppose that λ were to increase to 4:

```
lambda<-4
x<-numeric(40)
x[1]<-0.6
for (t in 2 : 40) x[t] <- lambda * x[t-1] * (1 - x[t-1])
plot(1:40,x,type="l",ylim=c(0,1),ylab="population",xlab="time",main="4.0")
```

Now this is really interesting. The dynamics to not repeat in any easily described pattern. They are said to be **chaotic** because the pattern shows extreme sensitivity to initial conditions: tiny changes in initial conditions can have huge consequences on numbers at a given time in the future.

Investigating the route to chaos

We have seen four snapshots of the relationship between λ and population dynamics. To investigate this more fully, we should write a function to describe the dynamics as a function of λ, and extract a set of (say, 20) sequential population densities, after any transients have died away (after, say, 380 iterations):

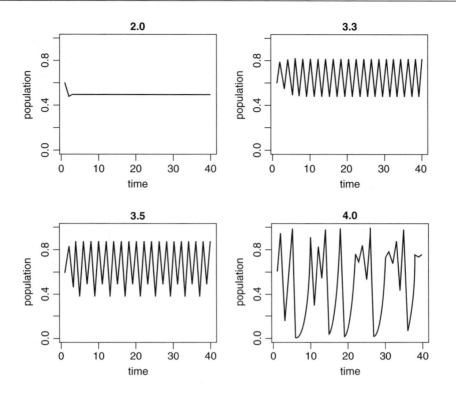

```
numbers<-function (lambda) {
x<-numeric(400)
x[1]<-0.6
for (t in 2 : 400) x[t] <- lambda * x[t-1] * (1 - x[t-1])
x[381:400] }
```

The idea is to plot these 20 values on the y axis of a new graph, against the value of λ that produced them. A stable point equilibrium will be represented by a single point, because all 20 values of y will be identical. Two-point cycles will show up as two points, four-point cycles as four points, but chaotic systems will appear as many points. Start with a blank graph:

```
par(mfrow=c(1,1))
plot(c(2,4),c(0,1),type="n",xlab="lambda",ylab="population")
```

Now, simulate using a for loop a wide range of values for λ between 2 and 4 (the range we investigated earlier), and use the function sapply to apply our function to the current value of λ, then use points to add these results to the graph:

```
for(lam in seq(2,4,0.01))
    points(rep(lam,20),sapply(lam,numbers),pch=16,cex=0.5)
```

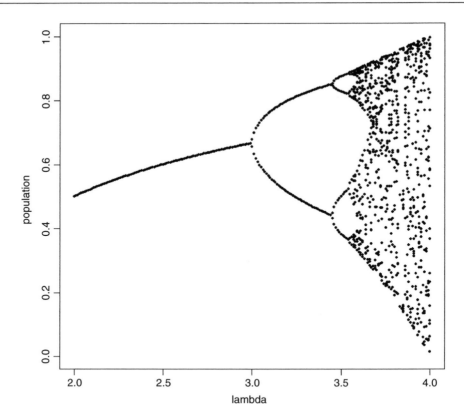

This graph shows what is called 'the period-doubling route to chaos'; see May (1976) for details.

Temporal and Spatial Dynamics: a Simulated Random Walk in Two Dimensions

The idea is to follow an individual as it staggers its way around a two-dimensional random walk, starting at the point (50, 50) and leaving a trail of lines on a square surface which scales from 0 to 100. First, we need to define what we mean by our random walk. Suppose that in the x direction the individual could move one step to the left in a given time period, stay exactly where it is for the whole time period, or move one step to the right. We need to specify the probabilities of these three outcomes. Likewise, in the y direction the individual could move one step up in a given time period, stay exactly where it is for the whole time period, or move one step down. Again, we need to specify probabilities. In R, the three movement options are c(1,0,-1) for each of the types of motion (left, stay or right, and up, stay or down) and we might as well say that each of the three motions is equally likely. We need to select one of the three motions at random independently for the x and y directions at each time period. In R we use the **sample** function for this:

```
sample(c(1,0,-1),1)
```

which selects one value (the last argument is 1) with equal probability from the three listed options ($+1$, 0 or -1). Out of 99 repeats of this procedure, we should expect an average of 33 ups and 33 downs, 33 lefts and 33 rights.

We begin by defining the axes and drawing the start position in red:

```
plot(0:100,0:100,type="n",xlab="",ylab="")
x<-y<-50
points(50,50,pch=16,col="red",cex=1.5)
```

Now simulate the spatial dynamics of the random walk with up to 10 000 time steps:

```
for (i in 1:10000){
xi<-sample(c(1,0,-1),1)
yi<-sample(c(1,0,-1),1)
lines(c(x,x+xi),c(y,y+yi))
x<-x+xi
y<-y+yi
if (x>100 | x<0 | y>100 | y<0) break
}
```

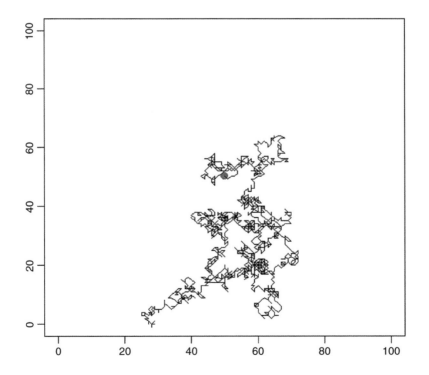

You could make the walk more sophisticated by providing wrap-around margins (see below). On average, of course, the random walk should stay in the middle, where it started, but as you will see by running this model repeatedly, most random walks do nothing of the sort. Instead, they wander off and fall over one of the edges in more or less short order.

Spatial Simulation Models

There are two broad types of spatial simulation models:

- spatial models that are divided but not spatially explicit;
- spatially explicit models where neighbours can be important.

Metapopulation dynamics is a classic example of a spatial model that is not spatially explicit. Patches are occupied or not, but the fate of a patch is not related to the fate of its immediate neighbours but rather by the global supply of propagules generated by all the occupied patches in the entire metapopulation.

Metapopulation dynamics

The theory is very simple. The world is divided up into many patches, all of which are potentially habitable. Populations on inhabited patches go extinct with a density-independent probability, e. Occupied patches all contain the same population density, and produce migrants (**propagules**) at a rate m per patch. Empty patches are colonized at a rate proportional to the total density of propagules and the availability of empty patches that are suitable for colonization. The response variable is the proportion of patches that are occupied, p. The dynamics of p, therefore, are just gains minus losses, so

$$\frac{\mathrm{d}p}{\mathrm{d}t} = p(1-p)m - ep.$$

At equilibrium $\mathrm{d}p/\mathrm{d}t = 0$, and so

$$p(1-p)m = ep,$$

giving the equilibrium proportion of occupied patches p^* as

$$p^* = 1 - \frac{e}{m}.$$

This draws attention to a critical result: there is a threshold migration rate $(m = e)$ below which the metapopulation cannot persist, and the proportion of occupied patches will drift inexorably to zero. Above this threshold, the metapopulation persists in dynamic equilibrium with patches continually going extinct (the mean lifetime of a patch is $1/e$) and other patches becoming colonized by immigrant propagules.

   The simulation produces a moving cartoon of the occupied (light colour) and empty patches (dark colour). We begin by setting the parameter values

```
m<-0.15
e<-0.1
```

We create a square universe of 10 000 patches in a 100×100 array, but this is not a spatially explicit model, and so the map-like aspects of the image should be ignored. The response variable is just the proportion of all patches that are occupied. Here are the initial conditions, placing occupied 100 patches at random in a sea of unoccupied patches:

```
s<-(1-e)
N<-matrix(rep(0,10000),nrow=100)
xs<-sample(1:100)
ys<-sample(1:100)
for (i in 1:100){
N[xs[i],ys[i]]<-1 }
image(1:100,1:100,N)
```

We want the simulation to run over 1000 generations:

```
for (t in 1:1000){
```

First we model the survival (or not) of occupied patches. Each cell of the universe gets an independent random number from a uniform distribution (a real number between 0 and 1). If the random number is bigger than or equal to the survival rate s ($= 1 - e$, above) then the patch survives for another generation. If the random number is less than s, then the patch goes extinct and N is set to zero:

```
S <-matrix(runif(10000),nrow=100)
N<-N*(S<s)
```

Note that this one statement updates the whole matrix of 10 000 patches. Next, we work out the production of propagules, im, by the surviving patches (the rate per patch is m):

```
im<-floor(sum( N*m))
```

We assume that the settlement of the propagules is random, some falling in empty patches but others being 'wasted' by falling in already occupied patches:

```
placed<-matrix(sample(c(rep(1,im) ,rep (0,10000-im))),nrow=100)
N<-N+placed
N<-apply(N,2,function(x) ifelse(x>1,1,x))
```

The last line is necessary to keep the values of N just 0 (empty) or 1 (occupied) because our algorithm gives $N = 2$ when a propagule falls in an occupied patch. Now we can draw the map of the occupied patches

```
image(1:100,1:100,N)
box(col="red")
}
```

Because the migration rate ($m = 0.15$) exceeds the extinction rate ($e = 0.1$) the metapopulation is predicted to persist. The analytical solution for the long term proportion of patches occupied is one-third of patches ($1 - 0.1/0.15 = 0.333$). At any particular time of stopping the simulation, you can work out the actual proportion occupancy as

```
sum(N)/length(N)
```

```
[1]  0.268
```

because there were 2680 occupied patches in this map:

Remember that a metapopulation model is *not* spatially explicit, so you should not read anything into any the apparent neighbour relations in this graph (the occupied patches should be distributed at random over the surface).

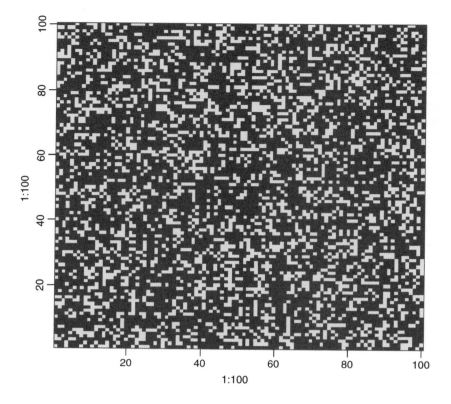

Coexistence resulting from spatially explicit (local) density dependence

We have two species which would not coexist in a well-mixed environment because the fecundity of species A is greater than the fecundity of species B, and this would lead, sooner or later, to the competitive exclusion of species B and the persistence of a monoculture of species A. The idea is to see whether the introduction of local neighbourhood density dependence is sufficient to prevent competitive exclusion and allow long-term coexistence of the two species.

The kind of mechanism that might allow such an outcome is the build-up of specialist natural enemies such as insect herbivores or fungal pathogens in the vicinity of groups of adults of species A, that might prevent recruitment by species A when there were more than a threshold number, say T, of individuals of species A in a neighbourhood.

The problem with spatially explicit models is that we have to model what happens at the edges of the universe. All locations need to have the same numbers of neighbours in the model, but patches on the edge have fewer neighbours than those in the middle. The simplest solution is to model the universe as having 'wrap-around margins' in which the left-hand edge is assumed to have the right-hand edge as its left-hand neighbour (and vice versa), while the top edge is assumed to have the bottom edge as its neighbour above (and vice versa). The four corners of the universe are assumed to be reciprocal diagonal neighbours.

We need to define who is a neighbour of whom. The simplest method, adopted here, is to assume a square grid in which a central cell has eight neighbours – three above, three below and one to either side:

```
plot(c(0,1),c(0,1),xaxt="n",yaxt="n",type="n",xlab="",ylab="")
abline("v"=c(1/3,2/3))
abline("h"=c(1/3,2/3))
xs<-c(.15,.5,.85,.15,.85,.15,.5,.85)
ys<-c(.85,.85,.85,.5,.5,.15,.15,.15)
for (i in 1:8) text(xs[i],ys[i],as.character(i))
text(.5,.5,"target cell")
```

This code produces a plot showing a target cell in the centre of a matrix, and the numbers in the other cells indicate its 'first-order neighbours':

1	2	3
4	target cell	5
6	7	8

We need to write a function to define the margins for cells on the top, bottom and edge of our universe, N, and which determines all the neighbours of the four corner cells. Our universe is 100×100 cells and so the matrix containing all the neighbours will be 102×102. Note the use of subscripts (see p. 20 for revision of this):

```
margins<-function(N){
edges<-matrix(rep(0,10404),nrow=102)
edges[2:101,2:101]<-N
edges[1,2:101]<-N[100,]
edges[102,2:101]<-N[1,]
edges[2:101,1]<-N[,100]
edges[2:101,102]<-N[,1]
edges[1,1]<-N[100,100]
edges[102,102]<-N[1,1]
edges[1,102]<-N[100,1]
edges[102,1]<-N[1,100]
edges}
```

Next, we need to write a function to count the number of species A in the eight neighbouring cells, for any cell *i*, *j*:

```
nhood<-function(X,j,i) sum(X[(j-1):(j+1),(i-1):(i+1)]==1)
```

Now we can set the parameter values: the reproductive rates of species A and B, the death rate of adults (which determines the space freed up for recruitment) and the threshold number, *T*, of species A (out of the eight neighbours) above which recruitment cannot occur:

```
Ra<-3
Rb<-2.0
D<-0.25
s<-(1-D)
T<-6
```

The initial conditions fill one half of the universe with species A and the other half with species B, so that we can watch any spatial pattern as it emerges:

```
N<-matrix(c(rep(1,5000),rep(2,5000)),nrow=100)
image(1:100,1:100,N)
```

We run the simulation for 1000 time steps:

```
for (t in 1:1000){
```

First, we need to see if the occupant of a cell survives or dies. For this, we compare a uniformly distributed random number between 0 and 1 with the specified survival rate $s = 1 - D$. If the random number is less than *s* the occupant survives, if it is greater than *s* it dies:

```
S <-1*(matrix(runif(10000),nrow=100)<s)
```

We kill the necessary number of cells to open up space for recruitment:

```
N<-N*S
space<-10000-sum(S)
```

Next, we need to compute the neighbourhood density of A for every cell (using the wrap-around margins):

```
nt<-margins(N)
```

```
tots<-matrix(rep(0,10000),nrow=100)
for (a in 2:101) {
for (b in 2:101) {
tots[a-1,b-1]<-nhood(nt,a,b)
}}
```

The survivors produce seeds as follows:

```
seedsA<- sum(N==1)*Ra
seedsB<- sum(N==2)*Rb
all.seeds<-seedsA+seedsB
fA=seedsA/all.seeds
fB=1-fA
```

Seeds settle over the universe at random.

```
setA<-ceiling(10000*fA)
placed<-matrix(sample(c(rep(1,setA) ,rep (2,10000-setA))),nrow=100)
```

Seeds only produce recruits in empty cells $N[i,j] == 0$. If the winner of an empty cell (placed) is species B, then species B gets that cell: $if(placed[i,j]== 2)$ $N[i,j]<-2$. If species A is supposed to win a cell, then we need to check that it has fewer than T neighbours of species A. If so, species A gets the cell. If not, the cell is forfeited to species B: if $(tots[i,j]>=T)$ $N[i,j]<-2$.

```
for (i in 1:100){
for(j in 1:100){
if (N[i,j] == 0 )
if(placed[i,j]== 2) N[i,j]<-2
else
if (tots[i,j]>=T) N[i,j]<-2
else N[i,j]<-1
}}
```

Finally, we can draw the map, showing species A in red and species B in white:

```
image(1:100,1:100,N)
box(col="red")}
```

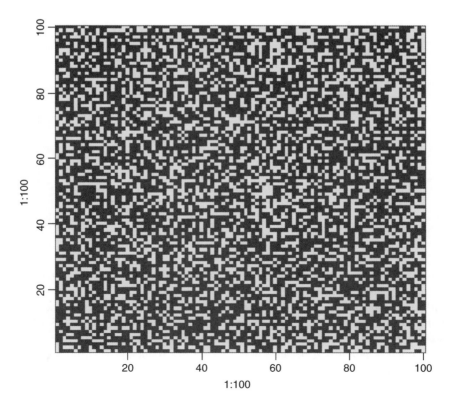

You can watch as the initial half-and-half pattern breaks down, and species A increases in frequency at the expense of species B. Eventually, however, species A gets to the point where most of the cells have six or more neighbouring cells containing species A, and its recruitment begins to fail. At equilibrium, species B persists in isolated cells or in small (white) patches, where the cells have six or more occupants that belong to species A.

If you set the threshold $T = 9$, you can watch as species A drives species B to extinction. If you want to turn the tables, and see species B take over the universe, set $T = 0$.

Pattern Generation Resulting from Dynamic Interactions

In this section we look at an example of an ecological interaction between a species and its parasite. The interaction is unstable in a non-spatial model, with increasing oscillations in numbers leading quickly to extinction of the host species and then, in the next generation, its parasite. The non-spatial dynamics look like this:

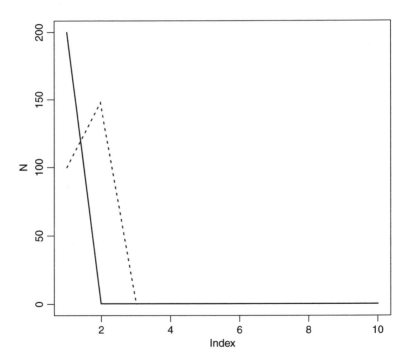

The parasite increases in generation number 1 and drives the host to extinction in generation 2, subsequently going extinct itself.

In a spatial model, we allow that hosts and parasites can move from the location in which they were born to any one of the eight first-order neighbouring cells (p. 819). For the purposes of dispersal, the universe is assumed have wrap-around margins for both species. The interaction is interesting because it is capable of producing beautiful spatial patterns that fluctuate with host and parasite abundance. We begin by setting the parameter values for the dynamics of the host (r) and the parasite (a) and the migration rates of the host

(Hmr = 0.1) and parasite (Pmr = 0.9): in this case the hosts are relatively sedentary and the parasites are highly mobile:

```
r<-0.4
a<-0.1
Hmr<-0.1
Pmr<-0.9
```

Next, we set up the matrices of host (N) and parasite (P) abundance. These will form what is termed a coupled map lattice:

```
N<-matrix(rep(0,10000),nrow=100)
P<-matrix(rep(0,10000),nrow=100)
```

The simulation is seeded by introducing 200 hosts and 100 parasites into a single cell at location (33,33):

```
N[33,33]<-200
P[33,33]<-100
```

We need to define a function called host to calculate the next host population as a function of current numbers of hosts and parasites (N and P), and another function called parasite to calculate the next parasite population as a function of N and P – this is called a Nicholson–Bailey model:

```
host<-function(N,P) N*exp(r-a*P)
parasite<-function(N,P) N*(1-exp(-a*P))
```

Both species need a definition of their wrap-around margins for defining the destinations of migrants from each cell:

```
host.edges<-function(N){
Hedges<-matrix(rep(0,10404),nrow=102)
Hedges[2:101,2:101]<-N
Hedges[1,2:101]<-N[100,]
Hedges[102,2:101]<-N[1,]
Hedges[2:101,1]<-N[,100]
Hedges[2:101,102]<-N[,1]
Hedges[1,1]<-N[100,100]
Hedges[102,102]<-N[1,1]
Hedges[1,102]<-N[100,1]
Hedges[102,1]<-N[1,100]
Hedges}

parasite.edges<-function(P){
Pedges<-matrix(rep(0,10404),nrow=102)
Pedges[2:101,2:101]<-P
Pedges[1,2:101]<-P[100,]
Pedges[102,2:101]<-P[1,]
Pedges[2:101,1]<-P[,100]
Pedges[2:101,102]<-P[,1]
Pedges[1,1]<-P[100,100]
Pedges[102,102]<-P[1,1]
```

```
Pedges[1,102]<-P[100,1]
Pedges[102,1]<-P[1,100]
Pedges}
```

A function is needed to define the eight cells that comprise the neighbourhood of any cell and add up the total number of neighbouring individuals:

```
nhood<-function(X,j,i) sum(X[(j-1):(j+1),(i-1):(i+1)])
```

The number of host migrants arriving in every cell is calculated as follows:

```
h.migration<-function(Hedges){
Hmigs<-matrix(rep(0,10000),nrow=100)
for (a in 2:101) {
for (b in 2:101) {
Hmigs[a-1,b-1]<-nhood(Hedges,a,b)
}}
Hmigs}
```

The number of parasites migrants is given by:

```
p.migration<-function(Pedges){
Pmigs<-matrix(rep(0,10000),nrow=100)
for (a in 2:101) {
for (b in 2:101) {
Pmigs[a-1,b-1]<-nhood(Pedges,a,b)
}}
Pmigs}
```

The simulation begins here, and runs for 600 generations:

```
for (t in 1:600){

he<-host.edges(N)
pe<-parasite.edges(P)

Hmigs<-h.migration(he)
Pmigs<-p.migration(pe)

N<-N-Hmr*N+Hmr*Hmigs/9
P<-P-Pmr*P+Pmr*Pmigs/9

Ni<-host(N,P)
P<-parasite(N,P)
N<-Ni

image(1:100,1:100,N)
}
```

You can watch as the initial introduction at (33,33) spreads out and both host and parasite populations pulse in abundance. Eventually, the wave of migration reaches the margin and appears on the right hand edge. The fun starts when the two waves meet one another. The pattern below is typical of the structure that emerges towards the middle of a simulation run:

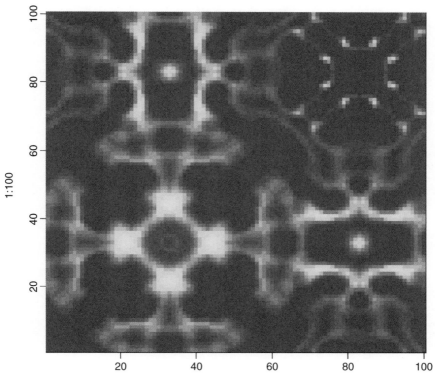

27

Changing the Look of Graphics

Many of the changes that you want to make to the look of your graphics will involve the use of the graphics parameters function, par. Other changes, however, can be made through alterations to the arguments to high-level functions such as plot, points, lines, axis, title and text (these are shown with an asterisk in Table 27.1).

Graphs for Publication

The most likely changes you will be asked to make are to the orientation of the numbers on the tick marks, and to the sizes of the plotting symbols and text labels on the axes. There are four functions involved here:

las   determines the orientation of the numbers on the tick marks;

cex   determines the size of plotting characters (pch);

cex.lab   determines the size of the text labels on the axes;

cex.axis determines the size of the numbers on the tick marks.

Here we show four different combinations of options. You should pick the settings that look best for your particular graph.

```
par(mfrow=c(2,2))
x<-seq(0,150,10)
y<-16+x*0.4+rnorm(length(x),0,6)
plot(x,y,pch=16,xlab="label for x axis",ylab="label for y axis")
plot(x,y,pch=16,xlab="label for x axis",ylab="label for y axis",
                las=1,cex.lab=1.2, cex.axis=1.1)
plot(x,y,pch=16,xlab="label for x axis",ylab="label for y axis",
                las=2,cex=1.5)
plot(x,y,pch=16,xlab="label for x axis",ylab="label for y axis",
                las=3,cex=0.7,cex.lab=1.3, cex.axis=1.3)
```

The top left-hand graph uses all the default settings:

las = 0, cex = 1, cex.lab = 1

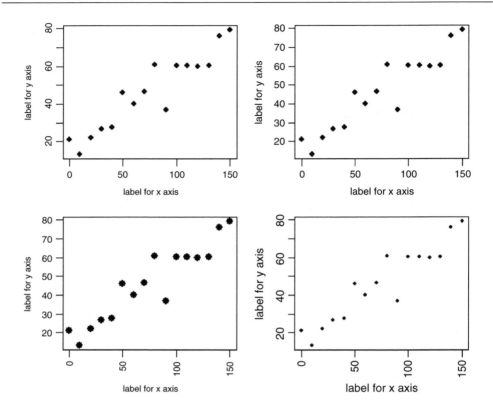

In the top right-hand graph the numbers have been rotated so that they are all vertical (las = 1), the label size has been increased by 20% and the numbers by 10%:

las = 1, cex = 1, cex.lab = 1.2, cex.axis=1.1

In the bottom left-hand graph the plotting symbol (pch = 16) has been increased in size by 50% and the numbers on both axes are parallel with their axes (las = 2):

las = 2, cex = 1.5, cex.lab = 1

Finally, in the bottom right-hand graph the label size has been increased by 30%, the plotting symbols reduced by 30% and the axes numbers are all at 90 degrees (las = 3)

las = 3, cex = 0.7, cex.lab = 1.3, cex.axis = 1.3

My favourite is the top right-hand graph with slightly larger text and numbers and with vertical numbering on the *y* axis.

Shading

You can control five aspects of the shading: the density of the lines, the angle of the shading, the border of the shaded region, the colour of the lines and the line type. Here are their default values:

density = NULL
angle = 45

```
border = NULL
col = NA
lty = par("lty"), ...)
```

Other graphical parameters such as xpd, lend, ljoin and lmitre (Table 27.1) can be given as arguments.

The following data come from a long-term study of the botanical composition of a pasture, where the response variable is the dry mass of a grass species called *Festuca rubra* (FR), and the two explanatory variables are total hay biomass and soil pH:

```
data<-read.table("c:\\temp\\pgr.txt",header=T)
attach(data)
names(data)
```

```
[1]  "FR"  "hay"  "pH"
```

The idea is to draw polygons to represent the convex hulls for the abundance of *Festuca* in the space defined by hay biomass and soil pH. The polygon is to be red for *Festuca* > 5, green for *Festuca* > 10 and cross-hatched in blue for *Festuca* > 20. After all of the solid objects have been drawn, the data are to be overlaid as a scatterplot with pch = 16:

```
plot(hay,pH)
x<-hay[FR>5]
y<-pH[FR>5]
polygon(x[chull(x,y)],y[chull(x,y)],col="red")
```

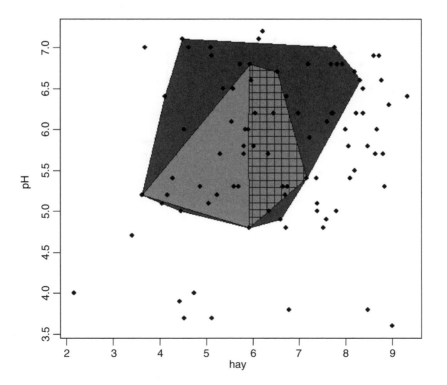

```
x<-hay[FR>10]
y<-pH[FR>10]
polygon(x[chull(x,y)],y[chull(x,y)],col="green")
x<-hay[FR>20]
y<-pH[FR>20]
polygon(x[chull(x,y)],y[chull(x,y)],density=10,angle=90,col="blue")
polygon(x[chull(x,y)],y[chull(x,y)],density=10,angle=0,col="blue")
points(hay,pH,pch=16)
```

The issue of transparency (i.e. what you can see 'through' what) is described in the help files for ?polygon and ?rgb. If in doubt, use points, lines and polygons in sequence, so that objects ('on top') that you want to be visible in the final image are drawn last.

Logarithmic Axes

You can transform the variables inside the plot function (e.g. plot(log(y)~x)) or you can plot the untransformed variables on logarithmically scaled axes (e.g. log="x").

```
data<-read.table("c:\\temp\\logplots.txt",header=T)
attach(data)
names(data)
```

```
[1]  "x"  "y"
```

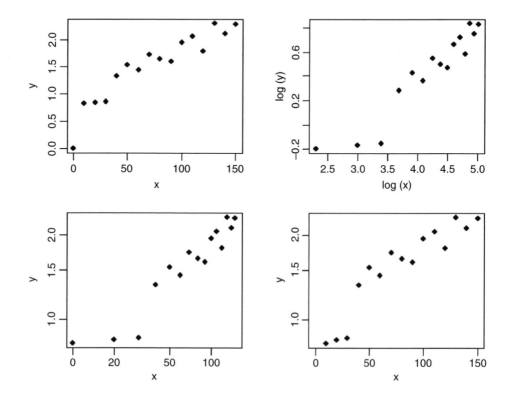

```
par(mfrow=c(2,2))
```

```
plot(x,y,pch=16)
plot(log(x),log(y),pch=16)
plot(x,y,pch=16,log="xy")
plot(x,y,pch=16,log="y")
```

The untransformed data are in the top left-hand graph, and both x and y are transformed to logs before plotting in the upper right. The bottom left-hand plot shows both axes log-transformed, while the bottom right shows the data with only the y axis log-transformed. Note that all logs in R are to the base e by default (not base 10). It is important to understand that when R is asked to plot the log of zero it simply omits any such points without comment (compare the top left-hand graph with a point at $(0, 0)$ with the other three graphs).

Axis Labels Containing Subscripts and Superscripts

The default xlab and ylab do not allow subscripts like r2 or superscripts or xi. For these you need to master the **expression** function. In R, the operator for superscripts is 'hat' (or, more correctly, 'caret') so 'r squared' is written r^2. Likewise subscripts in R involve square brackets [] so x_i is written x[i]. Suppose that we want r^2 to be the label on the y axis and x_i to be the label on the x axis. The **expression** function turns R code into text like this:

```
plot(1:10,1:10, ylab=expression(r^2), xlab=expression(x[i]),type="n")
```

Different font families for text

To change the typeface used for plotted text, change the name of a font family. Standard values are family = "serif", "sans" (the default font), "mono", and "symbol", and the Hershey font families are also available. Some devices will ignore this setting completely. Text drawn onto the plotting region is controlled using **par** like this:

```
par(family="sans")
text(5,8,"This is the default font")
par(family="serif")
text(5,6,"This is the serif font")
par(family="mono")
text(5,4,"This is the mono font")
par(family="symbol")
text(5,2,"This is the symbol font")
par(family="sans")
```

Don't forget to turn the family back to "sans", otherwise you may get some very unexpected symbols in your next **text**. To write the results of calculations using **text**, it is necessary to use **substitute** with **expression**. Here, the coefficient of determination (cd) was calculated earlier and we want to write its value on the plot, labelled with '$r^2 =$':

```
cd<- 0.63
...
text(locator(1),as.expression(substitute(r^2 == cd,list(cd=cd))))
```

Just click when the cursor is where you want the text to appear. Note the use of 'double equals'.

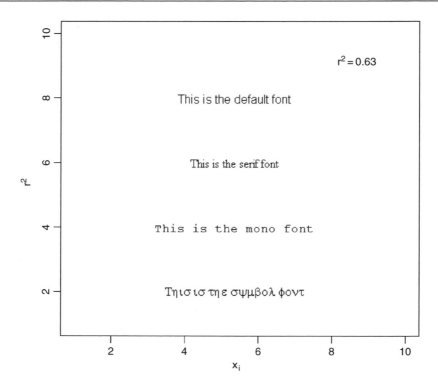

Mathematical Symbols on Plots

To write on plots using more intricate symbols such as mathematical symbols or Greek letters we use **expression** or **substitute**. Here are some examples of their use. First, we produce a plot of sin ϕ against the phase angle ϕ over the range $-\pi$ to $+\pi$ radians:

```
x <- seq(-4, 4, len = 101)
plot(x,sin(x),type="l",xaxt="n",
        xlab=expression(paste("Phase Angle ",phi)),
        ylab=expression("sin "*phi))
axis(1, at = c(-pi, -pi/2, 0, pi/2, pi),
        lab = expression(-pi, -pi/2, 0, pi/2, pi))
```

Note the use of **xaxt = "n"** to suppress the default labelling of the x axis, and the use of **expression** in the labels for the x and y axes to obtain mathematical symbols such as phi (ϕ) and pi (π). The more intricate values for the tick marks on the x axis are obtained by the **axis** function, specifying 1 (the x ('bottom') axis is axis no. 1), then using the **at** function to say where the labels and tick marks are to appear, and **lab** with **expression** to say what the labels are to be.

Suppose you wanted to add $\chi^2 = 24.5$ to this graph at location $(-\pi/2, 0.5)$. You use **text** with **substitute**, like this:

```
text(-pi/2,0.5,substitute(chi^2=="24.5"))
```

Note the use of 'double equals' to print a single equals sign. You can write quite complicated formulae on plots using paste to join together the elements of an equation. Here is the density function of the normal written on the plot at location $(\pi/2, -0.5)$:

```
text(pi/2, -0.5, expression(paste(frac(1, sigma*sqrt(2*pi)), " ",
    e^{frac(-(x-mu)^2, 2*sigma^2)})))
```

Note the use of frac to obtain individual fractions: the first argument is the text for the numerator, the second the text for the denominator. Most of the arithmetic operators have obvious formats $(+, -, /, *, \hat{\ },$ etc.); the only non-intuitive symbol that is commonly used is 'plus or minus' $\pm$; this is written as $\% + -\%$ like this:

```
text(pi/2,0,expression(hat(y) %+-% se))
```

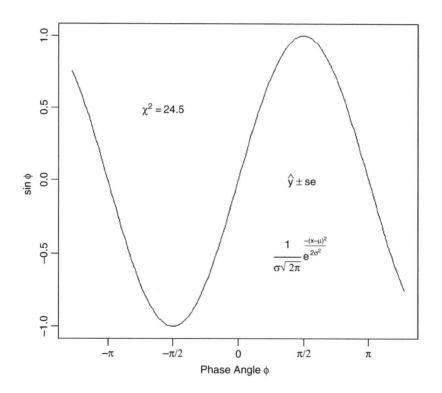

Phase Planes

Suppose that we have two competing species (named 1 and 2) and we are interested in modelling the dynamics of the numbers of each species (N_1 and N_2). We want to draw a phase plane showing the behaviour of the system close to equilibrium. Setting the derivates to zero and dividing both sides by $r_i N_i$ we get

$$0 = 1 - \alpha_{11}N_1 - \alpha_{12}N_2,$$

which is called the isocline for species 1. It is linear in N_1 and N_2 and we want to draw it on a phase plane with N_2 on the y axis and N_1 on the x axis. The intercept on the y axis shows the abundance of N_2 when $N_1 = 0$: this is $1/\alpha_{12}$. Likewise, when $N_2 = 0$ we can see that $N_1 = 1/\alpha_{11}$ (the value of its single-species equilibrium). Similarly,

$$0 = 1 - \alpha_{21} N_1 - \alpha_{22} N_2$$

describes the isocline for species 2. The intercept on the y axis is $1/\alpha_{22}$ and the value of N_1 when $N_2 = 0$ is $1/\alpha_{21}$. Now we draw a phase plane with both isoclines, and label the ends of the lines appropriately. We might as well scale the axes from 0 to 1 but we want to suppress the default tick marks:

```
plot(c(0,1),c(0,1),ylab="",xlab="",xaxt="n",yaxt="n",type="n")
abline(0.8,-1.5)
abline(0.6,-0.8,lty=2)
```

The solid line shows the isocline for species 1 and the dotted line shows species 2.

Now for the labels. We use at to locate the tick marks – first the x axis (axis = 1),

```
axis(1, at = 0.805, lab = expression(1/alpha[21]))
axis(1, at = 0.56, lab = expression(1/alpha[11]))
```

and now the y axis (axis = 2),

```
axis(2, at = 0.86, lab = expression(1/alpha[12]),las=1)
axis(2, at = 0.63, lab = expression(1/alpha[22]),las=1)
```

Note the use of las=1 to turn the labels through 90 degrees to the horizontal. Now label the lines to show which species isocline is which. Note the use of the function fract to print fractions and square brackets (outside the quotes) for subscripts:

```
text(0.05,0.85, expression(paste(frac("d N"[1],"dt"), " = 0" )))
text(0.78,0.07, expression(paste(frac("d N"[2],"dt"), " = 0" )))
```

We need to draw phase plane trajectories to show the dynamics. Species will increase when they are at low densities (i.e. 'below' their isoclines) and decrease at high densities (i.e. 'above' their isoclines). Species 1 increasing is a horizontal arrow pointing to the right. Species 2 declining is a vertical arrow pointing downwards. The resultant motion shows how both species' abundances change through time from a given point on the phase plane.

```
arrows(-0.02,0.72,0.05,0.72,length=0.1)
arrows(-0.02,0.72,-0.02,0.65,length=0.1)
arrows(-0.02,0.72,0.05,0.65,length=0.1)

arrows(0.65,-0.02,0.65,0.05,length=0.1)
arrows(0.65,-0.02,0.58,-0.02,length=0.1)
arrows(0.65,-0.02,0.58,0.05,length=0.1)

arrows(0.15,0.25,0.15,0.32,length=0.1)
arrows(0.15,0.25,0.22,0.25,length=0.1)
arrows(0.15,0.25,0.22,0.32,length=0.1)

arrows(.42,.53,.42,.46,length=0.1)
arrows(.42,.53,.35,.53,length=0.1)
arrows(.42,.53,.35,.46,length=0.1)
```

All the motions converge, so the point is a stable equilibrium and the two species would coexist. All other configurations of the isoclines lead to competitive exclusion of one of the two species. Finally, label the axes with the species' identities:

```
axis(1, at = 1, lab = expression(N[1]))
axis(2, at = 1, lab = expression(N[2]),las=1)
```

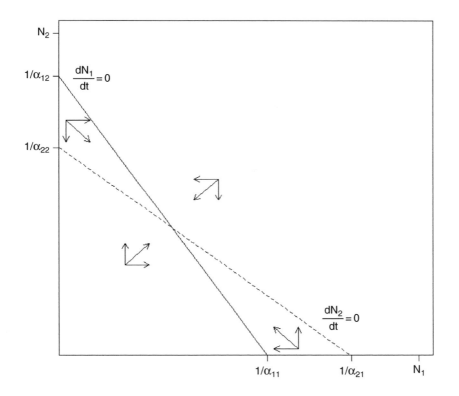

Fat Arrows

You often want to add arrows to plots in order to draw attention to particular features. Here is a function called **fat.arrows** that uses **locator(1)** to identify the bottom of the point of a vertical fat arrow. You can modify the function to draw the arrow at any specified angle to the clicked point of its arrowhead. The default widths and heights of the arrow are 0.5 scaled *x* or *y* units and the default colour is red:

```
fat.arrow<-function(size.x=0.5,size.y=0.5,ar.col="red"){
size.x<-size.x*(par("usr")[2]-par("usr")[1])*0.1
size.y<-size.y*(par("usr")[4]-par("usr")[3])*0.1
pos<-locator(1)
xc<-c(0,1,0.5,0.5,-0.5,-0.5,-1,0)
yc<-c(0,1,1,6,6,1,1,0)
polygon(pos$x+size.x*xc,pos$y+size.y*yc,col=ar.col) }
```

We will use this function later in this chapter (p. 857).

Trellis Plots

You need to load the **lattice** package and set the background colour to white. You can read the details in ?trellis.device.

library(lattice)

The most commonly use trellis plot is **xyplot**, which produces conditional scatterplots where the response, y, is plotted against a continuous explanatory variable x, for different levels of a conditioning factor, or different values of the shingles of a conditioning variable. This is the standard plotting method that is used for linear mixed-effects models and in cases where there are nested random effects (i.e. with **groupedData** see p. 668). The structure of the plot is typically controlled by the formula; for example

xyplot(y ~ x | subject)

where a separate graph of y against x is produced for each level of subject (the vertical bar | is read as 'given'). If there are no conditioning variables, **xyplot(y ~ x)**, the plot produced consists of a single panel. All arguments that are passed to a high-level trellis function like **xyplot** that are not recognized by it are passed through to the **panel** function. It is thus generally good practice when defining panel functions to allow a . . . argument. Such extra arguments typically control graphical parameters.

Panels are by default drawn starting from the bottom left-hand corner, going right and then up, unless **as.table = TRUE**, in which case panels are drawn from the top left-hand corner, going right and then down. Both of these orders can be modified using the **index.cond** and **perm.cond** arguments. There are some grid-compatible replacements for commonly used base R graphics functions: for example, **lines** can be replaced by **llines** (or equivalently, **panel.lines**). Note that base R graphics functions like **lines** will not work in a lattice panel function. The following example is concerned with root growth measured over time, as repeated measures on 12 individual plants:

results< -read.table("c:\\temp\\fertilizer.txt",header=T)
attach(results)
names(results)

```
[1]  "root"  "week"  "plant"  "fertilizer"
```

Panel scatterplots

Panel plots are very easy to use. Here is a set of 12 scatterplots, showing **root ~ week** with one panel for each plant like this: | plant

xyplot(root ~ week | plant)

By default, the panels are shown in alphabetical order by plant name from bottom left (ID1) to top right (ID9). If you want to change things like the plotting symbol you can do this within the **xyplot** function,

xyplot(root ~ week | plant,pch=16)

but if you want to make more involved changes, you should use a **panel** function. Suppose we want to fit a separate linear regression for each individual plant. We write

```
xyplot(root ~ week | plant ,
          panel = function(x, y) {
          panel.xyplot(x, y, pch=16)
          panel.abline(lm(y ~ x))
       })
```

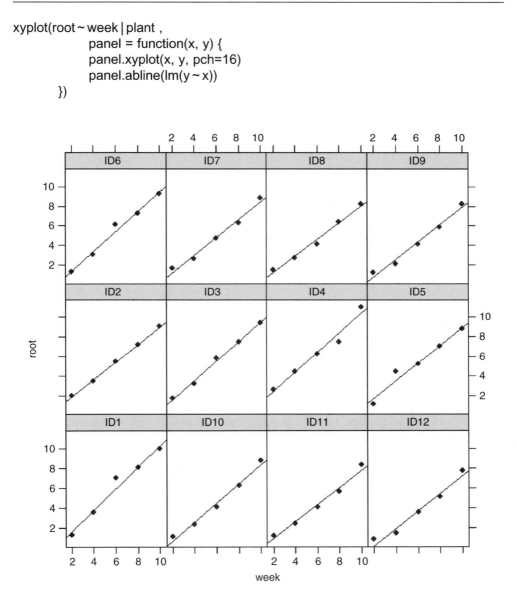

Panel boxplots

Here is the basic box-and-whisker trellis plot for the Daphnia data:

```
data< -read.table("c:\\temp\\Daphnia.txt",header=T)
attach(data)
names(data)
```

```
[1] "Growth.rate" "Water" "Detergent" "Daphnia"
```

```
bwplot(Growth.rate ~ Detergent | Daphnia, xlab = "detergent" )
```

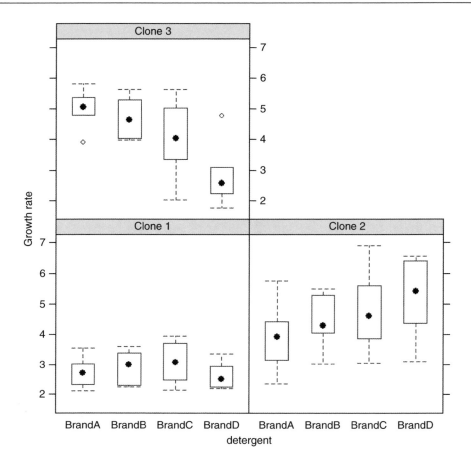

A separate box-and-whisker is produced for each level of detergent within each clone, and a separate panel is produced for each level of Daphnia.

Panel barplots

The following example shows the use of the trellis version of barchart with the barley data. The data are shown separately for each year (groups = year) and the bars are stacked for each year (stack = TRUE) in different shades of blue (col=c("cornflowerblue","blue")): The barcharts are produced in three rows of two plots each (layout = c(2,3)). Note the use of scales to rotate the long labels on the x axis through 45 degrees:

```
barchart(yield ~ variety | site, data = barley,
         groups = year, layout = c(2,3), stack = TRUE,
         col=c("cornflowerblue","blue"),
         ylab = "Barley Yield (bushels/acre)",
         scales = list(x = list(rot = 45)))
```

Panels for conditioning plots

In this example we put each of the panels side by side (layout=c(9,1)) on the basis of an equal-count split of the variable called *E* within the ethanol dataframe:

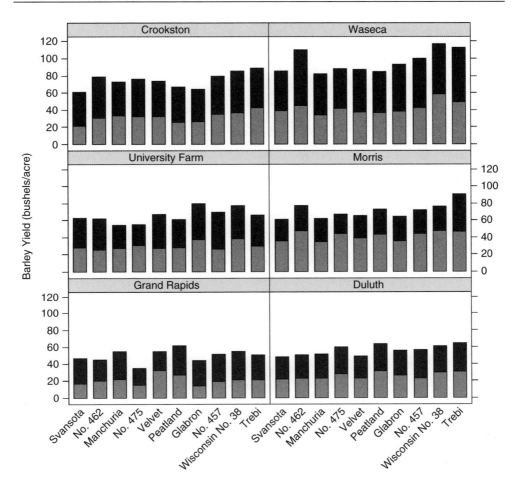

EE <- equal.count(ethanol$E, number=9, overlap=1/4)

Within each panel defined by *EE* we draw a grid (panel.grid(h=-1, v= 2)), create a scatterplot of NOx against *C* (panel.xyplot(x, y)) and draw an individual linear regression (panel.abline(lm(y ~ x))):

```
xyplot(NOx ~ C | EE, data = ethanol,layout=c(9,1),
      panel = function(x, y) {
        panel.grid(h=-1, v= 2)
        panel.xyplot(x, y)
        panel.abline(lm(y ~ x))
      })
```

This is an excellent way of illustrating that the correlation between NOx and C is positive for all levels of *EE* except the highest one, and that the relationship is steepest for values of *EE* just below the median (i.e. in the third panel from the left).

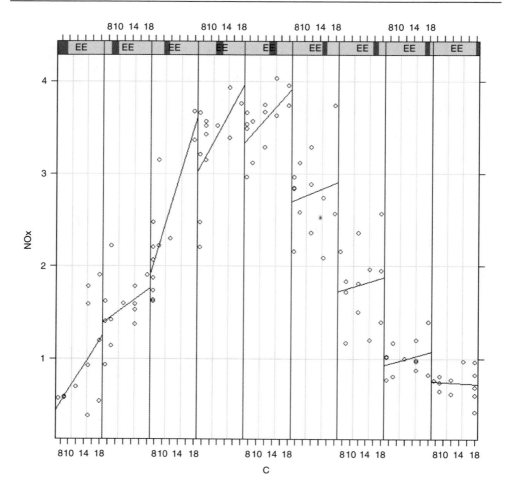

Panel histograms

The task is to use the Silwood weather data to draw a panel of histograms, one for each month of the year, showing the number of years in which there were 0, 1, 2, . . . rainy days per month during the period 1987–2005. We use the built-in function month.abb to generate a vector of labels for the panels.

```
months < -month.abb
data < -read.table("c: \\ temp \\ SilwoodWeather.txt",header=T)
attach(data)
names(data)
```

```
[1] "upper" "lower" "rain" "month" "yr"
```

We define a rainy day as a day when there was measurable rain

```
wet < -(rain > 0)
```

then create a vector, *wd*, containing the total number of wet days in each month in each of the years in the dataframe, along with a vector of the same length, mos, indicating the month, expressed as a factor:

```
wd <-as.vector(tapply(wet,list(yr,month),sum))
mos <-factor(as.vector(tapply(month,list(yr,month),mean)))
```

The panel histogram is drawn using the histogram function which takes a model formula *without a response variable* ~wd|mos as its first argument. We want integer bins so that we can see the number of days with no rain at all, breaks=-0.5:28.5, and we want the strips labelled with the months of the year (rather than the variable name) using strip=strip.custom(factor.levels=months)):

```
histogram( ~ wd | mos,type="count",xlab="rainy days",ylab="frequency",
    breaks=-0.5:28.5,strip=strip.custom(factor.levels=months))
```

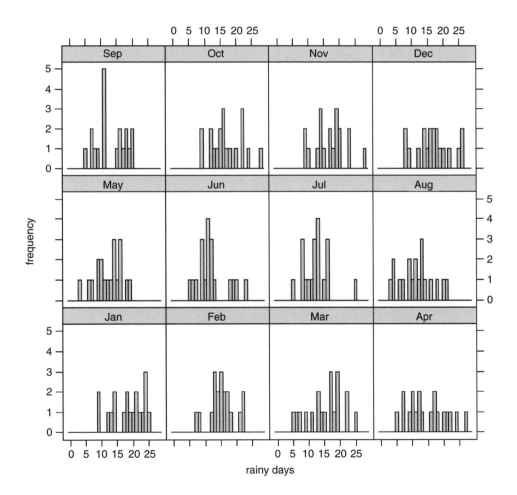

You can see at once that there is rather little seasonality in rainfall at Silwood, and that most months have had at least 20 wet days in at least one year since 1987. No months have been entirely rain-free, but May and August have had just 3 days with rain (in 1990 and 1995, respectively). The rainiest months in terms of number of wet days were October 2000 and November 2002, when there were 28 days with rain.

More panel functions

Plots can be transformed by specifying the grouping (groups = rowpos), indicating that each group should be drawn in a different colour (panel = "panel.superpose"), or by specifying that the dots should be joined by lines for each member of the group (panel.groups = "panel.linejoin"). Here are the orchard spray data with each row shown in a different colour and the treatment means joined together by lines. This example also shows how to use auto.key to locate a key to the groups on the right of the plot, showing lines rather than points:

xyplot(decrease ~ treatment, OrchardSprays, groups = rowpos,
      type="a",
      auto.key =
      list(space = "right", points = FALSE, lines = TRUE))

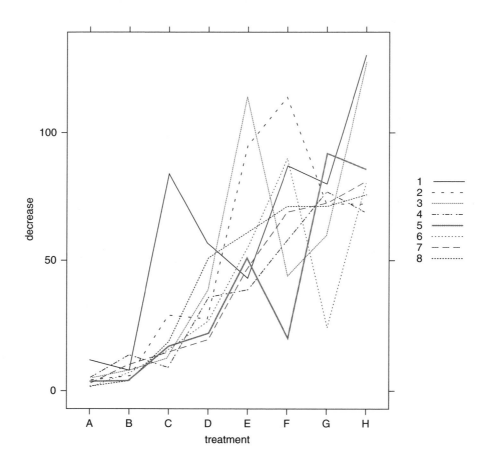

Three-Dimensional Plots

When there are two continuous explanatory variables, it is often useful to plot the response as a contour map. In this example, the biomass of one plant species (the response variable) is plotted against soil pH and total community biomass. The species is a grass called *Festuca rubra* that peaks in abundance in communities of intermediate total biomass:

```
data < -read.table("c: \\ temp \\ pgr.txt",header=T)
attach(data)
names(data)
```

```
[1] "FR" "hay" "pH"
```

You need the library called akima in order to implement bivariate interpolation onto a grid for irregularly spaced input data like these, using the function interp:

```
install.packages("akima")
library(akima)
```

The two explanatory variables are presented first (hay and pH in this case), with the response variable (the 'height' of the topography), which is FR in this case, third:

```
zz < -interp(hay,pH,FR)
```

The list called *zz* can now be used in any of the four functions contour, filled.contour, image or persp. We start by using contour and image together. Rather than the red and yellows of heat.colors we choose the cooler blues and greens of topo.colors:

```
image(zz,col = topo.colors(12),xlab="biomass",ylab="pH")
contour(zz,add=T)
```

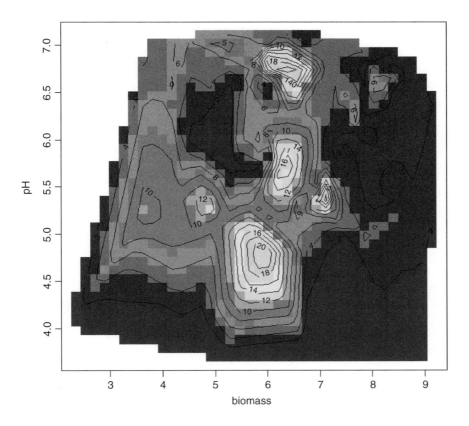

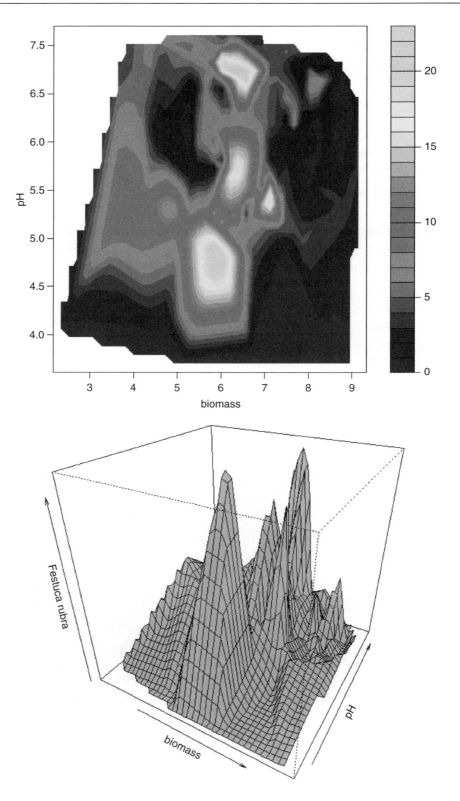

Alternatively, you can use the filled.contour function,

```
filled.contour(zz,col = topo.colors(24),xlab="biomass",ylab="pH")
```

which provides a useful colour key to the abundance of *Festuca*. Evidently the grass peaks in abundance at intermediate biomass, but it also occurs at lower biomasses on soils of intermediate pH (5.0 to 6.0). It is found in only trace amounts in communities where the biomass is above 7.5 tonnes per hectare, except where soil pH is *c*.6.6.

The function persp allows an angled view of a 3D-like object, rather than the map-like views of contour and image. The angles theta and phi define the viewing direction: theta gives the azimuthal direction and phi gives the colatitude.

```
persp(zz,xlab="biomass",ylab="pH",zlab="Festuca rubra",
                  theta = 30, phi = 30,col="lightblue")
```

It is straightforward to create 3D images of mathematical functions from regularly spaced grids produced by the outer function without using interp. First create a series of values for the *x* and *y* axis (the base of the plot):

```
x <-seq(0,10,0.1)
y <-seq(0,10,0.1)
```

Now write a function to predict the height of the graph (the response variable, *z*) as a function of the two explanatory variables *x* and *y*:

```
func <-function(x,y) 3 * x * exp(0.1*x) * sin(y*exp(-0.5*x))
```

Now use the outer function to evaluate the function over the complete grid of points defined by *x* and *y*:

```
image(x,y,outer(x,y,func))
contour(x,y,outer(x,y,func),add=T)
```

Complex 3D plots with wireframe

If you want to create really fancy 3D graphics you will want to master the wireframe function, which allows you to specify the location of the notional light source that illuminates your object (and hence creates the shadows). Here are two examples from demo(trellis) that produce pleasing 3D objects. In the first case, the surface is based on data (in the dataframe called volcano), whereas in the second case (strips on a globe) the graphic is based on an equation ($z \sim x * y$). It is in library(lattice). This is how wireframe is invoked:

```
wireframe(volcano, shade = TRUE, aspect = c(61/87,
     0.4), screen = list(z = -120, x = -45), light.source = c(0,
     0, 10), distance = 0.2, shade.colors = function(irr, ref,
     height, w = 0.5) grey(w * irr + (1 - w) * (1 - (1 - ref)^0.4)))
```

Next, we see a shaded globe with the surface turned into strips by leaving out every other pair of coloured orbits by setting their values to NA.

```
n <- 50
tx <- matrix(seq(-pi, pi, len = 2 * n), 2 * n, n)
ty <- matrix(seq(-pi, pi, len = n)/2, 2 * n, n, byrow = T)
xx <- cos(tx) * cos(ty)
yy <- sin(tx) * cos(ty)
```

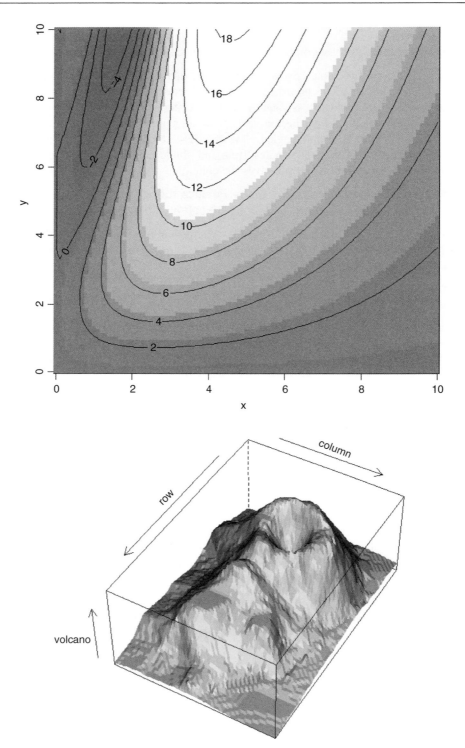

```
zz <- sin(ty)
zzz <- zz
zzz[, 1:12 * 4] <- NA
```

Now draw the globe and shade the front and back surfaces appropriately:

```
wireframe(zzz ~ xx * yy, shade = TRUE, light.source = c(3,3,3))
```

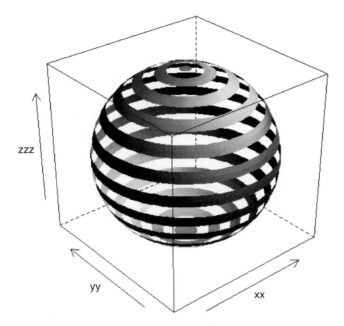

An Alphabetical Tour of the Graphics Parameters

Beginners cannot be expected to know which graphics attributes are changed with the par function, which can be changed inside the plot function, and which stand alone. This section therefore unites all the various kinds of graphics control into a single list (see Table 27.1): properties that are altered by a call to the par function are shown as par(name), while properties that can be altered inside a plot function are shown in that context; other graphics functions that stand alone (such as axis) are not shown in the table.

When writing functions, you need to know things about the current plotting region. For instance to find out the limits of the current axes, use

```
par("usr")
```

```
[1] 1947.92 2004.08 -80.00 2080.00
```

which shows the minimum x value par("usr")[1], the maximum x value par("usr")[2], the minimum y value par("usr")[3] and the maximum y value par("usr")[4] of the current plotting region for the gales data (p. 859).

If you need to use par, then the graphics parameters should be altered *before* you use the first plot function. It is a good idea to save a copy of the default parameter settings so that they can be changed back at the end of the session to their default values:

Table 27.1. Graphical parameters and their default values. Each of the functions is illustrated in detail in the text. The column headed 'In plot?' indicates with an asterisk whether this parameter can be changed as an argument to the plot, points or lines functions.

Parameter	In plot?	Default value	Meaning
adj	*	centred	Justification of text
ann	*	TRUE	Annotate plots with axis and overall titles?
ask		FALSE	Pause before new graph?
bg	*	"transparent"	Background style or colour
bty		full box	Type of box drawn around the graph
cex	*	1	Character expansion: enlarge if > 1, reduce if < 1
cex.axis	*	1	Magnification for axis notation
cex.lab	*	1	Magnification for label notation
cex.main	*	1.2	Main title character size
cex.sub	*	1	Sub-title character size
cin		0.1354167 0.1875000	Character size (width, height) in inches
col	*	"black"	colors() to see range of colours
col.axis		"black"	Colour for graph axes
col.lab	*	"black"	Colour for graph labels
col.main	*	"black"	Colour for main heading
col.sub	*	"black"	Colour for sub-heading
cra		13 18	Character size (width, height) in rasters (pixels)
crt		0	Rotation of single characters in degrees (see srt)
csi		0.1875	Character height in inches
cxy		0.02255379 0.03452245	Character size (width, height) in user-defined units
din		7.166666 7.156249	Size of the graphic device (width, height) in inches (the window is bigger than this)
family	*	"sans"	Font style: from "serif", "sans", "mono" and "symbol" (and see font, below)
fg		"black"	Colour for objects such as axes and boxes in the foreground
fig		0 1 0 1	Coordinates of the figure region within the display region: c(x1, x2, y1, y2)
fin		7.166666 7.156249	Dimensions of the figure region (width, height) in inches
font	*	1	Font (regular $= 1$, bold $= 2$ or italics $= 3$) in which text is written (and see family, above)
font.axis	*	1	Font in which axis is numbered
font.lab	*	1	Font in which labels are written
font.main	*	1	Font for main heading
font.sub	*	1	Font for sub-heading
gamma		1	Correction for hsv colours

hsv		1 1 1	Values (range [0, 1]) for hue, saturation and value of colour
lab		5 5 7	Number of tick marks on the x axis, y axis and size of labels
las		0	Orientation of axis numbers: use las = 1 for publication
lend		"round"	Style for the ends of lines; could be "square" or "butt"
lheight		1	Height of a line of text used to vertically space multi-line text
ljoin		"round"	Style for joining two lines; could be "mitre" or "bevel"
lmitre		10	Controls when mitred line joins are automatically converted into bevelled line joins
log	*	neither	Which axes to log: "log=x", "log=y" or "log=xy"
lty	*	"solid"	Line type (e.g. dashed: lty=2)
lwd	*	1	Width of lines on a graph
mai		0.95625 0.76875 0.76875 0.39375	Margin sizes in inches for c(bottom, left, top, right)
mar		5.1 4.1 4.1 2.1	Margin sizes in numbers of lines for c(bottom, left, top, right)
mex		1	Margin expansion specifies the size of font used to convert between "mar" and "mai", and between "oma" and "omi"
mfcol		1 1	Number of graphs per page (same layout as mfrow (see below), but graphs produced columnwise)
mfg		1 1 1 1	Which figure in an array of figures is to be drawn next (if setting) or is being drawn (if enquiring); the array must already have been set by mfcol or mfrow
mfrow		1 1	Multiple graphs per page (first number = rows, second number = columns): mfrow = c(2,3) gives graphs in two rows each with three columns, drawn row-wise
mgp		3 1 0	Margin line (in mex units) for the axis title, axis labels and axis line
new		FALSE	To draw another plot on top of the existing plot, set new=TRUE so that plot does not wipe the slate clean
oma		0 0 0 0	Size of the outer margins in lines of text c(bottom, left, top, right)
omd		0 1 0 1	Size of the outer margins in normalized device coordinate (NDC) units, expressed as a fraction (in [0,1]) of the device region c(bottom, left, top, right)
omi		0 0 0 0	Size of the outer margins in inches c(bottom, left, top, right)
pch	*	1	Plotting symbol; e.g. pch=16

Table 27.1. (Continued)

Parameter	In plot?	Default value	Meaning
pin		6.004166 5.431249	Current plot dimensions (width, height), in inches
plt		0.1072675 0.9450581 0.1336245 0.8925764	Coordinates of the plot region as fractions of the current figure region c(x1, x2, y1, y2)
ps		12	Point size of text and symbols
pty		"m"	Type of plot region to be used: pty="s" generates a square plotting region, "m" stands for maximal.
srt	*	0	String rotation in degrees
tck		tcl=-0.5	Big tick marks (grid-lines); to use this set tcl=NA
tcl		−0.5	Tick marks outside the frame
tmag		1.2	Enlargement of text of the main title relative to the other annotating text of the plot
type	*	"p"	Plot type: e.g. type="n" to produce blank axes
usr		set by the last plot function	Extremes of the user-defined coordinates of the plotting region c(xmin, xmax, ymin, ymax)
xaxp		0 1 5	Tick marks for log axes: xmin, xmax and number of intervals
xaxs		"r"	Pretty x axis intervals
xaxt		"s"	x axis type: use xaxt = "n" to set up the axis but not plot it
xlab	*	label for the x axis	xlab="label for x axis"
xlim	*	pretty	User control of x axis scaling: xlim=c(0,1)
xlog		FALSE	Is the x axis on a log scale? If TRUE, a logarithmic scale is in use; e.g. following plot(y~x, log ="x")
xpd		FALSE	The way plotting is clipped: if FALSE, all plotting is clipped to the plot region; if TRUE, all plotting is clipped to the figure region; and if NA, all plotting is clipped to the device region
yaxp		0 1 5	Tick marks for log axes: ymin, ymax and number of intervals
yaxs		"r"	Pretty y axis intervals
yaxt		"s"	y axis type: use yaxt = "n" to set up the axis but not plot it
ylab	*	label for the y axis	ylab="label for y axis"
ylim	*	pretty	User control of y axis scaling: ylim=c(0,100)
ylog		FALSE	Is the y axis on a log scale? If TRUE, a logarithmic scale is in use; e.g. following plot(y~x, log ="xy")

```
default.parameters <- par(no.readonly = TRUE)
...
par(...)
...
par(default.parameters)
```

To inspect the current values of any of the graphics parameters (**par**), type the name of the option in double quotes: thus, to see the current limits of the x and y axes, type

```
par("usr")
```

```
[1] 1947.92 2004.08 -80.00 2080.00
```

and to see the sizes of the margins (for the gales data on p. 859),

```
par("mar")
```

```
[1] 5.1 4.1 4.1 2.1
```

Text justification, adj

To alter the justification of text strings, run the **par** function like this:

```
par(adj=0)
```

The parameter **adj=0** produces left-justified text, **adj=0.5** centred text (the default) and **adj=1** right-justified text. For the **text** function you can vary justification in the x and y directions independently like this:

```
adj=c(1,0)
```

Annotation of graphs, ann

If you want to switch off the annotation from a plot (i.e. leave the numbers on the tick marks but not to write the x and y axis labels or print any titles on the graph), then set **ann = FALSE**.

Delay moving on to the next in a series of plots, ask

Setting **ask = TRUE** means that the user is asked for input, before the next figure is drawn.

Control over the axes, axis

The attributes of four sides of the graph ($1 =$ bottom (the x axis); $2 =$ left (the y axis); $3 =$ above and $4 =$ right) are controlled by the **axis** function.

When you want to put two graphs with different y scales on the same plot, you will likely want to scale the right axis (**axis = 4**) differently from the usual y axis on the left (see below).

Again, you may want to label the tick marks on the axis with letters (rather than the usual numbers) and this, too, is controlled by the axis function.

First, draw the graph with no axes at all using **plot** with the **axes=FALSE** option:

```
plot(1:10, 10:1, type="n", axes=FALSE,xlab="",ylab="")
```

For the purposes of illustration only, we use different styles on each of the four axes.

```
axis(1, 1:10, LETTERS[1:10], col.axis = "blue")
axis(2, 1:10, letters[10:1], col.axis = "red")
axis(3, lwd=3, col.axis = "green")
axis(4, at=c(2,5,8), labels=c("one","two","three"))
```

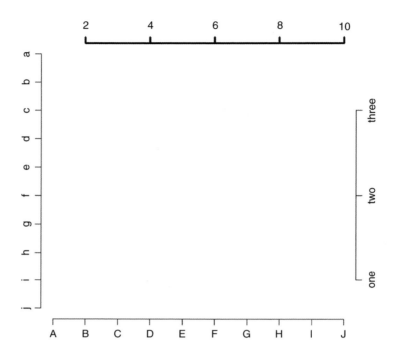

On axis 1 there are upper-case letters in place of the default numbers 1 to 10 with blue rather than black lettering. On axis 2 there are lower-case letters in reverse sequence in red on each of the 10 tick marks. On axis 3 (the top of the graph) there is green lettering for the default numbers (2 to 10 in steps of 2) and an extra thick black line for the axis itself (lwd = 3). On axis 4 we have overwritten the default number and location of the tick marks using at, and provided our own labels for each tick mark (note that the vectors of at locations and labels must be the same length).

Because we did not use box() there are gaps between the ends of each of the four axes.

Background colour for plots, bg

The colour to be used for the background of plots is set by the bg function like this:

par(bg="cornsilk")

The default setting is par(bg="transparent").

Boxes around plots, bty

Boxes are altered with the bty parameter, and bty="n" suppresses the box. If the character is one of "o", "l", (lower-case L not numeral 1), "7", "c", "u", or "]" the resulting box resembles the corresponding *upper* case letter. Here are six options:

```
par(mfrow=c(3,2))
plot(1:10,10:1)
plot(1:10,10:1,bty="n")
plot(1:10,10:1,bty="]")
plot(1:10,10:1,bty="c")
plot(1:10,10:1,bty="u")
plot(1:10,10:1,bty="7")
par(mfrow=c(1,1))
```

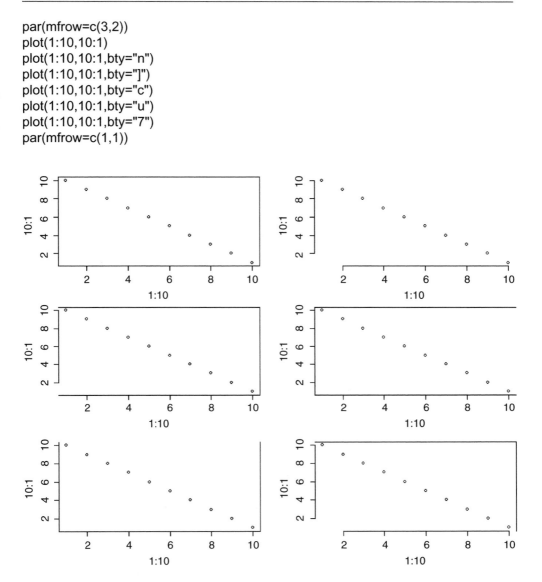

Size of plotting symbols using the character expansion function, cex

You can use points with cex to create 'bubbles' of different sizes. You need to specify the
x, y coordinates of the centre of the bubble, then use cex = value to alter the diameter of
the bubble (in multiples of the default character size: cex stands for character expansion).

```
plot(0:10,0:10,type="n",xlab="",ylab="")
for (i in 1:10) points(2,i,cex=i)
for (i in 1:10) points(6,i,cex=(10+(2*i)))
```

The left column shows points of size 1, 2, 3, 4, etc. (cex = i) and the big circles on the
right are in sequence cex = 12, 14, 16, etc. (cex=(10+(2*i))).

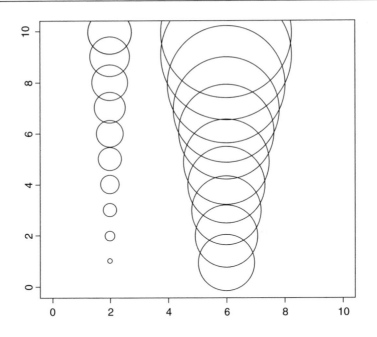

Colour specification

Colours are specified in R in one of three ways:

- by colour name (e.g. "red" as an element of colors());
- by a hexadecimal string of the form #rrggbb;
- by an integer subscript i, meaning palette()[i].

To see all 657 colours available in R (note the US spelling of colors in R), type

colors()

```
  [1]  "white"            "aliceblue"        "antiquewhite"
  [4]  "antiquewhite1"    "antiquewhite2"    "antiquewhite3"
  [7]  "antiquewhite4"    "aquamarine"       "aquamarine1"
 [10]  "aquamarine2"      "aquamarine3"      "aquamarine4"
 [13]  "azure"            "azure1"           "azure2"
 [16]  "azure3"           "azure4"           "beige"
 [19]  "bisque"           "bisque1"          "bisque2"
 [22]  "bisque3"          "bisque4"          "black"
 [25]  "blanchedalmond"   "blue"             "blue1"
...
[640]  "violet"           "violetred"        "violetred1"
[643]  "violetred2"       "violetred3"       "violetred4"
[646]  "wheat"            "wheat1"           "wheat2"
[649]  "wheat3"           "wheat4"           "whitesmoke"
[652]  "yellow"           "yellow1"          "yellow2"
[655]  "yellow3"          "yellow4"          "yellowgreen"
```

The simplest way to specify a colour is with a character string giving the colour name (e.g. col = "red"). Alternatively, colours can be specified directly in terms of their red/green/blue (RGB) components with a string of the form "#RRGGBB" where each of the pairs RR, GG, BB consists of two hexadecimal digits giving a value in the range 00 to FF. Colours can also be specified by giving an index into a small table of colours, known as the **palette**. This provides compatibility with S. The functions rgb (red–green–blue) and hsv (hue–saturation–value) provide additional ways of generating colours (see the relevant help ?rgb and ?hsv). This code demonstrates the effect of varying gamma in red colours:

```
n <- 20
y <- -sin(3*pi*((1:n)-1/2)/n)
par(mfrow=c(3,2),mar=rep(1.5,4))
for(gamma in c(.4, .6, .8, 1, 1.2, 1.5))
plot(y, axes = FALSE, frame.plot = TRUE,
     xlab = "", ylab = "", pch = 21, cex = 30,
     bg = rainbow(n, start=.85, end=.1, gamma = gamma),
     main = paste("Red tones; gamma=",format(gamma)))
```

Note the use of bg within the plot function to colour the different discs and the use of paste with format to get different titles for the six different plots.

Palettes

There are several built-in palettes: here are four of them

```
pie(rep(1,12),col=gray(seq(0.1,.8,length=12)),main="gray")
pie(rep(1,12),col=rainbow(12),main="rainbow")
pie(rep(1,12),col=terrain.colors(12),main="terrain.colors")
pie(rep(1,12),col=heat.colors(12),main="heat.colors")
```

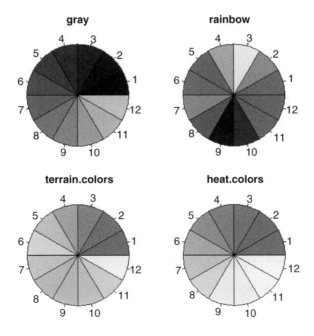

You can create your own customized palette either by colour name (as below) or by RGB levels (e.g. #FF0000, #00FF00, #0000FF):

```
palette(c("wheat1","wheat2","wheat3","wheat4","whitesmoke","beige",
          "bisque1","bisque2","bisque3","bisque4","yellow1",
          "yellow2","yellow3", "yellow4","yellowgreen"))
pie(1:15,col=palette())
```

To reset the palette back to the default use

```
palette("default")
```

The **RColorBrewer** package

This is a very useful package of tried and tested colour schemes:

```
install.packages("RColorBrewer")
library(RColorBrewer)
?brewer.pal
```

The function called colour.pics produces a square with a specified number (x) of striped colours: the default colours are drawn from mypalette which is set outside the function, using different options from brewer.pal.

```
colour.pics<-function(x){
image(1:x,1,as.matrix(1:x),col=mypalette,xlab="",
ylab="",xaxt="n",yaxt="n",bty="n") }
```

You can change the number of colours in your palette and the colour scheme from which they are to be extracted. Here are three schemes with 7, 9 and 11 colours, respectively:

```
mypalette<-brewer.pal(7,"Spectral")
colour.pics(7)
Sys.sleep(3)
mypalette<-brewer.pal(9,"Greens")
colour.pics(9)
Sys.sleep(3)
mypalette<-brewer.pal(11,"BrBG")
colour.pics(11)
```

Note the use of Sys.sleep(3) to create a pause of 3 seconds between the appearance of each of the palettes, as in a slide show.

Different colours and font styles for different parts of the graph

The colours for different parts of the graph are specified as follows:

col.axis   is the colour to be used for axis annotation;

col.lab   is the colour to be used for x and y labels;

col.main   is the colour to be used for plot main titles;

col.sub   is the colour to be used for plot sub-titles.

The font functions change text from normal (1 = plain text) to **bold** (2 = **bold face**), *italic* (3 = *italic* and 4 = ***bold italic***). You can control the font separately for the axis (tick mark numbers) with font.axis, for the axes labels with font.lab, for the main graph title with font.main and for the sub-title with font.sub.

```
plot(1:10,1:10, xlab="x axis label", ylab="y axis label", pch=16, col="orange",
      col.lab="green4",col.axis="blue",col.main="red",main="TITLE",
      col.sub="navy",sub="Subtitle",
      las=1,font.axis=3,font.lab=2,font.main=4,font.sub=3)
```

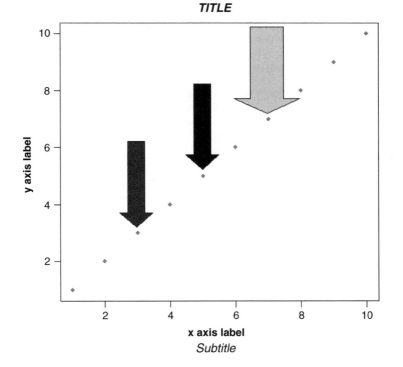

We add three fat arrows using locator(1) (p. 835) to draw attention to selected points:

```
fat.arrow()
fat.arrow(ar.col="blue")
fat.arrow(size.x=1,ar.col="green")
```

Foreground colours, fg

Changing the colour of such things as axes and boxes around plots uses the 'foreground' parameter, fg:

```
par(mfrow=c(2,2))
plot(1:10,1:10,xlab="x label",ylab="y label")
plot(1:10,1:10,xlab="x label",ylab="y label",fg="blue")
plot(1:10,1:10,xlab="x label",ylab="y label",fg="red")
plot(1:10,1:10,xlab="x label",ylab="y label",fg="green")
par(mfrow=c(1,1))
```

Colour with histograms

Let's produce a histogram based on 1000 random numbers from a normal distribution with mean 0 and standard deviation 1:

```
x <- rnorm(1000)
```

We shall draw the histogram on cornsilk coloured paper

```
par(bg = "cornsilk")
```

with the bars of the histogram in a subtle shade of lavender:

```
hist(x, col = "lavender", main = "")
```

The purpose of main = "" is to suppress the graph title. See what happens if you leave this out.

Changing the shape of the plotting region, plt

Suppose that you wanted to draw a map that was 30 m along the x axis and 15 m along the y axis. The standard plot would have roughly twice the scale on the y axis as the x. What you want to do is reduce the height of the plotting region by half while retaining the full width of the x axis so that the scales on the two axes are the same. You achieve this with the plt option, which allows you to specify the coordinates of the plot region as fractions of the current figure region. Here we are using the full screen for one figure so we want to use only the central 40% of the region (from $y = 0.3$ to 0.7):

```
par(plt=c(0.15,0.94,0.3,0.7))
plot(c(0,3000),c(0,1500),type="n",ylab="y",xlab="x")
```

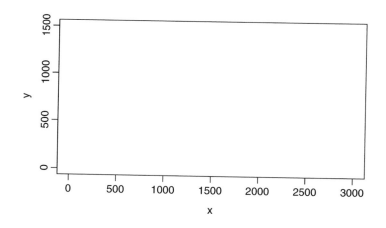

Locating multiple graphs in non-standard layouts using fig

Generally, you would use mfrow to get multiple plots on the same graphic screen (see p. 152); for instance, mfrow=c(3,2) would give six plots in three rows of two columns each. Sometimes, however, you want a non-standard layout, and fig is the function to use in this case. Suppose we want to have two graphs, one in the bottom left-hand corner of the screen and one in the top right-hand corner. What you need to know is that fig considers that the whole plotting region is scaled from (0,0) in the bottom left-hand corner to (1,1) in the top right-hand corner. So we want our bottom left-hand plot to lie within the space $x = c(0, 0.5)$ and $y = (0, 0.5)$, while our top right-hand plot is to lie within the space $x = c(0.5, 1)$ and $y = (0.5, 1)$. Here is how to plot the two graphs: fig is like a new plot function and the second use of fig would normally wipe the slate clean, so we need to specify that new=TRUE in the second par function to stop this from happening:

```
par(fig=c(0.5,1,0.5,1))
plot(0:10,25*exp(-0.1*(0:10)),type="l",ylab="remaining",xlab="time")
par(fig=c(0,0.5,0,0.5),new=T)
plot(0:100,0.5*(0:100)^0.5,type="l",xlab="amount",ylab="rate")
```

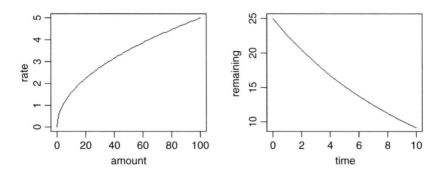

Two graphs with a common x scale but different y scales using fig

The idea here is to draw to graphs with the same x axis, one directly above the other, but with different scales on the two y axes (see also plot.ts on p. 718). Here are the data:

```
data<-read.table("c:\\temp\\gales.txt",header=T)
attach(data)
names(data)
```

```
[1] "year" "number" "February"
```

We use fig to split the plotting area into an upper figure (where number will be drawn first) and a lower figure (for February gales, to be drawn second but on the same page, so new=T). The whole plotting area scales from (0,0) in the bottom left-hand corner to (1,1) in the top right-hand corner, so

```
par(fig=c(0,1,0.5,1))
```

Now think about the margins for the top graph. We want to label the y axis, and we want a normal border above the graph and to the right, but we want the plot to sit right on top of the lower graph, so we set the bottom margin to zero (the first argument):

```
par(mar=c(0,5,2,2))
```

Now we plot the top graph, leaving off the x axis label and the x axis tick marks:

```
plot(year,number,xlab="",xaxt="n",type="b",ylim=c(0,2000),ylab="Population")
```

Next, we define the lower plotting region and declare that new=T:

```
par(fig=c(0,1,0,0.5),new=T)
```

For this graph we *do* want a bottom margin, because we want to label the common *x* axes (Year), but we want the top of the second graph to be flush with the bottom of the first graph, so we set the upper margin to zero (argument 3):

```
par(mar=c(5,5,0,2))
plot(year,February,xlab="Year",type="h",ylab="February gales")
```

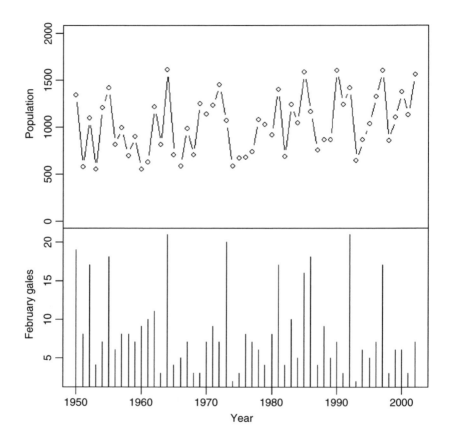

Contrast this with the overlaid plots on p. 868.

The layout function

If you do not want to use **mfrow** (p. 152) or **fig** (p. 858) to configure your multiple plots, then **layout** might be the function you need. This function allows you to alter both the location and shape of multiple plotting regions independently. The **layout** function is used like this:

```
layout(matrix, widths = ws, heights = hs, respect = FALSE)
```

where **matrix** is a matrix object specifying the location of the next *n* figures on the output device (see below), *ws* is a vector of column widths (with length = **ncol(matrix)**) and *hs* is a vector of row heights (with length = **nrow(matrix)**). Each value in the matrix must be 0 or a positive integer. If *n* is the largest positive integer in the matrix, then the integers

$\{1, \ldots, n-1\}$ must also appear at least once in the matrix. Use 0 to indicate locations where you do not want to put a graph. The respect argument controls whether a unit column width is the same physical measurement on the device as a unit row height and is either a logical value or a matrix object. If it is a matrix, then it must have the same dimensions as matrix and each value in the matrix must be either 0 or 1. Each figure is allocated a region composed from a subset of these rows and columns, based on the rows and columns in which the figure number occurs in matrix. The function layout.show(n) plots the outlines of the next n figures.

Here is an example of the kind of task for which layout might be used. We want to produce a scatterplot with histograms on the upper and right-hand axes indicating the frequency of points within vertical and horizontal strips of the scatterplot (see the result below). This is example was written by Paul R. Murrell. Here are the data:

```
x <- pmin(3, pmax(-3, rnorm(50)))
y <- pmin(3, pmax(-3, rnorm(50)))
xhist <- hist(x, breaks=seq(-3,3,0.5), plot=FALSE)
yhist <- hist(y, breaks=seq(-3,3,0.5), plot=FALSE)
```

We need to find the ranges of values within x and y and the two histograms lie:

```
top <- max(c(xhist$counts, yhist$counts))
xrange <- c(-3,3)
yrange <- c(-3,3)
```

Now the layout function defines the location of the three figures: Fig. 1 is the scatterplot which we want to locate in the lower left of four boxes, Fig. 2 is the top histogram which is to be in the upper left box, and Fig. 3 is the side histogram which is to be drawn in the lower right location (the top right location is empty), Thus, the matrix is specified as matrix(c(2,0,1,3),2,2,byrow=TRUE):

```
nf <- layout(matrix(c(2,0,1,3),2,2,byrow=TRUE), c(3,1), c(1,3), TRUE)
layout.show(nf)
```

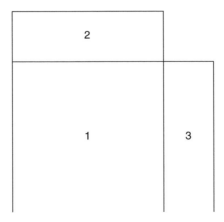

The figures in the first (left) column of the matrix (Figs 1 and 2) are of width 3 while the figure in the second column (Fig. 3) is of width 1, hence c(3,1) is the second argument.

The heights of the figures in the first column of the matrix (Figs 2 and 1) are 1 and 3 respectively, hence c(1,3) is the third argument. The missing figure is 1 by 1 (top right).

```
par(mar=c(3,3,1,1))
plot(x, y, xlim=xrange, ylim=yrange, xlab="", ylab="")
par(mar=c(0,3,1,1))
barplot(xhist$counts, axes=FALSE, ylim=c(0, top), space=0)
par(mar=c(3,0,1,1))
barplot(yhist$counts, axes=FALSE, xlim=c(0, top), space=0, horiz=TRUE)
```

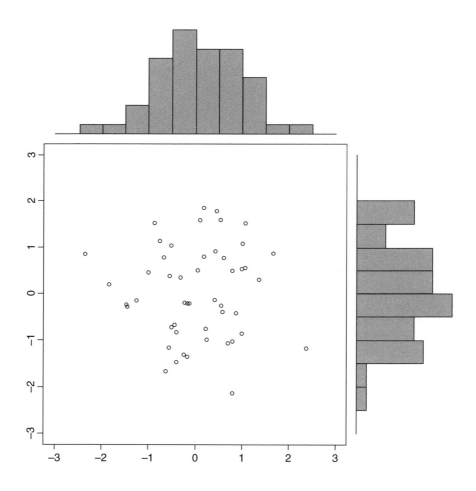

Note the way that the margins for the three figures are controlled, and how the horiz=TRUE option is specified for the histogram on the right-hand margin of the plot.

Creating and controlling multiple screens on a single device

The function split.screen defines a number of regions within the current device which can be treated as if they were separate graphics devices. It is useful for generating multiple plots on a single device (see also mfrow and layout). Screens can themselves be split, allowing for quite complex arrangements of plots. The function **screen** is used to select which screen to draw in, and **erase.screen** is used to clear a single screen, which it does by filling with the background colour, while **close.screen** removes the specified screen definition(s) and split-screen mode is exited by **close.screen(all = TRUE)**. You should complete each graph before moving on to the graph in the next screen (returning to a screen can create problems).

You can create a matrix in which each row describes a screen with values for the left, right, bottom, and top of the screen (in that order) in normalized device coordinate (NDC) units, that is, 0 at the lower left-hand corner of the device surface, and 1 at the upper right-hand corner (see **fig**, above)

First, set up the matrix to define the corners of each of the plots. We want a long, narrow plot on the top of the screen as Fig. 1, then a tall rectangular plot on the bottom left as Fig. 2 then two small square plots on the bottom right as Figs 3 and 4. The dataframe called **gales** is read on p. 859. Here is the matrix:

```
fig.mat<-c(0,0,.5,.5,1,.5,1,1,.7,0,.35,0,1,.7,.7,.35)
fig.mat<-matrix(fig.mat,nrow=4)
fig.mat
      [,1] [,2] [,3] [,4]
[1,]   0.0  1.0 0.70 1.00
[2,]   0.0  0.5 0.00 0.70
[3,]   0.5  1.0 0.35 0.70
[4,]   0.5  1.0 0.00 0.35
```

Now we can draw the four graphs:

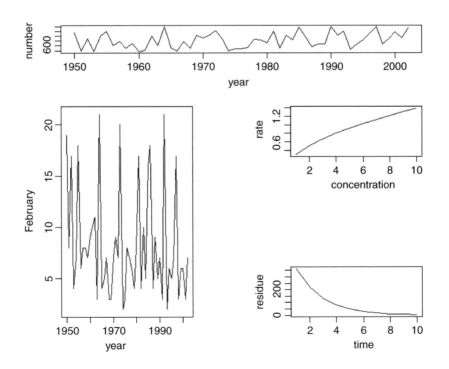

```
split.screen(fig.mat)
[1]  1  2  3  4

screen(1)
plot(year,number,type="l")
screen(2)
plot(year,February,type="l")
screen(3)
plot(1:10,0.5*(1:10)^0.5,xlab="concentration",ylab="rate",type="l")
screen(4)
plot(1:10,600*exp(-0.5*(1:10)),xlab="time",ylab="residue",type="l")
```

Orientation of numbers on the tick marks, las

Many journals require that the numbers used to label the y axis must be horizontal. To change from the default, use las:

las=0   always parallel to the axis (the default)

las=1   always horizontal (preferred by many journals)

las=2   always perpendicular to the axis

las=3   always vertical.

Note that you cannot use character or string rotation for this. Examples are shown on p. 828.

Shapes for the ends of lines, lend

The default is that the bare ends of lines should be rounded (see also **arrows** if you want pointed ends). You can change this to "butt" or "square".

```
par(mfrow=c(3,1))
plot(1:10,1:10,type="n",axes=F,ann=F)
lines(c(2,9),c(5,5),lwd=8)
text(5,1,"rounded ends")
par(lend="square")
plot(1:10,1:10,type="n",axes=F,ann=F)
lines(c(2,9),c(5,5),lwd=8)
text(5,1,"square ends")
par(lend="butt")
plot(1:10,1:10,type="n",axes=F,ann=F)
lines(c(2,9),c(5,5),lwd=8)
text(5,1,"butt ends")
```

Line types, lty

Line types (like solid or dashed) are changed with the line-type parameter lty:

lty = 1   solid

lty = 2   dashed

lty = 3   dotted

lty = 4   dot-dash

lty = 5   long-dash

lty = 6   two-dash

Invisible lines are drawn if lty=0 (i.e. the line is not drawn). Alternatively, you can use text to specify the line types with one of the following character strings: "blank", "solid", "dashed", "dotted", "dotdash", "longdash" or "twodash" (see below).

Line widths, lwd

To increase the widths of the plotted lines use lwd = 2 (or greater; the default is lwd=1). The interpretation is device-specific, and some devices do not implement line widths less than 1. The function abline is so called because it has two arguments: the first is the intercept (a) and the second is the slope (b) of a linear relationship $y = a + bx$ (see p. 136 for background):

```
plot(1:10,1:10,xlim=c(0,10),ylim=c(0,10),xlab="x label",ylab="y label",type="n")
abline(-4,1,lty=1)
abline(-2,1,lty=2)
abline(0,1,lty=3)
abline(2,1,lty=4)
abline(4,1,lty=5)
abline(6,1,lty=6)
abline(8,1,lty=7)
abline(-6,1,lty=1,lwd=4)
abline(-8,1,lty=1,lwd=8)
for( i in 1:5) text(5,2*i-1,as.character(i))
```

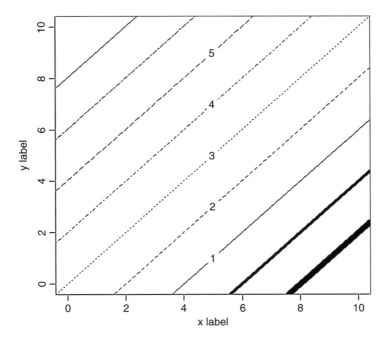

The numerals indicate the line types 1 to 5. In the bottom right-hand corner are two solid lines lty = 1 of widths 4 and 8.

Several graphs on the same page, mfrow

Multiple graph panels on the same graphics device are controlled by par(mfrow), par(mfcol), par(fig), par(split.screen) and par(layout), but par(mfrow) is much the most frequently used. You specify the number of rows of graphs (first argument) and number of columns of graphs per row (second argument) like this:

par(mfrow=c(1,1))   the default of one plot per screen

par(mfrow=c(1,2))   one row of two columns of plots

par(mfrow=c(2,1))   two rows of one column of plots

par(mfrow=c(2,2))   four plots in two rows of two columns each

par(mfrow=c(3,2))   six plots in three rows of two columns each

In a layout with exactly two rows and columns the base value of cex is reduced by a factor of 0.83; if there are three or more of either rows or columns, the reduction factor is 0.66. Consider the alternatives, layout and split.screen. Remember to set par back to par(mfrow=c(1,1)) when you have finished with multiple plots. For examples, see the Index.

Margins around the plotting area, mar

You need to control the size of the margins when you intend to use large symbols or long labels for your axes. The four margins of the plot are defined by integers 1 to 4 as follows:

$1 =$ bottom (the x axis),

$2 =$ left (the y axis),

$3 =$ top,

$4 =$ right.

The sizes of the margins of the plot are measured in lines of text. The four arguments to the mar function are given in the sequence bottom, left, top, right. The default is

par(mar=(c(5, 4, 4, 2) + 0.1))

with more spaces on the bottom (5.1) than on the top (4.1) to make room for a subtitle (if you should want one), and more space on the left (4.1) than on the right (2) on the assumption that you will not want to label the right-hand axis. Suppose that you do want to put a label on the right-hand axis, then you would need to increase the size of the fourth number, for instance like this:

par(mar=(c(5, 4, 4, 4) + 0.1))

Plotting more than one graph on the same axes, new

The new parameter is a logical variable, defaulting to new=FALSE. If it is set to new=TRUE, the next high-level plotting command (like plot(~x)) does *not* wipe the slate clean in the default way. This allows one plot to be placed on top of another.

Two graphs on the same plot with different scales for their *y* axes

```
gales<-read.table("c:\\temp\\gales.txt",header=T)
attach(gales)
names(gales)
```

```
[1] "year"  "number"  "February"
```

In this example we want to plot the number of animals in a wild population as a time series over the years 1950–2000 with the scale of animal numbers on the left-hand axis (numbers fluctuate between about 600 and 1600). Then, on top of this, we want to overlay the number of gales in February each year. This number varies between 1 and 22, and we want to put a scale for this on the right-hand axis (**axis = 4**). First we need to make room in the right-hand margin for labelling the axis with the information on February gales:

```
par(mar=c(5,4,4,4)+0.1)
```

Now draw the time series using a thicker-than-usual line (**lwd=2**) for emphasis:

```
plot(year,number,type="l",lwd=2,las=1)
```

Next, we need to indicate that the next graph will be overlaid on the present one:

```
par(new=T)
```

Now plot the graph of gales against years. This is to be displayed as vertical (**type="h"**) dashed lines (**lty=2**) in blue:

```
plot(year,February,type="h",axes=F,ylab="",lty=2,col="blue")
```

and it is drawn with its own scale (with ticks from 5 to 20, as we shall see). The right-hand axis is ticked and labelled as follows. First use **axis(4)** to create the tick marks and scaling information, then use the **mtext** function to produce the axis label (the name stands for 'margin text').

```
axis(4,las=1)
mtext(side=4,line=2.5,"February gales")
```

It looks as if unusually severe February gales are associated with the steepest population crashes (contrast this with the separate plots on p. 860).

Outer margins, oma

There is an area outside the margins of the plotting area called the **outer margin**. Its default size is zero, oma=c(0,0,0,0), but if you want to create an outer margin you use the function oma. Here is the function to produce an outer margin big enough to accommodate two lines of text on the bottom and left-hand sides of the plotting region:

```
par(oma=c(2,2,0,0))
```

Packing graphs closer together

In this example we want to create nine closely spaced plots in a 3×3 pattern without any tick marks, and to label only the outer central plot on the *x* and *y* axes. We need to take care of four things:

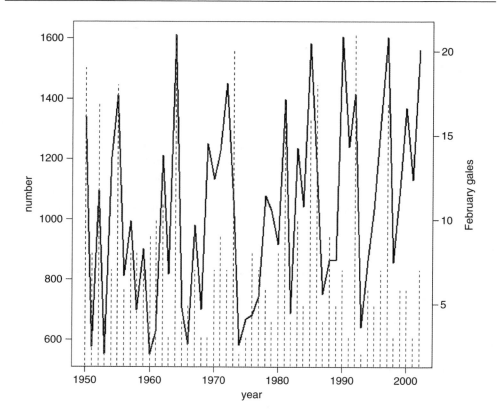

- mfrow=c(3,3) to get the nine plots in a 3×3 pattern;
- mar=c(0.2,0.2,0.2,0.2) to leave a narrow strip (0.2 lines looks best for tightly packed plots) between each graph;
- oma=c(5,5,0,0) to create an outer margin on the bottom and left for labels;
- outer = T in title to write the titles in the outer margin.

The plots consist of 100 pairs of ranked uniform random numbers sort(runif(100)), and we shall plot the nine graphs with a for loop:

```
par(mfrow=c(3,3))
par(mar=c(0.2,0.2,0.2,0.2))
par(oma=c(5,5,0,0))
for (i in 1:9) plot(sort(runif(100)),sort(runif(100)),xaxt="n",yaxt="n")
title(xlab="time",ylab="distance",outer=T,cex.lab=2)
```

Square plotting region, pty

If you want to have a square plotting region (e.g. when producing a map or a grid with true squares on it), then use the pty = "s" option. The option pty = "m" generates the maximal plotting region which is not square on most devices.

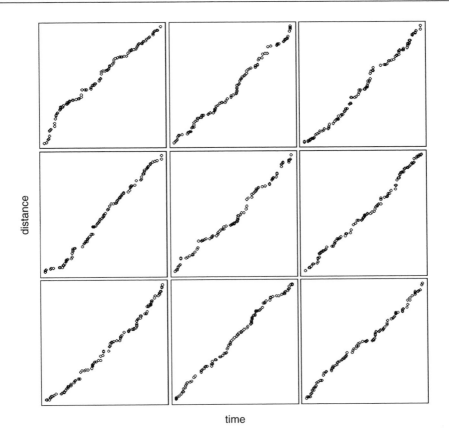

distance

time

Character rotation, srt

To rotate characters in the plotting plane use srt (which stands for 'string rotation'). The argument to the function is in degrees of counter-clockwise rotation:

```
plot(1:10,1:10,type="n",xlab="",ylab="")
for (i in 1:10) text (i,i,LETTERS[i],srt=(20*i))
for (i in 1:10) text (10-i+1,i,letters[i],srt=(20*i))
```

Observe how the letters i and I have been turned upside down (srt = 180).

Rotating the axis labels

When you have long text labels (e.g. for bars on a barplot) it is a good idea to rotate them through 45 degrees so that all the labels are printed, and all are easy to read.

```
spending<-read.csv("c:\\temp\\spending.csv",header=T)
attach(spending)
names(spending)
```

```
[1] "spend"  "country"
```

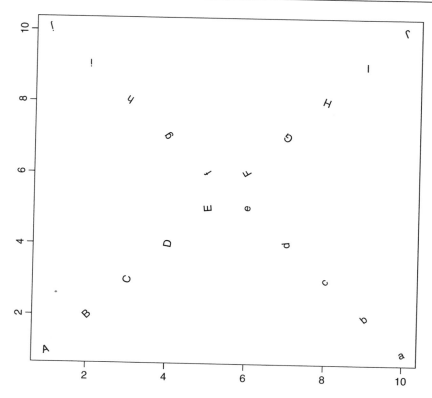

There are three steps involved:

- Make the bottom margin big enough to take the long labels (mar).
- Find the x coordinates of the centres of the bars (xvals) with usr.
- Use **text** with srt = 45 to rotate the labels.

```
par(mar = c(7, 4, 4, 2) + 0.1)
xvals<-barplot(spend,ylab="spending")
text(xvals, par("usr")[3] - 0.25, srt = 45, adj = 1,labels = country, xpd = TRUE)
```

Note the use of xpd = TRUE to allow for text outside the plotting region, and adj = 1 to place the right-hand end of text at the centre of the bars. The vertical location of the labels is set by par("usr")[3] - 0.25 and you can adjust the value of the offset (here 0.25) as required to move the axis labels up or down relative to the x axis.

Tick marks on the axes

The functions tck and tcl control the length and location of the tick marks. Negative values put the tick marks *outside* the box (tcl = -0.5 is the default setting in R, as you can see above). tcl gives the length of tick marks as a fraction of the height of a line of text.

    The default setting for tck is tck = NA but you can use this for drawing grid lines: tck=0 means no tick marks, while tck = 1 means fill the whole frame (i.e. the tick marks make a grid). The tick is given as a fraction of the frame width (they are 0.03 in the bottom

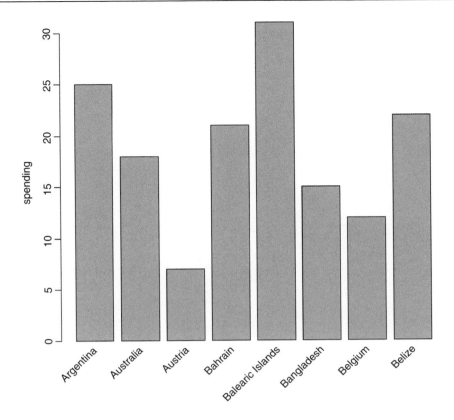

right-hand graph, so are internal to the plotting region). Note the use of line type = "b", which means draw both lines and symbols with the lines *not* passing through the symbols (compared with type="o" where lines *do* pass through the symbols).

```
par(mfrow=c(2,2))
plot(1:10,1:10,type="b",xlab="",ylab="")
plot(1:10,1:10,type="b",xlab="",ylab="",tck=1)
plot(1:10,1:10,type="b",xlab="",ylab="",tck=0)
plot(1:10,1:10,type="b",xlab="",ylab="",tck=0.03)
```

Axis styles

There are three functions that you need to distinguish:

axis   select one of the four sides of the plot to work with;

xaxs   intervals for the tick marks;

xaxt   suppress production of the axis with xaxt="n".

   The axis function has been described on pp. 834 and 852.

   The xaxs function is used infrequently: style "r" (regular) first extends the data range by 4% and then finds an axis with pretty labels that fits within the range; style "i" (internal) just finds an axis with pretty labels that fits within the original data range.

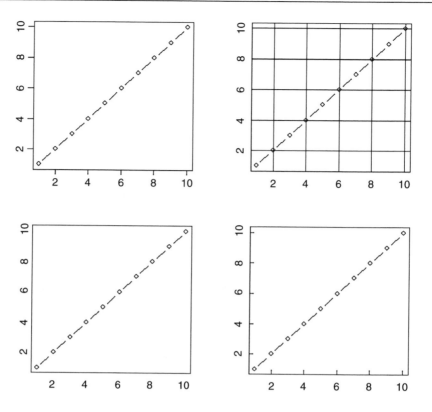

Finally **xaxt** is used when you want to specify your own kind of axes with different locations for the tick marks and/or different labelling. To suppress the tick marks and value labels, specify **xaxt="n"** and/or **yaxt="n"** (see p. 146).

References and Further Reading

Agresti, A. (1990) *Categorical Data Analysis*. New York: John Wiley.

Aitkin, M., Anderson, D., Francis, B. and Hinde, J. (1989) *Statistical Modelling in GLIM*. Oxford: Clarendon Press.

Atkinson, A.C. (1985) *Plots, Transformations, and Regression*. Oxford: Clarendon Press.

Bishop, Y.M.M., Fienberg, S.J. and Holland, P.W. (1980) *Discrete Multivariate Analysis: Theory and Practice*. New York: John Wiley.

Box, G.E.P. and Cox, D.R. (1964) An analysis of transformations. *Journal of the Royal Statistical Society Series B*, **26**, 211–246.

Box, G.E.P., Hunter, W.G. and Hunter, J.S. (1978) *Statistics for Experimenters: An Introduction to Design, Data Analysis and Model Building*. New York: John Wiley.

Box, G.E.P. and Jenkins, G.M. (1976) *Time Series Analysis: Forecasting and Control*. Oakland, CA: Holden-Day.

Breiman, L., Friedman, L.H., Olshen, R.A. and Stone, C.J. (1984) *Classification and Regression Trees*. Belmont, CA: Wadsworth International Group.

Caroll, R.J. and Ruppert, D. (1988) *Transformation and Weighting in Regression*. New York: Chapman and Hall.

Casella, G. and Berger, R.L. (1990) *Statistical Inference*. Pacific Grove, CA: Wadsworth and Brooks/Cole.

Chambers, J.M., Cleveland, W.S., Kleiner, B. and Tukey, P.A. (1983) *Graphical Methods for Data Analysis*. Belmont, CA: Wadsworth.

Chambers, J.M. and Hastie, T.J. (1992) *Statistical Models in S. Pacific Grove*. California: Wadsworth and Brooks Cole.

Chatfield, C. (1989) *The Analysis of Time Series: An Introduction*. London: Chapman and Hall.

Clark, P.J. and Evans, F.C. (1954) Distance to nearest neighbour as a measure of spatial relationships in populations. *Ecology*, **35**, 445–453 .

Cleveland, W.S. (1993) *Visualizing Data*. Summit, NJ: Hobart Press.

Cochran, W.G. and Cox, G.M. (1957) *Experimental Designs*. New York: John Wiley.

Collett, D. (1991) *Modelling Binary Data*. London: Chapman and Hall.

Conover, W.J. (1980) *Practical Nonparametric Statistics*. New York: John Wiley.

Conover, W.J., Johnson, M.E. & Johnson, M.M. (1981) A comparative study of tests for homogeneity of variances, with appplications to the outer continental shelf bidding data. *Technometrics*, **23**, 351–361.

Cook, R.D. and Weisberg, S. (1982) *Residuals and Influence in Regression*. New York: Chapman and Hall.

Cox, D.R. (1972) Regression models and life-tables (with discussion). *Journal of the Royal Statistical Society B*, **34**, 187–220?

Cox, D.R. and Hinkley, D.V. (1974) *Theoretical Statistics*. London: Chapman and Hall.

Cox, D.R. and Oakes, D. (1984) *Analysis of Survival Data*. London: Chapman and Hall.

Cox, D.R. and Snell, E.J. (1989) *Analysis of Binary Data*. London: Chapman and Hall.

Crawley, M.J. (2002) *Statistical Computing: An Introduction to Data Analysis using S-PLUS*. Chichester: John Wiley.

Crawley, M.J., Johnston, A.E., Silvertown, J., Dodd, M., de Mazancourt, C., Heard, M.S., Henman, D.F. and Edwards, G.R. (2005) Determinants of species richness in the Park Grass Experiment. *American Naturalist*, **165**, 179–192.

Cressie, N.A.C. (1991) *Statistics for Spatial Data*. New York: John Wiley.

Crowder, M.J. and Hand, D.J. (1990) *Analysis of Repeated Measures*. London: Chapman and Hall.

Dalgaard, P. (2002) *Introductory Statistics with R*. New York: Springer-Verlag.

Davidian, M. and Giltinan, D.M. (1995) *Nonlinear Models for Repeated Measurement Data*. London: Chapman and Hall.

Davison, A.C. and Hinkley, D.V. (1997) *Bootstrap Methods and Their Application*. Cambridge University Press.

Diggle, P.J. (1983) *Statistical Analysis of Spatial Point Patterns*. London: Academic Press.

Diggle, P.J., Liang, K.-Y. and Zeger, S.L. (1994) *Analysis of Longitudinal Data*. Oxford: Clarendon Press.

Dobson, A.J. (1990) *An Introduction to Generalized Linear Models*. London: Chapman and Hall.

Draper, N.R. and Smith, H. (1981) *Applied Regression Analysis*. New York: John Wiley.

Edwards, A.W.F. (1972) *Likelihood*. Cambridge: Cambridge University Press.

Efron, B. and Tibshirani, R.J. (1993) *An Introduction to the Bootstrap*. San Francisco: Chapman and Hall.

Eisenhart, C. (1947) The assumptions underlying the analysis of variance. *Biometrics*, **3**, 1–21.

Everitt, B.S. (1994) *Handbook of Statistical Analyses Using S-PLUS*. New York: Chapman and Hall / CRC Statistics and Mathematics.

Ferguson, T.S. (1996) *A Course in Large Sample Theory*. London: Chapman and Hall.

Fisher, L.D. and Van Belle, G. (1993) *Biostatistics*. New York: John Wiley.

Fisher, R.A. (1954) *Design of Experiments*. Edinburgh: Oliver and Boyd.

Fleming, T. and Harrington, D. (1991) *Counting Processes and Survival Analysis*. New York: John Wiley.

Fox, J. (2002) *An R and S-Plus Companion to Applied Regression*. Thousand Oaks, CA: Sage.

Gordon, A.E. (1981) *Classification: Methods for the Exploratory Analysis of Multivariate Data*. New York: Chapman and Hall.

Gosset, W.S. (1908) writing under the pseudonym "Student". The probable error of a mean. *Biometrika*, **6**(1), 1–25.

Grimmett, G.R. and Stirzaker, D.R. (1992) *Probability and Random Processes*. Oxford: Clarendon Press.

Hairston, N.G. (1989) *Ecological Experiments: Purpose, Design and Execution*. Cambridge: Cambridge University Press.

Hampel, F.R., Ronchetti, E.M., Rousseeuw, P.J. and Stahel, W.A. (1986) *Robust Statistics: The Approach Based on Influence Functions*. New York: John Wiley.

Harman, H.H. (1976) *Modern Factor Analysis*. Chicago: University of Chicago Press.

Hastie, T. and Tibshirani, R. (1990) *Generalized Additive Models*. London: Chapman and Hall.

Hicks, C.R. (1973) *Fundamental Concepts in the Design of Experiments*. New York, Holt: Rinehart and Winston.

Hoaglin, D.C., Mosteller, F. and Tukey, J.W. (1983) *Understanding Robust and Exploratory Data Analysis*. New York: John Wiley.

Hochberg, Y. and Tamhane, A.C. (1987) *Multiple Comparison Procedures*. New York: John Wiley.

Hosmer, D.W. and Lemeshow, S. (2000) *Applied Logistic Regression*. 2nd Edition. New York: John Wiley.

Hsu, J.C. (1996) *Multiple Comparisons: Theory and Methods*. London: Chapman and Hall.

Huber, P.J. (1981) *Robust Statistics*. New York: John Wiley.

Huitema, B.E. (1980) *The Analysis of Covariance and Alternatives*. New York: John Wiley.

Hurlbert, S.H. (1984) Pseudoreplication and the design of ecological field experiments. *Ecological Monographs*, **54**, 187–211.

Johnson, N.L. and Kotz, S. (1970) *Continuous Univariate Distributions. Volume 2*. New York: John Wiley.

Kalbfleisch, J. and Prentice, R.L. (1980) *The Statistical Analysis of Failure Time Data*. New York: John Wiley.

Kaluzny, S.P., Vega, S.C., Cardoso, T.P. and Shelly, A.A. (1998) *S+ Spatial Stats*. New York, Springer-Verlag:

Kendall, M.G. and Stewart, A. (1979) *The Advanced Theory of Statistics*. Oxford: Oxford University Press.

Keppel, G. (1991) *Design and Analysis: A Researcher's Handbook*. Upper Saddle River, N.J: Prentice Hall.

Khuri, A.I., Mathew, T. and Sinha, B.K. (1998) *Statistical Tests for Mixed Linear Models*. New York: John Wiley.

Krause, A. and Olson, M. (2000) *The Basics of S and S-PLUS*. New York: Springer-Verlag.

Lee, P.M. (1997) *Bayesian Statistics: An Introduction*. London: Arnold.

Lehmann, E.L. (1986) *Testing Statistical Hypotheses*. New York: John Wiley.

Mandelbrot, B.B. (1977) *Fractals, Form, Chance and Dimension*. San Francisco: Freeman.

Mardia, K.V., Kent, J.T. and Bibby, J.M. (1979) *Multivariate Statistics*. London: Academic Press.

May, R.M. (1976) Simple mathematical models with very complicated dynamics. *Nature*, **261**, 459–467.

McCullagh, P. and Nelder, J.A. (1989) *Generalized Linear Models. 2nd* Edition. London: Chapman and Hall.

McCulloch, C.E. and Searle, S.R. (2001) *Generalized, Linear and Mixed Models*. New York: John Wiley.

Michelson, A.A. (1880) Experimental determination of the velocity of light made at the U.S. Naval Academy, Annapolis. *Astronomical Papers*, 1: 109–145.

Millard, S.P. and Krause, A. (2001) *Using S-PLUS in the Pharmaceutical Industry*. New York: Springer-Verlag.

Miller, R.G. (1981) *Survival Analysis*. New York: John Wiley.

Miller, R.G. (1997) *Beyond ANOVA: Basics of Applied Statistics*. London: Chapman and Hall.

Mosteller, F. and Tukey, J.W. (1977) *Data Analysis and Regression*. Reading, MA: Addison-Wesley.

Nelder, J.A. and Wedderburn, R.W.M. (1972) Generalized linear models. *Journal of the Royal Statistical Society, Series A*, **135**, 37–384.

Neter, J., Kutner, M., Nachstheim, C. and Wasserman, W. (1996) *Applied Linear Statistical Models*. New York: McGraw-Hill.

Neter, J., Wasserman, W. and Kutner, M.H. (1985) *Applied Linear Regression Models*. Homewood, IL: Irwin.

OED (2004) *Oxford English Dictionary*. Oxford: Oxford University Press.

O'Hagen, A. (1988) *Probability: Methods and Measurement*. London: Chapman and Hall.

Pinheiro, J.C. and Bates, D.M. (2000) *Mixed-effects Models in S and S-PLUS*. New York: Springer-Verlag.

Platt, J.R. (1964) Strong inference. *Science*, **146**: 347–353.

Priestley, M.B. (1981) *Spectral Analysis and Time Series*. London: Academic Press.

Rao, P.S.R.S. (1997) *Variance Components Estimation: Mixed Models, Methodologies and Applications*. London: Chapman and Hall.

Riordan, J. (1978) *An Introduction to Combinatorial Analysis*. Princeton, NJ: Princeton University Press.

Ripley, B.D. (1996) *Pattern Recognition and Neural Networks*. Cambridge: Cambridge University Press.

Robert, C.P. and Casella, G. (1999) *Monte Carlo Statistical Methods*. New York: Springer-Verlag.

Rosner, B. (1990) *Fundamentals of Biostatistics*. Boston: PWS-Kent.

Ross, G.J.S. (1990) *Nonlinear Estimation*-. New York: Springer-Verlag.

Santer, T.J. and Duffy, D.E. (1990) *The Statistical Analysis of Discrete Data*. New York: Springer-Verlag.

Scott, D.W. (1992) *Multivariate Density Estimation. Theory, Practice and Visualization*. New York: John Wiley & Sons, Inc.

Searle, S.R., Casella, G. and McCulloch, C.E. (1992) *Variance Components*. New York: John Wiley.

Shao, J. and Tu, D. (1995) *The Jacknife and Bootstrap*. New York: Springer-Verlag.

Shumway, R.H. (1988) *Applied Statistical Time Series Analysis*. Englewood Cliffs, NJ: Prentice Hall.

Silverman, B.W. (1986) *Density Estimation*. London: Chapman & Hall.

Silvey, S.D. (1970) *Statistical Inference*. London: Chapman and Hall.

Snedecor, G.W. and Cochran, W.G. (1980) *Statistical Methods*. Ames, IA: Iowa State University Press.

Sokal, R.R. and Rohlf, F.J. (1995) *Biometry: The Principles and Practice of Statistics in Biological Research*. San Francisco, W.H. Freeman and Company.

Sprent, P. (1989) *Applied Nonparametric Statistical Methods*. London: Chapman and Hall.

Taylor, L.R. (1961) Aggregation, variance and the mean. *Nature*, **189**, 732–735.

Upton, G. and Fingleton, B. (1985) *Spatial Data Analysis by Example*. Chichester: John Wiley.

Venables, W.N. and Ripley, B.D. (2002) *Modern Applied Statistics with S-PLUS*. 4*th* Edition. New York: Springer-Verlag.

Venables, W.N., Smith, D.M. and the R Development Core Team (1999) *An Introduction to R*. Bristol: Network Theory Limited.

Wedderburn, R.W.M. (1974) Quasi-likelihood functions, generalized linear models and the Gauss–Newton method. *Biometrika*, **61**, 439–447.

Weisberg, S. (1985) *Applied Linear Regression*. New York: John Wiley.

Wetherill, G.B., Duncombe, P., Kenward, M., Kollerstrom, J., Paul, S.R. and Vowden, B.J. (1986) *Regression Analysis with Applications*. London: Chapman and Hall.

Winer, B.J., Brown, D.R. and Michels, K.M. (1991) *Statistical Principles in Experimental Design*. New York: McGraw-Hill.

Wood, S.N. (2000) Modelling and smoothing parameter estimation with multiple quadratic penalties. *Journal of the Royal Statistical Society B*, 62, 413–428.

Wood, S.N. (2003) Thin plate regression splines. *Journal of the Royal Statistical Society B*, 65, 95–114.

Wood, S.N. (2004) Stable and efficient multiple smoothing parameter estimation for generalized additive models. *Journal of the American Statistical Association*, 99, 673–686.

Zar, J.H. (1999) *Biostatistical Analysis*. Englewood Cliffs, NJ: Prentice Hall.

Index